Sediment Transport
and Depositional Processes

Sediment Transport and Depositional Processes

Edited by
KENNETH PYE
Reader in Sedimentology
Postgraduate Research Institute for Sedimentology
University of Reading

OXFORD
BLACKWELL SCIENTIFIC PUBLICATIONS
EDINBURGH LONDON BOSTON
MELBOURNE PARIS BERLIN VIENNA

© 1994 by
Blackwell Scientific Publications
Editorial Offices:
Osney Mead, Oxford OX2 0EL
25 John Street, London WC1N 2BL
23 Ainslie Place, Edinburgh EH3 6AJ
238 Main Street, Cambridge
   Massachusetts 02142, USA
54 University Street, Carlton
   Victoria 3053, Australia

Other Editorial Offices:
Librairie Arnette SA
1, rue de Lille
75007 Paris
France

Blackwell Wissenschafts-Verlag GmbH
Düsseldorfer Str. 38
D-10707 Berlin
Germany

Blackwell MZV
Feldgasse 13
A-1238 Wien
Austria

First published 1994

Set by Semantic Graphics, Singapore
Printed and bound in Great Britain
at the University Press, Cambridge

DISTRIBUTORS

Marston Book Services Ltd
PO Box 87
Oxford OX2 0DT
(*Orders*: Tel: 0865 791155
        Fax: 0865 791927
        Telex: 837515)

USA
Blackwell Scientific Publications, Inc.
238 Main Street
Cambridge, MA 02142
(*Orders*: Tel: 800 759-6102
          617 876-7000)

Canada
Oxford University Press
70 Wynford Drive
Don Mills
Ontario M3C 1J9
(*Orders*: Tel: 416 441-2941)

Australia
Blackwell Scientific Publications Pty Ltd
54 University Street
Carlton, Victoria 3053
(*Orders*: Tel: 03 347-5552)

A catalogue record for this title
is available from the British Library

ISBN 0-632-03112-3

Library of Congress
Cataloging in Publication Data

Sediment transport and depositional processes/
edited by Kenneth Pye.
        p.   cm.
   Includes bibliographical references
   and index.
   ISBN 0-632-03112-3
   1. Sedimentation and deposition.
2. Sediment transport. I. Pye, Kenneth.
QE571.S415 1994
551.3'03—dc20

Contents

List of contributors

JOHN R.L. ALLEN Postgraduate Research Institute for Sedimentology, University of Reading, Whiteknights, Reading RG6 2AB, UK

KEITH R. DYER Institute of Marine Studies, University of Plymouth, Drake Circus, Plymouth, Devon PI4 8AA, UK

RICHARD V. FISHER Department of Geological Sciences, University of California at Santa Barbara, CA 93106, USA

LYNNE E. FROSTICK Postgraduate Research Institute for Sedimentology, University of Reading, Whiteknights, Reading RG6 2AB, UK

JACK HARDISTY School of Geography and Earth Resources, University of Hull, Hull HU6 7RX, UK

WILLIAM G. NICKLING Department of Geography, University of Guelph, Guelph, Ontario N1G 2W1, Canada

KENNETH PYE Postgraduate Research Institute for Sedimentology, University of Reading, Whiteknights, Reading RG6 2AB, UK

IAN REID Department of Geography, Loughborough University of Technology, Loughborough LE11 3TU, UK

HANS-ULRICH SCHMINCKE GEOMAR Institute, Wischof Strasse, D-2300 Kiel 14, Germany

MICHAEL J. SELBY Department of Earth Sciences, University of Waikato, Private Bag 3105, Hamilton, New Zealand

PETER G. SLY Rawson Academy of Aquatic Science, 404, 1 Nicholas Street, Ottawa, Ontario K1N 7B7, Canada

DORRIK A.V. STOW Department of Geology, University of Southampton, Southampton SO9 5NH, UK

Preface

The processes involved in the transport and deposition of sediment grains are of considerable interest to a broad spectrum of scientists, including sedimentologists, oceanographers, geomorphologists, fluid mechanicists and engineers. These processes are not merely of academic interest, but are of basic importance to an understanding of many pressing environmental problems including soil erosion, siltation in reservoirs and harbours, coastal erosion, flood defence and waste disposal. The past two decades have seen an explosion of research interest in this area, and there is now a vast multi-disciplinary literature on the subject, written from a variety of specialist perspectives.

The principal aim of this book is to provide a summary of the fundamental principles concerning sediment transport and deposition. The emphasis of the book is on active processes and recent sediments, although references to the sedimentary record are made where appropriate. Although a wide range of issues is addressed, inevitably it has not been possible to give detailed consideration to all sediment transport processes and to every environment of deposition. Attention has been deliberately focused on some of the more common processes of sediment transport in air and water.

The individual chapters have been written by an invited team of internationally-respected specialists who are well acquainted with recent developments in their fields. It is intended that the resulting volume will provide a useful introduction and subsequent reference source for more advanced undergraduate students, for research workers and for professional scientists who are looking to broaden their knowledge or develop new research interests. A limited degree of background knowledge is assumed on the part of the reader, but for the benefit of those with little previous knowledge of sedimentology or fluid mechanics, introductions to the basic properties of sediment grains and fluids are presented in the first two chapters. Although numerical parameters and equations are widely used in the book, the balance of overall treatment is conceptual rather than mathematical.

The book is intended to supplement, rather than to replace, existing general texts on sedimentology, sedimentary facies and sediment transport mechanics. It has been written primarily with an audience of earth and environmental scientists in mind, but it is also likely to be of interest to engineers and fluid mechanicists concerned with applied problems of hydraulics and two-phase flow, and to life scientists who have interests in the relationships between sediments and biota.

The wide ranging nature of the subject matter has necessarily required some flexibility of approach in the individual chapters. Contributors have been encouraged to draw attention to those aspects which they consider to be of particular importance in their chosen subject areas. However, extensive referencing should allow interested readers to follow up points of particular concern in the specialist literature.

K. PYE
Reading

Acknowledgements

The undertaking of a project of this nature requires sustained commitment and patience on behalf of the authors, editor and publisher. In this respect I particularly wish to acknowledge the early contribution of Navin Sullivan, who first persuaded me to take on the task of editing this volume on behalf of Blackwell Scientific Publications Ltd, and the subsequent assistance and perseverance of Simon Rallison and Anna Illingworth in bringing the book to a successful conclusion.

References to the sources of previously published material are given in the text, where appropriate.

This publication represents University of Reading PRIS Contribution No. 264.

1 Properties of sediment particles

K. PYE

1.1 Introduction

The entrainment, transport and deposition of sediment particles by any fluid is controlled partly by the physical and chemical properties of the particles themselves and partly by those of the fluid in question. Entrainment of grains from the bed is controlled not only by individual grain characteristics but also by the bulk sediment properties which include the grain size distribution, sorting, grain orientation, packing arrangement, porosity, and degree of cohesion or cementation. During transport, sediment particles are frequently sorted according to size, shape and density, and changes in particle characteristics may be brought about through intergrain and grain–bed collisions. During extended transport, mineral grains with relatively low hardness or susceptibility to fracturing may be significantly reduced in size, modified in shape or destroyed altogether, while relatively hard, resistant minerals such as monocrystalline quartz are modified to a much lesser extent and become concentrated in the residual sediment. Analysis of the particle size, shape and compositional characteristics may therefore provide important clues to the sediment provenance, transport history and depositional conditions (see, for example, Krumbein, 1941; Passega, 1957; Dobkins & Folk, 1970; Sparks, 1976; Boulton, 1978; Friedman, 1979; Ballantyne, 1982; Ibbeken, 1983; Sheridan *et al.*, 1987, Bui *et al.*, 1990; Lirer & Vinci, 1991).

University of Reading PRIS Contribution 068.

1.2 Grain characteristics

1.2.1 Concepts of grain size

Grain size can be specified in a number of different ways, for example by measuring the external caliper dimensions of a particle (most commonly the orthogonal long, intermediate and short axes), by determining its volume or mass, by determining its settling velocity, a property which is influenced by grain density and shape as well as by external dimensions, or its equivalent spherical diameter. Different methods of particle size analysis describe different aspects of 'size'. For example, sieving, using a series of progressively finer square-mesh sieves, separates particles principally on the basis of their intermediate axial diameter (Sahu, 1965; Kennedy *et al.*, 1985), whereas electro-optical methods such as laser diffraction express size in terms of equivalent spherical diameter based on assumptions about the optical properties of the particles. All methods of particle size analysis are to some extent influenced by variations in grain shape (e.g. Komar & Cui, 1984; Kennedy *et al.*, 1985), and the results of some analysis methods are also influenced by grain density and optical properties. For this reason, the results obtained using different methods may not be directly comparable, and it can be difficult to assimilate grain size distribution data obtained using more than one method.

Among the most widely used concepts of particle size in sedimentology and related disciplines are the *maximum* and *intermediate caliper diameters*, which are usually determined by direct measurement, *sieve diameter*, determined by dry or wet sieving, and *equivalent spherical diameter*, determined by settling

tube or electro-optical methods. Measured in one of
these ways, the maximum diameter of natural sediment particles ranges from more than 1 m to less than
0.1 μm. Most sedimentologists have adopted the
Udden–Wentworth grade scale (Udden, 1914; Wentworth, 1922), or a modified version of it, which
identifies a series of grain size classes in which the
boundaries between successive size classes differ by a
factor of two (Table 1.1). Although the original
Udden–Wentworth scheme placed the boundary between silt and clay at 0.004 mm (4 μm), most workers
now accept the convention which places the boundary at 0.002 mm (2 μm).

In order to facilitate graphical presentation and
statistical manipulation of grain size frequency data,
Krumbein (1934) proposed that the grade scale
boundaries should be logarithmically transformed
into phi (φ) values, using the expression:

$$\phi = -\log_2 d, \qquad (1.1)$$

where d is the diameter in millimetres.

Since phi units are dimensionless, it is strictly more
correct to state that:

$$\phi = -\log_2 (d/d_o), \qquad (1.2)$$

where d_o is the standard grain size of 1 mm (Mc-
Manus, 1963).

With the development of computerized methods of
data processing and analysis, some of the original
advantages associated with use of the phi notation no
longer apply, particularly where sediments contain
only particles smaller than 1 mm. In such cases, size
data are usually expressed in millimetres or micrometres (μm). The equivalent phi, mm and μm values of
the principal sediment size class boundaries are
shown in Table 1.1.

Following Krumbein's (1934) observation that the
cumulative frequency distributions of many sediments approximate a log-normal distribution, many
sedimentologists have conventionally based their interpretations on comparisons between actual distributions and the ideal log-normal distribution (e.g.
Folk & Ward, 1957; Visher, 1969; Middleton, 1976).
Statistical parameters that describe the mean size,
sorting, skewness and kurtosis of the distribution in
comparison with the log-normal model have been
either derived graphically or computed numerically.
Moment measures (Friedman, 1961) have generally
been regarded as more sensitive than graphical parameters (Folk & Ward, 1957) on account of the fact
that they incorporate information from the whole of

the distribution rather than that part of it which lies
between the 5th and 95th percentiles.

A minority of workers (e.g. Bagnold, 1937; Bagnold & Barndorff-Nielsen, 1980) has maintained that
additional useful information can be gained from the
tails of the particle size distribution if both the grain
size and grain frequency scales are logarithmically
transformed (Bagnold used logarithms to base 10).
When the data are plotted on a log–log diagram, the
resulting frequency distribution plots as a hyper-
bola. A number of different parameters have been
proposed to characterize such log-hyperbolic distributions (Bagnold & Barndorff-Nielsen, 1980;
Barndorff-Nielsen et al., 1982; Barndorff-Nielsen &
Christiansen, 1988), which it has been suggested
yield more environmentally sensitive information
than conventional log-normal parameters (Christiansen, 1984; Hartmann & Christiansen, 1988,
1992; Barndorff-Nielsen & Christiansen, 1988; Hartmann, 1991). However, a majority of sedimentologists has not yet been convinced that the additional
information gained provides any significantly greater
insight into the processes involved (e.g. Wyrwoll &
Smith, 1985, 1988). Both the log-normal and the
log-hyperbolic models suffer from an inability to
model bimodal or trimodal distributions which are
not uncommon in natural sediments.

1.2.2 Grain mass and density

The behaviour of a grain when acted on by a fluid is
often controlled as much by its mass as by its external
dimensions and shape. *Mass* represents a measure of
the inertia of a body; that is, the resistance that the
body offers to having its velocity or position changed
by an applied force. The mass of a body is constant
throughout space, whereas *weight* varies with gravity.
Mass (m) is related to weight (w) by the expression:

$$m = w/g, \qquad (1.3)$$

where g is the acceleration due to gravity.

For sedimentological purposes, the gravitational
force can be regarded as constant over the Earth's
surface.

In the case of spherical particles, mass varies as the
third power of the radius. Consequently a 10-mm
diameter sphere is five times larger than a 2-mm
diameter sphere in terms of diameter, but 125 times
larger in terms of mass. Other things being equal, the
shear stress required to initiate movement of the
10-mm diameter particle should therefore be 125

Table 1.1 The Udden–Wentworth grain size scale, with class terminology modifications proposed by Friedman and Sanders (1978)

mm	μm	phi	Sediment size class terminology of Wentworth (1922)	Sediment size class terminology of Friedman and Sanders (1978)	
2048		−11		Very large boulders	
1024		−10		Large boulders	
512		−9	Cobbles	Medium boulders	
256		−8		Small boulders	
128		−7		Large cobbles	
64		−6		Small cobbles	Gravels
32		−5		Very coarse pebbles	
16		−4	Pebbles	Coarse pebbles	
8		−3		Medium pebbles	
4		−2		Fine pebbles	
2	2000	−1	Granules	Very fine pebbles	
1	1000	0	Very coarse sand	Very coarse sand	
0.5	500	1	Coarse sand	Coarse sand	
0.25	250	2	Medium sand	Medium sand	Sand
0.125	125	3	Fine sand	Fine sand	
0.063	63	4	Very fine sand	Very fine sand	
0.031	31	5		Very coarse silt	
0.016	16	6		Coarse silt	
0.008	8	7	Silt	Medium silt	Silt
0.004	4	8		Fine silt	
0.002	2	9		Very fine silt	
			Clay	Clay	Clay

Table 1.2 The densities of some common minerals found in sediments

Mineral	Composition	Density (kg m^{-3})
Light minerals		
Quartz	SiO_2	2650
Albite	$NaAlSi_3O_8$	2620
Labradorite	$(Ca,Na)(Al,Si)AlSi_2O_8$	2700
Anorthite	$CaAl_2Si_2O_8$	2750
Orthoclase	$KAlSi_3O_8$	2560
Microcline	$KAlSi_3O_8$	2560
Calcite	$CaCO_3$	2710
Aragonite	$CaCO_3$	2930
Dolomite	$CaMg(CO_3)_2$	2870
Gypsum	$CaSO_4.2H_2O$	2320
Halite	$NaCl$	2160
Anhydrite	$CaSO_4$	2890–2980
Heavy minerals		
Pyroxenes	$(Ca,Mg,Fe)_2(Si,Al)_2O_6$	3200–3550
Hornblende	$NaCa_2(Mg,Fe,Al)_5(Si,Al)_8O_{22}(OH)_2$	3000–3470
Garnet	$(Fe,Al,Mg,Mn,Ca)_5(SiO_4)_3$	3560–4320
Epidote	$Ca_2(Al,Fe)_3O_{12}(OH)$	3250–3500
Olivine	$(Mg,Fe)_2SiO_4$	3210–4390
Staurolite	$FeAl_4Si_2O_{10}(OH)_2$	3700
Kyanite	Al_2SiO_5	3690
Andalusite	Al_2SiO_5	3160–3200
Sillimanite	Al_2SiO_5	3230–3270
Zircon	$ZrSiO_4$	4670
Rutile	TiO_2	4250
Anatase	TiO_2	3900
Apatite	$Ca_5(PO_4)_3(F,Cl,OH)$	3100–3250
Tourmaline	$Na(Mg,Fe)_3Al_6(BO_3)_3(Si_6O_{18})(OH)_4$	3030–3100
Monazite	$(Ce,La,Y,Th)PO_4$	5270
Clay minerals and micas		
Muscovite	$KAl_2(AlSi_3O_{10})(OH)_2$	2800–2900
Biotite	$K(Mg,Fe)_3(AlSi_3O_{10})(OH)_2$	2800–3400
Chlorite	$(Mg,Fe,Al)_6(Al,Si)_4O_{10}(OH)_8$	2600–3300
Kaolinite	$Al_4SI_2O_5(OH)_4$	2600–2630
Illite	$KAl_2(Al,Si_3O_{10})(OH)_2$	2600–2700
Palygorskite	$(Mg,Al)_5(Si,Al)_8O_{20}.4H_2O(OH)_2$	2200–2360
Montmorillonite	$Na(Al_3Mg)(Si_8O_{20})(OH)_4.H_2O$	2000–2300
Ice	H_2O	920

times greater than that required to move the 2-mm diameter particle.

Density represents the mass per unit volume of a substance, expressed in kg m^{-3}. Mass is therefore equal to the volume of a particle multiplied by its density, while volume is equal to the mass divided by the density. *Relative density* is a dimensionless number which indicates the relationship between the density of a unit volume of a material relative to that of an equal volume of water at a temperature of 4°C (relative density value of 1). The densities of some solid materials commonly found in sediments are given in Table 1.2.

1.2.3 Effects of grain size and density on settling velocity and selective transport

The hydraulic behaviour of a sediment particle is significantly influenced by its *settling* or *fall velocity*. This, in turn, is dependent on the size, shape and density of the individual particle, on the concentration of particles in the fluid, on the fluid viscosity and on the turbulence intensity. For a particle to settle at all, it must have a higher mean density than the fluid in which it is immersed. When first released into the fluid the particle will accelerate rapidly until it attains a constant velocity, termed the *terminal fall velocity*, at which point the viscous resistance force opposing the settlement of the sphere through the fluid just equals the net downward force (the weight of the particle minus the buoyant force of the fluid). The rate of settling of quartz spheres smaller than approximately 50 µm diameter is accurately predicted by Stokes' Law (Stokes, 1851; see Chapter 2).

Other things being equal, a spherical particle with a density of >2.65 kg m^{-3} may be expected to have the same settling velocity as a larger quartz sphere; that is, the two differently sized particles will demonstrate *hydraulic equivalence* (Rubey, 1933a,b). However, mineral grains that are hydraulically equivalent in terms of settling velocity often show differential behaviour during entrainment, transport and deposition (Lowright *et al.*, 1972; Slingerland, 1977; Steidtmann, 1982; Trask & Hand, 1985).

As discussed further in Chapter 2, several theoretical and experimental studies have demonstrated that the magnitude of the critical fluid drag velocity and bed stress required to initiate the movement of particles is dependent, amongst other factors, on the size and density of the particles (e.g. Bagnold, 1941,

1979; Miller *et al.*, 1977). In practice, the magnitude of the critical stresses required to entrain a particle depends not only on the characteristics of the particle itself, and on the properties of the fluid in question, but also on the bulk sediment properties (especially the particle size distribution, packing arrangement and effectiveness of interparticle cohesion.

A grain of a given size at rest on a bed composed only of finer grains requires a significantly lower critical velocity for entrainment than an equivalent grain resting on a bed of similar sized grains. Conversely, a small particle sitting on a coarser bed will require a higher critical velocity for entrainment than one that sits on a bed of particles of the same size (Komar, 1987). Figure 1.1 shows three different sizes of sphere resting on a bed of uniformly sized spheres. The effect of the combined fluid drag and lift forces is to rotate the exposed particles on the surface of the bed about their forward point of contact with the underlying grains (the *pivot, p*). The angle φ, referred to as the *pivot angle*, reflects the ease with which a grain may be rotated about the pivot point. The magnitude of the pivot angle is inversely related to the grain diameter (Komar & Li, 1986, 1988), implying that grains that are larger than the mean diameter of the bed are more easily entrained than particles that are smaller than the mean diameter of the bed. As a result, in a bed that contains a wide range of grain sizes, selective entrainment may take place, with preferential forward transport of the larger

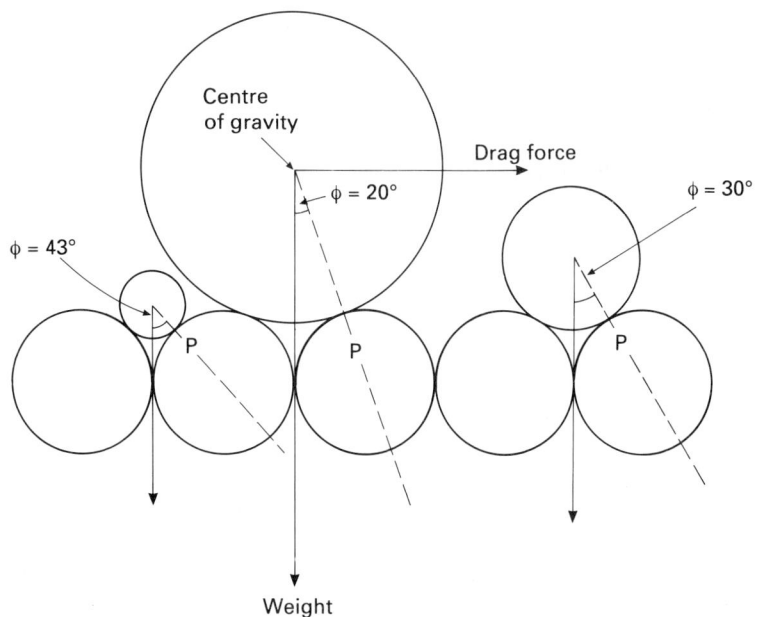

Fig. 1.1 Schematic two-dimensional diagram showing the dependence of the pivot angle (φ) on the relationship between the size of a spherical grain and the size of the grains on which it rests. P is the pivot point.

grains while the smallest particles, which to some extent are 'sheltered' by the larger grains, remain as a lag. Such selective entrainment of larger, lower density grains has been suggested to be a major process involved in the formation of some heavy mineral placer deposits (Slingerland, 1977, 1984; Komar & Wang, 1984; Trask & Hand, 1985; Slingerland & Smith, 1986; Komar, 1989; Frihy & Komar, 1991; Li & Komar, 1992a,b).

Field and experimental data suggest that selective entrainment and subsequent transport of larger grains are only likely to occur if the median grain size of the sediment lies in the medium to fine sand range. In the case of coarser sediments, where the median size lies in the range of medium sand to gravel, selective entrainment of the finer sizes is frequently observed (Komar, 1987). The precise nature of the hydraulic sorting in any given situation will reflect the local balance between the grain size characteristics of the bed (especially size distribution and sorting), and the energy regime associated with the operative transport process. For this reason, a spatial sequence of sediments in the direction of transport may become either coarser and better sorted or finer and better sorted (see McLaren & Bowles, 1985, 1991; Gao & Collins, 1991, for discussion). The skewness of the distribution can also become either more positive or more negative in the direction of transport, depending on the balance of forces.

1.3 Grain shape

1.3.1 Definitions of grain shape

The terms grain 'shape' and grain 'form' have been used in different senses by different authors (see, e.g., Whalley, 1972; Barrett, 1980; Winkelmolen, 1982; Willetts & Rice, 1983), but as defined here *shape* refers to all aspects of the external morphology of a particle, including the gross *form* (equidimensionality or sphericity), the roundness (sharpness of edges and corners), and the surface texture (small-scale roughness or smoothness).

Some grain shapes can be described qualitatively in terms of resemblance to readily recognizable geometric shapes or organic analogues, such as spheres, spheroids, ellipsoids, discoids, blades and rods (Fig. 1.2). However, such descriptive terms are subjective and are of little assistance when grains have no clearly identifiable form. Quantitative description and statistical comparison of the shapes of grain populations can only be achieved by the use of numerical shape parameters.

1.3.2 Grain form

Form indices provide a measure of the equidimensionality of a grain and have traditionally been obtained in one of two main ways: (1) by measuring the orthogonal long (L), intermediate (I) and short (S) axes of a grain and by calculating their ratios, or (2) by assessing the degree of deviation from a geometrical standard form, such as a sphere or ellipsoid.

Wadell (1933) first defined grain sphericity, ψ, as:

$$\psi = s/S, \qquad (1.4)$$

where s is the surface area of a sphere of the same volume of the particle and S is the actual surface area of the particle. However, the surface area of most sediment particles is difficult to measure directly, and Wadell's shape factor has not proved satisfactory for predicting the settling behaviour of grains. Consequently many workers have preferred to use shape factors based on the ratios of the axial dimensions, parameters that are more easily determined.

The ratio of the length of the long axes, L, to the intermediate axis I, provides a measure of particle elongation, while the ratios S/L and S/I describe the degree of particle flattening. Corey (1949) proposed a shape factor (CSF), widely used in experimental studies, which reflects the particle flatness:

$$CSF = S/(LI)^{1/2}. \qquad (1.5)$$

Numerical values of the formula range from 0 for a perfectly flat disc to 1 for a perfect sphere.

For analysis of two-dimensional grain outlines, the maximum projection sphericity, ψ_p, can be determined (Sneed & Folk, 1958):

$$\psi_p = (S^2/LI)^{1/3}. \qquad (1.6)$$

Sneed & Folk argued that this two-dimensional measure of sphericity is of direct relevance to the hydraulic behaviour of grains, since most grains settling in still water tend to orientate themselves with the maximum projected area normal to the direction of movement. It is not actually necessary to calculate values of ψ_p, since the ratios of the orthogonal axis dimensions can simply be plotted on a ternary diagram (Sneed & Folk, 1958; Hockey, 1970). Although some authors have advocated the use of advanced statistical techniques to extract the maximum environmentally sensitive information from grain-form

Fig. 1.2 Four pebbles of contrasting morphology: (a) oblate spheroid; (b) prolate spheroid; (c) highly prolate spheroid; (d) elongate, sub-angular pebble. (a), (b) and (c) are composed of metamorphosed mudrock; (d) is vein quartz. All were collected from a marine gravel beach at Towyn, west Wales.

data (e.g. Illenberger, 1991), in many instances sufficient information can be gained by plotting the axial ratios or basic formulae derived from them (Benn & Ballantyne, 1992).

1.3.3 Grain roundness

Wadell (1933) proposed a measure of grain roundness which was based on measurements of the radius of curvature of individual grain corners. However, owing to the time-consuming nature of such measurements, a majority of workers has preferred to estimate grain roundness by reference to a two-dimensional visual comparator such as that published by Powers (1953). Numerical values can be assigned to each Powers roundness class using Wadell's (1933) formula and the rho-scale conversion

of Folk (1955) (Table 1.3). However, this procedure is subject to a degree of imprecision and operator variation, and may not provide the sensitivity necessary to discriminate between the more subtle effects of shape on grain settling velocity and transport behaviour (e.g. Baba & Komar, 1981; Goldbery & Richardson, 1989).

Natural sediment grains vary greatly in shape depending on their composition, provenance and transport history (Figs 1.3, 1.4, 1.5, & 1.6). First cycle weathering debris, and sediment grains that have been transported only short distances from their source areas, are typically angular to sub-angular, whereas grains which have experienced a longer transport history, perhaps involving more than one cycle of erosion and deposition, show a much higher degree of rounding. The rate of rounding by abrasion

Table 1.3 (a) Numerical values for different Powers' (1953) roundness classes according to Wadell's (1933) formula and the rho-scale transformation proposed by Folk (1955). (b) grain sphericity and roundness classes recognized in the Power's (1953) visual comparator.

(a)

Powers' roundness class name	Corresponding Wadell (1933) class intervals	Corresponding values of Folk's rho scale (Folk, 1955)
Very angular	0.12–0.17	0–1.0
Angular	0.17–0.25	1.0–2.0
Sub-angular	0.25–0.35	2.0–3.0
Sub-rounded	0.35–0.49	3.0–4.0
Rounded	0.49–0.70	4.0–5.0
Well-rounded	0.70–1.00	5.0–6.0

(b)

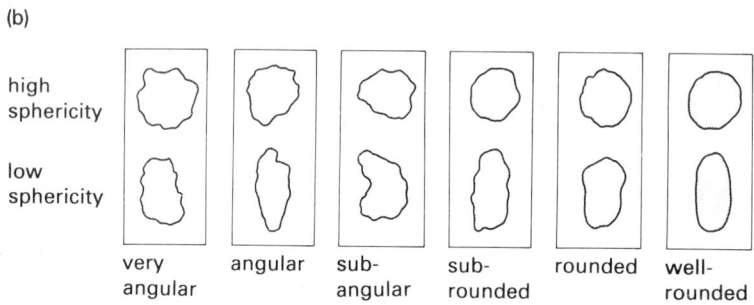

Fig. (b): high sphericity / low sphericity rows across roundness classes: very angular, angular, sub-angular, sub-rounded, rounded, well-rounded.

(a)

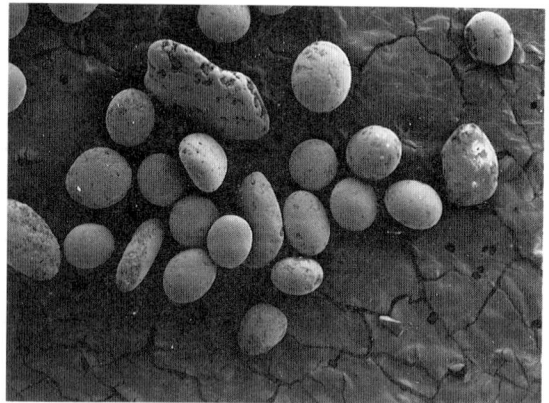

(b)

Fig. 1.3 Scanning electron micrographs of: (a) a mixture or angular, sub-angular and sub-rounded sand grains from the Stovepipe Wells dunefield, Death Valley, California (picture width 1200 μm); and (b) a mixture of spherical carbonate ooids and more irregular but well-rounded skeletal fragments, Andros Island, Bahamas (picture width = 500 μm).

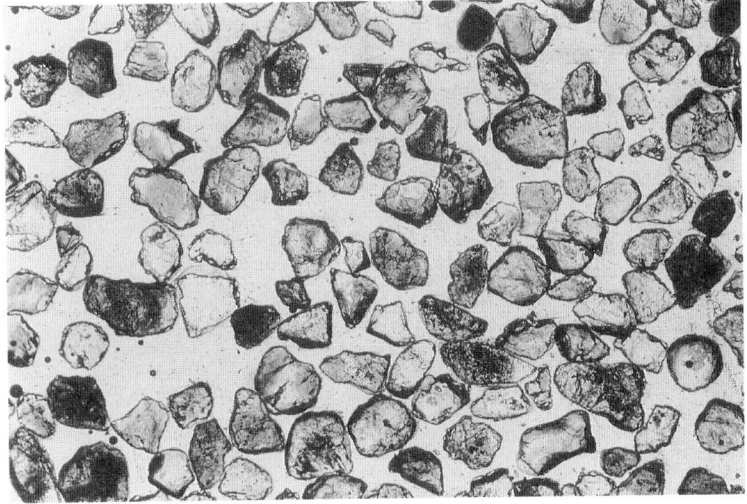

Fig. 1.4 Optical micrograph showing mixture of sub-rounded and sub-angular quartz grains from a marine beach deposit, North Queensland. The semi-opaque grains have a surface coating of iron oxyhydroxide. Picture width = 2.262 mm.

Fig. 1.5 Scanning electron micrograph showing needle-shaped crystals of authigenic aragonite from a sub-tropical carbonate mud. Picture width = 29.2 μm.

during transport is to a large extent governed by the hardness of the minerals concerned (Fig. 1.7)

1.3.4. Grain surface texture

In simple terms, the surface texture of a grain can be regarded as the degree of surface roughness or smoothness (i.e. degree of microrelief development). Under favourable circumstances, specific surface textural features may provide clues to the provenance, transport history and postdepositional alteration of the grains (e.g. Figs 1.8, & 1.9). Some surface textural features, such as polish or frosting, can be seen with the naked eye or with the aid of a hand lens, but others can only be identified using a binocular microscope or scanning electron microscope. Polish is related to the quality of light reflection. Grains that have been gently abraded, or that possess a surface coating of, for example, secondary silica, typically possess a smooth surface and an apparently high degree of polish (e.g. Folk, 1978). Frosting, on the other hand, is related to the scattering of light due to the presence of closely spaced surface irregularities which may be formed by violent abrasion or by

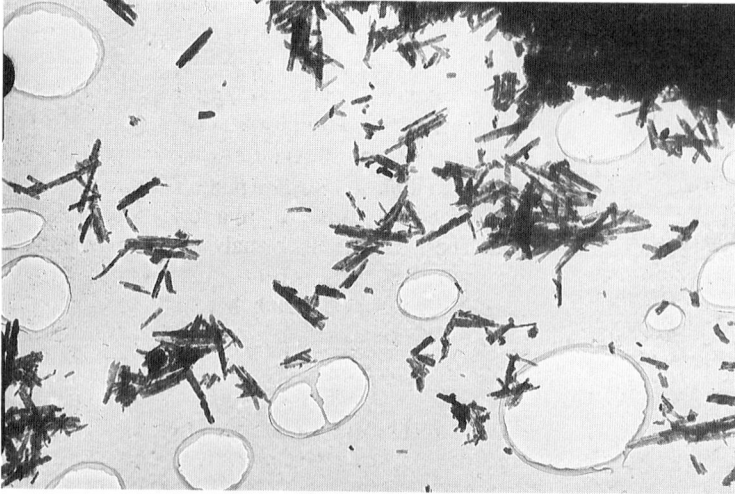

Fig. 1.6 Transmission electron micrograph of small iron oxide (maghaemite) particles from a tropical soil. Picture width = 4.8 μm.

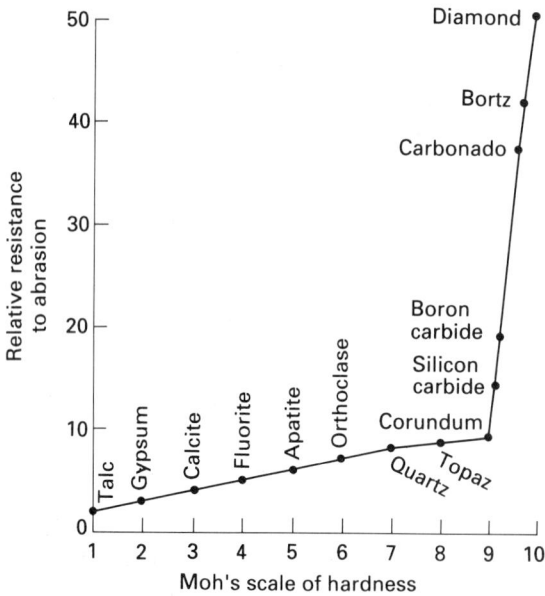

Fig. 1.7 Relationship between Moh's scale of hardness and relative resistance to abrasion of different minerals. Modified after Dana and Harlbut (1959).

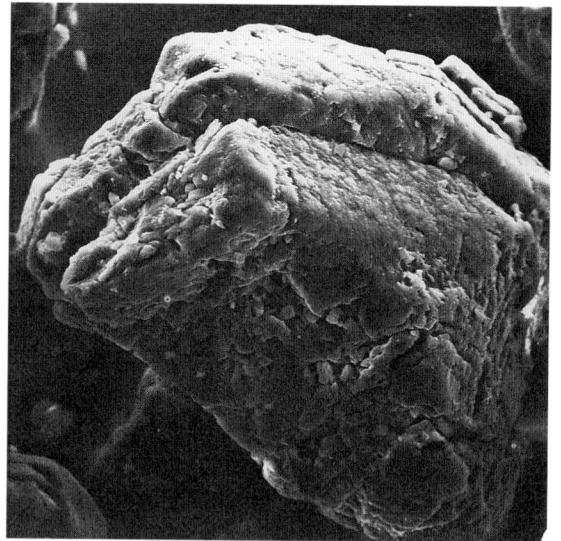

Fig. 1.8 Scanning electron micrograph of a highly weathered quartz grain from a podzolic soil A horizon, North Queensland, showing small-scale angularity and 'rough' surface texture produced by silica dissolution and break-up of the surface. Picture width = 162.5 μm.

chemical etching (cf. Kuenen & Perdok, 1962). Detailed analysis of the nature of the surface texture usually requires scanning electron microscopic examination in conjunction with X-ray microanalysis (e.g. Krinsley & Doornkamp, 1973; Culver *et al.*, 1983; Elzenga *et al.*, 1987).

1.3.5 Holistic characterization of grain shape

With the development of automatic image analysers and associated computer software, it has become easier to analyse the shape characteristics of large numbers of grains, at least in the sand and coarse- to

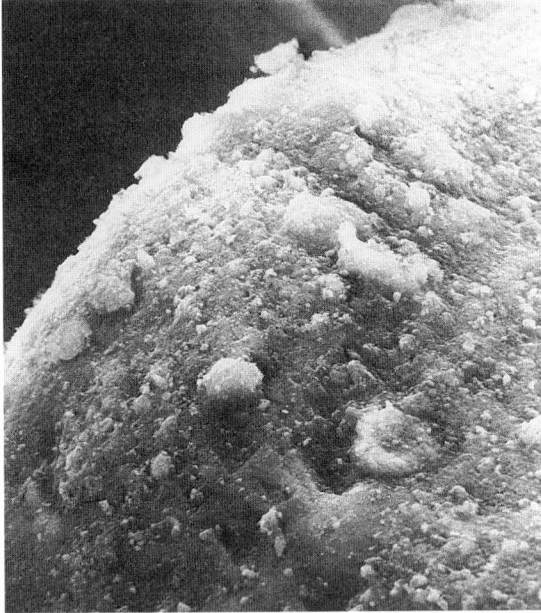

Fig. 1.9 Scanning electron micrograph of a sand grain from a podzolic B horizon, North Queensland, showing 'rough' surface texture produced by secondary silica precipitation. Picture width = 19.5 µm.

medium-silt size ranges. The large size of pebbles and cobbles makes it easier to measure their axial dimensions by hand, although photographic images of such particles can be processed automatically. Fine silt

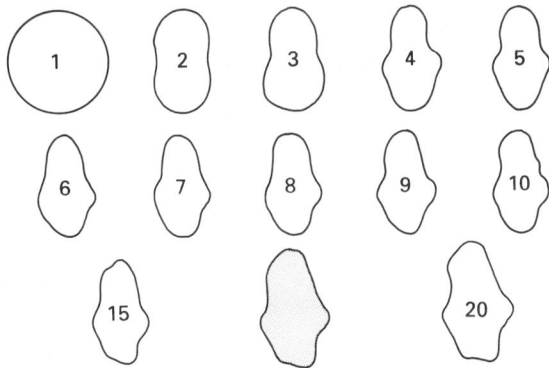

Fig. 1.10 Regeneration of a natural grain shape (stippled) by addition of successive harmonics computed by the Fourier series. The lower order harmonics reflect the gross morphology of the particle and the higher order harmonics add increasingly finer detail. After Ehrlich *et al.* (1980).

and clay particles need to be carefully dispersed and photographed using an electron microscope before image analysis can be undertaken properly.

Most automated shape analyses carried out to date have been performed on two-dimensional projected 'silhouette' images of sand or coarse-silt grains scattered on an illuminated glass plate. The grain outlines are digitized and converted to a series of x–y coordinates. These data are then analysed using techniques such as Fourier analysis (Ehrlich & Weinberg, 1970; Ehrlich *et al.*, 1974, 1980; Mazzullo *et al.*, 1986, 1992) and fractal analysis (Mandelbrot, 1967, 1977; Orford & Whalley, 1983, 1991; Clark, 1987).

In Fourier analysis, the maximum projected grain profile is compartmentalized into a series of standard shape components (harmonics) that converge to reproduce the natural grain shape (Fig. 1.10). The grain perimeter, $R(\theta)$, is expressed as a Fourier series expansion of the grain radius as a function of the polar angle about the centre of gravity of the grain:

$$R(\theta) = R_o + R_n \cos(n\theta - \theta_n), \qquad (1.7)$$

where R_n is the harmonic amplitude, θ is the polar angle, R_o the grain radius, n the harmonic number, and θ_n is the phase angle.

The lower order harmonics (1–5) reflect the broad form characteristics of the grain, while the higher order harmonics (usually up to 23) provide information about grain roundness and, to a limited extent, surface texture. Data may be presented graphically by plotting the frequency of occurrences as a function of each harmonic amplitude. The interval boundaries in these shape–frequency histograms are then defined by the maximum entropy concept, and the most informative harmonics identified by relative entropy analysis (Full *et al.*, 1984).

In routine Fourier analysis, 200 grains are typically taken from a standard grain size fraction which has been pretreated to remove carbonates, clay and organic coatings. For accurate results, analysis should be restricted to a single mineral (most commonly quartz).

Although the results of two-dimensional Fourier grain-shape analysis are to some extent influenced by preferred grain orientation (Tillman, 1973), the technique has been applied successfully in a wide range of provenance and sediment transport studies (e.g. Mazzullo *et al.*, 1983, 1986; Kennedy & Ehrlich, 1985; Dowdeswell *et al.*, 1985; Haines & Mazzullo, 1988). However, difficulties are encountered when grain outlines show a very high degree of irregularity, for example in the case of highly weathered or diagenetically altered grains which are characterized by well-developed re-entrants. In these circumstances, fractal analysis may provide a more suitable alternative (Orford & Whalley, 1983, 1991; Kennedy & Lin, 1991).

1.3.6 Effects of grain shape on settling velocity and selective transport

Several laboratory experimental studies have demonstrated that grain shape has a significant influence on the hydraulic behaviour of sediment grains (e.g. Lane, 1938; McNown & Malaika, 1950; Allen, 1969; Carrigy, 1970; Komar & Reimers, 1978; Hallermeier, 1981; Willetts, 1983; Willetts & Rice, 1983; Li & Komar, 1992a,b; Cui et al., 1983). Generally, the greater the departure of a grain from a spherical shape, the greater is the reduction in its settling velocity and the more irregular is its motion during settling. Highly angular particles such as volcanic glass shards may show markedly retarded settling behaviour compared with more rounded and blocky grains (Wilson & Huang, 1979). However, grains of the same shape but different size (and Reynolds number) may show quite different settling behaviour (see Allen, 1985, pp. 46–53, for detailed discussion). The effect of shape variations on settling velocities diminishes with decreasing grain size, since the differences between the drag on spheres and other shapes is progressively reduced with decreasing Reynolds number (Fig. 1.11 & 1.12; see also Chapter 2).

Experimental studies have shown that particle flatness, as measured by the Corey shape factor, is an important grain form characteristic that has a significant influence on settling rates right across the particle size range, although the effect is relatively greater for larger grains (Fig. 1.13). The more particles are flattened, the slower they will settle compared with a sphere of the same weight and density. This can be explained partly by the large cross-sectional area of discoid particles relative to their volume, and hence the higher flow resistance. Additionally, the sharply curved edges of discoid particles produce

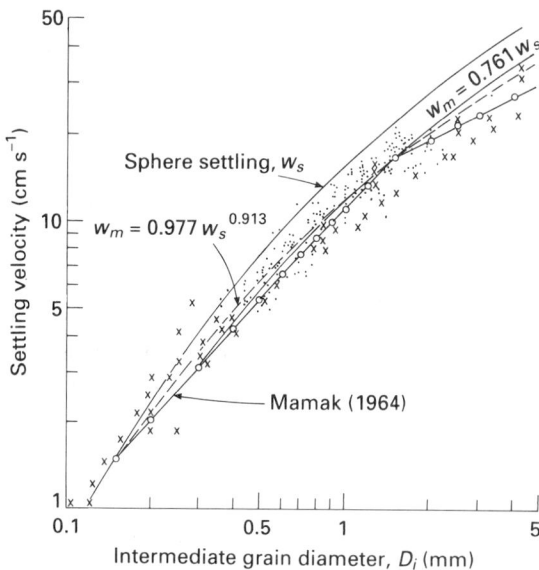

Fig. 1.11 The measured settling velocities of natural sand grains in water as a function of intermediate grain diameter (D_i) (after Baba & Komar, 1981). The upper solid curve represents the predicted settling velocity of quartz spheres (w_s). Also shown are the data of Lane (1938) (x) and Mamak (1964) (o).

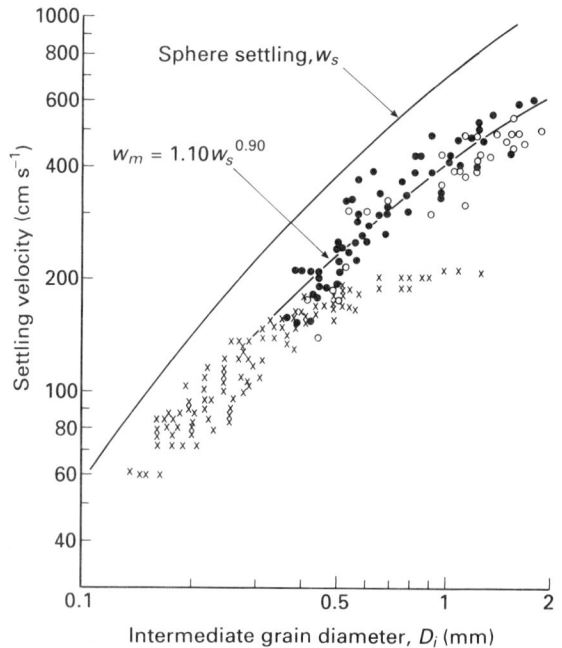

Fig. 1.12 The measured settling velocities of natural quartz sand grains in air (w_m) as a function of intermediate grain diameter (D_i). The open circles denote the measurements obtained in a stairwell while the solid dots represent data obtained using a 1120-cm settling tube. Also shown are the experimental data (x) of Bagnold (1941). After Cui et al. (1983).

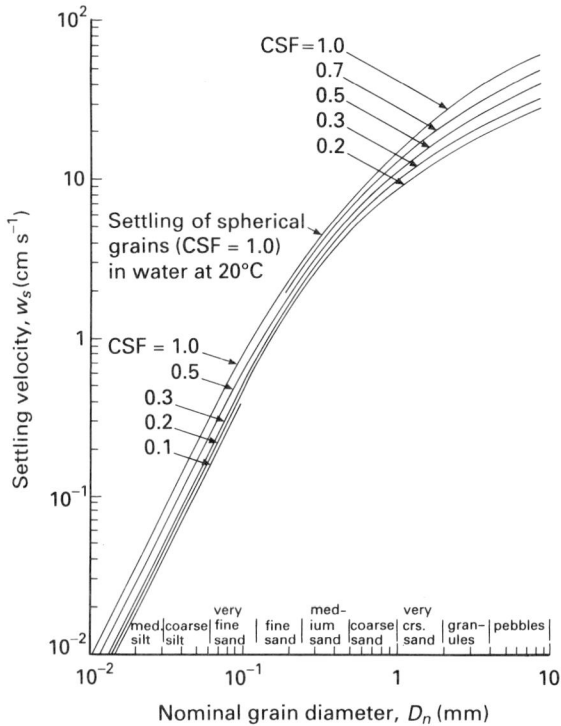

Fig. 1.13 Settling velocities of quartz density non-spherical grains in water as a function of their nominal diameter (D_n) and Corey shape factor (CSF). After Komar and Reimers (1978).

flow separation at much lower Reynolds numbers than is the case for spherical particles. Finally, marked flattening can induce instabilities in the settling of a particle, leading to possible rotation, tumbling and oscillating motion which will reduce the overall settling rate (Stringham *et al.*, 1969; Allen, 1985).

Particles with even more irregular shapes, such as the shells of bivalve molluscs, show widely varied settling behaviour, although a distinction between unsteady and steady settling behaviour can be drawn using a simple form index based on the long and intermediate axial ratios (Allen, 1984a,b).

Roundness and surface texture variations have less effect on settling velocity than does overall grain form (Williams, 1966; Baba & Komar, 1981), although their effects are measurable under laboratory experimental conditions (Goldbery & Richardson, 1989). Surface irregularities, including sharp edges and corners, will encourage unstable settling behaviour, localized flow separation and eddy generation. Al-

though small-scale surface roughness appears to have a relatively minor effect on settling behaviour (Williams, 1966), it can have a significant influence on such properties as angle of internal friction and propensity to avalanching (Allen, 1969, 1985).

The combined effect of grain flattening and roundness was investigated by Goossens (1989), based on Deitrich's (1982) compilation of the experimental data obtained by 18 previous authors who characterized their grains using the Corey shape factor and the Folk–Powers roundness index. The results show that the combined effect of flattening and roundness is greater than that of either characteristic in isolation, and that the nature of the effects is broadly similar in both air and water. For settling of small particles (<200 μm) in air, the relative contribution of particle rounding to the combined effect of combined flattening and rounding is apparently small, even for very slightly flattened grains. For grains larger than 500 μm, the effect of rounding assumes greater significance but is still secondary to particle flattening.

The roundness of particles is perhaps of greater significance in controlling the entrainment and transport of grains. For example, a single spherical (perfectly rounded) grain on a flat surface is much more easily entrained and kept in motion by a fluid than a highly angular particle of the same weight. This is embraced in Winkelmolen's (1971) concept of 'rollability', a behavioural property that can be measured directly as the time taken for grains to travel down the length of a rotating inclined cylinder.

Experimental results obtained by Komar & Li (1986, 1988) showed that angular particles of crushed gravel have larger pivoting angles (for uniform-sized grains) than either spherical particles or ellipsoids of the same size. The difference in pivoting angle between the angular and more rounded particles was found to become larger with increasing grain size (Fig. 1.14).

Once entrained, shape also has a significant effect on transport rate. For example, Willetts (1983) determined in wind tunnel studies that low sphericity (the maximum projection sphericity of Sneed & Folk, 1958) depresses the transport rate, apparently because such grains are dislodged less frequently and are transported by the wind less efficiently than more spherical grains. Field studies have also shown that more spherical, and to some extent more rounded, grains are preferentially transported during aeolian saltation (e.g. Mazzullo *et al.*, 1986).

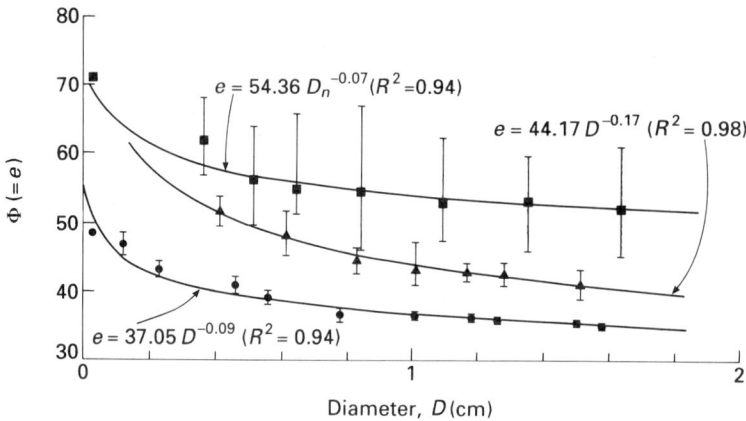

Fig. 1.14 Pivoting angles (ϕ) for uniform-size grains as a function of grain diameter (D) and grain shape: ●, spheres; ▲, ellipsoids; ■, angular. (After Li & Komar, 1986). ϕ is equal to coefficient e for uniform size grains (i.e. $D/K = 1$), where D is the diameter of the pivoting grain and K is the diameter of the base grains.

1.4   Tendency for grain aggregation and flocculation

1.4.1   Electrostatic charging and grain aggregation in air

Sedimentary particles of silt and clay size sometimes show a natural tendency to form aggregates owing to the build-up of electrostatic or electrochemical charges on their surfaces. Airborne particles can become charged in a number of ways (Greeley, 1979), including: (1) contact electrification, resulting from different materials coming into contact or from contact of the same materials that have different surface properties, (2) frictional electrification, resulting from the rubbing of one surface over another, (3) piezoelectric charge build-up, resulting from pressure, (4) cleavage electrification, resulting from break-up of certain minerals, (5) electrification resulting from freezing and thawing of water, and (6) photoelectric charging. Charging of particles during sand and dust storms is well documented (Gill, 1948; Jordan, 1954; Stow, 1969; Kamra, 1972), and is apparently a viable mechanism for the formation of suspended dust aggregates (Greeley & Leach, 1978; Marshall et al., 1981). Grain aggregation due to electrostatic forces has also been suggested to be important in the sedimentation of volcaniclastic particles (Brazier et al., 1983).

1.4.2   Particle flocculation in water

Aggregation of particles in water, or flocculation, is also an important process (Powers, 1957; Kranck, 1973, 1975; Gibbs, 1977, 1983). In the case of small particles, particularly platy minerals such as clays which have a large surface area relative to their volume, the interparticle attractive forces are relatively large in comparison with the gravitational forces causing the particles to settle. However, the attractive forces are significant over very short distances and particles must be brought into contact physically before they can have an effect. Processes that may bring suspended particles into contact include Brownian motion (random movement of the molecules of the fluid; effective only for sub-micron sized particles), fluid turbulence, and differential settling of the particles. Once in contact, the magnitude of the attractive forces depends on the mineralogy of the particles, in particular the charge-state of their surfaces, on the electrochemical nature of the suspending medium, and on the presence of any organic material.

The surfaces of clay particles are usually negatively charged due to two main factors: (1) vacancies or unbalanced ionic substitutions in the crystal lattice, and (2) the adsorption of ions where there are broken bonds, particularly on the faces (rather than the edges) of the particles. This net negative charge is generally balanced by a layer of positive ions (counter ions) in the immediately surrounding electrolyte. The negative charges on the particle surface and the adjacent zone of counter positive ions together form an *electrical double layer*. The concentration of counter ions decreases exponentially with distance outward from the particle surface. Since clay mineral particles of any one type have a similar electrical double-layer, they repel one another when brought sufficiently close together, provided that the thickness of the layer of positively charged ions is greater than the distance over which the interparticle attractive (principally van der Waals) forces are effective. In dilute suspensions (e.g. fresh water), the ion cloud is generally large and the repulsive forces operate over too large a distance for the attractive forces to have any effect. Consequently if grains are brought close together they immediately repel one another. However, at high electrolyte concentrations (e.g. sea water), the thickness of the neutralizing cloud of positive counter ions is much reduced, allowing particles to come sufficiently close together that van der Waals attractive forces become operative and the positively charged edges of some clay particles are attracted to the negatively charged surfaces of others. The electrical double layer can also be compressed if higher valence cations are exchanged for those already in the counter layer. The flocculating power of bivalent ions is approximately 20–80 times that of univalent ions and that of trivalent ions is 10–100 times that of bivalent ions (Weaver, 1989).

A measure of the particle repulsion is provided by the *Zeta potential*, measured in millivolts, which is the electrical potential at the interface between the double layer and the surrounding electrolyte. If the Zeta potential is large (either positive or negative), particle repulsion will occur; as the ion concentration increases and the Zeta potential falls towards zero, flocculation takes place.

Flocculation is a process of major importance when fluvially transported clay minerals enter a higher salinity estuarine or coastal environment (see Chapter 6). Owing to differences in size and surface charge, different clay minerals show differing tendencies to flocculate at a given salinity. On theoretical grounds, illite should show the greatest tendency to flocculate, followed by kaolinite and smectite. However, in practice, the relationship is complicated by variations in temperature, turbulence and particle collision efficiency, pH and the presence of hydroxides and organic compounds that influence the flocculation process (Whitehouse *et al.*, 1960; Kranck, 1981).

The size distribution of flocculated sediment aggregates (flocs) must be determined in the field since removal of the sample of the laboratory causes changes in the size distribution (Krone, 1972; Bale & Morris, 1987). Laboratory determinations of median settling velocity of flocs can be up to an order of magnitude lower than those measured in the field.

The median settling velocity of fine-grained sediment is closely dependent on the suspended sediment concentration. At concentrations of between 2 and 10 kg m^{-3} the settling flocs interfere with the flow of surrounding fluid, a condition known as *hindered settling* (Krone, 1972; Burt & Stevenson, 1983; Delo & Ockendon, 1992). Under field conditions, hindered settling may begin at suspended sediment concentrations as low as 2 g l^{-1}. Field-determined median settling velocity of flocs typically increases with suspended sediment concentration up to approximately 10 kg m^{-3}, but declines again at higher concentrations (Delo & Ockendon, 1992).

1.5 Bulk sediment properties

The principal bulk sediment characteristics that influence entrainment and transport are grain size distribution, orientation and packing arrangement, porosity and moisture content. These characteristics, in turn, govern the properties of bulk density, cohesion, internal friction, elasticity, permeability and shear strength.

1.5.1 Grain size distribution

The range of particle sizes present in a sediment has a direct influence on the porosity, permeability and the degree of cohesion. In general, a poorly sorted sediment with a wide range of particle sizes will require a higher critical velocity for entrainment than a sediment of the same median size but containing a narrower range of grain sizes (i.e. better sorted). Poorly sorted sediments in which fine particles infill the voids between larger grains will have a lower porosity and permeability than well-sorted sediments with near-uniform grain size (Beard & Weyl, 1973). Measures of the spread of the particle size distribution include the interquartile range (75th percentile–25th percentile, measured in mm, μm or phi units), and the standard deviation of the distribution (Folk & Ward, 1957; Friedman, 1961).

1.5.2 Voids ratio and porosity

The *voids ratio*, e, is the ratio of volume of voids (V_v) to volume of solids (V_s):

$$e = V_v/V_s. \tag{1.8}$$

This is expressed as a pure number (normally a decimal, although values can exceed 1).

The *porosity* (n) of a sediment is defined as the ratio of the pore volume (V_v) to the bulk volume (V), expressed as a percentage:

$$n = V_v/V = V_v/(V_v + V_s). \qquad (1.9)$$

The relationship between voids ratio and porosity is given by:

$$n = e/(1 + e) \text{ and } e = n/(1 - n). \qquad (1.10)$$

1.5.3 Grain orientation and packing arrangement

The grain packing arrangement has a strong influence on the porosity and voids ratio, as well as the ease with which individual grains may be entrained by a fluid. Theoretically, spherical grains may display several different packing arrangements (Fig. 1.15). A cubic packing arrangement gives rise to the loosest packing and a maximum porosity value of 48%, while a rhombohedral packing arrangement gives the tightest packing with a corresponding porosity value of 26% (Graton & Fraser, 1935). In general, the tighter the packing arrangement, the greater is the shear stress required to entrain the surface grains.

1.5.4 Moisture content

Moisture content (w) is the mass of water which can be removed from the sediment or soil by heating at 105°C (usually for 24 h), expressed as a percentage of the dry mass:

$$w = \text{loss of moisture/dry mass} \times 100\%.$$

Values for sands and muds typically lie in the range 3–60% but may exceed 100% for very absorbent material such as peats.

1.5.5 Bulk density and dry density

Bulk density is defined as the mass of a sediment or soil, including water and air contained in the pore spaces, per unit volume. Measurements of bulk density are often made at natural water content. *Dry density* refers to the mass of solids per unit volume after drying at 105°C, while *saturated density* is the density of solids plus water that completely fills the voids in a unit volume. Dry density, in particular, often provides a useful guide to the erosion resistance of a sediment (especially cohesive sediment).

1.5.6 Cohesion

Cohesion is the intermolecular attractive force acting between two adjacent portions of a substance. The attractive force that operates between two dissimilar substances is called *adhesion*. The attractive forces of cohesion and adhesion operate only over very short distances and vary in magnitude with the substances concerned. The relative importance of cohesive forces varies with particle size, being relatively large for clay-size particles and of virtually no significance in the case of coarse silts and sands. For this reason, granular solids such as sands are referred to as *cohesionless* while sediments containing a significant proportion of clay are referred to as *cohesive*. In practice, the properties and behaviour of cohesive sediments are highly variable, reflecting the interplay of physical, chemical and biological factors.

1.5.7 Angle of internal friction

The forces opposing the motion of dry, cohesionless, granular solids are principally frictional. A measure of the ability of a granular solid to resist failure under an applied stress is provided by the *angle of internal friction*, otherwise known as the *angle of initial yield* (ϕ_i). This angle is governed by the grain size, shape, packing arrangement and by the surface texture of the grains (Rowe, 1962; Carrigy, 1970; Allen, 1970; 1985, p. 37). If a slope becomes steepened beyond the angle of internal friction, failure will take place, typically by avalanching. During this process the original packing arrangement is disturbed and there is an expansion of the flowing grain mass, a process

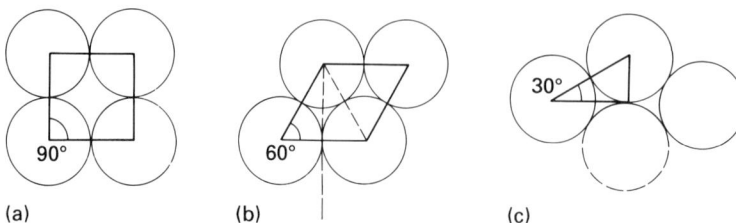

Fig. 1.15 Three alternative packing arrangements for uniformly sized spheres: (a) cubic packing; (b) orthorhombic packing; (c) tilted rhombic packing. Modified after Graton and Fraser (1935).

known as *dilation*, in which energy is expended
(Bagnold, 1956). When the grains finally come to rest
they assume a new angle, known as the *angle of repose*
(ϕ_r), which is typically several degrees smaller than
the angle of initial yield. In the case of sands, angles
of initial yield are typically 34–37° while angles of
repose are typically 32–34°. Angular gravel-sized
particles on talus and similar slopes may possess
higher angles of repose (typically up to 45°).

1.5.8 Deformability and elasticity

The deformability and elasticity characteristics of
sediments are particularly important for an understanding of their behaviour under conditions of
oscillating stress, as, for example, associated with
wave action on intertidal and sub-tidal muddy sediments (Hamilton, 1971, 1979). When a *stress* is
applied to a body, it usually responds by exhibiting
strain (change in unit length). The ratio of applied
stress to the resulting strain is defined by *Young's
modulus*. In general, strain increases in response to
increasing stress until the *yield stress* or *plastic limit* is
reached, at which point the material ruptures. A
more realistic picture of stress–strain behaviour is
provided by *Poisson's ratio*, which takes into account
the fact that during compression a material may
experience a decrease in length in one direction but
an increase in dimension perpendicular to the applied stress. Poisson's ratio is the non-dimensional
ratio of the strains in each of the two directions. In
practice, the stress–strain behaviour of natural materials is complex, being governed by moisture content,
temperature, pore fluid pressure and chemistry, rate
of load application and previous stress history.

1.5.9 Permeability

Permeability is the ability of a sediment, soil or rock
to discharge fluids under a hydraulic gradient. *Darcy's Law* relates volume of flow per unit time,
normally 1 s, (q) to the cross-sectional area (A) of a
sample, the coefficient of permeability (k) and the
hydraulic gradient (i):

$$q = Aki. \quad (1.11)$$

Permeability is normally measured directly in the
laboratory or in the field using a permeameter which
may be of either constant head or falling head type.
The *coefficient of permeability*, k, is then calculated
using the equation:

$$k = \eta q l / A(P_1 - P_o), \quad (1.12)$$

where η is the viscosity of the fluid; q is the volume of
fluid passing through the specimen in 1 s; l is the
length of the test specimen; A is the cross-sectional
area of the specimen perpendicular to the flow; P_1 is
the absolute pressure at the point of entrance to the
specimen (atmospheric pressure); and P_o is the absolute pressure at the point of exit from the specimen
(normally atmospheric pressure).

1.5.10 Shear strength

The fundamental relationship governing the dynamic
response of a material to an applied stress is given by
the *Mohr–Coulomb equation*:

$$\tau_f = C + \sigma \tan \phi_i, \quad (1.13)$$

where τ_f is the shear stress at failure, C is the cohesion
due to the presence of clays (in Pa), ϕ_i is the angle of
internal friction, and σ is the normal shear stress
(in Pa).

A wide variety of geotechnical methods is available
to help define the parameters necessary to solve
Eq. 1.13 for any given sediment or soil. One of the
most common methods is the determination of *shear
strength*, which allows determination of the cohesion
and angle of internal friction at a given stress level.

Several factors influence the shear strength of
undisturbed sediments under particular loading conditions, but the most important are the particle size
distribution, moisture content, and whether the sediment has previously been dried out or subject to
loading (Amos *et al.*, 1988). The shear-strength of
water-saturated muddy sediments decreases very
rapidly as a function of water content. The *erosion
shear strength* (τ_e) of a cohesive sediment bed is also
related to the dry density (HR Wallingford, 1989).
Erosion of the bed can only occur if the *constant fluid
bed shear stress* (τ_b) exceeds τ_e.

1.6 Types and origins of sediment grains

Sediment grains can be divided into four major groups
based on their nature, source and mode of origin
(Table 1.4). In terms of global significance, quartz and
silicate grains released by weathering and erosion of
crustal rocks are of by far the greatest significance, but
at regional and local scales sediments may consist
largely or entirely of biogenic skeletal debris, authigenic sediment grains, or organic matter.

Table 1.4 Major types and sources of sediment grains

Inorganic grains produced by weathering of igneous,
metamorphic and sedimentary rocks
Rock fragments
Quartz (monocrystalline and polycrystalline)
Feldspars
Heavy minerals (oxides and silicates)
Layer silicates (mainly micas)
Limestone and dolomite fragments
Clay minerals produced by chemical weathering

Inorganic grains produced by volcaniclastic processes
Shards, pumice, tephra and bombs formed during
  eruptions (glass, feldspars, rare amphiboles, pyroxenes
  olivines)
Particles produced by gas and liquid to solid conversions
  (e.g. sulphate particles)
Grain aggregates formed by electrostatic charging and
  interparticle collisions

Inorganic grains produced by physical and chemical
processes in near-surface environments
Authigenic crystals (gypsum, dolomite)
Clay pellets formed by wind erosion of muds
Clay floccules formed in water
Authigenic clay and ferriferous pellets, pisoliths, etc.
  formed in aqueous environments and soils (glauconite,
  berthierine, goethite)
Carbonate ooids and peloids formed in shallow marine
  environments (aragonite and high Mg-calcite)

Biogenic and biogenically mediated grains
Skeletal carbonate grains (calcified algae, foram tests,
  echinoderm plates and spicules, coral fragments,
  molluscs and gastropod shells, etc.)
Faecal pellets
Siliceous tests (diatoms)
Opal phytoliths from vegetation
Skeletal phosphatic grains (bones, some shells, etc.)
Organic matter

1.6.1 Sediment grains produced by weathering

The great majority of quartz and feldspar grains were
derived originally from plutonic igneous or metamorphic rocks, or from older sandstones. Quartz and
potassium-rich feldspar are especially abundant in
plutonic rocks of granitic or granodioritic composition, whereas these minerals are extremely rare in
basic plutonic rocks whose mineral composition is
dominated by plagioclase feldspars, pyroxenes and
amphiboles (Table 1.5). Metamorphic

rocks such as granitoid gneiss and migmatites are also
important sources of quartz and feldspar.

Both the range of particle sizes and the mineral
composition of weathering products derived from
igneous and metamorphic rocks are strongly controlled by the nature of the weathering processes in
the source area. These, in turn, are influenced principally by climate and tectonics, particularly the rate of
uplift and erosion (Basu, 1985). For example, the
amount of sand and the quartz/feldspar ratio in
weathering products derived from any granitoid rock
is dependent on the relative effectiveness of granular
disintegration and chemical decomposition (Pye,
1985). In humid climates, the chemical decomposition of feldspars, biotite and amphiboles is rapid, and
the resulting weathering products have a lower sand/
silt plus clay ratio than those formed in arid and
semi-arid climates under similar conditions of relief
and rates of surface stripping. Granular disintegration can be brought about either by mechanical
weathering processes, such as frost action and salt
weathering, or by chemical weathering processes that
affect only the areas adjacent to microfractures and
intercrystalline boundaries (Fig. 1.16). Temperature
also exercises an important control on weathering
rates and the nature of weathering products. Other
things being equal, chemical weathering is more
rapid in humid tropical areas than in humid temperate zones (see Ollier, 1984, and papers in Drever,
1984).

The weathering products derived from granitoid
rocks are typically enriched in monocrystalline
quartz relative to polycrystalline quartz and feldspar
in the source rock (Blatt, 1967; Basu, 1976). An
increase in the ratio of K-feldspar to plagioclase
feldspar is also commonly observed on account of the
slightly greater weathering resistance of the former
mineral (Pye *et al.*, 1985). Where chemical alteration
of the feldspars and mafic minerals is pronounced,
clay minerals represent an important component of
the weathering residues. The mineralogical composition of the clays is influenced by that of the source
rocks, by the climate and by local hydrological
conditions. However, at a very broad scale it is
possible to identify a pattern of kaolinite dominance
in humid areas, smectite dominance in arid areas,
and illite dominance in intermediate areas (see
Weaver (1989) for more detailed discussion).

Weathering of sandstones and sandy conglomerates also provides an important source of sand- and
gravel-sized sediment, while mudrocks act as an

Table 1.5 Proportions of different minerals found in plutonic igneous rocks. Data from Wedepohl (1969)

Mineral	Volume % of:					
	Granite	Grandiorite	Quartz diorite	Diorite	Gabbro	Upper crust average
Plagioclase	30	46	53	63	56	41
Quartz	27	21	22	2	—	21
K-feldspar	35	15	6	3	—	21
Amphibole	1	13	12	12	1	6
Biotite	5	3	5	5	—	4
Orthopyroxene	—	—	—	3	16	2
Clinopyroxene	—	—	—	8	16	2
Olivine	—	—	—	—	5	0.6
Magnetite, ilmenite	2	2	2	3	4	2
Apatite	0.5	0.5	0.5	0.8	0.6	0.5

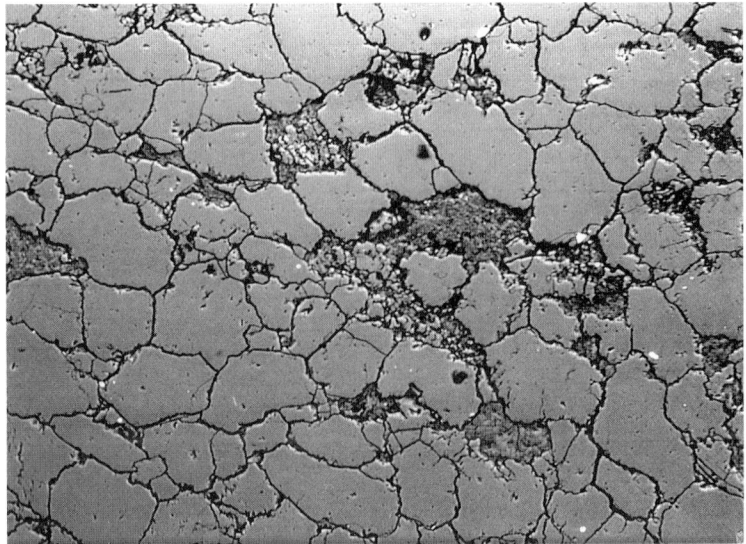

Fig. 1.16 Backscattered scanning electron micrograph of a polished section of quartzite undergoing granular disintegration. The grain boundary cracks and larger intergranular voids are filled with kaolinte formed by feldspar alteration. Sample from Mount Roraima, Venezuela. Picture width = 1.378 mm.

important source of silt and clay. The 'average' sandstone contains about 65% quartz compared with about 27% quartz in the 'average' granite (Blatt, 1970). The corresponding figures for feldspar are 12% and 65%. Consequently the sediments derived from sandstone source areas show a high quartz enrichment factor and are typified by low sand/silt plus clay ratios.

Mudrocks typically contain < 10% sand, 30–60% silt, and 30–70% clay. The sand and much of the coarse and medium silt is typically composed of quartz, whereas the very fine silt and clay fraction is dominated by micas and clay minerals. Mudrocks vary greatly in their degree of induration, reflecting variations in degree of compaction during burial, cementation and metamorphism. Consequently during weathering and erosion a wide range of clast sizes may be produced with greatly differing physical properties.

In arid continental areas, especially those that are tectonically active and of moderate to high relief, physical weathering of limestones, dolomites and calcretes can produce clastic material ranging in size from boulders to silt (Fig. 1.17). However, in more humid climates and in areas of lower relief, weathering is dominated by solution and only small amounts of insoluble residue (mainly quartz and clays) are produced.

Fig. 1.17 Break-down of a limestone boulder due to salt weathering on the surface of an alluvial fan, western side of the Dead Sea Rift Valley, Israel.

1.6.2 Inorganic grains formed in near-surface environments

In addition to clay formation during weathering, grains may be produced by authigenic processes in a wide range of surface and near-surface environments. Examples include the formation of gypsum, aragonite and dolomite crystals in the muds around playas, coastal sabkhas, and carbonate banks (Fig. 1.5; Jones, 1938; Eardley & Stringham, 1952; McKee & Moiola, 1975), the formation of ooids and pisolites in certain lakes, weathering profiles, and shallow marine areas (e.g. Kahle, 1974; Davies *et al.*, 1978; Ferguson *et al.*, 1978). Although some ooids appear to be formed entirely by inorganic precipitation of calcium carbonate, iron hydroxide or clay minerals (e.g. glauconite), the formation of others involves the activity of algae or bacterially mediated diagenetic alteration of faecal pellets and other biogenic debris. Vast numbers of faecal pellets are formed in sediments and soils that contain an active biota. Pellets of broadly similar size and appearance can also be formed during wind erosion of clay-rich sediments (Huffman & Price, 1949; Price, 1958; Bowler, 1973).

1.6.3 Volcaniclastic sediment particles

In areas of active volcanic activity, sediment grains may be produced both during eruptions and as a result of the subsequent weathering and break-down of lavas (see Chapter 10 for detailed discussion of volcaniclastic processes and products). In addition to forming extensive ashfall deposits, such particles may be reworked by fluvial, aeolian and marine processes to form alluvial, dune and beach deposits. Many of the beaches in the Hawaiian Islands, for example, consist of black sands composed of basaltic fragments, mafic minerals, plagioclase feldspars and glass shards (e.g. Moberley *et al.*, 1965).

1.6.4 Biogenic sediment grains

Many living organisms possess hard parts composed of aragonite, high-Mg calcite or low-Mg calcite. Among the more important sources of carbonate sediment are calcareous algae, foraminfera, echinoderms, corals, gastropods and bivalves (Bathurst, 1975; Scoffin, 1987; Tucker & Wright, 1990). In addition, some soft-bodied organisms such as worms are important formers of calcified tubes. Fish and mammals contain bones that are principally composed of calcium phosphate. On the death of these organisms, the calcified hard parts are gradually broken up both by physical and biological processes, forming a potentially wide range of particle sizes and shapes. Production rates of biogenic carbonate sand are particularly high in warm, shallow, tropical seas that receive relatively little sediment input from terrigenous sources.

1.7 **References**

Allen J.R.L. 1969. The maximum slope angle attainable by surfaces underlain by bulked equal spheroids with variable dimensional ordering. *Geol. Soc. Am. Bull.*, **80**, 1923–1930.

Allen J.R.L. 1970. The avalanching of granular solids on dune and similar slopes. *J. Geol.*, **78**, 326–351.

Allen J.R.L. 1984a. Experiments on the settling, overturning and entrainment of bivalve shells and related models. *Sedimentology*, **31**, 227–250.

Allen J.R.L. 1984b. Experiments on the terminal fall of the valves of bivalve molluscs loaded with sand trapped from dispersion. *Sedim. Geol*, **39**, 197–209.

Allen J.R.L. 1985. *Principles of Physical Sedimentology*. Allen & Unwin, London.

Amos C.L., Wagoner N.A. & Daborn G.R. 1988. The influence of subaerial exposure on the bulk properties of fine-grained intertidal sediment from the Minas Basin, Bay of Fundy. *Est. Coast Shelf. Sci.*, **21**, 1–13.

Baba J. & Komar P.D. 1981. Measurements and analysis of settling velocities of natural quartz sand grains. *J. Sedim. Petrol.*, **51**, 631–640.

Bagnold R.A. 1937. The size grading of sand by wind. *Proc. R. Soc. Lond. A*, **163**, 250–264.

Bagnold R.A. 1941. *The Physics of Blown Sand and Desert Dunes*. Methuen, London.

Bagnold R.A. 1956. Flow of cohesionless grains in fluids. *Phil. Trans. R. Soc. Lond. A*, **249**, 235–297.

Bagnold R.A. 1979. Sediment transport by wind and water. *Nordic Hydrol.*, **10**, 309–322.

Bagnold R.A. & Barndorff-Nielsen O.E. 1980. The pattern of natural grain size distributions. *Sedimentology*, **27**, 199–207.

Bale A. J. & Morris A.W. 1987. *In situ* measurements of particle size in estuarine waters. *Est. Coast. Shelf. Sci.*, **24**, 253–263.

Ballantyne C.K. 1982. Aggregate clast form characteristics of deposits near the margins of four glaciers in the Jotunheim Massif, Norway. *Norsk Geog. Tidsskr.*, **36**, 103–113.

Barndorff-Nielsen O.E. & Christiansen C. 1988. Erosion, deposition and size-distributions of sand. *Proc. R. Soc. Lond. A*, **417**, 335–352.

Barndorff-Nielsen O.E., Dalsgard K., Halgreen C., Kuhlman H., Moller J. T. & Schou G. 1982. Variation in particle size over a small dune. *Sedimentology*, **29**, 53–65.

Barrett P.J. 1980. The shape of rock particles, a critical review. *Sedimentology*, **27**, 291–303.

Basu A. 1976. Petrology of Holocene fluvial sand derived from plutonic source rocks: implications to palaeoclimatic interpretation. *J. Sedim. Petrol.*, **46**, 649–709.

Basu A. 1985. Influence of climate and relief on compositions of sands released at source areas. In: *Provenance of Arenites* (Ed. by G.G. Zuffa), pp. 1–15. Reidel, Dordrecht.

Bathurst R.G.C. 1975. *Carbonate Sediments and Their Diagenesis*, 2nd edn. Elsevier, Amsterdam.

Beard D.C. & Weyl P.K. 1973. Influence of texture on porosity and permeability of unconsolidated sand. *Bull. Am. Ass. Petrol. Geol.*, **51**, 349–369.

Benn D.I. & Ballantyne C.K. 1992. Pebble shape (and size)—discussion. *J. Sedim. Petrol.*, **62**, 1147–1150.

Blatt H. 1967. Original characteristics of quartz sand grains. *J. Sedim. Petrol.*, **37**, 401–424.

Blatt H. 1970. Determination of mean sediment thickness in the crust: a sedimentologic method. *Bull. Geol. Soc. Am.*, **81**, 255–262.

Boulton G.S. 1978. Boulder shapes and grain size distributions of debris as indicators of transport paths through a glacier and till genesis. *Sedimentology*, **25**, 773–799.

Bowler, J.M. 1973. Clay dunes; their occurrence, formation and environmental significance. *Earth Sci. Rev.*, **9**, 315–338.

Brazier S., Sparks R.S.J., Carey S.N., Sigurdsson S.N. & Westgate J.G. 1983. Bimodal grain size distribution and secondary thickening in air-fall ash layers. *Nature*, **301**, 115–119.

Bui E.N., Mazullo J. & Wilding L.P. 1990. Using quartz grain size and shape analysis to distinguish between aeolian and fluvial deposits in the Dallol Bosso of Niger (West Africa). *Earth Surf. Proc. Landf.*, **14**, 157–166.

Burt T.N. & Stevenson J.R. 1983. *Field settling velocity of Thames mud.* Hydraulics Research Ltd, Wallingford, Report IT 251.

Carrigy M.A. 1970. Experiments on the angle of repose of granular material. *Sedimentology*, **14**, 147–158.

Christiansen C. 1984. *A Comparison of Sediment Parameters from Log-Probability Plots and Log–Log Plots of the Same Sediments.* Dept Geology, University of Aarhus, Geoskrifter 20.

Clark N.N. 1987. A new scheme for particle shape characterization based on fractal harmonics and fractal dimension. *Powder Technol.*, **51**, 243–249.

Corey A.T. 1949. *Influence of shape on the fall velocity of sand grains.* Unpublished MS Thesis, A&M College, Colorado.

Cui B., Komar P.D. & Baba J. 1983. Settling velocities of natural sand grains in air. *J. Sedim. Petrol.*, **53**, 1205–1211.

Culver S.J., Bull P.A., Campbell S., Shakesby R.A. & Whalley W.B. 1983. Environmental discrimination based on quartz grain surface textures: a statistical investigation. *Sedimentology*, **30**, 129–136.

Dana J.D. & Harlbut C.S. 1959. *Manual of Mineralogy*, 7th edn. Wiley, New York.

Davies P.J., Bubela B. & Ferguson J. 1978. The formation of ooids. *Sedimentology*, **25**, 703–730.

Deitrich W.E. 1982. Settling velocity of natural particles. *Wat. Res.*, **18**, 1615–1626.

Delo E.A. & Ockenden M. C. 1992. *Estuarine muds manual.*

Hydraulics Research Ltd, Wallingford, Report SR 309.

Dobkins J.E. & Folk R.L. 1970. Shape development on Tahiti-Nui. *J. Sedim. Petrol.*, **40**, 1167–1203.

Dowdeswell J.A., Osterman L.E. & Andrews J.T. 1985. Quartz sand grain shape and other criteria used to distinguish glacial and non-glacial events in a marine core from Frobisher Bay, Baffin Island, NWT, Canada. *Sedimentology*, **32**, 119–132.

Drever J.I. (ed.) 1984. *The Chemistry of Weathering. NATO ASI Series, Ser. C*, **149**. Reidel, Dordrecht.

Eardley A.J. & Stringham B. 1952. Selenite crystals in the clays of the Great Salt Lake. *J. Sedim. Petrol.*, **22**, 234–238.

Ehrlich R. & Weinberg B. 1970. An exact method for characterization of grain shape. *J. Sedim. Petrol.*, **40**, 205–212.

Ehrlich R., Brown P.J., Yarus J.M. & Przygocki R.S. 1980. The origin of shape frequency distributions and the relationship between size and shape. *J. Sedim. Petrol.*, **50**, 475–483.

Ehrlich R., Orzeck J. & Weinberg B. 1974. Detrital quartz as a natural tracer—Fourier grain shape analysis. *J. Sedim. Petrol.*, **44**, 145–150.

Elzenga W., Schwan J., Baumfalk T.A., Vendenbergh J. & Krook L. 1987. Grain surface characteristics of periglacial aeolian and fluvial sands. *Geol. Mijn.*, **65**, 273–286.

Ferguson J.B., Bubela B. & Davies P.J. 1978. Synthesis and possible mechanism of formation of radial carbonate ooids. *Chem. Geol.*, **22**, 285–308.

Folk R.L. 1955. Student operator error in determination of roundness, sphericity and grain size. *J. Sedim. Petrol.*, **25**, 297–301.

Folk R.L. 1978. Angularity and silica coatings of Simpson Desert sand grains, Northern Territory. *J. Sedim. Petrol.*, **52**, 93–101.

Folk R.L. & Ward W.C. 1957. Brazos River bar: a study in the significance of grain size parameters. *J. Sedim. Petrol.*, **27**, 3–26.

Friedman G.M. 1961. Distinction between dune, beach and river sands from their textural characteristics. *J. Sedim. Petrol.*, **31**, 514–529.

Friedman G.M. 1979. Differences in size distributions of populations of particles among sands of various origins. *Sedimentology*, **26**, 3–32.

Friedman G.M. & Sanders J.E. 1978. *Principles of Sedimentology*. Wiley, New York.

Frihy O.E. & Komar P.D. 1991. Patterns of beach sand sorting and shoreline erosion on the Nile delta. *J. Sedim. Petrol.*, **61**, 544–550.

Full W., Ehrlich R. & Kennedy S.K. 1984. Optimal configuration and information content of sample data generally displayed as histograms or frequency plots. *J. Sedim. Petrol.*, **54**, 117–126.

Gao S. & Collins M. 1991. A critique of the 'McLaren Method' for defining sediment transport paths — discussion. *J. Sedim. Petrol.*, **61**, 143–146.

Gibbs R.J. 1977. Clay mineral segregation in the marine environment. *J. Sedim. Petrol.*, **47**, 237–243.

Gibbs R.J. 1983. Coagulation rates of clay minerals and natural sediments. *J. Sedim. Petrol.* **53**, 1193–1203.

Gill E.W.B. 1948. Frictional electrification of sand. *Nature*, **162**, 568.

Goldbery R. & Richardson D. 1989. The influence of bulk shape factors on settling velocities of natural sand-sized sedimentary suites. *Sedimentology*, **36**, 125–136.

Goossens D. 1989. Interference phenomena between particle flattening and particle rounding in free vertical sedimentation. *Sedimentology*, **34**, 155–168.

Graton L.C. & Fraser H.J. 1935. Systematic packing of spheres with particular relation to porosity and permeability. *J. Geol.*, **43**, 785–909.

Greeley R. 1979. Silt–clay aggregates on Mars. *J. Geophys. Res.*, **84**, (B11), 6248–6254.

Greeley R. & Leach R. 1978. A preliminary assessment of the effects of electrostatics on aeolian processes. *NASA Tech. Mem.*, **79729**, 236–237.

Haines J. & Mazzullo J.M. 1988. The original shapes of quartz silt grains. *Mar. Geol.*, **78**, 227–240.

Hallermeier R.J. 1981. Terminal settling velocity of commonly occurring sand grains. *Sedimentology*, **28**, 859–865.

Hamilton E.L. 1971. Elastic properties of marine sediments. *J. Geophys. Res.*, **76**, 579–604.

Hamilton E.L. 1979. V_p/V_s and Poisson's ratios in marine sediment and rocks. *J. Acoust. Soc. Am.*, **66**, 1093–1101.

Hartmann D. 1991. Cross-shore selective sorting processes and grain size distributional shape. *Acta Mech. Suppl.*, **2**, 49–62.

Hartmann D. & Christiansen C. 1988. Settling velocity distributions and sorting processes on a seif dune: a case study. *Earth Surf. Proc. Landf.*, **13**, 649–656.

Hartmann D. & Christiansen C. 1992. The hyperbolic shape triangle as a tool for discriminating populations of sediment samples of closely connected origin. *Sedimentology*, **39**, 697–708.

Hockey B. 1970. An improved coordinate system for particle shape representation. *J. Sedim. Petrol.*, **40**, 1054–1056.

HR Wallingford 1989. *Grangemouth mud properties*. Hydraulics Research Ltd, Wallingford, Report SR 197.

Huffman G.G. & Price W.A. 1949. Clay dune formation near Corpus Christi, Texas. *J. Sedim. Petrol.*, **19**, 118–127.

Ibbeken H. 1983. Jointed source rock and fluvial gravels controlled by Rosin's Law: a grain size study in Calabria, South Italy. *J. Sedim. Petrol.*, **53**, 1213–1231.

Illenberger W.K. 1991. Pebble shape (and size!) *J. Sedim. Petrol.*, **61**, 756–767.

Jones D. J. 1938. Gypsum–oolite dunes, Great Salt Lake Desert, Utah. *Bull. Am. Ass. Petrol. Geol.*, **37**, 2530–2538.

Jordan D.W. 1954. The adhesion of dust particles. *Brit. J. App. Phys.* **5**, 5194–5197.

Kahle C.F. 1974. Ooids from Great Salt Lake, Utah, as an analogue for the genesis and diagenesis of ooids in marine limestones. *J. Sedim. Petrol.*, **44**, 30–39.

Kamra A.K. 1972. Measurement of the electrical properties of duststorms. *J. Geophys. Res.*, **77**, 5856–5869.

Kennedy S.K. & Ehrlich R. 1985. Origin of shape changes of sand and silt in a high-gradient stream system. *J. Sedim. Petrol.*, **55**, 57–64.

Kennedy S.K. & Lin W.H. 1991. A comparison of Fourier and Fractal techniques in the analysis of closed forms. *J. Sedim. Petrol.*, **62**, 842–848.

Kennedy S.K., Meloy T.P. & Gurney T.E. 1985. Sieve data—size and shape information. *J. Sedim. Petrol.*, **55**, 356–360.

Komar P.D. 1987. Selective grain entrainment by a current from a bed of mixed sizes: a re-analysis. *J. Sedim. Petrol.*, **57**, 203–211.

Komar P.D. 1989. Physical processes of waves and currents and the formation of marine placers. *Reviews Aquat. Sci.*, **1**, 393–423.

Komar P.D. & Cui B. 1984. The analysis of grain size measurements by settling tube techniques. *J. Sedim. Petrol.*, **54**, 603–614.

Komar P.D. & Li Z. 1986. Pivoting analysis of selective entrainment of sediments by shape and size with application to gravel threshold. *Sedimentology*, **33**, 425–436.

Komar P.D. & Li Z. 1988. Applications of grain pivoting and sliding analyses to selective entrainment of gravel and to flow competence evaluations. *Sedimentology*, **35**, 681–695.

Komar P.D. & Reimers C.E. 1978. Grain shape effects on settling rates. *J. Geol.*, **86**, 193–209.

Komar P.D. & Wang C. 1984. Processes of selective grain transport and the formation of placers on beaches. *J. Geol.*, **92**, 637–655.

Kranck K. 1973. Flocculation of suspended sediment in the sea. *Nature*, **246**, 348–350.

Kranck K. 1975. Sediment deposition from flocculated suspensions. *Sedimentology*, **22**, 111–123.

Kranck K. 1981. Particulate matter, grain size characteristics and flocculation in a partially mixed estuary. *Sedimentology*, **28**, 107–114.

Krinsley D.H. & Doornkamp J.C. 1973. *Atlas of Quartz Sand Grain Surface Textures.* Cambridge University Press, Cambridge.

Krone R.B. 1972. A field study of flocculation as a factor in estuarial shoaling processes. *Technical Bulletin*, **19**. US Army Corps of Engineers, Committee Tidal Hydraulics.

Krumbein W.C. 1934. Size frequency distributions of sediments. *J. Sedim. Petrol.*, **4**, 65–77.

Krumbein W.C. 1941. Measurement and geologic significance of shape and roundness of sedimentary particles. *J. Sedim. Petrol.*, **11**, 64–72.

Kuenen P.H. & Perdok W.G. 1962. Experimental abrasion. 5. Frosting and defrosting of quartz grains. *J. Geol.*, **70**, 648–658.

Lane E.W. 1938. Notes on the formation of sand. *Trans. Am. Geophys. Un.*, **18**, 505–508.

Li M.Z. & Komar P.D. 1992a. Longshore grain sorting and beach placer formation adjacent to the Columbia River. *J. Sedim. Petrol.*, **62**, 429–441.

Li M.Z. & Komar P.D. 1992b. Selective entrainment and transport of mixed size and density sands: flume experiments simulating the formation of black sand placers. *J. Sedim. Petrol.*, **62**, 584–590.

Lirer L. & Vinci A. 1991. Grain size distributions of pyroclastic deposits. *Sedimentology*, **38**, 1075–1084.

Lowright R., Williams E.G. & Dachille F. 1972. An analysis of factors controlling deviations in hydraulic equivalence in some modern sands. *J. Sedim. Petrol.*, **42**, 634–645.

Mandelbrot B.B. 1967. How long is the coastline of Britain? Statistical self-similarity and fractional dimension. *Science*, **155**, 636–638.

Mandelbrot B.B. 1977. *Fractals: Form, Chance and Dimension.* W.H. Freeman, San Francisco.

Marshall J.R. Krinsley D.H. & Greeley R. 1981. An experimental study of the behaviour of electrostatically-charged fine particles in atmospheric suspension. *NASA Tech. Mem. 84211*, 208–210.

Mazzullo J.M., Ehrlich R. & Pilkey O.H. 1983. Local and distal origin of sands in the Hatteras Abyssal Plain. *Mar. Geol.*, **48**, 75–88.

Mazzullo J.M., Sims D. & Cunningham D. 1986. The effects of eolian sorting and abrasion upon the shapes of fine quartz sand grains. *J. Sedim. Petrol.*, **56**, 45–56.

Mazzullo J.M., Alexander A., Tieh T. & Menglin D. 1992. The effects of wind transport on the shapes of quartz silt grains. *J. Sedim. Petrol.*, **62**, 961–971.

McKee E.D. & Moiola J. 1975. Geometry and growth of the Whitesands dune field, New Mexico. *J. Res. U.S. Geol. Surv.*, **3**, 59–66.

McLaren P. & Bowles D. 1985. The effects of sediment transport on grain size distributions. *J. Sedim. Petrol.*, **55**, 457–470.

McLaren P. & Bowles D. 1991. A critique of the 'McLaren method' for defining sediment transport paths—reply. *J. Sedim. Petrol.*, **61**, 147.

McManus D.A. 1963. A criticism of certain usage of the phi notation. *J. Sedim. Petrol.*, **35**, 792–796.

McNown J.S. & Malaika J. 1950. Effects of particle shape on settling velocity at low Reynolds numbers. *Trans. Am. Geophys. Un.*, **31**, 74–82.

Middleton G.V. 1976. Hydraulic interpretation of sand size distributions. *J. Geol.*, **84**, 405–426.

Miller M.C., McCave I.N. & Komar P.D. 1977. Threshold of sediment motion under unidirectional currents. *Sedimentology*, **24**, 507–527.

Moberley R., Bauer L.D. & Morrison A. 1965. Source and variation of Hawaiian littoral sand. *J. Sedim. Petrol.*, **35**, 589–598.

Ollier C.D. 1984. *Weathering.* Arnold, London.

Orford J.D. & Whalley W.B. 1983. The use of fractal dimensions to quantify the morphology of irregular-shaped particles. *Sedimentology*, 30, 655–668.

Orford J.D. & Whalley W.B. 1991. Quantitative grain form analysis. In: Syvitski J. (ed.) *Principles, Methods and Application of Particle Size Analysis*, pp. 88–108. Cambridge University Press, Cambridge.

Passega R. 1957. Texture as characteristic of clastic deposition. *Am. Ass. Petrol. Geol. Bull.*, 41, 1952–1984.

Powers M.C. 1953. A new roundness scale for sedimentary particles. *J. Sedim. Petrol.*, 23, 117–119.

Powers M.C. 1957. Adjustment of land-derived clays to the marine environment. *J. Sedim. Petrol.*, 27, 355–372.

Price W.A. 1958. Sedimentology and Quaternary geomorphology of south Texas. *Trans. Gulf. Coast. Ass. Geol. Socs.*, 8, 410–475.

Pye K. 1985. Granular disintegration of gneiss and migmatites. *Catena*, 12, 191–199.

Pye K., Goudie A.S. & Watson A. 1985. An introduction to the physical geography of the Kora area of central Kenya. *Geog. J.*, 151, 168–181.

Rowe P.W. 1962. The stress–dilatancy relation for static equilibrium of an assembly of particles in contact. *Phil. Trans. R. Soc. Lond. A*, 269, 500–527.

Rubey W.W. 1933a. Settling velocities of gravel, sand and silt particles. *Am. J. Sci.*, 25, 325–338.

Rubey W.W. 1933b. The size distribution of heavy minerals within a water-laid sandstone. *J. Sedim. Petrol.*, 3, 3–29.

Sahu B.K. 1965. Theory of sieving. *J. Sedim. Petrol.*, 35, 750–753.

Scoffin T.P. 1987. *An Introduction to Carbonate Sediments and Rocks*. Blackie, Glasgow.

Sheridan M.F., Wohletz K.H. & Dehn J. 1987. Discrimination of grain-size sub-populations in pyroclastic deposits. *Geology*, 15, 367–370.

Slingerland R. 1984. Role of hydraulic sorting in the origin of fluvial placers. *J. Sedim. Petrol.*, 54, 37–50.

Slingerland R. & Smith N.D. 1986. Occurrence and formation of water-laid placers. *Ann. Rev. Earth Planet. Sci.*, 14, 113–147.

Slingerland R.L. 1977. The effects of entrainment on the hydraulic equivalence relationships of light and heavy minerals in sand. *J. Sedim. Petrol.*, 54, 137–150.

Sneed E.D. & Folk R.L. 1958. Pebbles in the lower Colorado River, Texas: a study in particle morphogenesis. *J. Geol.*, 66, 114–150.

Sparks R.S.J. 1976. Grain size variations in ignimbrites and implications for transport of pyroclastic flows. *Sedimentology*, 23, 147–188.

Steidtmann J.R. 1982. Size–density sorting of sand-size spheres during deposition from bedload transport and implications concerning hydraulic equivalence. *Sedimentology*, 29, 877–883.

Stokes G.G. 1851. On the effect of the internal friction of fluids on the motion of pendulums. *Trans. Cambridge Phil. Soc.*, 9, (2), 8–106.

Stow C.D. 1969. Dust and sand storm electrification. *Weather*, 24, 134–140.

Stringham G.E., Simons D.B. & Guy H.P. 1969. The behaviour of large particles falling in quiescent liquids. *U.S. Geol. Surv. Prof. Pap.*, 562-C.

Tillman S.E. 1973. The effect of grain orientation on Fourier shape analysis. *J. Sedim. Petrol.*, 43, 867–869.

Trask C.B. & Hand B.M. 1985. Differential transport of fall-equivalent sand grains, Lake Ontario, New York. *J. Sedim. Petrol.*, 55, 226–234.

Tucker M.E. & Wright V.P. 1990. *Carbonate Sedimentology*. Blackwell Scientific Publications, Oxford.

Udden J.A. 1914. *Mechanical Composition of Some Clastic Sediments*. Publications of the Augustana Library No. 1., USA

Visher G.S. 1969. Grain size distributions and depositional processes. *J. Sedim. Petrol.*, 39, 1074–1106.

Wadell H. 1933. Sphericity and roundness of rock particles. *J. Geol.*, 41, 310–331.

Weaver C.E. 1989. *Clays, Muds and Shales*. Elsevier, Amsterdam.

Wedepohl K.H. (ed.) 1969. *Handbook of Geochemistry*, Vol. Springer, Heidelberg.

Wentworth C.K. 1922. A scale of grade and class terms for clastic sediments. *J. Geol.*, 30, 377–392.

Whalley W.B. 1972. The description and measurement of sedimentary particles and the concept of form. *J. Sedim. Petrol.*, 42, 961–965.

Whitehouse U.G., Jeffrey L.M. & Debrecht J.D. 1960. Differential settling tendencies of clay minerals in saline waters. *Clays Clay Min.*, 7, 1–80.

Willetts B.B. 1983. Transport by wind of granular materials of different shapes and densities. *Sedimentology*, 30, 669–679.

Willetts B.B. & Rice A. 1983. Practical representation of characteristic gain shape of sands: a comparison of methods. *Sedimentology*, 30, 557–565.

Williams G.P. 1966. Particle roundness and surface texture effects on fall velocity. *J. Sedim. Petrol.*, 36, 255–259.

Wilson L. & Huang T.C. 1979. The influence of shape on the atmospheric settling velocity of volcanic ash particles. *Earth Planet. Sci Lett.*, 44, 311–324.

Winkelmolen A.M. 1971. Rollability, a functional shape property of sand grains. *J. Sedim. Petrol.*, 41, 703–714.

Winkelmolen A.M. 1982. Critical remarks on grain parameters with special emphasis on shape. *Sedimentology*, 29, 255–265.

Wyrwoll K.H. & Smith G.K. 1985. On using the log-hyperbolic distribution to describe the textural characteristics of eolian sediments. *J. Sedim. Petrol.*, 55, 471–478.

Wyrwoll K.H. & Smith G.K. 1988. On using the log-hyperbolic distribution to describe the textural characteristics of eolian sediments: reply. *J. Sedim. Petrol.*, 58, 161–162.

2 Fundamental properties of fluids and their relation to sediment transport processes

J.R.L. ALLEN

2.1 Introduction

The beginning of a scientific understanding of sediment transport and depositional processes, as they operate at the Earth's surface, lies in a grasp of the character and motion of natural fluids and in an appreciation of the interactions that can occur between these fluids, either stationary or in motion, and the sorts of particles discussed in Chapter 1. These three themes — the character and motion of natural fluids and fluid–particle interactions — are the central subjects of this chapter.

Atmospheric air, and water, either fresh or salt, are the natural fluids that chiefly interest sedimentologists. However, it is important to stress that materials which in bulk can be treated as fluids will arise naturally whenever sediment particles become mixed with either water or air. Some familiar examples of these multiphase, fluid-like materials are liquid mud and the dense mixtures of sand grains and intergranular air that avalanche down the fronts of desert dunes. It is also worth remembering that lava is fluid when erupted, and that comparatively dense mixtures of gases and solid particles capable of flow over the Earth's surface arise during certain kinds of volcanic activity.

Fluid–particle interactions in natural sedimentary systems occur on a wide range of spatio-temporal scales. Those at the larger scale underlie general sediment transport and the formation of such familiar sedimentary structures as the bedforms of sandy rivers, tidal shoals and desert sand seas. It is the smaller scale interactions that we examine in the present chapter. These relate to individual sediment

University of Reading PRIS Contribution No. 066.

particles and to numbers of particles on a scale generally smaller than that of a bedform or of some layer of sediment in a state of transport. The most important cases are the initiation of motion of particles by currents, the behaviour of particles during transport, and the settling of particles through and from fluids that are seldom at rest.

2.2 Physical properties of fluids

2.2.1 Definition of a fluid

The term *fluid* refers to a particular state of matter, and most experienced people would claim to be able to recognize one. However, it is not easy to provide a rigorous definition of the fluid state and, in a sedimentological context, a commonsense approach is to be preferred. Generally speaking, fluids are less 'weighty' than solids and, again generally speaking, they flow readily (i.e. deform) under the action of relatively small forces (e.g. their own weight), in contrast to solids which tend to retain the shapes given them. An especially striking property of a fluid is its ability to assume the shape, wholly or in part, of a vessel into which it is introduced. The extent to which this is possible provides us with a basis for dividing fluids into two classes. *Liquids* are those fluids which, on being poured into a vessel, form a horizontal free surface normal to the gravity field. Liquids (e.g. water) prove to be the weightier fluids; another property allying them with solids is their possession in some cases, intermittently and locally, of ordered, crystalline structures, as shown by X-ray diffraction studies. By contrast, *gases* (e.g. atmospheric air) are much less weighty than solids and, unlike liquids, do not form a free surface but spread uniformly through the whole space provided for them. A gas is by implication readily compressible, in contrast with solids and liquids which for practical purposes are incompressible.

It may be objected from everyday experience that some of these states of matter are not in practice rigidly bounded and that many real materials can be assigned to more than one state. Dry sand will flow under gravity and assume the shape of a jar into which it is poured, making it a liquid, but like a solid will support a substantial weight and stand at a slope. Liquid mud can behave similarly, especially if the solids content is large. These and similar materials are conveniently treated as a special class of fluids (non-Newtonian fluids), which possess properties additional to those of true fluids (Newtonian fluids).

2.2.2 Fluid density and specific gravity

The 'weightiness' or density of a fluid or solid refers to the amount of matter present per unit volume of space. It is measured in kilograms per cubic metre (kg m^{-3}) and has the dimensions ML^{-3}, where M stands for mass and L for length. The small Greek letter rho (ρ) is ordinarily used to represent fluid density in formulae and equations; for the density of a solid, the small Greek letter sigma (σ) is similarly employed. The specific gravity of a solid or fluid is its density

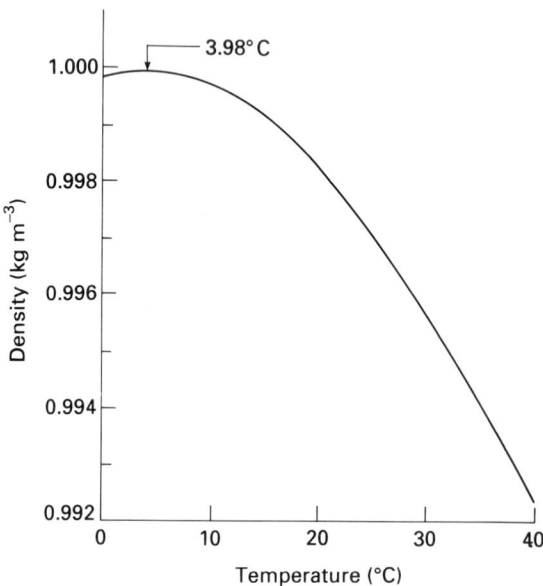

Fig. 2.1 Variation with temperature of the density of pure air-free water at one atmosphere pressure.

relative to that of water and, with the dimensions ML^{-3}/ML^{-3}, is simply a number without units (i.e. 'non-dimensional').

The density of water can for many practical purposes be assumed constant at 1000 kg m^{-3}, making it less dense than mineral sediment particles (Chapter 1). However, with varying temperature, the density of water changes in a complex manner of great importance to understanding the stratification and currents of such large water bodies as oceans and lakes (Fig. 2.1). Note that there is a density maximum at a temperature of about 4°C (Weast & Astle, 1980). Water is an excellent solvent, and its density is increased by the presence of dissolved solids. The sodium chloride and other salts dissolved in sea water give it a density of about 1028 kg m^{-3}. The overall or bulk density of water can also be raised by the presence of suspended mineral solids, and if these are sufficiently fine grained as to remain suspended for long periods, the increase can be semi-permanent. A useful formula for the bulk density of a solids–water solution or mixture is:

$$\rho_m = \sigma C + \rho(1 - C), \qquad (2.1)$$

where ρ_m is the density of the mixture, σ the density of the substance, and C the fractional volume concentration of the dissolved or admixed substance.

The density of air, under ordinary conditions, is three orders of magnitude smaller than that of water, and is meaningful as a numerical quantity only if the temperature and pressure are also indicated. At sea level and ordinary temperatures, that is, under full atmospheric pressure, air is of density 1.3 kg m^{-3}. With decreasing pressure, the density of air also declines.

Because the density of mineral solids relative to water and relative to air is so different, we may expect a substantial contrast in the behaviour of particles transported in the two media (see below).

2.2.3 Molecular (dynamic) viscosity

Although fluids are generally speaking readily deformed, it is not true to say that they present no resistance to deformation. The air resists the passage of the bicyclist and a sensible effort has to be made to stir such liquids as honey, syrup, engine lubricating oil and even water. Fluids may therefore be described as variously 'thin' or 'thick', depending on the degree to which they oppose an applied force. This intrinsic property of a fluid, deriving from the strength of the

Moving plate

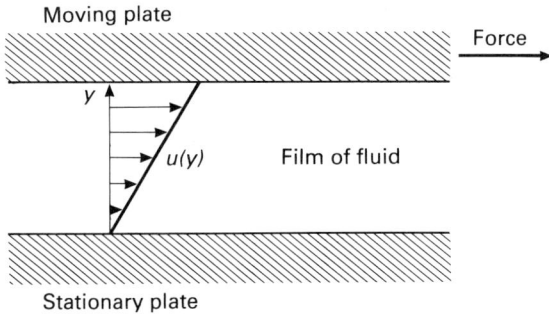

Fig. **2.2** A simple experiment to measure the molecular viscosity of a fluid.

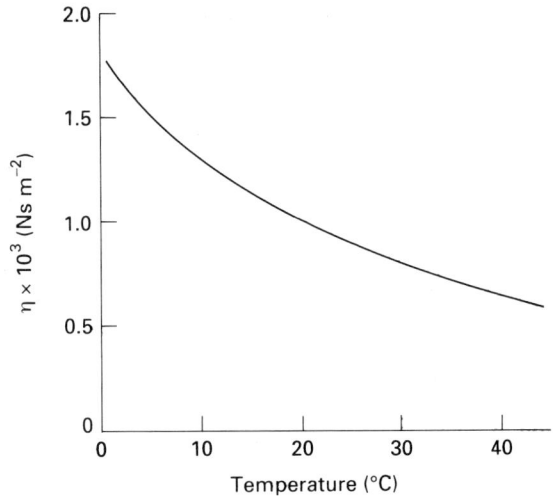

Fig. **2.3** Variation with temperature of the molecular viscosity of pure water.

forces operating between its molecules, is called the *molecular viscosity* and varies in value from fluid to fluid. The term *dynamic viscosity* is sometimes used instead, indicating that the property of resisting deformation becomes manifest only when there is relative movement between different parts of the fluid.

The molecular viscosity of a fluid is commonly defined in terms of a simple experiment in which a thin but extensive film of the fluid is held between two parallel smooth plates, one of which is free to move steadily in its own plane under the action of a constant applied tangential force (Fig. 2.2). Fluid molecules adhere without slipping to each plate, and a linear velocity profile consequently arises within the entrapped film, the shear applied to the moving plate being transmitted through the fluid to the static one. We then have the relation:

$$\tau = \eta \frac{du}{dy}, \qquad (2.2)$$

on which τ (tau) is the shear stress, u the fluid velocity, y the normal distance from the stationary plate, and η (eta) the molecular viscosity. The velocity is measured in metres per second (m s^{-1}) and has the dimensions LT^{-1}, where T stands for time. The shear stress is the (tangential) force per unit area and is measured in Newtons per square metre (N m^{-2}), with the dimensions $(ML/T^2)/L^2 = ML^{-1}T^{-2}$. The quantity du/dy is called the velocity gradient or shear rate; as it has the dimensions of velocity (LT^{-1}) divided by distance (L), it is measured per second (s^{-1}). The molecular viscosity is therefore a force per unit area per unit velocity gradient, with the dimensions $ML^{-1}T^{-1}$, and is measured in Newton-seconds per square metre (Ns m^{-2}).

The molecular viscosity of water (Weast & Astle, 1980) is of the order 1×10^{-3} Ns m^{-2} and is remarkably sensitive to changes of temperature, declining by about a half between, for example, polar and tropical oceans (Fig. 2.3). Temperature therefore substantially influences the behaviour of particles settling in water and the textural properties of aqueous suspensions. There is little variation with temperature in the case of atmospheric air, with a molecular viscosity of 1.8×10^{-5} Ns m^{-2} at sea level and 20°C. Thus air is significantly 'thinner' than water and both are much thinner than, say, honey or engine lubricating oil.

2.2.4 Kinematic viscosity

Workers in the field of loose-boundary hydraulics and fluid mechanics commonly use another definition of viscosity. This is the *kinematic viscosity*, the ratio of the molecular viscosity of the fluid to its density (η/ρ), represented by the small Greek letter v (nu). The kinematic vicosity has the dimensions L^2T^{-1} and is measured in m^2 s^{-1}. The kinematic viscosities of water and air vary with temperature in a similar fashion to the corresponding molecular viscosities.

The physical significance of the kinematic viscosity of a fluid is as follows. Since unit fluid force is the rate of change of momentum per unit volume, that is, the mass per unit volume times the velocity (the unit-volume momentum) times the same velocity, the shear stress in Eq. 2.2 is equivalent to the flux of

momentum to wetted unit area on the stationary
plate. Eq 2.2 then becomes:

$$\tau = \frac{\eta}{\rho}\frac{d(\rho u)}{dy}, \qquad (2.3)$$

in which $d(\rho u)/dy$ is the momentum gradient and
$\eta/\rho = \nu$ the coefficient linking the flux to the gradient.
Eq. 2.3 is a specific example of an important general
relationship, in which the flux of a property is related
through a coefficient called the diffusivity to the
'concentration' gradient of that property. The kinematic viscosity thus emerges as the diffusivity of
momentum. Relations identical in form to Eq. 2.3
can be used to describe the transfer of heat (e.g. from
a sun-heated desert surface) or of a soluble substance
(e.g. limestone or gypsum in a karst terrain) into a
fluid stream. An analogy exists between the processes
of momentum, heat and mass transfer, all of which
are important naturally, and the outcome of one of
the processes may often be predicted in terms of
another (review in Allen, 1982).

2.2.5 Newtonian and non-Newtonian fluids

Eq. 2.2 is a general description of the behaviour
during deformation of the important class of fluids
described as *Newtonian*, to which water and air
belong. In this class of (true) fluids, the relationship
between the shear stress and the shear rate is linear
(Fig. 2.4a), and the molecular viscosity, which is the
slope of the straight line connecting the two, is a
constant at each temperature and pressure and independent of the rate of shear.

There are other kinds of fluid of practical importance for which the shear stress and shear rate do not
bear a simple linear relationship (Fig. 2.4a,b). These
are grouped into the general class called *non-
Newtonian* fluids, extensively reviewed by Wilkinson
(1960). One important sub-class is the *Bingham
plastic*, which behaves as an elastic solid up to a
characteristic yield strength, but at greater stresses as
a Newtonian liquid with a constant apparent viscosity. Many water–sediment mixtures approximate to
Bingham plastics. The *power-law fluids* form a second
important sub-class. The stress–shear rate relationship for these is non-linear and a simple power
function, with the form of the curve depending on the
value of the exponent n. In this sub-class the value of
the apparent viscosity changes with the shear rate.
For dilatant power-law fluids ($n > 1$), the apparent
viscosity increases with shear rate, but for the

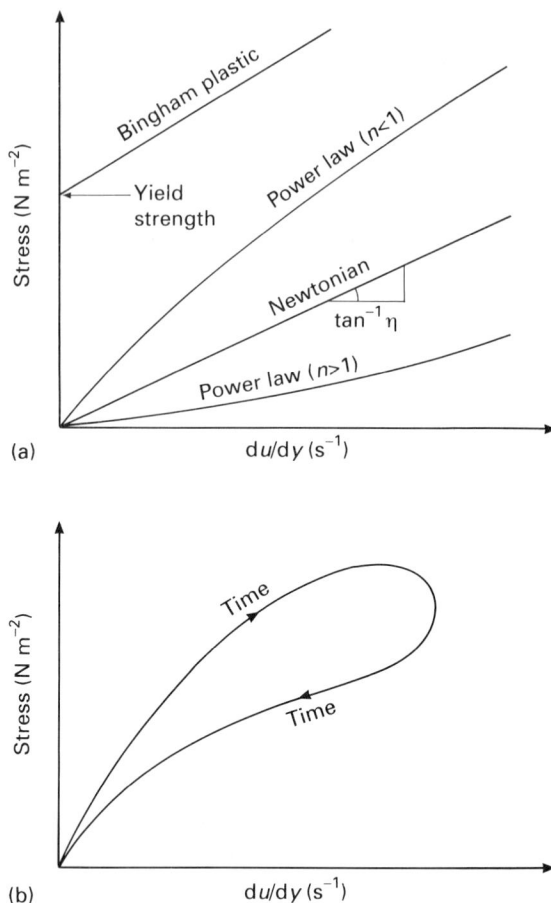

(a)

(b)

Fig. 2.4 Stress-shear rate variation of fluids
(a) Newtonian and a variety of non-Newtonian fluids.
(b) Response of a thixotropic (shear-thinning)
non-Newtonian fluid to a cycle of stress variation.

pseudo-plastic ones ($n < 1$) it declines as the shear
rate rises. The effect of continued shear on many
fluid–sediment mixtures, especially those involving
clay mineral particles, is to cause the apparent viscosity gradually to decrease as the structural relationships of the particles become increasingly disturbed
and ruptured. Such materials, called *thixotropic* or
shear-thinning, afford stress–shear rate curves of the
kind shown in Fig. 2.4b when deformed initially from
rest over a cycle of first increasing then decreasing
shear rate. Many muddy sediments prove to be
shear-thinning on being disturbed. Finally, we should
mention *visco-elastic* or 'memory' fluids. These substances, although viscous, have a certain elasticity of
shape, and so can store or release elastic strain as the

shear conditions change. The behaviour of emulsions, certain dilute suspensions of solid particles, and even some unconsolidated sediments and rocks can be modelled in terms of a visco-elastic fluid.

2.2.6 Mixtures of solid particles and fluids

As we have just seen, sediment particles and either air or water tend to form fluid-like mixtures that are non-Newtonian. A layer of particles moving as a bed load in a river is an example, and others are afforded by liquid mud and by debris flows, and by recently deposited, water-logged sediments liquefied by earthquake shocks. It is difficult to either predict or measure the relevant properties of these complicated materials. However, it is easy to grasp that their apparent viscosity will in some manner increase as the fractional volume concentration of the solids increases, for the more concentrated the grains the more particle collisions there will be during motion and the more the flow of the intergranular fluid will be constrained. A dilute mixture of either sand or clay minerals in water is very much easier to stir than a mixture of equal volumetric proportions, and there comes a sufficiently large particle concentration when the mixture becomes effectively a solid. The apparent viscosity of a particle–fluid mixture should therefore range from the molecular viscosity of the pure fluid phase up to an infinitely large value at the maximum possible particle concentration.

Much theoretical work, with supporting laboratory studies, has been done on the apparent viscosity of particle–fluid mixtures (review in Allen, 1982). A successful relationship for low and intermediate particle concentrations is the polynomial:

$$\frac{\eta_a}{\eta} = 1 + k_1 C + k_2 C^2 + k_3 C^3 \dots, \qquad (2.4)$$

in which η is the molecular viscosity of the (Newtonian) fluid phase, η_a the apparent viscosity of the mixture, and C the particle fractional volume concentration. The coefficients k_1, k_2, k_3... must generally be obtained empirically but can under some circumstances be predicted. At high concentrations:

$$\frac{\eta_a}{\eta} = \frac{k}{1 - (C/C_{max})^{1/3}} \qquad (2.5)$$

is useful, in which k is again a coefficient and C_{max} the maximum possible particle concentration, between 0.6 and 0.7 for many natural particle–fluid combinations. At concentrations within a few per

cent of the maximum, for example, η_a is of the order of 10^3 times η.

2.3   Fluids in motion

2.3.1 Fluid flow and its characterization

A fluid flow is a quantity of fluid in motion. We may characterize and classify the flow in terms of its context or boundary conditions, pattern, rate and changeability.

Most fluid flows of sedimentological interest involve some kind of solid boundary, typically of loose or potentially mobile sediment, for example, the sandy floor of a desert, the bed and banks of a river, or the deep ocean floor. When the solid boundary partly or wholly confines the motion, as in a tidal channel or a conduit forming part of a cave system, the flow may be described as confined or *internal*. The appelation 'free-surface' or 'open-channel' is applied to an internal flow only partly bounded by a solid bed, as the remainder of the boundary is formed of a deformable interface with another fluid. On the other hand, an *external* or unconfined flow is one that embraces the whole of the solid boundary, as with a sediment particle suspended in or sinking through a current.

An appreciation of fluid flow is enormously enhanced if the motion can be mapped or in some other manner rendered visible, as by the calculation of streamlines (Francis, 1975) or their visualization (Van Dyke, 1982) using dye, smoke or streamers (Fig. 2.5). A *streamline* is an imaginary curve that, at a given instant, is tangent to the directions of motion of all fluid particles lying on it. A pattern of streamlines indicates how the direction of flow changes from one part of the flow field to another. It also shows changes in the strength of flow, for the closer the streamlines the faster is the current. A flow pattern can also be depicted by plotting *pathlines*, where a pathline is the actual path traversed by a given fluid particle. When dye or hydrogen bubbles are introduced at a fixed point in a stream of water, or smoke is released from a fire into the wind, a streakline results to render the pattern of motion visible. A *streakline* is defined as the curve connecting all fluid particles passing through a given point in the flow. We shall see below that streamlines, pathlines and streaklines coincide only under restricted flow conditions.

Rate of fluid flow embraces the notion of amount as well as of strength. Now that streamlines have been

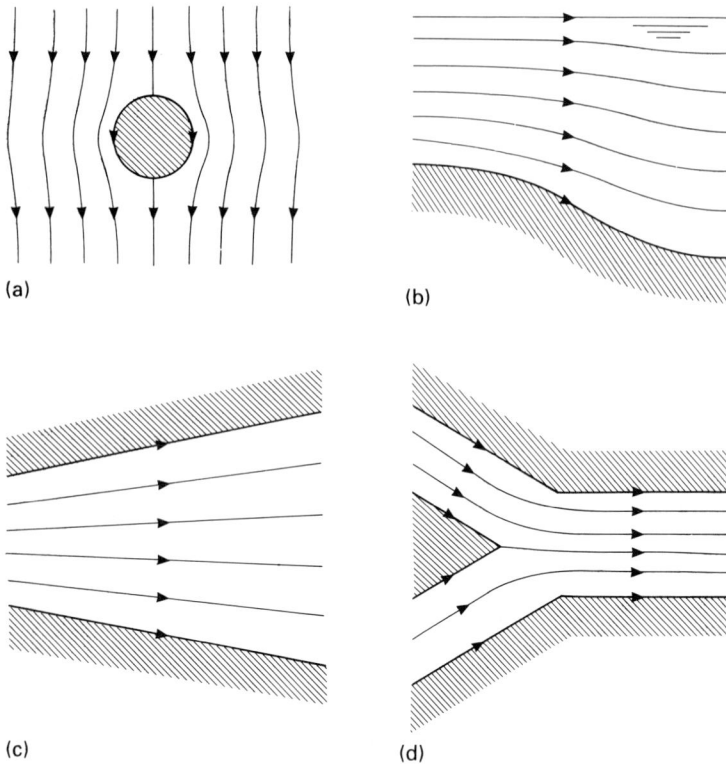

Fig. 2.5 Patterns of streamlines characteristic of different kinds of fluid flow. (a) Steady uniform flow round a sphere or cylinder (a case of external flow). (b) Steady streamlines in the streamwise plane for non-uniform channel flow (a partly confined, internal flow). (c) Streamlines for steady, non-uniform flow in an expanding pipe or channel (internal flow). (d) Streamlines for steady, non-uniform flow at the confluence between two pipes or channels (internal flow).

introduced, it is worth noting that velocity — the measure of strength — is a vector quantity, having both magnitude (speed) and direction. The amount of flow is measured by the *discharge*, namely, the product of the velocity and the normal cross-sectional area to which the value of the velocity applies. Discharge is measured in cubic metres per second ($m^3 s^{-1}$), sometimes abbreviated in river studies to cumecs (dimensions $L^3 T^{-1}$). The velocity is numerically equal, then, to the discharge per unit normal area.

Natural currents change in both space and time. A flow is called *steady* if the velocity measured at a fixed point remains constant in direction and magnitude, and it is only in these flows that streamlines, pathlines and streaklines coincide and remain unchanged over time. When the velocity u measured at a point varies with time t, the flow is described as *unsteady* and the fluid particles are subject to a temporal acceleration, written as du/dt and measured in metres per second ($m s^{-2}$; dimensions LT^{-2}). The pattern of streamlines now changes from moment to moment, and any given pattern has only an instantaneous significance. River flow, for example, is unsteady

because there are seasonal and other variations in discharge. The flow of a river also varies in space as the current passes from the deep cross-sections typical of bends to the wide shallow ones between. Such a flow is called *non-uniform*, further examples appearing in Fig. 2.5. Fluid particles in non-uniform flows experience a spatial acceleration $u.du/dx$ ($m s^{-2}$), in which x is downstream distance, on account of the change of velocity from point to point along the flow. However, in a *uniform* flow, there are no velocity changes in the current direction and the streamlines are parallel. Thus the terms steady and unsteady, and uniform and non-uniform, permit in terms of spatio-temporal change the recognition of four classes of flow. Only in steady, uniform flows are there no accelerations whatsoever.

2.3.2 Limitations of inviscid flow theory

Any fluid motion can in principle be described in terms of a threefold group of general equations, called the Navier–Stokes equations after their discoverers. It is not necessary to set out these equations here, but it is important to note that they specify the motion in terms of the forces due to the pressure in the fluid, the inertia of the fluid, and the fluid viscosity. However, the equations are complex and solutions cannot always be achieved.

One condition under which a solution is generally possible is the assumption that the fluid possesses no viscosity and is *inviscid* or *ideal*. It is, for example, a comparatively straightforward matter to calculate and map streamlines for motions involving ideal fluids (Francis, 1975). The resulting flows are often strikingly close to reality, as measured in the field or from a laboratory experiment. The differences, which under certain circumstances are serious, lie in the fact that, although the velocity in an ideal fluid may vary perpendicular to the streamlines in response to changes of flow geometry, the flow can exert no frictional drag on adjoining solid boundaries because of the assumption of an inviscid fluid. This is a severe or even insuperable limitation in the treatment of many sedimentological problems.

2.3.3 Boundary layer theory

The concept of the *boundary layer*, introduced in 1904 by the German fluid dynamicist Ludwig Prandtl, allows many of the limitations of inviscid fluids to be overcome. According to this simple but revolutionary idea, the friction arising between a real fluid and a solid with which it is in relative motion is restricted to a comparatively thin layer, called the boundary layer, adjacent to the solid. Within the boundary layer there is a gradient of velocity, equivalent to a rate of shear, because of the viscous or equivalent stress that arises between the solid boundary and the fluid (see Eq. 2.2). There are considered to be no stresses in the flow outside the layer, which thus behaves inviscidly, although its streamlines may be distorted by the presence of the boundary and there may consequently be transverse changes of velocity. Thus the momentum of the fast, inviscid, outer flow is transmitted through the viscous fluid of the boundary layer as a stress applied to the solid boundary.

2.3.4 Conservation of matter (law of continuity)

Whether a fluid flow is treated as inviscid or viscous, or as a combination of the two, the law of conservation of matter, or continuity, must be satisfied throughout the field of flow. In the case of gases, which are compressible, and so may vary significantly

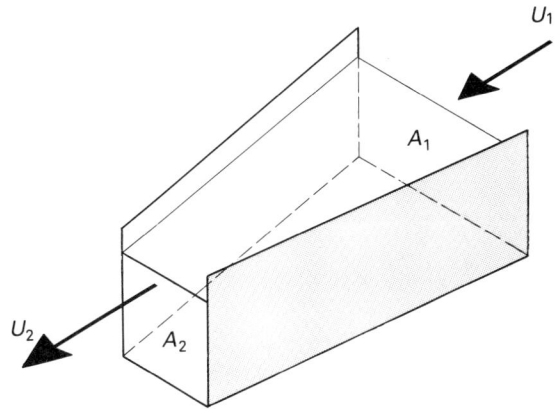

Fig. 2.6 The principle of conservation of matter, illustrated by the flow of a liquid in an open channel.

in density from place to place within the flow, it may be necessary to evaluate the motion in terms of the mass flow-rate. The flow of liquids normally presents fewer difficulties, because they can be assumed incompressible. It is then sufficient to conserve the volume flow-rate or discharge.

As a simple example, consider the flow of water in an impermeable channel varying in cross-sectional area from A_1 to A_2 between two stations along the flow (Fig. 2.6). If U_1 and U_2 are the respective overall mean velocities, continuity is satisfied if the discharge $Q = A_1 U_1 = A_2 U_2$. The same formula applies to a flow of water filling a pipe or cave conduit. There is more difficulty if the problem concerns, say, storm currents in a large water body with a shoreline, which acts as a barrier to flow. If the wind-driven current, always confined to the upper layers of the water body, has an onshore component of velocity, then in the lower layers there must be a compensating current with an offshore component of flow. Recalling that velocity is equivalent to unit discharge, continuity is satisfied in this case if the component velocity is zero when integrated from the bottom to the water surface at every place.

2.3.5 Conservation of energy (Bernoulli equation)

A conservation law of particular importance in fluid flow concerns energy, stating that the total energy on a streamline is a constant, varying from streamline to streamline. The most general form of the law is the equation, named after Daniel Bernoulli, its 18th century Swiss discoverer:

$$\tfrac{1}{2}\rho u^2 + \rho g y + p + E_{\text{loss}} = \text{constant},\qquad (2.6)$$

in which ρ is the fluid density, u the flow velocity at a point on the streamline, g the acceleration due to gravity (9.81 m s^{-2}), y the elevation of the point on the streamline above an arbitrary horizontal datum, p the pressure, and E_{loss} the frictional heat loss. Allen (1985) gives a proof.

Each term represents a particular kind of energy per unit volume of flow, measured in Joules per cubic metre (J m^{-3}) with the final dimensions $ML^{-1}T^{-2}$, the same as a stress. The first term is the kinetic energy per unit volume, that is, the energy of motion. The second is the potential energy per unit volume, that is, the energy due to the position of the fluid element relative to the Earth. The pressure energy is represented by the third term. The fourth term should be included when the fluid is viscous and under shear, but can often be safely neglected.

The fact that these four terms sum to a *constant* total energy on each streamline implies that the form of the energy may change as we follow the streamline from point to point in the flow. Potential energy may be exchanged for kinetic energy, as happens when water flows down the bed of a river. Another inference is that a high pressure can be achieved by slowing down the flow, provided that the potential energy is unchanged. The energy dissipated as heat will be particularly high if a flow can in some way be significantly slowed while its potential energy is reduced, as at the base of a spillway.

2.3.6 Froude number

The Bernoulli equation is especially helpful in exploring free-surface flows, that is, flows that involve a deformable interface between two fluids in relative motion, an example of which is provided by a river. Using the Bernoulli equation, and assuming no energy losses, it is readily proved (see Allen, 1985) that an open-channel flow of a given discharge and total energy may exist in either of two states, each specified by a particular value of:

$$\frac{q^2}{gh^3} = \frac{U}{(gh)^{1/2}},\qquad (2.7)$$

called the *Froude number* (*Fr*), after an English naval architect. In this formula, which is non-dimensional, q is another unit discharge, in this instance the discharge per unit width of flow (cubic metres per second per metre or m^3 m^{-1}s^{-1}; dimensions L^2T^{-1})

and h the flow depth. Note that the mean velocity $U = q/h$.

The two alternative flow states are called *sub-critical* and *supercritical* flow. Sub-critical flow, specified by a Froude number less than one, is typified by a water surface that is smooth and on which waves caused by, for example, throwing in a stone, are quickly damped out. Supercritical flows, typically fast and of small depth, have a Froude number greater than one and they can sustain large surface waves and related disturbances. The most impressive of these are antidunes, which typically break noisily and reform in a regular cycle. The boundary between the two states of flow is defined by $Fr = 1$, describing a critical flow. The total flow energy is at a minimum for a given unit discharge when $Fr = 1$, and is otherwise in excess of the minimum value.

The Froude number is also a statement about the balance between certain of the forces acting in open-channel flow. The velocity forming the numerator implies the inertial force, whereas the denominator is indicative of a body or gravitational force. The Froude number is therefore the ratio of the inertial to the gravitational force operating in the flow. As such, it defines the *stability* of the deformable water surface. In sub-critical flow, for example, the gravitational force dominates, and so the free surface tends to be smooth and flat. The gravitational force is insufficient to stabilize the surface when the flow is supercritical. These flows are subject to a variety of instabilities.

2.3.7 Laminar and turbulent flow

These are also different states of flow, but of a different kind than the sub-critical and supercritical flows just discussed. Laminar and turbulent motions are recognized in both internal and external flows, and in unconfined as well as confined settings.

The streamlines representing a *laminar flow* at an instant are either straight or smoothly curved and, depending on whether they record a uniform or a non-uniform flow, either parallel or sub-parallel. The streamlines are never intertwined but at all times highly ordered in appearance.

In the case of *turbulent flows*, the streamlines as depicted at each instant are intertwined in a very complicated way, their pattern (if it can be so described) changing dramatically from one instant to another. Only the pattern of streamlines averaged over a sufficiently long time has the ordered appearance of the streamlines of laminar motion. It is the presence of turbulent eddies, changing in size, shape and velocity structure from instant to instant, that creates the intricate streamline arrangements characteristic of turbulent flows. Most natural currents are turbulent and expressions of turbulence are everyday experiences. The gustiness of a strong wind, occasionally visualized in the form of flurries of debris, records the presence of huge eddies in the hurrying air. In river and tidal currents, surface features called 'boils' are generated as the larger eddies within the flow surge upward to interact with the water surface (Fig. 2.7). The water surface within each boil is elevated slightly above the general level, and the surface current is seen to be directed radially outward, both features proving the upthrust of fluid from the depths. These eddies take on a scale that is comparable to the flow depth itself.

Physically, the turbulence of a fluid means that mixing is going on at a macroscopic scale, irrespective of whether the fluid as a whole is in translational motion. The mixing is brought about by the eddies, through their growth and decay. Because of this action of the eddies, properties carried by the fluid, such as momentum, heat, and dissolved or suspended materials, tend to become quickly and uniformly spread throughout the entire flow field. However, an actually uniform spread is only achieved under special conditions, as other factors that are present tend to constrain the process. By contrast, in a laminar flow, mixing takes place only on a molecular scale, and thus very slowly.

For example, compare the rate at which sugar dissolves into a cup of coffee on stirring the cup, which creates turbulence, with the rate at which it passes into solution on leaving the cup unstirred (normally an unappealing overnight wait)! Mixing in turbulent flows ordinarily takes place orders of magnitude faster than in laminar ones, a fact that accounts for the much steeper near-bed velocity profiles of turbulent boundary layers than laminar ones (see below), a feature implying the much faster transfer of momentum to the boundary of a turbulent flow.

2.3.8 Reynolds number

The parameter now called the *Reynolds number* serves for laminar and turbulent flows in the same way as the Froude number for sub-critical and supercritical ones, namely, it is a non-dimensional quantity which distinguishes the two states of flow and which defines the degree to which one of the states is realized by any particular flow. The Reynolds number was introduced in 1883 and 1894 by the British mathematician and experimental physicist Osborne Reynolds, in the course of his seminal work on states of flow and their mathematical analysis. In its most general form, the Reynolds number (Re) may be written:

$$Re = \frac{\rho U L}{\eta} = \frac{UL}{\nu}, \qquad (2.8)$$

in which ρ is the fluid density, U a characteristic velocity, L a characteristic length, η the fluid molecular viscosity, and ν the fluid kinematic viscosity. From earlier explanations, it will be clear that the Reynolds number has the dimensions $ML^{-1}T^{-1}/ML^{-1}T^{-1}$. Physically, it describes the balance between the inertial flow forces, represented by the product in the numerator, and the viscous forces as described by the molecular viscosity. Rather like the Froude number, it is a criterion of stability, for turbulent motions may be regarded as unstable in comparison to laminar ones with their high degree of internal order.

A Reynolds number can be defined for any flow, for it is only a question of choosing an appropriate velocity and length. In the case of a river — a partly confined internal flow with a free surface — it is customary to take the mean flow velocity and the mean flow depth. The Reynolds number for a boundary layer may be framed in terms of the boundary-layer thickness and the velocity of the undisturbed flow. If our interest lies in the behaviour of sediment particles — a case of external flow — it is sensible to

Fig. 2.7 Boils 1–2 m across on the surface of a vigorous flood tidal current, Severn Estuary. Flow from right to left.

choose either the particle radius or diameter and the terminal settling velocity. When the Reynolds number is small, viscous forces dominate the motion, and inertial ones can be ignored. At high enough Reynolds numbers, on the other hand, inertial forces dominate and it is often possible to ignore viscosity, except close to the flow boundaries.

2.4   Characteristics of turbulent flows

2.4.1 Description and statistical treatment of turbulence

Great efforts have been devoted over many decades to the question of the description and analysis of turbulent flows, which are the fluid motions most often encountered naturally. Traditionally, turbulent flow has been thought of as essentially chaotic and consequently as amenable only to statistical description and analysis. The more modern view, as will be seen, envisages turbulent flow as partly deterministic, because it is now clear that it involves coherent flow structures, the size, shape and lifespan of which depend on overall flow properties (Cantwell, 1981).

The traditional statistical approach begins with the measurement over time of appropriate properties of the moving fluid at selected fixed points in the flow. For example, the flow velocity at each point can be measured in three mutually perpendicular directions using hot-wire or hot-film anemometers coupled to signal-processors (Fig. 2.8). It is customary, in the case, say, of a uniform, steady, turbulent channel flow, to designate the mean flow direction as the positive x-direction, the upward direction normal to the bed as the positive y-direction, and the bed-parallel direction to the right of the positive x-axis as the positive z-direction. The anemometer would then measure an instantaneous unsteady velocity u parallel with x, an instantaneous unsteady velocity v parallel with y, and an instantaneous unsteady velocity w parallel with z. By making the measurements over a sufficiently long period and then averaging, we obtain the time-averaged velocities $U > 0$, $V = 0$ and $W = 0$. The instantaneous and time-averaged turbulent velocities are related through:

$$u = u' + U$$
$$v = v' + V \qquad (2.9)$$
$$w = w' + W,$$

in which u', v' and w' are called the fluctuating turbulent velocity components. These fluctuating components are implicitly either positive or negative, but it is sometimes helpful to identify them explicitly as to direction using appropriate sub-scripts, for example, $u'_{forward}$, $u'_{backward}$, v'_{up}, v'_{down}, w'_{right}, and w'_{left}.

The values of the fluctuating velocity components can be used in a variety of ways to describe and analyse the turbulence. By calculating the standard deviation of each component, that is, the root-mean-square values $(\overline{u'^2})^{1/2}$, $(\overline{v'^2})^{1/2}$ and $(\overline{w'^2})^{1/2}$, where the bar implies the time-averaged value, we can estimate the strength or intensity of the turbulence and also its degree of anisotropy. Turbulence intensity is usually given as the ratio of the root-mean-square value of the streamwise fluctuating velocity to the time-averaged streamwise velocity. The turbulence may only be described as *isotropic* if the root-mean-square values of the three component fluctuating velocities prove to be the same. When different values are found, the turbulence is *anisotropic*. The degree of symmetry of the turbulence in any one of the principal directions may be explored by comparing the average magnitudes and root-mean-square values of the positive and negative parts of the appropriate fluctuating velocity, for example, v'_{up} and v'_{down}. Perfect symmetry requires that the two quantities are the same in the up and down directions. Isotropic turbulence is almost invariably symmetrical in all three principal directions, but anisotropic turbulence is normally asymmetrical parallel to either the y-direction or to both the x-axis and the y-axis. Turbulent channel flow is an especially important case of asymmetrical anisotropic turbu-

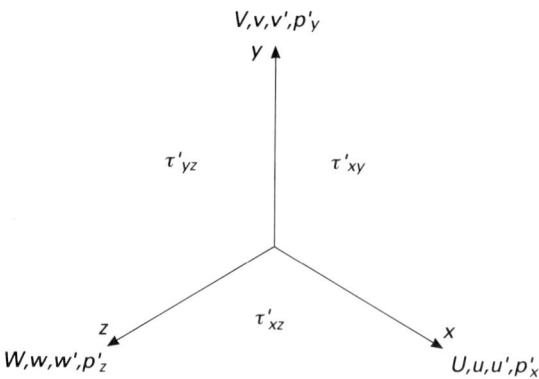

Fig. 2.8 Variables ascertainable from turbulent flows in an orthogonal system of coordinate axes in which the x-direction is parallel with the mean flow.

lence, for it turns out in this type of flow that v'_{up} exceeds v'_{down} in average magnitude. This is equivalent to saying that the vertical components of the turbulence exert a net upward fluid force (proportional to the velocity squared), which appears to explain why dense mineral particles can be transported in suspension by a turbulent wind or river current.

2.4.2 Reynolds stresses and eddy viscosity

The description and statistical treatment of turbulence just outlined is wholly kinematic and makes no reference to the forces either operating within the flow or exerted by the flow on its solid boundaries. When it comes to understanding the response of sedimentary boundaries to turbulent flows, it is important not only to be able to describe the turbulence itself, but also to know about the steady and fluctuating pressure and shear forces due to the turbulence that influence the bed.

We saw above that resistance to flow during laminar motion arises at a molecular scale because of the forces gluing the molecules together. In a turbulent flow, intermolecular friction is also present but is vastly exceeded by resistance to motion on a macroscopic scale, arising from friction between the constantly reforming and decaying eddies in relative motion that are present. The resisting forces associated with the eddying are called the Reynolds stresses. Listing the steady time-averaged values, they comprise the three components of the pressure:

$$\overline{p'_x} = \rho\,\overline{u'^2}, \; \overline{p'_y} = \rho\,\overline{v'^2}, \; p'_z = \rho\,\overline{w'^2}, \quad (2.10)$$

where the sub-scripts imply the pressure parallel with the respective axes (see Fig. 2.8), and the three components of the shear stress:

$$\overline{\tau'_{xy}} = -\rho\,\overline{u'v'}, \; \overline{\tau'_{yz}} = -\rho\,\overline{v'w'}, \; \overline{\tau'_{xz}} = -\rho\,\overline{u'w'} \quad (2.11)$$

where the sub-scripts xy, yz and xz indicate respectively the xy-plane, the yz-plane and the xz-plane. The primes again imply fluctuating values and the bars the average over time. The corresponding instantaneous values of the Reynolds stresses can be represented by removing the bars from the expressions appearing in Eqs 2.10 and 2.11.

In analogy with Eq. 2.2 above, written for the viscosity due to molecular forces, we can proceed to write for each of the principal directions in a turbulent flow an expression for a macroscopic or 'eddy' viscosity. Choosing for illustration the x-direction:

$$\tau_{xy} = \rho\,\overline{u'v'} = \eta_{eddy}\frac{dU}{dy}, \quad (2.12)$$

in which η_{eddy} is the eddy viscosity and U the point value of the time-averaged streamwise velocity. Unlike the molecular viscosity, the eddy viscosity is neither a fluid property nor a constant, but varies with position and direction within a given flow, as well as with Reynolds number and with the overall character of the flow. It is not readily predicted, but on measurement tends to be orders of magnitude greater than the molecular viscosity.

As far as sediment beds are concerned, the quantities whose behaviour at the bed is of greatest interest are the fluctuating and steady streamwise stresses, and the corresponding pressure values parallel with flow and perpendicular to the bed. These quantities together determine the rapidly changing drag and lift forces acting on either some small area of the bed or a single sediment grain exposed at the surface.

2.4.3 Mean velocity profile of a turbulent boundary layer

A very rapid increase in the time-averaged streamwise velocity with increasing distance from the bed is observed experimentally in turbulent boundary layers developed in wide channels or on flat surfaces (review in Schlichting, 1960). A formula of the form:

$$\frac{U}{U_{max}} = \left(\frac{y}{\delta}\right)^n, \quad (2.13)$$

describes these profiles quite well, where U is the time-averaged velocity at a distance y from the boundary, U_{max} the velocity at the outer edge of the boundary layer, δ the boundary-layer thickness (equal to the flow depth in a wide channel), and n an exponent. The exponent varies from about 1/5 at Reynolds numbers placing the flow just in the turbulent range to about 1/7 at the high Reynolds numbers typical of, say, most rivers (Fig. 2.9). The all-important time-averaged shear stress τ_0 at the bed is given by the definition:

$$\tau_0 = \frac{f}{8}\rho\,U_{mean}, \quad (2.14)$$

in which f is a non-dimensional coefficient called the Darcy–Weisbach friction coefficient, ρ the fluid density, and U_{mean} the mean of the local time-averaged

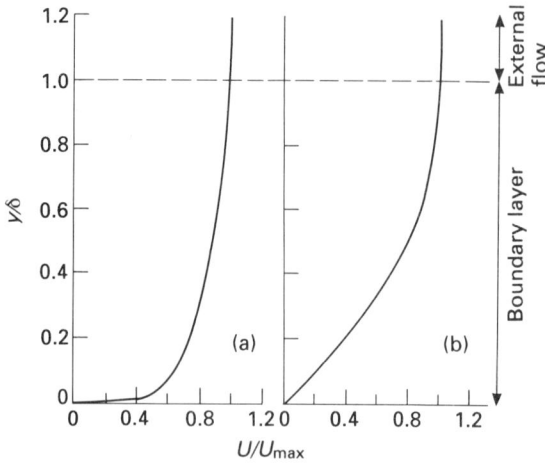

Fig. 2.9 A comparison of (non-dimensional) velocity profiles in (a) turbulent, and (b) laminar boundary layers.

velocity over the whole boundary layer. The Darcy–Weisbach friction coefficient, which is very small or small, can be predicted only under special conditions, and is generally assigned a value empirically on the basis of the Reynolds number and the character of the flow boundary.

It is interesting to compare the laminar boundary layer corresponding to the turbulent boundary layer just discussed (review in Schlichting, 1960). The laminar velocity profile is given by:

$$U = U_{max} - \frac{\tau_0}{2\eta\delta}(\delta - y)^2, \qquad (2.15)$$

where U is the streamwise velocity at a distance y from the bed, U_{max} the velocity at the outer limit of the boundary layer, τ_0 the bed shear stress, η the molecular viscosity, and δ the boundary layer thickness. The profile is a parabola and much less steep at the bed than in the turbulent case (Fig. 2.9). The formula:

$$\tau_0 = \frac{2\eta U_{max}}{\delta}, \qquad (2.16)$$

gives the value of the shear stress at the bed in the laminar case. Clearly, the turbulent boundary layer exerts by far the larger bed stress.

Equation 2.13 is physically correct in giving $U = 0$ at $y = 0$ but does not describe accurately the profile of velocity near to the bed. Provided that the boundary is sufficiently even, it is found experimentally that adjoining the bed there is a very thin zone in which

the velocity profile is essentially linear. Such a pattern in a thin layer points to laminar flow and to a situation in which the viscous forces are so dominant that turbulent eddying is suppressed. The zone of linear change of velocity is called the laminar or, preferably, *viscous sub-layer*, and is of a thickness given experimentally by:

$$\delta_{viscous} = 11.5 \frac{\eta}{\rho(\tau_0/\rho)^{1/2}}, \qquad (2.17)$$

in which the quantity $(\tau_0/\rho)^{1/2}$ is called the *shear velocity* and is measured in m s^{-1}. The shear velocity, with the dimensions of $(L^4M/L^2MT^2)^{1/2} = LT^{-1}$, is a quantity often quoted instead of the shear stress itself in discussions involving turbulent boundary layers and their effects. Equation 2.17 shows that the viscous sub-layer increases in thickness with fluid viscosity but thins as the shear stress grows larger.

A better description of the velocity profile, to within about two viscous sub-layer thicknesses from the flow boundary, is given by the so-called universal velocity distribution (Schlichting, 1960). This logarithmic law takes the general form:

$$U = \left(\frac{\tau_0}{\rho}\right)^{1/2} \frac{1}{k} \ln \frac{y}{y_0}, \qquad (2.18)$$

in which the shear velocity again appears, k is the von Kármán constant, ln stands for natural logarithms and, because the formula is logarithmic, y_0 is the notional distance from the flow boundary at which the time-averaged velocity apparently becomes zero. The universal velocity distribution is very convenient for the estimation of the shear stress from either field or laboratory measurements, for it is necessary only to graph on appropriate scales the observed time-averaged velocity against distance from the flow boundary (Fig. 2.10). The notional height y_0 is given as an intercept on extrapolating the scatter of measured points, and the shear stress follows by measuring the slope $(1/k)(\tau_0/\rho)^{1/2}$. The von Kármán constant is normally taken to equal 0.4, but varies slightly with the Reynolds number and the amount of sediment present (McCave, 1973; Nouh, 1989).

The presence of the viscous sub-layer, with a linear velocity profile, affords a natural explanation for the paradoxical prediction by the universal law of a zero velocity at a non-zero distance from the flow boundary. In naturally occurring turbulent boundary layers, much of the unevenness of the bed is provided by the sediment particles forming the bottom. If these are so small that they lie enclosed within the viscous sub-

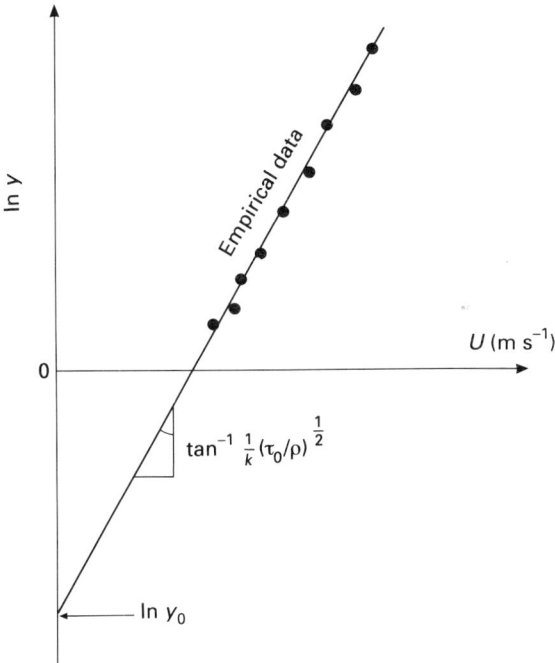

Fig. 2.10 Practical application of the logarithmic (universal) velocity profile for the turbulent boundary layer.

layer, the near-bed flow is dominated by viscous forces and the flow can be described as *hydraulically smooth* (Fig. 2.11). Smooth boundaries are afforded by most mud beds and by the finer grades of sand. Under these

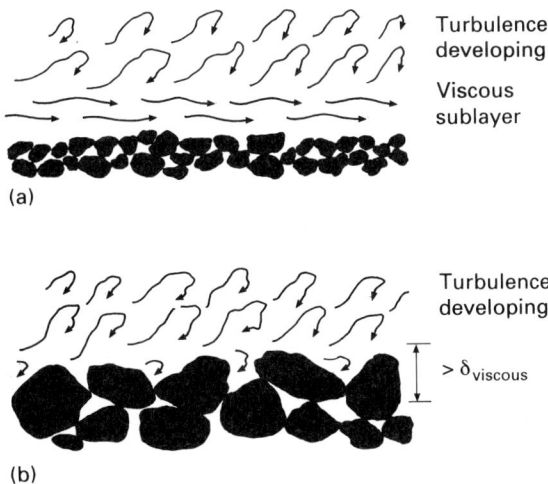

Fig. 2.11 Hydraulic classification of granular flow boundaries. (a) Hydraulically smooth. (b) Hydraulically rough.

conditions, the Darcy–Weisbach coefficient is calculable and is found to decline slowly with increasing boundary-layer Reynolds number. A so-called *hydraulically rough* boundary arises when the bed grains are so large in size that they exceed the thickness of the (notional) viscous sub-layer and so protrude into the turbulent portion of the flow (Fig. 2.11). The intercept y_0 given by the universal law then increases with the scale of the bed unevenness and is usually called the roughness length. Experiments show that the Darcy–Weisbach coefficient for a rough boundary takes a value that is independent of the Reynolds number but which increases with the roughness length and with the ratio of particle size to boundary-layer thickness. Gravel particles and the coarser grades of sand typically afford hydraulically rough beds.

2.4.4 Turbulent eddies and the production of turbulence

Turbulent boundary layers have proved over the last 20 years to be far less chaotic than they seemed (Cantwell, 1981), chiefly because of advances in techniques of flow visualization, and the application of the signal-processing power of modern computers to the output from arrays of sophisticated hot-wire and hot-film anemometers. The structural picture these studies suggest is of (1) relatively persistent, streamwise, corkscrew vortices in a thin zone adjoining the flow boundary, which are (2) periodically disrupted and lifted up (bursting) into the outer parts of the flow as steeply inclined, horseshoe or hairpin vortices, that (3) combine together to form larger eddies that grow over a significant period to a scale comparable with the boundary-layer thickness itself. Studies exploring all or substantial parts of this scheme are described by Falco (1977, 1991), Nakagawa and Nezu (1981), Head and Bandyopadhyay (1981), Thomas and Bull (1983), Kim and Moin (1986), Utami and Ueno (1987), and Grass *et al.* (1991). An appreciation of these organized, coherent structures of turbulence is becoming increasingly important to an understanding of sediment transport and the origin of sedimentary structures.

The streamwise corkscrew vortices that are present in a thin wall zone several viscous sub-layers thick express themselves as a 'streaky' structure (Figs. 2.12, 2.13) in visual studies exploiting soluble dyes, hydrogen bubbles or sand grains as markers (Kline *et al.*, 1967; Grass, 1971; Nakagawa & Nezu, 1981; Smith & Metzler, 1983; Jang *et al.*, 1986). The vortices are

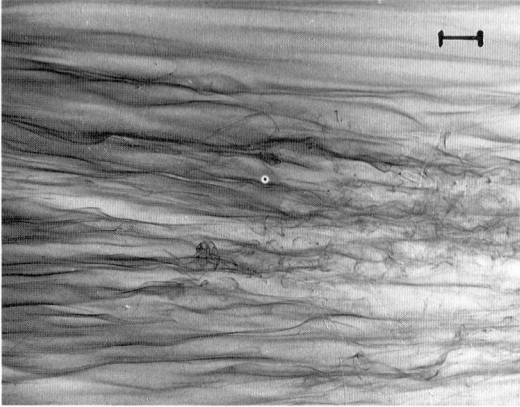

Fig. 2.12 Streaky structure in the wall-region (*xz*-plane) of a turbulent boundary layer, visualized using dye released from wall. Current from left to right. Scale bar 0.01 m.

oppositely rotating in pairs, with a typical transverse scale on hydraulically smooth boundaries of about 8.6 viscous sub-layer thicknesses, Eq. 2.17, as set by a resonance effect. They are not individually structures of the flow that are permanent, but slowly grow up and then decay as they are convected downstream. As may be judged from Fig. 2.12, the vortices tend to waver or meander as they drift along. Their streamwise length at maturity is very large compared to the transverse and vertical scales. Flow markers such as dye or sand grains become concentrated in the zones of convergent bottom flow, called 'low-speed streaks'

(Fig. 2.13). In these zones of upward movement the fluid momentum is relatively low and the velocity gradient weak. The zones emptied of flow markers, termed 'high-speed streaks', are characterized by high-momentum fluid and a steep velocity gradient.

Periodically, the corkscrew vortices drifting past any chosen position on the bed are found to experience a sudden disruption, during which a blob of low-momentum fluid is drawn outward away from the bed into the faster portion of the boundary layer, in the form of a hairpin-shaped vortex with legs (Head & Bandyopadhyay, 1981; Perry & Chong, 1982; Kim & Moin, 1986; Utami & Ueno, 1987). A series of these hairpin vortices, which may be likened to a forest, is photographed in the vertical streamwise plane in Fig. 2.14, and in Fig. 2.15 an attempt is made to represent them schematically. The width of a vortex is comparable to that of a pair of corkscrew vortices beneath and its legs lean downstream at an angle of 40–50° from the bed. On hydraulically rough boundaries, the vortices appear to scale as the size of the roughness elements (Grass *et al.*, 1991). The motion within a typical hairpin vortex is complex but, in a reference frame moving at the average speed of the vortex, is like the movements described by a sequence of spots of paint spaced along each of a series of elastic threads parallel to the bed and at right angles to the current (i.e. vortex lines), when that series is gathered up at a place, stretched, and twisted like a ball in top spin. The data from hot-wire studies are similarly difficult to interpret in terms of the details of the motion within the

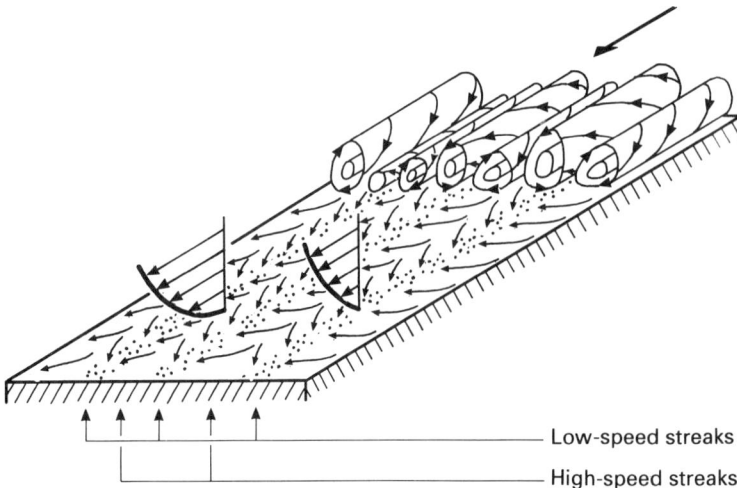

Low-speed streaks
High-speed streaks

Fig. 2.13 Flow pattern associated with the streaky structure of the boundary layer wall-region.

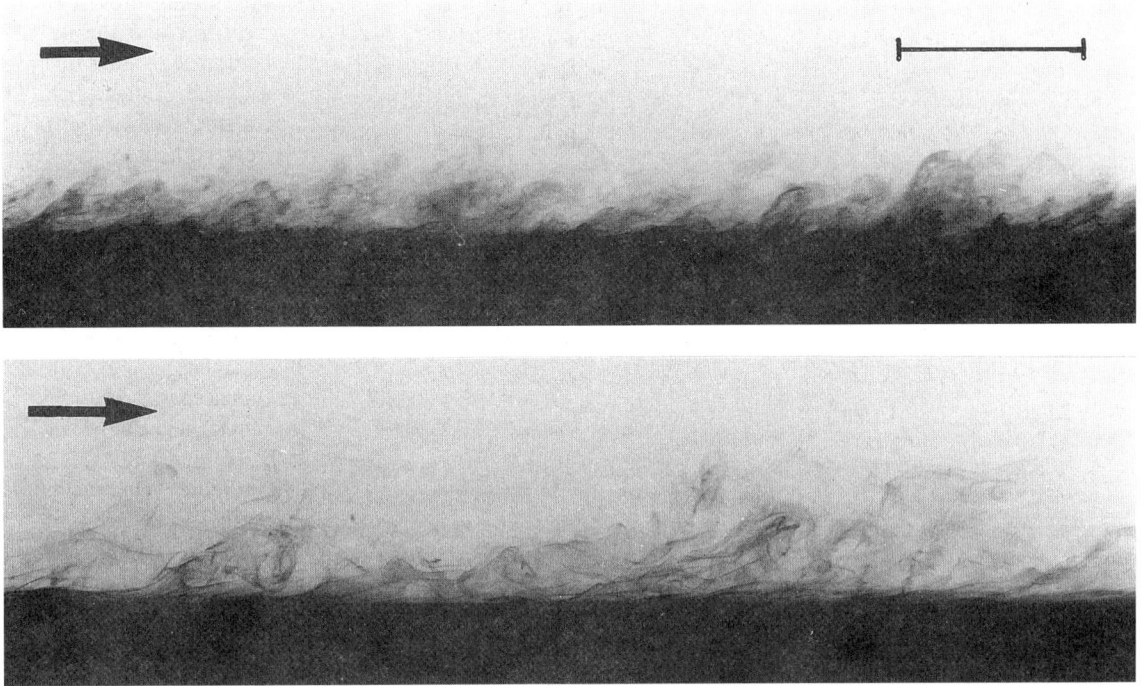

Fig. 2.14 Two views in the vertical streamwise plane (xy-plane) of hairpin vortices rising from the wall-region of a turbulent boundary layer, visualized by dye released from the wall. Current from left to right. Scale bar 0.05 m.

vortices. However, different parts of the vortex yield characteristic anemometer signals, especially when the instrument is measuring the velocity field in the plane of symmetry of the vortex, that is, in the dividing xy-plane. Where the fluctuating components are either $v' < 0$ and $u'v' < 0$, or $v' < 0$ and $u'v' > 0$, the motion of the fluid is toward the bed (called a sweep) and the front of the vortex is passing the instrument. The hot wire lies on the trailing side of the vortex when the fluctuating components read either $v' > 0$ and $u'v' < 0$, or $v' > 0$ and $u'v' > 0$. Here the motion is away from the wall (called an ejection).

Visualization studies (Head & Bandyopadhyay, 1981; Utami & Ueno, 1987) indicate that the hairpin vortices in the forest rising from the bed become organized, and possibly twisted rope-like together, into a series of large horseshoes ranging to roller-shaped vortices inclined downstream at about 20° from the bed. These vortices, convected at a speed slightly less than the mean boundary layer velocity, probably create the 'boils' visible on the surface of a turbulent river (Fig. 2.7). The vortex streamwise spacing appears to be a

few times the flow or boundary-layer thickness, and their width about half this value. It seems that the 'bursting' of boundary layer streaks to form hairpin vortices is controlled and triggered by the passage of these larger vortices in the outer parts of the flow (Thomas & Bull, 1983; Aubrey *et al.*, 1988). Figure 2.16 suggests the form and internal motion of the eddies and gives the observed timing of bursting and streak-formation relative to their passage. The instantaneous bed shear stress fluctuates substantially in relation to position beneath the larger eddies, especially in association with bursting.

2.5 An introduction to types of fluid flow

2.5.1 Open-channel flow

This term is applied to the flow of a liquid confined in a channel and with a free surface exposed to the atmosphere. To the sedimentologist, river flow is implied. Chow (1959) thoroughly reviews open-channel flow.

Open-channel flows are driven by gravity, that is,

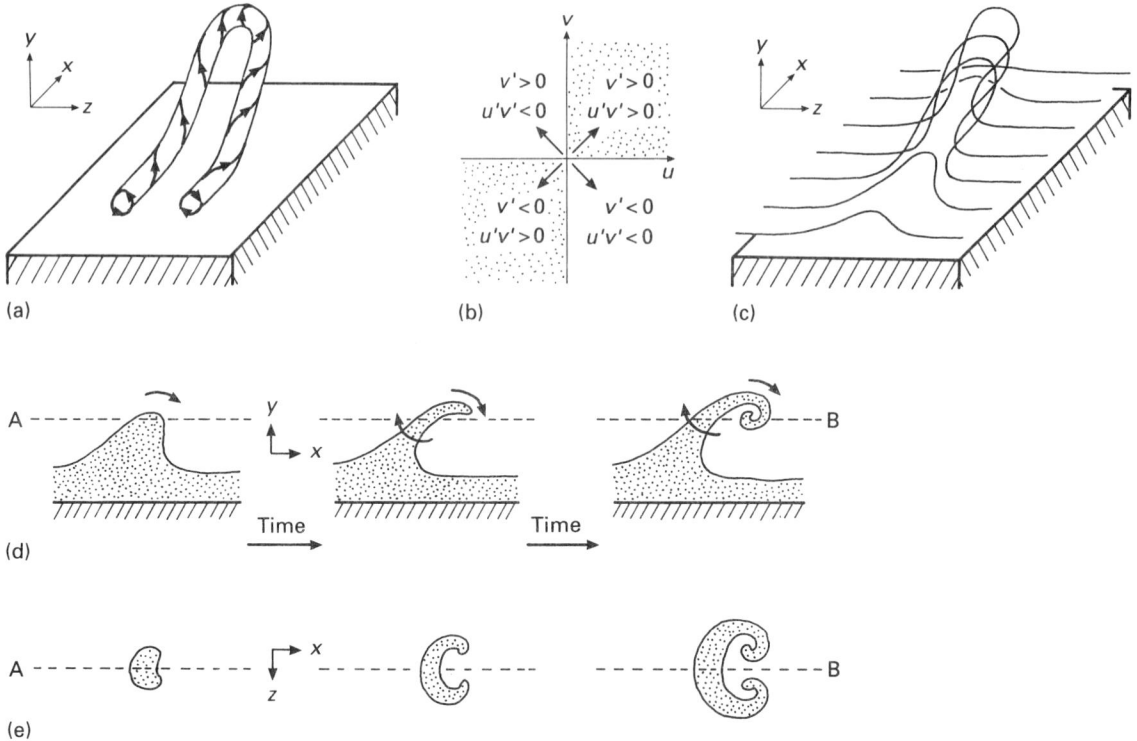

Fig. 2.15 Schematic representations of a horseshoe/hairpin vortex in the wall-region of a turbulent boundary layer. (a) A vortex represented as a vortex tube. (b) Classes of anemometer signal in the streamwise plane, specified in terms of fluctuating velocity components. (c) A vortex represented by twisted vortex filaments. Adapted from Utami and Ueno (1987). (d, e) A vortex represented by the progressively distorting shape in the streamwise and horizontal planes of a low-momentum wall layer. Adapted from Perry and Chang (1982).

by the weight of the liquid contained in the channel acting along the sloping channel bed, the liquid thereby exchanging potential for kinetic energy. Opposing the motion is the frictional drag exerted on the flow by the wetted bed and banks. In analysing the motion in the practical case of a river, we can ignore the presence of the atmosphere, since air is of very small density compared to water.

Assuming that the flow is uniform and steady, and takes place in a straight channel of wide, rectangular cross-section, the balance between the driving and resisting forces leads to the simplified relationship, fully proved in Allen (1985):

$$\tau_0 = \rho g h S, \qquad (2.19)$$

in which τ_0 is the mean shear stress acting over the wetted perimeter, ρ the fluid density, g the acceleration due to gravity, h the flow depth, and S the slope of the parallel bed and water surfaces. The bed shear stress thus increases linearly with the depth and slope. Substituting for the stress from Eq. 2.14, we discover that:

$$U_{mean} = \left(\frac{8S}{f}\right)^{1/2} (gh)^{1/2}, \qquad (2.20)$$

and on further rearranging that:

$$\left(\frac{8S}{f}\right)^{1/2} = \frac{U_{mean}}{(gh)^{1/2}} = Fr, \qquad (2.21)$$

where U_{mean} is the overall mean flow velocity and Fr the Froude number. Unlike the shear stress, the velocity increases as the square root of the depth and slope, and inversely as the square root of the friction coefficient. The Froude number behaves similarly with respect to the slope and friction coefficient.

Four regimes of open-channel flow can be identified using the Froude number above and the Reynolds number (Eq. 2.8), calculated from the mean

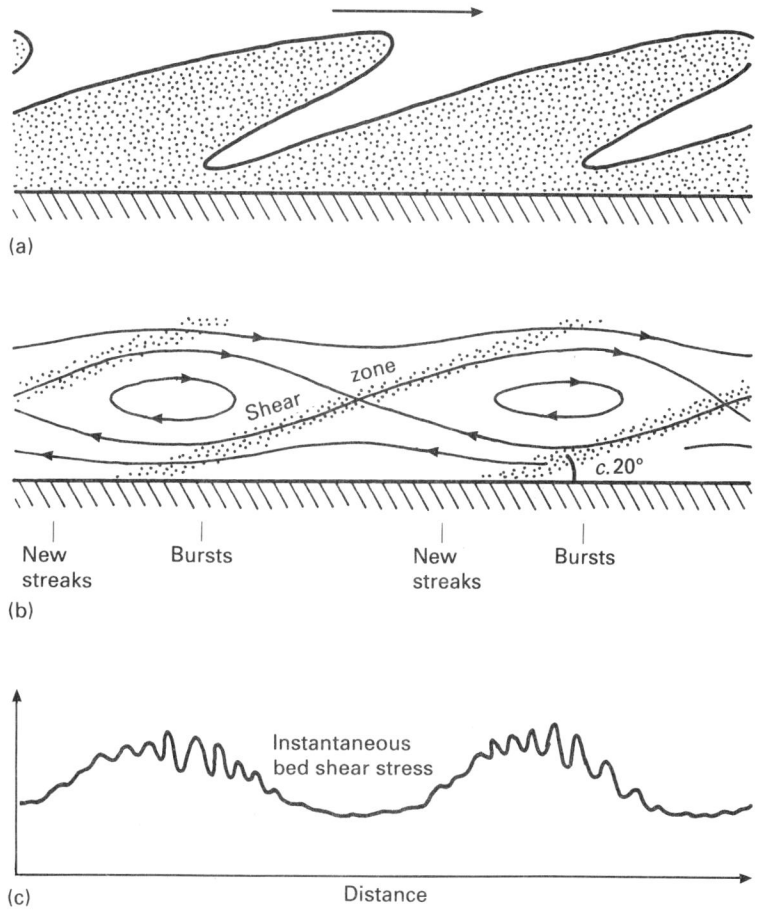

Fig. 2.16 Schematic representation (streamwise plane) of the larger coherent structures of the turbulent boundary layer and the relationship of streaks and bursts to them. Based largely on Thomas and Bull (1983). (a) Schematic instantaneous representation of the larger eddies. (b) Steady flow pattern within the eddies in a frame of reference carried with the eddies, and the relationship of streak creation and bursting to the eddies. (c) Spatial variation of the instantaneous bed shear stress beneath the eddies (see (b) for relationship to streaks and bursts).

flow velocity and flow depth (Fig. 2.17). The regimes are defined by the critical Froude number, $Fr = 1$, and $500 < Re < 2000$, the range for the laminar–turbulent transition. From the graph, it may be suggested that all rivers are turbulent and that many, especially when in flood, are supercritical. Natural laminar flows of open-channel type are probably rare, and chiefly confined to films of rain water running off rock surfaces or mudflats at low tide.

As a river flows down its sloping bed, exchanging potential for kinetic energy, power is made available which can be used to perform sedimentologically useful work, for example, erosion or sediment transport. The power of the stream, represented by the small Greek letter omega (ω), is the product of the mean bed shear stress with the mean flow velocity, that is, $\tau_0 U_{mean}$ or its equivalent. It has the dimensions $(ML^{-1}T^{-2})(LT^{-1}) = MT^{-3}$, and is measured in watts per square metre (W m^{-2} or N m s^{-1}m^{-2}).

2.5.2 Oscillatory flows due to waves and tides

A very important property of the free surface between two immiscible fluids, and of any sufficiently thin mixed zone between two miscible ones, is that travelling undulations or waves can be created on the interface by the application of suitable forces. The most important waves occurring naturally are due on the one hand to the wind (Wiegel, 1964; Sleath, 1984), as it blows over oceans, seas, lakes and rivers, and on the other to the pull of the Earth, Moon and Sun on the global ocean, which creates the tide (Pugh, 1987).

When the wind blows over water, it exerts on the surface both a tangential drag and, because of the turbulent eddies present, a pushing and pulling as the normal component of the local pressure changes. The disorderly hummocks and hollows that consequently appear on the surface are convected with the wind and those of a particular scale, set by the wind speed,

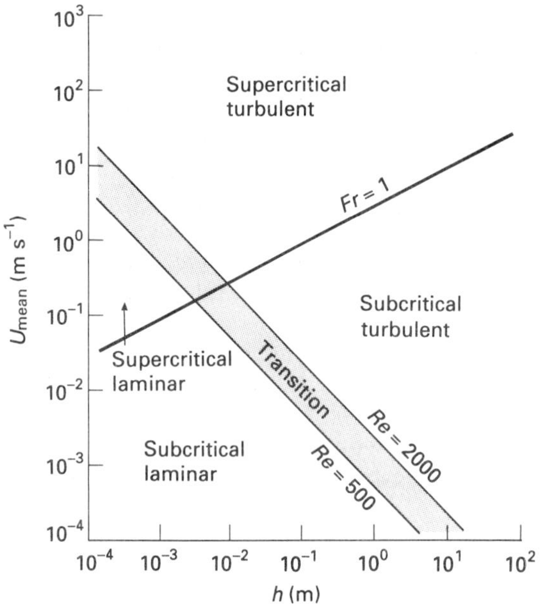

Fig. 2.17 Regimes of open-channel flow as defined by the critical Froude number (*Fr*) and the Reynolds number (*Re*) for transition to turbulence, where *h* is the mean flow depth and U_{mean} is the average flow velocity.

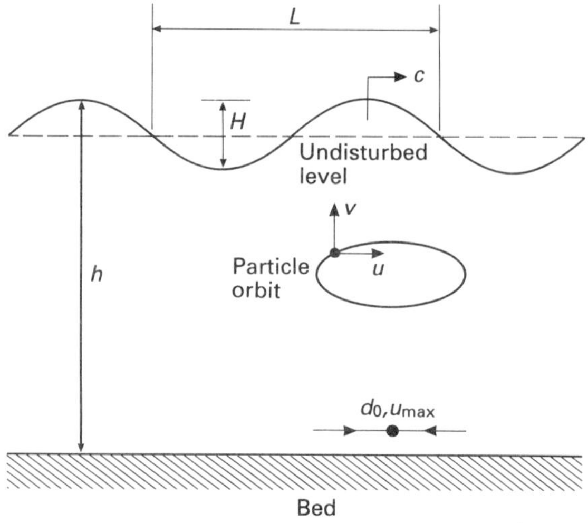

Fig. 2.18 Definition diagram for waves travelling over the surface of water (or over a similar interface): *c*, wave celerity; *h*, water depth; *H*, wave height; *L*, length.

tend to grow preferentially in amplitude, through the operation of a non-linear wavelength-selection mechanism. These preferred undulations subsequently come to dominate the observed spectrum of waves, provided that the wind remains steady and blows for long enough over a sufficient distance. Two forces prevent these waves from growing indefinitely in height. One is the gravity force, which is generally dominant, and the other is surface tension, effective only for very small waves between immiscible fluids.

As perceived by a stationary coastal observer, the tide is the vertical rise and fall of the sea, a movement accompanied by horizontal currents called tidal streams. The tide results from the combined effects of the masses of the Earth, Moon and Sun on the mobile water covering today nearly three-quarters of the surface of the Earth (Pugh, 1987). Consequently, the tide is moderated as well as driven by gravitational forces. However, the relative motion of the three heavenly bodies is complex, and where a detailed description is needed, the tide should be treated as a dominant wave modulated by a range of other periodicities. The basic wave has a period of 12 h and 26 min, set by the period of rotation of the Earth and

the period of the Moon's orbiting of the Earth, but significant effects on semi-monthly, monthly, semi-annual and even longer time scales (e.g. the lunar nodal cycle of 18.6 years) are also evident (Pugh, 1987). Sedimentologically, the semi-diurnal and semi-monthly or spring-neap periodicities are especially important, for they are detectable as bedding and textural patterns in tidal deposits.

Wave motion has proved to be a fruitful field of study by applied mathematicians for over a century, and many theories of waves, each approximating closer to the truth than its predecessors, have now been advanced (Sleath, 1984). The simplest, and in many respects the most useful, is Airy's theory, which assumes monochromatic, sinusoidal waves of very small amplitude on a still, inviscid fluid. Referring to the definitions in Fig. 2.18, the celerity *c* of waves of wavelength *L* and period *T* on water of depth *h* is given by:

$$c = \frac{gT}{2} \tanh\left(\frac{2\pi h}{L}\right), \qquad (2.22)$$

where *g* is the acceleration due to gravity and tanh stands for one of the extensively tabulated (see Wiegel, 1964) hyperbolic functions (the others are sinh and cosh). Tanh *x* varies such that, when *L* is very small compared to *h*, as is true for wind waves in the open ocean and during calms in shelf seas (so-called short or deep-water waves), Eq. 2.22 simplifies to:

$$c = \left(\frac{gL}{2\pi}\right)^{1/2}. \qquad (2.23)$$

Similarly, when L is large compared to h we find the simplification:

$$c = (gh)^{1/2}, \qquad (2.24)$$

characteristic of what are called long or shallow-water waves. The corresponding equations for wavelength are as follows. In water of any depth:

$$L = \frac{gT^2}{2\pi} \tanh\left(\frac{2\pi h}{L}\right), \qquad (2.25)$$

for deep-water conditions:

$$L = \frac{gT^2}{2\pi}, \qquad (2.26)$$

and for waves in shallow water:

$$L = (ghT^2)^{1/2}. \qquad (2.27)$$

Typically, wind waves have periods of the order of 1–10 s and wave lengths of the order of 1–100 m, and so all three equations are needed to describe them. On the other hand, the tide has a wavelength ideally equal to half the Earth's circumference, and so can invariably be treated as a long wave. Note that Eq. 2.24 is effectively the critical Froude number (see Eqs 2.7, 2.21), implying that the waves would be stationary relative to the ground if on water flowing opposite to the direction of wave propagation at a speed $U = c$.

When waves travel over a fluid interface, the fluid particles distant from the interface are set in orbital motion, an aspect of behaviour described by the Airy theory (Fig. 2.19). In the case of short or deep-water waves, the orbits are circles which in diameter decline exponentially downward. Shallow-water or long waves yield orbits of constant diameter but increasing ellipticity with depth. In the intermediate case, the orbits decline downward in size and also change downward from circular to increasingly elliptical. What is of interest sedimentologically is the character of the orbit at the bed, for it is the forces exerted here that affect sediment. In the general case (see Fig. 2.18), according to Airy, the maximum horizontal component of the velocity of a near-bed water particle due to a wave of height H is given by:

$$u_{max} = \pm \frac{\pi H}{T \sinh\left(\frac{2\pi h}{L}\right)}, \qquad (2.28)$$

and the orbital diameter by:

$$d_0 = \frac{H}{\sinh\left(\frac{2\pi h}{L}\right)}. \qquad (2.29)$$

For long waves, for example the tide, these equations simplify to:

$$u_{max} = \pm \frac{1}{2} H \left(\frac{g}{h}\right)^{1/2}, \qquad (2.30)$$

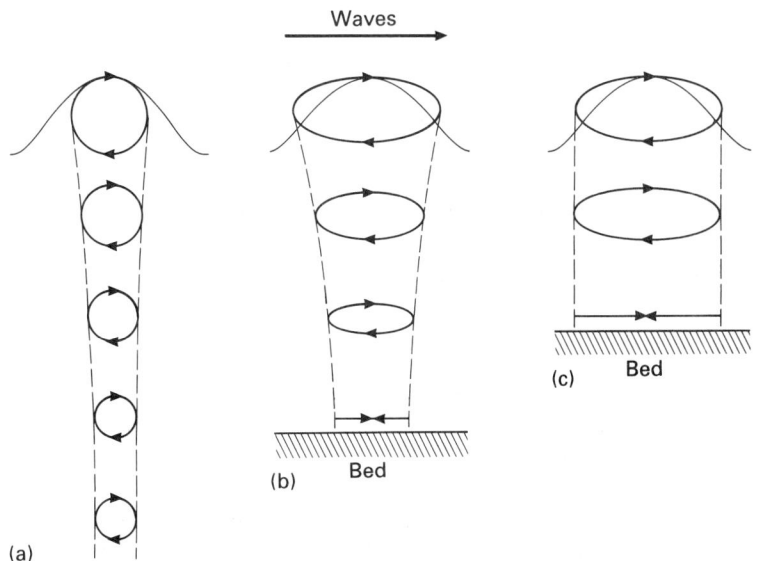

Fig. 2.19 Water–particle orbital motions beneath (a) short (deep-water), (b) intermediate, and (c) long (shallow-water) waves.

and:

$$d_0 = \frac{H}{2\pi\left(\dfrac{h}{L}\right)}. \tag{2.31}$$

This aspect of the Airy theory gives a very useful first-order description of wave currents, but suffers from the defects of its chief assumptions. Real waves have a finite amplitude and become distorted in shape, the crests sharpening relative to the troughs, in shallow water, causing an asymmetry in the pattern of the orbital velocity (forward stroke faster than backward stroke). Furthermore, water is viscous, and not the ideal fluid assumed, and effects are observed in the laboratory and field which are not encompassed by the Airy theory. One of these is a weak but significant drift or mass-transport current in the direction of wave propagation.

Again because of viscosity, the maximum orbital velocity given by Eqs 2.28 and 2.30 may in reality be expected to influence the bed through a thin boundary layer, across which the velocity profile changes in pattern and strength on the same period as the waves. The boundary-layer thickness is of order $(\eta T/\pi\rho)^{1/2}$, where η is the fluid molecular viscosity and ρ the density. Experiments show that transition to turbulence within the boundary layer occurs at a Reynolds number between about 10^4 and 10^5, calculated on the basis of u_{max} and the above thickness, depending on the bed roughness. Note that this transitional Reynolds number is one to two orders of magnitude greater than in the case of uniform, steady open-channel flow (Fig. 2.17). The presence at the bed of a turbulent wave boundary layer does not necessarily imply that the overlying water is also turbulent.

2.5.3 Stratified flows and gravity currents

Although river flow and the propagation of wind waves and the tide take place in the presence of the atmosphere, and therefore in a context of stably stratified fluids, effects due to the air can be safely ignored because of its very small density compared to water. This simplification is not justified where the fluid strata are similar in density, as in the oceans and in sufficiently large lakes, where small density differences can arise because of differences of temperature and content of dissolved salts or dispersed solids. These stratified flows can be driven by buoyancy as well as by pressure and gravity forces (Turner, 1973; Yih, 1980), and in natural environments involve

water or air masses on a thickness scale of 10^1–10^3 m.

Internal waves can be generated between gravitationally stable layers of miscible fluid similar in density in much the same way as on a water surface beneath the atmosphere. Where the mixing layer between two semi-infinite fluid strata is very thin, waves of length L will propagate over the interface at the celerity:

$$c = \left(\frac{gL}{2\pi}\frac{\rho_2-\rho_1}{\rho_2+\rho_1}\right)^{1/2}, \tag{2.32}$$

where ρ_1 and ρ_2 are the density respectively of the upper and lower fluid layers. Note how Eq. 2.32 collapses to Eq. 2.23 when ρ_2 is very much greater than ρ_1. Comparing the two equations, we see that the celerity of internal waves in an ocean or lake is much less than the speed of surface waves of a similar wavelength.

Internal waves can also arise where there is continuous change of density with depth in a thick mixing layer between gravitationally stable fluid strata. The period of these waves is:

$$T = 2\pi\left(-\frac{g}{\rho_r}\frac{d\rho}{dy}\right)^{-1/2}, \tag{2.33}$$

where ρ_r is a reference density characteristic of the mixing layer and $d\rho/dy$ the density gradient, y being measured positive upward.

It is often the case that the several layers in a stable stratification of fluids are moving relatively, as is normal in the atmosphere and the ocean. The stability of the interface between each pair of layers then becomes an important matter, as it governs the extent of mixing between the strata. Consider the simple case of a layer of fluid of thickness h and density ρ_2 travelling at a speed U over a bed beneath still fluid of a lower density ρ_1. Whether or not the equivalent of wind waves can propagate in the opposite direction to the current over the interface is determined by the value of:

$$Fr' = \frac{U}{\left(g\dfrac{\rho_2-\rho_1}{\rho_2}h\right)^{1/2}}, \tag{2.34}$$

where Fr' is a modified Froude number, called the densiometric Froude number. $Fr' = 1$ is its critical value. At greater values, the flow of the lower fluid is supercritical and, depending on whether the layers are immiscible or not, either large stationary waves or billows (vortices) likely to effect considerable mixing, can be sustained on its upper surface. There are in

Fig. 2.20 Turbidity current (aqueous suspension of clay) advancing from left to right into still water. Scale bar 0.25 m.

fact a number of kinds of instability that can arise at the interface between fluids in relative motion, depending on the particular combination of stabilizing and destabilizing forces (Turner, 1973). However, the criterion for instability can often be expressed as the inverse square of the densiometric Froude number, called the Richardson number.

A sedimentologically very important type of primarily horizontal stratified flow can be generated when a mass of fluid becomes more dense than its surroundings by acquiring dispersed sediment. These flows, belonging to the general class called gravity currents (Simpson, 1987), include turbidity currents in oceans and lakes, and the base surges and pyroclastic flows created during certain types of volcanic eruption. It is useful to link with these outflows of fresh water from river mouths (plumes) and a range of atmospheric phenomena — bores, thunderstorm outflows, and sea-breeze fronts — all which, although not primarily driven by dispersed sediment, are capable of entraining and/or transporting significant quantities of chiefly fine particles. Whether in the atmosphere or hydrosphere, the flows mentioned share common anatomical and dynamical features. One noteworthy common property is that they tend to be catastrophic.

The head of a catastrophic gravity current advancing over the ground has a characteristic appearance (Figs. 2.20, 2.21 & 2.22). It is wedge-shaped in streamwise profile, with a rounded nose raised a relatively small distance from the bed, and carries on the upper side a series of regularly spaced transverse billows which grow in size toward the rear. Experimentally, the velocity U_{head} at which the profile advances into a still fluid is:

$$U_{head} = 0.75 \left(g \frac{\rho_2 - \rho_1}{\rho_2} h \right)^{1/2},$$

(2.35)

where ρ_1 and ρ_2 are the density respectively of the ambient medium and the current, and h is the head height (Middleton, 1966a). Note that this formula resembles a Froude number. The transverse billows,

Fig. 2.21 Head of a turbidity current (aqueous suspension of clay) advancing through still water toward the observer. Head is 0.5 m wide between images reflected from walls of tank.

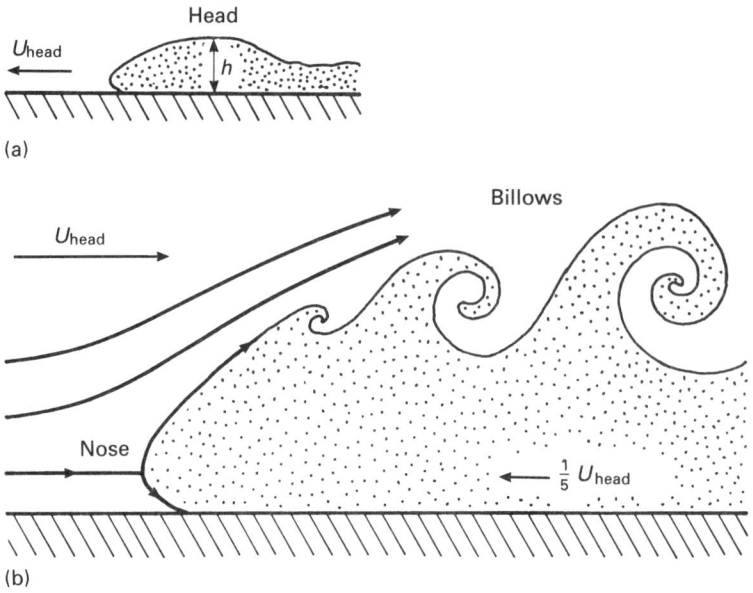

Fig. 2.22 The head of a gravity (e.g. turbidity) current in the streamwise plane. (a) General character of the flow relative to the ground h, head height; U_{head}, velocity of head. (b) Appearance of the head in a frame of reference in which the head is arrested relative to the ground by superimposing on the ambient medium a velocity equal and opposite to that of the head.

which are virtually stationary relative to the ground, serve to mix the current and the ambient medium together, creating a wake of intermediate density. The fact that such mixing takes place within the scope of the head means that the current entering the lower head from upstream has a velocity relative to the ground some 1.2 times greater than U_{head} itself (Fig. 2.22). It follows that a supply of fresh fluid is needed to sustain the head.

A striking and invariable feature of the front of the head (Simpson, 1987) is the overhanging nose (Fig. 2.20) and a transverse alternation of lobes and clefts (Fig. 2.21). At sufficiently large Reynolds numbers, the nose lies at an approximately constant 0.1 head heights above the ground. The lobes and clefts at the front of a substantial, naturally occurring current can take a hierarchy of scales, the largest of which appears to be set by the nose height itself. These curious structures, together with the overhanging nose, record the viscous resistance of the near-bed ambient fluid to the overriding current, combined with the fact that the fluid in the current is the more dense and thus gravitationally unstable (Allen, 1971a). The clefts lead far back into the head (Fig. 2.23), apparently along sites where the bottom flow in a series of paired streamwise, corkscrew vortices at the base of the head is convergent (Allen, 1985). The mixing due to the engulfment of the ambient medium at the clefts is much slower than in the billows on the top of the

head. However, the chief importance of the lobes and clefts is the large transverse variation expected in the value of the accompanying bed shear stress, which

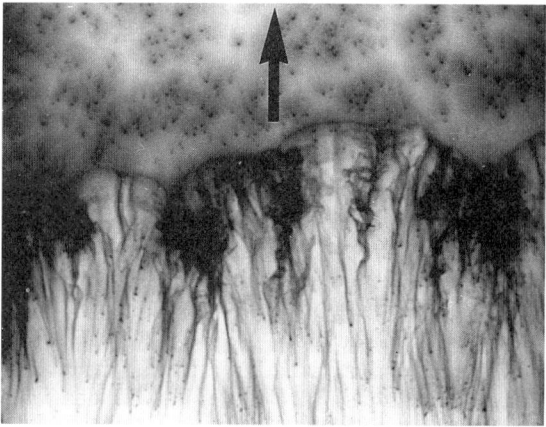

Fig. 2.23 View from below the head of a gravity current (salt solution, 0.5 m wide) advancing across the floor of a tank of still water on which crystals of potassium permanganate had been sprinkled and left to dissolve. The front of the head consists of a series of large lobes and clefts on which smaller lobes and clefts are superimposed. The dye has become concentrated in the clefts and reveals the presence of secondary currents within the large lobes and clefts. Bottom flow convergent behind clefts (low bed shear stress) but divergent behind lobes (high bed shear stress).

can be reflected in patterns of sedimentary textures, fabrics and structures in any deposits formed.

Where a gravity current can be sustained for a significant period, the suddenly appearing head is followed by a long body in which the flow is more uniform and steady. Whereas the head is normally driven by a combination of buoyancy and gravity forces, flow in the body depends on the downslope weight of the current as it moves over the ground or ocean bed. Assuming uniform, steady conditions, and recognizing the frictional resistance arising at the fluid interface, the motion may be analysed as in the case of a wide river, yielding:

$$\tau_0 + \tau_i = (\rho_2 - \rho_1)gh \sin \beta , \qquad (2.36)$$

where τ_0 is the shear stress at the bed and τ_i the value at the interface, ρ_2 and ρ_1 the density respectively of the current and the ambient medium, h the flow thickness, and $\sin \beta$ the slope of the bed. Introducing Eq. 2.14, the mean current velocity U_{body} becomes:

$$U_{body} = \left(\frac{8}{f_0 + f_i} \right)^{1/2} \left(\frac{\rho_2 - \rho_1}{\rho_2} g \right)^{1/2} (h \sin \beta)^{1/2} , \quad (2.37)$$

where f_0 and f_i are the friction coefficients respectively at the bed and interface. Experiments reveal that f_i is of the same order as f_0 (Middleton, 1966b). Because of the presence of the density term, uniform steady gravity currents (Eq. 2.37) are slower in the ocean than rivers (Eq. 2.20) on the same slopes. In practice, the two turn out to be similar in speed, because of the normally much greater thickness of naturally occurring aqueous gravity currents.

2.5.4 Separated flows

The near-bed fluid in a laminar or turbulent boundary layer has little momentum and, where it travels into a regime of adverse pressure gradient (see Eq. 2.6), may become halted. The faster fluid then breaks away or separates from the bed, isolating the low-momentum fluid within a closed region in which there is a sluggish recirculation. This region, the *separation bubble*, is divided from the faster, external stream by a zone of rapid velocity change, called a *free shear-layer*, in which there is normally intense mixing. Chang (1970) and Allen (1982) comprehensively review separated flows.

Typically, flow separation occurs where there is a sharp change in the attitude of the flow boundary, such that the current becomes directly partly toward the bed (Fig. 2.24). The resulting patterns of streamlines — time-averaged ones in the case of turbulent flows — include some which detach from the bed at what is called a separation point and which rejoin the surface at a point of attachment. These streamlines combine with the flow boundary to enclose the separation bubble. Many natural features, for example, ribs of rock which stretch across a river bed or the sea floor, resemble the steps shown in Fig. 2.24a and create a turbulent separated flow. A pebble or shell on the bed has a similar effect, generating a horseshoe vortex which curves from the front around the sides of the obstacle (Paola *et al.*, 1986). The downward steps of Fig. 2.24b are abundantly represented in natural environments by current ripples and dunes in river and tidal flows, the steep ripples with sharp crests fashioned by wave currents, and by transverse aeolian dunes. The pattern and dynamical properties of the turbulent separated flows associated with these structures have a significant part to play in determining the shape and motion of the mobile bed features. A significant role is also played by the separated flows that fill flute marks scoured into cohesive mud beds by sand-laden turbidity currents. In each case, the pattern of time-averaged motion in the separated flow is complicated, primarily because the bed features to which the flow is coupled have curved rather than straight crests (Allen, 1968, 1971b).

It is important with turbulent separated flows to distinguish between the instantaneous and time-averaged properties of the motion. In the case of the separated flows generated at current ripples, dunes and flute marks, the time-averaged length of the separation bubble is typically several times the height of the crest of the structure above the trough. The time-averaged velocity characteristic of the separation bubble is normally a small fraction of the mean velocity of the stream outside. However, the instantaneous properties of the motion point to a separated region that is highly disturbed relative to the external flow (Clark & Markland, 1972; Kiya & Sasaki, 1985). This is because the free shear-layer — the zone of steep velocity gradient between the separation bubble and the external stream — rolls up into a series of transverse, billow-like vortices which grow in size and become stretched into a series of horseshoe forms as they are convected along the layer toward the bed (Fig. 2.25a). Each vortex as it strikes the surface scours the bed like a gusting wind. Anemometers placed in the free shear-layer record fluctuating components of velocity two to three times greater than in the unseparated boundary layer, and Reynolds stresses in the streamwise plane

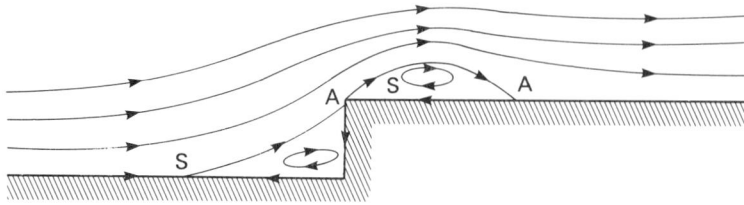

(a)

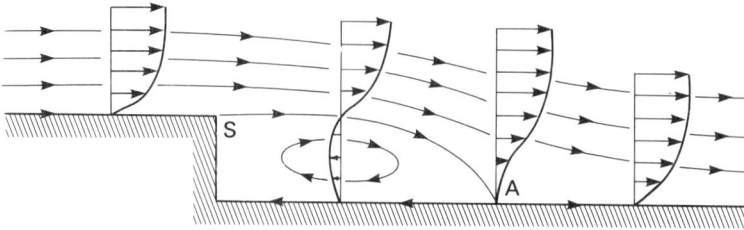

(b)

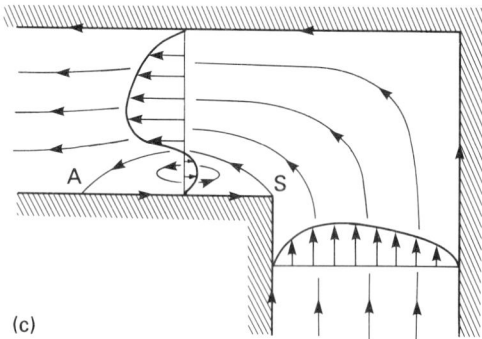

(c)

Fig. 2.24 Some examples of flow separation (S = separation point, A = attachment point) in steady currents, illustrated by time-averaged streamline patterns at (a) an upward transverse step, (b) a downward transverse step, and (c) a sharp bend in a conduit or open channel.

several times the ordinary values (Fig. 2.25b,c). The bed shear stresses experienced in the attachment zone are very variable and can reach exceptionally high instantaneous values in response to the arrival of eddies. Such high values have a significant effect on the entrainment of grains from the bed and on the erosion by suspended particles of a cohesive surface. A new boundary layer is gradually formed downstream from attachment, in which the turbulence intensity and Reynold stresses slowly fall to their ordinary values.

2.2.5 Secondary flows

This term decribes a mean flow in a channel or boundary layer with superimposed on it one or more streamwise, corkscrew vortices which, if more than two are present, are counter-rotating ordinarily in pairs. An example already described is the streaky structure in the wall-region of the turbulent boundary layer (Fig. 2.13). Secondary flows have many different causes, but only a few are important in natural environments (see review in Allen, 1982).

Essentially unconfined boundary-layer flows with secondary currents are encountered naturally in wide, shallow river channels, in the lower atmosphere, and in association with tidal streams. Figure 2.26 summarizes the characteristics of these flows. The paired vortices typically have a transverse wavelength of two to four flow or boundary-layer thicknesses. The maximum angle by which the streamlines deviate from the mean flow direction is small, generally between 5 and 10°. As with the streaky structure, the velocity profile is steepest, and the mean bed shear stress greatest, in the zones of flow divergence on the bed.

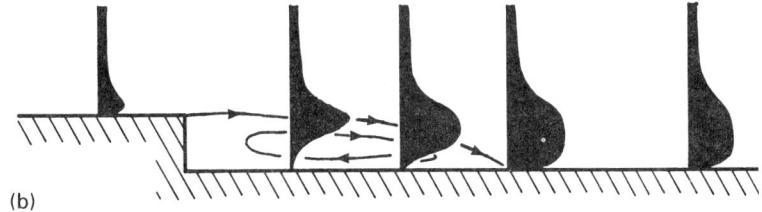

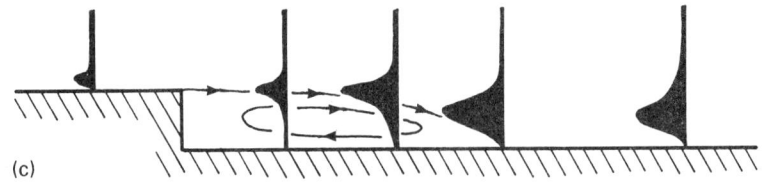

Fig. 2.25 Schematic representation of the properties of a separated flow at a transverse downward step. (a) Instantaneous impression of the transverse vortices developed by the rolling up of the shear layer (see Fig. 2.24) between the external stream (ornamented) and the separation bubble. (b) Variation in the relative intensity of turbulence $(u'^2)^{1/2}/U_{max}$. (c) Variation in the Reynolds stress $-\rho u'v'/U_{max}$.

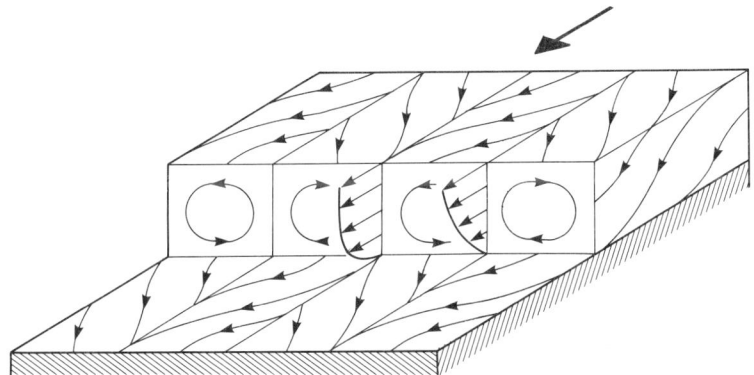

Fig. 2.26 Secondary flow consisting of pairs of oppositely rotating corkscrew vortices.

Extensive systems of well-organized corkscrew vortices readily arise in the atmospheric boundary layer where the wind can be heated from below and caused to rise, as in the desert during the day or above a relatively warm ocean (Brown, 1980). The vortices, commonly made evident by the development of parallel 'cloud streets', typically measure a few kilometres between pairs and can be shown by tracking balloons to have an internal spiralling motion. They afford one possible explanation for longitudinal desert dunes, which are thought to form in the zones of convergent bottom flow.

Secondary currents may also be responsible for streamwise erosional furrows observed from tidal environments and from the ocean floor (Allen, 1982; Viekman *et al.*, 1989), as well as for flow-parallel ribbons of different grades of sandy or gravelly sediment encountered in river and tidal currents and even beneath the wind (McClean, 1981; Pantin *et al.*, 1981; Allen, 1982). In some instances, these currents may arise because of turbulence anisotropy in the *yz*-plane. Another possibility is that transverse variations in shear stress are initiated during sediment transport by the chance clustering into patches of either exceptionally coarse or unusually fine particles, when they become self-perpetuating zones of contrasting roughness. Viekman *et al.* (1989) proved in the field the association of secondary flows with erosional mud furrows. The furrows in this and similar cases may record the presence, in the zones of convergent bottom flow over the initially smooth bed, of heightened concentrations of coarse debris which could act to erode the bed, either directly as tools or indirectly by locally enhancing turbulence.

Perhaps the most important naturally occurring secondary flows are those found in curved river and tidal channels (Fig. 2.27), as extensively reviewed in Allen (1982). These currents have great sedimentological importance and arise because of the retarding action of viscosity on the flow near the channel bed. When an element of water traces a curved path, it can be regarded as being acted on by a radially outward centrifugal force:

$$\text{centrifugal force} = \frac{\rho U^2}{r}, \qquad (2.38)$$

in which ρ is the fluid density, U the tangential velocity, and r the radius of curvature of the path. The mean centrifugal force acting at any radial distance in a curved channel is therefore the average of Eq. 2.38 over the streamwise velocity profile at that position. But the curved motion of the element can be steady only if there arises a force equal and opposite to the mean centrifugal force, that is, a force acting radially inward. Recalling that coffee stirred in a cup assumes a concave-up surface, the required force on the element is a pressure related to the radially inward hydraulic gradient, that is:

$$\text{pressure force} = \rho g \frac{dy}{dr}, \qquad (2.39)$$

in which g is the acceleration due to gravity, y the height of the water surface above an arbitrary datum, and dy/dr the hydraulic gradient. However, the centrifugal and pressure forces are generally unbalanced locally. In the low-momentum fluid near the bed, where the velocity is less than average, the pressure force is the greater, and tends to drive the fluid inward over the bed. Close to the free-surface, where the local velocity exceeds the mean, the centrifugal force exceeds the pressure force, driving the fluid outward. Hence a spiral motion, inward at the bed and outward at the surface, becomes superimposed on the mean flow in the bend.

It is only in relatively deep channels that a single spiral vortex arises. Several vortices are likely to form in a curved channel that is shallow compared to the width. Experiments show that the deviation of the streamlines in the vortex from the mean flow direction (Fig. 2.27) is given by:

$$\tan \alpha = 11 \frac{h}{r}, \qquad (2.40)$$

where α is the deviation angle. The deviation angle is

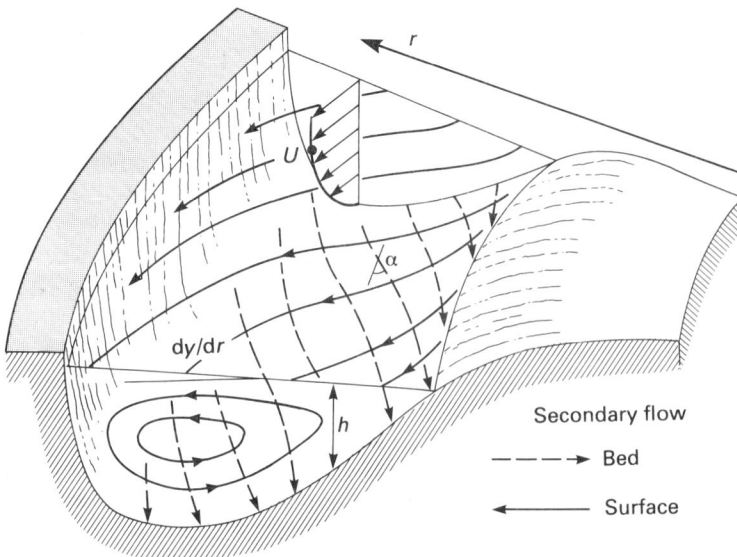

Fig. 2.27 Features of steady secondary flow in a curved open river or tidal channel.

likely to be very small or small in the case of natural channels, for the width-to-depth ratio generally lies between 10 and 100, and the minimum radius of curvature is ordinarily at least one to two channel widths.

2.6 Relation of fluid and flow properties to sediment transport

2.6.1 Forces acting on a stationary particle on a static bed

The forces exerted by a current on its bed are commonly sufficient to entrain and then transport stationary particles or particle aggregates from the bed. Grain, fluid, and flow properties combine to determine the entrainment threshold and the modes and rate of sediment transport.

Consider an idealized spherical particle, more dense than the fluid shearing steadily past it, perched on a bed of the same particles (Fig. 2.28). The forces acting on the particle include the fluid drag F_D and lift F_L, the particle immersed weight F_W, and the interparticle cohesion F_C at each grain contact. The first motion of the grain upon entrainment will be about the downstream pivot P, and the force-balance:

$$aF_L + bF_D = aF_W + cF_C, \qquad (2.41)$$

will be satisfied, where a, b and c are the lengths of the

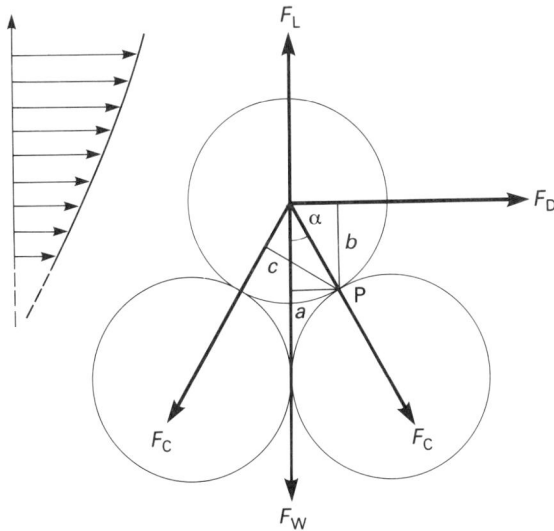

Fig. 2.28 Forces acting on a particle resting on a granular bed subject to a steady current.

moment arms from P (note that F_C acting through P has zero moment). It is assumed that the forces act through the grain centre.

These forces vary in relative importance with the circumstances of entrainment. For sand- and gravel-sized particles in water or air, interparticle cohesion can normally be ignored, leaving in consideration only the fluid drag and lift and the particle immersed weight. However, the magnitude of the lift force remains obscure, although there is evidence that for relatively large grains it is of the same order as the drag (Raudkivi, 1976; Greeley & Iversen, 1985). Interparticle cohesion can under some circumstances be important in both aqueous and aeolian environments. Especially in warm climates, and elsewhere during sunny weather, sand particles exposed at low tide or low river stage can acquire a sticky surface film of algal or bacterial mucilage (Grant & Gust, 1987; Vos *et al.*, 1988). Sand grains acted on by the wind may adhere because of either a surface film of moisture or a coating precipitate left after the evaporation of overnight dew. In each case, larger fluid forces would be necessary for particle entrainment than if the grains had possessed a clean surface. Other sources of interparticle cohesion — van der Waals and electrostatic forces — become increasingly significant relative to weight, drag and lift as grain size decreases through the silt and clay ranges, in air as well as water as the transporting medium. Water-laid clay mineral muds are cohesive sediments *par excellence*. Typically, mud beds yield to aqueous currents not grain-by-grain, but by the tearing of large particle aggregates from the surface (fluid stripping) (Allen, 1971b). The yield strength of the mud — a bulk and not a grain property — largely governs this process. Aggregates can also be lost from mud beds by the cutting action of larger particles either suspended or being rolled along in the current. This process, of natural sand-blasting, is called *corrasion* (Allen, 1971b).

The fluid drag in Eq. 2.41 can be specified either as the mean bed shear stress at the threshold of entrainment adjusted by the projection area of the grain, or through a definition of the form of Eq. 2.14, involving a drag coefficient and the mean velocity at the particle level, again applied over the projected particle area. The lift force is more difficult to define, but one possible source of lift on a stationary grain can be identified from the Bernoulli equation (Eq. 2.6). Consider the spherical particles represented in section in Fig. 2.29 and remember that there would be uninterrupted shear flow past the perched grain in nearby parallel sections. Bernoulli's equation predicts a lower

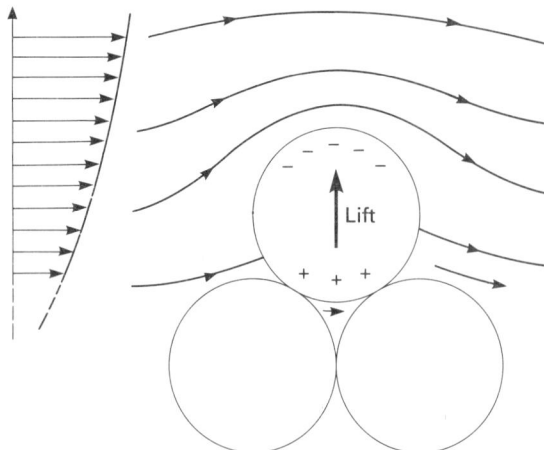

Fig. 2.29 Lift force due to the Bernoulli effect on a particle on a granular bed subject to fluid shear. The fluid pressure is greater on the underside of the particle (plus signs), where the fluid velocity is low, than over the upper surface (minus signs), where a high velocity obtains.

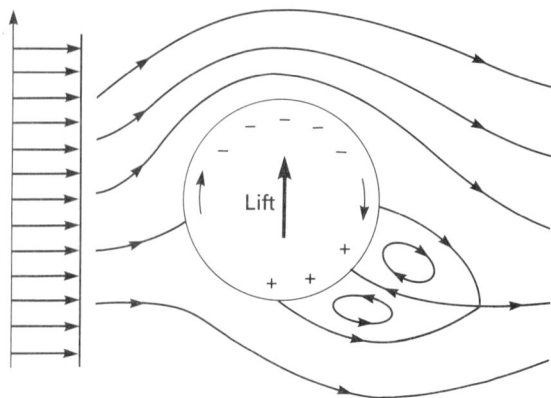

Fig. 2.30 Lift force due to the Bernoulli effect on a particle with top spin in a uniform current (no vertical change in velocity). The pressure is high (plus signs) and the velocity low on the underside of the particle, where the spin opposes the flow, but low (minus signs) on the upper surface, where the flow and particle surface are moving in the same direction.

pressure in the fast flow over the upper surface of the sphere than in the slower stream beneath. Hence the particle will experience an upward-acting lift. Through the operation of the Magnus effect (Fig. 2.30), a moving particle which had a sufficiently fast top spin in the near-bed shear zone could experience further upward lift on entering the outer flow in which the velocity was more uniform.

2.6.2 Threshold of entrainment

In attempts to predict the entrainment threshold for cohesionless grains, fluid lift is often ignored or regarded as uniquely and simply related to the drag. In the former case, and generalizing the problem to spherical grains of radius r on a bed of uniform spheres of radius R (Fig. 2.31), we may write:

$$\tau_{cr} = \frac{4}{3}(\sigma - \rho)g\frac{rR}{\{(r+R)^2 - R^2\}^{1/2}}, \qquad (2.42)$$

starting from Eq. 2.40 and the geometry of the configuration, where τ_{cr} is the critical or threshold mean bed shear stress for entrainment, σ the solids density, ρ the fluid density, and g the acceleration due to gravity. Theoretically, it follows that, when $r = R$, the threshold force increases linearly with the particle size and excess density. The threshold stress is sometimes conveniently represented by the equivalent threshold shear velocity (Section 2.4.3), when the entrainment condi-

tion varies as the square root of the grain size. Inspection of Eq. 2.42 will also show that, theoretically, the threshold stress is a maximum when $r = R$, implying that grains both larger and smaller than the bed particles are easier to entrain than grains of the same size as those in the static bed. This condition, applied over a protracted series of particle entrainments and depositions, should tend to associate or sort like particles together, as is true of most far-travelled water- and wind-laid sediments. It will act in addition to the spectrum of instantaneous turbulent bed shear stresses and

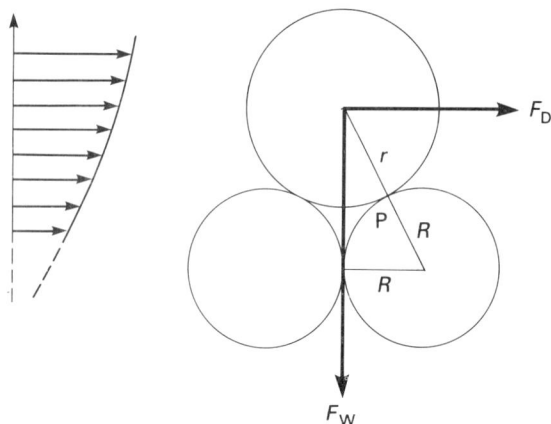

Fig. 2.31 Definition diagram for the entrainment of a particle on a granular bed affected by a steady current.

pressures, which also tend to ensure that like particles are associated together (Bridge, 1981).

The bed slope, as well as the relative size of the entrained and static grains, affects the entrainment threshold. If $\tau_{cr,0}$ is the threshold stress on a horizontal bed, and $\tau_{cr,\beta}$ the threshold on a bed of slope angle β measured positively downward with the flow, it is readily shown that:

$$\frac{\tau_{cr,\beta}}{\tau_{cr,0}} = \frac{\sin(\alpha - \beta)}{\sin\alpha}, \qquad (2.43)$$

where α is the angle subtended at the centre of the entrained particle between the normal to the bed and the pivotal point P (Fig. 2.28). As was recently confirmed (Whitehouse & Hardisty, 1988), a slope downward in the flow direction rapidly diminishes the threshold stress, whereas an upward slope demands an increased force.

For several reasons these theoretical considerations give only a limited insight into the entrainment conditions for natural sediment mixtures in real flows. Of particular importance is the fact the fluid forces acting on bed grains in turbulent flows are strongly time-dependent (Section 2.4). Not only does the instantaneous bed shear stress fluctuate rapidly and over a substantial range, but so also does the total lift force, contributed by fluctuations of the normal pressure and of the local velocity gradient. For a given sediment, even if perfectly sorted, entrainment is observed experimentally to occur over a significant range of flow conditions. It is consequently difficult to obtain universal agreement as to what constitutes entrainment. However, it is generally accepted that entrainment has occurred once the flow conditions permit a sustained and general grain motion.

In practice, threshold conditions are predicted from extensive laboratory experiments using, generally speaking, well-graded sediments on artificially smoothed beds. Figure 2.32 shows Bagnold's (1941) widely accepted graph for quartz particles in a uniform, steady air flow. The threshold is stated in terms of the critical shear velocity $U_{*cr} = (\tau_{cr}/\rho)^{1/2}$ and has a minimum value for a particle diameter of about 80 μm. At larger sizes, the theoretical relation, Eq. 2.42 and associated discussion, is obeyed. The other curve plotted is the so-called 'impact threshold', that is, the shear velocity at which the impact with the bed of particles already moving is sufficient to cause further entrainment. In the case of this aeolian threshold, it is clearly necessary to add an appropriate impact force to those already included in Fig. 2.28. A corresponding

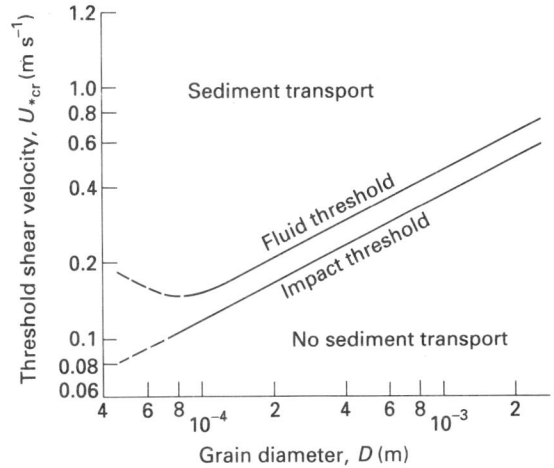

Fig. 2.32 Experimental threshold criterion for the entrainment of quartz particles in air, by the action of fluid forces alone (fluid threshold) and by wind-driven grains (impact threshold) where D is the particle diameter and u_{*cr} is the threshold shear velocity (Bagnold, 1941).

graph for water appears in Fig. 2.33, based on a careful selection of data intended to reduce scatter (Miller *et al.*, 1977). The threshold stress is here normalized by a quantity proportional to the grain immersed weight, and the grain size is stated as a grain Reynolds number. This graph has a certain elegance, in that both variables are non-dimensional, but is less practical than plots like Fig. 2.32 in which the particle diameter is explicit. There appears for uniform, steady water transport to be no detectable equivalent of the impact threshold in air. Komar and Miller (1973) and others (see review in Allen, 1982) give graphs for the entrainment of sediment into the highly unsteady boundary layers created by wind waves.

2.6.3 Settling of a single particle in a still fluid

The settling of a single grain in a still fluid of large extent compared to the particle is controlled by the fluid viscosity and by the particle excess density, size, shape and surface roughness. The particle size and excess density determine the weight force that acts during the motion, while the fluid viscosity and the particle shape and roughness, in combination with the size, control the fluid drag.

A smooth spherical particle of diameter D and density σ, settling at a steady velocity V_0 through a still fluid of density ρ, is acted on by a downward weight

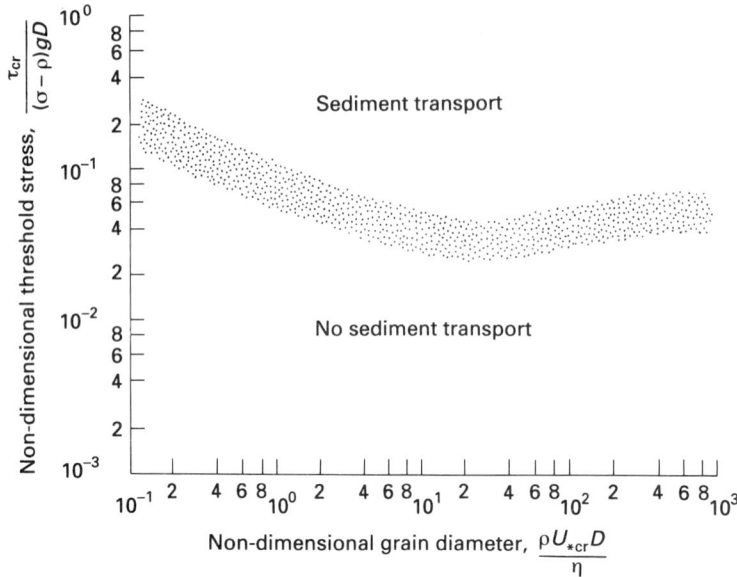

Fig. 2.33 Experimental threshold criterion for the entrainment of quartz-density solids by a water stream. Adapted from Miller *et al.* (1977).

force:

$$\frac{4}{3}\pi\left(\frac{D}{2}\right)^3(\sigma-\rho)\,g\,,\qquad(2.44)$$

and by an equal and opposite fluid drag:

$$C_{D,0}\,\pi\left(\frac{D}{2}\right)^2\frac{\rho V_0^2}{2}\,,\qquad(2.45)$$

where g is the acceleration due to gravity and $C_{D,0}$ a non-dimensional drag coefficient. The 19th century mathematician G.G. Stokes showed theoretically that, for small Reynolds number Re based on V_0 and D, when inertial forces can be neglected, the drag coefficient $C_{D,0} = 24/Re$. Substituting into Eq. 2.44 gives:

$$V_0 = \frac{1}{18}\frac{(\sigma-\rho)\,gD^2}{\eta}\,,\qquad(2.46)$$

where η is the fluid molecular viscosity, known as *Stokes Law* (Allen, 1985). By this law, the steady settling velocity varies as the square of the particle size and inversely as the fluid viscosity. Similar calculations have been made for various other regular shapes of particle at small Reynolds number (Happel & Brenner, 1965; Clift *et al.*, 1978).

Laboratory experiments show that Stokes Law begins to fail at $Re \approx 1$ and that at increasing Reynolds numbers the drag coefficient tends to a constant value in the range 0.1–1 (Fig. 2.34). Hence on substituting $C_{D,0}$ = constant into Eq. 2.44, we find that, for large enough Reynolds numbers, the steady settling

velocity varies as the square root of the particle size (the *'impact' law*). These changes in the drag and settling laws are linked to a progressive alteration in the pattern of flow around the settling particle (Fig. 2.34). With increasing Reynolds number, we see an increasingly large separated flow, then a region of regular vortex shedding, and finally the development of a turbulent boundary layer and wake.

As discussed in Chapter 1, naturally occurring sediment particles are neither spherical nor smooth, and their shape and surface roughness exert a strong influence on settling behaviour. The effect of a departure from a perfectly spherical form is to increase the drag coefficient and to reduce the settling velocity relative to the equivalent sphere; a variety of shape factors has been introduced into the settling laws to adjust for this effect. An increase in surface roughness has a similar but less marked result. With increasing distortion of form may come a change in the mode of settling, from descent on a simple, vertical path to complicated patterns of fall. Baba & Komar (1981) found experimentally that natural quartz sand grains, which are rough and more or less angular, although relatively equant, settled somewhat more slowly than the equivalent smooth spheres. Gravel (Alger & Simons, 1968; Komar & Reimers, 1978) and biogenic particles, such as molluscan shells (Allen, 1984), also sink more slowly than their spherical equivalents, while executing complicated descending motions. Such particles may oscillate, tumble, pitch

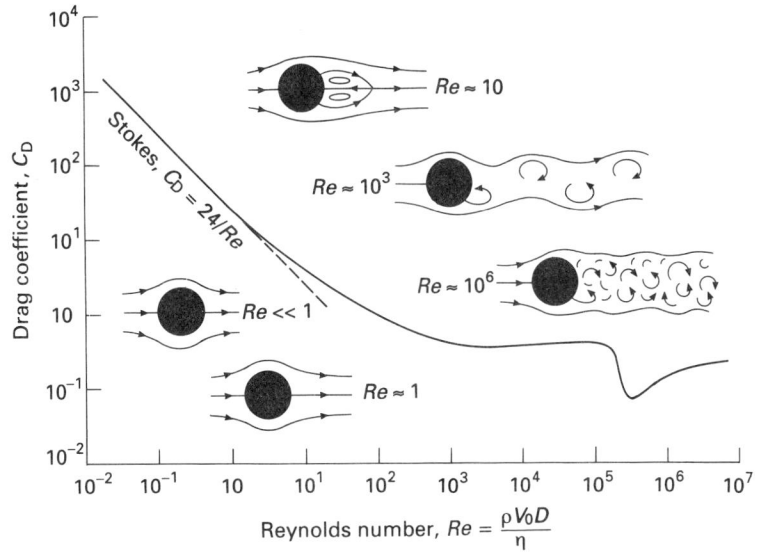

Fig. 2.34 Dependence of the flow pattern and drag coefficient on Reynolds number for the steady settling of a smooth, spherical particle. At Reynolds numbers in excess of the Stokes range, the relationship is based on experimental results.

or spiral during fall, depending on their Reynolds number, relative elongation, and degree of asymmetry (Chapter 1).

2.6.4 Settling of particles in bulk

Equations 2.44–2.46 give an invaluable insight into the general process of settling, but should not be expected to describe accurately the behaviour of grains settling in bulk, as is commonly the case in natural flows.

The fluid drag that opposes the immersed weight of a settling particle is the sum of skin friction and of pressure forces linked to the distortion of the fluid by the body. As a solitary sphere bends the streamlines representing a flow past it to an outward distance of the order of one to two diameters (Fig. 2.35), the drag coefficient for a grain settling together with other like grains should increase with the particle concentration, because of the additional distortion of the fluid caused by neighbours. Experimentally and theoretically (Maude & Whitmore, 1958; Allen, 1985):

$$C_D = \frac{C_{D,0}}{(1 - C)^{2n - 2}}, \qquad (2.47)$$

and

$$\frac{V}{V_0} = (1 - C)^n, \qquad (2.48)$$

where C_D is the drag coefficient for the grains in bulk, $C_{D,0}$ the coefficient for the solitary grain, C the

fractional volume concentration of the particles, V_0 the steady settling of the solitary grain, V the settling velocity of the grains in bulk, and $2.33 \leqslant n \leqslant 4.65$ an exponent dependent on the Reynolds number

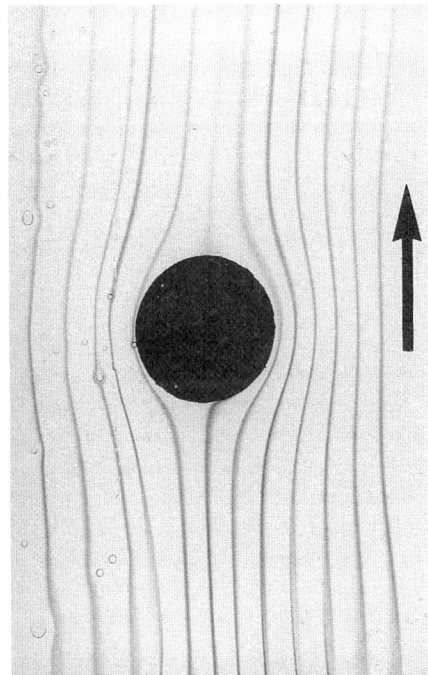

Fig. 2.35 Streamlines of the low-Reynolds number flow around an equatorial section through a sphere.

$Re = \rho D V_0/\eta$. The exponent takes its larger limit for $Re \leqslant 1$ and its smaller constant value when $Re \geqslant 10^3$. Thus the settling velocity of grains in bulk, as in turbidity currents and pyroclastic flows, is a steeply decreasing function of the concentration, especially at small Reynolds numbers (e.g. the finer grades of sand in a turbidity current).

2.6.5 Fluidization and liquefaction of granular beds

Reference was made in Sections 2.2.5 and 2.2.6 to the ability of mixtures of grains and either water or air to behave as a new and more viscous fluid with, generally speaking, non-Newtonian properties. Two processes either related or leading to settling — *fluidization* and *liquefaction* — are available naturally to change a deposited bed of clean silt or sand into a fluid-like state, provided the sediment is unlithified and not too consolidated.

A change only of the reference frame is required to convert the process of bulk grain settling into that of *fluidization* (Davidson & Harrison, 1963). When a fluid is passed vertically up through a fixed bed of grains, each particle experiences an upward fluid drag that counteracts its downward weight. On slowly increasing the fluid discharge, there will come a stage when these two forces just balance, whereupon the grains will just disengage from each other and the bed expand very slightly, as the support of one particle by another becomes unnecessary. This state of incipient fluidization may be specified in terms of the overall or superficial fluid velocity V_{fl} necessary to effect it:

$$V_{fl} = V_0(1 - C_{bed})^n, \qquad (2.49)$$

in which V_0 is the steady settling velocity of a solitary grain, C_{bed} the grain concentration in the bed, and n the previous exponent. As C_{bed} is generally 0.6–0.7, varying inversely as the rate of deposition, V_{fl} is very small compared to V_0.

Fluidized beds occur in two main regimes (Fig. 2.36), determined largely by the relative density of the solids and fluid involved. Particulate fluidization, marked by an essentially uniform grain distribution, is typical of mineral sands in water, as in the seeps at the toes of river levees and alluvial fans. The fluidization of mineral debris by gases, as in volcanic vents, is normally in the more varied aggregative regime.

It should be noted that the fluidized state can be maintained only for so long as there is an adequate upward flow of fluid. A sufficient reduction (or cessation) of flow will cause the bed to settle and resume its originally solid-like character.

Sands deposited rapidly acquire a loose, metastable grain packing which is readily destroyed by the application of modest stresses, especially if these are cyclical (e.g. earthquake shocks, passage of storm waves). The process of changing the state of a granular bed from solid-like to fluid-like in this way is called *liquefaction* (review in Allen, 1982). However,

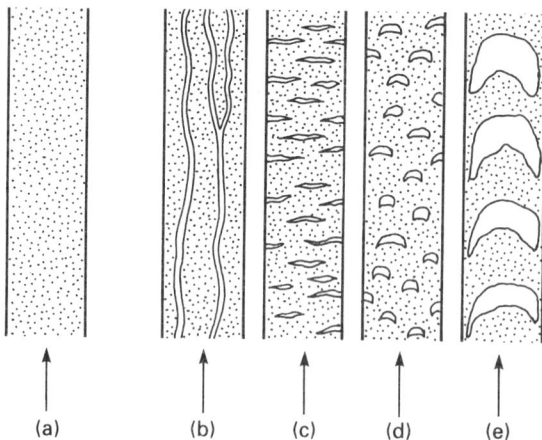

Fig. 2.36 Particulate fluidization (a) and (b–e) types of aggregative fluidization: (b) channelling; (c) parvoids; (d) bubbling; (e) slugging.

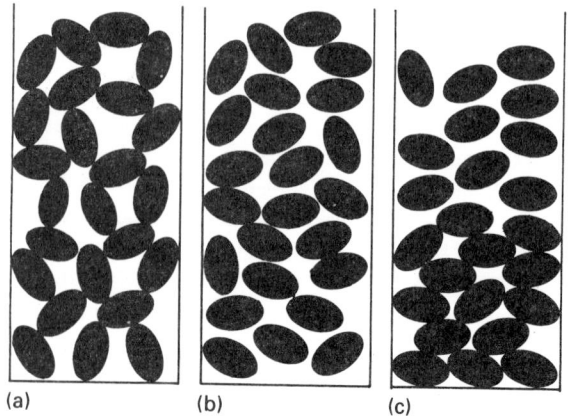

Fig. 2.37 A model for the liquefaction of a granular sediment. (a) Initial deposit with a metastable grain packing. (b) The mixture of particles and intergranular fluid at the moment of complete liquefaction. (c) The dispersion in the process of resedimenting after liquefaction.

the fluid-like state is short-lived, and its attainment is almost at once followed by resettling, the pattern of events being summarized in Fig. 2.37.

Liquefaction occurs in a closed system, unlike fluidization which demands an external source of fluid, and liquefied beds do not generally become much mixed internally during the change of state. Primary lamination therefore normally survives the process of liquefaction, although the layering may change its shape if external forces are sufficient to cause the liquefied sediment to flow. The degree to which primary lamination becomes deformed is chiefly controlled by the apparent viscosity of the liquefied sediment, (Eqs 2.4 & 2.5), and by the time T required for the liquefied bed to resettle fully. Assuming a uniform initial bed, it can be shown (Allen, 1985) that:

$$T = \frac{\Delta C \cdot h}{C_{disp} V_0 (1 - C_{bed})^n}, \qquad (2.50)$$

where ΔC is the concentration difference between the liquefied dispersion and resettled bed, C_{disp} the grain concentration in the liquefied bed, C_{bed} the grain concentration in the resettled bed, h the initial bed thickness, V_0 the steady settling velocity of a solitary grain, and n the exponent in Eqs 2.47 and 2.48. As ΔC is very small, T is of the order of 10–100 s for sand beds of ordinary thickness. This is long enough for laminae to become distorted significantly but too short for worse than a modest blurring of their appearance.

2.6.6 Modes of particle motion during sediment transport

Grains entrained from a bed travel in various ways with the current, depending on their size, shape and excess density, and the viscosity and speed of the transporting fluid. For two reasons sediment transport in air and water differ in important respects. Mineral-density particles are three orders of magnitude more dense than air, but of the same order of density as water. Air has a molecular viscosity two orders of magnitude less than that of water. Hence viscous forces should generally play the chief role in grain–grain and grain–fluid interactions during aqueous transport, whereas in air the domination should lie with inertial and impact forces.

The four modes of transport in water are *sliding, rolling, saltation* and *suspension* (Fig. 2.38a). Sliding

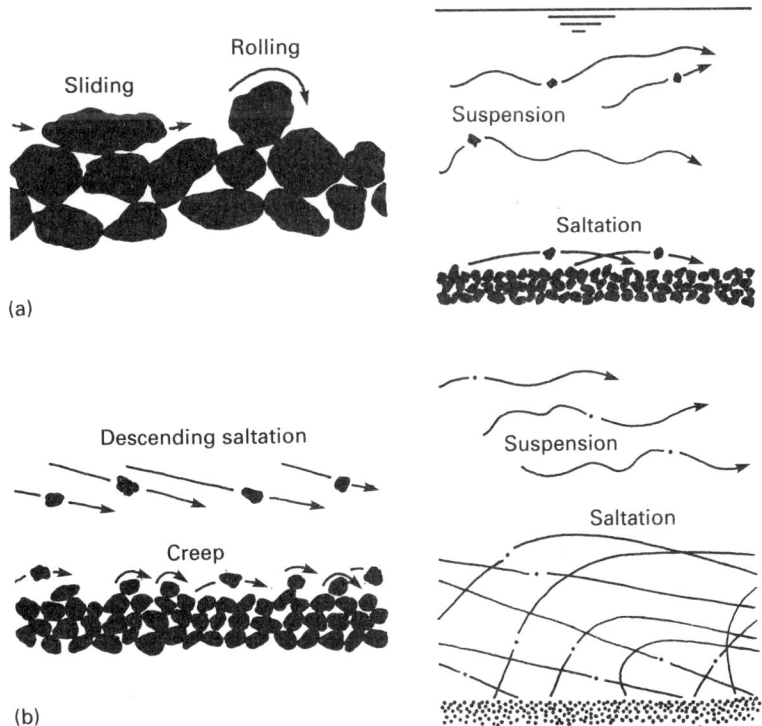

Fig. 2.38 Modes of sediment transport in (a) water, and (b) air. The general scale of each type of motion is indicated by the relative size of the particles involved.

particles remain in continuous contact with the bed and, while they may tip up or down slightly during travel, experience no consistent rotation about a flow-transverse axis. However, a rolling grain turns continuously about a flow-transverse axis while remaining essentially in contact with the bed. Cartwheeling is a spectacular form of rolling, preferred by well-rounded, discoidal particles. A grain that is saltating jumps over the bed on a series of low trajectories of the order of 10–20 particle diameters long and a few diameters high. Particles that are sliding, rolling or saltating are densely arrayed and collectively form the *bedload*. The *suspended load* is formed of suspended grains, that is, particles that follow irregular, lengthy paths within the fluid that practically never bring them into contact with the static bed.

Experimentally, although criteria for suspension have been advocated theoretically, there is no sharp division between the four modes of particle motion, and in the stronger flows the modes will occur simultaneously (Abbott & Francis, 1977). However, sliding and rolling are most prevalent at low transport stages, whereas saltation and suspension dominate at high stages. Saltation involves repeated collisions between moving and bed grains, and it is these that sustain the load represented by the saltating particles (Bagnold, 1966). The 'bursting' of boundary-layer streaks to form hairpin vortices is probably the chief mechanism urging bed grains up into the outer flow, where they can become suspended (Sumer & Deigaard, 1981). As grains once suspended rarely afterward make contact with the bed, the load they represent must be supported by a purely fluid force. This appears to originate in an asymmetry of the shear-related turbulence, expressed by upward fluctuating velocities more powerful than downward-acting ones (Bagnold, 1966; Brodkey *et al.*, 1974; Wei & Willmarth, 1991).

The modes of particle transport recognized during aeolian transport are *creep* or *reptation*, *saltation* and *suspension* (Fig. 2.37b). Creep or reptation (Bagnold, 1941; Mitha *et al.*, 1986), is the slow and intermittent forward movement of densely arrayed, near-bed grains caused by the impact of an intense rain of saltating particles. Sliding, rolling and hopping on short, flat trajectories are all involved in creep. The process of saltation is controlled by elastic impacts between the moving particles and stationary bed grains, as the former splash down into the surface (Bagnold, 1941; Anderson & Hallett, 1986; Werner & Haff, 1988). Sand grains saltating in the wind typically follow trajectories of the order of 1–2 m long and several deci-

metres in height, rising from the bed at speeds of a few metres per second and returning at the end of their leap at angles of the order of 10°.

2.6.7 Dispersed sediment–flow interactions

Dispersed particles and a fluid in motion interact in such a way as each to modify the properties and behaviour of the other. Several studies have been made showing that the general effect of turbulence is to lower the settling velocity of a particle relative to its value in a still fluid (review in Allen, 1982). It seems that the effect of the particles is to weaken the higher-frequency turbulence that would otherwise be present (Wang & Qian, 1989). Saltating and suspended grains, by exerting a drag on the fluid transporting them, extract some of its forward momentum. Consequently, the flow of the fluid is slower than in sediment-free conditions. The effect becomes increasingly marked as the grains become more concentrated on approaching closer to the bed (e.g. Owen, 1964). A flatter and more uniform near-bed velocity profile typically results.

2.7   **References**

Abbott J.E. & Francis J.R.D. 1977. Saltation and suspension trajectories of solid grains in water streams. *Phil. Trans. R. Soc.*, **A284**, 225–254.

Alger G.R. & Simons D.B. 1968. Fall velocity of irregular shaped particles. *J. Hydraul. Div. Am. Soc. Civ. Engrs*, **94**, 721–737.

Allen J.R.L. 1968. *Current Ripples*. North-Holland, Amsterdam.

Allen J.R.L. 1971a. Mixing at turbidity current heads, and its geological implications. *J. Sedim. Petrol.*, **41**, 97–113.

Allen J.R.L. 1971b. Transverse erosional marks of mud and rocks: their physical basis and geological significance. *Sediment. Geol.*, **5**, 167–385.

Allen J.R.L. 1982. *Sedimentary Structures, their Character and Physical Basis*, 2 Vols. Elsevier, Amsterdam.

Allen J.R.L. 1984. Experiments on the settling, overturning and entrainment of bivalve shells and related models. *Sedimentology*, **31**, 227–250.

Allen J.R. L. 1985. *Principles of Physical Sedimentology*. Allen & Unwin, London.

Anderson R.S. & Hallet B. 1986. Sediment transport by wind: toward a general model. *Bull. Geol. Soc. Am.*, **97**, 523–535.

Aubrey N., Holmes P., Lumley J.L. & Stone E. 1988. The dynamics of coherent structures in the wall region of a turbulent boundary layer. *J. Fluid Mech.*, **192**, 115–173.

Baba J & Komar P.D. 1981. Settling velocities of irregular

grains at low Reynolds numbers. *J. Sedim. Petrol.*, **51**, 121–128.

Bagnold R.A. 1941. *The Physics of Blown Sand and Desert Dunes.* Chapman & Hall, London.

Bagnold R.A. 1966. An approach to the sediment transport problem from general physics. *U.S. Geol. Surv. Prof. Pap.*, **422-I**.

Bridge J.S. 1981. Hydraulic interpretation of grain-size distributions using a physical model for bedload transport. *J. Sedim. Petrol.*, **51**, 1109–1124.

Brodkey R.S., Wallace J.M. & Eckelman H. 1974. Some properties of truncated turbulence signals in bounded shear flows. *J. Fluid Mech.*, **63**, 209–224.

Brown R.A. 1980. Longitudinal instabilities and secondary flows in the planetary boundary layer: a review. *Rev. Geophys. Space Phys.*, **18**, 683–697.

Cantwell B.J. 1981. Organized motion in turbulent flow. *Ann. Rev. Fluid Mech.*, **13**, 457–515.

Chang P.K. 1970. *Separation of Flow.* Pergamon, London.

Chow V.T. 1959. *Open Channel Hydraulics.* McGraw-Hill, New York.

Clark J.A. & Markland E. 1972. Flow visualization in free shear layers. *J. Hydraul. Div. Proc. Am. Soc. Civ. Engrs*, **99**, 1897–1913.

Clift R., Grace J.R. & Weber M.E. 1978. *Bubbles, Drops and Particles.* Academic Press, New York.

Davidson J.F. & Harrison D. 1963. *Fluidized Particles.* Cambridge University Press, Cambridge.

Falco R.E. 1977. Coherent motions in the outer regions of turbulent boundary layers. *Physics Fluids*, **20**(Suppl.), S124–S132.

Falco R.E. 1991. A coherent structure model of the turbulent boundary layer and its ability to predict Reynolds number dependence. *Phil. Trans. R. Soc. Lond.*, **A336**, 103–129.

Francis J.R.D. 1975. *Fluid Mechanics for Engineering Students*, 4th edn. Arnold, London.

Grant G. & Gust G. 1987. Prediction of coastal sediment stability from photopigment content of mats of purple sulphur bacteria. *Nature*, **330**, 244–246.

Grass A.J. 1971. Structural features of turbulent flow over smooth and rough boundaries. *J. Fluid Mech.*, **50**, 233–255.

Grass A.J., Stuart R.J. & Mansour-Tehrani M. 1991. Vortical structures and coherent motion in turbulent flow over smooth and rough boundaries. *Phil. Trans R. Soc. Lond.*, **A336**, 35–65.

Greeley R. & Iversen J.D. 1985. *Wind as a Geological Process.* Cambridge University Press, Cambridge.

Happel J. & Brenner H. 1965. *Low Reynolds Number Hydrodynamics.* Prentice-Hall, Englewood Cliffs.

Head M.R. & Bandyopadhyay P. 1981. New aspects of turbulent boundary-layer structure. *J. Fluid Mech.*, **107**, 297–338.

Jang P.S., Benney D.J. & Gran R.L. 1986. On the origin of streamwise vortices in a turbulent boundary layer. *J. Fluid Mech.*, **169**, 109–123.

Kim J. & Moin P. 1986. The structure of the vorticity found in turbulent channel flow. Part 2. Study of ensemble-averaged fields. *J. Fluid Mech.*, **162**, 339–363.

Kiya M. & Sasaki K. 1985. Structure of large-scale vortices and unsteady reverse flows in the reattaching zone of a turbulent separation bubble. *J. Fluid Mech.*, **154**, 463–491.

Kline S.J., Reynolds W.C., Schraub F.A. & Runstadler P.W. 1967. The structure of turbulent boundary layers. *J. Fluid Mech.*, **30**, 741–773.

Komar P.D. & Miller M.C. 1973. The threshold of sediment movement under oscillatory water waves. *J. Sedim. Petrol.*, **43**, 1101–1110.

Komar P.D. & Reimers C.E. 1978. Grain shape effects on settling rates. *J. Geol.*, **86**, 193–209.

Maude A.D. & Whitmore R.L. 1958. A generalized theory of sedimentation. *Br. J. Appl. Phys.*, **9**, 477–482.

McCave I.N. 1973. Some boundary layer characteristics of tidal currents bearing sand in suspension. *Mem. Soc. R. Sci. Liège*, **(6)6**, 187–206.

McLean S.R. 1981. The role of non-uniform roughness in the formation of sand ribbons. *Mar. Geol.*, **42**, 49–74.

Middleton G.V. 1966a. Experiments on density and turbidity currents. I. The motion of the head. *Can. J. Earth Sci.*, **3**, 523–546.

Middleton G.V. 1966b. Experiments on density and turbidity currents. II. Uniform flow of density currents. *Can. J. Earth Sci.*, **3**, 629–637.

Miller M.C., McCave I.N. & Komar P.D. 1977. Threshold of sediment motion under unidirectional currents. *Sedimentology*, **24**, 507–527.

Mitha S., Tran M.Q., Werner B.T. & Haff P.K. 1986. The grain–bed impact process in aeolian saltation. *Acta Mechanica* **63**, 267–278.

Nakagawa H. & Nezu I. 1981. Structure and space-time correlation of bursting phenomena in an open-channel flow. *J. Fluid Mech.*, **104**, 1–43.

Nouh M. 1989. The von Kármán coefficient in sediment-laden flow. *J. Hydraulic Res.* **27**, 477–499.

Owen P.R. 1964. Saltation of uniform grains in air. *J. Fluid Mech.*, **20**, 225–242.

Pantin H.M., Hamilton D. & Evans C.D.R. 1981. Secondary flow caused by differential roughness, Langmuir circulations, and their effect on the development of sand ribbons. *Geo. Mar. Lett.*, **1**, 255–260.

Paola C., Gust C. & Southard J.B. 1986. Skin friction behind isolated obstacles and the formation of obstacle marks. *Sedimentology* **33**, 279–293.

Perry A.E. & Chang M. S. 1982. On the mechanism of wall turbulence. *J. Fluid Mech.*, **119**, 173–217.

Pugh D.T. 1987. *Tides, Surges and Mean Sea-level.* Wiley, Chichester.

Raudkivi A.J. 1976. *Loose Boundary Hydraulics*, 2nd edn. Pergamon, Oxford.

Schlichting H. 1960. *Boundary Layer Theory*, 4th edn. McGraw-Hill, New York.

Simpson J.E. 1987. *Gravity Currents*. Chichester, Ellis Horwood.

Sleath J.F.A. 1984. *Sea Bed Mechanics*. Wiley, New York.

Smith C.R. & Metzler S.P. 1983. The characteristics of low-spread streaks in the near-wall region of a turbulent boundary layer. *J. Fluid Mech.*, **129**, 27–54.

Sumer M. & Deigaard R. 1981. Particle motion near the bottom in turbulent flow in an open channel. Part 2. *J. Fluid Mech.*, **109**, 311–337.

Thomas A.S.W. & Bull M.K. 1983. On the role of wall-pressure fluctuations in determining motions in the turbulent boundary layer. *J. Fluid Mech.*, **128**, 283–322.

Turner J.S. 1973. *Buoyancy Effects in Fluids*. Cambridge University Press, Cambridge.

Utami T. & Ueno T. 1987. Experimental study on the coherent structure of turbulent open-channel flow using visualization and picture processing. *J. Fluid Mech.*, **174**, 399–440.

Van Dyke M. 1982. *An Album of Fluid Motion*. Parabolic Press, Stanford.

Viekman B.E., Wimbush M. & Van Leer J.C. 1989. Secondary circulations in the bottom boundary layer over sedimentary furrows. *J. Geophys. Res.*, **94**, 9721–9730.

Vos P.C., De Boer P.L. & Misdorp P. 1988. Sediment stabilization by benthic diatoms in intertidal sandy shoals: qualitative and quantitative observation. In: De Boer P.L., Van Gelder A. & Nio S.D.(eds) *Tide-influenced Sedimentary Environments and Facies*, pp. 511–526. Reidel, Dordrecht.

Wang X. & Qian N. 1989. Turbulence characteristics of sediment-laden flow. *J. Hydraul. Engng*, **115**, 781–800.

Weast R.C. & Astle M.J. 1980. *CRC Handbook for Chemistry and Physics*, 60th edn. CRC Press, Boca Raton.

Wei T. & Willmarth W.W. 1991. Examination of v-velocity fluctuations in a turbulent channel flow in the context of sediment transport. *J. Fluid Mech.*, **223**, 241–252.

Werner B.T. & Haff P.K. (1988). The impact process in aeolian saltation: two-dimensional simulations. *Sedimentology*, **35**, 189–196.

Whitehouse R.J.S. & Hardisty J. 1988. Experimental assessment of two theories for the effect of bed slope on bedload transport. *Mar. Geol.*, **79**, 135–139.

Wiegel R.L. 1964. *Oceanographical Engineering*. Prentice-Hall, Englewood Cliffs.

Wilkinson W.L. 1960. *Non-Newtonian Fluids*. Pergamon, New York.

Yih C.S. 1980. *Stratified Fluids*, 2nd edn. Academic Press, New York.

3 Hillslope sediment transport and deposition

M.J. SELBY

3.1 Introduction

Hillslopes make up a large proportion of all continental surfaces, except those of Antarctica, and are consequently exposed to the full range of climatically and tectonically induced processes of weathering and erosion. These processes are moderated in their effectiveness by lithology; vegetation cover; availability of rock and soil particles for transport; slope length, steepness and roughness; and by the results of human activity. The processes are not constant in the energy or periodicity of their operation and the available resistance to them is also variable.

3.2 Variability of sediment yield and erosion processes in space and over time

Sediment yield studies are usually conducted at one of two scales: (1) *net sediment yields* are from whole river basins, and (2) *gross sediment yields* are from local landscape units and are the total of all rain splash, sheetwash, rill, gully and landslide erosion. Gross erosion represents a higher yield per unit area than does net erosion because, within large river basins, sediment may be stored in floodplains, terraces, fans, deltas and lakes for periods which may range up to hundreds of millions of years.

3.2.1 Variations of net sediment yield in space

Global studies of sediment yield have commonly sought relationships between yield and precipitation for the major climatic zones. Such studies have resulted in the publication of world maps or graphs that suggest direct correlations (Langbein & Schumm,

1958; Corbel, 1959; Fournier, 1960; Douglas, 1967; Strakhov, 1967; Wilson, 1973; Wolman & Gerson, 1978; Jansson, 1982; Ohmori, 1983; Walling & Webb, 1983; Dedkov & Mozzherin, 1984; Walling & Webb, 1987). However, such relationships have five major deficiencies:

1 The data are normally derived from the loads of suspended sediment carried by major rivers and few data are available for bedload and dissolved load.
2 Relief has a major effect upon sediment yield both because rainfall and erosion rates are often related to relief, and also because the rate of delivery of runoff to rivers is variably related to relief.
3 Land use is a major, and often dominant, control on local sediment yield (Table 3.1) yet its effect is subsumed in the generalized data.
4 Yields from whole catchments do not reflect local variations in rate of delivery of sediment to the major rivers.
5 Rare, but large magnitude, events such as storms and earthquakes, may have short- or long-term effects upon sediment yields, yet their influence may be unrecognized.

Table 3.1 Representative rates of erosion from various land uses in USA

Land use	Sediment yield (t km^{-2} year^{-1})	Relative to forest = 1
Forest	8.5	1
Grassland	85.0	10
Abandoned surface mines	850.0	100
Cropland	1700.0	200
Construction sites	17000.0	2000

Source: US Environmental Protection Agency (1973).

61

A critique of published studies by Jansson (1988) has demonstrated a number of significant points:

1 there are notable differences amongst the world maps and conclusions of the various authors,

2 there is considerable variation in the quality and representativeness of the data used,

3 the data show a large range in annual yields (expressed as t km^{-2}) in several climatic regions, especially where relief, lithology and deforestation have major influences on erosion rates,

4 there is no clear theoretical or statistical relationship between annual runoff and sediment yield, and

5 the use of mean values for sediment yield from major climatic zones obscures the extremes and the magnitude of the range of yields (Fig. 3.1).

3.2.2 Gross sediment yields from small sites

In small catchments, the inclinations of hillslopes provide the potential energy for runoff and therefore there is likely to be a close relationship between local sediment yield and the slope angle. However, numerous local studies have demonstrated that there are many other factors involved; discussion of these matters is the topic of Section 3.3.

Sediment volumes available for transport from small catchments over long periods are controlled by the natural rate at which new soil and regolith materials are created by weathering. Estimated rates of soil thickening around the world range from 0.01 to 7.7 mm $year^{-1}$ with an average of about 0.1 mm $year^{-1}$. This average thickening of the soil profile indicates the increase in mass of the soil profile by chemical weathering of bedrock to be 0.1 kg m^{-2} $year^{-1}$ (if the bulk density for soil is assumed to be 1.0 Mg m^{-3}) (Buol et al., 1973). Cultivation may increase the natural average rate of soil formation for a short period to about 1.1 kg m^{-2}, by mixing soil from the A and lower horizons and by the incorporation of organic matter and fertilizer; however, such effects are characteristic only of the first few years of cultivation (Hall et al., 1979).

Measured sediment yields from catchments indicate that many areas of the world have natural soil formation rates which are approximately in equilibrium with the local natural denudation rates. However, cultivation can cause wide departures from the natural rate. In a study by Freebairn and Wockner (1986), cropland in the Darling Downs, Australia, had a soil loss of 2.9–6.2 kg m^{-2} $year^{-1}$ on cultivated soils with no stubble retention after cropping. When stubble mulching was practised the loss was 0.5 kg m^{-2} $year^{-1}$ and in areas with zero tillage 0.2 kg m^{-2} $year^{-1}$.

Agricultural scientists are far from unanimous in their view of what erosion rates can be tolerated if soil fertility is to be maintained, but the figures used are often based upon maintaining fertility over 20–25 years and indicate a tolerance of annual soil losses of 0.7 to 1.1 kg m^{-2} (Wischmeier, 1970). Such losses imply the total loss of the soil in a few hundred years.

The figures quoted above assume that sediment is derived from soil profiles derived from a chemically weathered bedrock and that the rate of soil formation is in approximate equilibrium with the rate of erosion. This is not always the situation. Some landscapes have

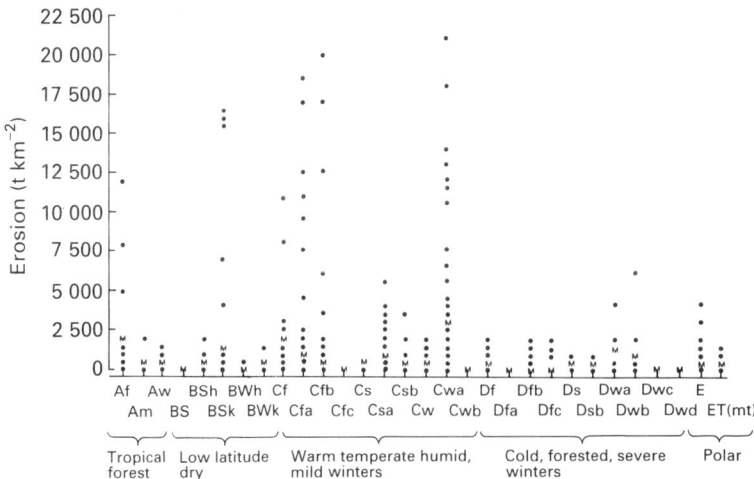

Fig. 3.1 Mean (M) sediment yield and extreme values in different climates. Climatic regimes are indicated by the nomenclature of the Köppen classification: A, tropical climates; B, dry climates; C, warm temperate climates; D, cold climates; E, polar climates. After Jansson (1988).

Fig. 3.2 A deeply weathered soil profile, with quartz veins indicating that it is *in situ*; Snowy Mountains, Australia.

been stripped to bedrock under a past environment, others have been remarkably stable and have experienced the accumulation of deep regoliths over very long periods (Fig. 3.2). At other sites, glacial, aeolian, fluvial and other deposits have been left by past processes and form a reservoir of sediment that may be entrained and transported in a period of dominant erosion. Such stores can be the sources of unusually large volumes of sediment that far exceed the normal 'background' rate of sediment yield.

3.2.3 Variations of net sediment yield over time

Climatic changes, occurring at scales of tens to hundreds of thousands of years, and past climatic regimes have left evidence in the terrestrial sedimentary record in the form of glacial and periglacial deposits in areas that now have temperate climates, sand dunes and loess deposits in areas now humid indicating earlier phases of aridity, and paleosols in areas now arid indicating former humidity.

On much shorter timescales, earthquakes and severe storms, or periods of storminess, may cause sudden and large-volume increases in erosion rates. The evidence for such events is seldom datable from the scars left by landslides or gullies, but major events create deposits on footslopes, fans, terraces and floodplains which can be ascribed an age by application of such methods as dendrochronology, tephrochronology, and radio-carbon dating of included wood.

A remarkably successful study of this kind was that of Grant (1985) who was able to recognize eight major periods of erosion and alluvial sedimentation in northern New Zealand over the last 1800 years. These periods have been dated. The earliest period may have been caused by heavy rainfalls induced by volcanic eruptions, but the others are ascribed to

increased northerly airflows and increased magnitudes of major rainstorms. Such changes are associated with temporary strengthening of meridional upper atmospheric circulation in the Southwest Pacific. Between the periods of increased erosion are intervals in which hillslope stability and soil formation on floodplain surfaces are characteristic.

The work of Wolman and Miller (1960) led most geomorphologists to recognize that climatically induced erosional events of moderate frequency and magnitude cause most of the erosion in river channels and on many hillslopes. In river channels the dominant event corresponds to the stage of bankfull discharge; an event that has a return period of between 1.33 and 2 years.

In areas where vegetation cover is limited, or the ground surface is disturbed by agriculture, there is substantial evidence that dominant erosional events are likely to be the result of storms that occur several times each year (Roose, 1967; Lal, 1976; Hudson, 1981; Morgan, 1986). In areas of complete vegetation cover, dominant erosional events on hillslopes are commonly found to be landslides with return periods of several years. Working in Tanzania, Temple and Rapp (1972) identified the 5-year return period storm as being dominant and, in New Zealand, Selby (1976) recognized the 30-year storm as being dominant on hillslopes with pasture-grass covers and the 100-year storm as being dominant on similar hills with a full forest cover.

Very large-magnitude storms or tectonic events may cause such large-scale erosional events that they create deposits which will not only survive in the landscape but will supply debris for fluvial transport for hundreds of years and so greatly increase long-term rates of mean denudation.

Large-scale urban construction, open-pit mining

and other examples of human activity may cause local increases in sediment yield over periods of tens of years (Wolman, 1967; US Environmental Protection Agency, 1973).

Recognition that the dominant processes on hillslopes may differ across the landscape and vary in magnitude and frequency as well as through time, has caused some workers to base field study on the concept of erosion–transport–accumulation (ETA) systems (Engelen & Venneker, 1988). Figure 3.3 is an attempt to identify characteristic ETA systems, operating and relict, on hillslopes of high and medium ranges of relative relief, at the scale of small drainage basins with areas of a few square kilometres.

The division of relief forms into just two classes, mountains and hills, is extremely crude, but it helps

Fig. 3.3 A representation of ETA systems that are common in mountain and hill country. Modern systems are overprinted on relicts of paleosystems.

make the major distinction between (1) mountains, in which steep-sided valleys have slopes developing as a result of large landslides, rockfalls and other types of mass movement that contribute to rapid rates of slope denudation and delivery of sediment, and (2) hillslopes that are generally stable against deep-seated landsliding, are evolving relatively slowly and have much lower rates of denudation and sediment delivery.

3.3 Erosion and transport processes on hillslopes by raindrops and overland flow

3.3.1 Empirical models of soil erosion

Erosion of soil from hillslopes has been a major subject of scientific investigation since the 1930s, when the magnitude of economic losses by soil erosion in many countries, and especially in the USA, was recognized (Bennett, 1939). A basic understanding of most of the factors affecting erosion was developed from many qualitative studies and measurements performed on experimental plots (Ayres, 1936). The importance of raindrop impact on bare agricultural soils was first recognized as a result of the studies of Laws (1940) and the analysis of the mechanical action of raindrops by Ellison (1947). Much of the work of Laws on the fall velocities of water drops of various diameters (Table 3.2) has been largely confirmed by recent work (Epema & Riezebos, 1983).

The use of erosion plots under experimental and field conditions permitted the development of a num-ber of empirical equations. Zingg (1940) related soil loss to slope steepness and length in the expression:

$$A = CS^m L^{n-1}, \qquad (3.1)$$

where A is the average soil loss per unit area from a land slope of unit width, C is a constant of variation, S is the degree of land slope, L is the horizontal length of land slope, and m, n are the exponents of degree and horizontal length of land slope, respectively, with values of 1.4 and 1.6.

In Eq. 3.1, the constant of variation, C, combines the effects of rainfall, crop cover, soil characteristics and land management.

The relationship of soil loss to ground conditions and rainfall intensity, for a number of stations, was expressed by Musgrave (1947):

$$E = (0.00527)IRS^{1.35}L^{0.35}P_{30}^{1.75}, \qquad (3.2)$$

where E represents the soil loss (mm year^{-1}), I the inherent erodibility of a soil at 10% slope and 22 m slope length (mm year^{-1}), R is a vegetation cover factor, S is the degree of slope (%), L is the length of slope (m), and P_{30} the maximum rainfall in 30 min.

A method for estimating soil loss from clay pan soils of Missouri was developed by Smith and Whitt (1947, 1948) in the form:

$$A = CSLKP, \qquad (3.3)$$

where A is the average soil loss, C the average annual rotation soil loss from plots, and S, L, K, and P are the multipliers to adjust the plot soil loss, C, for slope steepness, length, soil group and conservation practice.

Table 3.2 Raindrop velocity and kinetic energy

Rainfall type	Median diameter (mm)	Velocity of fall (m s^{-1})	Drops (m^2 s^{-1})	Intensity (mm h^{-1})	Kinetic energy (J m^{-2} per mm of rain)
Fog	0.01	0.0	67 000 000	0.05	0.52
Mist	0.10	0.2	27 000	0.13	4.14
Drizzle	0.96	4.1	150	0.25	6.61
Light rain	1.24	4.8	280	1.02	11.95
Moderate rain	1.60	5.7	500	3.81	16.94
Heavy rain	2.05	6.7	500	15.24	22.17
Very heavy rain	2.40	7.3	820	40.64	25.92
Torrential rain-1	2.85	7.9	1215	101.60	29.42
Torrential rain-2	4.00	8.9	440	101.60	29.42
Torrential rain-3	6.00	9.3	130	101.60	29.42

Sources: Basic data from Laws (1941); Laws and Parsons (1943); and Ellison (1947).

Soil erosion research in the USA was consolidated by the compilation and evaluation of over 8000 plot-years of data from 36 locations in 21 states. A new evaluation of the data and factors affecting soil loss led to the development of the prediction method called the Universal Soil Loss Equation (USLE), (Smith & Wischmeier, 1957; Wischmeier & Smith, 1958; Wischmeier *et al.*, 1958):

$$A = RKLSCP,\qquad(3.4)$$

where A is the soil loss, R is the rainfall erosivity factor, K is the soil erodibility factor, L is the slope length factor, S is the slope gradient factor, C is the cropping management factor, and P is the erosion control practice factor.

The USLE was devised for practical conservation purposes and, as stated by Wischmeier (1976), may be used to:
1 predict average annual loss of soil from a cultivated field with specific land use conditions,
2 guide the selection of cropping and management systems, and conservation practices for specific soils and slopes,
3 predict the change in soil loss that would result from a change in crops or land use on a specific field,
4 determine how conservation practices should be adjusted to allow higher crop yields,
5 estimate soil losses from areas that are not in agricultural use, and
6 provide estimates of soil losses for planners of conservation works.

An understanding of the importance of the USLE factors was gained from study of erosion on standard plots using cultivation practices appropriate to the USA. In order to apply its concepts more widely, modified versions have been developed and applied in other parts of the world: for example, Hudson (1961) developed a method for use in the African sub-tropics, Elwell (1977) a method for southern Africa, Stehlík (1975) a method for Czechoslovakia, and Morgan *et al.* (1984), a general method with refinements beyond those of the USLE. All of these methods are for use on fields or slopes of limited area and are concerned with annual sediment loss estimates; they cannot be used for studies of drainage basins, nor for sediment yields from individual storms. An account of an attempt to develop a watershed model, capable of simulating the results of proposed soil erosion control measures in part of Holland, has been presented by De Roo *et al.* (1980). Charts demonstrating how the USLE and Morgan *et*

al., methods can be applied are presented in Morgan (1986).

All of the methods listed above are founded on identifying statistically significant relationships between variables that were believed to be important from observational experience, or were found to be so in application of multivariate statistical testing.

3.3.2 Physical and physically-based models of erosion and transport

In order to obtain a better understanding of erosion processes for scientific and practical purposes, two approaches are in common use:
1 Physical models of reality are built in the field or in laboratories, using such devices as rainfall simulators, infiltrometers, soil splash trays and flumes. Such methods require a careful assessment of the similitude between models and the real world.
2 Physically-based models are mathematical equations formulated to describe processes by founding them on the laws of conservation of mass and energy.

Physically-based models have the advantage, over empirical models, that they can improve prediction as well as understanding of processes and their consequences. The US Department of Agriculture is currently undertaking a major study, called the USDA Water Erosion Prediction Project (WEPP), with the aim of developing a new methodology for erosion prediction based upon fundamental erosion mechanics. WEPP is intended to replace the USLE by the mid-1990s (Lane *et al.*, 1988).

WEPP is to apply to 'field-sized areas' ranging in size from single hillslopes up to a few hundred hectares in area. The procedures are limited to areas where overland flow and surface runoff occur and cannot be applied to areas where partial area hydrology and sub-surface flow dominate.

The basis of WEPP, like most fundamental models, is the empirical recognition that soil loss from the upper slope, and sediment yield at the base of the slope, is determined by the process of rainfall causing soil detachment (splash), with overland flowing entraining, transporting and depositing particles. Overland flow processes are usually conceptualized as a combination of broad sheet flow (called inter-rill flow) and concentrated flow (called rill flow).

Many reports of soil erosion phenomena have their value limited by uncertainties in the terminology used, consequently the key terms are defined here.

Raindrop erosion is recognized as being responsible for four effects:
1 disaggregation of soil units as a result of impact,
2 minor lateral displacement of soil particles (a process sometimes referred to as creep),
3 splashing of soil particles into the air (sometimes called saltation),
4 selection or sorting of soil particles by raindrop impact may occur as a result of two effects — (a) the forcing of fine-grained particles into soil voids causing the infiltration rate to be reduced, and (b) selective splashing of detached grains. Wash is the process in which soil particles are entrained and transported by the shallow sheet flow which is also called Hortonian overland flow. Rainwash is the combined effect from raindrops falling into a sheet flow.

3.3.3 Raindrop erosion

Splash erosion has been studied in many experiments carried out since the pioneer work of Laws (1941) and Ellison (1944) but soil is such a variable material that reliable generalizations, about the relative importance of the factors that influence sediment yield by this process, are few. However, a splash transport model is likely to involve an equation derived from the following general form (Poesen, 1985; De Ploey & Poesen, 1987):

$$q_s = f(E_k, R, D_{50}, \beta, B),\qquad (3.5)$$

where q_s is the net splash transport on a bare smooth, unchannelled surface (m³ m⁻¹ year⁻¹), E_k is the kinetic energy of rainfall (J m⁻² year⁻¹), R is the shearing resistance of soil (N m⁻²), D_{50} is the median grain size of the soil material (m), β is the slope angle (°), and B is the rainfall obliquity (°).

The kinetic energy of raindrops is derived from the expression:

$$E_k = 1/2\ m\ v^2,\qquad (3.6)$$

where m represents the mass (kg), and v the velocity (m s⁻¹).

Most of this energy is dissipated in friction so that as little as 0.2% of the energy may be available to cause erosion (Pearce, 1976). The effect of raindrop impact both compacts the soil surface and disperses from the crater of impact in lateral flow jets which have local velocities that are nearly double those of the raindrop at impact (Huang et al., 1982). These jets form water droplets that may contain soil particles.

Soil surface compaction by impact is further enhanced by selective dispersal of fine grains of soil which are moved into the soil pores to create a sealed crust about 0.1–3 mm thick (Farres, 1978). Aggregates at the soil surface are consequently broken down, but those below the crust are usually protected. A major effect of the crust is to reduce infiltration capacity. The percentage reduction is variable, depending upon the soil physical characteristics and the diameter, and therefore the energy, of the raindrops; recorded reductions are as high as 50% for a single storm and 1000% for a succession of storms (Morin et al., 1981; Hoogmoed & Stroosnijder, 1984). Soils with high clay contents and high organic matter contents are generally less prone to crusting than loams and sandy loams. The overall detachment of grains, as measured by mean weight of soil displaced, usually follows the order sand>silt>clay loam>clay (Quansah, 1985). Under most conditions gravel-size particles would be the least detachable and remain as a lag on the soil surface (Fig. 3.4). The term, D_{50}, is a useful approximation for practical purposes to the problem of characterizing soil texture.

The resistance of soil to detachment by raindrop impact depends upon its shear strength, that is its cohesion (c) and angle of friction (ϕ) (Cruse & Larson, 1977). It is difficult, in practice, to measure the appropriate values of c and ϕ for grains at the surface of a soil or soil crust, partly because of variability in the size, packing and shape of particles and partly because of the varying degrees of wetting and submergence of grains by water (see Poesen, 1985, p. 197, table 1). It has been suggested by Brunori et al. (1989) that many of the conflicts in published experimental data and conclusions are attributable to the lack of standard instruments and methods; after a comparison of penetrometers and shear vanes they concluded that pocket shear vanes (of the torsional type, such as the Soil Test Torvane) give the most reliable measure of soil resistance. This conclusion has been reinforced by studies of Crouch and Novruzi (1989) on rill initiation.

The inclination of the hillslope is important because on a sloping surface more water droplets and soil particles are splashed downslope than upslope. Experimental results indicate that splash detachment is greatest when slope angles fall in the range 10–20° with lower values at angles <10° and >20° (Froehlich, 1986). The relationship is usually a curvilinear one and is usually influenced by particle size (De Ploey & Savat, 1968; Meyer et al., 1975; Savat,

Fig. 3.4 Lags of quartz gravel produced by overland flow on slopes in central Australia.

1981). Soil detachment angles are usually in the range 11–18°; on low angles of slope nearly as many grains are projected upslope as downslope; at high angles of slope the proportion moving upslope declines above about 20° of slope.

Raindrops driven by wind-strike the soil surface at an angle from the vertical and so affect the splash detachment angle. Wind therefore affects both the proportions of upslope and downslope splash and increases the impact energy of raindrops. On hill-slopes into which raindrops are driven by wind, low angle slopes may experience net upslope splash and the maximum downslope transport will occur at slope angles >20° (Moeyersons, 1983).

A generalized and approximate formula for the calculation of sediment discharge by splash from vertically falling raindrops could take the form:

$$q_s = \frac{E_k \cdot \cos \beta}{R}(\sin \beta + D_{50}),\qquad (3.7)$$

and for wind-driven raindrops:

$$q_s = \frac{E_k \cdot \cos (\beta \pm B)}{R}((\sin \beta \pm B) + D_{50}),\quad (3.8)$$

where the terms are as given for Eq. 3.5 and B is the angle between a vertical line and a mean trajectory of raindrops. Then:

$(\beta + B)$ = rain directed to produce increased splash,
$(\beta - B)$ = rain directed to produce decreased splash.

A specific form of these equations has been presented by Poesen (1985) in which the resistance to splash is represented as a function of soil bulk density and the kinetic energy needed to detach 1 kg of material. The forms of the equations given above assume that R is a measure of total shearing resistance. It must be stressed that rainfall energy varies rapidly over time and the median particle size is a convenient, rather than fundamental, characterization of soil properties. Furthermore, soil from A horizons is often aggregated so that the effective particle size is different from that measured for individual grains. Because many aggregates breakdown under the impact of raindrops, by slaking, agitation and splashing, the effective particle size on a surface changes during a storm. Changes due to aggregate instability enhance those due to selection by splashing, washing into voids and transport in overland flow. The practical value of the generalized formulae of Eqs 3.7 and 3.8 may therefore be limited. Further discussion of these topics is presented by De Ploey and Poesen (1985).

3.3.4 Overland flow and sediment transport

During rainstorms, minor depressions and cracks in the soil are first filled and then water percolates into the soil at a rate which is at a maximum when the soil is saturated but has no barrier to continuing vertical or lateral drainage. That water which cannot infiltrate runs off the slope and is known variously as Hortonian overland flow, sheet flow, sheetwash or wash. The flow of water over natural ground is seldom as a sheet of uniform depth. Irregularities on the surface promote the formation of anastomosing threads of water that diverge and converge around pebbles and the stems of plants. All such obstructions, together with the ground surface, create frictional resistance to flow and promote turbulence. The Reynolds number (R_e) is the index of turbulence of the flow and the Froude number (F_e) describes the state of the flow or, more specifically, the ratio between the velocity of the flow and the force of gravity (see Chapter 2). Frictional resistance to flow may be described by coefficients such as Chezy's C, Manning's n and the Darcy–Weisbach friction factor f. The factor f is superior to C and n for studies of overland flow because it is dimensionless and applies to all states of flow:

$$f = \frac{8g R S_e}{v^2}, \qquad (3.9)$$

where g is the acceleration due to gravity, R is the hydraulic radius, S_e is the energy slope, and v is the mean flow velocity.

Because of the irregularity of most natural ground surfaces, in contrast to laboratory flumes and the like, flow depth and velocity may vary over very short distances and give rise to changes in the state of the flow (F_e) through laminar, turbulent and transitional forms. Furthermore, the depth of flow varies during storms, changing the immersion of obstacles and hence the values of f.

Many mathematical models of overland flow incorporate a relationship between f and R_e. It has been pointed out by Abrahams et al. (1986) that for desert hillslopes, irregularities of the surface are so extreme that no single expression of either factor, or relationship between them, can describe flow conditions. The surfaces of cultivated soils may be as irregular as those of deserts, especially after cultivation or harvesting.

When the Froude number value is less than 1.0 the flow is described as tranquil or sub-critical; values greater than 1.0 denote super-critical or rapid flow, which is more erosive. Published values of F_e for overland flow range from 15 for a plane-bed flume to 0.1 for some field studies (Pearce, 1976; Savat, 1977; Morgan, 1980); indicating that super-critical flow does occur. However, a study of Govers (1989) suggests the transporting capacity of overland flow is not strongly dependent on flow regime.

Reynolds numbers above 2000 indicate turbulent flow conditions, and less than 500 indicate laminar flow, the range 500–2000 is a transition; published values of R_e for overland flow are usually less than

100 on cultivated soils but on desert hillslopes are commonly in the range of 1200–3000.

For erosion to occur, a threshold velocity of flow has to be exceeded and that velocity is related to particle size (Hjulström, 1935). Once entrained the particle will remain in suspension until a depositional velocity occurs. Recorded water velocities of overland flow are commonly in the range of 0.015–0.3 m s^{-1}, which is great enough to move silts and fine sands.

Because of variability in ground roughness and ground slope; rainfall intensity; supply of splashed particles; and particle density, shape and size, the hydraulic conditions of flow vary greatly over short distances and ground surfaces show patterns of alternating scours and sediment fans (Moss & Walker, 1978). Consequently, no single, simple, mathematical relationship is likely to describe all relationships between sediment discharge, flow and soil particles. Relationships have to be determined from experimental studies, but will take the general form:

$$q_s = \frac{k\, Q_w^a \sin \beta^b}{r^c\, D_{50}}, \qquad (3.10)$$

where Q_w = water discharge, r = resistance to entrainment, q_s, β, D_{50} are as in previous equations; and k, a, b, c have empirically derived values.

Komura (1976) and Morgan (1980) use forms of the above equation and have determined values for the exponents.

Because overland flow water-depths are usually limited to a few millimetres, raindrops falling into the flow may create splash and enhanced entrainment. Moss and Green (1983) and Palmer (1965) found that the rate of transport of soil by this mechanism increases with increasing depth of water to a maximum at a depth that is about one to three times the diameter of the raindrops, and is consequently at a maximum when flow depth is in the range of 3–10 mm. However, it is possible that at depths of flow greater than one diameter the increased soil detachment is due to the turbulence in the flow created by the raindrops. At depths greater than three diameters water is likely to be turbulent anyway and the raindrop impact energy is dissipated in the flow.

In discussing the nature of overland flow, Horton (1945) described flow as resulting from the rainfall intensity exceeding the infiltration capacity of the soil and suggested that it occupied up to two-thirds or more of hillslope surfaces during the peak period of a storm. Horton postulated a belt of no erosion at the

Fig. 3.5 Rills, developed on mudrocks, showing the effects of channel and plungepool erosion. Note the variation in spacing of rills down the slope.

top of a hill and, further down a slope, a zone in which sufficient water accumulates for flow to begin, yet towards the base of the slope the flow depth increases until flow becomes channelled into rills (Fig. 3.5).

The formation of overland flow of the Hortonian kind in semi-arid areas is not questioned, but in well-vegetated areas such flow occurs infrequently and seldom covers two-thirds of the ground surface. In areas of relatively deep soils and complete vegetation cover saturated zones are likely to develop in hollows, at the heads of ephemeral channels and alongside channels, and saturated zone overland flow is more likely to occur than infiltration-excess overland flow; only in areas with very intense rainfalls is the Hortonian type of flow likely to occur. Saturated zone overland flow in vegetated areas rarely entrains much sediment, unless from bare soil patches or forest floors. In areas with bare, cultivated, soil overland flow in association with splash can account for large proportions of the total soil loss — up to 95% (van Asch, 1983; Morgan, 1986).

Sub-surface flow and pipe flow are not known to contribute more than a few per cent to sediment discharge.

3.3.5 Rill processes

Most studies of rills have been carried out either on agricultural soils or on field or laboratory plots established in agricultural soils. Studies related to geomorphic development are fewer and have been concentrated in semi-arid environments or specific environments such as those with calcareous and smectite clay dominated soils (see Bryan, 1987, for a review). The agricultural importance of rills is indicated in the most commonly used definition: rills are 'microchannels . . . small enough to be removed by normal tillage operations' (FAO, 1965). As rills exist in areas that are not cultivated, a more general definition is needed.

Rills are small channels with cross-sectional dimensions of a few centimetres to a few tens of centimetres. They are usually discontinuous; may have no connection to a stream channel system; are often obliterated between one storm and the next, or even during a storm when the supply of sediment from splash on inter-rill areas, or collapse of rill walls, or liquefaction of the bed and walls, exceeds the transporting capacity of the rill flow; and they occur on slopes steeper than 2–3° (De Ploey, 1983; Dunne & Aubry, 1986; Bryan & Poesen, 1989).

It is commonly stated that rills do not become progressively enlarged downslope, but instances are recognized of enlargement (Schumm, 1956) even to the size of gullies, and rills may form the heads of natural drainage systems. By contrast, rills may become wider, shallower, and feed into a braided wash downslope.

Rill extension can extend upslope by the development of headcuts and downslope by channel erosion. Lateral changes can occur by channel-wall collapse, micropiracy and cross-grading with formation of a master-rill which drains an increasing area of a slope until it is both a permanent feature and too large to be defined as a rill.

Fig. 3.8 A gully head formed in pumiceous infill of a valley, central North Island, New Zealand. The dark organic soil of the valley floor has lenses of pumice gravels within it and a thin pumice fan deposit lies on the valley floor. The fan deposit is derived from a small gully further up the valley.

terials, and boulders may occur in sites below steep slopes with bedrock outcrops.

Soil A horizons and total soil profile depths tend to thicken in downslope directions across the colluvial footslope unit. Where deposition has been episodic, with periods of upslope erosion punctuated by periods of slope stability with predominant soil formation, paleosols may form in the colluvium. Paleosols are particularly common where there has been an unusual period of vegetation destruction, as by natural fires or human activity causing deforestation or cultivation. In Fig. 3.8 the dark soil of the valley floor, above the gully head, is composed of organic-rich A horizon material interbedded with two lenses of pumiceous gravel. These deposits record episodes of slope erosion producing the dark deposits and valley-floor transport of pumice during periods of erosion between 1945 and 1965.

Fan deposits, of colluvial footslopes, are usually of small dimensions — a few tens of centimetres to a few metres in length — because catchments and rills are also of limited dimensions. Particle sizes are commonly in the silt to sand range with some organic material and remnants of aggregates which may be predominantly of clay particles. As with wash-derived sediments, the deposits from rills are the products of erosion and transport processes. Where rills are large, steep, or have occasional high storm flows through them, they may carry coarse debris which will be incorporated in the fan.

Because of the short distance of transport, rill and wash deposits rarely have distinct bedforms, but there may be lenses of alternating coarser and finer material related to fluctuations in discharge during storms, or variations in the supply of material.

Whatever the texture, colluvium is usually deposited in such thin units, and at such low rates, that bioturbation by plants and animals, such as earthworms, causes bedding to be destroyed as pedological processes incorporate the sediment into the thickening soil. Rates of colluvial soil thickening over the last 1000 years in Iceland, where soil erosion has been periodically very active and where tephra is a readily eroded material, have been assessed as being in the range of $0.7–5.7$ mm year^{-1} for slope wash and $1.3–10$ mm year^{-1} for fans (Gerrard, 1985, p. 93); most colluvial sites have rates of soil thickening that are of much smaller magnitude.

3.5 Gully processes

Gullies are erosional features found in many parts of the world, and variously known by regional names, for example, 'arroyo' in the southwestern United

Fig. 3.9 A gully in the Transkei, southern Africa, resulting from overland flow from the slopes in the background; the gully is cut into old colluvial valley infills.

States (Schumm & Hadley, 1957; Cooke & Reeves, 1976), and 'lavaka' in Madagascar (Riquier, 1958; Hurault, 1971). The chief features of gullies are their steep walls, flat floors, and their incision into weak sediments and soils. They may be continuous from a steep head downvalley to a depositional feature such as an alluvial fan, a delta, or to a river, or they may be discontinuous with incised sections shallowing downvalley to a fan of sediment.

A major factor in the consideration of gullies is their incision into weak materials. In many areas these are fine-grained, Late Cenozoic, terrestrial sediments that have been subjected to little or no consolidation by overburdens and few or no lithification processes. Alluvial fans, volcanic flow and avalanche deposits, loess, and sand alluvia may be entrenched by gullies (see Fig. 3.8) (Blong, 1970; Stocking, 1980). A second general class of gullies is of those that develop in weak argillic rocks such as argillites, phyllites, mudrocks and shales. These gullies are often channels through which debris flows and mudflows, with very high sediment discharges, drain unstable catchments (Schouten, 1984; Sidle *et al.*, 1985; Li Jian & Wang Jingrung, 1986) (Figs 3.9 & 3.10).

The causes of gully erosion are seldom simple,

Fig. 3.10 Rills cut into playa deposits, derived from reworked tephra, on the margin of the Salar de Atacama, Chile. The rills drain into the gully in the middle-ground.

being a group of inter-related factors, but can usually be related to some environmental change which involves one or more of: (1) a local increase in channel slope; (2) concentration of water flow; and (3) removal of vegetation cover.

Local increases of slope may be due to such occurrences as creation of a headcut through migration of a main stream with undercutting of a tributary which is flowing over weak sediments, ground subsidence above a natural pipe (Gibbs, 1945; Swanson *et al*, 1989), deposition of a sediment fan on a valley floor or ponding behind a fence. Concentration of flow may be caused by runoff from roads, broken drainage ditches or stream diversion. Removal of vegetation may be by fires, over-grazing, or cultivation. Whatever the direct causes there is an increase in exposure of the weak sediment to a tractive stress which may have been increased by greater discharges. Gully widening, deepening and lengthening may then occur by plunge-pool erosion, bank collapse, sapping, piping and surface flow.

3.5.1 Erosive power of surface flows

One of the oldest methods used for understanding the operations of factors influencing the erosive power of flows in open channels uses the mean velocity of flow as a surrogate for erosiveness. Conservators are often concerned to restrict the velocity of flow to that which is below the threshold at which a non-eroding flow is transformed into an erosive flow. The Manning's equation is a useful guide for estimating the mean velocity of turbulent flow in open channels:

$$\bar{v} = \frac{1}{n} R^{2/3} S^{1/2}, \qquad (3.11)$$

where $\bar{v}$ is the mean velocity of flow (m s^{-1}), S is the average gradient of the channel (m m^{-1}), R is the hydraulic radius (m), and n is Manning's roughness coefficient.

Therefore, increased erosiveness through valley floors could be achieved by one or more of three related changes: increase of slope, increase of hydraulic radius, and reduction of surface roughness.

The hydraulic roughness is approximately equal to flow depth, so reducing the concentration of flow by reducing the contributing area for runoff or increasing water losses by infiltration, or take up by plants, could reduce flow depths.

Surface roughness in valley floors is largely caused by vegetation, so increasing vegetation cover by appropriate species will decrease velocities. In the southwestern USA, flow velocities which are 'permissible' (i.e. will not be great enough to cause incision) on erosion-resistant soils are in the range 2.5–0.9 m s^{-1}, and on easily eroded soils are 1.8–0.7 m s^{-1} (Ogrosky & Mockus, 1964).

The Manning's equation has its deficiencies when used to calculate permissible velocities, primarily because the factors are to some extent interdependent.

3.5.2 Deposits from gully erosion

The material eroded from gullies is usually carried into main rivers or lakes where it becomes part of another system (see Chapters 4, 5, 6). The material deposited within the gully system has characteristics that are largely controlled by the original source with only slight modification by sorting and comminution during transport, primarily because the distance of transport is usually small.

Fan deposits are usually spread widely and thinly over a valley floor, unless erosion up-valley has been catastrophic (see Fig. 3.8, and Selby, 1982, p. 111). It is sometimes possible to recognize the presence of thin lenses of sediment, resulting from locally varying discharges, and hence sorting by texture, However, the primary control on texture of sediments is the texture of materials in the source area; most gully deposits are therefore fine-grained, but in areas of shattered bedrock may be of gravel and sand size and, in the unusual situation where rhyolitic low-density pumice is being transported, they may be predominantly of gravel and cobble sizes (Fig. 3.11).

3.6 **Mass wasting and materials**

The distinction between a soil mass and a rock mass is of major importance when considering the origin of material that is involved in mass wasting processes. 'Soil' is used in the geotechnical sense of:

1 naturally occurring loose or soft deposit formed at or near the surface of Earth,

2 it is weakened or softened by immersion in water,

3 it may be the result of physical, chemical and biological processes acting to produce an organic-rich material,

4 it may be formed from weathering of harder rock or older soils; it may be formed on the site in which it currently occurs,

5 it may be of transported material, and

Fig. 3.11 A fan of argillite fragments derived from a gully in the background. Southern New Zealand.

6 it may be deposited as a weak geological formation.

Whatever its origin, however, a soil mass is essentially a continuous material with few large-scale joints or fissures. Rock, by contrast, is an intact material of mineral grains cemented together; it is a hard, elastic substance that does not significantly soften in water and in a mass it is a discontinuous material with joints and fractures that separate the mass into discrete blocks (Selby, 1985, p. 172).

The importance of the distinction is in (1) the effect of water to create softening; (2) the shear strength of the material; and (3) the presence of fractures. Soil may fail in blocks but in the process of transport down a hillslope the blocks will break up and, if moist, behave as a plastic substance or viscous fluid. Rock may fail as blocks along pre-existing discontinuities and may break into small blocks in transport. The behaviour of the particulate rock material in transport will depend upon the velocity of the displacement and the nature of the matrix material and fluids in the voids between the rock particles (Figs 3.12 & 3.13).

3.6.1 Mass wasting in soils

Processes of mass wasting by slumping and translational sliding in soils have been intensively studied and reported in the technical literature (e.g. Brunsden & Prior, 1984; Anderson & Richards, 1987). Compared with the mechanisms that initiate movements, the processes of transport and deposition have been given little attention. However, some general conclusions can be offered:

1 Sediment volumes from mass wasting depend almost entirely upon the volume of material removed from the site of the landslide, with a general increase in volume being possible as a result of (a) dilation of the material in transport; (b) the opening of voids in the moving mass; and (c) accretion of debris in the failure path; however, there may be a loss in volume from the formation of levees and disintegration during movement.

2 The deeper the failure plane, the greater is the chance that the materials will be of varied structure and texture.

3 The distance of travel of the transported material will increase with (a) water content of the moving mass; (b) steepness of the slope; (c) length of slope; fineness of texture of the material; and (d) volume of the material.

4 The transported material will come to rest as a result of decrease in slope, loss of water and decrease in volume of the mass.

5 Remoulding of soil in transport will break down

Fig. 3.12 Unsorted sandstone boulders derived from a rockslide in the background. Tararua Range, New Zealand.

Fig. 3.13. A small debris flow deposit derived from translational landslides in the background. The material is unsorted and ranges in size from coarse gravel to clay. Note the levees. Hapuakohe Range, North Island, New Zealand.

structural units in the moving mass and, for the same water content, increase the fluidity of the mass.

6 Moving soil masses may incorporate water, from overland flow or channels, as they move and so become less viscous.

7 Moving masses may either erode the materials they travel over or leave them unmodified (Figs 3.14 & 3.15).

8 Deposits which come to rest on landsurfaces may be recognized from their lack of pedological features — horizons, and structure — and from the buried paleosol on which they lie. If the upper surface has not been modified by rainwash, cultivation or other processes, it may preserve an irregular relief, pressure ridges and flowlines.

3.6.2 Debris flows

Debris flows are a principal form of mass movement

in semi-arid zones and some more humid environments with a prolonged dry season, such as some 'mediterranean' climatic zones. They also occur in many mountains.

Debris flows often evolve into fluvial phenomena in debris torrents and on alluvial fans, and the fluvial part of the feature results in important modifications of the debris flow deposits. Furthermore, the change from debris flow to fluvial forms is often gradational (Johnson & Rodine, 1984; Hooke, 1987). Debris flows also share many features and mechanics of processes with mudslides, mudflows, lahars and mud torrents (Brunsden, 1984). The literature has been reviewed by Costa (1984) and papers that have been published subsequently to that review include Hungr *et al.*, (1984); Van Dine (1985); Nash *et al.*

Fig. 3.14 Translational landslides developing into mudflows downslope. The material is a clay–silt–fine sand slurry derived from a loess-rich soil. Wairarapa, New Zealand.

(1985); Osterkamp *et al.* (1986) and Van Steijn *et al.* (1988).

Debris flow materials have densities in the range 1.6–2.4 Mg m^{-3}; with the lower values usually applying to materials with a high clay content, such as mudslides; the higher values apply to debris flows containing large proportions of boulders of dense rock. The dynamic viscosities of flows range from about 100 to 700 Ns m^{-2}, compared with water

Fig. 3.15 A small translation slide, which became a flow over a short distance, has flowed over the grass without eroding it. Hawkes Bay, New Zealand.

at 0.001 Ns m^{-2}. The strength of debris at rest can be described by the Coulomb law, as modified by Terzaghi:

$$\tau_f = c + (\sigma_n - u) \tan \phi. \qquad (3.12)$$

The strength of mobile debris is more appropriately described by inclusion of a term that describes the viscosity of the flow and the rate of strain (Johnson & Rodine, 1984):

$$\tau_c = c' + \sigma'_n \tan \phi' + \eta(\mathrm{d}w/\mathrm{d}y), \qquad (3.13)$$

where τ_f is the resistance to shearing, τ_c is the yield strength, c, c' are cohesion and effective cohesion respectively, σ_n, σ'_n represent normal stress and effective normal stress, ϕ, ϕ' are friction and effective friction, u is pore water pressure, η is the coefficient of viscosity, and $(\mathrm{d}w/\mathrm{d}y)$ is the rate of shear strain.

For such materials, no flow occurs until the yield strength is exceeded, but then the viscosity decreases gradually with increasing applied stress. The pore fluid in a debris flow is not water but a water–clay–silt–sand mixture. The pore fluid both separates coarse clasts and holds them in suspension with the result that interlocking of large clasts is reduced and frictional resistance to flow is relatively low. Debris flows with 80–90% granular solids by weight can move in sheets about 1 m thick over surfaces with slopes of 3–10° (Johnson & Rodine, 1984, pp. 278, 288).

The main body of a debris flow is a raft, or plug, which has a critical thickness. As the flow slows, for example, as a result of a decrease in slope angle, the plug thickens, and if the thickness of the plug equals the total thickness of the flow, the flow stops. Debris flows also stop when the shear stress on the bed no longer exceeds the yield strength of the mud at the base of the plug.

Debris flows are initiated in one of two major ways, both of which involve the availability of abundant water:

1 As a landslide mass that cracks, rotates and churns into a material looking like wet concrete within 2–3 m of movement. The churning is probably essential to the incorporation of water and the formation of a mixed pore fluid.

2 By formation of a slurry, primarily of clay and water, that erodes its channel and thereby increases its load until the concentration of granular solids approaches 80–90%.

In transport, little sorting occurs except that waves of debris may form with the leading face of the wave having a concentration of large rock fragments. Because the density of the mass is so close to that of the source rock, clasts with dimensions nearly as great as the depth of flow may be transported.

Because of the origin and transport mechanisms, debris flow deposits are largely unsorted, unbedded, have a large range of particle sizes, and are predominantly granular with voids between large clasts filled with fine-grained materials. There is little or no rounding of clasts during transport so clasts are angular. Very large clasts may occur within the flow. Very dense flows commonly result in matrix-supported beds, and more fluid flows may leave clast-supported bodies of sediment. The base of a debris flow makes an abrupt contact with underlying material. Many of these features are evident in Fig. 3.16. Experimental studies, carried out in flumes, indicate that debris flows may have downslope particle orientation dominant in lateral parts of flows, whereas the terminal lobes show more or less random orientations (Van Steijn, 1988; Van Steijn & Coutard, 1989) this has yet to be verified in field observations of large features.

The shape of the debris deposit is that of a thin, elongated lobate mass, usually sited on a flat surface down-slope from the channel along which it travelled. Up-valley, deposits left behind form levees alongside the channel (see Fig. 3.13).

Mudflows also move as plugs in regular waves, and form lobate deposits on flat, low-angle surfaces. Unlike debris flows, they may incorporate so much water that they become slurries and are transformed into turbulent mudflows. The resulting deposits are fans rather than lobes. Fan deposits may have some sorting and show limited bedding.

3.6.3 Talus mantles and deposits

Many attempts have been made to develop simple models for the form, development, and deposits of talus sheets and talus cones. However, simplicity is not a feature of talus accumulations for reasons that include the following:

Fig. 3.16 A fan deposit in the Dudh Kosi, Nepal. The unsorted deposit on the left is interpreted as being of debris flow origin, with a lense of fluvial deposits within it. The main body of material to the right is weakly bedded and sorted and is interpreted as being produced by overland flow in the monsoon season.

1 Talus deposits found in formerly glaciated valleys have accumulated during the 10 ky of Holocene time, in which there have been many fluctuations of climate that have influenced both the supply of debris and the processes acting to modify the surfaces of talus deposits (e.g. Kotarba & Strömquist, 1984; Francou, 1988a,b).

2 In mountains in the tropics and sub-tropics, the dominant processes may have been, and may still be, substantially different from those acting in glacial or periglacial zones (Francou, 1988a).

3 In any locality, microclimatic variations may cause considerable variation in rates of supply to talus surfaces and in the importance of such processes as snow avalanches, nivation, interstitial ice and creep created by that ice (Olyphant, 1983; Åkerman, 1984).

4 The material supplied to a talus may be (a) of mixed size, or relatively equidimensional; (b) dominantly cubic, or tabular; (c) delivered as single clasts, or by catastrophic events; or (d) part of a dry mass of material, a wet slush avalanche, a powder snow avalanche, part of a debris flow or debris torrent.

5 The processes acting to modify the talus surface may include one of, or a combination of (a) creep and rolling of particles caused by collisions; (b) creep caused by needle ice; (c) subsidence caused by melting of buried snow and ice; and (d) progressive weathering of talus materials.

In spite of the great range of influences on talus surfaces, it is possible to make some observations that are generally valid and apply to many talus features:

1 Talus sheets develop against the bases of cliffs and form generally rectilinear slopes on deposits which have varying thickness; reported ranges are from 1 to 35 m (Rapp, 1960).

2 Where accumulation develops not under influences distributed more or less evenly along the cliff face, but under the influence of chutes through which debris is delivered, then the accumulation takes the form of a talus cone which may grow upwards until there is continuity between the talus mantle and the debris in the floor of the chute (Fig. 3.17). This example indicates that there may be a change over time from a delivery dominated by rockfall to one that is dominated by rolling, avalanche and debris flow processes.

3 The long profiles of talus mantles may take many forms: rectilinear; upper concavity → rectilinear → basal concavity; upper convexity → rectilinear → basal concavity; upper concavity → greater basal concavity.

Attempts have been made to explain slope profile angles in terms of the properties of a granular cohesionless material:

1 It was suggested by Ward (1945) that the commonly observed, relatively straight 35° 'angle of repose' of talus materials is also the angle of shearing resistance of the loose material. This view is in conflict with the analysis of Chandler (1973) who showed that the minimum angle of shearing resistance of most talus materials is likely to be about 39–40°. He also showed that the inclination of well-drained talus slopes will reflect the angle of

Fig. 3.17 Debris cones below schist slopes near Cho Oyu, Nepal. The sequence shows increasing growth, from left to right, into the chutes which supply the debris, and increasing fluvial or debris flow influence.

shearing resistance only if (a) the slopes are being rapidly eroded; or (b) deposition rates are high compared with the degradation rates.

2 A model developed by Kirkby & Statham (1975) emphasizes the accumulation of individual particles that roll, bounce and slide downwards until they reach a zone where particles of similar size to their own form the surface. Large particles will roll over smaller ones and small particles will be trapped in the spaces between large ones, creating sorting. Consequently small particles will come to dominate upper slopes while large particles will dominate lower slopes. The lower concavity will be created by two effects: (a) some particles will travel farther than others, creating a thinning tail, and (b) large blocks falling from a high cliff will have high kinetic energy that will carry them across the toe of the talus mantle. As the cliff diminishes, this last effect will cease and the talus should become straighter.

Four main kinds of fabric are commonly recognized in talus deposits:

1 An open-work fabric, in which there are few small clasts, is supported on the point contacts between clasts and usually results from fall of individual fragments or from small rockfalls in which the rock fragments are of similar size. An alternative source is from the accumulation of such materials in a chute, or high on the talus, to form a lobe which when over-steepened, or pushed by other debris, will form a dry flow that moves down the slope as a sheet. Such dry flows have been observed in the Karakoram.

2 A partly open-work fabric may result from infilling of some of the voids, in an open-work, by washing down or fall of small grains.

3 A closed clast-supported fabric has all of the voids filled with fine-grained material; this is usually the result of washing of fines into an original open work.

4 A matrix-supported fabric is usually the product of debris flows.

Where there are several processes acting on a slope, a section through a talus may show varied fabrics (see Fig. 3.16): surface individual grain movements, caused by dry or wet sheet flows, create fine laminations or beds; channelized water creates distinct bedded units of limited width; and debris flows create unsorted mixed-size and matrix-supported masses.

Stratified deposits, of different dominant fabrics and particle sizes, indicate alternations of processes in the accumulation of the talus mantle. A widely recognized form of such alternate beds is the type known as grèze-litée. This phenomenon consists of layers of angular rock fragments embedded in a finer matrix alternating with layers of equally angular material with an open work structure. Such materials have been regarded as being produced by alternations of gelifluction, which is responsible for the material with a matrix, and sliding of coarse angular clasts over snow, which gives the open-work fabric (Van Steijn et al., 1984). Alternative hypotheses involve slope wash, eluviation to create the open-work fabrics, snow creep, rockfall and nivation (see Van Steijn et al. (1984) for discussion and references), and miniature debris flows (Van Steijn & Coutard, 1989).

In spite of the comments above emphasizing the great range of processes and sedimentary features of talus deposits, some mantles have remarkably uniform fabrics and lack of structure. The example in Fig. 3.18 is of a talus formed in a cliff in a 'mediterranean' climate. In the dry season individual blocks fall and roll and come to rest on the talus surface. In the wet season slope wash carries fines and large clasts, incorporating all particles into a predomi-

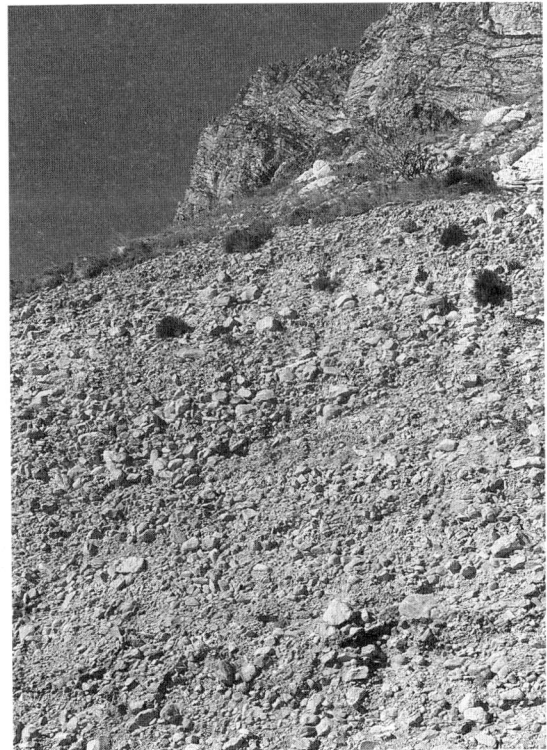

Fig. 3.18 Section through a fan deposit. Cape Mountains, South Africa.

nantly clast-supported deposit devoid of bedding or significant variations in structure.

3.6.4 Rock avalanches (sturzstrom)

Rock avalanches deserve special mention because, although they are rare events and confined to high mountains, they are producers of the most voluminous and far dispersed debris derived from slopes (Fig. 3.19).

The largest known rock avalanche is the Saidmarreh landslide in southwestern Iran (Harrison & Falcon, 1937). It involved a segment of an anticlinal ridge, 15 km long and 5 km wide with a thickness of 300 m. The whole mass slid off the mountain and had sufficient momentum to cross a valley, rise over a 600 m high ridge and come to rest in the next valley 20 km from source. The mass of rock involved was about 20 km^3, and the debris covered an area of

Fig. 3.19 Debris from a small rock avalanche. Langtang Valley, Nepal.

166 km^2 to a maximum depth of 300 m, and an average depth of 130 m. The edges of the deposits are sharp fronts at least 50 m high.

Other well-known rock avalanches are the Turtle Mountain (McConnell & Brock, 1904) and Huascarán (Plafker & Ericksen, 1978) failures. A recent study by Evans *et al.* (1989) has a substantial bibliography. The first fundamental paper on the subject was published by Heim in 1882, who reported the Elm rock avalanche in Switzerland (Hsü, 1978).

Heim recognized that most rock avalanches start as rockfalls and rockslides on steep slopes. They achieve high velocities, 90–350 km h^{-1} have been calculated, and appear to flow as flexible debris sheets that cover a large area in comparison with the area of the source. The material of the avalanche is mostly bedrock although snow, ice, water and some debris may be incorporated in transport. The failed material is highly fractured and forms a chaotic flow in which there is a general absence of sorting. Large blocks may be either fitted together or separated by a fine-grained matrix. The debris sheet is commonly thin, being a few tens of metres thick; it usually has a well-defined distal rim, lateral ridges and transverse surface patterns of hummocks, and ridges. Very large boulders, up to 16 kt mass, have been noted on the surfaces of some avalanche deposits.

Although there have been a number of hypotheses developed to explain the fluidity of rock avalanches, the original view of Heim, restated in modern terms, is generally regarded as the most probable mechanism.

Heim noted that an individual block travels in a zig-zag, bounding path through elastic impacts with its surroundings. A large aggregate of smaller blocks behaves quite differently because each block is confined to bouncing back and forth between its neighbours, and only the outer blocks may fly away. Thus kinetic energy is exchanged between particles by elastic collisions, and the same energy keeps the particles separated during countless elastic contacts. The mass therefore behaves as a fluid with very low internal friction.

The presence of fine particles increases the frequency of collisions, and also the dispersion of large blocks which may then pass one another. Rock avalanches thus have high fluid viscosities when the mass is thick (perhaps around 4 MNs m^{-2} at a thickness of 2–3 m) but as little as 100 kNs m^{-2} when the flow is less than this in thickness. High fluid viscosities prevent internal turbulent mixing and internal deformation, so the mass moves as a thin flexible sheet. At rest, the notable feature of the debris is the lack of attrition of particles, the matrix-supported fabric, the presence of very large boulders on or in the mass, and the lack of sorting. The features that distinguish rock avalanches from debris flows are their greater thicknesses, larger boulders, greater volumes, lack of channels, and the fact that debris in rock avalanches can travel up opposing slopes (up to 600 m has been recorded).

3.7 Synthesis

Hillslope deposits are rarely discussed in books on sedimentary environments and processes. Some brief concluding statements may help to explain why this has been so.

1 Hillslope sediments are seldom preserved in the geological record because they are subject to erosion and removal by fluvial, glacial and, less commonly, by aeolian processes. Only if they are laid down in subsiding tectonic basins are they likely to be buried, lithified and preserved.

2 Bioturbation and pedological processes destroy thin colluvial deposits as distinct units, and incorporate them in soils.

3 The distances of transport from sites of erosion to those of deposition are short, often only a few metres, consequently hillslope sediments are likely to retain the mixture of particle sizes derived from the erosion site with little or no attrition, rounding, sorting, bedding or bedforms being evident in the ultimate deposit.

4 Rainsplash and wash are, compared with most sub-aerial processes, low-energy phenomena and consequently they transport only fine-grained materials. Increasing power and erosional and transport effectiveness occurs in the range: rill < gully < translational landslide < debris flow < rock avalanche. Consequently, the capacity to transport large clasts increases in that order also.

5 Gully and alluvial fan processes are increasingly fluvial in nature as the depth and regularity of flow increases away from source, and are best treated in that context (see Chapter 4).

6 Landslide and avalanche processes have a range of distances of travel, up to a few tens of kilometres for the largest debris flows and rock avalanches. The high densities and viscosities of the moving masses largely prevent turbulence and sorting, with the result that the particle sizes cover the widest possible range and are similar in their lack of sorting and bedding to deposits laid down by glaciers. Distinguishing between hillslope and glacial deposits may then depend more upon circumstantial evidence, such as striations, related deposits and landforms, and local and regional environments, than upon unique sedimentary features.

3.8 References

Abrahams A.D., Parsons A.J. & Luk S-H. 1986. Resistance to overland flow on desert hillslopes. *J. Hydrol.*, **88**, 343–363.

Åkerman H.J. 1984. Notes on talus morphology and processes in Spitsbergen. *Geogr. Ann.*, **66A**, 267–284.

Anderson M.G. & Richards K.S. (Eds) 1987. *Slope Stability*. Wiley, Chichester.

Ayres Q.C. 1936. *Soil Erosion and its Control*. McGraw-Hill, New York.

Bennett H.H. 1939. *Soil Conservation*. McGraw-Hill, New York.

Beverage J.P. & Culbertson J.K. 1964. Hyperconcentrations of suspended sediment. *J. Hydraul. Div. Am. Soc. Civ. Engnrs*, **90**, (HY6), 117–128.

Blong R.J. 1970. The development of discontinuous gullies in a pumice catchment. *Am. J. Sci.*, **268**, 369–383.

Brunori F., Penzo M.C. & Torri D. 1989. Soil shear strength: its measurement and soil detachability. *Catena*, **16**, 59–71.

Brunsden D. 1984. Mudslides. In: Brunsden D. & Prior D.B. (eds) *Slope Instability*, pp. 363–418. Wiley, Chichester.

Brunsden D. & Prior D.B. (eds) 1984. *Slope Instability*. Wiley, Chichester.

Bryan R.B. 1987. Processes and significance of rill development. *Catena*, **8**, (Suppl.) 1–15.

Bryan R.B. & Poesen J. 1989. Laboratory experiments on the influence of slope length on runoff, percolation and rill development. *Earth Surf. Proc. Landf.*, **14**, 211–231.

Bryan R.B., Yair A. & Hodges W.K. 1978. Factors controlling the initiation of runoff and piping in Dinosaur Provincial Park badlands, Alberta, Canada. *Z. Geomorph. Suppl. Bd*, **29**, 151–168.

Buol S.W., Hole F.D. & McCracken R.J. 1973. *Soil Genesis and Classification*. Iowa State University Press, Iowa.

Chandler R.J. 1973. The inclination of talus, arctic talus terraces, and other slopes composed of granular materials. *J. Geol.*, **81**, 1–14.

Conacher A.J. & Dalrymple J.B. 1977. The nine unit landsurface model: an approach to pedogeomorphic research. *Geoderma*, **18**, 1–154.

Cooke R.U. & Reeves R.W. 1976. *Arroyos and Environmental Change in the American South-West*. Clarendon Press, Oxford.

Corbel J. 1959. Vitesse de l'érosion. *Z. Geomorph. N.F.*, **3**, 1–28.

Costa J.E. 1984. Physical geomorphology of debris flows. In: Costa J.E. & Fleischer P.J. (eds) *Developments and Applications of Geomorphology*, pp. 268–317. Springer-Verlag, Berlin.

Crouch R.J. & Novruzi T. 1989. Threshold conditions for rill initiation on a vertisol, Gunnedah N.S.W., Australia. *Catena*, **16**, 101–110.

Cruse R.M. & Larson W.E. 1977. Effect of soil shear strength on soil detachment due to raindrop impact. *Soil Sci. Soc. Am. J.*, **41**, 777–781.

Dalrymple J.B., Blong R.J. & Conacher A.J. 1968. A hypothetical nine unit landsurface model. *Z. Geomorph. N.F.*, **12**, 60–76.

Dedkov A.P. & Mozzherin V.I. 1984. Eroziya i Stok Nanosov na Zemle. *Izdatelstvo Kazanskogo Universiteta.*

De Ploey J. (1983) Runoff and rill generation on sandy and loamy topsoils. *Z. Geomorph. Suppl. Bd.*, **46**, 15–23.

De Ploey J. & Poesen J. 1985. Aggregate stability, runoff generation and interrill erosion. In: Richards K.S., Arnett R.R. & Ellis S. (eds) *Geomorphology and Soils*, pp. 99–120. Allen and Unwin, London.

De Ploey J. & Poesen J. 1987. Some reflections on modelling hillslope processes. *Catena*, **10** (Suppl.), 67–72.

De Ploey J. & Savat J. 1968. Contribution à l'étude de l'érosion par le splash. *Z. Geomorph. N.F.*, **12**, 174–193.

De Roo A.P.J., Hazelhoff L. & Burrough P.A. 1989. Soil erosion modelling using 'ANSWERS' and geographical information systems. *Earth Surf. Proc. Landf.*, **14**, 517–532.

Douglas I. 1967. Man, vegetation and sediment yield of rivers. *Nature*, **215**, 925–928.

Dunne T. & Aubry B.F. 1986. Evaluation of Horton's theory of sheetwash and rill erosion on the basis of field experiments. In: Abrahams A.D. (ed.) *Hillslope Processes*, pp. 31–53. Allen and Unwin, Boston.

Ellison W.D. 1947. Soil erosion studies. *Agric. Engng.*, **28**, 145–146, 197–201, 245–248, 297–300, 349–351, 402–405, 442–450.

Elwell H.A. 1977. Soil loss estimation system for southern Africa. *Dept of Conservation and Extension, Research Bulletin*, **22**. Salisbury, Rhodesia.

Engelen G.B. & Venneker R.G.W. 1988. ETA (Erosion, Transport, Accumulation) systems, their classification, mapping and management. *I.A.S.H. Publ.*, **174**, 397–412.

Epema G.F. & Riezebos H. Th. 1983. Fall velocity of waterdrops at different heights as a factor influencing erosivity of simulated rain. *Catena*, **4** (Suppl.), 1–17.

Evans R. 1980. Mechanics of water erosion and their spatial and temporal controls: an empirical viewpoint. In: Kirkby M.J. & Morgan R.P.C. (eds) *Soil Erosion*, pp. 110–128. Wiley, Chichester.

Evans S.G., Clague J.J., Woodsworth G.J. & Hungr O. 1989. The Pandemonium Creek rock avalanche, British Columbia. *Can. Geotech. J.*, **26**, 427–446.

FAO 1965. *Soil Erosion by Water: Some Measures for its Control on Cultivated Lands.* FAO/UNESCO, Rome.

Farres P. 1978. The role of time and aggregate size in the crusting process. *Earth Surf. Proc.*, **3**, 243–254.

Ford D.G. & Lundberg J. 1987. A review of dissolutional rills in limestone and other soluble rocks. *Catena*, **8** (Suppl.), 119–139.

Foster G.R. & Meyer L.D. 1975. Mathematical simulation of upland erosion by fundamental erosion mechanics. Present and prospective technology for predicting sediment yields and sources. *USDA Agric. Res. Serv. Pub.*, ARS-S-40, 190–207.

Fournier F. 1960. *Climat et Erosion.* Presses Universitaires de France, Paris.

Francou B. 1988a. Talus formation in high mountain environments, Alps and tropical Andes (abstract). Centre de Geomorphologie du CNRS, Caen.

Francou B. 1988b. Eboulis statifiés dans les Hautes Andes Centrales du Pérou. *Z. Geomorph. N.F.*, **32**, 47–76.

Freebairn D.M. & Wockner G.H. 1986. A study of soil erosion on vertisols of the eastern Darling Downs, Queensland. 1. Effects of surface conditions on soil movement within contour bay catchments. *Austr. J. Soil Res.*, **24**, 135–158.

Froehlich W. 1986. Influence of the slope gradient and supply area on splash, scope of the problem. *Z. Geomorph. Suppl. Bd*, **60**, 105–114.

Gerits J., Imeson A.C., Verstraten J.M. & Bryan R.B. 1987. Rill development and badland regolith properties. *Catena*, **8** (Suppl.), 141–160.

Gerrard A.J. 1985. Soil erosion and landscape stability in southern Iceland: a tephrochronological approach. In: Richards K.S., Arnett R.R. and Ellis S. (eds.) *Geomorphology and Soils*, pp. 78–94. Allen and Unwin, London.

Gibbs H.S. 1945. Tunnel-gully erosion in the Wither Hills, Marlborough. *N.Z. J. Sci. Technol.*, **A27**, 135–146.

Govers G. 1989. Grain velocities in overland flow: a laboratory study. *Earth Surf. Proc. Landf.*, **14**, 481–498.

Grant P.J. 1985. Major periods of erosion and alluvial sedimentation in New Zealand during the Late Holocene. *J. Roy. Soc. N.Z.*, **15**, 67–121.

Hall G.F., Daniels R.B. & Foss J.E. 1979. Soil formation and renewal rates in the US. Symposium on Determinants of Soil Loss Tolerance, Soil Science Society of America, Annual Meeting, Fort Collins, Colorado.

Harrison J.V. & Falcon N.L. 1937. The Saidmarreh landslip, southwest Iran. *Geogr. J.*, **89**, 42–47.

Hjulström F. 1935. Studies of the morphological activity of rivers as illustrated by the River Fyris. *Bull. Geol. Inst. Univ. Uppsala*, **25**, 221–527.

Hodges W.K. 1982. Hydrologic characteristics of a badland pseudopediment slope system during simulated rainstorm experiments. In: Bryan R.B. and Yair A. (eds) *Badland Geomorphology and Piping*, pp. 127–152. Geo-Books, Norwich.

Hodges W.K. & Bryan R.B. 1982. The influence of material behaviour on runoff initiation in the Dinosaur Badlands, Canada. In: Bryan R.B. and Yair A. (eds) *Badland Geomorphology and Piping*, pp. 13–46. GeoBooks, Norwich.

Hoogmoed W.B. & Stroosnijder L. 1984. Crust formation on sandy soils in the Sahel. I. Rainfall and infiltration. *Soil and Tillage Res.*, **4**, 5–24.

Hooke R. LeB. 1987. Mass movement in semi-arid environments and the morphology of alluvial fans. In: Richards K.S. and Anderson M.G. (eds), *Slope Stability*, pp. 505–529. Wiley, Chichester.

Horton R.E. 1945. Erosional development of streams and their drainage basins: hydrophysical approach to quantitative morphology. *Bull. Geol. Soc. Am.*, **56**, 275–370.

Hsü K.J. 1978. Albert Heim: observations on landslides and relevance to modern interpretations. In: Voight B. (ed.) *Rockslides and Avalanches*, pp. 71–93. Elsevier, Amsterdam.

Huang C., Bradford J.M. & Cushman J.H. 1982. A numerical study of raindrop impact phenomena: the rigid case. *Soil Sci. Soc. Am. J.*, **46**, 14–19.

Hudson N.W. 1961. An introduction to the mechanics of soil erosion under conditions of subtropical rainfall. *Rhodesia Sci. Assoc. Proc.*, **49**, 14–25.

Hudson N.W. 1981. *Soil Conservation*. Batsford, London.

Hungr O., Morgan G.C. & Kellerhals R. 1984. Quantitative analysis of debris torrent hazards for design of remedial measures. *Can. Geotech. J.*, **21**, 663–677.

Hurault J. 1971. La significance morphologique des lavaka. *Rev. Geomorph. Dyn.*, **19**, 121–128.

Imeson A.C. & Verstraten J.M. 1988. Rills on badland slopes: a physico-chemically controlled phenomenon. *Catena*, **12** (Suppl.), 139–150.

Jansson M.B. 1982. Land erosion by water in different climates. *U.N.G.I. Rapp.* **57**.

Jansson M.B. 1988. A global survey of sediment yield. *Geogr. Ann.*, **70A**, 81–98.

Johnson A.M. & Rodine J.R. 1984. Debris flow. In: Brunsden D. & Prior D.B. (eds.) *Slope Instability*, pp. 257–361. Wiley, Chichester.

Kirkby M.J. & Statham I. 1975. Surface stone movement and scree formation. *J. Geol.*, **83**, 349–362.

Komura S. 1976. Hydraulics of slope erosion by overland flow. *J. Hydraul. Div. Am. Soc. Civ. Engnrs*, **102**, 1573–1586.

Kotarba A. & Strömquist L. 1984. Transport, sorting and deposition processes of alpine debris slope deposits in the Polish Tatra Mountains. *Geogr. Ann.*, **66A**, 285–294.

Lal R. 1976. Soil erosion problems on an alfisol in western Nigeria and their control. *IITA Monograph* **1**.

Lane L.J., Schertz D.L., Alberts E.E., Laflen J.M. & Lopes V.L. 1988. The US national project to develop improved erosion prediction technology to replace the USLE. *IAHS Publ.*, **174**, 473–481.

Langbein W.B. & Schumm S.A. 1958. Yield of sediment in relation to mean annual precipitation. *Trans. Am. Geophys. Union*, **39**, 1076–1084.

Laws J.O. 1940. Recent studies in raindrops and erosion. *Agric. Engng*, **21**, 431–433.

Laws J.O. 1941. Measurement of fall-velocity of water-drops and rain-drops. *Trans. Am. Geophys. Union*, **22**, 709–721.

Laws J.O. & Parsons A. 1943. The relation of raindrop size to intensity. *Trans. Am. Geophys Union*,

Li Jian & Wang Jingrung 1986. The mudflows of Xiaojiang Basin. *Z. Geomorph. Suppl. Bd.*, **58**, 155–164.

McConnell R.G. & Brock R.W. 1904. The great landslide at Frank Alberta. *Canada Department of the Interior Annual Report 1902–1903*, Part 8, appendix.

Meyer L.D., Foster G.R. & Romkens P. 1975. Source of soil eroded by water from upland slopes. *US Agric. Res. Service Rept*, **ARS-S-40**, 177–189.

Moeyersons J. 1983. Measurements of splash–saltation fluxes under oblique rain. *Catena*, **4** (Suppl.), 19–31.

Morgan R.P.C. 1980. Field studies of sediment transport by overland flow. *Earth Surf. Proc.*, **5**, 307–316.

Morgan R.P.C. 1986. *Soil Erosion and Conservation*. Longman, Harlow.

Morgan R.P.C., Morgan D.D.V. & Finney H.J. 1984. A predictive model for the assessment of soil erosion risk. *J. Agric. Engng Res.*, **30**, 245–253.

Morin J., Benyamini Y. & Michaeli A. 1981. The effect of raindrop impact on the dynamics of soil surface crusting and water movement in the profile. *J. Hydrol.*, **52**, 321–336.

Moss A.J. & Green P. 1983. Movement of solids in air and water by raindrop impact. Effects of drop-size and water-depth variations. *Aust. J. Soil Res.*, **21**, 257–269.

Moss A.J. & Walker P.H. 1978. Particle transport by continental water flows in relation to erosion, deposition, soils and human activities. *Sediment. Geol.*, **20**, 81–139.

Moss A.J., Green P. & Hutka J. 1982. Small channels: their formation, nature and significance. *Earth Surf. Proc. Landf.*, **7**, 401–415.

Musgrave G.W. 1947. The quantitative evaluation of factors in water erosion — a first approximation. *J. Soil Water Conserv.*, **2**, 133–138.

Nash D.F.T., Brunsden D.K., Hughes R.E., Jones D.K.C & Whalley B.F. 1985. A catastrophic debris flow near Gupis, Northern areas, Pakistan. *Proc. 11th Int. Conf. Soil Mech. Found. Eng.*, **3**, 1163–1166.

Ogrosky H.O. & Mockus V. 1964. Hydrology of agricultural lands. In: Chow V.T. (ed.) *Handbook of Applied Hydrology*, Section 21. McGraw-Hill, New York.

Ohmori H. 1983. Characteristics of the erosion rate in Japanese mountains from the viewpoint of climatic geomorphology. *Z. Geomorph. Suppl. Bd*, **46**, 1–14.

Olyphant G.A. 1983. Analysis of the factors controlling cliff burial by talus within Blanca Massif, southern Colorado, USA. *Arc. Alp. Res.*, **15**, 65–75.

Osterkamp W.R., Hupp C.R. & Blodgett J.C. 1986. Magnitude and frequence of debris flows, and areas of hazard on Mount Shasta, northern California. *US Geol. Surv. Prof. Pap.*, **1396-C**, C1–C21.

Palmer R.S. 1965. Water drop impact forces. *Proc. Am. Soc. Agric. Engnrs*, **8**, 70–72.

Pearce A.J. 1976. Magnitude and frequency of erosion by Hortonian overland flow. *J. Geol.*, **84**, 65–80.

Plafker G. & Ericksen G.E. 1978. Nevados Huascáran avalanches, Peru. In: Voight B. (ed.) *Rockslides and Avalanches*, pp. 277–314. Elsevier, Amsterdam.

Poesen J. 1985. An improved splash transport model. *Z. Geomorph. N.F.*, **29**, 193–211.

Poesen J. 1987. Transport of rock fragments by rill flow — a field study. *Catena*, **8** (Suppl.), 35–54.

Quansah K. 1985. The effect of soil type, slope, flow rate and their interactions on detachment by overland flow with and without rain. *Catena*, **6** (Suppl.), 19–28.

Rapp A. 1960. Talus slopes and mountain walls at Tempelfjorden, Spitsbergen. *Norsk Polarinstitutt*, Skr. Nr. 119.

Rauws G. 1987. The initiation of rills on plane beds of non-cohesive sediments. *Catena*, **8** (Suppl.), 107–118.

Riquier J. 1958. Le 'lavaka' de Madagascar. *Bull. Soc. Geogr. Marseilles*, **69**, 181–191.

Roose E.J. 1967. Dix années de mesure de l'érosion et du ruissellement au Sénégal. *L'Agron. Trop.*, **22**, 123–152.

Savat J. 1976. Discharge velocities and total erosion of a calcareous loess: a comparison between pluvial and terminal runoff. *Rev. Géomorph. Dyn.*, **24**, 113–122.

Savat J. 1977. The hydraulics of sheet flow on a smooth surface and the effect of simulated rainfall. *Earth Surf. Proc.*, **2**, 125–140.

Savat J. 1980. Resistance to flow in rough supercritical sheetflow. *Earth Surf. Proc.*, **5**, 103–122.

Savat J. 1981. Work done by splash: laboratory experiments. *Earth Surf. Proc.*, **6**, 275–283.

Savat J. & De Ploey J. 1982. Sheetwash and rill development by surface flow. In: Bryan R.B. & Yair A. (eds) *Badland Geomorphology and Piping*, pp. 113–126. Geo-Books, Norwich.

Schouten C.J. 1984. Measurement of gully erosion and the effects of soil conservation techniques in Puketurua Experimental Basin (New Zealand). *Z. Geomorph. Suppl. Bd*, **49**, 151–164.

Schumm S.A. 1956. Evolution of drainage systems and slopes in badlands at Perth Amboy, New Jersey. *Bull. Geol. Soc. Am.*, **67**, 597–646.

Schumm S.A. & Hadley R.F. 1957. Arroyos and the semi-arid cycle of erosion. *Am. J. Sci.*, **225**, 161–174.

Selby M.J. 1976. Slope erosion due to extreme rainfall: a case study from New Zealand. *Geogr. Ann.*, **58A**, 131–138.

Selby M.J. 1982. *Hillslope Materials and Processes.* Oxford University Press, Oxford.

Selby M.J. 1985. *Earth's Changing Surface.* Clarendon, Oxford.

Sidle R.C., Pearce A.J. & O'Loughlin C.L. 1985. Hillslope Stability and Land Use. *Water Resources Monograph Series*, **11**, Am. Geophys. Union.

Smith D.D. & Whitt D.M. 1947. Estimating soil losses from field area of claypan soil. *Proc. Soil Sci. Soc. Am.*, **12**, 485–490.

Smith D.D. & Whitt D.M. 1948. Evaluating soil losses from field areas. *Agric. Engng*, **29**, 394–396.

Smith D.D. & Wischmeier W.H. 1957. Factors affecting sheet and rill erosion. *Trans. Am. Geophy. Union*, **38**, 889–896.

Stehlík O. 1975. Potenciálni eroze pudy proudíci vodou na území CSR. *Studia Geographica*, **42**. Brno.

Stocking M.A. 1980. Examination of the factors controlling gully growth. In: De Boodt M. & Gabriels D. (eds) *Assessment of Erosion*, pp. 505–520. Wiley, Chichester.

Strakhov N.M. 1967. *Principles of Lithogenesis*, Vol. 1. Oliver and Boyd, Edinburgh.

Swanson M.L., Kondolf G.M. & Boison P.J. 1989. An example of rapid gully initiation and extension by subsurface erosion: coastal San Mateo County, California. *Geomorphology*, **2**, 393–403.

Temple P.H. & Rapp A. 1972. Landslides in the Mgeta area, western Uluguru Mountains, Tanzania. *Geogr. Ann.*, **55A**, 157–193.

US Environmental Protection Agency 1973. *Methods for Identifying and Evaluating the Nature and Extent of Nonpoint Sources of Pollutants.* Washington, DC.

Van Asch Th.W.J. 1983. Water erosion on slopes in some land units in a Mediterranean area. *Catena*, **4** (Suppl.), 129–140.

Van Dine D.F. 1985. Debris flows and debris torrents in the southern Canadian Cordillera. *Can. Geotech. J.*, **22**, 44–68.

Van Steijn H. 1988. Debris flows involved in the development of Pleistocene stratified slope deposits. *Z. Geomorph. Suppl. Bd.*, **71**, 45–58.

Van Steijn H. & Coutard, J-P. 1989. Laboratory experiments with small debris flows: physical properties related to sedimentary characteristics. *Earth Surf. Proc. Landf.*, **14**, 587–596.

Van Steijn H., Van Brederode L.E. & Goedheer G.J. 1984. Stratified slope deposits of the grèze-litée type in the Ardèche region in the south of France. *Geogr. Ann.*, **66A**, 295–305.

Van Steijn H., Ruig J. De, & Hoozemans F. 1988. Morphological and mechanical aspects of debris flows in parts of the French Alps. *Z. Geomorph. N.F.*, **32**, 143–161.

Walling D.E. & Webb B.W. 1983. Patterns of sediment yield. In: Gregory K.J. (ed.) *Background to Paleohydrology*, pp. 69–100. Wiley, Chichester.

Walling D.E. & Webb B.W. 1987. Material transport by the world's rivers: evolving perspectives. *IAHS Publ.*, **164**, 313–329.

Ward W.H. 1945. The stability of natural slopes. *Geogr. J.*, **105**, 170–197.

Wilson L.J. 1973. Variations in mean annual sediment yield as a function of mean annual precipitation. *Am. J. Sci.*, **273B**, 335–347.

Wischmeier W.H. 1970. Relation of soil erosion to crop and soil management. *Proc. Int. Water Erosion Symp., Int. Comm. Irrig. Drainage, Prague*, **2**, 201–220.

Wischmeier W.H. 1976. Use and misuse of the universal soil loss equation. *J. Soil Water Conserv.*, **31**, 5–9.

Wischmeier W.H. & Smith D.D. 1958. Rainfall energy and its relationship to soil loss. *Trans. Am. Geophys. Union*, **39**, 285–291.

Wischmeier W.H., Smith D.D. & Uhland R.E. 1958. Evaluation of factors in the soil-loss equation. *Agric. Engng*, **39**, 458–462.

Wolman M.G. 1967. A cycle of sedimentation and erosion in urban river channels. *Geogr. Ann.*, **49A**, 385–395.

Wolman M.G. & Gerson R. 1978. Relative scales of time and effectiveness of climate in watershed geomorphology. *Earth Surf. Proc.*, **3**, 189–208.

Wolman M.G. & Miller J.P. 1960. Magnitude and frequency of forces in geomorphic processes. *J. Geol.*, **68**, 54–74.

Zingg A.W. 1940. Degree and length of land slope as it affects soil loss in runoff. *Agric. Engng*, **21**, 59–64.

4 Fluvial sediment transport and deposition

I. REID & L.E. FROSTICK

4.1 Introduction

Rivers are the landscape's self-formed gutter system. Their primary purpose is the conveyance of rainfall and snow-melt towards topographic sinks. The transport of sediment is entirely incidental in that it arises because momentum is transferred from the flow to the granular solids that form its boundary. However, this movement of sediment acts as a regulator, so that a river's character and geometry — its planform, cross-section and long-profile — are controlled not only by hydrologic or hydraulic considerations, but also by erosion and deposition at a number of scales from highly local to basinal.

It might be argued that rivers are the primary cause of sedimentary cycling despite the fact that the clastic element of the world's stratigraphic column is dominated by sediments which are classified as marine. Indeed, rivers are often merely the transfer conduits for material that has originated on the hillslopes of water catchments and is on its way to the oceans. In these circumstances, accumulated river sediments may be comparatively thin and are often ephemeral. However, where the topographic sink is continental, as in the present Lake Eyre basin of Australia, the Chad basin of the Sahel and the rifts of East Africa, great thicknesses of riverine sediments can accumulate and be preserved. Even at the continental margins, where river meets ocean, hybridized sediments of mixed process parentage can accumulate in great thicknesses, as in the large deltas of the Nile, the Niger, the Mackenzie and the Ganga.

However, rivers are keenly attuned to their environment. They are also extremely responsive to environ-

University of Reading PRIS Contribution No. 068.

mental change. As a result, the character of the drainage network may differ significantly from place to place and through time. So, for example, the rainfall–runoff relationship in arid zones encourages higher drainage densities and this leads in turn to a channel form that differs significantly from that typical of humid environments. In a temporal context, rapid climatic fluctuation may cause a change in the rainfall–runoff relationship. This may, in turn, mean adjustments in channel geometry that are brought about by, but also have consequences for, patterns of erosion and sedimentation. It is therefore important to understand the process of river flood generation in order to be able to assess the hydraulic environment and, thus, the potential mobility of sediment.

4.2 River floods

4.2.1 Flood frequency and magnitude

Rain-fed floods are a response to a permutation of regulating variables that are both endogenous and exogenous to the water catchment of a river. These include soil conditions and vegetation, storm intensity and duration, among many others. As a result, flood generation can be regarded as a stochastic process that will provide a spectrum of peak water discharges. Over a period of time, the permutation of regulating variables will produce a large number of moderate floods and a small number of large floods so that a frequency distribution of peak flood discharges will be positively skewed (Fig. 4.1). The longer the period of record, the more representative will be the sample of all possible floods.

The fact that a set of individual flood discharges can be arranged to form a continuous distribution

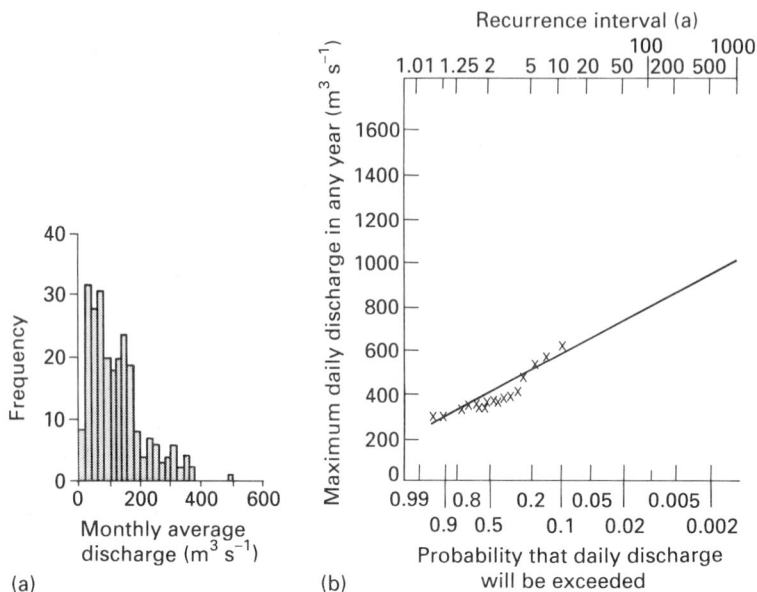

Fig. 4.1 (a) Frequency distribution of monthly average discharge in a humid temperate region river over a return period of 20 years, showing the moderately positive skew expected of flows in this morphoclimatic setting. (b) Ranked maximum annual flood (annual flood series) plotted against the estimated probability of being exceeded. The line is a least-squares best-fit. Recurrence interval is the reciprocal of probability and indicates the likely interval between floods of the same magnitude.

is useful in that it allows a probabilistic approach to flood prediction which revolves around the *magnitude–frequency* concept. The magnitude of each flood can be assigned a probability of occurrence that can be used to determine its likely *recurrence interval*, or, in other words, the likely time between floods of the same size. The probability of occurrence decreases as floods increase in size because the chance of experiencing an appropriate permutation of the regulating variables decreases as flood magnitude increases (Fig. 4.1). So, for instance, it is comparatively rare for a storm of both high intensity and long duration to fall on soil with low infiltration capacity (perhaps because of a well-developed surface crust or because snow-melt or previous rain has brought it to field capacity). As a result, the streamflow generated by such a storm would be expected only infrequently and the flood peak discharge would have a much larger recurrence interval than that associated with a storm of moderate intensity and duration which had fallen on a catchment whose soil were in the same prestorm condition.

An analysis of river flow records suggests that bankfull discharge has a recurrence interval of about 1.5 years (Leopold & Maddock, 1953). Overbank flow can therefore be expected to occur with about the same frequency. Because channel and floodplain sediments are distinctly different, this point on the magnitude–frequency plot has considerable significance. The fact that rivers of considerably different character utilize their floodplains with about the same frequency suggests that channel geometry is adjusted to cope with floods of *moderate* magnitude. Indeed, fluvial geomorphologists have argued that bankfull flow is channel formative, being both frequent enough and powerful enough to erode banks and arrange bed sediments in such a way that channel form is largely unaffected by lesser flows, while the rarity of greater flows reduces their importance, at least in the short term (Kirkby, 1977; Pickup & Warner, 1976).

In fact, much larger floods may not have been recorded if only because the development of suitable instrumentation and an interest in documenting flood history are both largely a product of the 20th century. (The network of 'Nilometers' established by the Egyptian Pharaoh Menes more than three millennia ago was, of course, a notable exception.) In this circumstance, the magnitude — frequency approach to flood prediction can be of assistance in estimating the likely size of an event of large recurrence interval.

Assume that there is a need to assess the likely discharge of the *1000-year flood* (i.e. a flow with a probability of recurring once in 1000 years — potentially devastating in engineering terms and damaging geomorphologically). The record of flow for the river in question (the *return period*) is analysed and the highest discharge of each successive day is extracted. This provides a dataset referred to as the *diurnal flood series* because it amalgamates single values of flow from each day. A *monthly flood series* would consists of the highest flows recorded in successive months of the return period. An *annual* flood series can also be extracted, but its value will depend upon the length of the return period, that is, on the number of individual variates that will make up the dataset. The flows are ranked in order of decreasing magnitude, and assigned a probability of occurrence. A number of methods have been proposed for this, all of which have similar characteristics. The exact method chosen may depend upon the nature of the flood dataset, but will take the general form:

$$p = 1 - M/(N - 1), \qquad (4.1)$$

where p is the level of probability, M is the rank order, and N is the number of events in the flood series. The event probability is then plotted against event magnitude (Fig. 4.1).

Since most hydrological datasets have a frequency distribution which is highly skewed, and since the range of discharge values is often large, the plot is normally made on either a log-probability or a linear-probability graticule. The linearity that this achieves allows the relationship to be extrapolated beyond the limits of the plotted coordinates, especially in the direction of high flows, which are unlikely to have been measured. A second scale is constructed to lie alongside that of probability. This indicates the recurrence-interval and is the reciprocal of probability. Finally, a least-squares regression line is added to the graph and extrapolated beyond the plotted data.

It is now possible to estimate the magnitude of the 1000-year flood. By intersecting the regression line at a recurrence interval of 1000 years, it can be seen that a discharge of just over $1000 \, \mathrm{m^3 \, s^{-1}}$ might be expected in the example given. For a variety of reasons, not least of which is the fact that the extrapolation procedure involves making considerable assumptions about the rainfall–runoff processes that lead to higher and higher flood discharges, the results of this type of analysis are usually treated with due caution. Nevertheless, the magnitude–frequency method offers not only a means of assessing likely flood size, but, if peak discharge can be estimated, also of assessing the likely recurrence interval of damaging flood events of the past.

4.2.2 Catastrophism versus gradualism

A consideration of extreme flows leads to the contro-

versy over whether moderate but frequent events are as significant in determining the landform and in rearranging sediments as has previously been thought (Graf, 1982). It was as long ago as 1923 that Bretz attempted to convince a cynical earth science community that the Washington Scablands were the product of a flood deluge of gigantic proportions, albeit that the cause of the flow was the sudden release of meltwater from the Pleistocene Lake Missoula. It was not until Baker's (1973) comprehensive palaeohydrological investigation half a century later that this controversy was finally to be dispelled.

In fact, Baker and others have gone on to establish and assess records of river flow for periods that predate recorded histories at a number of locations around the globe. Deeply incised channels such as those of the Katherine Gorge of Australia's Northern Territories and the Guadalupe River of Texas have been particularly useful in confining large palaeo-floods, while tributary re-entrants in the valley wall have provided the location for separated flow and the deposition of tell-tale fine sediments as markers of peak flood stage (Patton *et al.*, 1982; Baker *et al.*, 1983; Fig. 4.2). The erosional havoc wrought by these large floods is thought to have been channel formative. Outside the thalweg itself, Nanson (1986) has shown that periodic *floodplain stripping* accounts for the incongruous landforms and cut-and-fill stratigraphy at decametre scale that characterizes some of the floodplains of the coastal rivers in New South Wales.

The impact of these large floods is such that the river may be considered as *underfit* during the intervening periods of 'normal' flood flows. The implication of this is that river sedimentation is disequilibrated for much of the time in these circumstances. The movement of individual clasts of boulder size and bigger is beyond the competence of even large moderate-sized floods, and these channel sediments become part of a transient storage with a residence time that is dependent on the recurrence interval of the high magnitude flows (Fig. 4.3).

In fact, even large floods with recurrence intervals of much less than 1000 years are known to leave morphological legacies that may control erosion and sedimentation for considerable periods of time. For example, Schumm and Lichty (1963) have shown that the 1914 flood on the Cimarron River in Kansas led to a 24-fold widening of the channel from an average of 15–365 m in the period to 1942, after which it narrowed as the floodplain was re-established.

This has led some to postulate that the erosional

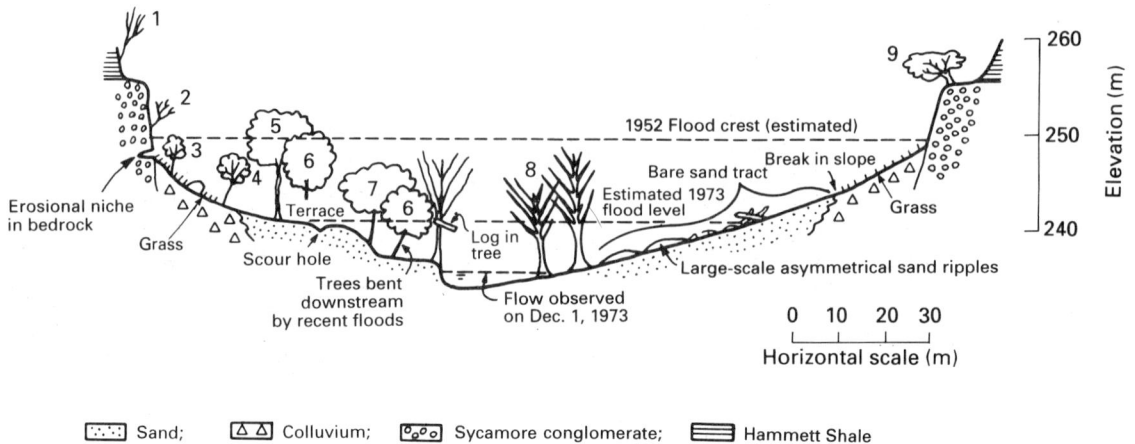

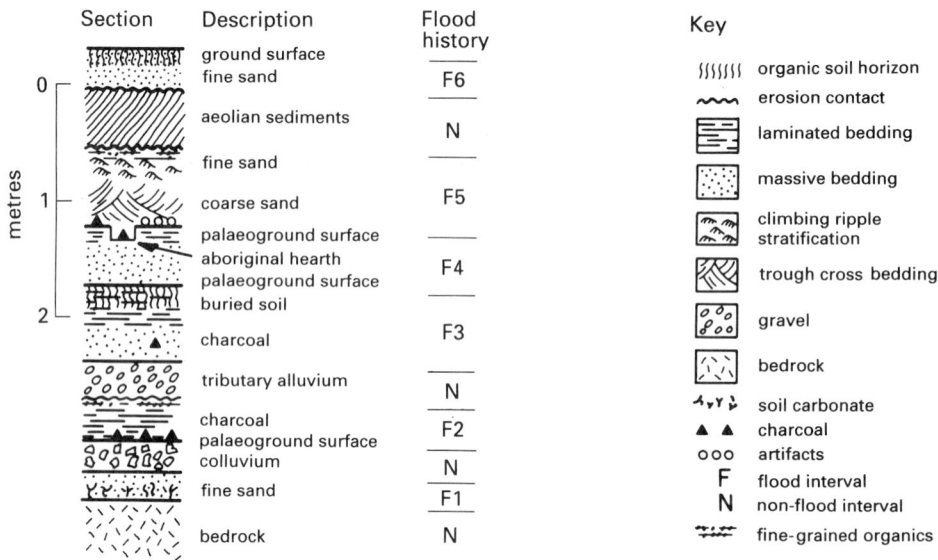

(b)

Fig. 4.2 (a) Tell-tale indicators of recent high-magnitude/low-frequency flood stage on the Pedernales River, Texas.
1, mesquite; 2, eastern red cedar; 3, myrtle oak; 4, hackberry; 5, walnut; 6, pecan; 7, American sycamore; 8, bald cypress;
9, live oak. After Patton *et al.* (1982). (b) Schematic section through a fluvial sequence showing how sedimentary
parameters can be employed in the interpretation of palaeoflood history. After Baker (1989).

damage or sedimentary accumulation that arises from
extreme events is gradually mended through the action of more frequent flood events of lower magnitude
as the system struggles to regain its former equilibrium. Wolman and Gerson (1978) have argued that
the length of this *relaxation period* may depend upon
the local environment. In their view, channel recovery

in humid environments is likely to be more rapid than
in arid-zone counterparts where floods are few and far
between (Fig. 4.4). However, this presumes that fluvial systems are elastic and are likely to recover once
the stress is relaxed. It may be that long-term changes
in climate or tectonism dictate a drift in the equilibrium condition which is being sought by the river

Fig. 4.3 (a) Giant boulder deposit of a high-magnitude/low-frequency flood event in the ephemeral Torrente de Pareis, Sierra Tramuntana, Isle of Mallorca. Individual clasts have diameters up to 15 m producing high values of grain roughness! (b) Giant boulder 'cluster' — one of several — on a Holocene terrace cut into the Upper Pleistocene Zeelim alluvial fan complex, Dead Sea trough.

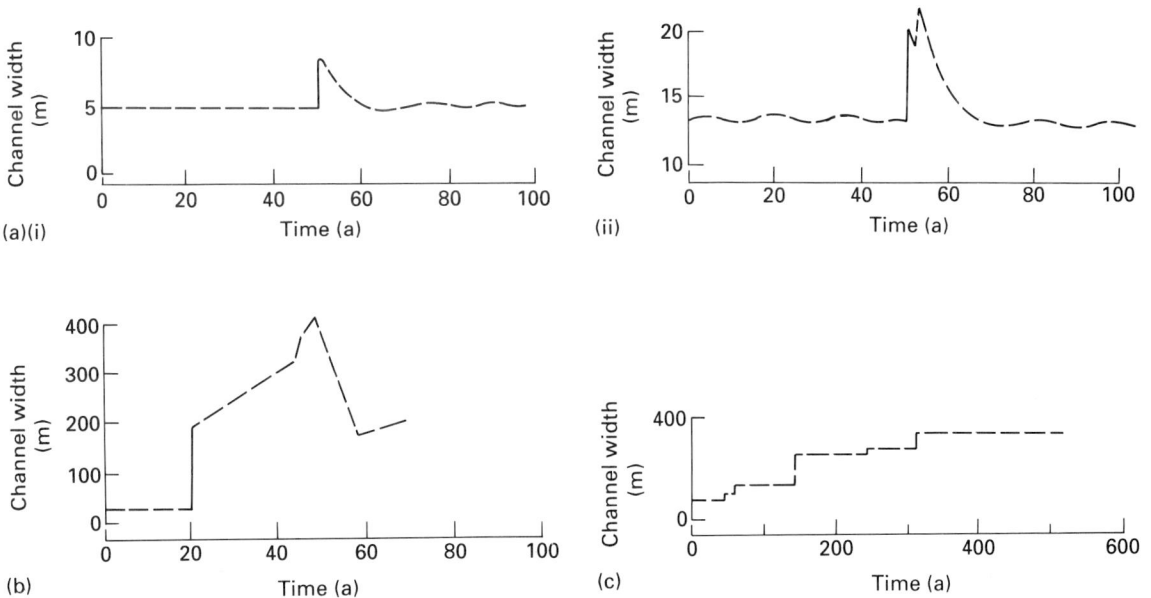

Fig. 4.4 Relaxation period following the channel changes brought about by high-magnitude/low-frequency floods. The differences between one environmental setting and another is a product of the frequency of amending flood events of moderate size and the potential for vegetational recolonization and sediment entrapment. (a) Humid: (i) Baisman Run; (ii) Patuxent Run. (b) Semi-arid: Cimarron River. (c) Arid: Negev and Sinai (schematic). After Wolman and Gerson (1978).

system. The river may not be able to mend the damage *and* keep pace with changing circumstances.

The impact of catastrophic floods is far from ubiquitous, and much fluvial activity is far less spectacular. However, there is a need for caution in assessing sedimentary processes that are more gradual especially if these are to be used as analogues in interpreting ancient river sediments. In the first instance, the advent of advanced technologies such as those of Iron-age Man and his successors led to deforestation and a mobilization of sediment in excess of that of 'natural' landscapes (Douglas, 1967a). Secondly, the climatic fluctuations of the Pleistocene and the glaciation of high latitudes have left a legacy that has had important consequences for Holocene sediment transport in regions such as these. Church and Slaymaker (1989) have pointed out that the sediment yields of British Columbian drainage basins follow a trend which opposes those of unglaciated regions in that the unit efflux of sediment *increases* with increasing basin size. They rationalize that this reflects the availability of erodible sediment in the form of moraine and outwash valley fills.

Needless to say, some care has to be exercised in selecting sediment transport data as representative of natural environments that are assumed to be in long-term equilibrium. Indeed, the climatic fluctuations of the Pleistocene together with the present ubiquity of Mankind might be grounds for excluding practically all datasets!

4.3 River sediment load

4.3.1 Sources of sediment

River water is rarely free of sediment, although at low flows or where gradient and boundary roughness are such that turbulence is suppressed, the solids being transported may be entirely organic. As a general, though by no means universal, rule, the transport of mineral sediment increases with increasing flow. Fluvial sediments in transit have been classified as *bed material load* and *wash load* in order to distinguish them according to their origin. Bed material load is derived by disturbing the channel sediments and is generally thought of as comparatively coarse in size. The degree of disturbance may differ between perennial and ephemeral rivers. Leopold *et al.* (1966) reported a fairly well-defined direct relationship between flood flow and scour depth down to *c.* 0.3 m in a gravelly sand-bed ephemeral stream in New Mexico (Fig. 4.5). Foley (1978) noted a similar pattern for a sand-bed river in California, but related scour depth to the amplitude of antidunes believed to have formed during flood flows. In the Negev Desert, Israel, Schick *et al.* (1987) have detected scour to 0.4 m in a gravel-bed ephemeral stream. In contrast,

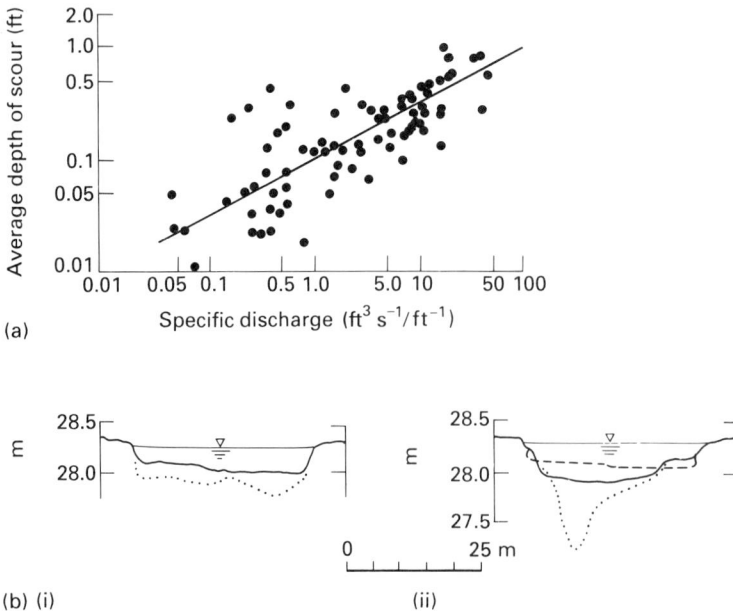

Fig. 4.5 (a) Generalized relationship between channel scour-depth and flood peak discharge in the gravelly sand bed Arroyo de los Frijoles, New Mexico. After Leopold *et al.* (1966). (b) Scour and fill arising from two floods of different magnitude on the ephemeral sand bed of Quatal Creek, California. (i) $\overline{Y}$, 0.23 m; $\bar{u}$, 2.9–4.4 m s^{-1}. —, pre- and post-flow bed; ....., maximum scour. (ii) $\overline{Y}$, 0.34 m; $\bar{u}$, 3.8–5.7 m s^{-1}. - - -, pre-flow bed; —, post-flow bed; ....., maximum scour. After Foley (1978).

scour appears to be less significant in perennial gravel-bed rivers, where the bed seems to be better armoured. So, for example, Jackson and Beschta (1982) and Parker *et al.* (1982) indicate that the partial disruption of this armour during flood flows allows the smaller calibre sub-armour layer to be entrained, but there is no deep scour. Undoubtedly, some of these differences arise from the flashier nature of the flood hydrograph in ephemeral streams. However, scour and fill are not necessarily synchronized from reach to reach, and scour may be occurring in one stretch of the channel while another stretch is aggrading, as shown for individual rain-fed floods by Lane and Borland (1953) and over the much more extended duration of seasonal snowmelt by Andrews (1979). Nevertheless, despite such complications, the bed material does provide material of widely varying size.

The wash load is generally finer and has traditionally been associated with hillslope source areas. However, there are few studies that have been able to trace the pathway from slope to river. The size distribution of wash load particles has been used to establish a connection, but this has been largely inferential after comparison with the size distribution of slope soils (Peart & Walling, 1982; Frostick *et al.*, 1983; Fig. 4.6). The same can be said of the use of other natural properties of the sediment such as its remanent magnetism (Oldfield *et al.*, 1979). Ironically, the fallout of radionuclides in NW Europe after the Chernobyl nuclear reactor accident in 1986 labelled enough soil clay particles with adsorbed [137]Caesium to make a tracing programme feasible (Walling *et al.*, 1989). Because it consists in part of clay particles with settling velocities of less than 10^{-6} ms^{-1}, the wash load is often assumed to remain suspended in the flow. This is undoubtedly the case where the channel bed consists of 'clean' sands, but gravel beds offer a framework of large clasts with interstitial spaces in which fine particles can and do settle (Carling & Reader, 1982; Frostick *et al.*, 1984). As a result, when disturbed, a gravel bed can act as an additional source of material traditionally classified as wash load.

In reality then, the source of sediment is more complex than envisaged by a twofold classification. One additional source of considerable importance that has often been overlooked is the channel bank (Thorne & Lewin, 1982). Several studies have attempted to isolate and measure its contribution. Hansen (1971) concluded from a 3-year study of the Pine River, Michigan, that 45% of the sediment discharge could be

attributed directly to bank erosion. Murgatroyd and Ternan (1983) have shown that riparian landuse is one determinant of the rate of bank collapse; the dense root mat of a grass sod is efficient at strengthening the bank material while the lack of ground flora under closed-canopy coniferous woodland leaves the bank susceptible to erosion. However, the contribution of material by stream banks will depend also upon cohesion and, therefore, upon the clay content of the channel wall. Schumm (1961) revealed an inverse relationship between the fractional amount of silt and clay and the width : depth ratio of ephemeral channels in four western states of the USA. In cases where the banks were more competent (i.e. have higher shear strength through higher clay content) the channels were found to be less wide and more incised.

4.3.2 Modes of sediment motion

Whatever the source of sediment, once it is entrained by the flow it moves in one of three ways depending upon the relative strength of the flow and the immersed weight of the clasts.

The *bedload* includes those grains which roll or slide along the bed essentially in contact at all times. The actual mode of movement depends upon the transport rate. At low rates, a step-and-rest pattern appears to dominate (Rathbun & Guy, 1967; Andrews, 1983). This may be both spasmodic and sporadic and involve a burst of activity by a number of neighbouring particles followed by a period of corporate rest (Reid *et al.*, 1984; Custer *et al.*, 1987; Fig. 4.7). At higher transport rates, motion may still involve discrete steps, but the rest period may be minimal. However, so many particles are now in motion that the bed surface may resemble a mobile carpet. This is a pattern that is more familiar on sand beds than on gravels.

The *suspended load* includes those grains whose settling velocity is more than matched by the upward component of turbulence and so remain within the flow, having no contact with the bed for an unspecifiable but significant fraction of time. The size of suspended sediment can range more widely than the wash load, and has been shown to include coarse sand (Nordin, 1963; Reid & Frostick, 1987). However, Colby (1963) illustrates the conflict between turbulence and settling with a vertical profile through the Mississippi: medium and coarse sand is thrown only 3 m into suspension while fine sand can be found to be well mixed within the flow, reaching 8 m

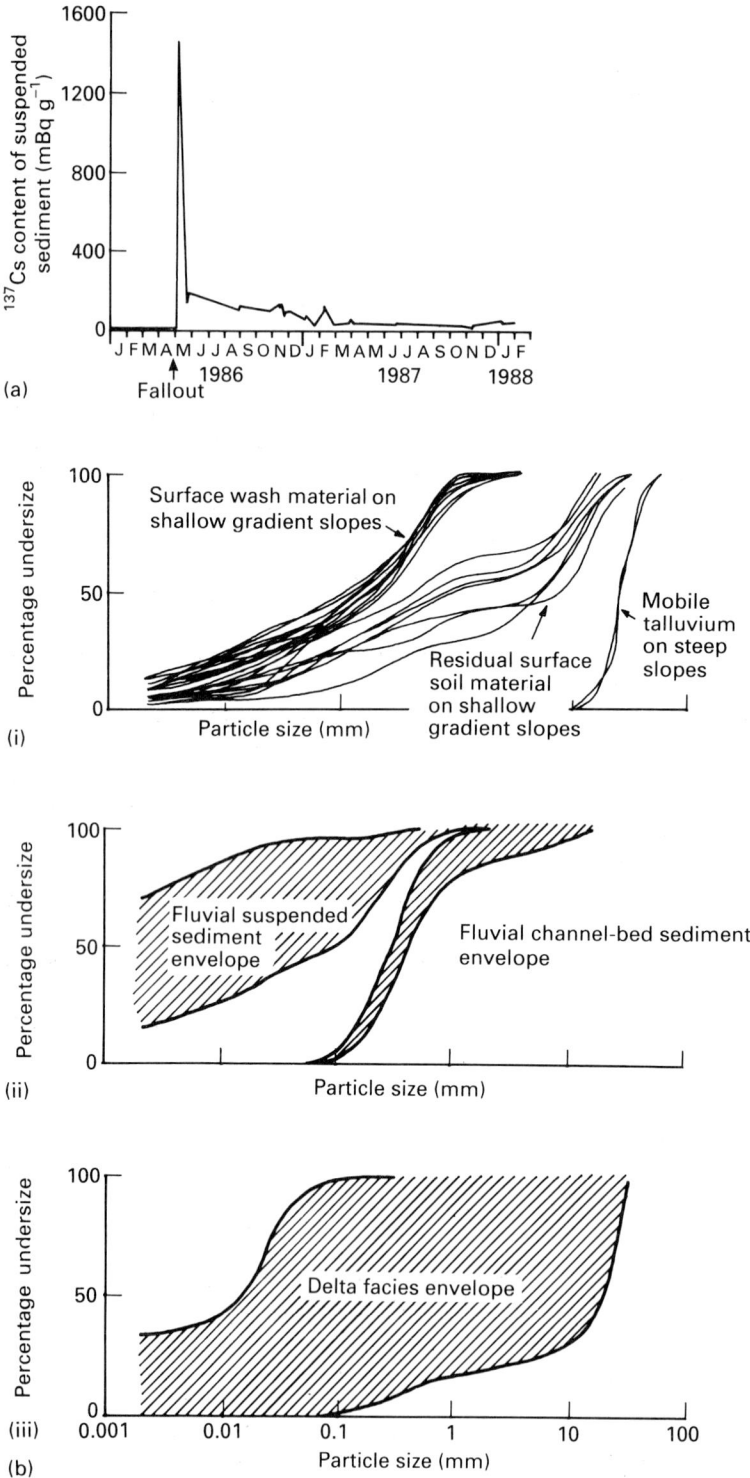

Fig. 4.6 (a) Radio-caesium in suspended sediment samples from the River Severn, England, following fallout from the nuclear reactor accident at Chernobyl in the Ukraine on 26 April 1986. After Walling *et al.*, 1989.
(b) Size-distribution curves for sediments in various parts of a semi-arid drainage system, the Tulu Bor, northern Kenya. The lake delta contains facies that represent all the sediments that are mobile in the river catchment. However, the fluvial suspension is largely derived from fine clasts that make up the lowest two quartiles of the hillslope-wash size-distribution, while this calibre material is almost absent from the channel bed of these sand-bed streams. (i) Hillslope–source; (ii) river channel–conveyance; (iii) delta–deposit. After Reid and Frostick (1986b).

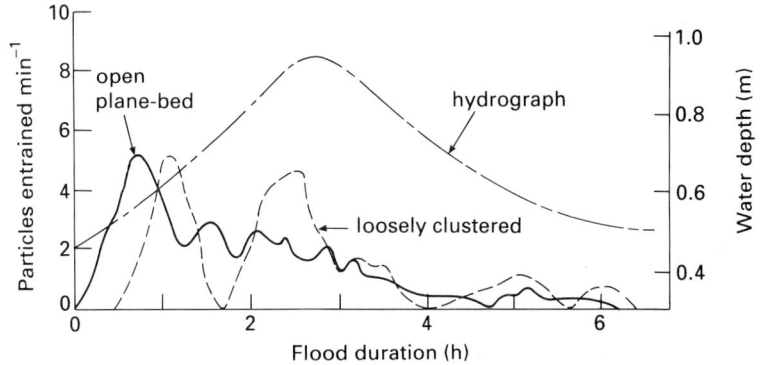

Fig. 4.7 Spasmodic entrainment of metal-tagged clasts seeded in open plane bed and loosely clustered configurations on the gravel bed of Turkey Brook, England. After Brayshaw (1985).

from the bed on the particular occasion he depicts.

The *saltation load* involves a hybrid mode of transport that consists of momentary contact with the bed and a series of asymmetric downstream trajectories within the boundary layer of the flow. Francis (1973) captured the motion with a multi-exposure film technique and reported that grain motion appeared to be ballistic, that is, the trajectory was concave to the bed throughout the saltation hop, reaching a maximum height of between two and four grain diameters. Clasts moving as part of the saltation load will have an immersed mass which is intermediate between those moving as part of either the bedload or the suspended load.

Although it is convenient to distinguish these three modes of transport, it is important to bear in mind that particles which are carried as bedload in one reach or at one flow may become suspended either downstream, where flow conditions are different, or in a single reach as a flood wave waxes. However, as a general rule, clays and fine to medium silts (particles < 32 µm) will almost always act as suspended load, and pebbles and coarser gravels (> 4 mm) will generally travel as bedload. Coarse silts, sands and granules (32 µm–4 mm) may switch from one mode to another depending upon local flow conditions.

4.4 Fluvial bedload

Sediment transport as bedload has long fascinated engineers, geomorphologists and sedimentologists alike. To one, it represents a potential hazard as revetments and bridge piers are undermined by erosion, or reservoirs are filled by sediment. To the others, it is seen as a major factor in forming and changing the character of river channels and, therefore, the style of fluvial sedimentation. However, it should not be forgotten that best estimates of the

contribution that bedload makes to the total sediment yield of a drainage basin lie between 0 and 50%, depending upon whether the local environment is humid or arid and whether the channel bed is composed of sand or gravel. It may have a significance beyond its relative contribution in that it is channel-forming, but its denudational role is generally inferior to that of suspended sediment.

Serious interest in the prediction of bedload flux rates goes back to Du Boys (1879). It was Du Boys who developed the idea that a fluid in motion exerts a shearing force on the stream bed and that this causes clasts to be displaced in the direction of the energy gradient. This shear force can be calculated from the properties and dimensions of the fluid. The bulk transport rate was conceptualized as the sum displacement of a number of layers, each of finite but small thickness, that shear against each other progressively less and less with increasing depth below the bed surface down to a layer that is static (Fig. 4.8). At the point of initial entrainment — incipient motion

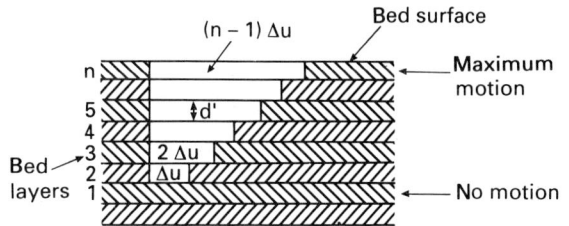

Fig. 4.8 Du Boys' (1879) portrayal of shear within a stream bed yielding bedload. Each layer thickness (d') is notional, but might be envisaged as equivalent to the diameter of a single grain. At incipient motion, only the surface layer is poised for imminent motion. Δu is the difference in velocity between any one layer and that immediately below; n is the total number of layers.

— only the topmost layer is mobile. Du Boys' bedload equation is:

$$q_b = A \tau_o (\tau - \tau_o), \qquad (4.2)$$

where q_b is the specific bedload flux rate (in subaerial mass terms); A is a coefficient characteristic of the sediment and varies inversely with grain diameter, D, such that $A = 0.17/D^{0.75}$; τ is the shear stress at the bed; and τ_o is the critical shear stress for clast entrainment in the uppermost layer of Du Boys' schematic representation of the bed surface.

Subsequently, the empirical observations of the American pioneering geologist Gilbert (1914), both in a natural sand-bed stream and in an experimental flow channel, were to confirm the fact that a hydraulic parameter could be used to predict bedload transport rates, albeit that these early experiments were not as controlled as might have been desirable.

Attempts to find a solution to the prediction of bedload flux rates now took several directions, and can be classified as *empirical* or *semi-theoretical*.

4.4.1 Empirical bedload equations

The empirical bedload equations generally follow a format that resembles Du Boys'. Bedload flux rate is a function of excess shear (defined as average shear minus shear at incipient motion) and bed sediment character (grain size and density, etc.). The relationships may or may not include a consideration of flow resistance due to bedform or grain roughness, and constants and exponents which calibrate them are derived from field or laboratory flume data.

Perhaps the most widely used of the empirical bedload relationships is that of Meyer-Peter and Muller (1948), the calibration of which rests on a large body of laboratory flume data obtained for a wide range of hydraulic conditions and sedimentary variables:

$$q_b = \frac{\gamma_s}{\gamma_s - \gamma} \cdot \left[\frac{(K_b/K_G)^{3/2} YS - 0.047\{(\gamma_s - \gamma)/\gamma\}D_m}{(0.25/\gamma)(\gamma/g)^{1/3}} \right]^{3/2},$$

$$(4.3)$$

where q_b is specific bedload flux rate in sub-aerial mass terms; γ is the specific weight of the fluid; γ_s is the specific weight of the sediment; K_b specifies the bed roughness as $u/Y^{2/3}S^{1/2}$, where u is mean water velocity; K_G specifies grain roughness as $26/D_{90}^{1/6}$, in which D_{90} is the ninetieth percentile of the bed surface grain-size distribution; Y is water depth and

which should be substituted by hydraulic radius in narrow channels in order to account for the loss of energy due to sidewall drag; S is the slope of the water surface and approximates the energy gradient; D_m is the 'effective' diameter of the surface bed sediment defined as $\Sigma D_i P_i/100$, where D_i and P_i are the average diameter and percentage fraction by weight of the ith size fraction, respectively, and which often approximates D_{64} in gravel-bed streams, for which the formula was originally devised; and g is the acceleration of gravity.

4.4.2 Semi-theoretical bedload equations

The semi-theoretical equations adopt various approaches to the problem of predicting bedload. Perhaps the most unusual is that of Einstein (1942) who argued against a clearly defined entrainment threshold and excess shear. Einstein's approach was probabilistic. He argued that, because all grains on a bed of uniformly sized grains do not move at a single point in time as the stress is increased to a critical level, the critical tractive stress approach is inappropriate. Instead, he rationalized that entrainment occurs when the local instantaneous lift force exceeds the immersed weight of an individual particle. The probability of movement is calculated for a number of narrow particle size classes after noting the fractional contribution that each makes to the bed sediment. Einstein recognized that some particles are shielded from the flow by larger upstream neighbours, and so he devised a 'hiding factor' that progressively reduces the lift force and, therefore, the chance of being moved as particle size decreases. Once mobile, the step-length of each size grade is assumed to take on an average value regardless of flow conditions. Einstein had observed this to be about one hundred times the particle diameter for sands.

The computation of bedload using the Einstein formula in either its original or modified forms (Einstein, 1950) is complex and can be found in either the original papers or in many hydraulics texts such as those of Garde and Ranga Raju (1977) or Chang (1988).

A different theoretical approach was adopted by Bagnold (1966). He rationalized that the rate of work (as represented by sediment transport) should be related to the rate of energy expenditure. The total energy available was defined as stream power, which for unit channel width, ω, is given by:

$$\omega = \rho g Y S u, \qquad (4.4)$$

where ρ is fluid density; g is gravitational acceleration; Y is water depth; S is water surface slope; and u is mean water velocity.

In early versions of his bedload equation, Bagnold incorporated an efficiency term which recognized that most energy is expended in moving the water rather than sediment. Reid and Frostick (1986a) have shown that the level of efficiency for two American and one British gravel-bed rivers lies at a level of less than 1% and that the modal value is 0.05%. Typical values for the East Fork River in Wyoming, which largely transports sand, lie around 5% although Gilbert's (1914) flume data had previously led Bagnold to believe that efficiencies lay at about 10% for sand. The problem of establishing the efficiency of the 'engine', together with the fact that field data were emerging which were reasonably reliable and against which the theoretical model could be tested, finally led Bagnold (1980, 1986) to empiricism. The last version of his equation is:

$$i_{\mathrm{b}} \propto (\omega - \omega_{\mathrm{o}})^{3/2}(Y/Y_{\mathrm{r}})^{-2/3}(D/D_{\mathrm{r}})^{-1/2}, \quad (4.5)$$

where i_{b} is specific bedload flux rate, ω is specific stream power, ω_{o} is specific stream power at incipient sediment motion, Y is water depth, D is bed sediment diameter and taken as D_{50}, and Y_{r} and D_{r} are arbitrary reference values of water depth and sediment size, introduced by Bagnold as a convenient way of collapsing data from different sources in order to demonstrate the generality of the trend between energy expenditure and rate of work, that is, bedload transport.

A useful working version of the equation, in which specific bedload flux rate is given in subaerial mass terms, is:

$$i_{\mathrm{b}} = \frac{\gamma_{\mathrm{s}}}{\gamma_{\mathrm{s}} - \gamma} i_{\mathrm{br}} \left[\frac{\omega - \omega_{\mathrm{o}}}{(\omega - \omega_{\mathrm{o}})_{\mathrm{r}}} \right]^{3/2} (Y/Y_{\mathrm{r}})^{-2/3}(D/D_{\mathrm{r}})^{-1/2}, \quad (4.6)$$

where $\omega = \rho g Y S u$, W m^{-2}, $\omega_{\mathrm{o}} = 5.75 [0.04(\gamma_{\mathrm{s}} - \gamma)\rho]^{3/2}$ $(g/\rho)^{1/2}D^{3/2} \log (12Y/D)$, W m^{-2}, $D = D_{50}$ for unimodal bed sediment size distributions, m, $i_{\mathrm{br}} = 0.01$ kg s^{-1} m^{-1} and is an arbitrary reference value for bedload flux rate, $(\omega - \omega_{\mathrm{o}})_{\mathrm{r}} = 0.5$ W m^{-2} and is a reference value for excess specific stream power, $Y_{\mathrm{r}} = 0.1$ m, and $D_{\mathrm{r}} = 0.0011$ m.

4.4.3 The bedload prediction problem

A successful solution to the prediction of bedload

transport has yet to be found despite more than a century of work on the subject. Formulae have proliferated because the processes that are involved are not only fundamental to present-day engineering problems, but are also the basis for understanding long-term rates of continental denudation and sedimentation. Interest in finding an equation that has universal application remains high and has been passed from generation to generation of hydraulic engineers, geomorphologists and sedimentologists alike. Carson and Griffiths (1987) offer an excellent summary of this quest and appraise the reasons why a solution remains tantalizingly elusive.

The problem lies in the variable nature of rivers. The size- and shape-distributions of bed sediments change across a single bedform and from channel bar to thalweg. They also change from reach to reach in a single river, let alone from one drainage system to another. Add to this the changes in hydraulic environment that arise as a flood wave waxes and wanes and the complexity of the problem starts to become evident. Despite this, several early datasets appeared to suggest that the fluvial system reacts in a fairly simple fashion. The field data of the sand-bed Niobrara River, Nebraska, are an illustration (Colby & Hembree, 1955), although the fact that, in this case, the bedload was sampled as suspended load where the channel is severely constricted may have helped to reduce the scatter in the relationship between flow and 'bedload' flux rate. Nevertheless, because these datasets exist, an orderly response may have been, and may still be, expected from a system that is inherently complex, if not disorganized. Perhaps because of this, and because formulae have often been devised around, or calibrated by, data collected in a specific natural or artificial channel and therefore appear to perform tolerably, there has been a tendency to accept bedload equations as solutions to the problem of predicting sediment flux.

However, Parker *et al.* (1982) have drawn attention to the poor performance of bedload transport equations when they are rated against data from rivers other than those for which they were devised (Fig. 4.9). More recently, Gomez and Church (1989) have reviewed the predictive power of equations against data that were selected deliberately to represent steady flow conditions, so eliminating, as far as is possible, the complication of rapidly changing hydraulic conditions. Their conclusions must only apply to gravel-bed rivers since this is the source of the data. However, they are of the opinion that none

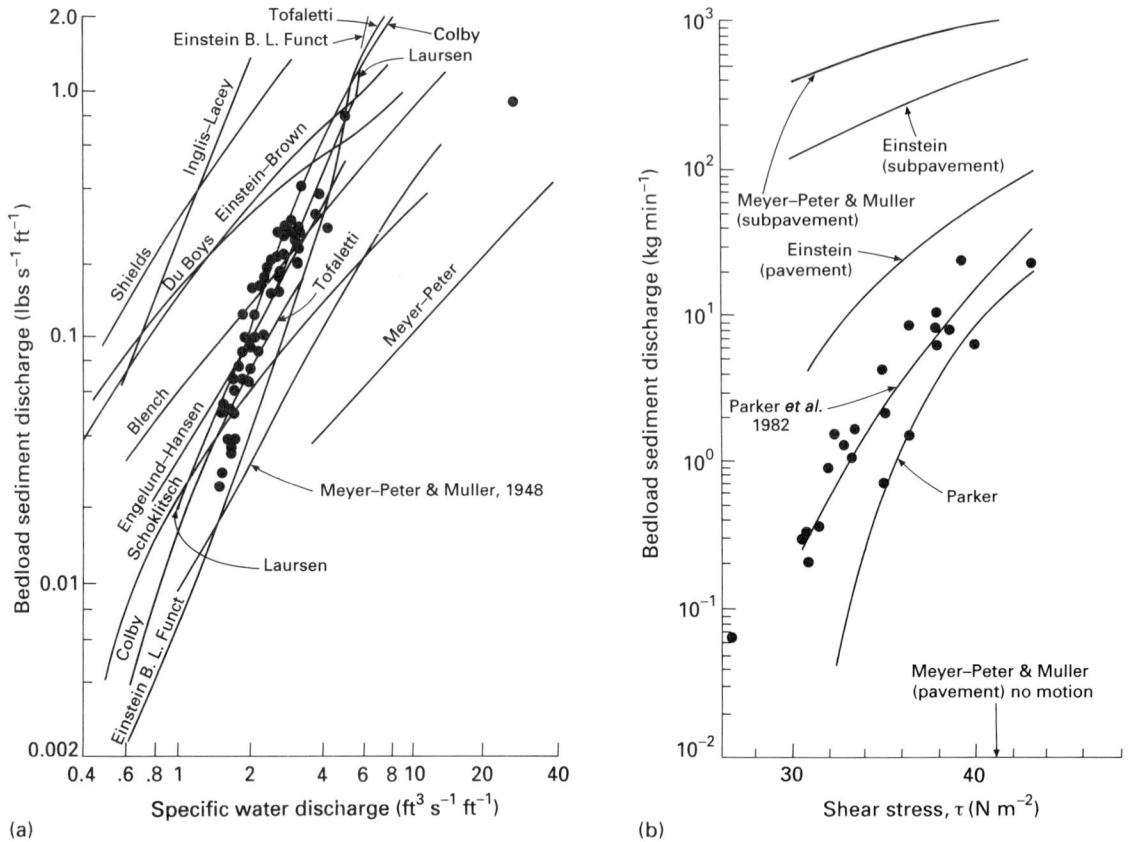

Fig. 4.9 Bedload transport according to predictive bedload equations and measured bedload for (a) the sand bed Niobrara River, Nebraska, after Vanoni (1975); and (b) the gravel bed Oak Creek, Oregon. ●, Observed. After Parker *et al.* (1982).

of the formulae that they have tested is capable of a satisfactory general prediction of bedload transport. Despite this declaration, Bagnold's 1980 equation provides a best fit for the field and flume data, and they consider that the approach to the problem that it represents will be worthy of pursuing in the future (Fig. 4.10).

Much effort has been devoted to establishing the reasons behind the poor performance of predictive bedload equations. This has led to a greater understanding of the mechanics of sediment transport even if, as is often the case, newly found processes are too complex to build into the formulae.

The armour layer in gravel-bed rivers

The gravel beds of perennial rivers consist of two active vertical layers. The *armour layer* forms the surface, is no more than one grain thick and overlies the *sub-armour layer* which may be of variable thickness (Andrews & Parker, 1987). The term *pavement* has been used to describe a virtually immobile surface layer (Kellerhals, 1967), but has been abandoned in favour of *armour* as a general descriptor. The grain size of the armour layer is usually coarser than its substrate, with typical values of the armour/ sub-armour size ratio lying in the region of 1.5–2, though Dunkerley (1990) gives an interesting example in the Tambo River of Victoria where the ratio reaches values as high as 25. The reason for relative coarseness is most probably the selective winnowing of small particles (Proffitt & Sutherland, 1983; Sutherland, 1987) and Carling and Reader (1982) have certainly shown that the matrix of fine particles can be removed from the interstices between the larger framework clasts to form a *censored armour*

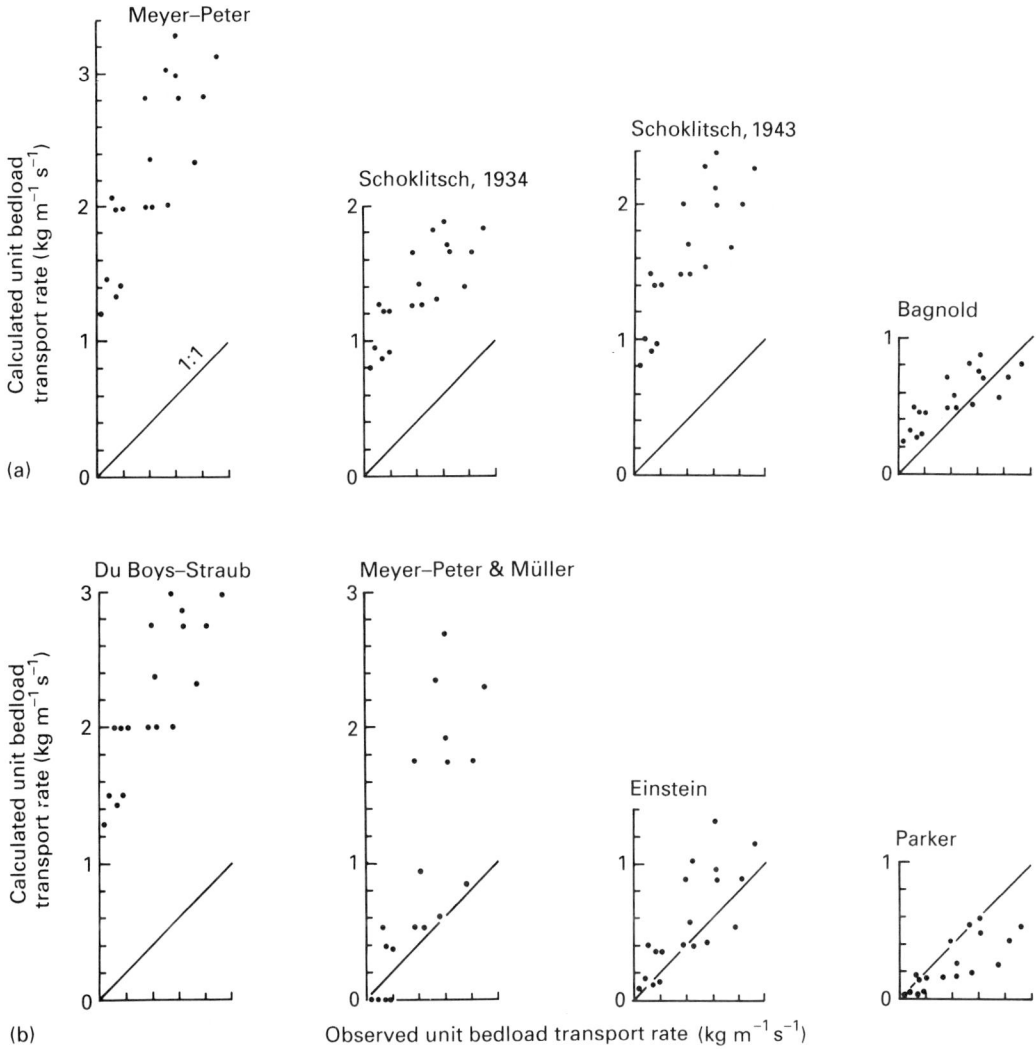

Fig. 4.10 Bedload transport predicted by stream power and tractive force formulae rated against bedload transport observed in the Elbow River, Alaska. (a) Streampower formulae; (b) tractive force formulae. After Gomez and Church (1989).

layer down to about 0.08 m. However, the significance of the armour layer for sediment transport had not been appreciated until recently. Parker *et al.* (1982) working with Milhous' data (1973) for Oak Creek, Oregon, commented on the fact that the size distribution of the bedload bore more resemblance to the sub-armour layer than to the armour. They surmise that the partial breaching of the armour allows the smaller sub-armour layer to be ravaged. Reid and Frostick (1986a) have noted the same in

Turkey Brook, England, while Jackson and Beschta (1982), working on another Oregon stream, propose a two-phase bedload transport model that involves a change in transport efficiency before and after the armour layer is breached and takes account of the changing size of clasts exposed to the flow. Gomez (1983) has drawn attention to the reduction in bedload flux rate that accompanies the re-establishment of an armour layer during waning flood flows.

Flow resistance and small-scale bedforms in sand-bed rivers

Einstein's original bedload equation did not allow for form resistance probably because it was developed for gravel-bed rivers in which repeating microforms have only recently been recognized. However, it was soon acknowledged that the changing boundary roughness associated with small-scale bedforms would affect stream efficiency, and hence bedload. As a result, an attempt was made later to apportion roughness between the grains composing the bed and the bedforms (Einstein & Barbarossa, 1952). This was done by dividing the hydraulic radius into components representing the two sources of rough-

ness. However, the form of the bed changes with the strength of the flow. Guy *et al.* (1966) showed a rise in the Manning roughness coefficient from around 0.02 to 0.035 as flow increases through the lower flow regime and the bedforms change from ripples to dunes. As the dunes wash out and the upper flow regime becomes established, first over a plane bed and then over antidunes, the roughness coefficient drops abruptly to a level between 0.015 and 0.02 (Fig. 4.11).

Microform flow resistance on gravel beds

It is only recently that repeating microforms have been recognized in modern gravel-bed rivers, even

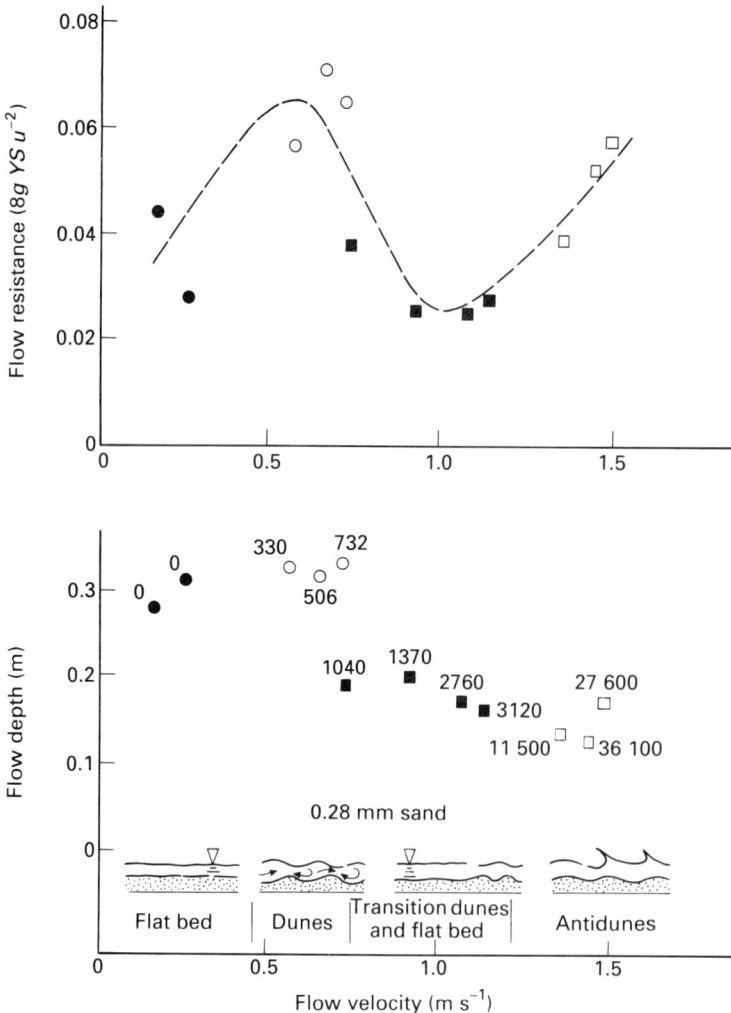

Fig. 4.11 Changes in small-scale bedform and suspended sediment concentrations and, consequently, in flow resistance with increasing flow velocity. Numbers attached to symbols refer to suspended sediment load in parts per million. ■, Transition dunes and flat beds; □ antidunes; ●, flat beds; ○, dunes. After Guy *et al.* (1966).

though they had been noted some time previously as part of the fabric of ancient river sediments (Dal Cin, 1968a; Teisseyre, 1977). One difficulty is undoubtedly the variety of shape that arises because, unlike in sand beds, individual clasts, themselves ranging widely in size and shape, can make up a large fraction of the bedform. Some gravel-bed microforms have a structure that lies orthogonal to the flow such as *clast dams* (Bluck, 1987) and *steps* (Whittaker & Jaeggi, 1982). Others are streamlined such as *pebble clusters*.

Koster (1978) has provided a fairly comprehensive analysis of discrete *transverse ribs* of coarser than average clasts in braided rivers. He suggests that they are formed at times when flow is supercritical, that is, when the Froude number exceeds a value of one. They are equivalent, therefore, to antidunes in sand-bed streams, the wavelength reflecting the dimensions of the standing waves responsible for their formation. But because supercritical flow is transitory, and more often than not highly localized, ribs are not ubiquitous, tending to form locally along the thalweg and not on bar shoulders, etc.

The other repeating microform recognized as a constituent of gravel-bed surfaces is the pebble cluster (Brayshaw, 1985). Unlike transverse ribs, clusters are streamlined and discrete in both planform and longitudinal section. An 'obstacle' clast forms the nucleus and initiates cluster formation by distorting the flow, on the one hand, and by impeding the progress of other clasts, on the other. Obstacle clasts are often among the largest particles — Brayshaw (1984) indicates that, on average, they fall in the largest decile in British streams that have a wide range of clast lithotype. However, even comparatively small clasts can impede the downstream progress of others by increasing the pivot angle of entrainment. Against the obstacle on its upstream stoss side, an imbricate structure will develop to a point where the addition of one more clast produces an unstable condition. These particles may correspond to the 60th percentile of the size distribution, but can range up to the 90th percentile where lithology encourages platey clasts. In the wake of the obstacle, flow is separated from the free stream and the net drag force on clasts in this region may be 'negative', that is, the vector may be upstream (Brayshaw *et al.*, 1983). A streamlined wake tail develops as clasts enter, but do not leave, the separation bubble. Generally, these are fine particles, often

corresponding to the smallest decile of the armour layer size distribution. A wide spectrum of cluster forms can be identified, and all of the constituent parts are not necessarily represented in each case. Nevertheless, they are prevalent, covering as much as 10–20% of the bed, but having a hydrodynamic influence which extends beyond this.

Pebble clusters have been shown to delay sediment entrainment, and to be responsible, in part, for a four- to five-fold increase in the value of the dimensionless critical shear (the Shields' parameter) relative to the experimental values derived for 'plane' bed conditions (Reid & Frostick, 1984). Because clusters have a mechanical and hyrodynamic strength, not only do they contribute proportionately less to the bedload than the intervening plane bed, but because entrainment is delayed, particles emanating from them have been shown to travel less far during a single flood wave. More recently, cluster density has been shown to affect flow resistance in a manner similar to sand bed microforms (ripples and dunes) and, therefore, to control rates of bulk bed-load transport (Hassan & Reid, 1990).

Transverse ribs and pebble clusters are recognizable and repeating microforms. The intervening bed may be more or less organized, and Laronne and Carson (1976) have drawn attention to the wide spectrum of niche types facing a particle mobilized by the flow. The spectral range will depend in large part on the size and degree of sorting of the armour layer sediment. Steep upland streams possess the highest level of apparent disorder and Bathurst (1978) has shown that flow resistance is difficult to parameterize where the roughness elements are variably swamped at any single flood stage, that is, where some boulders are submerged while others remain partially exposed. Notwithstanding the difficulty of establishing appropriate values for flow resistance in order to estimate residual levels of transporting energy, the inverse relationship expected between flow depth and resistance can be demonstrated with field data. However, since total flow resistance emanates from both bed and bank roughness, Carling (1983) has drawn attention to the difference between wide and narrow channels. All other things equal, unit rise in flood stage will mean a larger reduction in resistance in a wide channel where the wall drag is proportionately less important (Fig. 4.12). As a result, sediment transport rates might be expected to rise faster as the flood wave grows in a wide channel.

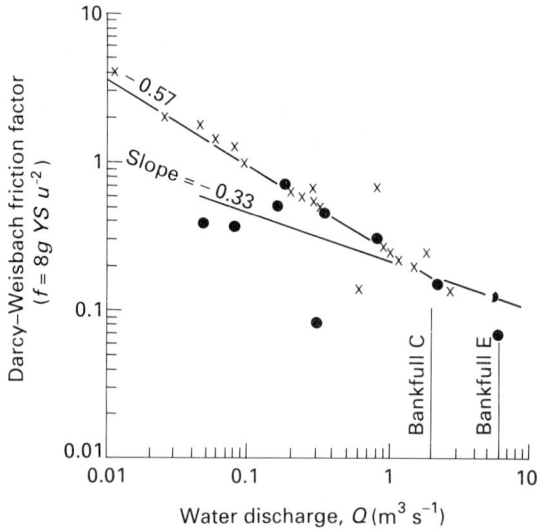

Fig. 4.12 Changes in flow resistance as roughness elements are drowned out with increasing water discharge in narrow (x, C: Carl Beck) and wide (●, E: Gt Eggleshope Beck) upland gravel-bed streams. After Carling (1983).

Entrainment thresholds

A simplified analysis of entrainment mechanics such as that of White (1940) in which lift and drag forces are balanced against the body force at incipient motion indicates that clast size is the main controlling variable. A number of compilations of entrainment threshold data, some combining flume and field studies (Shields, 1936; Meyer-Peter & Muller, 1948), have lent empirical support to the notion. This has led to a commonly held view that a knowledge of bed material size will allow the estimation of excess shear

force and, therefore, bedload transport flux rates. In very broad terms and for situations where bed material is uniform in size, this may be a reasonable assumption — sands *will* have a lower entrainment threshold than gravels.

Indeed, this is reflected by the fact that Shields' entrainment function, θ, which describes shear at incipient motion in dimensionless form:

$$\theta = \frac{u_*^2}{gD(\rho_s - \rho)},\qquad(4.7)$$

(where u_* is shear velocity, which can be defined by $(\tau\rho)^{0.5}$; τ is shear stress at the bed; g is gravitational acceleration; D is grain diameter and usually taken as D_{50}; ρ_s is particle density; and ρ is fluid density) is more or less constant at a value of 0.045 over a sediment size-range that has been extended to include coarse gravel by Miller *et al.* (1977).

Fenton and Abbott (1977) were able to demonstrate that Shields' entrainment condition is appropriate for planar bed conditions — in fact, the very conditions contrived in each of the experimental runs that formed the basis of Shields' data. However, particles that sit proud of the bed and are fully exposed to the flow are entrained at a dimensionless shear as low as 0.01 (Fig. 4.13). In fact, Andrews (1983) has argued, using field observations of bedload in the East Fork River, Wyoming, and the Snake and Clearwater Rivers, Idaho, that an appropriate value of the entrainment function lies between 0.02 and 0.025 for much of the bed material. This would predict an earlier onset of bedload and greater excess shear than use of the traditional value of 0.06 (Shields, 1936) or that of 0.045 (Miller *et al.*, 1977).

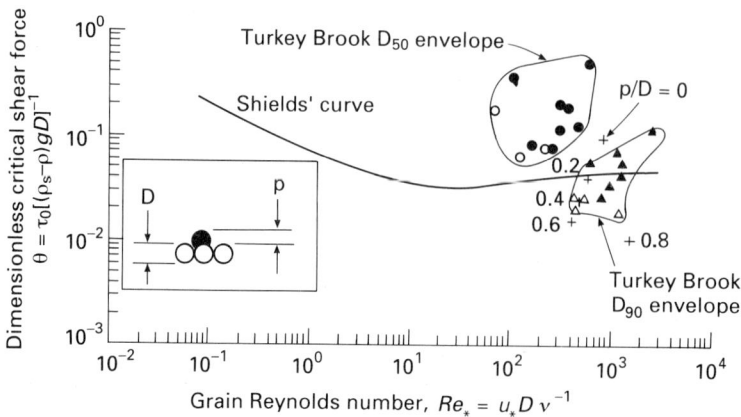

Fig. 4.13 Dimensionless critical shear stress plotted against grain Reynolds' number. Shields' (1936) curve derives from flume experiments with 'plane' beds. The Turkey Brook data show the effect of pebble clusters and other factors in strengthening the bed structure and delaying transport (after Reid & Frostick, 1984), while Fenton and Abbott's (1977) flume and initial motion data (see box) illustrate the effect of relative grain protrusion in delaying or encouraging transport relative to that expected of a plane bed. Turkey Brook: ●▲, initial motion: ○△, final motion; +, Fenton and Abbot's flume data.

Conversely, data for Turkey Brook, England, suggest that initial sediment motion occurs when the entrainment function reaches values in excess of either 0.06 or 0.045. It has been argued here that the presence of pebble clusters in natural channels both strengthens the bed and increases flow resistance, so that critical shear values are higher than those determined under laboratory conditions (Reid & Frostick, 1984). This would mean a delay in the onset of bedload.

This apparent conflict may, in part, reflect the fact that the field data used by Andrews (and similar data used by others) describe the competence of the flow to keep the largest clast in an established traction carpet mobile during sustained high flows (bedload data were established with a portable sampler), whereas the Turkey Brook data are actually for *first* entrainment of the bed sediments by discrete rain-fed floods, as sampled with a recording and permanently installed pit-sampler.

One problem encountered in establishing entrainment thresholds is the difficulty of arriving at a satisfactory definition of initial sediment movement. A single grain might take a step and then rest, but no one would accept this as the threshold condition since minor adjustments of the bed might simply reflect a shift towards greater consolidation. This is a common experience in laboratory flume experiments where preparation of the bed is more often than not artificial. Vanoni (1964) observed that particles moved in bursts and considered a burst period of between 1 and 2 s as representing the fact that the traction threshold had been reached. Neill (1968) has cautioned that visual criteria tend to be more severe as particle size decreases because the area of bed under observation tends to remain the same regardless of bed material size, that is, the number of potentially mobile grains in the field of view increases as size decreases.

There are few field determinations because observation of the bed of a stream in spate is usually precluded by the turbid nature of the flow. In order to get around this problem, Helley (1969) trapped floats under cobbles in Blue Creek, California, which, when released, signified that the cobbles had been entrained. Reid *et al.* (1984) developed an electromagnetic sensor to establish the time of entrainment of metal-tagged clasts. However, both of these methods provide data for the movement of individual clasts and not the onset of general bedload.

The continuous slot samplers permanently installed in the bed of Turkey Brook have, however, provided information on the first general movement of the bed material, and this throws some light on the problems of predicting bedload discharge. There are, in fact, two conditions of 'entrainment': the first is that which is traditionally associated with first motion and involves, among other things, a consideration of the inertia of a static bed sediment; the second, is that which is associated with transportation and sedimentation rather than first motion and which reduces the consideration of inertia (Francis, 1973). The magnitude of the first might be derived by observing the initial motion of the bed sediment as a flood wave develops; the magnitude of the second might be assumed to be associated with the last flux of bedload as the flood wave wanes. The bedload data for Turkey Brook shows that, on average, the *depositional* threshold is only 35% of the *entrainment* threshold (Reid & Frostick, 1986a).

The recognition of multiple entrainment thresholds is not new. Although for different mechanical reasons, Bagnold (1941) was able to describe a 'fluid threshold' and an 'impact threshold' for aeolian sediment transport. However, all the fluvial sediment transport equations that relate sediment discharge to excess shear incorporate only a single threshold. The fact that these lie between the appropriate entrainment and depositional thresholds because of the way they have been derived may be the reason why predicted bedload transport rates fall at least within one or two orders of magnitude of measured values. However, the fact that the need for a varying threshold has yet to be recognized may be a major cause of the disappointment over the performance of equations that are otherwise well-founded.

Bed consolidation

Undoubtedly one of the difficulties in arriving at a satisfactory assessment of entrainment conditions is the fact that the bed of a natural channel may be in various states of consolidation. Church (1978) considered that freshly deposited sediments are 'underloose' and that, in this condition, they closely resemble the state of laboratory flume beds. Subtle rearrangement of particles during low flows increases the interlock between neighbouring grains so that the strength of the bed increases with time. Church dubbed this condition 'normally loose'. The bed may move towards a 'consolidated' state with further time. During this period, the movement of fine sediment into the voids between the framework clasts produces a matrix (Beschta & Jackson, 1979; Carling, 1984; Frostick *et al.*, 1984), and this helps to increase bed strength.

Where floods are separated by variable periods of time, the shear force required to initiate sediment transport can be expected to vary. Although there are considerable technical difficulties associated with attempting to measure this, it has been shown that the shear stress at initial entrainment can be up to three times the average when the first flood of the winter season follows summer months with no bed disturbance (Reid *et al.*, 1985).

Clast shape

The entrainment function of Shields (1936) was based on flume data derived for sediments ranging widely in their composition and, therefore, their density. Although the use of these different materials will have meant a variety of clast shapes, the strength of the argument embodied in the entrainment function was that there was no apparent segregation according to lithotype on the Shields diagram. Subsequent analysis has shown that this is too convenient a simplification (Li & Komar, 1986), and that lithological control over clast shape has important consequences for the entrainment threshold (Fig. 4.14). As might be expected, angular clasts sitting proud on a bed of similar clasts require greater shear to move than eliptical equivalents, while spheres will move before elipses. The reason is that the pivot angle through which a grain passes in order to remove itself from an interstitial niche in the bed is greater the further the grain deviates in shape from spherical. In addition to this, since imbrication is better developed as a packing structure with platey grains and that this further increases the pivot angle, all other things equal, the entrainment of river sediments in some metamorphic terrains might be expected less fre-

quently than in regions where the country rock produces clasts of more equant shape.

Pulses in bedload transport rates

A simple mechanical view of bedload transport — inevitably characteristic of almost all predictive equations — leads to an assumption that bedload discharge will remain approximately constant if flow is steady and exceeds the traction threshold. Equally, if flow were to increase and decrease, as during the passage of a flood wave, it would be expected that rates of bedload transport would mimic the changing pattern of water discharge. However, it has been known for some time that bedload fluctuations occur in a way that often appears independent of flow conditions. Some early records on the Rivers Danube, Inn and Rhine (Ehrenberger, 1931; Muhlhoffer, 1933; Einstein, 1937) had suggested a regular periodicity of sediment flux under more or less steady flows, but there had been doubts expressed about the efficiency of the samplers being used. More recently, there have been numerous reports of unsteady bedload transport. Gomez *et al.* (1989) have given a useful summary from which it can be seen that the sediment transport pulse can have a period ranging from seconds up to several months (Table 4.1). In some cases there is no doubt about sampler efficiency, and this has increased the level of confidence in the reality of the phenomenon (Hayward, 1979; Reid *et al.*, 1985; Custer *et al.*, 1987; Tacconi & Billi, 1987; Fig. 4.15).

An explanation of unsteady bedload discharge usually invokes the migration of bedforms — large, intermediate or small. If measurements have been made in a natural channel, more often than not the

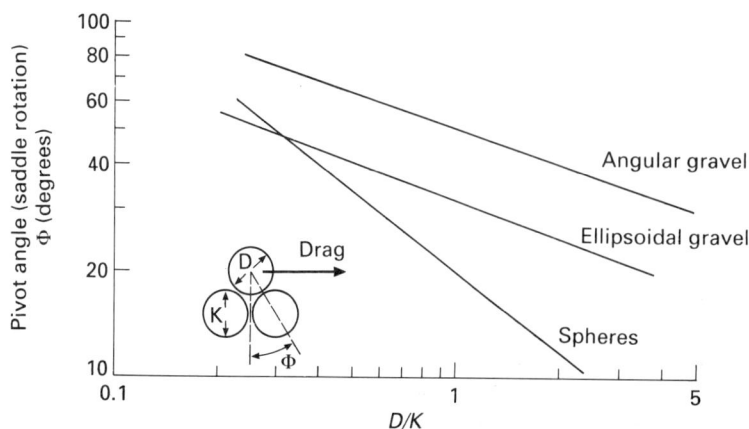

Fig. 4.14 The pivot angle of a grain with diameter D moving from a rest position within a bed niche through the saddle between two subjacent downstream grains of diameter K. After Li and Komar (1986).

Table 4.1 Timescales of and mechanisms responsible for generating temporal variations in bedload transport rates that are independent of variations in water discharge. Figures in italics indicate approximate period of perturbation in hours. After Gomez *et al.* (1989)

Timescale	Probable cause of temporal variability	Author	
		Field observations	*Laboratory or theoretical model*
Long/ intermediate-term	Wave-like translation of bed material through a reach	Gilbert (1913) Hayward and Sutherland (1974) Ashida *et al.* (1976) Mosley (1978) Griffiths (1979) Hayward (1979) Arkell *et al.* (1983) Meade (1985) Nakamura (1986) Roberts and Church (1986)	Whittaker and Davies (1982) Pickup *et al.* (1983) Ashmore (1987) (*4 to 8*) Kelsey *et al.* (1987) Whittaker (1987)
	Intraseasonal exhaustion	Nanson (1974) Leopold and Emmett (1976) Knott *et al.* (1987)	
Short-term	Intraevent exhaustion Scour and fill	Károlyi (1957) Andrews (1979) Jackson and Beschta (1982) Leopold and Emmett (1984)	
	Movement of bedforms (e.g. bars, dunes, sheets, particle clusters)	Ehrenberger (1931) (*0.3*) Muhlhofer (1933) (*0.1*) Nesper (1937) (*0.75*) Swiss Federal Authority (1939) (*1.0*) Zanen (1967) (*3.5*) Solov'ev (1969) (*0.2*) Emmett (1975) (*0.4*) Leopold and Emmett (1977) Beschta (1981) Lekach and Schick (1983) (*0.5*) Ying (1983) (*0.45*) Carey (1985) (*0.5*) Reid *et al.* (1985) (*1.7*) Custer *et al.* (1987) (*<0.3*) Tacconi and Billi (1987) (*0.5*) Whiting *et al.* (1988) (*0.25*)	Einstein (1937) (*0.16 to 20*) Skibinski (1968) Gibbs and Neill (1972) Mizuyama (1977) Hamamori (1962) Ikeda (1983) Hubbell *et al.* (1987) (*0.7 to 2.0*) Iseya and Ikeda (1987) Naden (1987) Kuhnle and Southard (1988) (*0.05 to 0.42*)
	Armouring	Milhous and Klingeman (1973) Graf and Pazis (1977) Gomez (1983)	Harrison (1950) Parker *et al.* (1982) (*1.0*)
Instantaneous	Mechanics of particle entrainment and motion	Sayre and Hubbell (1965)	Einstein (1937) Kalinske (1947) Stelczer (1981)

cause is conjectural because direct observation of the bed has been precluded by the turbid nature of the flow. A notable exception can be found in the East Fork River, Wyoming, where excellent bedload transport data have been matched by a continual survey of sand waves which migrate downstream in response to the seasonal snow-melt runoff regime (Leopold & Emmett, 1976; Meade *et al.*, 1981). Other direct field measurements confirm the spasmodic nature of bedload at much smaller scale. In these instances, un-

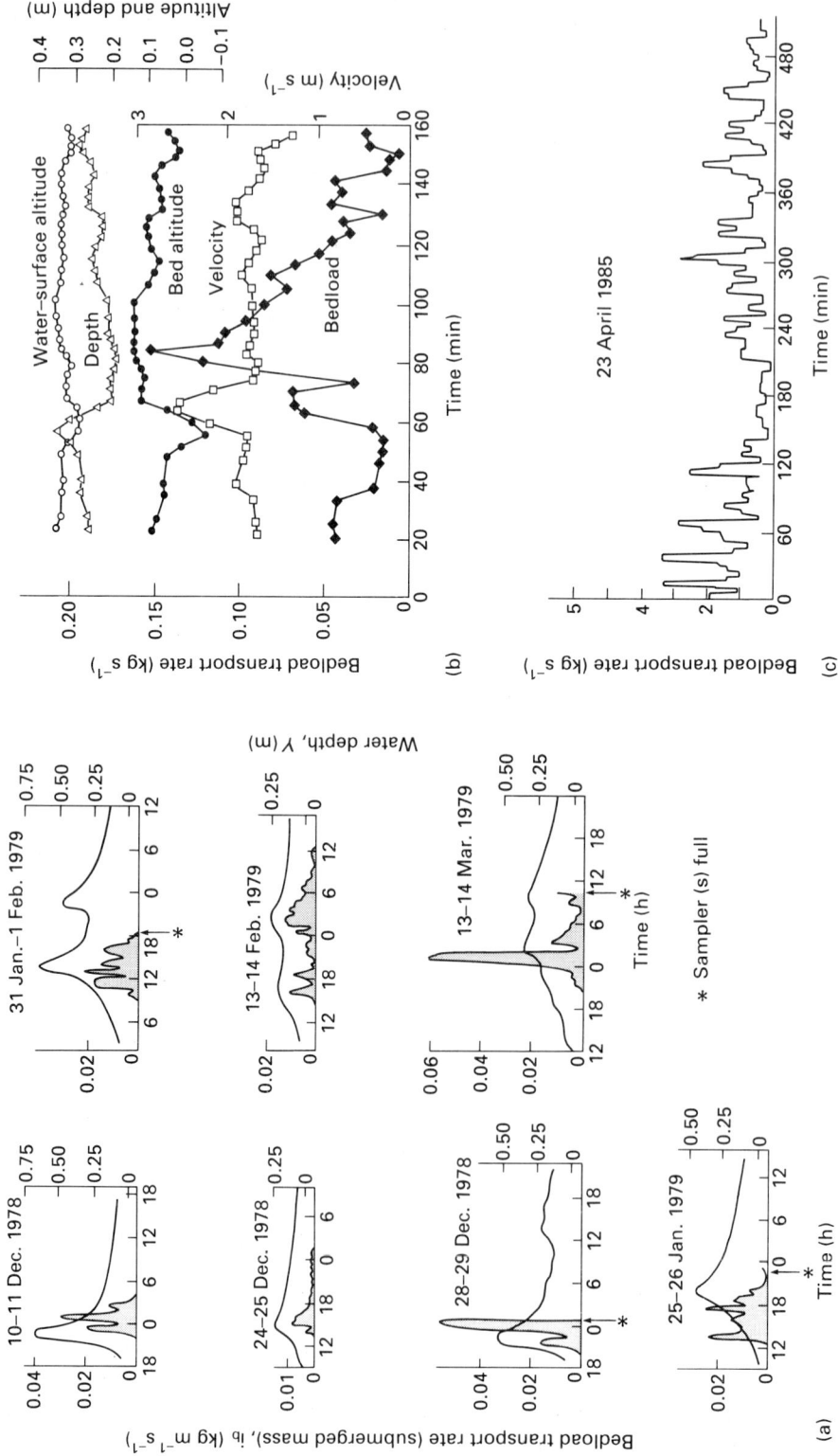

Fig. 4.15 Pulses in bedload transport in gravel-bed streams. (a) Turkey Brook, England. After Reid *et al.* (1985). (b) Slate Creek, Idaho. After Emmett (1975). (c) Virginio Creek, Italy. After Tacconi and Billi (1987).

steady transport rates are attributed to either the burst-sweep of eddies impinging on the bed (Sayre & Hubbell, 1965; Ergenzinger & Custer, 1983; Custer *et al.*, 1987) or the break-up of pebble clusters (Reid *et al.*, 1984). As yet, there is no directly observed explanation for the cause of fluctuations in bedload at intermediate timescales, that is, with periods of between 0.5 and a few hours. However, recent laboratory flume studies have identified 'dune' migration as one possible cause in gravel beds (Iseya & Ikeda, 1987; Kuhnle & Southard, 1988), a factor which plays a more obvious role on sand beds under lower flow regime conditions. Attention has been paid also to low-amplitude *gravel sheets* which appear to be congregations of marginally coarser bed material that can be distinguished from intervening streets of finer sediment and which have axes that lie transverse to the flow (Whiting *et al.*, 1988).

4.5   Fluvial suspended load

Of the three major transport loads carried by rivers — bedload, suspended, and dissolved — suspended sediment is by far the largest. One of the reasons for its supremacy is the fact that non-cohesive fine sediments of silt and fine sand have low thresholds of entrainment. So, for example, fine sand with an average diameter of 0.2 mm may be entrained when shear stress reaches a value of around 0.2 Pa, whereas gravel with an average diameter of 20 mm may not move until shear reaches around 2 Pa. Still finer particles of clay size have much higher traction thresholds when bound to other clays by cohesive forces and especially when consolidated either through desiccation or overburden pressure. However, the fact that they dominate suspended sediment load is a testimony to the efficacy of processes other than those associated with fluvial traction.

In fact, clay particles or floccules are often made available by the weathering and erosional processes that operate on the hillslopes of a drainage basin and are carried to the channel system by overland flow. Much of the suspended load may be in train, therefore, before water enters the river's first-order tributaries. Sediment concentrations in overland flow have been reported as high as 60% (Gerson, 1977), though values as high as this may reflect, in part, the hand of Man in accelerating the erosion process and tend to occur in semi-arid environments. Besides hillslope sources, the channel bed and banks also provide fine particles. In fact, the suspended sediment load is significant, if not necessarily spectacular, even at low

flows, and concentrations of several hundred mg l⁻¹ (< 0.1%) are often reported for perennial streams in humid temperate environments during the long periods that intervene between floods. Unlike bedload, which may contribute to total sediment transport for much less than 1% of the time in some streams, suspended sediment load almost never falls to zero values, at least in perennial systems.

4.5.1   Mechanism of suspension

It used to be thought that the lift force generated by pressure differences on the top and bottom sides of a particle originally at rest on the bed of the stream was primarily responsible for moving clasts up into the flow (Jeffreys, 1929). However, there is considerable scepticism as to the precise role of the lift force in entrainment mechanics, some believing that it merely assists the drag force by partially counteracting the particle body force. In fact, Laursen (1958) considers suspension as an accidental consequence of transport as bedload. Particles are envisaged as losing contact with the bed momentarily as they move over an irregularity (e.g. a dune) thereafter being swept up into the free stream. Sutherland (1967) proposes a different mechanism based on detailed observations of the flow. He is of the opinion that suspension arises as turbulent eddies penetrate the laminar sublayer to rake the bed. Grass (1971) was able to provide a vizualisation of the process which he dubbed *burst–sweep* and which involved, first, the sweep of an eddy along the bed and its subsequent burst away from the boundary. Obviously, if turbulence increases in intensity with flow velocity, the frequency of such encounters will increase as flow increases. All other things equal, the amount of material suspended from the bed should increase as flow increases.

Once suspended, the fate of a particle, whether a single grain or a floccule, will depend upon its fall velocity. This will be controlled largely by its submerged mass, but will be affected also by its shape. Set against this is the vertical component of flow associated with each turbulent eddy. However, the balance of forces means that there should be a sediment concentration gradient, at least in deep flows. The mass of sediment per unit volume will be much higher close to the bed and will decline rapidly with height in the flow in an inverse image of the velocity profile. This has been confirmed by sampling on large rivers such as the Mississippi (Fig. 4.16). In shallower, highly turbulent flows, the concentration

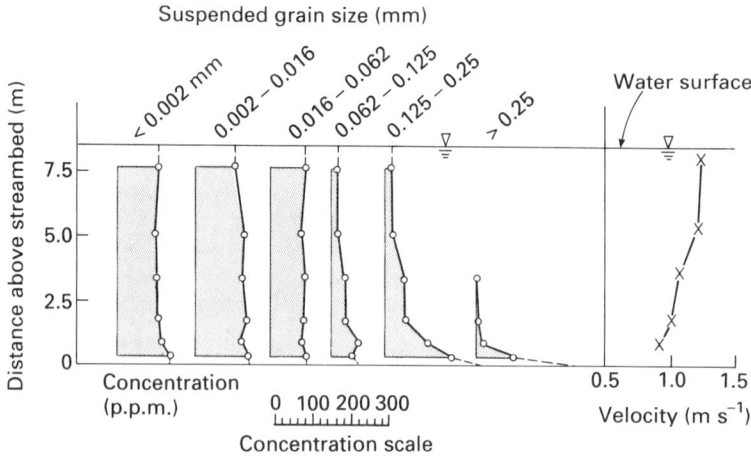

Fig. 4.16 Concentration of suspended sediment size fractions and water velocity in a vertical section through the River Mississippi at St Louis, Missouri, 24 April 1956. p.p.m., Parts per million. After Colby (1963).

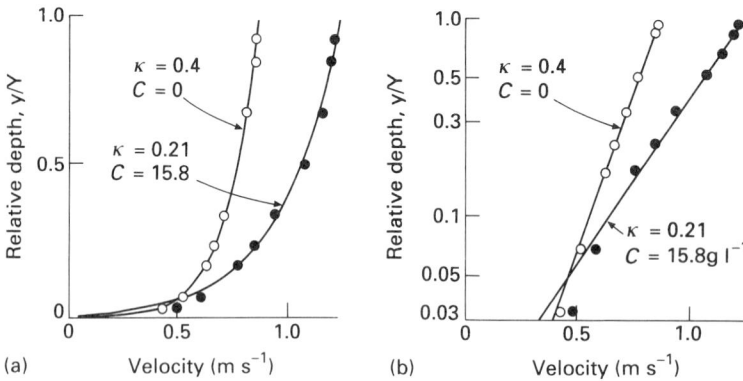

Fig. 4.17 Effect of suspended sediment concentration on the shape of the water velocity profile. κ is the von Karman constant, and C is the average sediment concentration. After Vanoni (1953).

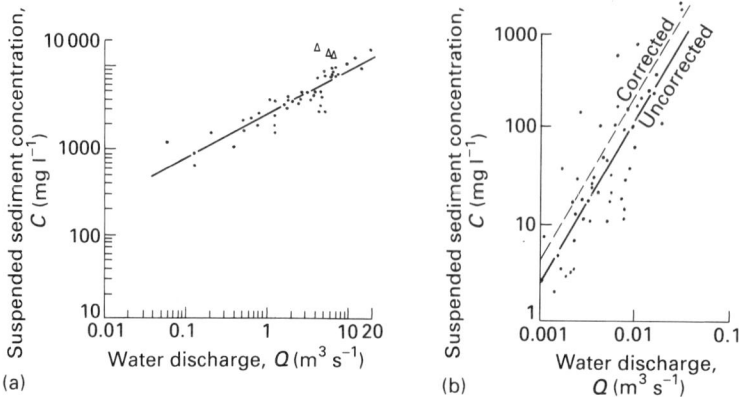

Fig. 4.18 (a) Rating of suspended sediment concentration against water discharge in the flash flood regime of the ephemeral Il Kimere, northern Kenya. △, Flood bore. After Frostick *et al.* (1983). (b) Suspended sediment rating curves for the Merevale Stream, England. That labelled *uncorrected* is the standard least-squares regression of log C on log Q; that labelled *corrected* is derived after applying a statistical correction that allows for the bias introduced by the use of logarithmic values of C. After Ferguson (1986).

Table 4.2 Relationship between total concentration of suspended sediment, C, and water discharge, Q, $C = aQ^b$, in arid-zone ephemeral and temperate zone perennial rivers. After Frostick *et al.* (1983)

Reference	River	Environment	Const. a	Exp. b
Frostick *et al.* (1983)	Il Kimere, N. Kenya	Arid	2570	0.512
Nordin and Beverage (1965)	R. Grande, N. Mexico	Arid	100	0.700
Negev (1969)	N. Qishon, Israel	Arid	4217	0.159
Nordin (1963)	R. Puerco, N. Mexico	Arid	80 000	0.200
Müller and Forstner (1968)	Alpenrhein, Austria	Temperate, humid	0.004	2.200
Jordan (1965)	Mississippi, Missouri	Temperate, sub-humid	0.01	1.600
Fahnestock (1963)	White R., Washington	Temperate, humid	40	2.500
Bauer and Tille (1967)	Helbe, Germany	Temperate, humid	31	1.391

gradient may be less pronounced due to greater mixing.

The mechanics of suspension are amenable to theoretical treatment as shown by Vanoni (1946) and Laursen (1958), among many others. In reality, however, many factors are either unpredictable or too complex to model without excessive simplification. For example, the generation of turbulent eddies is controlled in part by changes in bulk properties of the flow which may themselves reflect changes in suspended sediment concentration. For instance, the gradient of the velocity profile appears to be steepened near the bed when suspended loads are high, that is, when the density of the suspension rises appreciably above that of clear water (Vanoni, 1953; Fig. 4.17). Turbulence is dampened as kinematic viscosity (η/ρ) increases. Just as turbulence varies in intensity, so the fall velocity of a single grain varies as it descends through a water column, the density of which is changing as a function of differences in sediment concentration. Besides these and many other complications, the fact that much suspended material originates outside the channel, that it is delivered as wash load and, therefore, without regard for local hydraulic conditions, means that predicted concentrations of suspended sediment often fail to match actual concentrations that have been obtained as a result of a field sampling programme.

Because of this, much effort has been devoted to establishing empirical correlations between sediment concentration and water discharge, river by river. Discrete values of the two variables are plotted after logarithmic transformation and a least-squares regression analysis is carried out to provide a *sediment rating curve* which takes the form:

$$C = aQ^b, \qquad (4.8)$$

where C is sediment concentration, and Q is water discharge; a is a constant and b an exponent, both derived empirically (Fig. 4.18). Exponent b has been shown to range narrowly around a value of 2, although Leopold and Maddock (1953) found that it rose to a value of 3 for several North American rivers. Levels such as these may describe the behaviour of perennial rivers, but there is some evidence that ephemeral streams differ significantly (Frostick *et al.*, 1983). Data are sparse because few have sampled flash floods in arid regions. Nevertheless, it would appear that, here, exponent b takes on a value of less than unity, that is, the slope of the sediment rating curve is much less than its perennial stream counterpart. On the other hand, the value of constant a is also very different: where flow is perennial, it lies at a level < 50 and, in many cases, much lower than this; for desert flash floods, it may lie two or three orders of magnitude higher at several thousands or several tens of thousands (Table 4.2).

4.5.2 Size of suspended sediment

Much less interest has been shown in assessing the size of suspended sediment than in describing its concentration or its downstream flux. This may have arisen because most sampling programmes have been established on perennial rivers in temperate latitudes where comparatively low concentrations mean that the amount of material obtained from a manageable water sample of a few litres or less makes conventional size analysis inappropriate. Despite this, a number of discrete sampling programmes appear to have confirmed the cross-sectional or vertical patterns that might be expected from a theoretical consideration of the balance of forces. So, for example, on the

Mississippi at St Louis, the average size of the suspension decreases away from the bed because greater fall velocity ensure that sand is carried upward less far by turbulent eddies (Colby, 1963). In a shallow sand bed such as the Niobrara in Nebraska, where water depth is less than 1 m, Colby and Hembree (1955) were able to show that the average size of the suspension mirrored the cross-sectional pattern of water velocity, increasing towards the main flow cell.

In contrast, samples taken over a wide range of discharges during a gauging programme produce an ambiguous relationship between suspended sediment size and flow, at least in perennial streams. Some studies such as those on the River Clyde in Scotland (Fleming & Poodle, 1970), the Broad River in Georgia (Kennedy, 1964) and the River Exe in England (Walling & Kane, 1982) show no significant change in size with water discharge (Fig. 4.19). The records for the Niobrara River, Nebraska (Colby & Hembree, 1955) indicate that the median size falls with increasing discharge. In this case the decrease in sediment size is thought to reflect the infux of clay originating on the slopes of the water catchment and brought to the river channel by overland flow during large storm events.

The pattern appears to be clearer in the case of ephemeral streams, although, again, it should be remembered that the number of studies is small, perhaps because the duration of flash floods and their unpredictability reduces the likelihood of being on-site during flood flows. Desert streams appear to behave in a fashion that would be predicted from a theoretical consideration of the forces involved, at least in the case of sand- and silt-sized particles. The behaviour of clay particles seems to be less determinate even here, probably because clays are largely brought to the stream by overland flow on adjoining hillslopes. However, the average size of the suspended load generally increases and decreases in sympathy with flow (Nordin, 1963; Lekach & Schick, 1982). In the case of the sand bed Il Kimere in northern Kenya, the average size of the suspension has been measured as 0.002 mm at low flows, rising to 0.032 mm as sand is added, much of this from scour of the bed (Reid & Frostick, 1987).

A note of caution should be added here since the hydrodynamics of the suspension depend on the particle fall velocity, which is determined by grain size in the absence of significant differences in grain

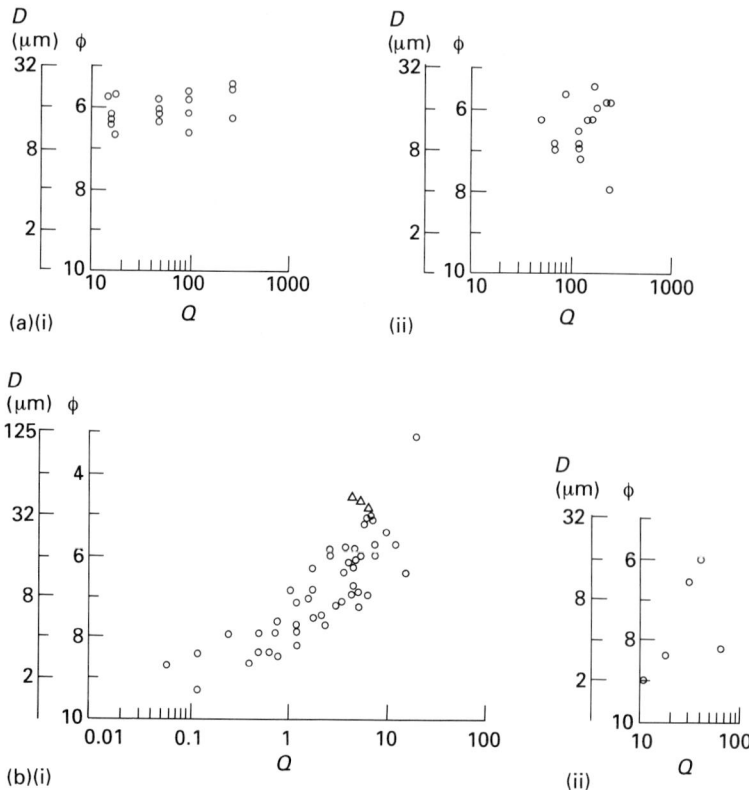

Fig. 4.19 Size of suspended sediment as a function of flow showing a contrast between (a) perennial rivers of temperate humid regions and (b) ephemeral rivers of sub-tropical and tropical semi-deserts. (a) Perennial rivers: (i) River Clyde, $r = 0.389$; (ii) Broad River, $r = 0.134$. (b) Ephemeral rivers: (i) Il Kimere, $r = 0.826$; (ii) Rio Puerco, $r = 0.503$. $\triangle$, Flood bore; Q, water discharge ($m^3 s^{-1}$); D, mean particle size.

shape. Analytical procedures aim specifically at segregating primary particles, yet the cohesive properties of clays mean that they probably travel as floccules with greater fall velocity than their individual constituent particles. Indeed, Nanson *et al.* (1986) have drawn attention to mud deposits of Cooper Creek in the Lake Eyre basin, Australia, that consist of soil peds whose size dictates that they are more than likely to be transported as bedload rather than in suspension. However, although suspended 'clays' may be carried as floccules that act as small silt-sized particles, present analytical procedures make it difficult to determine the exact proportion of material classified as of clay size that is involved — a problem that has been pointed out by Flint (1972) and Walling and Moorehead (1987).

4.5.3 The suspended sediment load prediction problem

Just as with bedload, our understanding of suspended load dynamics has been improved greatly by investigating the processes that complicate the relationship between sediment flux and water flow. Indeed, a high degree of the scatter of coordinates on a sediment rating curve might now be expected as normal, and analysis will be directed at partitioning the dataset in order to establish a number of sediment rating curves that express, for example, seasonal differences in suspended sediment concentration (Walling, 1977; Fig. 4.20).

Non-uniqueness of the sediment load/water discharge relationship

Walling (1974) was among the first to champion frequent sampling for suspended sediment, in part because he had found a repeating pattern of hysteresis between sediment concentration and water discharge during floods on a number of perennial rivers with drainage basins of small to moderate size (i.e. less than 10^3 km^2). In general, sediment concentrations are higher on the rising limb of a flood hydrograph than at corresponding levels of discharge on the falling limb, the difference reaching as much as two orders of magnitude (Fig. 4.20). It is believed that sediment is flushed from accessible storage sites (bed, banks, and the riparian zone) in the early stages of a flood event, after which it is surmised that sediment availability becomes a limiting factor on the amount of material in suspension.

In fact, Negev (1969) has pointed to declining sediment supply as the cause of drift in the sediment rating curve from one flood to the next in ephemeral streams of northern Israel. This progressive exhaustion of available sediment has also been noted as a possible cause of scatter in the lumped data sets of small perennial streams (Wood, 1977). On an even longer timescale, a seasonal shift in the rating curve can be detected. Summer sediment concentrations tend to be higher than those of winter at the same level of water discharge (Walling, 1974).

Flood wave celerity

In larger drainage basins, the sediment concentration wave may become progressively out of phase with the flood wave as each passes downstream. Heidel (1956) illustrates this phenomenon for the Bighorn River which flows from Wyoming to Montana. In one flood, discharge and sediment concentration maxima are more or less coincident at Manderson, but at Bighorn some 270 river-kilometres downstream, they are 13 h apart. On this particular occasion, the celerity of the flood wave was 1.6 m s^{-1}, while the sediment wave representing flow velocity appeared to be travelling at 0.8 m s^{-1}, or about half the speed (Fig. 4.21). The pattern is complicated at other times by the influx of sediment-laden waters from downstream tributaries. However, the general lesson is clear: the higher the proportion of suspended sediment consisting of wash load, that is, originating outside the river channel, and the larger the drainage basin, the lower the likelihood of establishing a relationship between sediment flux rate and a local hydraulic parameter such as water discharge.

4.5.4 High and hyperconcentrations of suspended sediment

In perennial streams, the concentration of suspended sediment rarely exceeds a few thousand milligrams per litre unless there has been severe disturbance of the vegetation or careless agriculture that has led to accelerated soil erosion. In contrast, ephemeral streams almost invariably carry high sediment loads in suspension. Even during low flows on the recession limb of the flood hydrograph, concentrations may reach several thousand milligrams per litre (> 0.1%). This accounts for the high values of constant *a* in the sediment rating curve (Eq. 4.8). One of the reasons for this disparate behaviour is a significant difference in the path taken by the majority of rainwater.

Overland flow with its erosive potential may be

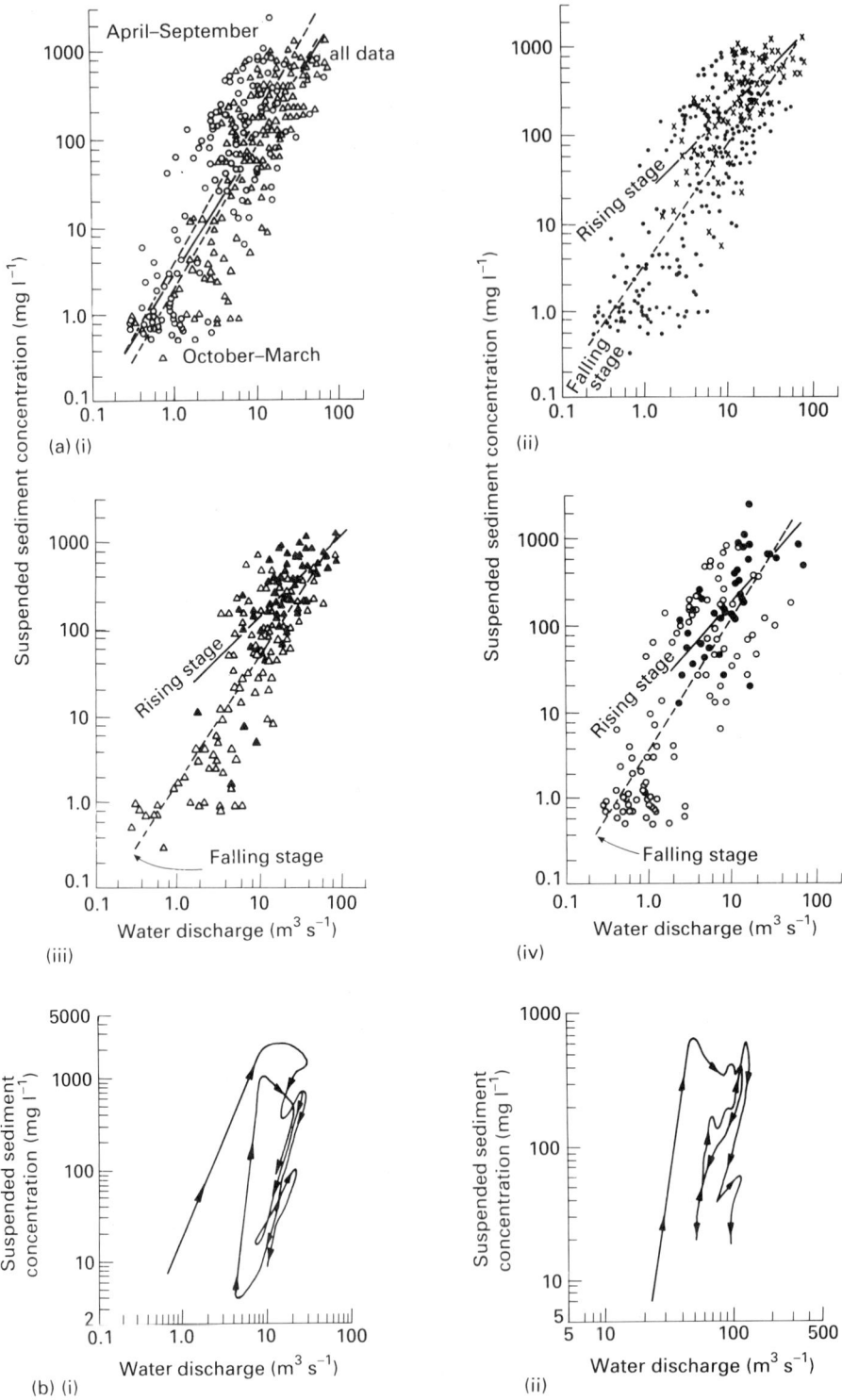

Fig. 4.20 (a) Suspended sediment rating curves for the River Creedy, England, showing the effect of separating data (a) (i) according to season. ○, April–September; △, October–March. Waxing (ii) and waning flood stage. x, rising stage; ●, falling stage. (iii) Flood stage in winter (October–March). ▲, rising stage; △, falling stage. (iv) Flood stage in summer (April–September). ●, rising stage; ○, falling stage. (b) Hysteretic relationship in the suspended sediment/water discharge relationship during single flood events on the Rivers Dart (i) and Exe (ii), England. After Walling (1977).

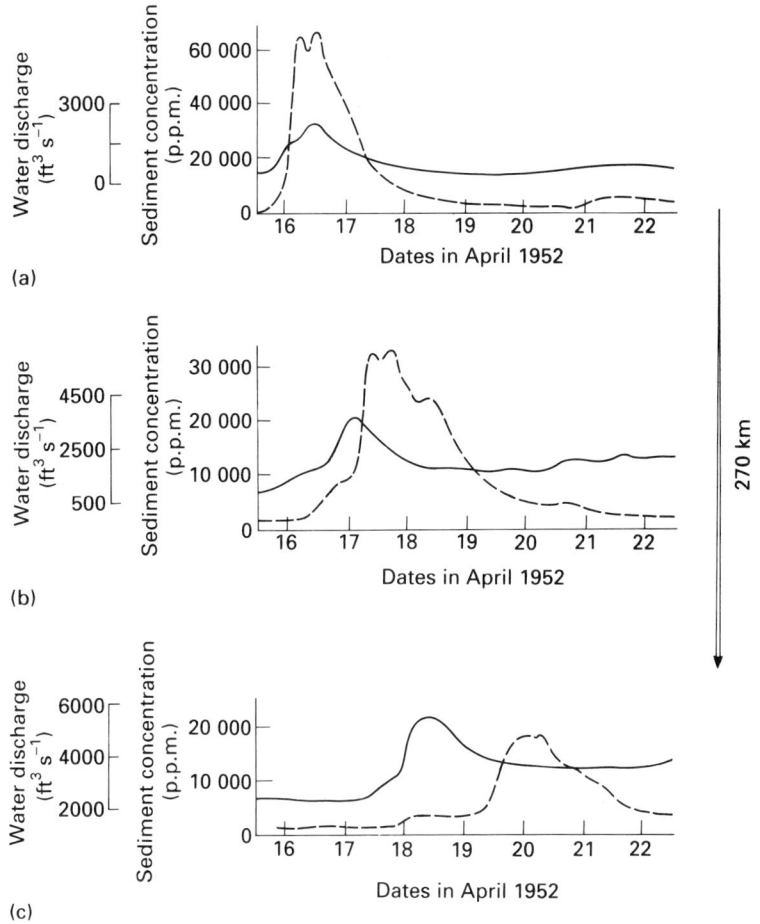

Fig. 4.21 Water discharge and suspended sediment concentrations at three stations along the Bighorn River, Wyoming and Montana, during a headwater-generated flood event. (a) Manderson, Wyoming; (b) Kane, Wyoming; (c) Bighorn, Montana. – – –, Sediment concentration; —, water discharge. After Heidel (1956).

as frequent as rainfall in arid zones, but the high infiltration rates characteristic of well-structured soils in humid zones ensures that, here, water is generally transmitted much less destructively as *soil interflow.*

Concentrations of suspended sediment may regularly reach several ten thousand parts per million as a flash flood waxes. However, much higher loads have been reported. Lekach and Schick (1982) measure a rising stage value of 28.5% (285 000 p.p.m.) for a small stream in the southern Negev Desert. This is exceeded by a value of 41% for the Paria River of Arizona (Rainwater, 1962) and 68% for the Rio Puerco of New Mexico (Bondurant, 1951). In fact, Beverage & Culbertson (1964) have argued that sediment concentrations in excess of 40% but less than *c.* 80% should be dubbed *hyperconcentrated* since the hydraulic properties of such a sediment–water mixture differ from those of both clear water and flows with lower concentrations. Above 80%, the

fluid becomes non-Newtonian in character and starts to behave as a pseudo-plastic typical of mudflows and debris flows.

There is some controversy over the exact nature of hyperconcentrated flows. Some suggest that the high viscosity which develops because of the increase in shear force required to move one parcel of sediment–water mixture against another dampens turbulence completely. Such flows have been observed and described as 'oily' (Pierce, 1917). However, Hino (1963) considers that the flow remains turbulent but that its viscosity leads to a decrease in the lifetime of individual eddies. This view is supported by Bondurant's observations of the Rio Puerco where surface boils and other indications of turbulence were noted even when suspended sediment reached hyperconcentrated levels.

Whatever the nature of turbulence during hyperconcentrated flows, Bagnold (1954) has demonstrated that the dispersive stress acting between grains is itself partially capable of supporting the suspended load. This, together with the reduction in density contrast between fluid and sediment, is undoubtedly one reason why the sand fraction of the suspended load increases with increasing concentration (Lekach & Schick, 1982), and why the range of particle size carried by ephemeral flows is greater than those of perennial streams (Reid & Frostick, 1987).

4.6 Sediment yield

Many geomorphologists have equated sediment yield

with the efflux of suspended sediment from a drainage basin. This reflects, in part, the comparative ease of establishing the concentration of suspended sediment in a sampling programme and, consequently, the availability of a large number of records for a wide variety of rivers. The other component of sediment yield — bedload — has often been ignored unless estimates have been derived from repeated surveys of reservoir sedimentation. However, evidence suggests that suspended sediment accounts for more than 90% of the total load in perennial streams (where most investigations have been conducted). Fortuitously, therefore, there is some validity in the presumption that the flux of suspended sediment can

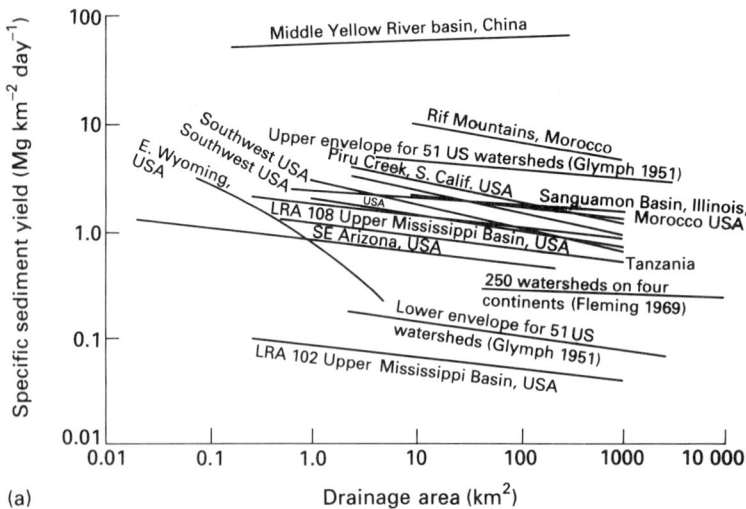

(a)

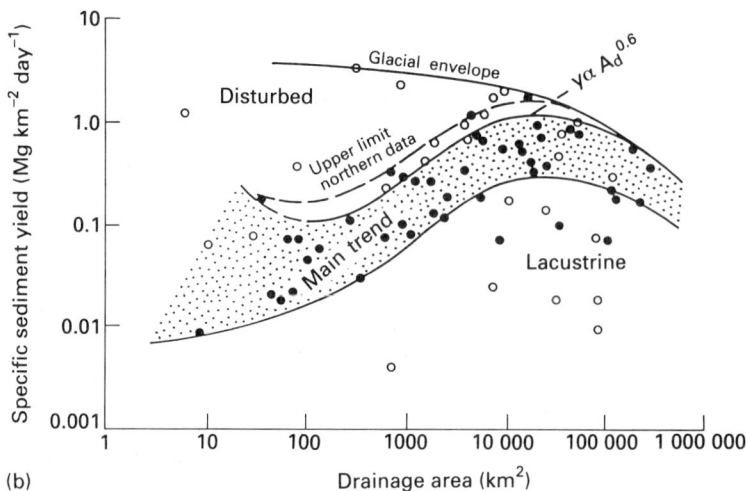

(b)

Fig. 4.22 (a) Compilation of clastic sediment yield curves showing the expected decrease in yield with increasing water catchment area. (b) Specific sediment yields of water catchments in British Columbia showing the impact of a glacial legacy and a trend which runs counter to that derived in non-glacierized terrain. After Church *et al.* (1989).

be used as a measure of sediment yield in this environment. In contrast, ephemeral streams in arid zones may shift large quantities of sand and gravel as bedload. Where this is supplied to an exotic perennial crossing a desert region, the sediment yield will be higher than anticipated from measurements of the suspended load alone. This can lead to serious engineering problems. The rapid siltation of Lake Mead on the Colorado River (Vetter, 1940) and the Elephant Butte reservoir on the Rio Grande (Vanoni, 1975) are celebrated examples. More recently, the huge Tarbella Reservoir on the Indus River has been shown to have lost *c.* 6% of its capacity in the first 5 years of its life as a result of sedimentation (Ackers & Thompson, 1987).

Specific sediment yield — the efflux of sediment from unit area per unit time — varies widely from one environment to another. If the hyper-arid and arid deserts are ignored because runoff here is extremely rare, there is a strong inverse relationship between sediment yield and rainfall (Langbein & Schumm, 1958; Fig. 4.22). Ironically, the greatest yields are derived from semi-arid terrain despite the infrequence of flash floods, while temperate forests only produce about one third as much sediment despite the much higher frequency of storm events and the fact that flow is perennial. The role of vegetation is multifold. However, even in a single climatic zone, comparative studies of the water balance in coniferous forest and grassland indicate that 40% less rainfall passes through to the river system where a tree canopy is effective in intercepting rainfall before it reaches the soil (Calder & Newson, 1979). Interestingly enough, undisturbed tropical rain forest yields sediment at about the same low level as vegetated temperate drainage basins.

The significance of vegetation has been highlighted by the effect of clearing vast tracts of forest, especially in equatorial regions (Douglas, 1967b). In this context, it should be remembered that 'benchmark' rates of sediment yield can rarely be established for truly natural terrain, untouched by human endeavour. It is difficult, therefore, to gauge how artificial our estimates of sediment yield are. For example, the Hwang Ho, feeding on the easily ravaged loess of western China, often carries hyperconcentrations of suspended sediment and can produce a daily yield as high as 1.2×10^9 t (Todd & Eliassen, 1938), but much of this reflects accelerated soil erosion due to agricultural landuse.

Besides this, Church *et al.* (1989) have drawn attention to the fact that some areas in middle, and almost all in high, latitudes suffer from the legacy of Pleistocene glaciation. This has left an accessible store of material in the form of valley deposits that are easily mobilized to give anomalously high values

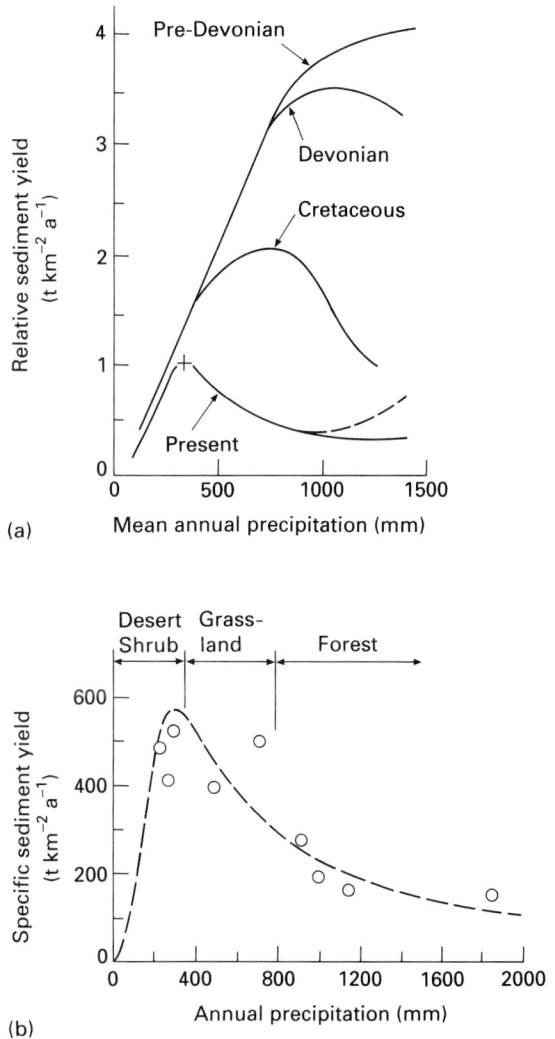

(a)

(b)

Fig. 4.23 (a) Postulated trends in sediment yield as a function of annual precipitation before significant colonization of the continents by plants (pre-Devonian), during the early stages of higher land plant evolution (Devonian), at the inception of the angiosperms, especially the grasses (Cretaceous), and under present conditions. +, Reference value. After Schumm (1968). (b) Present-day specific sediment yield in different bioclimatic zones of North America. After Langbein and Schumm (1958).

of sediment yield when judged against unglaciated drainage basins (Fig. 4.22).

On top of all these complications, Ferguson (1986) has drawn attention to the fact that the statistical method commonly used in establishing a sediment rating curve leads to an underestimation of the sediment load by as much as 50%. The degree of underestimation depends upon the amount of scatter in the bivariate plot of sediment concentration and water discharge (Fig. 4.18). However, it does mean that present estimates of continental erosion rates may be extremely conservative.

Nevertheless, looking forward to a time when global temperatures may be higher as a function of anthropogenically induced climatic change, affecting, as this may, the water balance and hence the zonal latitudinal pattern of vegetation, the areas yielding most sediment may shift towards what are currently the world's grasslands as the desert cores expand and as the grasslands change in character towards semi-desert.

Looking back in time, and adopting a cautious uniformitarian approach, it is interesting to speculate that turnover rates in the form of global sedimentary recycling might have been far higher before the evolution of the grasses (i.e. pre-Cretaceous) and much higher still before the advent of a significant number of land plants (i.e. pre-Devonian). In fact, sediment yields might have been universally closer to those of the present-day desert fringes during the Lower Palaeozoic (Schumm, 1968; Fig. 4.23).

4.7 Downstream changes in sediment calibre

As a general rule, the calibre of channel-bed material decreases with progressive distance from headwater reaches (Fig. 4.24). There is also some indication that suspended sediment decreases in size (Colby et al., 1956; Mapes, 1969). On the other hand, size sorting tends to increase in the same direction. It might be argued that elucidation of these downstream patterns

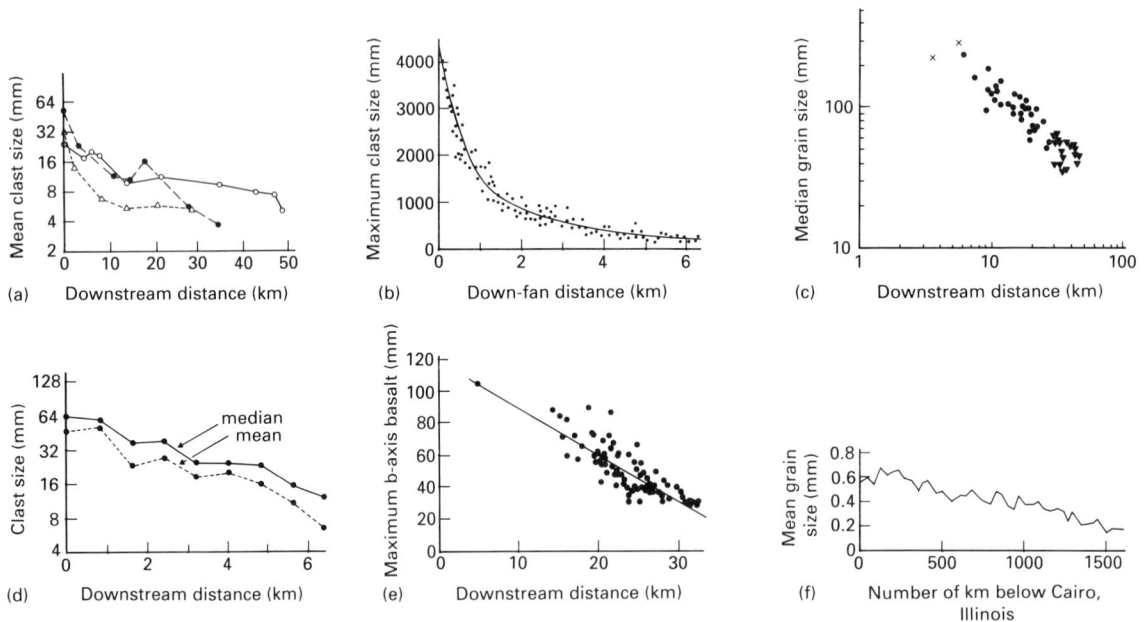

Fig. 4.24 Downstream decline in clast size in single-thread and braided gravel- and sand-bed channels. (a) Single-thread gravel bed: three tributaries of the Cheyenne River, Black Hills, South Dakota. After Plumley (1948). (b) An alluvial fan (braided gravel bed) of the Santa Catalina Mts, Arizona. After Blissenbach (1952). (c) Variable planform gravel bed; x, canyon; ●, braided; ▲, meandering. The Squamish River, British Columbia. After Brierley and Hickin (1985). (d) A braided gravel-bed reach of the Upper Kicking Horse River, British Columbia. After Smith (1974). (e) Braided gravel-bed: the Koobi Fora Plateau Gravel, northern Kenya. y = 119.528 − 2.910 x; r = − 0.840. After Frostick and Reid (1980). (f) Single-thread sand bed: the Mississippi River below Cairo, Illinois. After Mississippi River Commission (1935).

represents the most important single contribution that sedimentology has made to earth science since it provides an indication of both local fluvial environment and provenance in ancient river sediments that are now contorted or incomplete because of erosion. It also provides clues about past tectonic landforms.

Sternberg (1875), working on the River Rhine, was the first to formalize the relationship between downstream distance and bed sediment calibre in what has become dubbed 'Sternberg's Law':

$$W_L = W_0 e^{-CL}, \qquad (4.9)$$

where L is transport distance, W_L is clast weight at L, W_0 is the weight of the clast at $L = 0$, and C is a constant known as the coefficient of abrasion.

Subsequent work by others on a variety of rivers has confirmed the general pattern of downstream 'fining', although Brierly and Hickin (1985) consider that a power function is more appropriate as a statistical descriptor than the exponential relationship used by Sternberg. However, as with all grey or black boxes, there is considerable controversy over the relative importance of the processes involved. The gradual attrition of clasts, or their selective entrainment by variably competent flood flows have all been implicated. However, even after a plethora of studies, there is still uncertainty over the importance of each process.

4.7.1 Clast comminution

Mineralogy confers a range of different strengths on rock particles that are subject to mechanical abrasion and weathering during transport. Plumley (1948) was able to demonstrate the susceptibility of feldspar to comminution through abrasion in the Cheyenne River of Dakota, while Bradley (1970) considered that the weathering of clasts while stored in channel bars to be the primary cause of the disintegration of granitic material in the Colorado River of Texas. However, the processes producing downstream fining in these and other studies are almost always inferred. It was Keunen (1956) who demonstrated through a series of laboratory flume experiments that abrasion produced a progressive rounding and comminution. He also showed that there are considerable differences between different lithotypes, with, for example, gabbro surviving longer than limestone. The general form of the patterns found by Keunen in his 'endless' circular flume has been confirmed since by studies of actual river sediments. Among them, Dal Cin's

(1968b) analysis of the Piave River, an Alpine tributary of the Po in Italy, produced a downstream pattern of increasing clast roundness that was remarkably close to those of Keunen (Fig. 4.25).

4.7.2 Stream incompetence

In addition to the effects of attrition is the notion of 'stream competence' or, rather, 'incompetence'. River gradients decline from values of > 0.01 to < 0.0001 as they pass from their mountain headwaters to the ocean or to inland basins, and, although water depth and discharge generally increase downstream, the shear force acting on the river bed is more sensitive to the changes in longitudinal slope. It has been argued that the calibre of local bed material will reflect local competence among other things (Kirkby, 1977), and that the downstream reduction in slope, and hence shear, will involve a concomitant reduction in bed-material size.

Where a decrease in slope is accompanied by a decrease in the discharge carried by individual channels, as in a braided stream, the downstream reduction in competence may be even more dramatic. For example, Smith (1974) shows that median grain size falls at a rate of 6 mm km^{-1} on a braided reach of the Upper Kicking Horse River of British Columbia. Blissenbach's (1952) celebrated study of the alluvial fans of Arizona reveals a grain-size gradient of 215 mm km^{-1}, while Bluck (1964) gives gradients of 200 mm km^{-1} for a number of fans in Nevada. These can be compared, if only loosely, with the grain-size gradients reported for single-thread rivers. For example, Sternberg (1875) gives a value of 0.26 mm km^{-1} for the Rhine, while the Mississippi River Commission (1935) reports a gradient of 0.0001 mm km^{-1} for the lower Mississippi.

4.7.3 Size selective entrainment and transport

Inherent to the concept of (in)competence as a determinant of downstream fining is the notion that the probability of entrainment and the transport velocity of clasts are both related inversely to their size. The subject is currently being debated again among the engineering and geomorphological community. The difficulty of finding a resolution of the argument is bound up inevitably with the need to acquire incontrovertible field data and with the fact that the data available at present are contradictory.

Parker *et al.* (1982) and Andrews (1983) have

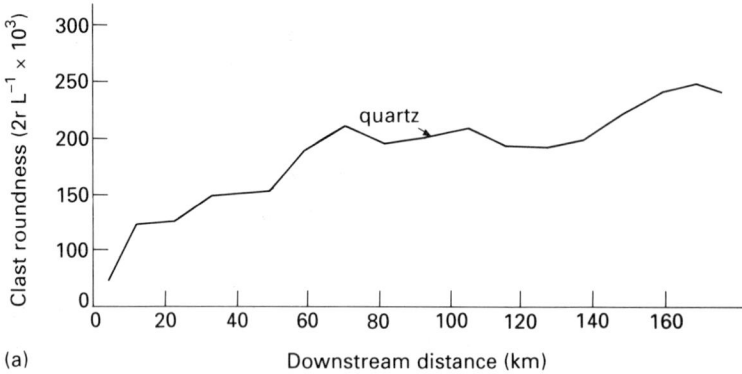

(a)

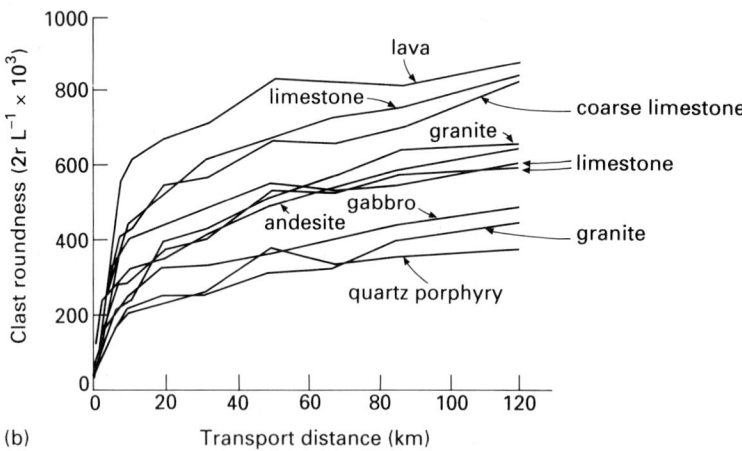

(b)

Fig. 4.25 Downstream rounding of clasts. (a) The Piave River, northern Italy. After Dal Cin (1968b). (b) Experiments in a circular flume channel. After Keunen (1956).

argued for equal or near-equal mobility of clasts in gravel-bed streams regardless of clast size. The rationale for this argument is that the greater body force of larger particles is offset by greater protrusion above the general level of the bed so that the fluid drag acting on them is commensurately increased. At the same time, the angle through which a grain must pivot in order to vacate a bed niche decreases with increasing grain size so that the turning moment is reduced and entrainment requires less stress. In contrast, smaller particles are sheltered from the flow to a greater or lesser degree by larger clasts so that they are moved less easily than a consideration of body weight alone would suggest. The net effect should be a bedload size-distribution that approximates that of the bed sediment.

Komar (1987a), in particular, has pointed to the fact that those datasets that are available to test the theory, indicate a departure from equal mobility at entrainment, and his arguments have been supported by Ashworth and Ferguson (1989) (Fig. 4.26). In mixed-sized sediments ranging from medium sands through to gravels, the larger the diameter of an individual grain, the larger is the stress required for its entrainment. Obviously, this promises to be more in keeping with the downstream fining pattern found in the majority of rivers.

However, the picture is far from clear-cut. Komar and Shih (1991) have noted that the size distribution of bedload in the gravel bed Oak Creek, Oregon, gets coarser, moving closer to that of the bed material, with increasing bed shear. This suggests a drift towards a condition of equal mobility with increasing flood discharge. Indeed, on the gravelly sandbed of the Arroyo de los Frijoles, Leopold et al. (1966) showed a tendency towards less size dependence in the flood dispersal of gravel with increasing flood peak discharge. Were this a general relationship, it would be tempting to conclude, albeit from a limited number of observations, that the downstream fining

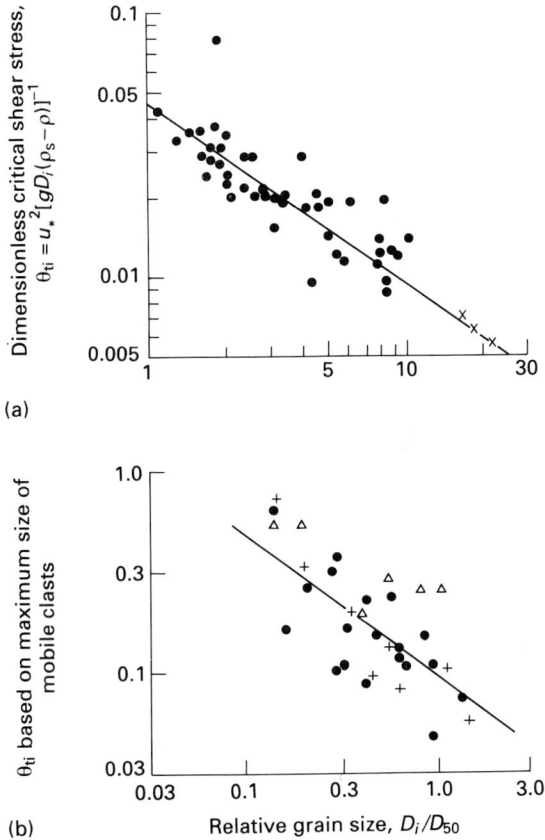

(a)

(b)

Relative grain size, D_i/D_{50}

Fig. 4.26 Dimensionless critical shear stress for clasts of different size (D_i) relative to the median diameter of the bed material (D_{50}) showing selective entrainment. A slope of -1 in the least-squares relationship would indicate equal clast mobility regardless of size. (a) Bedload data from the gravel bed Great Eggleshope Beck, England, where D_i is maximum size of moved clast and θ_{ti} is assumed to correspond to maximum flood stage.
Slope $= -0.68$; ●, bedload trap; x, tagged boulders.
After Komar (1987; data from Carling, 1983) (b) Bedload data derived with use of a portable Helley-Smith sampler with θ_{ti} based on contemporary flow parameters in two Scottish and one Norwegian braided gravel-bed rivers.
Slope $= -0.74$; △, Lyngsdalselva; ●, Dubhaig;
$+$, Feshie. After Ashworth and Ferguson (1989).

pattern found in most river systems is established by floods of moderate to high frequency and moderate to low magnitude. Indeed, by amalgamating data-sets from several rivers — some perennial, some ephemeral — Hassan and Church (1992) have shown that the transport distance of unconstrained clasts in gravel-bed rivers is very size dependent (at least for particles larger than the median size of the bed material) when transport rates are comparatively low.

This debate will undoubtedly continue for some time.

4.7.4 Complications to downstream fining

Not all studies of sediment calibre have reported downstream fining. In fact, a survey of studies covering 36 rivers of wide-ranging length and sediment texture and draining a variety of terrains indicates that just under 30 % show either downstream coarsening or no significant change in calibre (Frostick & Reid, 1979). Although no universal explanations can be adduced for either of these patterns, several factors are known to complicate the more usual downstream fining. Among these are the addition of material by tributaries which may have higher gradients and, hence, greater competence than the trunk stream at the point of confluence (Knighton, 1980; Ichim & Radoane, 1990); and the local reduction in trunk stream gradient and, hence, competence, by the outward growth of tributary fans (Dawson, 1988).

4.8 Fluvial placers

Fluvial processes have been responsible for producing many of the world's economic deposits of heavy minerals and precious metals. These range widely in age. For example, the gold palaeoplacers of the Witwatersrand are Proterozoic (Minter, 1978), the diamond placers of Swaziland are Triassic (Turner & Minter, 1985), the cassiterite placers of Indonesia, now submerged by the Holocene rise in sea level, are Pleistocene (Aleva, 1985; Fig. 4.27) as are the gold placers of the Yukon (Morison & Hein, 1987), while the diamond placers of the Central African Republic lie in Recent river terraces and modern river channels (Sutherland, 1984).

4.8.1 Concentration processes

It might be argued that the processes involved in placer formation are known, at least in broad terms. So, for example, Sutherland (1982) has drawn attention to the downstream fining of diamonds in the Central African Republic, while the flow separation associated with bedrock scour-holes appears to lead to preferential accumulation of diamonds (Hall *et al.*, 1985). However, assessments of the mechanical concentrating processes have been largely inferential and

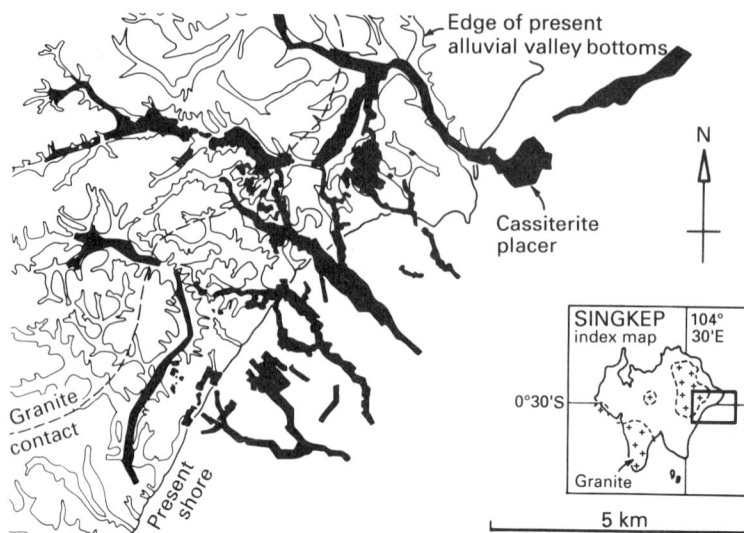

Fig. 4.27 On- and offshore distribution of the fluvial cassiterite placers of Singkep, Indonesia, showing the effects of sea-level change during the Pleistocene and Holocene. After Aleva (1985).

generally based on the patterns of occurrence commonly observed in exploration and exploitation.

This allowed an ascendency of the concept of *hydraulic equivalence*, developed by Rubey (1933) to account for the coexistence of grains of different density and size in single laminae of a deposit. Hydraulic equivalence is probably best dubbed *settling equivalence*, if only because, as Rubey himself pointed out, many factors control the selective entrainment or deposition of variously dense grains. Slingerland (1984) has drawn attention particularly to the interplay of relative grain exposure (with its implications for fluid drag) and the variable degree of inertia conferred on different grains by density contrasts. In an individual lamination, heavy minerals are often too large for settling equivalence with neighbouring quartz grains: the equivalent quartz grains, even if initially present, would have been too protrusive and would have been removed downstream by selective entrainment, leaving the placer grains alongside non-placer grains that are only marginally larger (Reid & Frostick, 1985). In this circumstance, it is undoubtedly more appropriate to consider the placer and non-placer grains to have *entrainment equivalence*, or, perhaps even more appropriately, *inertial equivalence*.

However, placer grains are often much smaller than either settling or entrainment equivalence would allow. For instance, Smith and Minter (1980) have demonstrated the apparent affinity of gold whose grain diameter may be only a few tens of

micrometres for pebble-sized gravels and the relative dearth of gold in the interbedded sands of the Witwatersrand palaeoplacers (Fig. 4.28). This fre-

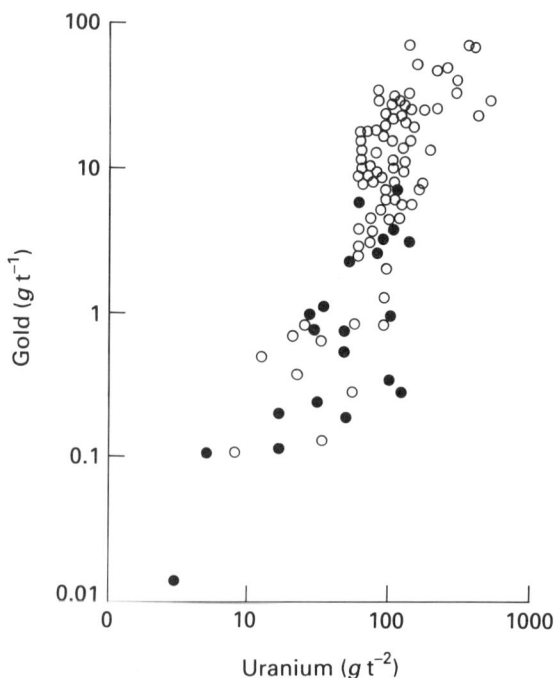

Fig. 4.28 Crossplot showing the affinity of gold and uranium for coarse fluvial sediments in the Witwatersrand placers, South Africa. ○, Conglomerate; ●, sandstone. After Smith and Minter (1980).

quent association of placer minerals with coarse gravels and conglomerates suggests that other concentrating processes are important. In this context, Minter and Toens (1970) have shown experimentally that the ingress of fine grains into the interstices of a gravel bed can enrich a traction carpet consisting of placer and non-placer sand. They speculate that this enrichment produces a matrix of economic significance further downstream and argue that this might be the reason why midfan rather than fanhead sediments are more important sites for placers in the Witwatersrand reefs.

4.8.2 Channel morphology and placer formation

Grain-scale processes obviously exercise considerable fundamental control over placer concentration. However, the fact that particular bedforms or other channel features form type-sites for placers in a wide range of rivers means that these parts of a river bed are conditioned to entrap heavy minerals and that hydraulic conditions recur frequently enough at these locations to ensure greater influx than efflux of the placer mineral. So, for instance, it may be that convergence or divergence of flow produces a local increase in stream competence and a concomitant increase in the average grain size of the bed material. Best and Brayshaw (1985) show that flow separation at river confluences produces a vortex train along the edge of the separation zone bar just downstream of the stream junction and that this leads to heavy mineral concentration (Fig. 4.29). Mosley and Schumm (1977) and Smith and Beukes (1983) highlight a similar process, but, here, flow is converging, in the one case, downstream of a confluence, and, in the second, at the tail end of a point bar. Nami and James (1987) point to the energy loss that occurs at the interface between channel and floodplain during overbank flows and, taking this into account, they develop a numerical model that mimics gold placer distribution in some of the Witwatersrand reefs. In this case, the placer is thought to have settled out of suspension from flood flows along narrow strips that border the low-flow channels (Fig. 4.30).

4.9 Depositional bedforms

The interplay of flow turbulence and bedforms exerts a strong controlling influence on both sediment transport and deposition. Leeder (1983) envisaged continuous feedback in the relationships between

turbulence, sediment transport and bedform development, each of the three affecting the other two in a very direct way (Fig. 4.31). For example, sediment transport has a dampening effect on turbulence and, thereby, affects flow separation and bedform generation indirectly; sediment transport rates vary across individual bedforms as a function of form-induced fluid acceleration and deceleration; and the development of bedforms influences flow structure and, therefore, turbulence, etc. No consideration of sediment transport and deposition can thus ignore the influence of bedforms. To do so would run the risk of providing an inadequate explanation of the processes involved.

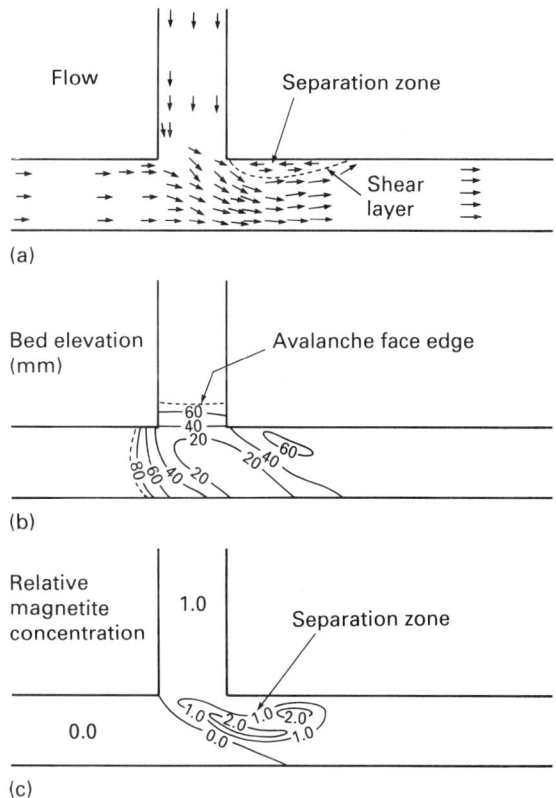

Fig. 4.29 Flow pattern (a), bed elevation (b) and 'placer' concentration (c) at the right-angle junction of an experimental laboratory flume tributary channel that is carrying a placer mineral, and where tributary discharge has been set at 1.5 times that of the trunk stream. After Best and Brayshaw (1985).

(a)

(b)

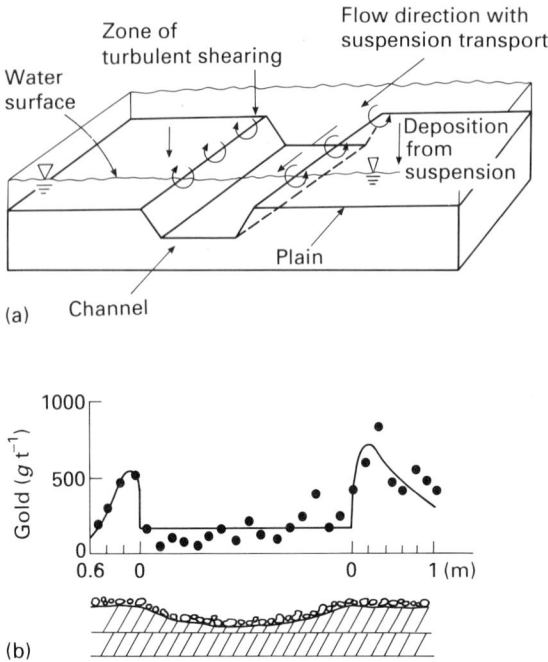

Fig. 4.30 (a) Schematic representation of hydraulic conditions at flood stages involving overbank flow showing regions of energy loss in the levee shear zones. (b) Cross-sectional distribution of gold placer concentration in a braid-stream complex of the Carbon Leader, Witwatersrand, South Africa. ●, Measured; ——, simulated. After Nami and James (1987).

4.9.1 Bedform generation and turbulent flow

The work of Grass (1971) and Jackson (1976) did much to establish the relationship between boundary-layer structure, sediment transport and bedforms in rivers. Central to their ideas is the premise that 'burst–sweep' cycles of turbulence characterize stream flows (see Chapter 2). In essence, they divided the flow into two layers: (1) a near-boundary 'inner' layer which contains streaks of high and low velocity fluid, the disposition of which is controlled in part by processes operating in (2) the outer layer. Parcels of low-velocity fluid burst upwards within the outer layer to be replaced by quantities of high velocity fluid which sweep downwards towards the bed, disrupting the laminar sub-layer (Fig. 4.32). The complex structure of flow within the outer layer has largely been established through flow visualization experiments (e.g. Fielder & Head, 1966; Dimotakis & Brown, 1976; Bradshaw, 1981). Rao *et al.* (1971) proposed a direct link between processes in the inner and outer layers as being necessary to maintain the energy balance within the turbulent boundary layer. This link has also been confirmed through flow visualization, low and high velocity streaks being associated with bursts and sweeps respectively Best (1992).

Burst–sweep cycles are largely responsible for the initiation of sediment transport since sweeps towards the bed penetrate the inner layer and exert high instantaneous shear stresses on bed sediments while bursts carry suspended sediment away from the bed. Because bed disturbance is not uniform, they therefore play a critical role in initiating and maintaining bedforms. This is evident, for example, from experimental work that links particularly strong burst activity with the lee side of dunes (Jackson, 1976). The feedback between flow, sediment transport and bedforms is obviously of great significance. It is perhaps surprising, therefore, that the effects of bed roughness elements such as bedforms on turbulence is still poorly understood. The few data that do exist suggest that roughness has a large effect close to the bed, reducing the longitudinal and increasing the vertical dimensions of turbulent structure. Even at some distance from the boundary, surface boils are

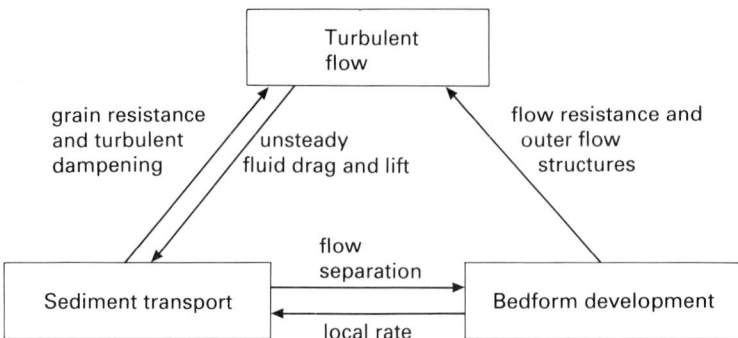

Fig. 4.31 Interrelationship between flow characteristics and bedform development in channelized flow. After Leeder (1983).

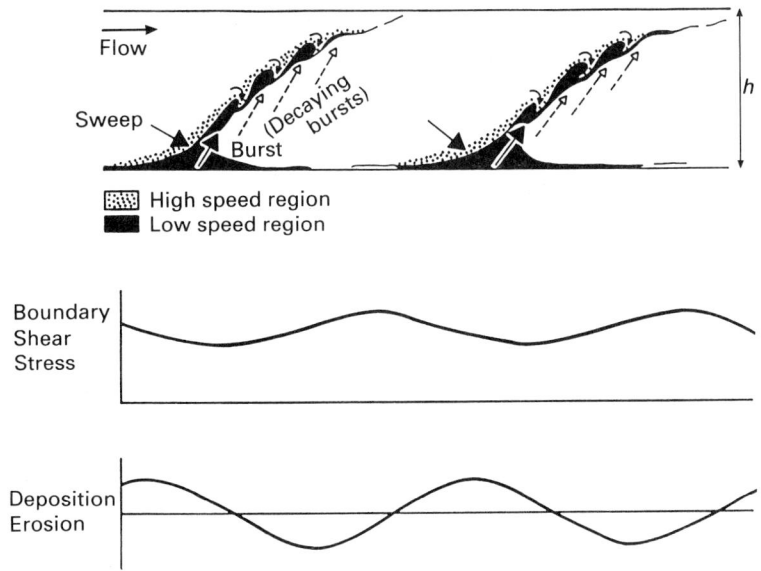

Fig. 4.32 Burst–sweep cycles of turbulence and associated patterns of shear stress, erosion and deposition. After Allen (1983) and Leeder (1983).

commonly seen in rivers where large dunes or other bed protruberances have formed (Znamenskaya, 1964; Jackson, 1976).

4.10 Character of fluvial bedforms

The range of bedforms that occurs in a river system is wide and varies not only downstream but within a single reach if flood character is highly variable or sediment supply changes, for example, as a product of seasonal exhaustion. Treatises on alluvial bedforms and the internal primary structures that are preserved are numerous (e.g. Middleton, 1965a; Conybeare & Crook, 1968; Collinson & Thompson, 1982; Allen, 1982) and give a firm basis for understanding the significance of bedforms in the construction of river deposits.

The range, size and character of alluvial bedforms are direct products of the contemporary balance between erosion and deposition at different points on the bed. It is significant that several authors have noted the variability of sediment transport rates with the passage of bedforms at a variety of scales (Davoren & Mosley, 1986; Ashmore, 1987; Bridge & Best, 1988; Gomez et al., 1989). In addition, migrating bedforms may leave an interpretable record of their movement under conditions of net aggradation. Such sedimentary structures are often used to infer palaeohydraulic parameters associated with ancient deposits despite our incomplete understanding of the

processes governing bedform generation in modern streams (Tanner, 1967; Leeder, 1980; Allen, 1983).

Bedforms occur at a variety of scales ranging from ripples through dunes to riffles and bars. Small-scale bedforms occur at discrete, though often channel-extensive, sites on the bed and are dependent upon highly localized flow conditions (flow velocity, water depth, etc.). Larger scale bedforms are a product of channel planform and can be directly related to patterns of secondary flow (Thorne et al., 1985).

4.10.1 Small-scale bedforms

Sediment size is probably the most important determinant of bedform development. Channels with sand-sized beds have the potential for developing a wide range of small-scale bedforms and these have been widely studied in attempts to link flow conditions with bedform type (Simons & Richardson, 1961; Simons et al., 1965; Southard, 1971). In contrast, small-scale bedforms in gravel-bed streams are less well-known and poorly studied. Indeed, there is a school of thought which believes that properly organized bedforms cannot exist within coarse deposits as a result of the strong vertical turbulence caused by the large bed roughness. Pebble imbrication has long been the only organized structure readily identified in both modern and ancient deposits (Harms et al., 1975). But more recently, careful observation of gravel beds of widely differing character has led to an identification of streamlined pebble clusters as an important type of small-scale bedform (Teisseyre, 1977; Brayshaw, 1984, 1985; Reid et al., 1992; de Jong, 1991). At a slightly larger scale, ribs, steps, and boulder jams have also gained acceptance as bedforms typical of coarser alluvial deposits (Wertz, 1966; Scott & Gravlee, 1968; Bowman, 1977).

Sand-bed streams

Small-scale bedforms in sand-bed streams range from ripples and dunes through plane beds to antidunes. Bedform phase diagrams — explaining bedform character in terms of sediment size and some function of

flow, for example, average velocity, shear or stream power — are numerous and well known (Harms et al., 1975; Leeder, 1980; Allen, 1982; Fig. 4.33). Such diagrams largely rely on data derived from controlled flume experiments and leave the user with the problem of deciding the extent to which such results can be applied to natural streams where flow varies dramatically both with location on the stream bed at any instant and over longer periods with the passage of a flood wave. However, despite reservations about the representativeness of bedforms, phase diagrams do offer an opportunity for both forward and reverse modelling: either flow and bed material size can be used to predict likely bedform character, or preserved sedimentary structures and abandoned bedforms can

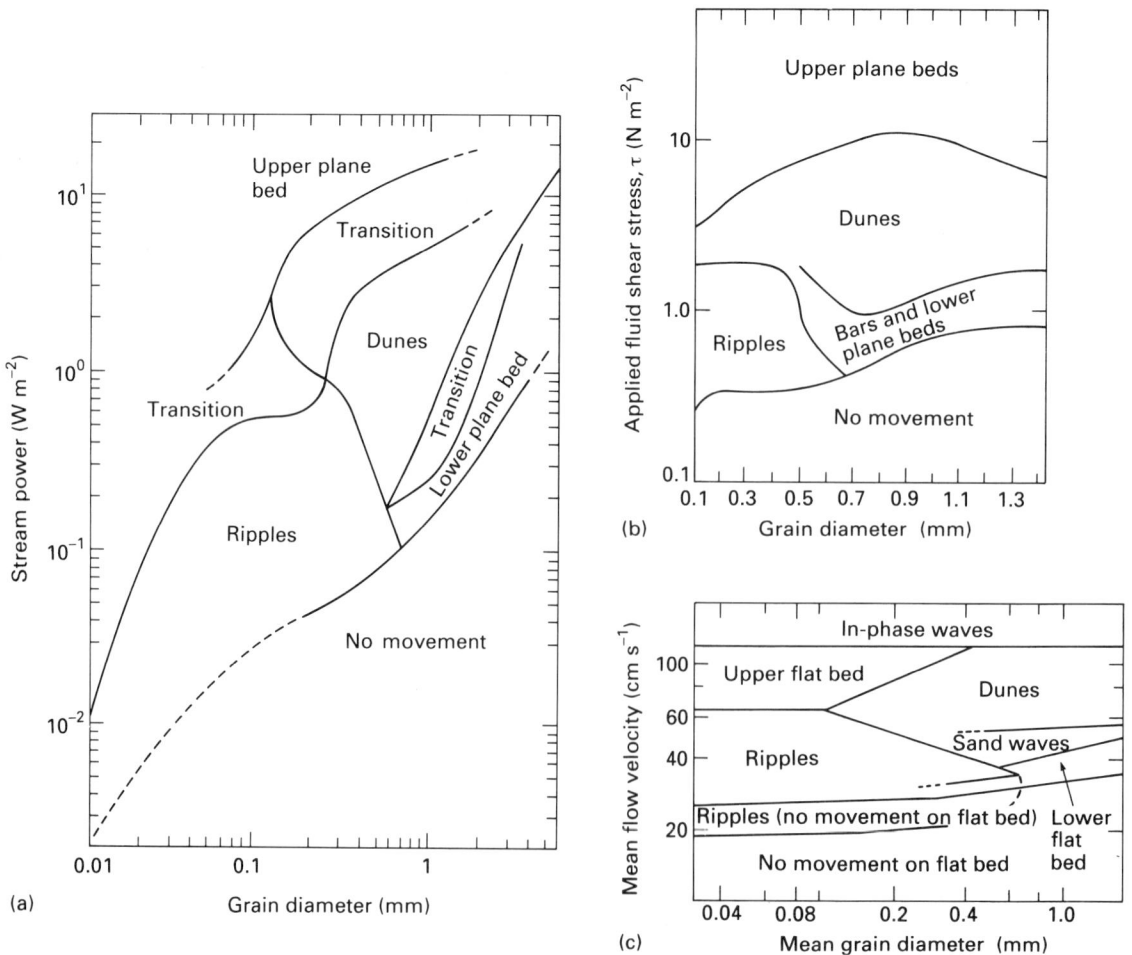

Fig. 4.33 Stability fields for different bedforms in relation to: (a) stream power, after Allen (1983); (b) shear stress, after Leeder (1983); and (c) flow velocity, after Harms et al. (1975).

be used to infer past flows. Such diagrams can be useful in interpreting alluvial deposits but their shortcomings must always be borne in mind.

Ripples are the smallest of the non-planar sand bedforms and can vary greatly in character. They are asymmetric in longitudinal cross-section with a steep lee face. Planforms range from straight, through wavy to linguoid and this variation has produced the variety of terms used to describe them (Fig. 4.34). Ripples migrate as a function of stoss side erosion and leeside deposition. This leads to sediment avalanching down the lee face and the grain size-segregation that occurs during this movement is responsible for the development of laminations inclined to the bedding surface, that is, cross bedding.

Ripples are initiated as burst–sweep phenomena disturb an initial plane bed forming irregularities. Initial defects are amplified and propagated downstream through lee side flow separation and reattachment (Allen, 1968, 1969). The character of ripples is unaffected by water depth but shows a weak relationship with flow velocity and bed shear. This is largely due to changes in the intensity of turbulence at reattachment, which in turn affects entrainment and scour depth, to changes in the downstream length of the separation bubble, and to the transport hop-

length of individual grains, all of which influence the wavelength of the bedform. Ripples only exist in sediments with a mean grain diameter of less than 0.66 mm. Above this size, grains protrude from the viscous sub-layer and the resulting increase in turbulence complicates the systematic enlargement and propagation of small bedform defects.

As flow velocity increases, small ripples may be overtaken by larger ones. This, when added to the increase in scour depth associated with the larger forms, eventually leads to the formation of dunes in sands finer than 0.51 mm. In coarser sediment, the initial bedforms which develop may have long straight crests and lower amplitude than true dunes (Costello & Southard, 1981). They appear to result from the fact that slower, larger grains hinder the progress of smaller, faster ones and are akin to the kinematic particle queues described by Langbein and Leopold (1968). Unfortunately these microforms have been referred to as 'bars', a term probably better reserved for the large bedforms associated with channel planforms, for example, point bars. Low-amplitude *pseudo-ripples* or *pseudo-dunes* might be a more appropriate descriptive term.

Dunes are common both on the beds of streams and in tidal channels. They are larger scale structures

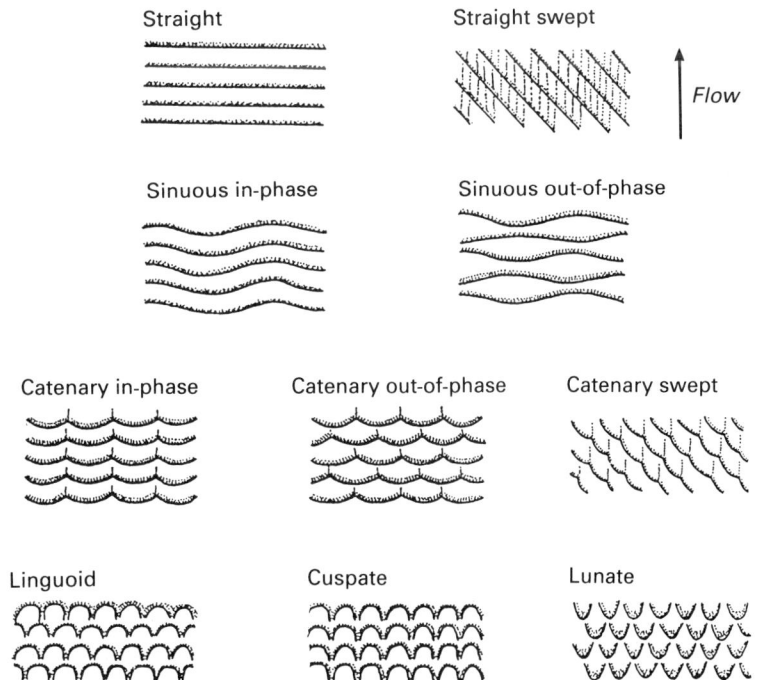

Fig. 4.34 Various planforms recognized for ripples. After Collinson and Thompson (1982).

than current ripples although many features of their shape and internal organization may be similar. However, in the fluvial environment, dunes tend to exhibit greater variation in longitudinal cross-section than do ripples, and may be symmetrical or asymmetrical, strongly convex or straight-backed depending on the prevailing local hydraulic conditions. They are generally distinguished from ripples by differences in both amplitude and wavelength (Fig. 4.35). Dunes are ripple-like structures over 1 m in wavelength and 0.1 m in height. However, they differ from ripples in that both height and wavelength are positively related to water depth (Allen, 1968; Jackson, 1976). Processes that lead to the formation and migration of dunes are associated with the macroturbulent structure of the 'outer layer' rather than to flow separation (Yalin, 1977). Dunes are sufficiently large to produce major leeside bursts which cause deeper scour and higher sediment transport rates. This increases bedform height and leads to greater acceleration of the flow up to that point at which the rate of sediment delivery to the crest matches the rate of erosion. This is an equilibrium condition, and bedform dimensions are commensurate with prevailing flow depth and velocity.

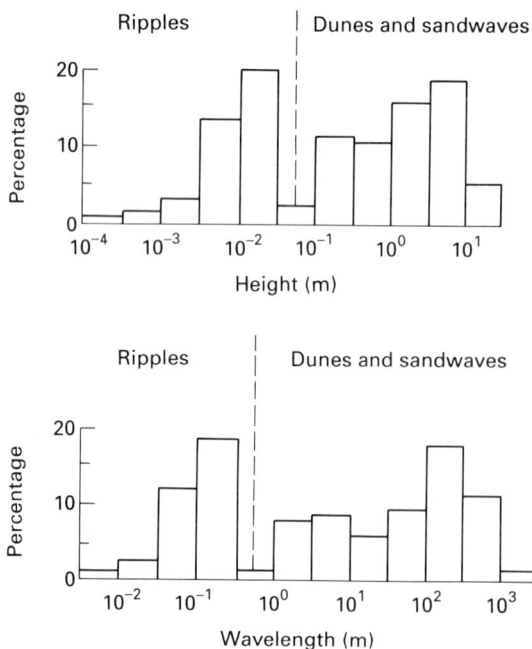

Fig. 4.35 Histograms of wavelength and height for ripples and dunes. After Allen (1968).

The term dune is generally applied to bedforms with limited crestal continuity. However, there are dune-like structures with straight and laterally persistent crests. These are often referred to as sandwaves (Collinson & Thompson, 1982). This can prove confusing, since sandwaves are widely documented in tidal deposits and are much larger banks of sediment. It may be appropriate to abandon the term sandwave when describing small-scale bedforms in rivers and substitute it with 'straight-crested dune'.

Plane beds occur both in flows close to the threshold of sediment transport and at velocities sufficiently high to wash out dune structures. In both cases the formation of plane beds involves erosion and deposition and can leave behind a lamination which may be preserved in ancient deposits. Upper-stage plane beds probably develop as a consequence of very high concentrations of sediment close to the bed which tend to dampen turbulence over any incipient bed topography. This prevents the patterns of erosion at reattachment and deposition downstream that are required for the formation of three-dimensional bedforms (Leeder, 1983). Linear ridges of coarser particles sometimes occur on the otherwise plane sediment surface. These are related to the occurrence of high and low velocity streaks and are known as current lineations. Lower-stage plane beds are a consequence of the strong vertical turbulence that occurs over a rough boundary at velocities close to the threshold for sediment transport. This tends to inhibit the development of flow separation so that bed irregularities are not amplified and propagated downstream (Leeder, 1980).

Standing waves can be observed in steep streams as the velocity of water flowing over an upper-stage plane bed increases (Simons & Richardson, 1961; Kennedy, 1963). The waveform that develops at the water surface is mirrored by the sediment below. The bedforms are referred to as *antidunes*. Standing waves remain more or less stationary and do not migrate downstream. However, field observations reveal that they are transitory, typically persisting for tens of seconds and then only in restricted parts of the flow where conditions are supercritical (Reid & Frostick, 1987). In fact, they periodically break upstream, leaving a highly turbulent but more or less planar water surface and (presumably) a bed surface to match. Antidunes probably produce upstream-dipping, low-angle cross-bedding and such structures in ancient deposits have been interpreted as resulting from their migration (Kennedy, 1961; Middleton

1965b). The main problem with such an interpretation arises from the fact that bedforms developed
under conditions of such rapid flow are largely
erosional and extremely transitory so that the
chances of both construction and preservation are
extremely low. In addition, the distribution of standing waves within the channel during flood flows is
discrete and any lamination that might be produced
would be highly localized and subject to wash-out as
upper-stage plane-bed conditions were re-established.

Gravel-bed streams

Gravel-bed streams certainly develop small-scale
bedforms but the fact that constituent clasts are often
of the same order of magnitude as the microform
leads to wide-ranging morphotypes and this makes
recognition less easy than in sand-bed channels.
Partly because of this, the sedimentology community
may yet need to be convinced of their ubiquity.

The commonest microform is the *pebble-cluster*,
now recognized by a growing number of studies in a
variety of rivers having different clast lithology and,
hence, shape (e.g. Laronne & Carson, 1976; Billi,
1988). Detailed analysis of these bedforms by
Brayshaw (1984) has shown that they comprise an
obstacle clast, normally equivalent to the D_{90} of
the bed sediment, and a collection of stoss and/or lee
side particles which accumulate in response to both
mechanical interlock with, and flow separation
brought about by, the obstacle (Brayshaw *et al.*, 1983;
Brayshaw, 1984; Fig. 4.36). They may be the gravel-
bed equivalent of ripples and dunes since their
longitudinal spacing appears to relate to maximum
flow resistance and minimum sediment transport flux
(Hassan & Reid, 1990).

Ribs, steps and boulder jams have also been recognized in a variety of gravel-bed streams (e.g. Krumbein, 1942; Scott & Gravlee, 1968; Bowman, 1977;
Bluck, 1987). They are interpreted variously as
queues (Scott & Gravlee, 1968), the coarse sediment
equivalent of antidunes (Koster, 1978) and as kinematic waves of sediment similar to those postulated
by Langbein and Leopold (1968). They comprise an
accumulation of the coarsest clasts within the bed
sediment and form themselves into cross-channel
'steps' with a steep downstream gradient.

4.10.2 Large-scale bedforms

Large-scale bedforms are related to channel-wide

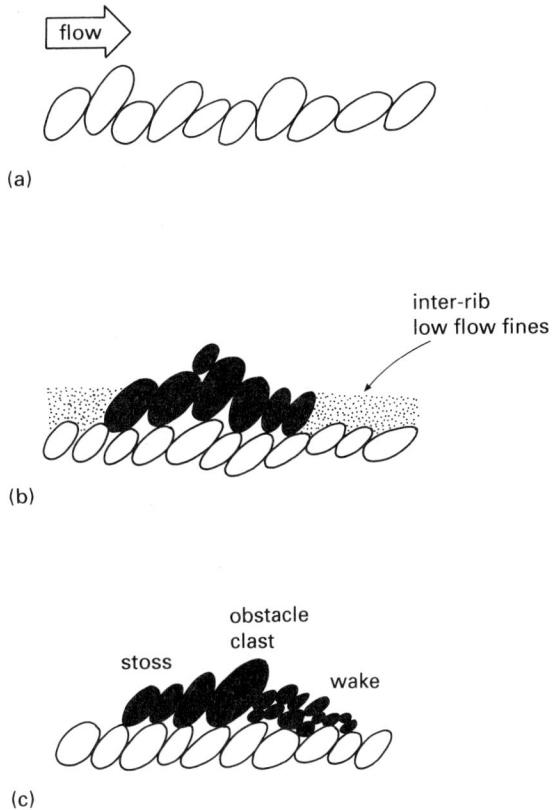

Fig. 4.36 Internal arrangement of clasts in (a) plane
beds; (b) ribs and (c) cluster bedforms in gravel-bed rivers
as seen in sections cut parallel to the flow.

patterns of accumulation and erosion which both
control and reflect channel planform. Channel planform relates directly to the hydrodynamics of flow
and the associated processes of sediment transfer and
energy dissipation. Channels attempt to achieve an
equilibrium between gradient, sediment supply and
water discharge. Adjustments of the channel pattern
are a product of changes in energy of the stream. A
shift from single thread (straight, sinuous and meandering) to multithread (braided and, less commonly,
anastomosing) will be in response to changes in the
balance between stream power and resistance to flow.
The range of channel patterns forms a continuum,
but, despite this, a number of attempts have been
made to discriminate between single-thread (straight
and meandering) and braided streams on the basis of
discharge and channel slope (Fig. 4.37). The validity
of simple bivariate discriminators has been questioned by later workers. For example, Carson (1984)

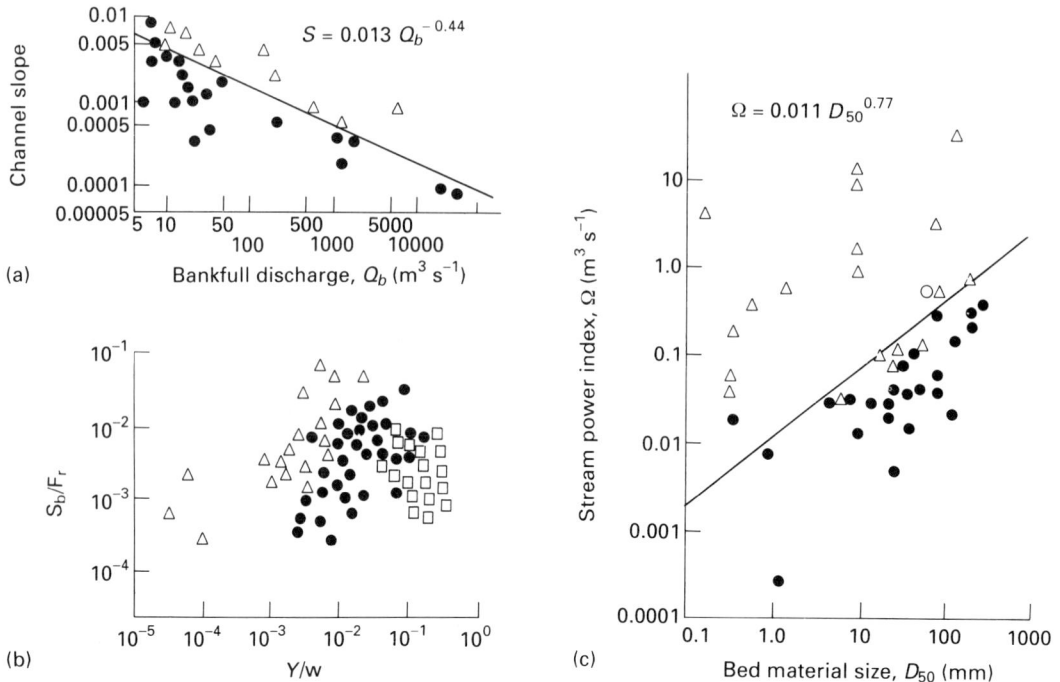

Fig. 4.37 Bivariate plots used to discriminate between single thread and braided streams on the basis of: (a) channel slope and bankfull discharge, after Leopold and Wolman (1957); (b) slope of stream bed/Froude number (Sb/Fr) and depth/width (Y/w) ratios, after Parker (1976); and (c) stream power index and median bed material size, after Parker (1976). △, Multithread; ●, single thread — meandering; □, single thread — straight; stream power index is the product of bankfull discharge and slope.

was critical of Leopold and Wolman's (1957) classical diagram showing different fields for channel types on a diagram of bankfull discharge and slope. He stressed the importance of including bed sediment size as a third factor, since gravel-bed streams will always plot higher on a discharge-slope diagram because of the higher power requirements for bed-load transport. However, generalizations such as those of Leopold and Wolman do offer an understandable and potentially useful basis for explaining river planform. For any one channel capacity and water discharge, they predict braids under conditions of high slope and large sediment load and single-thread streams where the channel has lower slope and smaller sediment load (e.g. Schumm & Khan, 1971; Parker, 1976).

Straight channels

Straight reaches tend to occur for only short lengths of a river — characteristically less than 10 times the channel width (Richards, 1982). Even though the

stream is straight the flow is not uniform in a downstream direction reflecting the development of riffles and pools. Riffles are topographical 'highs' on an undulating stream long profile. They are spaced typically at distances approximating five to seven channel widths and usually contain the coarsest components of the bed sediment (Leopold et al., 1964; Richards, 1978a). Pools, as their name suggests, are the corresponding low points which are often lined with marginally finer bed sediments, at least at low flows. At high flows, scour may expose a coarse 'lag' material (Lisle, 1979).

Observations of riffle–pool systems suggest that, during relatively low flow, riffles are zones of high flow velocity and steep water-surface gradient (Richards, 1978b). Backwelling from the riffle produces a corresponding pool upstream where the water is deep and slower flowing. However, this situation is believed to hold only at low discharges when the character of the flow is controlled by the gross geometry of the bed. As the discharge increases, it has been suggested that the pattern of relative velocity or

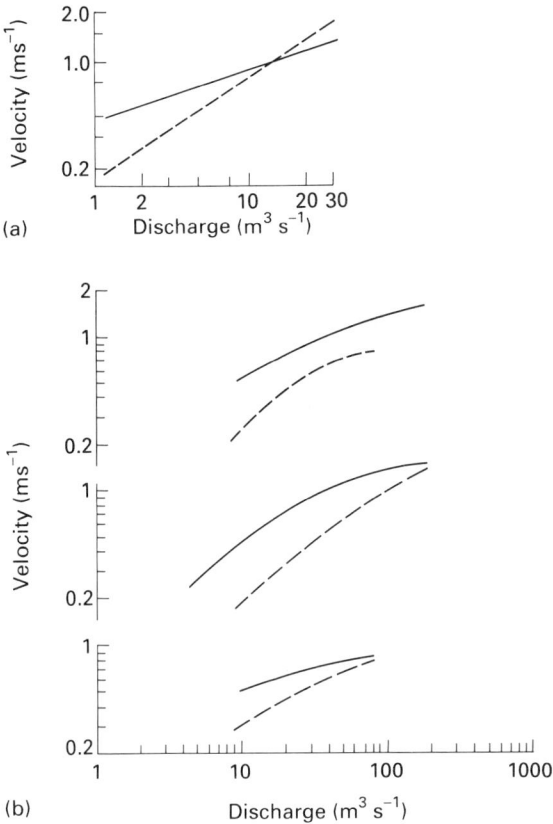

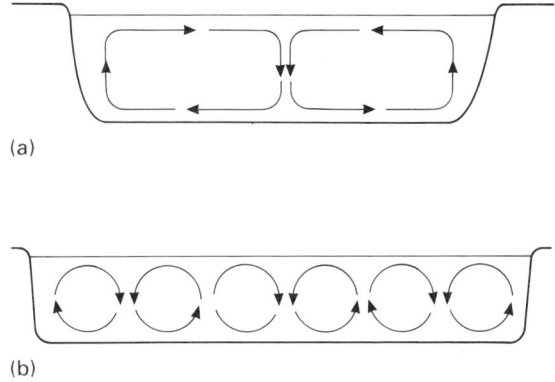

Fig. 4.38 (a) Velocity–discharge relationship showing velocity reversal for riffles and pools as discharge increases. After Lisle (1979). (b) Velocity–discharge relationships for three experimental reaches in the River Severn showing no velocity reversal. —, Riffle; – – –, pool. After Carling (1991).

bed shear stress is reversed (e.g. Keller, 1971; Andrews, 1979; Lisle, 1979; Fig. 4.38). Pools are thought to experience higher shear velocities and, therefore, to suffer greater scour. The eroded material is deposited on the downstream riffle which becomes a zone of net storage. At lower flows, finer sediment such as sands are swept across riffles by fast, shallow, flows and may be deposited in the pool in deltaic fashion. The degree of sediment sorting between riffles and pools is therefore thought to relate to recent flood history (Lisle 1979).

This idea of 'velocity reversal' is the basis for a variety of models of sedimentation and morphology in riffle–pool channels (Andrews, 1982; Hooke, 1986; Sidle, 1988). However, its universal applicability has been questioned recently, most notably by Carling

Fig. 4.39 Secondary flow cells for (a) paired cells in a narrow channel and (b) multiple cells in a broad channel. After Richards (1982).

(1991). He maintains that riffles must be significantly wider than adjacent pools before velocity reversal can occur. This is not the case in some rivers, especially larger rivers with strong banks. The controversy will run for some time, if only because of the logistical problems of obtaining reliable and meaningful data from streams in full flood.

Within all channels, both straight and sinuous, a feature of the flow is the development of a series of secondary circulation cells, the number depending largely on the width–depth ratio of the flow. Narrow deep streams may develop only two, whereas flow in wide shallow channels may compartmentalize into a larger number (Richards, 1982; Fig. 4.39). Keller and Melhorn (1973) examined the differences in circulation patterns in riffles and pools. They found that the secondary flow in pools is such that water moves down towards the bed at the centre increasing the potential for erosion. By contrast, on riffles, flow ascends in the centre tending to lift faster water away from the bed, reducing shear and favouring deposition (Fig. 4.40).

Meandering channels

Nature does not favour straight channels if only because of the inherent tendency for open-channel flows to propagate secondary circulation (Francis & Asfari, 1970; Thorne *et al.*, 1985). As a result, they are unstable, tending to transform themselves towards more sinuous planform with time. In a sinuous and meandering stream, the riffle–pool pattern shifts to conform with the channel plan, so that pools lie

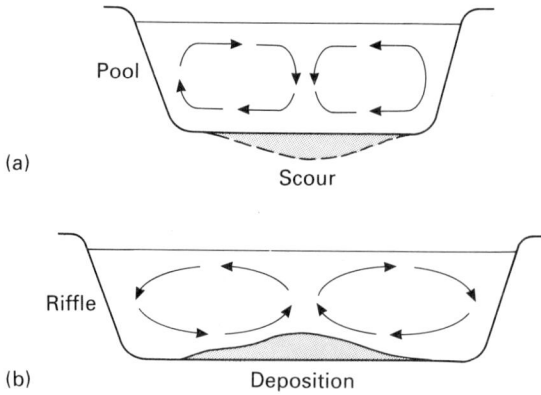

Fig. 4.40 Patterns of scour (a) and deposition (b) in relation to flow patterns in riffles and pools. After Richards (1982).

Fig. 4.41 Diagrammatic representation of a typical river reach showing locations of pools and riffles and the energy grade line for the same reach. After Leopold *et al.* (1964).

at bend apices and 'cross-over' riffles lie between adjacent pools.

The distinctive feature of meandering streams is the repeated, often large-scale, bending of the channel and, in unconfined situations, the migration of these bends both laterally across the floodplain and downstream. This results from selective erosion and undermining of the concave bank, and deposition of material in the form of a point bar on the convex bank of a bend (Thorne & Lewin, 1982). Rates of migration appear to be a function of meander curvature, stream power, outer bank height and its capacity to resist scour (Hickin & Nanson, 1984).

The main bed features of meandering rivers are scour pools, accretionary point bars and riffle bars between adjacent bends (Fig. 4.41). An additional accretionary feature of some rivers is the concave bench noted by Nanson and Page (1983) on the Murrumbidgee River (NSW, Australia). These tend to form in very acute bends as a result of flow separation upstream of the pool and on the concave bank of the bend.

Point and riffle bars frequently merge and become indistinguishable. Patterns of erosion and deposition are a function of the development of a distinct asymmetry in secondary flow cells as they move through the bend (Hey & Thorne, 1975; Bridge & Jarvis, 1982; Fig. 4.42). This compresses one of the cells against the outswing bank causing erosion. The resulting material can then either be deposited locally, displaced by the larger circulation cell onto the adjacent point bar, or carried downstream, according to particle size and flow conditions.

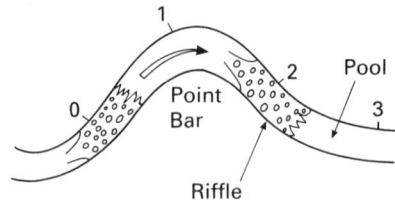

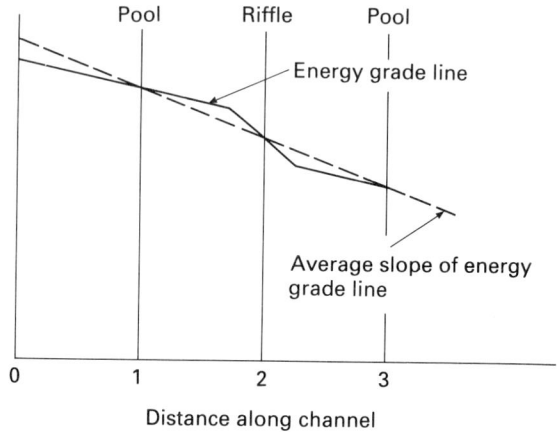

Meanders only develop in relatively deep rivers where width allows for the development of only two secondary cells. Unconfined meandering therefore tends to occur where the banks are fine-grained and cohesive, a characteristic encouraged where clay-rich suspended sediment dominates the load. Where the development of meander bends is controlled in some way, for example, by rock outcrops, the planform of the channel may be distorted in both shape and scale. However, in instances where meanders are incising an accommodating country rock, they may still maintain the classic plan symmetry of the best unconfined examples. Perhaps the best known case is that of the 'gooseneck' meanders on the San Juan River of southern Utah where it cuts the Pennsylvanian rocks of the Hermosa Group in its plunge towards the Grand Canyon.

Many attempts have been made to relate the scale of meanders to the factors known to be linked to their initial development, that is, discharge (either bankfull, dominant or mean annual) or channel width,

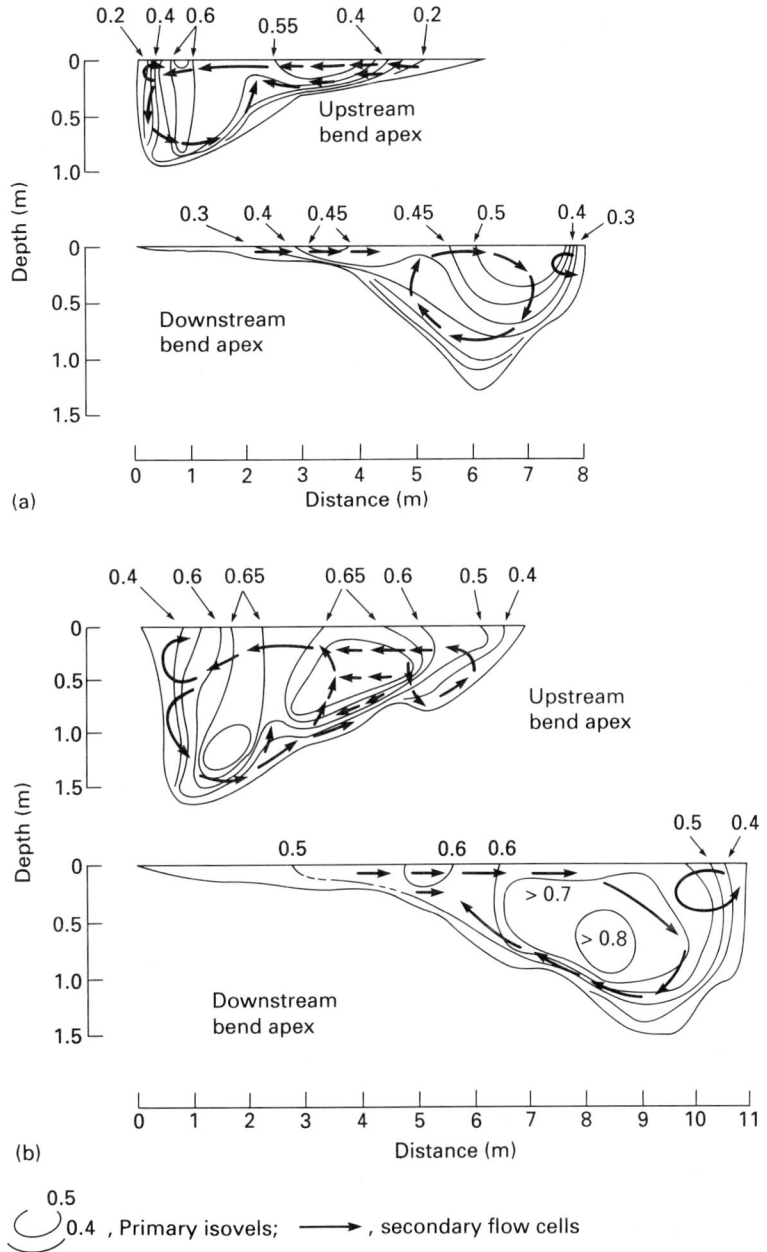

Fig. 4.42 Patterns of secondary flow in a meandering sand-bed river for two bends at discharges of (a) 1.7 and (b) 4.0 m^3 s^{-1}. After Thorne *et al.* (1985).

and silt–clay content of the banks (Table 4.3). These have met with some success and many of the relationships are in regular use by geomorphologists to infer either hydrological characteristics in areas where records are scanty or non-existent, or past climates and different rainfall–runoff relationships than those that exist in an area at present (Dury, 1964, 1976).

Point bars are the dominant depositional bedform in meanders. The calibre of the sediment and the potential for development of smaller scale bedforms on the surface of these bed features can be related to patterns of flow, and, therefore, to shear around the bend (Allen, 1970a; Bluck, 1971; Bridge & Jarvis, 1982; Fig. 4.43). Grain size tends to decrease from the pool onto the point bar. However, the upper surface of the bar is exposed at low flows and is

Table 4.3 Equations used by various authors to predict relationships between meander wavelength and other channel parameters

Predictive equation	Source
$L = 54.3Q_b^{0.5}$	Dury (1955)
$L = 7.32w^{1.01}$	Leopold and Wolman (1957, 1960)
$L = 12.13w^{1.09}$	Leopold and Wolman (1957, 1960)
$L = 4.59r_cw^{0.98}$	Leopold and Wolman (1957, 1960)
$L = 1935\,Q_m^{0.34}B^{-0.74}$	Schumm (1968)
$L = 618Q_b^{0.43}B^{-0.74}$	Schumm (1968)
$L = 62Q_d^{0.47}$	Ackers and Charlton (1970)
$L = 166Q_m^{0.46}$	Carlston (1965)
$L = 11w = 32.86Q_b^{0.55}$	Dury (1976)

L = meander wavelength; Q_b = bankfull discharge; Q_m = mean annual discharge; Q_d = dominant discharge; w = channel width; B = silt–clay index for banks; r_c = radius of curvature of bend.

submerged only periodically during floods. Deposits preserved on this surface are therefore the product of a wide variety of processes including sub-aerial ones (Nanson, 1980). Scroll and chute bars are commonly formed as a result of flow expansion downstream and soil-forming processes act on all exposed material during interflood periods.

Multithread channels

Multithread channels can be divided into two distinct types with contrasting planforms and processes — anastomosed and braided rivers. Anastomosed systems are characterized by rapid aggradation, low gradients, and well-defined low sinuosity sand- or gravel-bed channels that have irregularly branching planform (Smith, 1983). Channels are generally separated by marsh or other wetland areas and differ from regular single-thread sinuous channels by virtue of their branching nature and their very rapid aggradation. They are considered by some to be a feature of tectonically active areas where the sediment load is dominated by clays which lead to the development of cohesive and competent banks (Smith & Putnam, 1980).

By contrast, braided streams have unstable, shallow channels separated by actively migrating and accretionary braid bars. Braiding is favoured where valley gradients are high, where discharge is large and highly variable, where banks are non-cohesive and lack the stabilizing influence of vegetation, and where bedload discharge is high. In the wide, shallow channels that form, the secondary circulation cells that are a quasi-stable feature of flood flows in narrower, deeper streams become unstable and break up into a larger number of smaller cells. This leads to accretion where flows converge and rise from the bed and development of braid bars (Ashmore, 1982). Diversion of the flow around the emerging bars causes bank erosion and increases the channel width. This in turn leads to an increase in the number of secondary circulation cells, the development of further bars, and so on. Bars are emergent only under low flow conditions when they may be subjected to dissection, become vegetated and develop soils. During large floods they may be completely submerged and the flood waters may carry both coarse and fine bedload sediments onto the upper surface. Larger particles tend to become lodged at the proximal end of each bar and a strong negative size gradient may develop towards the distal end (Ashworth & Ferguson, 1986).

Braid bars have been classified in a variety of ways (e.g. Church, 1972; Smith, 1974; Ferguson & Werrity, 1983). Any categorization inevitably ignores the fact that they are constantly altering in shape and position and may well evolve back and forth from one form to another. Nevertheless, three main bar forms can be distinguished: longitudinal, transverse and diagonal (Fig. 4.44). Bar tops are generally dominated by the coarser fraction of the sediment carried by the stream. However, there is an overall tendency for material to become finer downstream, and the bar often terminates in a slipface at the upstream end of a scour hole that develops where branching channels confluence (Best, 1988).

Braids are a common feature of alluvial fans from the humid tropics to arid zones (Hooke, 1967; Nilsen, 1982; Kochel & Johnson, 1984). They are characteristic of glacial outwash plains and desert bajadas (Bluck, 1974; Smith, 1974; Hein & Walker, 1977) and have even been reported in submarine channels (Hein, 1984). One interesting example of a braided system which appears to defy the drainage 'laws' is that of Cooper Creek, southwest Queensland (Nanson *et al.*, 1986). Here, a mud-dominated braid plain is 'dissected' by contemporaneous sand-bed anastomosing channels. Nanson *et al.* propose that

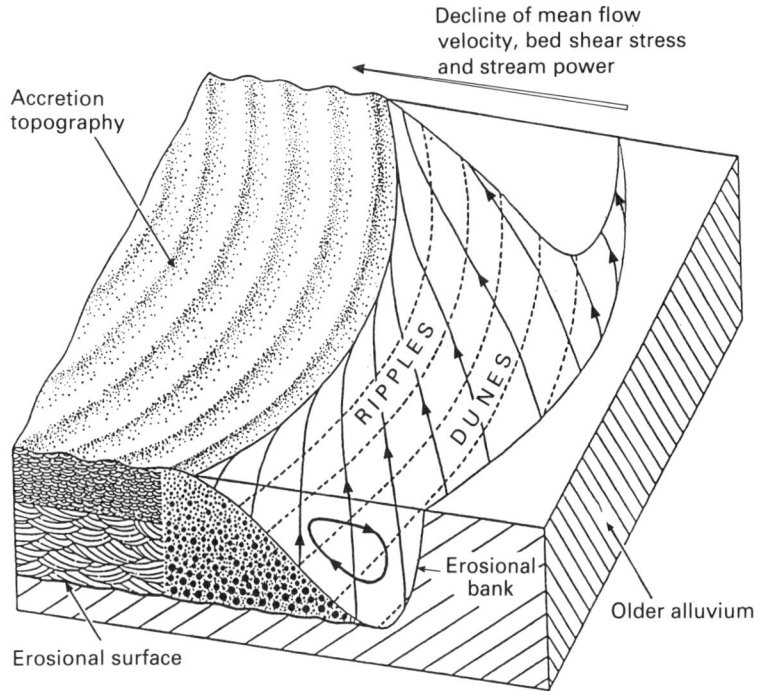

Fig. 4.43 Idealized patterns of deposition around a hypothetical channel bend in relation to fluid motion. After Allen (1970a).

▨ , Cross-lamination; ▨ , cross-bedding; ▨ , decreasing grain size; ⎯⎯ , skin-friction line; ⎯ ⎯ ⎯ , constant stream power.

the braid channels are formed and used by exceptionally high flood discharges. However, the clay peds which dominate the bedload at these times consolidate to form the coherent banks of the incised anastomosed channels that accommodate lesser more frequent floods.

4.11 Sedimentary structures in preserved fluvial deposits

The migration of bedforms at all scales leaves traces in the bed material that can be used in hydrological and hydraulic interpretation of both recently abandoned and more ancient alluvia (Baker, 1973; Bridge, 1978; Ethridge & Schumm, 1978; Bridge & Diemer, 1983; Allen, 1985). Frequently, the internal structures that result from the passage of small-scale bedforms are the most obvious, since they can be seen in their entirety in a relatively restricted outcrop or exposure. Larger structures are often too extensive

to be appreciated without both careful logging of sequences in adjacent exposures and intelligent interpolation where intervening information is missing or unexposed.

4.11.1 Small-scale sedimentary structures

Ripples and dunes

The migration of both ripples and dunes results in cross-bedding which varies in character according to the pattern of the original bedform (Allen, 1969; Collinson & Thompson, 1982). Several attempts have been made to classify cross-bedding in terms of the geometry of individual lamina and the relationship between 'sets' of laminae (Allen, 1968) but the complexity of the classifications has tended to discourage their use on a day-to-day basis. Most commonly, cross-bedding is defined as just trough-shaped or tabular.

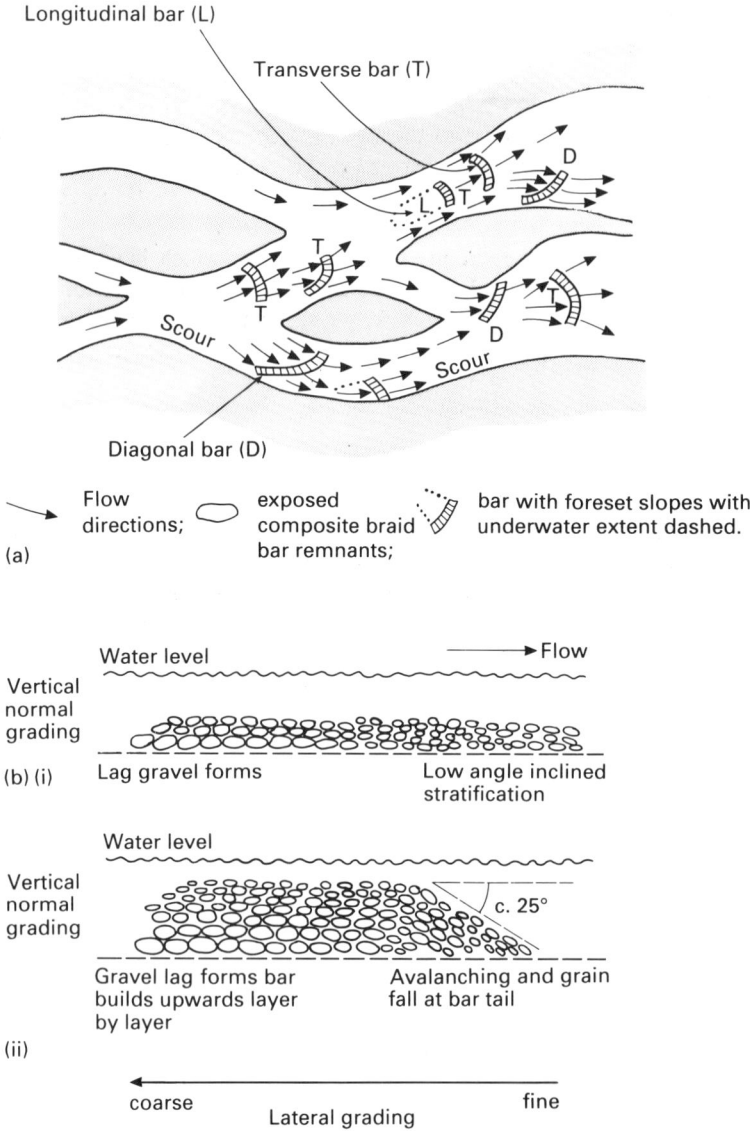

Fig. 4.44 (a) Terminology, morphology and characteristic deposits of braid bars. (b) (i) Longitudinal and diagonal bars; (ii) transverse bar. After Collinson and Thompson (1982).

In general, ripple migration leaves behind units or sets of cross laminae 3–4 cm thick, each made up of inclined layers which are concave upwards with tangential lower, and sharp, truncated, upper contacts. Commonly, the sets are trough-shaped when seen in a section cut transverse to the mean foreset dip. However, some examples may be tabular, resulting from the migration of ripples with straighter crests. As a rule, linguoid ripples tend to produce trough cross-sets and straight ripples more tabular sets.

Dunes also generate cross-bedding, but at a larger scale and with greater scope for small-scale complexities. Straight-crested dunes tend to generate tabular cross-sets, whereas dunes which are more variable in planform leave behind a variety of trough cross-bedding. Trough sets seldom exceed 1.5 m in thickness, have widths of up to a few metres and can be tens of metres long. Generally they have concave-up foresets with bases that are tangential to the underlying deposit — often another set. Tabular sets vary more widely in their dimensions, but are generally less than 1 m thick. They often extend laterally for several tens of metres and may have either angular or tangential bases.

Complications in the structure arise, for example, from large fluctuations in flow conditions during the passage of a flood wave. Erosion of the crest may truncate the lee face. A smaller dune, migrating across the back of a larger one that has become inactive will result in the superimposition of smaller cross-sets on larger ones and separated from them by an erosional surface. In addition, small-scale burst features often develop on the stoss sides of dunes. These may generate diminutive bedforms which migrate up the stoss face and leave behind cross-bedding to mark their passage.

Changes in dune morphotype occur when flow moves from lower- to upper-stage flow conditions during the transition to plane bed. This produces a variety of forms from asymmetrical through symmetrical to humpback which leave behind different patterns of cross-bedding. A synopsis of these relationships is given by Saunderson and Lockett (1983) and is shown in Fig. 4.45.

Plane beds

Plane beds leave a clear record of their development in the form of planar lamination. These are a notable feature of ephemeral stream deposits and theories concerning their formation range from the passage of low-relief bedwaves (Bridge & Best, 1988) through the turbulent bursting process (Paola *et al.*, 1989) to the strong pulsations in flow which have been observed in arid-zone rivers (Frostick & Reid, 1977; Reid & Frostick, 1989). Planar laminations are often no more than 5–20 grain diameters in thickness and there is often grain-size or grain-density sorting between adjacent laminae.

Small-scale structures in coarse-grained deposits

In gravel-bed rivers, the main record of former bedforms is preserved in the arrangement of clasts and the overall fabric of the deposit. Pebble clusters may be recognized by the presence of a large obstacle clast against which slightly smaller particles may be stacked on its stoss side (Teisseyre, 1977; Brayshaw,

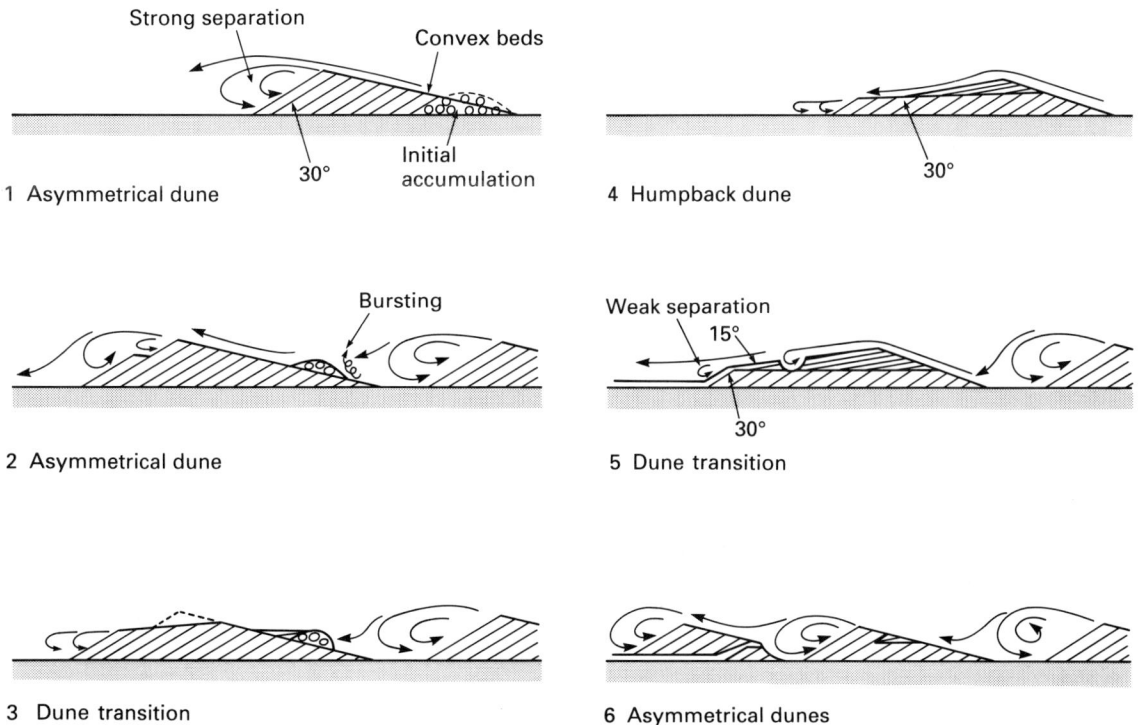

Fig. 4.45 Schematic diagram to show transitions from asymmetric to humpback dunes observed in a flume study by Saunderson and Lockett (1983).

1984; Fig. 4.36). In addition, the lee side often comprises much finer sediment strung out in a tail and orientated with the prevailing flow direction like a sedimentary wind sock.

Clast dams and ribs may be more difficult to diagnose, if only because, in longitudinal section, they may only appear to be complex tabular sheets (Fig. 4.36).

4.11.2 Large-scale structures

Large-scale sedimentary structures are varied and tend to reflect the differences in process–form relationships that differentiate single- and multithread streams.

Meandering streams

Meandering stream deposits have been the subject of intensive study and several models have been suggested for their development (Allen 1970a,b; Bridge, 1975, 1977). The channel plan is inevitably simplified and a sinusoidall wave-form is adopted in order to avoid complicating an already complex flow structure. The pattern of sedimentation is assumed to be associated largely with bankfull discharge when the cross-channel bed shear of secondary currents is expected to have fully developed within the bend. The theoretical patterns of flow and the sedimentary consequences are shown in Fig. 4.43. Lateral and down-valley bend migration together with accretion on the point bar result in the development of a tabular sediment body, often presumed to be a sand body, the thickness of which is a function of channel depth and the lateral extent of which depends on maintenance of the channel form and an absence of avulsion (Campbell, 1976; Bridge & Leeder, 1979; Miall, 1988). The (sand) body rests on an erosional bounding surface and a lag gravel may be postulated if clasts of this calibre are deemed to be present at the point in the drainage system that is of interest. The sorting process built into the model dictates that particle size progressively decreases up the point bar and the set size of any cross-bedding associated with superimposed small-scale bedforms also diminishes. The process of lateral accretion leaves behind a record of successive point bar faces in the form of large-scale sigmoidal surfaces generally orthogonal to the local primary flow direction (Willis, 1989; Miall, 1988).

Such models, though useful, only approximate

natural patterns of sedimentation since the planform and flow in field prototypes is usually far more complex while some of the premises surrounding sedimentary processes may be inappropriate or at least unproven. For example, in many streams — perhaps most (Pickup & Warner, 1976) — dominant discharge lies below that of bankfull so that most bedload is transported at discharges below that of channel capacity. The point bars of perennial meandering gravel-bed rivers are characterized by an armour layer (i.e. inverse grading at the surface), while the coarsest clasts are often found on the bar platform, contrasting with the simple upward-fining sequence predicted by the models. Chutes and scrolls often occur on the upper surface of the point bar. This may result in large-scale tabular and trough cross-bedding at the top of the sequence, etc., but the result is a deposit that is, therefore, very complex with many internal erosional bounding surfaces and a variety of deposits superimposed on each other (e.g. Wolman & Leopold, 1957).

Anastomosing channels

Patterns of sedimentation in anastomosing channels are less well documented than those of either meandering or braided streams. The primary feature of these rivers is the stability of the channels, with strong banks often fixed by vegetation, and rapid vertical aggradation. As a result, the sand bodies created have a shoe-string character — well-defined and relatively thick but limited in width. The internal structure of the deposits has received little attention, but Smith's (1983) study of the Columbia and Saskatchewan Rivers in Canada suggests that the dominant bedforms are straight-crested dunes. Consequently, the main structures of these channel-fills are likely to be tabular cross-sets arranged in multi-storey cycles. The lateral accretion deposits characteristic of meandering streams are almost entirely absent.

Braided streams

Deposition in braided systems is dominated by the various bar forms which separate the shallow, wide branches of the channel. Clast disposition and internal structures vary according to the dominant grain-size of the bed material. In coarse sediments, longitudinal and diagonal bars show a characteristic reduction of sediment size downstream (Ashworth &

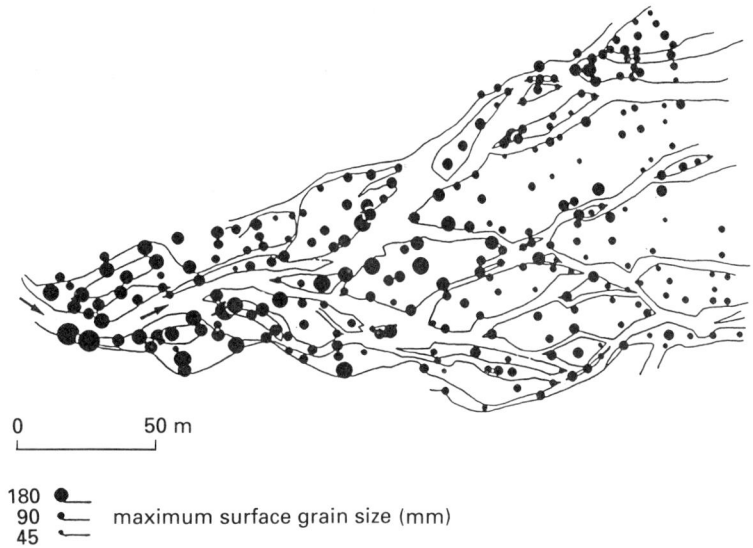

Fig. 4.46 Maximum surface grain size on bars within a braided reach of the Lyngsdalselva River. After Ashworth and Ferguson (1986).

0   50 m

180 ●—
90 ●—   maximum surface grain size (mm)
45 —

Ferguson, 1986; Fig. 4.46). Internally they comprise massive or horizontally bedded gravels, often well imbricated. Layers of matrix-filled and openwork gravels frequently alternate and may represent fluctuations in discharge (Smith, 1974). Cross-bedding will develop if an avalanche slope is present at the downstream extremity of the bar. Transverse bars, by comparison, tend to produce extensive sheets of cross-bedded deposits with numerous erosional surfaces (Fig. 4.44). In sand-dominated braided streams, the dominant bar type is the transverse bar which may take on a number of forms. These leave behind tabular cross-bedded sand units as a consequence of foreset avalanching and the migration of the downstream slipfaces. In addition, sandflats are a common and characteristic feature of sandy braided streams (Cant & Walker, 1978). In these cases, there are no slipfaces and they grade imperceptibly into adjacent channel material. In some channels, the flats are very stable features and may spread laterally by accretion. They form by amalgamation with other, smaller bars and for this reason, the internal structures are complex, comprising both tabular and trough cross-bedding at a variety of scales and with widely diverging orientations.

In ancient deposits, bedforms provide a key to palaeohydraulic interpretation. In present-day streams, they provide a basis for inferring details of recent flood history as well as interpreting the overall hydraulic and hydrological regime.

4.12 The floodplain

The proportion of the total floodplain occupied by active channel varies with river type and planform. Single-thread rivers often occupy only a small proportion of the valley width, though instances have been recorded in semi-arid settings where large recurrence interval floods have expanded the percentage of bottomland given over to channel to near 100% albeit until the system recovers its former character (Schumm & Lichty, 1963; Burkham, 1972). Braided channels can occupy close to 100% of the floodplain. Glacial outwash in particular often leads to broad sandurs as a result of abundant and available sediment, the calibre of which dictates transportation as bedload (Collinson, 1970; Church, 1978). Those parts of a valley not currently occupied by an active channel system may be inundated periodically, depending on the frequency of overbank flow. This means that sedimentation is even more spasmodic than that in the channel system. Floodplain topography also dictates that sedimentation is highly sporadic.

4.12.1 Channel–floodplain interaction

The processes and depositional products of overbank flows differ markedly from adjacent channel sequences. They contain less evidence of migrating bedforms and are generally finer, only rarely achieving grain sizes that are coarser than fine sand and then only locally, close to active channels (Kessel *et al.*, 1974; Nanson & Young, 1981). When a channelized flow spreads beyond its banks to encompass the broad flat areas of the floodplain there is a strong interaction between the flows in the two areas. In general, there is a transfer of momentum between the deep, fast flow of the channel and the shallow, slow flow of the floodplain. This results in a decrease in flow velocity and in shear in the channel and a concomitant increase in both of these hydraulic parameters in the immediate overbank area (James, 1985). The energy transfer is accompanied by a flux of suspended sediment from the channel onto the floodplain. Here rapid settling occurs as flow velocity and transport capacity diminish in response to lower gradients and increased flow resistance (Pizutto, 1986). Patterns of flow and rates of deceleration depend on floodplain topography and vegetation (Kessel *et al.*, 1974). As a result, the distribution of overbank deposits and variations in their grain-size characteristics will reflect not only the channel size and planform and the character of the suspended load, but also the topography of the floodplain (Lambert & Walling, 1987; Walling & Bradley, 1989).

4.12.2 Vertical and lateral accretion processes

Butzer (1976) divided floodplains into two categories, *convex* types formed by vertical accretion and *flat* types formed by the lateral accretion of channel deposits. However, both before and since Butzer declared his classification there had and has been much debate over the relative importance of these two main processes and their relative contribution to floodplain deposits. Most authors emphasize the role played by lateral accretion in floodplain development. This view was first established over 50 years ago in a seminal paper by Mackin (1937) and was reinforced by the work of Wolman and Leopold (1957). Both proposed a depositional model dominated by the migration of channels across the floodplain and a resulting sequence of laterally accreted point bar deposits characterized by a fining-upwards signature (Allen, 1970a; Fig. 4.43). Vertical accretion from overbank flows produces only a thin, fine-grained veneer on the floodplain surface. This model has gained widespread acceptance and is given prominence in many reviews (Leopold *et al.*, 1964; Douglas, 1977; Collinson, 1986). But its universal

application has not gone unchallenged. Nanson (1986) describes the floodplain deposits of the coastal rivers in New South Wales and concludes that the main formative process is vertical accretion. Other authors have suggested a similar dominance of overbank deposition for the southern Mississippi (Kessel *et al.*, 1974), for several rivers in Papua New Guinea (Speight, 1965; Blake & Ollier, 1971) and for various rivers in the Broads of eastern England (Lambert *et al.*, 1960; Lewin, 1981).

Of particular interest is the dominance of vertical accretion on the floodplains of low gradient anastomose streams (Smith & Putnam, 1980). Here, channels are fixed in position as a function of both low flow velocities and cohesive banks, both of which restrict bank erosion. The main mechanism of channel movement is therefore abrupt avulsion rather than progressive lateral migration.

The case studies that have been published to date suggest widely varying contributions to floodplain sediments from channel and overbank sources. The balance between vertical and lateral accretion is a function of channel stability which in turn reflects sediment load and bank character. Vertical accretion may be more characteristic of perennial rivers in humid environments than it is of ephemeral arid-zone streams (Reid, 1994).

The accumulation of layer upon layer of overbank deposits relies upon the passage of successive flood waves of sufficient magnitude to overtop the channel banks. Vertical accretion of itself is self-limiting since vertical growth of the floodplain will eventually take it beyond the reach of all but very infrequent events. However, erosion of the floodplain either by laterally migrating channels or during large flood events can strip off overbank deposits and reset the system for a further period of deposition. In river systems where vertical accretion is the dominant floodplain process, this periodic floodplain stripping can lead to complex cut-and-fill structures within the deposit (Nanson, 1986).

Vertical accretion can persist over longer periods under conditions of active aggradation. If base-level rises, the river may deposit much of its load. Increased sediment supply as a result of tectonic or climatic modification of the hinterland can produce a wave of sediment that passes down the system and results in aggradation (Fisk, 1939; Tinkler 1971; Maizels, 1979). Of course, vertical accretion can also be curtailed, for example, by uplift or a fall in base-level, in which cases the channel will tend to incise, leaving the former floodplain beyond the reach of all but the largest floods.

Nanson (1986) proposes a general theory for floodplain formation which encompasses both vertical and lateral accretion processes. He suggests that floodplains represent a balance between erosional energy and alluvial resistance. Upland rivers are high-energy systems which cut narrow gorges that preclude the development of a floodplain. Where gorges are wider, only coarse-grained alluvium accumulates between large and infrequent flood events (Baker, 1977; Nanson & Hean, 1985). In wider, less steep valleys where the river is still relatively energetic but where migration is restricted for some reason, for example by bedrock outcrop, floodplains will form by vertical accretion but will be eroded episodically. This has been documented for some arid-zone rivers (Schumm & Lichty, 1963) as well as in more temperate settings (Nanson, 1986). In this part of the river system, the most important flood events are of high magnitude and low frequency; these strip out the accumulated deposits and reset the depositional cycle. In even less steep valleys where lateral migration is possible, floodplains will be dominated by lateral accretion. Typical fining-upwards sequences of gravel and sand will develop, capped by a veneer of even finer grained overbank deposits. The formative flood events for this type of deposit are of moderate magnitude and higher frequency, since these continually erode the base of the outswing banks, causing bankfalls and achieving channel migration. In that part of the system where gradient is very low indeed, stream power may be insufficient to cause bank erosion. Here, the channel is unable to migrate and vertical accretion becomes the main depositional process on the floodplain (Smith & Smith, 1980; Nanson & Young, 1981). Deposits in this case reflect the extreme stability of the river. However, where stream power is insufficient to maintain the channel, aggradation may reduce its capacity. As a consequence, overbank flows may increase in frequency and vertical accretion may be accelerated.

4.12.3 Floodplain deposits

The surface of a floodplain is by no means featureless. A variety of depositional environments can be recognized including abandoned sections of channel often preserved as small lakes or sloughs, levees, crevasse splays and chutes (Coleman, 1969; Singh, 1972; Lewis & Lewin, 1983; Fig. 4.47). All of these features can contribute to the evolving stratigraphy. In addition, they give the floodplain a microtopography which deflects and retards overbank flows and can promote deposition or encourage local reworking.

Levees develop on the margins of the channel as a result of rapid fall-out of the coarser components of the suspended load. Consequently, there is a rapid decrease in both grain size and sediment thickness away from the source channel (Pizzuto, 1986). Thicker sequences of sands and silts are deposited in levees, which accrete more rapidly than the flood basin beyond, so that they become areas of positive relief (Kessel et al., 1974).

Levees may be breached locally by crevasses. Rising flood waters first overtop the levees at topographic low points which are lowered further by erosion. The load entrained by the confined flow in the crevasse is immediately deposited as the waters spread out on the floodplain and form lobate bodies

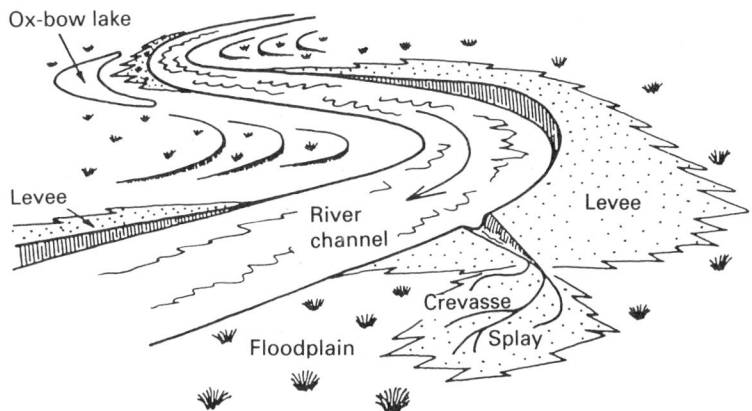

Fig. 4.47 Schematic diagram showing typical surface morphology of a river floodplain.

of sediment referred to as crevasse splays. These deposits frequently extend beyond the distal fringe of the levee and become interbedded with the finer deposits of the floodplain. Generally, they are sand wedges which show internal evidence of deposition under waning flow conditions, including graded beds and bedforms ranging from dunes through ripples to plane beds. Indeed, crevasse splays and levees are the only sites within overbank sediments where depositional bedforms are important. Levees often exhibit ripple drift lamination, evidence of the very rapid nature of deposition. The preservation potential of deposits so close to the channel is relatively low. If they do survive, the preservation of original internal structures in a recognizable form will depend largely on the prevailing climate. If conditions dictate colonization by plants and bioturbation by soil fauna, the floodplain deposit will tend to be homogenized and sedimentary structure will be lost.

The overbank deposits laid down away from the main channels and beyond any levees that may have developed are predominantly fine silts and clays. Generally, they show a proximal to distal fining and are laminated, each layer the deposit of a single flood event and its thickness varying spatially within the floodplain and in comparison with other layers. Few documented flood laminae exceed a few centimetres and most are only a few millimetres thick (Brown, 1983; Lambert & Walling, 1987; Walling *et al.*, 1992), although an interesting example of floodplain deposits in excess of 1 m thick and arising from one large flash flood event is given for Bijou Creek, Kansas, by McKee *et al.*, (1967).

Vegetation on the floodplain makes a variable contribution to overbank sedimentation. The amount and character of organic matter in the deposits is a function of climate and proximity to local or regional base-level (the ocean or perhaps a lake in closed drainage basins). In humid areas, swamps may occupy the interchannel sites, as, for example, in the Atchafalaya River basin (Coleman, 1966); organic productivity is high and thick peats can accumulate. Under more arid conditions, organic material is either sparse or absent. Desiccation features such as mudcracks are often found and there may be evidence of aeolian reworking and accumulation around isolated patches of vegetation or individual plants (Cooke & Warren, 1973; Goldsmith, 1973). Indeed, considerable thicknesses of windblown silt can accumulate on floodplains (Lambrick, 1967) and dune deposits are often found interbedded with those of

ephemeral streams, for example in the Permian and Triassic sediments of the southern North Sea and Moray Firth basins (Glennie, 1986; Clemmenson, 1987; Frostick *et al.*, 1988).

4.12.4 Rates of overbank deposition

Comparatively little is known about contemporary rates of overbank deposition and the significance of floodplains in the overall sediment budgets of river systems (Walling, 1989). Yet this information is of great environmental significance since many sediment-related pollutants travel with the fine fraction of the suspended load and may accumulate within floodplain deposits Macklin *et al.* (1992).

The methods used to establish rates of deposition fall into two groups, those that yield data on a very short-term basis, often for single flood events, and those that look at longer term trends. Data for individual floods are frequently obtained either by post-event examination of the most recent layer of sediment or by trapping material on the surface while deposition is in progress (Gretener & Stromquist, 1987; Walling & Bradley, 1989; Marriott, 1992). This information, though extremely useful, usually suffers from the logistical problems of obtaining sufficient spatial information in order to characterize such a variable phenomenon. There is also the ever-present problem that short-term records may be biased by the unrepresentative nature of the flood events occurring during the period of record. Some authors have attempted to establish sediment budgets for individual events by measuring suspended sediment losses from reach to reach and ascribing these losses to overbank deposition (e.g. Gretener & Stromquist, 1987). This approach overcomes the problems of post-flood sampling of the flood deposit but can give only average rates of accumulation for a stretch of floodplain between gauging stations. In addition, the accuracy of the estimate is only as good as that of the suspended sediment measurements used to balance the budget, besides which, it assumes that the only sink for fine material is the floodplain.

Longer-term trends can be established either by monitoring individual river systems over prolonged periods, normally considered not to be cost effective, or by establishing datable marker horizons within the overbank stratigraphy and calculating an average rate of deposition for the time interval. Methods used to date horizons include identification of artefacts (Costa, 1975), relating trace metals within a profile to a known history of mining upstream (Lewin & Macklin, 1987; Popp *et al.*, 1988) and using the fallout of radionuclides, particularly ^{137}Caesium, either from atomic testing which began in 1954 or from more dramatic singular events such as the atmospheric discharge of radioactive materials from the Chernobyl reactor in 1986 (Walling *et al.*, 1992).

Rates of overbank deposition calculated using the various methods outlined above vary between fractions of a millimetre to a few centimetres per year. For example, Shotton (1978) reports average rates of deposition of 5 mm year^{-1} and Brown (1987) 1.4 mm year^{-1} for the Avon and lower Severn valleys during the past 3 and 10 ka respectively. Walling and his co-workers have collected data for the Culm in Devon, the Severn near Tewkesbury,

both in the UK, and for the Leira in southern Norway. Their research has focused upon deposition over a period of 35 years between the beginning of atomic testing and 1989. They show considerable variation in accumulation rates that can be related to the microtopography of the floodplain (Walling & Bradley, 1989; Walling *et al.*, 1992; Fig. 4.48). For the Culm, rates have varied between 0 and 7 mm year^{-1}, for the Severn the range is 0–10 mm year^{-1}, while for the Leira rates have been between 3 and 40 mm year^{-1}. The higher rates of sedimentation on the Norwegian floodplain are attributed to landuse changes upstream. In all cases, deposition is greatest close to the channel, especially where the floodwaters are constrained by microtopography and flow is directed into ponded areas on the floodplain. In these

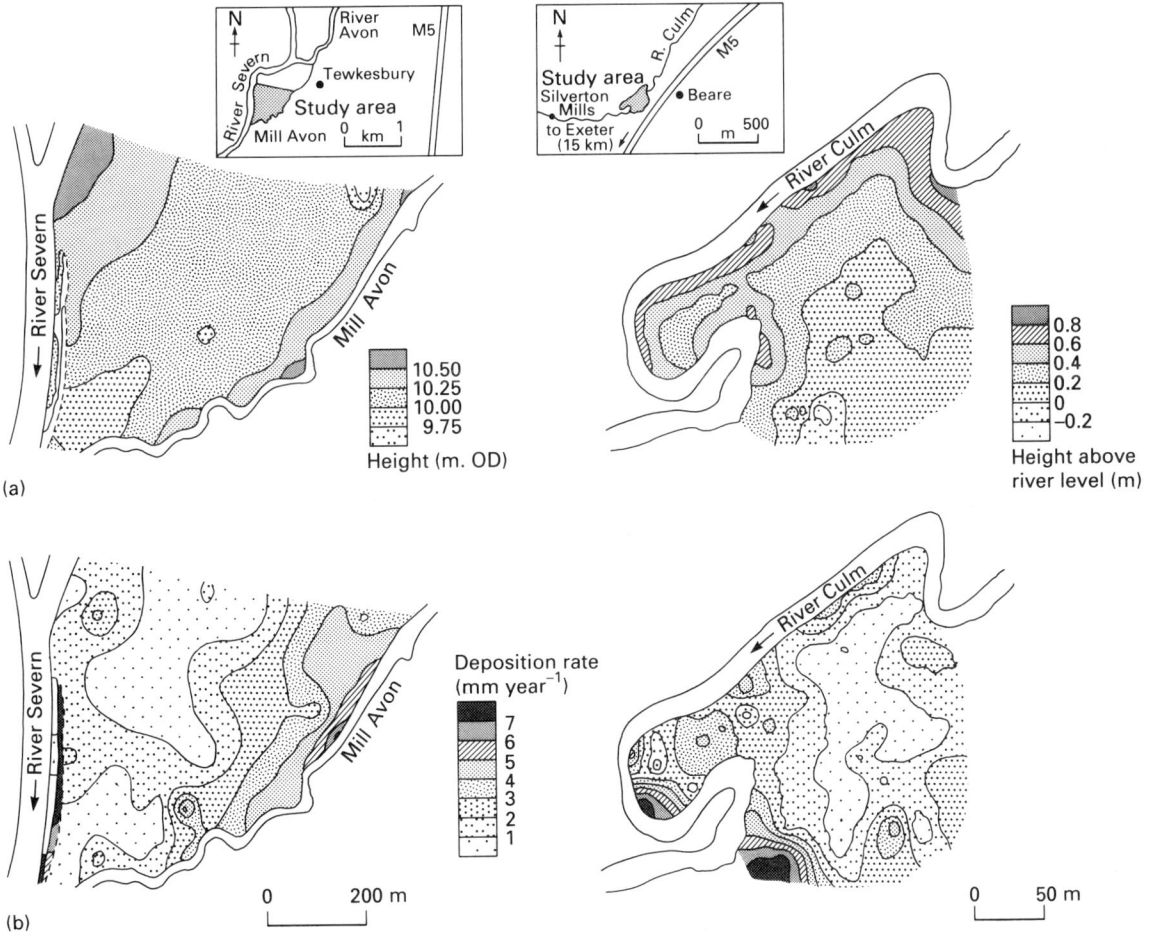

Fig. 4.48 Floodplain topography (a) and rates of sediment deposition (b) for the lower Severn and Culm river valleys. OD, ordnance datum (Newlyn). After Walling *et al.* (1992).

areas, the deposit is not only thicker but also coarser, containing up to twice the average percentage of fine sand (30% compared with 10–15%).

4.12.5 Pedogenic alteration of floodplain deposits

Soil development is a significant process controlling deposit character in interchannel areas. It is interesting to note that many ancient alluvial sequences contain large numbers of readily identified palaeosols, for example the Eocene Willwood Formation of Wyoming contains between 500 and 1200 palaeosols (Kraus & Bown, 1986) and the Lower Old Red Sandstone of southern Britain contains 600 calcretized horizons in 3 km of sediment (Allen, 1986). The degree of pedogenic alteration depends both on the amount of time the deposit is undisturbed at or near the surface and the rate of soil formation. Wright (1992) calculated that a floodplain with a mean rate of deposition of 2–3 mm year^{-1} will remain within the pedogenic zone for up to 10^3 years. It is therefore to be expected that most floodplain sediments will have undergone some pedogenic modification, although conditions of progressive burial are not conducive to well-developed soil profiles. Indeed, the abundance of recognizable soils in ancient alluvial sequences suggests sporadic deposition rather than ubiquitous gradual accretion.

The extent of pedogenetic modification varies across the floodplain, mirroring deposition rates. All else being equal, areas close to the channel where accumulation is more rapid will have more rudimentary soil profiles than those in distal locations. This fact can be used to infer palaeoposition relative to the active channel in ancient fluvial sequences (Kraus & Bown, 1988; Wright, 1992). Pedogenesis will also vary with relief and soil drainage so that floodplain microtopography can lead to the development of catenary relationships. For instance, better drained levee deposits may produce entisols whilst lower lying areas may develop gley or pseudogley soils under humid climatic conditions and vertisols in a more seasonally dry setting. As the floodplain aggrades, drainage frequently improves resulting in a suite of soils which varies through time.

A detailed analysis of soil types is beyond the scope of this chapter. The type of soil that may develop in any floodplain setting will depend ultimately on climate and organic productivity as well as on local controls such as relief and drainage. From this point of view, it is interesting to note that the most widely recognized floodplain soil in pre-Quaternary deposits is the semi-arid calcrete. This may in part reflect the lessening role of vegetation in more distant geological times but also suggests that geologists have failed to recognize more subtle palaeosols that were developed on the floodplains of more humid environments (Besley & Tumer, 1988).

4.13 **References**

Ackers P. & Charton F.G. 1970. Meander geometry arising from varying flows. *J. Hydrol.*, **11**, 230–252.

Ackers P. & Thompson G. 1987. Reservoir sedimentation and influence of flushing. In: Thorne C.R., Bathurst J.C. & Hey R.D. (eds) *Sediment Transport in Gravel-bed Rivers*, pp. 845–861. Wiley, Chichester.

Aleva G.J.J. 1985. Indonesian fluvial cassiterite placers and their genetic environment. *J. Geol. Soc. Lond.*, **142**, 815–836.

Allen J.R.L. 1968. *Current Ripples and Their Relation to Patterns of Water Flow and Sediment Motion.* North Holland, Amsterdam.

Allen J.R.L. 1969. Some recent advances in the physics of sedimentation. *Proc. Geol. Ass.*, **80**, 1–42.

Allen J.R.L. 1970a. A quantitative model of grain size and sedimentary structures in lateral deposits. *Geol. J.*, **7**, 129–146.

Allen J.R.L. 1970b. Studies in fluviatile sedimentation: a comparison of fining-upwards cyclothems with special reference to coarse-member composition and interpretation. *J. Sedim. Petrol.*, **40**, 298–323.

Allen J.R.L. 1982. *Sedimentary Structures: Their Character and Physical Basis. Developments in Sedimentology*, **30**, Elsevier, Amsterdam.

Allen J.R.L. 1983. River bedforms: progress and problems. In: Collinson J.D. & Lewin J. (eds) *Modern and Ancient Fluvial Systems. International Association Sedimentologists Special Publication*, **6**, 19–33. Blackwell Scientific Publications, Oxford.

Allen J.R.L. 1985. Loose boundary hydraulics and fluid mechanics: selected advances since 1961. In: Brenchley P.J. & Williams B.P.J. (eds) *Sedimentology, Recent Developments and Applied Aspects*, pp. 7–28. Blackwell Scientific Publications, Oxford.

Allen J.R.L. 1986. Pedogenic calcrete in the Old Red Sandstone facies (Late Silurian–Early Carboniferous) of the Anglo-Welsh area, southern Britain. In: Wright V.P. (ed.) *Palaeosols: Their Recognition and Interpretation*, pp. 58–86. Blackwell Scientific Publications, Oxford.

Andrews E.D. 1979. Scour and fill in a stream channel, East Fork River, western Wyoming. *U.S. Geol. Surv. Prof. Pap.*, **1117**.

Andrews E.D. 1982. Bank stability and channel width adjustment, East Fork River, Wyoming. *Wat. Res. Res.*, **18**, 1184–92.

Andrews E.D. 1983. Entrainment of gravel from naturally sorted riverbed material. *Geol. Soc. Am. Bull.*, **94**, 1225–1231.

Andrews E.D. & Parker G. 1987. Formation of a coarse surface layer as the response to gravel mobility. In: Thorne C.R., Bathurst J.C. & Hey R.D. (eds) *Sediment Transport in Gravel-bed Rivers*, pp. 269–300. Wiley, Chichester.

Arkell B., Leeks G., Newson M. & Oldfield F. 1983. Trapping and tracing: some recent observations of supply and transport of coarse sediment from upland Wales. *International Association of Sedimentologists Special Publications*, **6**, 107–119. Blackwell Scientific Publications, Oxford.

Ashida K., Takahashi T. & Sawada T. 1976. Sediment yield and transport on a mountainous watershed, *Bull. Dis. Prev. Res. Inst., University of Kyoto*, **26**, 119–144.

Ashmore P. 1987. Bedload transfer and channel morpholgy in braided streams. *Int. Ass. Sci. Hydrol. Publ.*, **165**, 333–341.

Ashworth P.J. & Ferguson R.I. 1986. Interrelationships of channel processes, changes and sediments in a proglacial river. *Geog. Ann.*, **68A**, 361–371.

Ashworth P.J. & Ferguson R.I. 1989. Size-selective entrainment of bed load in gravel bed streams. *Wat. Res. Res.*, **25**, 627–634.

Bagnold R.A. 1941. *The Physics of Blown Sand and Desert Dunes*. Methuen, London.

Bagnold R.A. 1954. Experiments on a gravity free dispersion of large solid spheres in a Newtonian fluid under shear. *Proc. R. Soc. Lond.*, **225A**, 49–63.

Bagnold R.A. 1966. An approach to the sediment transport problem from general physics. *U.S. Geol. Surv. Prof. Pap.*, **422-I**.

Bagnold R.A. 1980. An empirical correlation of bedload transport rates in flumes and natural rivers. *Proc. R. Soc. Lond. A*, **372**, 453–473.

Bagnold R.A. 1986. Transport of solids by natural water flow: evidence for a world-wide correlation. *Proc. R. Soc. Lond. A*, **405**, 369–374.

Baker V.R. 1973. Paleohydrology and sedimentology of Lake Missoula flooding in eastern Washington. *Geol. Soc. Am. Spec. Pap.*, **144**.

Baker V.R. 1977. Stream channel response to floods with examples from central Texas. *Bull. Geol. Soc. Am.*, **88**, 1057–1071.

Baker V.R. 1989. Magnitude and frequency of palaeofloods. In: Beven K. & Carling P. (eds) *Floods*, pp. 171–183. Wiley, Chichester.

Baker V.R., Kochel R.G., Patton P.C. & Pickup G. 1983. Palaeohydrologic analysis of Holocene flood slack-water sediments. In: Collinson J.D. & Lewin J. (eds) *Modern and Ancient Fluvial Systems. International Association Sedimentologists Special Publication*, **6**, 229–239. Blackwell Scientific Publications, Oxford.

Bathurst J.C. 1978. Flow resistance of large-scale roughness. *Proc. Am. Soc. Civ. Engnrs. J. Hyd. Div.*, **104**, 1587–1603.

Bauer L. & Tille W. 1967. Regional differentiations of the suspended sediment transport in Thuringia and their relation to soil erosion. *Int. Assoc. Sci. Hydrol. Publ.*, **75**, 367–377.

Beschta R.L. 1981. Patterns of sediment and organic matter transport in Oregon Coast Ranges stream. *Int. Ass. Sci. Hydrol. Publ.*, **132**, 179–188.

Beschta R.L. & Jackson W.L. 1979. The intrusion of fine sediments into a stable gravel bed. *J. Fish. Res. Bd Can.*, **36**, 204–210.

Besley B.M. & Turner P. 1983. Origin of red beds in a moist tropical climate (Etruvia Formation, Upper Carboniferous, UK). In: Wilson R.L.L. (ed.) Residual Deposits: Surface-Related Weathering Processes and Materials. Geological Society of London Special Publication, **11**, 131–147.

Best J.L. 1988. Sediment transport and bed morphology at river channel confluences. *Sedimentology*, **35**, 481–495.

Best J.L. 1992. On the entrainments of sediment and initiation of motion bed defects: insights from recent developments within turbulent boundary layer research. *Sedimentology*, **39**, 797–811.

Best J.L. & Brayshaw A.C. 1985. Flow separation — a physical process for the concentration of heavy minerals within alluvial channels. *J. Geol. Soc. Lond.*, **142**, 747–755.

Beverage J.P. & Culbertson J.K. 1964. Hyperconcentrations of suspended sediment. *Proc. Am. Soc. Civ. Engrs, J. Hydr. Div.*, **90**, 117–128.

Billi P. 1988. A note on cluster bedform behaviour in a gravel-bed river. *Catena*, **15**, 473–481.

Blake D.H. & Ollier C.D. 1971. Alluvial plains of the Fly River, Papua. *Zeit. Geomorph. Supp. Bd*, **12**, 1–17.

Blissenbach E. 1952. Relation of surface angle distribution to particle-size distribution on alluvial fans. *J. Sedim. Petrol.*, **22**, 25–28.

Bluck B.J. 1964. Sedimentation of an alluvial fan in southern Nevada. *J. Sedim. Petrol.*, **34**, 395–400.

Bluck B.J. 1971. Sedimentation in the meandering River Endrick. *Scott. J. Geol.*, **7**, 93–138.

Bluck, B.J. 1974. Structure and directional properties of some valley sandur deposits in southern Iceland. *Sedimentology*, **21**, 533–544.

Bluck B.J. 1987. Bedforms and clast size changes in gravel-bed rivers. In: Richards K.S. (ed.) *River Channels: Environment and Process. Institute of British Geographers Special Publication*, 159–178.

Bondurant D.C. 1951. Sediment studies at Concash River in New Mexico. *Trans. Am. Soc. Civ. Engnrs*, **116**, 1283–1295.

Bowman D. 1977. Stepped bed morphology in arid gravelly channels. *Bull. Geol. Soc. Am.*, **88**, 291–298.

Bradley W.C. 1970. Effect of weathering on abrasion of

granitic gravel, Colorado River, Texas. *Geol. Soc. Am. Bull.*, **81**, 61–80.

Bradshaw P. 1981. Turbulence. *Sci. Progr.*, **67**, 185–204.

Brayshaw A.C. 1984. Characteristics and origin of cluster bedforms in coarse-grained alluvial channels. In: Koster E.H. & Steel R.J. (eds) *Sedimentology of Gravels and Conglomerates. Memoir Canadian Society Petroleum Geologist*, **10**, 77–85.

Brayshaw A.C. 1985. Bed microtopography and entrainment thresholds in gravel-bed rivers. *Geol. Soc. Am. Bull.*, **96**, 218–223.

Brayshaw A.C., Frostick L.E. & Reid I. 1983. The hydrodynamics of particle clusters and sediment entrainment in coarse alluvial channels. *Sedimentology*, **30**, 137–143.

Bretz J.H. 1923. The channeled scablands of the Columbia Plateau. *J. Geol.*, **31**, 617–649.

Bridge J.S. 1975. Computer simulation of sedimentation in meandering streams. *Sedimentology*, **22**, 3–43.

Bridge J.S. 1977. Flow, bed topography, grain size and sedimentary structures in open channel bends: a three-dimensional model. *Earth Surf. Proc.*, **2**, 401–416.

Bridge J.S. 1978. Palaeohydraulic interpretation using mathematical models of contemporary flow in meandering channels. In: Miall A.D. (ed.) *Fluvial Sedimentology. Memoir Canadian Society Petroleum Geologists*, **5**, 723–742.

Bridge J.S. & Best J.L. 1988. Flow, sediment transport and bedform dynamics over the transition from dunes to upper-stage plane beds: implications for the formation of planar laminae. *Sedimentology*, **35**, 753–763.

Bridge J.S. & Diemer J.A. 1983. Quantitative interpretation of an evolving ancient river system. *Sedimentology*, **30**, 599–623.

Bridge J.S. & Jarvis J. 1982. The dynamics of a river bend: a study in flow and sedimentary processes. *Sedimentology*, **29**, 499–541.

Bridge J.S. & Leeder M.R. 1979. A simulation model of alluvial stratigraphy. *Sedimentology*, **26**, 617–644.

Brierley G.J. & Hickin E.J. 1985. The downstream gradation of particle sizes in the Squamish River, British Columbia. *Earth Surf. Proc. Landf.*, **10**, 597–606.

Brown A.G. 1983. An analysis of overbank deposits of a flood at Blandford Forum, Dorset, England. *Rev. Geomorph. Dyn.*, **32**, 95–99.

Brown A.G., 1987. Holocene floodplain sedimentation and channel response of the lower River Severn, UK. *Zeit. Geomorph.* **31**, 293–310.

Burkham D.E. 1972. Channel changes of the Gila River in Safford Valley, Arizona, 1846–1970. *U.S. Geol. Surv. Prof. Pap.* **655G**.

Butzer K.W. 1976. *Geomorphology from the Earth.* Harper & Row, New York.

Calder I.R. & Newson M.D. 1979. Land use and upland water resources in Britain — a strategic look. *Wat. Res. Bull.*, **15**, 1628–1639.

Campbell C.V. 1976. Reservoir geometry of a fluvial sheet sandstone. *Bull. Am. Ass. Petrol. Geol.*, **60**, 1009–1020.

Cant D.J. & Walker R.G. 1978. Fluvial processes and facies sequences in sandy braided South Saskatchewan River, Canada. *Sedimentology*, **25**, 625–648.

Carey W.P. 1985. Variability in measured bedload transport rates. *Am. Wat. Res. Ass. Bull.*, **21**, 39–48.

Carling P.A. 1983. Threshold of coarse sediment transport in broad and narrow rivers. *Earth Surf. Proc. Landf.*, **8**, 1–18.

Carling P.A. 1984. Deposition of fine and coarse sand in an open-work gravel bed. *Can. J. Fish. Aquat. Sci.*, **41**, 263–270.

Carling P.A. 1991. An appraisal of the velocity-reversal hypothesis for stable pool–riffle sequences in the River Severn, England. *Earth Surf. Proc. Landf.*, **16**, 19–31.

Carling, P.A. and Reader, N.A. 1982. Structure, composition and bulk properties of upland stream gravels. *Earth Surf. Proc. Landf.* **7**, 349–366.

Carlston C.W. 1965. The relation of a free meander geometry to stream discharge and its geomorphic implications. *Am. J. Sci.*, **263**, 864–885.

Carson M.A. 1984. The meandering-braided river threshold: a reappraisal. *J. Hydrol.*, **73**, 315–334.

Carson M.A. & Griffiths G.A. 1987. Bedload transport in gravel channels. *J. Hydrol. (New Zealand)*, **26**, 1–151.

Chang H.H. 1988. *Fluvial Processes in River Engineering.* Wiley, New York.

Church M. 1972. Baffin Island sandurs: a study of Arctic fluvial processes. *Geological Survey of Canada Bulletin*, **216**.

Church M. 1978. Palaeohydrological reconstructions from a Holocene valley fill. In: Miall A.D. (ed.) *Fluvial Sedimentology. Canadian Society Petroleum Geologists Memoir*, **5**, 743–772.

Church M. & Slaymaker O. 1989. Disequilibrium of Holocene sediment yield in glaciated British Columbia. *Nature*, **337**, 452–454.

Church M., Kellerhalls R. & Day T.J. 1989. Regional clastic sediment yields in British Columbia. *Can. J. Earth Sci.*, **26**, 31–45.

Clemmensen L.B. 1987. Complex star dunes and associated aeolian bedforms, Hopeman Sandstone (Permo-Triassic) Moray Firth Basin, Scotland. In: Frostick L.E. & Reid I. (eds) *Desert Sediments: Ancient and Modern. Geological Society Special Publication*, **35**, 213–231. Blackwell Scientific Publications, Oxford.

Colby B.R. 1963. Fluvial sediments — a summary of source, transportation, deposition, and measurement of sediment discharge. *U.S. Geol. Surv. Bull.*, **1181-A**.

Colby B.R. & Hembree C.H. 1955. Computations of total sediment discharge, Niobrara River near Cody, Nebraska. *U.S. Geol. Surv. Wat. Supp. Pap.*, **1357**.

Colby B.R., Hembree C.H. & Rainwater F.H. 1956. Sedimentation and chemical quality of surface waters in the

Wind River Basin, Wyoming. *U.S. Geol. Surv. Wat. Supp. Pap.*, **1373**.

Coleman J.M. 1966. Ecological changes in a massive freshwater clay sequence. *Trans. Gulf Coast Geol. Soc.*, **16**, 159–174.

Coleman J.M. 1969. Brahmaputra River: channel processes and sedimentation. *Sed. Geol.*, **3**, 129–139.

Collinson J.D. 1970. Bedforms of the Tana River. *Geogr. Ann.*, **52A**, 31–56.

Collinson J.D. 1986. Alluvial sediments. In: Reading H.G. (ed.) *Sedimentary Environments and Facies*, pp. 20–62. Blackwell Scientific Publications, Oxford.

Collinson J.D. & Thompson D.B. 1982. *Sedimentary Structures*. Allen & Unwin, London.

Conybeare C.E.B. & Crook K.A.W. 1968. Manual of sedimentary structures. *Bur. Min. Res., Geol. Geophys. Austr. Bull.*, **102**.

Cooke R.U. & Warren A. 1973. *Geomorphology in Deserts*. Batsford, London.

Costa J.E. 1975. The effects of agriculture on erosion and sedimentation in Piedmont Province, Maryland. *Bull. Geol. Soc. Am.*, **86**, 1281–1286.

Costello W.R. & Southard J.B. 1981. Flume experiments on lower-flow-regime bedforms in coarse sand. *J. Sedim. Petrol.*, **51**, 849–864.

Custer S.G., Bugosh N., Ergenzinger P.E. & Anderson B.C. 1987. Electromagnetic detection of pebble transport in streams: a method for measurement of sediment-transport waves. In: Ethridge F.G., Flores R.M. & Harvey M.D. (eds) *Recent Developments in Fluvial Sedimentology. Society of Economic Paleontologists and Mineralogists Special Publication, 39*, 21–26. SEPM, Tulsa.

Dal Cin R. 1968a. Pebble clusters: their origin and utilization in the study of palaeoccurents. *Sed. Geol.*, **2**, 233–241.

Dal Cin R. 1968b. Climatic significance of roundness and percentage of quartz in conglomerates. *J. Sedim. Petrol.*, **38**, 1094–1099.

Davoren A. & Mosley M.P. 1986. Observations of bedload movement, bar development and sediment supply in the braided Ohau River. *Earth Surf. Proc. Landf.*, **11**, 643–652.

Dawson M. 1988. Sediment size variation in a braided reach of the Sunwapta River, Alberta, Canada. *Eath Surf. Proc. Landf.*, **13**, 599–618.

de Jong C. 1991. A reappraisal of the signifiance of obstacle clasts in cluster bedform dispersal. *Earth Surf. Proc. Londf.*, **16**, 737–744.

Dimotakis P.E. & Brown G.L. 1976. The mixing layer at high Reynolds number: large structure dynamics and entrainment. *J. Fluid Mech.*, **78**, 535–560.

Douglas I. 1967a. Man, vegetation and the sediment yields of rivers. *Nature,* **215**, 925–928.

Douglas I. 1967b. Natural and man-made erosion in the humid tropics of Australia, Malaysia and Singapore. *Int.*

Ass. Sci. Hydrol. Pub., **75**, 17–29.

Douglas I. 1977. *Humid Landforms.* Australian University Press, Canberra.

Du Boys P. 1879. Le Rhone et les rivieres a lit affouillable. *Ann. Ponts et Chauss.*, **18**, 141–195.

Dunkerley D.L. 1990. The development of armour in the Tambo River, Victoria, Australia. *Earth Surf. Proc. Landf.*, **15**, 405–412.

Dury G.H. 1955. Bed-width and wave-length in meandering valleys. *Nature*, **176**, 31.

Dury G.H. 1964. Principles of underfit streams. *U.S. Geol. Surv. Prof. Pap.*, **452A**.

Dury G.H. 1976. Discharge prediction present and former, from channel dimensions. *J. Hydrol.*, **30**, 219–245.

Ehrenberger R. 1931. Direkte Geschiebemessungen an der Donau bei Wein und deren bischerige Ergehuisse. *Die Wasserwirtschaft*, **34**, 1–9.

Einstein, H.A. 1937. Die Eichung des im Rhein verwendeten Geschiebefangers. *Schweizer Bauzeitung*, **110**, 29–32.

Einstein H.A. 1942. Formulas for the transportation of bed load. *Trans. Am. Soc. Civ. Engnrs*, **107**, 561–573.

Einstein H.A. 1950. The bed-load function for sediment transportation in open channel flows. *U.S. Dept. Agric. Tech. Bull.*, **1026**.

Einstein H.A. & Barbarossa N.L. 1952. River channel roughness. *Trans. Am. Soc. Civ. Engnrs*, **117**, 1121–1146.

Emmett W.W. 1975. The channels and waters of the Upper Salmon River area, Idaho. *U.S. Geol. Surv. Prof. Pap.*, **870-A**.

Ergenzinger P.J. & Custer S.G. 1983. Determination of bedload transport using naturally magnetic tracers: first experiences at Squaw Creek, Gallatin County, Montana. *Wat. Res. Res.*, **19**, 187–193.

Ethridge F.G. & Schumm S.A. 1978. Reconstructing paleochannel morphologic and flow characteristics: methodology limitations and assessment. In: Miall A.D. (ed.) *Fluvial Sedimentology. Canadian Society Petroleum Geologists Memoir*, **5**, 703–721.

Fahnestock P.K. 1963. Morphology and hydrology of a glacial stream — White River, Mount Rainier, Washington. *US Geol. Surv., Prof. Dap.*, **422-A**.

Fenton J.D. & Abbott J.E. 1977. Initial movement of grains on a stream bed: the effect of relative protrusion. *Proc. R. Soc. Lond. A*, **352**, 523–537.

Ferguson R.I. 1986. River loads underestimated by rating curves. *Wat. Res. Res.*, **22**, 74–76.

Ferguson R.I. & Werritty A. 1983. Bar development and channel changes in the gravelly River Feshie, Scotland. In: Collinson J.D. & Lewin J. (eds) *Modern and Ancient Fluvial Systems. International Association Sedimentologists Special Publication*, **6**, 181–193. Blackwell Scientific Publications, Oxford.

Fielder H. & Head M.R. 1966. Intermittency measurements in the turbulent boundary layer. *J. Fluid Mech.*, **25**, 719–735.

Fisk H.N. 1939. Depositional terrace slopes in Louisiana. *J. Geomorph.*, **2**, 385–410.

Fleming G. & Poodle T. 1970. Particle size of river sediments. *Proc. Am. Soc. Civ. Engnrs, J. Hydr. Div.*, **96**, 431–440.

Flint R.F. 1972. Fluvial sediments in Hocking River sub-watershed 1 (N Branch, Hunters Run), Ohio. *U.S. Geol. Surv. Wat. Supp. Pap.*, **1798-I**.

Foley M.G. 1978. Scour and fill in steep, sand-bed ephemeral streams. *Geol. Soc. Am. Bull.*, **89**, 559–570.

Francis J.R.D. 1973. Experiments on the motion of solitary grains along the bed of a water-stream. *Proc. R. Soc. Lond. A*, **332**, 443–471.

Francis J.R.D. & Asfari A.F. 1970. Velocity distributions in wide, curved open channel flows. *J. Hydr. Res.*, **9**, 73–90.

Frostick L.E. & Reid I. 1977. The origin of horizontal laminae in ephemeral stream channel-fill. *Sedimentology*, **24**, 1–9.

Frostick L.E. & Reid I. 1979. Drainage-net control of sedimentary parameters in sand-bed ephemeral streams. In: Pitty A.F. (ed.) *Geographical Approaches to Fluvial Processes*, pp. 173–201. Geo Abstracts, Norwich.

Frostick L.E. & Reid I. 1980. Sorting mechanisms in coarse-grained alluvial sediments: fresh evidence from a basalt plateau gravel, Kenya. *J. Geol. Soc. Lond.*, **137**, 431–441.

Frostick L.E., Lucas P.M. & Reid I. 1984. The infiltration of fine matrices into coarse-grained alluvial sediments and its implications for stratigraphical interpretation. *J. Geol. Soc. Lond.*, **141**, 955–965.

Frostick L.E., Reid I. & Layman J.T. 1983. Changing size distribution of suspended sediment in arid-zone flash floods. In: Collinson J.D. & Lewin J. (eds) *Modern and Ancient Fluvial Systems. International Association Sedimentologists Special Publication*, **6**, 97–106. Blackwell Scientific Publications, Oxford.

Frostick L.E., Reid I., Jarvis J. & Eardley H. 1988. Triassic sediments of the Inner Moray Firth, Scotland: early rift deposits. *J. Geol. Soc. Lond.*, **145**, 235–248.

Garde R.J. & Ranga Raju K.G. 1977. *Mechanics of Sediment Transportation and Alluvial Stream Problems*. Wiley Eastern, New Delhi.

Gerson R. 1977. Sediment transport for desert watersheds in erodible materials. *Earth Surf. Proc.*, **2**, 343–361.

Gibbs C.J. & Neill C.R. 1972. Interim report on laboratory study of basket-type bedload samplers, *Research Council of Alberta Report*, **REH/72/2**.

Gilbert G.K. 1913. Hydraulic mining debris in the Sierra Nevada, *US Geol. Surv. Prof. Pap.*, **105**.

Gilbert G.K. 1914. The transportation of debris by running water. *U.S. Geol. Surv. Prof. Pap.*, **86**.

Glennie K.W. 1986. Early Permian Rotliegend. In: Glennie K.W. (ed.) *Introduction to the Petroleum Geology of the North Sea*, pp. 63–86. Blackwell Scientific Publications, Oxford.

Goldsmith V. 1973. Internal geometry and origin of vegetated coastal sand dunes. *J. Sedim. Petrol.*, **43**, 1128–1142.

Gomez B. 1983. Temporal variations in bedload transport rates: the effect of progressive bed armouring. *Earth Surf. Proc. Landf.* **8**, 41–54.

Gomez B. & Church M. 1989. An assessment of bed load sediment transport formulae for gravel bed rivers. *Wat. Res. Res.*, **25**, 1161–1186.

Gomez B., Naff R.L. & Hubbell D.W. 1989. Temporary variations in bedload transport rates associated with the migration of bedforms. *Earth Surf. Proc. Landf.*, **14**, 135–156.

Graf W.L. 1982. Catastrophe theory as a model for change in fluvial systems. In: Rhodes D.D. & Williams G.P. (eds) *Adjustments of the Fluvial System*, pp. 13–32. Allen & Unwin, London.

Graf W.L. & Pazis G.C. 1977. Les phénomèns des dépositions et d'érosion dans une canal alluvionnaire (Deposition and erosion in an alluvial channel). *J. Hydraul. Res.*, **15**, 151–166.

Grass A.J. 1971. Structural features of turbulent flow over smooth and rough boundaries. *J. Fluid Mech.*, **50**, 233–255.

Gretener B. & Stromquist L. 1987. Overbank sedimentation rates of fine grained sediments. A study of the recent deposition in the lower River Fyrisan. *Geog. Ann.*, **69A**, 139–146.

Griffiths G.A. 1979. Recent sedimentary history of the Waimakiriri River, New Zealand. *J. Hydrol. N.Z.*, **18**, 6–28.

Guy H.P., Simons D.B. & Richardson E.V. 1966. Summary of alluvial channel data from flume experiments. *U.S. Geol. Surv. Prof. Pap.*, **462-I**.

Hall A.M., Thomas M.F. & Thorp M.B. 1985. Late Quaternary alluvial placer development in the humid tropics: the case of the Birim Diamond Placer, Ghana. *J. Geol. Soc. Lond.*, **142**, 777–787.

Hamamori A. 1962. A theoretical investigation on the fluctuations of bedload transport, *Delft Hydraulics Laboratory Report*, **R4**.

Hansen E.A. 1971. Sediment in a Michigan trout stream. *U.S. Dept. Agric. Forest Serv. Res. Pap.*, **NC-59**.

Harms J.C., Southard J.B., Spearing D.R. & Walker R.G. (eds.) 1975. *Depositional Environments as Interpreted from Primary Sedimentary Structures and Stratification Sequences. Society Economic Paleontologists & Mineralogists Short Course Notes*, **2**, SEPM, Tulsa.

Harrison A.S. 1950. Report on special investigation of bed sediment segregation in a degrading bed. *Calif. Inst. Engr. Res. Ser.*, **33**, 1.1–1.205.

Hassan M.A. & Church M. 1992. The movement of individual grains on the streambed. In: Billi P., Hey R.D., Thorne C.R. & Tacconi P. (eds) *Dynamics of Gravel-bed Rivers*, pp. 159–173. Wiley & Sons, Chichester.

Hassan M.A. & Reid I. 1990. The influence of microform bed roughness elements on flow and sediment transport in gravel bed rivers. *Earth Surf. Proc. Landf.*, **15**, 739–750.

Hayward J.A. 1979. Mountain stream sediments. In: Murray D.L. & Ackroyd P. (eds) *Physical Hydrology*, pp. 193–212. New Zealand Hydrological Society, Wellington North.

Hayward J.A. & Sutherland A.J. 1974. The Torless stream vortex-tube sediment trap. *J. Hydrol. N.Z.*, **13**, 41–53.

Heidel S.G. 1956. The progressive lag of sediment concentration with flood waves. *Trans. Am. Geophys. Un.*, **37**, 56–66.

Hein F.J. 1984. Deep sea and fluvial braided channel conglomerates: a comparison of two case studies. In: Koster E.H. & Steel R.J. (eds) *Sedimentology of Gravels and Conglomerates. Memoir Canadian Society Petroleum Geologists*, **10**, 33–50.

Hein F.J. & Walker R.G. 1977. Bar evolution and development of stratification in the gravelly braided Kicking Horse River, British Columbia. *Can. J. Earth Sci.*, **14**, 562–570.

Helley E.J. 1969. Field measurements of the initiation of large bed particle motion in Blue Creek near Klamath, California. *U.S. Geol. Surv. Prof. Pap.*, **562-G**.

Hey R.D. & Thorne C.R. 1975. Secondary flows in river channels. *Area*, **7**, 191–195.

Hickin E.J. & Nanson G.C. 1984. Lateral migration rates of river bends. *Am. Soc. Civ. Engnrs, J. Hydr. Div.*, **110**, 1557–1567.

Hino M. 1963. Turbulent flow with suspended particles. *Proc. Am. Soc. Civ. Engnrs, J. Hydr. Div.*, **89**, 161–185.

Hooke J.M. 1986. The significance of mid-channel bars in an active meandering river. *Sedimentary*, **33**, 839–850.

Hooke R. LeB. 1967. Processes in arid region alluvial fans. *J. Geol.*, **75**, 438–460.

Ichim I. & Radoane M. 1990. Channel sediment variability along a river: a case study of the Siret River (Romania). *Earth Surf. Proc. Landf.*, **15**, 211–225.

Ikeda H. 1983. *Experiments on bedload transport, bed forms and sedimentary structures using fine gravel in the 4-metre-wide flume*. Environmental Research Centre, University of Tsukuba, Paper, No. 2, 78 pp.

Iseya F. & Ikeda H. 1987. Pulsations in bedload transport rates induced by longitudinal sediment sorting: a flume study using sand and gravel mixtures. *Geog. Ann.*, **69A**, 15–27.

Jackson R.G. 1976. Sedimentological and fluid dynamic implications of the turbulent bursting phenomenon in geophysical flows. *J. Fluid Mech.*, **77**, 531–560.

Jackson W.L. & Beschta R.L. 1982. A model of two-phase bed load transport in an Oregon Coast Range stream. *Earth Surf. Proc. Landf.*, **7**, 517–527.

James C.S. 1985. Sediment transfer to overbank sections. *J. Hydr. Res.*, **23**, 435–452.

Jeffreys H. 1929. On the transport of sediments in streams. *Proc. Cambridge Philo. Soc.*, **25**.

Jordan P.R. 1965. Fluvial sediment of the Mississippi River at St Louis, Missouri. *US Geol. Surv. Wat. Sup. Pap.*, 1802.

Kalinske A.A. 1947. Movement of sediment as bedload in rivers, *Trans. Am. Geophys. Uni.*, **28**, 615–620.

Károlyi Z. 1957. A study into inconsistencies in bedload transport on the basis of measurements in Hungary, *Int. Ass. Sci. Hydrol., Proc. General Assembly, Toronto*, **1**, 286–299.

Keller E.A. 1971. Areal sorting of bedload material: the hypothesis of velocity reversal. *Bull. Geol. Soc. Am.*, **82**, 753–756.

Keller E.A. & Melhorn W. 1973. Bedforms and fluvial processes in alluvial channels: selected observations. In: Morisawa M. (ed.) *Fluvial Geomorphology*, pp. 253–283. Binghampton Publications in Geomorphology, SUNY Binghampton.

Kellerhals R. 1967. Stable channels with gravel-paved beds. *Proc. Am. Soc. Civ. Engnrs, J. Water. Harb. Div.*, **93**, 63–84.

Kelsey H.M., Lamberson R. & Madej M.A. 1987. Stochastic model for the long-term transport of stored sediment in a river channel. *Wat. Res. Res.*, **23**, 1738–1750.

Kennedy J.F. 1961. *Stationary wave antidunes in an alluvial channel*. Report W.M. Keck Laboratory of Hydraulics and Water Resources, California Institute of Technology, Pasadena.

Kennedy J.F. 1963. The mechanics of dunes and antidunes in erodible-bed channels. *J. Fluid Mech.*, **16**, 521–544.

Kennedy V.C. 1964. Sediment transported by Georgia streams. *U.S. Geol. Surv. Prof. Pap.*, **1668**.

Kessel R.H., Dunn K.C., McDonald R.C. & Allison K.R. 1974. Lateral erosion and overbank deposition in the Mississippi River in Louisiana caused by the 1973 flooding. *Geology*, **2**, 461–464.

Keunen, Ph.H. 1956. Experimental abrasion of pebbles. 2, Rolling by current. *J. Geol.*, **64**, 336–368.

Kirkby M.J. 1977. Maximum sediment efficiency as a criterion for alluvial channels. In: Gregory K.J. (ed.) *River Channel Changes*, pp. 429–442. Wiley & Sons, Chichester.

Knighton A.D. 1980. Longitudinal changes in size and sorting of stream-bed material in four English rivers. *Geol. Soc. Am. Bull.*, **91**, 55–62.

Knott J.M., Lipscomb S.W. & Lewis T.W. 1987. Sediment transport characteristics of selected streams in the Susitna River basin, Alaska: Data for water year 1985 and trends in bedload discharge, 1981–85, *US Geol. Surv. Open-file Rept.*, **87–229**.

Kochel R.C. & Johnson R.A. 1984. Geomorphology and sedimentology of humid temperate alluvial fans, central Virginia. In: Koster E.H. & Steel R.J. (eds) *Sedimentology of Gravels and Conglomerates. Canadian Society Petro-*

leum Geologists Memoir, **10**, 109–122.

Komar P.D. 1987a. Selective grain entrainment by a current from a bed of mixed sizes: a reanalysis. *J. Sedim. Petrol.,* **57**, 203–211.

Komar P.D. 1987b. Selective gravel entrainment and the empirical evaluation of flow competence. *Sedimentology,* **34**, 1165–1176.

Komar P.D. & Shih, S-M. 1992. Equal grain mobility versus changing bedload grain sizes in gravel-bed streams. In: Billi P., Hey R.D., Thorne C.R. & Tacconi P. (eds) *Dynamics of Gravel-bed Rivers,* 73–93. Wiley & Sons, Chichester.

Koster E.H. 1978. Transverse ribs: their characteristics, origin and palaeohydraulic significance. In: Miall A.D. (ed.) *Fluvial Sedimentology. Canadian Society Petroleum Geologists Memoir,* **5**, 161–186.

Kraus M.J. & Bown T.M. 1986. Palaeosols and time resolution in alluvial stratigraphy. In: Wright V.P. (ed.) *Palaeosols: Their Recognition and Interpretation,* pp. 180–207. Blackwell Scientific Publications, Oxford.

Kraus M.J. & Brown T.M. 1988. Pedofacies analysis: a new approach to reconstructing fluvial sequences. *Geol. Soc. Am. Spec. Pap.,* **216**, 143–152.

Krumbein W.C. 1942. Flood deposits of the Arroyo Seco, Los Angeles County, California. *Bull. Geol. Soc. Am.,* **53**, 1355–1402.

Kuhnle R.A. & Southard J.B. 1988. Bed load transport fluctuations in a gravel bed laboratory channel. *Wat. Res. Res.,* **24**, 247–260.

Lambert C.P. & Walling D.E. 1987. Flood plain sedimentation: a preliminary investigation of contemporary deposition within the lower reaches of the River Culm, Devon, UK. *Geogr. Ann.,* **69A**, 47–59.

Lambert J.M., Jennings J.N., Smith C.T., Green C. & Hutchinson J.N. 1960. *The Making of the Broads.* Royal Geographical Society, London.

Lambrick H.T. 1967. The Indus floodplain and the 'Indus' civilization. *Geog. J.,* **133**, 483–495.

Lane E.W. & Borland W.M. 1953. River-bed scour during floods. *Trans. Am. Soc. Civ. Engnrs,* **254**, 1069–1079.

Langbein W.B. & Leopold L.B. 1968. River channel bars and dunes — theory of kinematic waves. *U.S. Geol. Surv. Prof. Pap.,* **422-L**.

Langbein W.B. & Schumm S.A. 1958. Yield of sediment in relation to mean annual precipitation. *Trans. Am. Geophys. Un.,* **39**, 1076–1084.

Laronne J.B. & Carson M.A. 1976. Interrelationships between bed morphology and bed material transport for a small gravel-bed channel. *Sedimentology,* **23**, 67–85.

Laursen E.M. 1958. The total sediment load of streams. *Proc. Am. Soc. Civ. Engnrs, J. Hydr. Div.,* **84**, 1530–1536.

Leeder M.R. 1980. On the stability of lower stage plane beds and the absence of current ripples in coarse sands. *J. Geol. Soc. Lond.,* **137**, 423–430.

Leeder M.R. 1983. On the interactions between turbulent flow, sediment transport and bedform mechanics in channelized flow. In: Collinson J.D. & Lewin J. (eds) *Modern and Ancient Fluvial Systems. International Association Sedimentologists Special Publication,* **6**, 5–18. Blackwell Scientific Publications, Oxford.

Lekach J. & Schick A.P. 1982. Suspended sediment in desert floods in small catchments. *Israel J. Earth Sci.,* **31**, 144–156.

Lekach J. & Schick A.P. 1983. 'Evidence for transport of bedload in waves: analysis of fluvial sediment samples in a small upland stream channel', *Catena,* **10**, 267–279.

Leopold L.B. & Emmett W.W. 1976. Bedload measurements, East Fork River, Wyoming. *Proc. Nat. Acad. Sci. USA,* **73**, 1000–1004.

Leopold L.B. & Emmett W.W. 1977. 1976 bedload measurements, East Fork River, Wyoming. *Proc. Natl. Acad. Sci. USA,* **74**, 2644–2648.

Leopold L.B. & Emmett W.W. 1984. Bedload movement and its relation to scour. In: *River Meandering* pp. 640–649. American Society of Civil Engineers, New York.

Leopold L.B. & Maddock T. 1953. The hydraulic geometry of stream channels and some physiographic implications. *U.S. Geol. Surv. Prof. Pap.,* **252**.

Leopold L.B. & Wolman M.G. 1957. River channel patterns — braiding, meandering and straight. *U.S. Geol. Surv. Prof. Pap.,* **282B**.

Leopold L.B. & Wolman M.G. 1960. River meanders. *Bull. Geol. Soc. Am.,* **71**, 769–794.

Leopold L.B., Emmett W.W. & Myrick R.M. 1966. Channel and hillslope processes in a semiarid area, New Mexico. *U.S. Geol. Surv. Prof. Pap.,* **352-G**.

Leopold L.B., Wolman M.G. & Miller J.P. 1964. *Fluvial Processes in Geomorphology.* Freeman, San Francisco.

Lewin J. 1981. Contemporary erosion and sedimentation. In: Lewin J. (ed.) *British Rivers,* pp. 34–58. Allen & Unwin, London.

Lewin J. & Macklin M.G. 1987. Metal mining and floodplain sedimentation in Britain. In: Gardner V. (ed.) *International Geomorphology 1986,* pp. 1009–1027. Wiley, Chichester.

Lewis G.W. & Lewin J. 1983. Alluvial cutoffs in Wales and the Borderlands. In: Collinson J.D. & Lewin J. (eds) *Modern and Ancient Fluvial Systems. International Association Sedimentologists Special Publication,* **6**, 145–154. Blackwell Scientific Publications, Oxford.

Li Z. & Komar P.D. 1986. Laboratory measurements of pivoting angles for applications to selective entrainment of gravel in a current. *Sedimentology,* **33**, 413–423.

Lisle T. 1979. A sorting mechanism for a riffle–pool sequence. *Bull. Geol. Soc. Am.,* **90**, 1142–1157.

Mackin J.H. 1937. Erosional history of the Bighorn Basin, Wyoming. *Bull. Geol. Soc. Am.,* **48**, 813–894.

Macklin M.G., Rumsby B.T. & Newson M.D. 1992. Historical floods and vertical accretion of fine-grained alluvium in the Lower Tyne Valley, northeast England. In: Billi P.,

Hey R.D., Thorne C.R. & Tacconi P. (eds) *Dynamics of Gravel-Bed Rivers*. Wiley & Sons, Chichester.

Maizels J.K. 1979. Proglacial aggradation and changes in braided channel patterns during a period of glacial advance: an Alpine example. *Geogr. Ann.*, **61A**, 87–101.

Mapes B.E. 1969. Sediment transport by streams in the Walla Walla River Basin, Washington and Oregon, July 1962–June 1965. *U.S. Geol. Surv. Wat. Supp. Pap.*, **1868**.

Marriott S. 1992. Textural analysis and modelling of a flood deposit: River Severn, UK. *Earth Surf. Proc. & Landf.*, **17**, 687–697.

McKee E.D., Crosby E.J. & Berryhill H.L. 1967. Flood deposits of Bijou Creek, Colorado, June 1965. *J. Sedim. Petrol.*, **37**, 829–851.

Meade R.H. 1985. Wave-like movement of bedload sediment, East Fork River, Wyoming. *Environ. Geol. Wat. Sci.*, **7**, 215–225.

Meade R.H., Emmett W.W. & Myrick R.M. 1981. Movement and storage of bed material during 1979 in East Fork River, Wyoming, USA. In: Davies T.R.H. & Pearce A.J. (eds) *Erosion and Sediment Transport in Pacific Rim Steeplands*. International Association Hydrological Sciences Publication, **132**, 225–235.

Meyer-Peter E. & Muller R. 1948. Formulas of bed-load transport. *International Association for Hydraulic Structures Research, Report of Second Meeting, Stockholm*, 39–64.

Miall A.D. 1988. Reservoir heterogeneities in fluvial sandstones: lessons from outcrop studies. *Bull. Am. Ass. Petrol.*, **72**, 682–697.

Middleton G.V. 1965a. *Primary Sedimentary Structures and Their Hydrodynamic Interpretation*. Society Economic Paleontologists and Mineralogists Special Publication, **12**. SEPM, Tulsa.

Middleton G.V. 1965b. Antidune cross-bedding in a large flume. *J. Sedim. Petrol.*, **35**, 923–927.

Milhous R.T. 1973 *Sediment transport in a gravel-bottomed stream*. Unpubl PhD Thesis, Oregon State University.

Milhous R.T. & Klingeman P.C. 1973. Sediment transport system in a gravel-bottomed stream. *Proc. 21st Ann. Speciality Conf. Hydraul. Div., A.S.C.E.*, 293–303.

Miller M.C., McCave I.N. & Komar P.D. 1977. Threshold of sediment motion under unidirectional currents. *Sedimentology*, **24**, 507–527.

Minter W.E.L. 1978. A sedimentological synthesis of placer gold, uranium and pyrite concentrations in Proterozoic Witwatersrand sediments. In: Miall A.D. (ed.) *Fluvial Sedimentology. Canadian Society Petroleum Geologists Memoir*, **5**, 801–829.

Minter W.E.L. & Toens P.D. 1970. Experimental simulation of gold deposition in gravel beds. *Trans. Geol. Soc. S. Afr.*, **73**, 89–99.

Mississippi River Commission 1935. Studies of river bed materials and their movement with special reference to

the lower Mississippi River. *United States Waterways Experiment Station Paper*, **17**.

Mizuyama T. 1977. *Bedload transport in steep channel*. Unpublished Ph.D. Thesis, Kyoto University.

Morison S.R. & Hein F.J. 1987. Sedimentology of the White Channel Gravels, Klondike Area, Yukon Territory: fluvial deposits of a confined valley. In: Ethridge F.G., Flores R.M. & Harvey M.D. (eds) *Recent Developments in Fluvial Sedimentology. Society Economic Paleontologists and Mineralogists Special Publication*, **39**, 206–216. SEPM, Tulsa.

Mosley M.P. 1978. Bed material transport in the Tamaki River, near Dannevirke, Nort Island, New Zealand. *N.Z. J. Sci.*, **21**, 619–626.

Mosley M.P. & Schumm S.A. 1977. Stream junctions: a probable location for bedrock placers. *Econ. Geol.*, **72**, 691–697.

Muhlhoffer L. 1933. Untersuchungen uber die Schwebstoff- und Geschiebefuhrung des Inn nachst Kirchbichl (Tirol). *Die Wasserwirtschaft*, **26**, 48–51.

Müller G. & Forstner U. 1968. General relationship between suspended sediment concentration and water discharge in the Alpenrhein and some other rivers. *Nature*, **217**, 244–245.

Murgatroyd A.L. & Ternan J.L. 1983. The impact of afforestation on stream bank erosion and channel form. *Earth Surf. Proc. Landf.*, **8**, 357–369.

Naden P. 1987. Modelling gravel-bed topography from sediment transport. *Earth Surf. Proc. Landf.*, **12**, 353–367.

Nakamura F. 1986. Analysis of storage and transport processes based on age distribution of sediment. *Trans Jap. Geomorph. Uni.*, **7**, 165–184.

Nami M. & James C.S. 1987. Numerical simulation of gold distribution in the Witwatersrand placers. In: Ethridge F.G., Flores R.M. & Harvey M.D. (eds) *Recent Developments in Fluvial Sedimentology. Society Economic Paleontologists and Mineralogists Special Publication*, **39**, 353–357. SEPM, Tulsa.

Nanson C.G. 1974. Bedload and suspended load transport in a small steep mountain stream. *Am. J. Sci.*, **274**, 471–486.

Nanson G.C. 1980. Point bar and floodplain formation of the meandering Beatton River, northeastern British Columbia, Canada. *Sedimentology*, **27**, 3–29.

Nanson G.C. 1986. Episodes of vertical accretion and catastrophic stripping: a model of disequilibrium floodplain development. *Geol. Soc. Am. Bull.*, **97**, 1467–1475.

Nanson G.C. & Hean D. 1985. The West Dapto flood of February 1984; rainfall characteristics and channel changes. *Austr. Geogr.*, **16**, 249–258.

Nanson G.C. & Page K. 1983. Lateral accretion of fine-grained concave benches on meandering rivers. In: Collinson J.D. & Lewin J. (eds) *Modern and Ancient Fluvial Systems. Special Publication International Association*

Sedimentologists, **6**, 133–143. Blackwell Scientific Publications, Oxford.

Nanson G.C. & Young R.W. 1981. Overbank deposition and floodplain formation on small coastal streams of New South Wales. *Zeit. Geomorph. N.F.* **25**, 332–347.

Nanson G.C., Rust B.R. & Taylor G. 1986. Coexistent mud braids and anastomosing channels in an arid-zone river: Cooper Creek, central Australia. *Geology*, **14**, 175–178.

Negev M. 1969. Analysis of data on suspended sediment discharge in several streams in Israel. *Israel Ministry of Agriculture, Water Commission, Hydrological Service, Hydrological Paper*, **12**.

Neill C.R. 1968. Note on initial movement of coarse uniform material. *Journal Hydraulic Research*, **6**, 173–176.

Nesper F. 1937. Ergebuisse der Messungen uber Gesschiebe- und Schlammfuhrung des Rheins an der Brugger Rheinbrucke. *Schweizer Bauzeitung*, **110**, 143–148, 161–164.

Nilsen T.V. 1982. Alluvial fan deposits. In: Scholle P.A. & Spearing D. (eds) *Sandstone Depositional Environments. American Association Petroleum Geologists Memoir*, **31**, 49–86.

Nordin C.F. 1963. A preliminary study of sediment transport parameters, Rio Puerco near Bernado, New Mexico. *U.S. Geol. Surv. Prof. Pap.*, **462-C**.

Nordin C.F. & Beverage J.P. 1965. Sediment transport in the Rio Grande, New Mexico. *US Geol. Surv. Prof. Pap.*, **462-F**.

Oldfield F., Rummery T.A., Thompson R. & Walling D.E. 1979. Identification of suspended sediment sources by means of magnetic measurements: some preliminary results. *Wat. Res. Res.*, **15**, 211–218.

Paola C., Wiele S.M. & Reinhart M.A. 1989. Upper regime parallel lamination as a result of turbulent sediment transport and low amplitude bedforms. *Sedimentology*, **36**, 47–59.

Parker G. 1976. On the cause and characteristic scale of meandering and braiding in rivers. *J. Fluid Mech.*, **76**, 459–480.

Parker G., Klingeman P.C. & McLean D.G. 1982. Bedload and size distribution in paved gravel-bed streams. *Proc. Am. Soc. Civ. Engnrs, J. Hydr. Div.*, **108**, 544–571.

Patton P.C., Baker V.R. & Kochel R.C. 1982. Slack-water deposits: a geomorphic technique for the interpretation of fluvial paleohydrology. In: Rhodes D.D. & Williams G.P. (eds) *Adjustments of the Fluvial System*, pp. 225–253. Allen & Unwin, London.

Peart M.R. & Walling D.E. 1982. Particle size characteristics of fluvial suspended sediment. *International Association Hydrological Sciences Publication*, **137**, 397–407.

Pickup G. & Warner W.A. 1976. Effects of hydrologic regime on magnitude and frequency of dominant discharge. *J. Hydrol.*, **29**, 51–75.

Pickup G., Higgins R.J. & Grant I. 1983. Modelling sediment transport as a moving wave — the transfer and

deposition of mining waste. *J. Hydrol.*, **60**, 281–301.

Pierce R.C. 1917. The measurement of silt-laden streams. *U.S. Geol. Surv. Wat. Supp. Pap.*, **400**.

Pizzuto J.E. 1986. Flow variability and the bankfull depth of sand-bed streams of the American MidWest. *Earth Surf. Proc. Landf.*, **11**, 441–450.

Plumley W.J. 1948. Black Hills terrace gravels: a study in sediment transport. *J. Geol.*, **56**, 526–577.

Popp C.L., Horley J.W., Love D.W. & Dehn M. 1988. Use of radiometric (^{137}Cs, ^{210}Pb), geomorphic and stratigraphic techniques to date recent oxbow sediments in the Rio Puerco drainage, Grants Uranium region, New Mexico. *Env. Geol. Wat. Sci.*, **11**, 253–269.

Proffitt G.T. & Sutherland A.J. 1983. Transport of non-uniform sediments. *J. Hydraul. Res.*, **21**, 33–43.

Rainwater F.H. 1962. Stream composition of the coterminous United States. *U.S. Geol. Surv. Hydrol. Invest. Atlas*, **HA-61**.

Rao K.N., Narasimha R. & Narayanan M.A.B. 1971. The bursting phenomenon in a turbulent boundary layer. *J. Fluid Mech.*, **48**, 339–352.

Rathbun R.E. & Guy H.P. 1967. Measurement of hydraulic and sediment transport variables in a small recirculating flume. *Wat. Res. Res.*, **3**, 107–122.

Reid I. 1994. River landforms and sediments: evidence of climatic change. In: Abrahams A.D. & Parsons A.J. (eds) *Geomorphology of Desert Environments*, 571–592. Harper Collins, London.

Reid I. & Frostick L.E. 1984. Particle interaction and its effect on the thresholds of initial and final bedload motion in coarse alluvial channels. In: Koster E.H. & Steel R.J. (eds) *Sedimentology of Gravels and Conglomerates. Canadian Society Petroleum Geologists Memoir*, **10**, 61–68.

Reid I. & Frostick L.E. 1985. Role of settling, entrainment and dispersive equivalence and of interstice trapping in placer formation. *J. Geol. Soc. Lond.*, **142**, 739–746.

Reid I. & Frostick L.E. 1986a. Dynamics of bedload transport in Turkey Brook, a coarse-grained alluvial channel. *Earth Surf. Proc. Landf.*, **11**, 143–155.

Reid I. & Frostick L.E. 1986b. Slope processes, sediment derivation and landform evolution in a rift valley basin, northern Kenya. In: Frostick L.E., Renaut R.W., Reid I. & Tiercelin, J-J. (eds) *Sedimentation in the African Rifts. Geological Society of London Special Publication*, **25**, 99–111. Blackwell Scientific Publications, Oxford.

Reid I. & Frostick L.E. 1989. Channel form, flows and sediments in deserts. In: Thomas D.S.G. (ed.) *Arid-Zone Geomorphology*, pp. 117–135, Belhaven Press, London.

Reid I. & Frostick L.E. 1987. Flow dynamics and suspended sediment properties in arid zone flash floods. *Hydrol. Proc.*, **1**, 239–253.

Reid I., Brayshaw A.C. & Frostick L.E. 1984. An electromagnetic device for automatic detection of bedload motion and its field applications. *Sedimentology*, **31**, 269–276.

Reid I., Frostick L.E. & Brayshaw A.C. 1992. Microform roughness elements and the selective entrainment and entrapment of particles in gravel-bed rivers. In: Billi P., Thorne C.R., Hey R.D. & Tacconi P. (eds) *Dynamics of Gravel-bed Rivers*, pp. 251–272. Wiley, Chichester.

Reid I., Frostick L.E. & Layman J.T. 1985. The incidence and nature of bedload transport during flood flows in coarse-grained alluvial channels. *Earth Surf. Proc. Landf.*, **10**, 33–44.

Richards K.S. 1978a. Simulation of flow geometry in a riffle–pool stream. *Earth Surf. Proc.*, **3**, 345–354.

Richards K.S. 1978b. Channel geometry in the riffle–pool sequence. *Geog. Ann.*, **60A**, 23–27.

Richards K.S. 1982. *Rivers: Form and Process in Alluvial Channels*. Methuen, London.

Roberts R.G. & Church M. 1986. The sediment budget of severely disturbed watersheds, Queen Charlotte Ranges, British Columbia, *Can. J. For. Res.*, **16**, 1092–1106.

Rubey W.W. 1933. The size-distribution of heavy minerals within a water-laid sandstone. *J. Sedim. Petrol.*, **3**, 3–29.

Saunderson H.C. & Lockett F.P.J. 1983. Flume experiments on bedforms and structures at the dune/plane bed transition. In: Collinson J.D. & Lewin J. (eds) *Modern and Ancient Fluvial Systems. International Association Sedimentologists Special Publication*, **6**, 49–60. Blackwell Scientific Publications, Oxford.

Sayre W.W. & Hubbell D.W. 1965. Transport and dispersion of labelled bed material, North Loup River, Nebraska. *U.S. Geol. Surv. Prof. Pap.*, **433-C**.

Schick A.P., Lekach J. & Hassan M.A. 1987. Vertical exchange of coarse bedload in desert streams. In: Frostick L.E. & Reid I. (eds) *Desert Sediments: Ancient and Modern. Geological Society of London Special Publication*, **35**, 7–16. Blackwell Scientific Publications, Oxford.

Schumm S.A. 1961. Effect of sediment characteristics on erosion and deposition in ephemeral-stream channels. *U.S. Geol. Surv. Prof. Pap.*, **352-C**.

Schumm S.A. 1968. Speculations concerning paleohydrologic controls of terrestrial sedimentation. *Geol. Soc. Am. Bull.*, **79**, 1573–1588.

Schumm S.A. & Khan H.R. 1971. Experimental study of channel patterns. *Nature*, **233**, 407–409.

Schumm S.A. & Lichty R.W. 1963. Channel widening and flood-plain construction along Cimarron River in southwestern Kansas. *U.S. Geol. Surv. Prof. Pap.*, **352-D**.

Scott K.M. & Gravlee G.C. 1968. Flood surge on the Rubicon River, California, hydrology, hydraulics and boulder transport. *U.S. Geol. Surv. Prof. Pap.*, **422M**.

Shields A. 1936. Anwendung der Aehnlichkeils mechanik, und der Turbulenzforschung auf die Geschiebebewegung. *Mitteilungen der Preussischen Versuchsanstalt für Wasserbau und Schiffbau*, **26**, 98–109.

Shotton F.W. 1978. Archaeological inferences from the study of alluvium in the lower Severn–Avon valley. In: Limbrey S. & Evans J.G. (eds) *Man's Effect on the Landscape: The Lowland Zone. Council for British Archaeology Research Report*, **21**, 27–32.

Sidle R.C. 1988. Bedload transport regime of a small forest stream. *Wat. Res. Res.*, **24**, 207–218.

Simons D.B. & Richardson E.V. 1961. Forms of bed roughness in alluvial channels. *Trans. Am. Soc. Civ. Engnrs*, **HY3 87**, 87–105.

Simons D.B. & Richardson E.V. 1962. Resistance to flow in alluvial channels. *Trans. Am. Soc. Civ. Engnrs*, **127**, 927–953.

Simons D.B., Richardson E.V. & Nordin C.F. 1965. Sedimentary structures generated by flow in alluvial channels. In: Middleton G.V. (ed.) *Primary Sedimentary Structures and Their Hydrodynamic Interpretation. Society Economic Paleontologists and Mineralogists Special Publication*, **12**, 34–52.

Singh I.B. 1972. On the bedding in the natural levee and point bar deposits of the Gompti River, Uttar Pradesh, India. *Sed. Geol.*, **7**, 309–317.

Skibinski J. 1968. Bedload transport at flood time. *Int. Ass. Sci. Hydrol. Publ.*, **75**, 41–47.

Slingerland R.L. 1984. Role of hydraulic sorting in the origin of fluvial placers. *J. Sedim. Petrol.*, **54**, 137–150.

Smith D.G. 1983. Anastomosed fluvial deposits: modern examples from western Canada. In: Collinson J.D. & Lewin J. (eds) *Modern and Ancient Fluvial Systems. International Association Sedimentologists Special Publication*, **6**, 155–168. Blackwell Scientific Publications, Oxford.

Smith D.G. & Putnam P.E. 1980. Anastomosed river deposits: modern and ancient examples in Alberta, Canada. *Can. J. Earth Sci.*, **17**, 1396–1406.

Smith D.G. & Smith N.D. 1980. Sedimentation in anastomosed river systems. Examples from alluvial valleys near Banff, Alberta. *J. Sedim. Petrol.*, **50**, 157–164.

Smith N.D. 1974. Sedimentology and bar formation in the Upper Kicking Horse River, a braided outwash stream. *J. Geol.*, **82**, 205–223.

Smith N.D. & Beukes N.J. 1983. Bar to bank flow convergence zones: a contribution to the origin of alluvial placers. *Econ. Geol.*, **78**, 1342–1349.

Smith N.D. & Minter W.E.L. 1980. Sedimentological controls of gold and uranium in Witwatersrand paleoplacers. *Econ. Geol.*, **75**, 1–14.

Solov'ev N.Ya. 1969. Pulsation in the movement of bedload in mountain streams. *Trans. State Hydrol. Inst. (TRUDY-GGI)*, **175**, 119–123.

Southard J.B. 1971. Representation of bed configuration in depth–velocity–size diagrams. *J. Sedim. Petrol.*, **41**, 903–915.

Speight J.G. 1965. Flow and channel characteristics of the Angabunga River, Papua. *J. Hydrol.*, **3**, 16–36.

Stelczer K. 1986. *Bedload Transport*, Water Resources Publications, Fort Collins, Colorado.

Sternberg H. 1875. Untersuchungen uber langen- und Querprofil gerschiebefuhrende Flusse. *Zeit. Bauwesen*, **25**, 483–506.

Sutherland A.J. 1967. Proposed mechanism for sediment entrainment by turbulent flows. *J. Geophys. Res.*, **72**, 6183–6194.

Sutherland A.J. 1987. Static armour layers by selective erosion. In: Thorne C.R., Bathurst J.C. & Hey R.D. (eds) *Sediment Transport in Gravel-bed Rivers*, pp. 243–260. Wiley, Chichester.

Sutherland D.G. 1982. The transport and sorting of diamonds by fluvial and marine processes. *Econ. Geol.*, **77**, 1613–1620.

Sutherland D.G. 1984. Geomorphology and mineral exploration: some examples from exploration for diamondiferous placer deposits. *Zeit. Geomorph. Supp.-Bd*, **51**, 95–108.

Swiss Federal Authority 1939. Untersuchungen in der Natur über Bettbildung Geschiebe- und Schwebestoffuhrung (Field research on bed formations, bedload and suspended load movement. *Mitteilung Nr 33 des Amtes für Wasserwirtschaft*, Switzerland.

Tacconi P. & Billi P. 1987. Bed load transport measurements by the vortex-tube trap on Virginio Creek, Italy. In: Thorne C.R., Bathurst J.C. & Hey R.D. (eds) *Sediment Transport in Gravel-bed Rivers*, pp. 583–606. Wiley, Chichester.

Tanner W.F. 1967. Ripple mark indices and their uses. *Sedimentology*, **9**, 89–104.

Teisseyre A.K. 1977. Pebble clusters as a directional structure in fluvial gravels: modern and ancient examples. *Geologische Sudetica*, **12**, 79–94.

Thorne C.R. & Lewin J. 1982. Bank processes, bed material movement and planform development in a meandering river. In: Rhodes D.D. & Williams G.P. (eds) *Adjustments of the Fluvial System*, pp. 117–137. Allen & Unwin, London.

Thorne C.R., Zevenbergen L.W., Pitick J.C., Rais J.B., Bradley J.B. & Julien P.Y. 1985. Direct measurements of secondary currents in a sand-bed river. *Nature*, **316**, 746–747.

Tinkler K.J. 1971. Active valley meanders in south-central Texas and their wider implications. *Bull. Geol. Soc. Am.*, **82**, 1873–1899.

Todd O.J. & Eliassen C.E. 1938. The Yellow River problem. *Proc. Am. Soc. Civ. Engnrs, J. Hydr. Div.*, **64**. 1921–1991.

Turner B.R. & Minter W.E.L. 1985. Diamond-bearing upper Karoo fluvial sediments in NE Swaziland. *J. Geol. Soc. Lond.*, **142**, 765–776.

Vanoni V.A. 1946. Transportation of suspended sediment by water. *Trans. Am. Soc. Civ. Engnrs*, **111**, 67–133.

Vanoni V.A. 1953. Some effects of suspended sediment on flow characteristics. *Proc. 5th Hydraul. Conf. State Univ. Iowa Bull.*, **34**, 137–158.

Vanoni V.A. 1964. Measurement of critical shear stress for entraining fine sediments in a boundary layer. *California Institute Technology Report*, **KH-R-7**.

Vanoni V.A. (ed.) 1975. *Sedimentation Engineering. American Society Civil Engineers Manual on Sedimentation*, **54**, American Society for Civil Engineers, New York.

Vetter C.P. 1940. Technical aspects of the silt problem on the Colorado River. *Civ. Engng*, **10**, 698–701.

Walling D.E. 1974. Suspended sediment and solute yields from a small catchment prior to urbanization. In: Gregory K.J. & Walling D.E. (eds) *Fluvial Processes in Instrumented Watersheds. Institute British Geographers Special Publication*, **6**, 169–192.

Walling D.E. 1977. Limitations on the rating curve technique for estimating suspended sediment loads, with particular reference to British rivers. *Int. Ass. Hydrol. Sci. Publ.*, **122**, 34–48.

Walling D.E. & Kane P. 1982. Temporal variation of suspended sediment properties. In: *Recent Developments in the Explanation and Prediction of Erosion and Sediment Yield. International Association of Publications*, **137**, 409–419.

Walling D.E. & Bradley S.B. 1989. Rates and patterns of contemporary floodplain sedimentation: a case study of the River Culm, Devon, UK. *GeoJournal*, **19**, 153–162.

Walling D.E. & Moorehead P.W. 1987. Spatial and temporal variations of particle-size characteristics of fluvial suspended sediment. *Geogr. Ann.*, **69A**, 47–59.

Walling D.E., Quine T.A. & He Q. 1992. Investigating contemporary floodplain sedimentation. In: Carling P.A. & Petts G.E. (eds) *Lowland Floodplain Rivers: Geomorphological Perspectives*, pp. 165–184. Wiley, Chichester.

Walling D.E., Rowan J.S. & Bradley S.B. 1989. Sediment-associated transport and redistribution of Chernobyl fallout radionuclides. *Int. Ass. Hydrol. Sci. Publ.*, **184**, 37–45.

Wertz J.B. 1966. The flood cycle of ephemeral streams in the South-Western United States. *Ann. Ass. Am. Geogr.*, **56**, 598–633.

White C.M. 1940. The equilibrium of grains on the bed of a stream. *Proc. R. Soc. Lond. A*, **174**, 322–328.

Whiting P.J., Dietrich W.E., Leopold L.B., Drake T.G. & Shreve R.L. 1988. Bedload sheets in heterogeneous sediment. *Geology*, **16**, 105–108.

Whittaker J.G. 1987. Sediment transport in step-pool streams. In: Thorne C.R., Bathurst J.C. & Hey R.D. (Eds) *Sediment Transport in Gravel-bed Rivers*, pp. 545–579. Wiley & Sons, Chichester.

Whittaker J.G. & Davies R.H. 1982. Erosion and sediment transport processes in step-pool torrents, *Int. Ass. Hydrol. Sci. Publ.*, **137**, 99–104.

Whittaker J.G. & Jaeggi M.N.R. 1982. Origin of step-pool systems in mountain streams. *Proc. Am. Soc. Civ. Engnrs, J. Hydr. Div.*, **108**, 758–773.

Willis B.J. 1989. Palaeochannel reconstructions from point bar deposits: a three-dimensional perspective. *Sedimentology*, **36**, 757–766.

Wolman M.G. & Gerson R. 1978. Relative scales of time and effectiveness of climate in watershed geomorphology. *Earth Surf. Proc.*, **3**, 189–208.

Wolman M.G. & Leopold L.B. 1957. River flood plains: some observations on their formation. *U.S. Geol. Surv. Prof. Pap.* **282-C**.

Wood P.A. 1977. Controls of variation in suspended sediment concentration in the River Rother, West Sussex, England. *Sedimentology*, **24**, 437–445.

Wright V.P. 1992. Palaeopedology: stratigraphic relationships and empirical models. In: Martini P. & Chesworth W. (eds) *Weathering, Soils and Palaeosols. Developments in Earth Surface Processes*, **2**, 475–499. Elsevier, Amsterdam.

Yalin M.S. 1977. *Mechanics of Sediment Transport*. Pergamon, Oxford.

Ying T. 1983. 'Intermittent surges of gravel transport in rivers', *Proc. 2nd Internatl. Symp. on River Sedimentation, Nanjing*, **3**, 368–377.

Zanen I.A. 1967. The results of continuous bedload measurements related to fluctuations of the river bed. *Int. Ass. Sci. Hydrol. Publ.*, **75**, 255–275.

Znamenskaya N.S. 1964. Experimental study of the dune movement of sediment. *Soviet Hydrol.*, **3**, 253–275.

5 Sedimentary processes in lakes

P.G. SLY

5.1 Introduction

The deposition, erosion, and transport of sediments in lakes is of interest in many fields of investigation, including lake productivity and nutrient dynamics, aquatic habitat, contaminant transport and pathways, sediment/water and sediment/biota interactions, and, of course, quantitative assessments related to shore erosion, longshore drift, resource extraction, dredging and water-storage capacity.

This chapter addresses three main themes. These are: the environmental setting of lakes and broadscale controls on limnological processes; the nature of sediment/hydraulic interactions; and the relationships between sediment distributions and limnological processes (including palaeoconditions). Attention is focused mainly on the physical aspects of particle behaviour and only limited consideration is given to the influence of biological and chemical factors on sedimentary processes. Most of the lakes used as examples are found in the northern hemisphere and much of the supporting data has been drawn from the Laurentian Great Lakes region of North America where extensive limnological studies have been ongoing for nearly three decades.

5.2 Environmental setting of lakes and limnological processes

5.2.1 Lakes, rivers and watersheds

From a geological perspective, lakes are ephemeral features of the landscape and only large lakes, persisting over long periods of time, provide conditions under which within-lake processes may be capable of eroding more resistant forms of bedrock. Hydraulic energies are usually sufficient only to rework unconsolidated shoreline materials or to selectively distribute the incoming sedimentary load. Thus, to a great extent, the characteristics and quantities of sediment that are subject to within-lake processes are a product of external and temporal factors. Temporal factors can be thought of as the extent to which some earlier stages in the evolution of a lake basin continue to influence the modern environment of a lake. Lakes and rivers represent opposite ends in a near-continuous spectrum of systems closure; rivers are the most open members of this spectrum and have the shortest retention times. Lakes are further characterized by the degree to which within-lake processes are dominated by edge effects. Small lakes are considerably more influenced by edge effects than large lakes.

Although lakes are an integral part of many drainage systems, a lake may be thought of as a receiving basin, partly or wholly downstream of a surrounding watershed (Fig. 5.1). The watershed acts as a collector of diffuse source inputs (mostly atmospheric) and its discharge, composed of water, bedload and suspended particulate materials, usually flows through a river channel towards the continental margin. Lakes modify this discharge and function both as a filter and a buffer. As a filter, lakes retain a portion of the particulate matter being discharged from the watershed. As a buffer, they modify both water and sediment quality in response to in-lake chemical and biochemical processes, and by integrating the inflow from different drainage basins they dampen the extremes of discharge.

Figure 5.2 provides a summary of the more important factors that directly and indirectly influence limnological conditions. Sedimentary processes are a

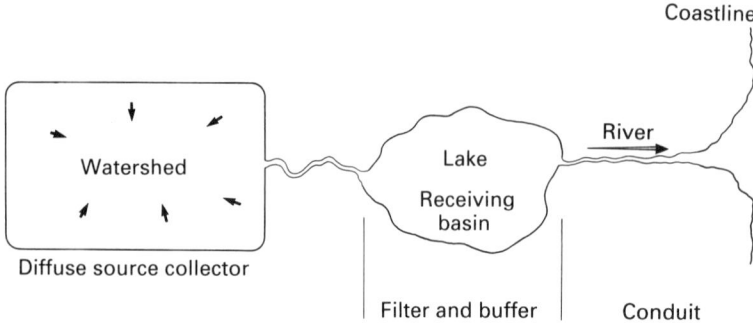

Fig. 5.1 Principal functions of watersheds, lakes, and rivers in the hydrologic cycle. Lakes act as a filter and buffer of materials in transit between source environments and receiving ocean waters.

major component of such conditions. Climate is seen to have both a direct and indirect influence on lakes (Wilson, 1977). Directly, it affects conditions over a lake surface. It also affects conditions throughout a surrounding watershed and thus, indirectly, the lake that receives discharge from that watershed. Climate is strongly influenced by air-mass circulation and, for most lakes, the scales of atmospheric structure are usually greater than that shared by the lakes and their immediate watersheds. Locally, however, there may be significant climatic variations across some very large lake basins. Further, the directions of winds blowing simultaneously across the surface of a large lake may not be the same over different parts of it. Also, spatial patterns in surface temperature structures can reflect more detailed meteorological conditions (Elder & Lane, 1970).

The characteristics of watersheds themselves also influence the quantity and quality of water and sediment that is discharged from them (Gregory & Walling, 1973; Mankiewicz *et al.*, 1975; Meybeck,

1976; Pearce, 1976; IAHS, 1977; Slaymaker, 1982; Walling & Moorehead, 1989). The size of many individual drainage basins may be less than the lakes into which they discharge. Area, slope and vegetation exert a major influence on the physical characteristics of watershed discharge. There is a close relationship between area denudation rates and suspended sediment concentrations in most drainage basins. For example, denudation rates in the Colorado River basin and the St Lawrence River basin differ by the same order of magnitude as the concentrations of their suspended loads (Sly, 1978).

5.2.2 Lake origins

As noted by Hutchinson (1957), large-scale tectonic movements have provided mechanisms for the formation of some of the world's largest lake basins. During the Miocene, vast areas of Eurasia were covered by inland seas; the Aral Sea, the Caspian Sea and the Black Sea (Hsu & Kelts, 1978) are modern remnants of the complex of earlier basins in this region. A smaller crustal sag underlies Lake Victoria in east Africa, Lake Titicaca (Richardson *et al.*, 1977) is structurally controlled and there are many other examples of rift valley lakes such as Lake Baikal (USSR), Lakes Tanganyika (Stoffers & Hecky, 1978) and Malawi (east Africa), Lake Tahoe (USA), and the Dead Sea. Lake basins formed as a result of volcanic processes are often deep and they are usually much smaller than those formed by tectonic warping. Crater Lake (USA), for example, was formed by a massive crater explosion and crater collapse, and Lake Kivu (east Africa) and the Sea of Galilee were both formed by lava dams.

Although ice-ponded lakes may be relatively small (e.g. proglacial Lake Malaspina, Alaska; Gustavson, 1975), many glacial and late-glacial lakes formed by

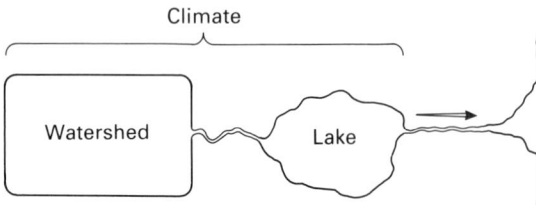

Fig. 5.2 Major factors having a direct or indirect influence on within-lake processes. *Watershed* — in which discharge and flow regime, TDS, and bed and suspended loads are additionally influenced by: area, slope, soil, vegetation, geology and relief. *Climate* — in which temperature and precipitation are largely influenced by: latitude, altitude and global air-mass circutation. *Lake* — in which residence, hydronamics and productivity are additionally influenced by: basin norphometry, size, orientation and basin volume/rate of inflow.

ice-front ponding were also of great size. During its lifetime, North American Lake Agassiz covered much of western Ontario, Manitoba, and parts of Saskatchewan, North Dakota and Minnesota (Teller *et al.*, 1983). Most of the large lakes that lie along the boundary of the Canadian Shield, including the Great Bear and Great Slave lakes, lakes Athabasca and Winnipeg, and the Laurentian Great Lakes, developed in response to a combination of factors that includes repeated glaciation, isostatic rebound, and both structural and lithological control. The huge Lake Ancylus (Hutchinson, 1957), that filled much of what is now the Baltic Sea, shared similar origins.

Glacial and late-glacial lakes include those formed by relatively shallow ice scour, and great numbers of these basins now cover the Canadian Sheild and the Scandinavian Shield. Valley glaciation has formed numerous long, narrow and deep lakes such as the lakes of the English Lake District and Alpine Europe, and fjord lakes. Many lakes, also, have been ponded behind moraine barriers (such as the Finger Lakes of northern New York State). At a much smaller scale, cirque lakes are formed by freeze–thaw effects at the head of glaciated valleys, and kettle lakes infill depressions which formerly held ice residuals in areas of glacial drift.

Numerous lakes form as a result of shifting river channels (oxbow lakes) and by deposition of sediment barriers in delta areas, for example levee lakes in estuarine deltas such as the Mississippi (Gould, 1970; Mail, 1976; Hatton *et al.*, 1983), and perched lakes in the Peace–Athabasca delta in Canada (Environment Canada, 1976). Sediment barriers also form lagoons and may similarly close embayments or valley mouths in both marine and freshwater coastal areas (Steers, 1953). Lakes may also form in wind-eroded deflation basins (Reeves, 1966) or behind barriers of wind-blown sand (e.g. Les Landes, southern France).

Other types of lakes may be formed in response to solution effects in karst areas, by ponding of organic matter, by permafrost melting, or even in meteorite craters (such as Lake Manicouagan, Quebec).

The largest man-made lakes have been formed most often by constructing barriers across existing drainage systems, for example, Lake Powell on the Colorado River, USA, and Lake Nasser on the River Nile, Egypt (Shalash, 1982). The IJsselmeer was formed by closure of the North Sea barrier, in 1932, and although polder lands are being progressively reclaimed, large areas of the western Netherlands are still covered by this shallow man-made lake. Many smaller lakes infill quarries and other forms of excavation.

Shallow lake basins, such as Lake Chad (Mothershill, 1975) are particularly responsive to climatic variability and lack consistent areal extent.

5.2.3 Lake response to physical forcing

Lake typology, based on mode of origin (Hutchinson, 1957), can be a useful guide to the sedimentary processes within lakes. Table 5.1 lists the lake and watershed areas, and depth-related values of selected lakes.

Lake response to physical forcing occurs through the integration of a number of different mechanisms (Sly, 1978). On a site-specific basis, individual mechanisms may dominate. Thus, for example, lakewide circulation patterns may be locally overridden by the effects of river discharge. Figure 5.3 shows the relationships between forcing functions, controlling factors, mechanisms and lake hydraulic response. Sedimentary processes reflect local and regional time-integration of hydraulic response.

The principal forcing functions consist of wind, river inflow, solar heating, barometric pressure and gravity. Winds cause wave generation, currents and seiche effects (Håkanson, 1977a,b). River inflows produce various types of plume structure and entrained flow. Solar heating results in stratification, and cooling may result in surface ice formation. Large variations in barometric pressure cause seiches, almost always in conjunction with wind effects. Gravitational forces result in tidal response but measurable tides occur only in large lakes; tidal currents may add to circulatory flow on a regional basis.

Controlling factors

Controlling factors are of two types, those that provide some form of fixed confinement (such as basin morphometry), and those that provide some form of variable or gradational confinement (such as stratification in a lake).

Fixed confinement. Lake morphometry is a key factor (Rawson, 1955) that, together with size and orientation, acts as a principal control on sedimentary processes in lakes (Håkanson & Jansson, 1983). In very shallow lakes, surface wave action may extend to the base of the water column and continually disturb

Table 5.1 Some physical characteristics of different lake types and their watersheds

Name	Surface area (km²)	Drainage area (km²)	Depth (m) Mean	Depth (m) Max.	Length (km)	Width (km)
Black Sea	410 000			2250	1100	600
Caspian Sea	374 000	1 400 000	182	1025	1207	483
Superior	82 100	128 000	149	407	563	259
Sea of Aral	64 500	625 000	16	67	430	280
Victoria	62 900	263 000	40	85	402	241
Huron (+ G.B.)	59 500	134 000	59	299	331	294
Michigan	57 800	118 000	85	282	494	190
Tanganyika	32 000	263 000	572	1471	676	48
Baikal	31 500	588 000	680	1741	635	78
Great Slave	28 600		74	614	456	225
Erie	25 700	58 800	19	64	338	92
Winnipeg	24 400		13	28	416	89
Malawi	22 500	65 000	273	706	579	80
Ontario	19 000	70 000	86	245	311	85
Ladoga	18 100		52	230	209	129
Chad	10 300–25 900 25 900	2 500 000	2–5	4–11	224	144
Titicaca	8 000		103	304	209	56
Vanern	5 600		32	98	145	80
Manitoba	4 600		4	28	198	48
Great Salt	4 400	21 000	4	15	121	80
Kivu	2 200		240	480	97	48
Vattern	1 900		39	128	137	19
St Clair (Laurentian Great Lakes region)	1 100	17 900	3	6	42	39
Dead Sea	1 000	12 000	149	433	79	16
Simcoe	700			41	47	29
Biwa	700		41	103	64	19
Balaton	600	5 200	4	11	77	13
Geneva	500		154	310	72	13
Tahoe	500		249	501	35	19
Crater	50			580		
Eyre	0–7 700	550 000	0–3	0–4	209	64

Sources include Hutchinson (1957), Environment Canada (1973), Golterman (1975), Wetzel (1975), Martin & Olver (1976), Sly (1978) and Herdendorf (1982).
Note: Lake Huron area includes Georgian Bay (GB).

the sediment/water interface. In such lakes, also, surface heating during the summer months may extend throughout the water column, to the extent that there is essentially no stratification. However, in deep lakes much of the water column may be below wave base (Sly, 1978; Håkanson, 1982), the depth below which there is negligible effect from wave motion. In most deep temperate lakes, the water column will be seasonally stratified, with much of it below the influence of seasonal surface water heating. Thus, depth is a major influence upon the extent to which wind stress and surface heating may influence enclosed bodies of water. Depth morphometry is further characterized by the relative proportions of the lake surface that overlie deep and shallow parts of the lake bed. A fjord-lake with a deep 'U' section would have very little of its total surface area over lake bed less than the mean depth. On the other hand, a shallow saucer-shaped depression would have a much greater proportion of its lake bed less than the mean depth. Hydraulically, there are major differences between these two extremes of lake form and they will be mirrored by the types of sedimentary processes occurring in such lakes.

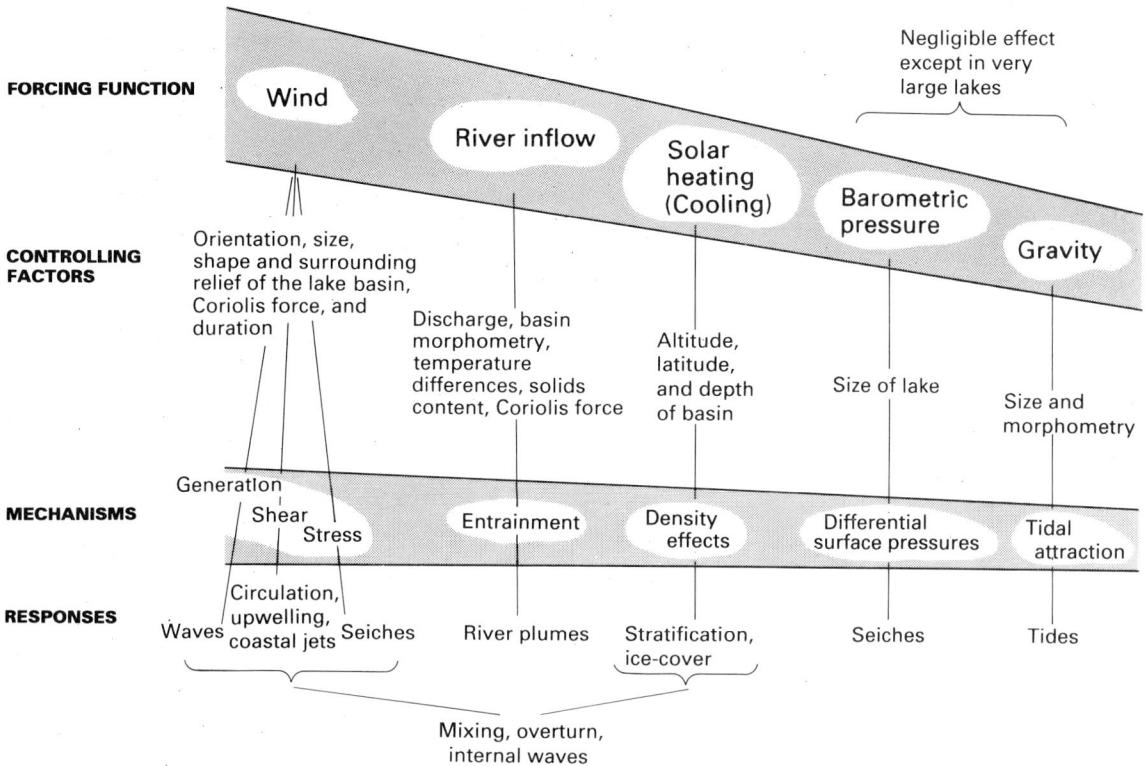

Fig. 5.3 Lake response to various forms of physical forcing functions. Modified after Sly (1978).

Shape is another morphometric characteristic which has an important influence on lake processes. The length to width ratio can be used to indicate the extent to which a lake is elongate or to record how closely this shape approximates circular form. *Complexity* reflects the extent to which a lake may be composed of a number of distinct basins or sub-basins. Basins are characterized by the near or partial closure of shoreline form, and sub-basins are usually related to the presence of sub-aqueous features within a lake. *Shoreline development* is a further description of macrorelief within a basin. It is a measure of the actual length of shoreline relative to a straight-line distance between shared end points. Values for shoreline development are very large for highly indented and island-fringed shorelines, they are negligible for stretches of featureless shoreline. However, it should be noted that shoreline development values will also vary depending upon the scale of survey chosen (Håkanson, 1981; Håkanson & Jansson, 1983).

The size of lakes varies over more than six orders of magnitude (Fig. 5.4), from features a few hectares in area to basins the size of the Caspian Sea which has an area of nearly 400 000 km² (Golterman, 1975). Lake size is particularly important because, to a large extent, it controls fetch (the length of the over-water wind path) and fetch is a major determinant of wave development in any body of water. The importance of fetch (as circular equivalent fetch) and wind speed, as determinants of the potential water depth of erosion, is illustrated in Fig. 5.4. This figure demonstrates that the effective depth of wave erosion (meaning the ability to initiate motion of the most easily eroded bed material, see later discussion) will reach maximum effect over a fetch of about 400 km for constant winds of about 5 m s⁻¹. A comparable maximum depth of effect for 10 m s⁻¹ winds would require a basin several times as large. It is unlikely that any enclosed lake basin is large enough to reach maximum depth of effect for winds of 15 m s⁻¹.

Orientation is also significant in that lakes may or may not be aligned with their long-axes parallel to

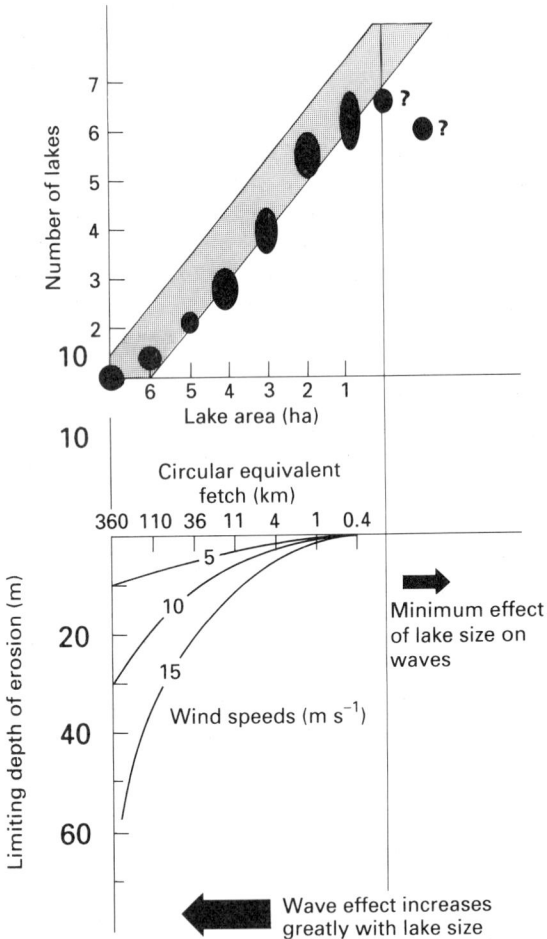

Fig. 5.4 General relationships between lake size and maximum water depth of wind–wave erosion. In the upper part of this figure, an idealized proportional relationship between lake size and frequency is shown by the screened plot. Data points are based on global (large lakes) and Canadian sources (see text). Upper data points (?) reflect temporary ponding and survey artefacts. In the lower part of this figure, lake bed erosion curves are for medium-fine sand (see Fig. 5.5) and for winds of unlimited duration. Modified after Sly (1987).

dominant wind directions. Lake shape, size and orientation form a collective influence on fetch and the development of wave climate regime. Shoreline development, on the other hand, provides some indication of the degree of shelter that can exist along different stretches of shoreline; it can also indicate the degree of complexity that may be expected in the distribution of nearshore sediments.

Variable or gradational confinement. Factors that control mechanisms of entrainment and thus the formation of river plumes include their discharge, and the relative densities of receiving and inflowing waters due to differences in temperature, dissolved solids and suspended solids (Sherman, 1953; Hutchinson, 1957; Hamblin & Carmack, 1978; Pharo & Carmack, 1979; Carmack *et al.*, 1986). In response to changes in density which are often seasonal in nature, the behaviour of features such as river plumes is often cyclical or episodic.

The Coriolis force induced by the Earth's rotation may also impart weak but consistent support to long-term current flows directed anticlockwise in the northern hemisphere and clockwise in the southern hemisphere (Csanady, 1978). The effect is most apparent in large lakes. The Coriolis effect is equally an influence on wind-induced circulation.

Although neither the altitude nor latitude of a lake basin changes in relation to the Earth's surface, solar elevation varies consistently with the seasons. Thus solar heating may be seen both as a long-term constant, and as a seasonal variable with least change at low altitude in equatorial latitudes. Passing generally, from low latitude to high latitude and based on thermal stratification, lakes have been classified (Hutchinson & Loffler, 1956; Wetzel, 1975) as oligomictic, monomictic, polymictic and amictic. Oligomictic lakes have water temperatures well above 4°C and are characterized by irregular or rare vertical circulation. Warm monomictic lakes do not drop below 4°C and circulate vertically once a year, for example, Lake Malawi (Talling, 1969) and Lake Windermere, UK (Jenkins, 1942). Cold polymictic lakes (such as Lake Joseph in Alberta; Hickman, 1979) have temperatures close to 4°C. Warm polymictic lakes have higher temperatures and are characterized by frequent vertical circulation; they are characteristic of climatic regions having small seasonal temperature change but a large daily temperature range that is often accompanied by strong winds. Lake George, in east Africa, is an example of this type (Ganf & Viner, 1973). Dimictic lakes such as Lake Baikal (Hutchinson, 1957) and Lake Mendota (Ragotzkie, 1978) circulate vertically twice a year. Warm water stratifies above cool water in summer, and cool water above cold water in winter. Most of the lakes in cool temperate climatic zones are dimictic. Cool monomictic lakes remain below 4°C but circulate once a year; Lake Stanwell-Fletcher (Rust & Coakley, 1970) and Char Lake (Schindler *et al.*,

1974) are of this type, Amictic lakes do not circulate in response to wind stress and they remain ice covered. Cool monomictic and amictic lakes such as Lakes Vanda (Goldman *et al.*, 1967; Ragotzkie & Likens, 1964), Fryxell (Vincent, 1981) and Bonney (Ragotzkie & Likens, 1964) are typical of sub-arctic and arctic climatic zones. More recently, a revised classification of lakes based on mixing has been presented by Lewis (1983).

Meromictic lakes remain generally stable and stratified and do not overturn (e.g. Lake Tanganyika; Carpart, 1952). However, these stabilities are controlled by gradients in dissolved solids or salinities rather than by thermal stratification.

On an episodic basis, extreme differences in barometric pressure across a lake surface may cause tilting of the water surface (seiching). Seiches can be considerable on large lakes; on Lake Erie, for example, the maximum range in water levels can exceed 2.5 m (Sly, 1978). This phenomenon is almost always coupled with wind stress effects. Whereas wind stress effects occur in lakes of all sizes, the coupling of wind stress and barometric effects is most noticeable in large lakes.

In summary, wind (and the accompanying effects of changing barometric pressure during storm events) is the dominant forcing function in almost all lakes having an exposed water surface. River inflows and entrainment may dominate on a site-specific or regional basis (Wright & Nydegger, 1980; Håkanson & Jansson, 1983), and can represent an important hydraulic influence in lakes of any size; river influence may be overriding in small lakes. River inflow may be the only source of high hydraulic energy in ice-covered lakes. Density stratification occurs in several different forms, depending particularly on latitude, and has a particular influence upon the movement of fine sediment in lakes.

Residence time, also, may be an important factor modifying sedimentary processes in lakes (Pharo & Carmack, 1979; Gilbert & Shaw, 1980), controlling the extent to which bed and suspended loads are retained, especially in small lakes. Residence times also have an important buffering effect on chemical fluxes from the bottom and from suspended sediments, thereby influencing water quality (Stiller & Imboden, 1986).

5.3 Sediment/hydraulic interactions

To understand sedimentary processes in lakes, it is also necessary to understand the sediment/hydraulic interactions that control them. Fundamentally different relationships exist between sediment/hydraulic interactions in erosional and depositional regimes such that it is possible to separate sediments into high and low hydraulic energy regimes. The divide between these two regimes occurs near the middle of the sand-size fraction (Sly, 1989a,b). Mechanisms that control sediment transport include material rolling and bouncing in contact with the bed, and suspension over more prolonged periods of time. Sediment transport occurs throughout both high and low energy regimes but coarse sediment transport is more or less limited to the high energy regime. For the purpose of the following discussion, the sand/silt and silt/clay boundaries are set at 4 phi (63 μm) and 8 phi (4 μm), respectively.

5.3.1 The divide between high and low energy regimes

In the lower part of Fig. 5.5, plots of erosion and sedimentation velocities are presented for particulate materials ranging in size from cobbles to clays. Whereas the plot for sedimentation shows a nearly constant relationship between particle size and flow velocity, the plot for near-bed erosional velocities shows a minimum value in the midrange of the sand-size fraction. Higher erosional velocities are associated with material both coarser and finer than this divide in the sand-size range. This means that as particle size decreases, the current velocities required to maintain particles in suspension also decrease. Also, it means that higher shear velocities are required to initiate particle motion in both very coarse (cobble and gravel) and very fine (clay) sediments. Hjulström's (1939) curves apply to cohesive fine sediments. Different curves apply to cohesionless high water content and organic-rich sediments (Postma, 1967; Fisher *et al.*, 1979).

Hjulström's (1939) observations are substantiated by the characteristics of size-frequency distributions in sediments that have a different mean particle size. In the upper part of Fig. 5.5 the envelope of standard deviation values based on more than 2000 marine and lacustrine samples is interpreted to characterize the mixing of particle sizes between end-member size populations in high energy (cobble–sand) and low energy (sand–clay) sedimentary regimes. Standard deviation is an expression of sediment sorting. High standard deviation values indicate that there is a

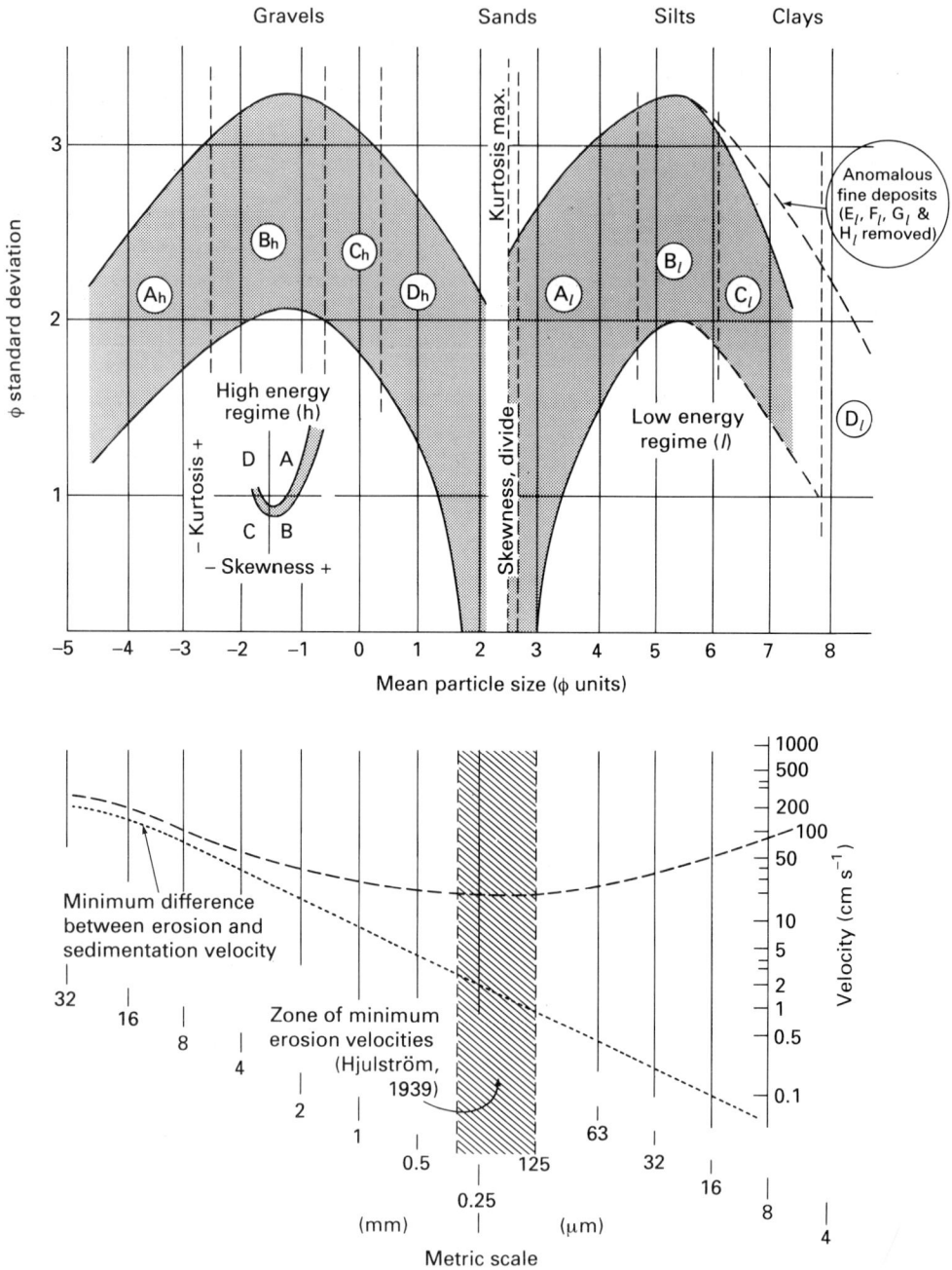

Fig. 5.5 Sediment standard deviation envelope for materials ranging in size from cobble to clay. In the lower part of this figure, flow velocities are shown in relation to particle erosion and sedimentation (Hjulström, 1939). Note, the erosion curves (Fig. 5.4) refer to materials from the zone of minimum erosion velocities (above). Particle size is given in both phi scale and metric units. Modified after Sly *et al.* (1982).

greater spread of particle size about the mean. This occurs midway between sands and cobbles, and midway between sands and clays. Sorting is very good in the mid sand-size range, coincident with the

divide between high and low energy regimes.

The third and fourth moment measures of skewness and kurtosis provide a further refinement in explanation of the particle size divide in the mid sand-size range (Sly, 1989a,b). Skewness is an expression of the symmetry of particle size sorting about the mean. Kurtosis is an expression of sorting in the tails of the size–frequency distribution relative to its central part. Changes in skewness/kurtosis relationships are particularly dramatic in the region of the high/low energy divide, as illustrated in Fig. 5.6. These changes imply that although sediment samples in the mid sand-size range have little tail to their distribution, they also require only very slight additions of coarse or fine material to make their skewness values negative (coarse-skewed) or positive (fine-skewed). The divide is therefore a sediment textural region that is extremely sensitive to slight compositional change. Sediments having a mean particle size at or near the divide respond to the minimum threshold velocities required to initiate particle motion and are highly mobile.

Since, under natural conditions, the divide reflects interactions between a number of factors, in particular the availability of materials of different particle size and changing flow conditions, the divide is best described as a balance point between the dominance of erosional and depositional conditions.

Under optimum conditions, presently active sediments would be expected to lie close to the outer boundary of the skewness/kurtosis curves shown in Fig. 5.6, exhibiting a more or less continuous compositional change from clay through silty-clay to silty-sand (low energy regime), and from sand through sandy-gravel to cobbles (high energy regime). However, actual sediment compositions, as evidenced by numerous samples, often depart from this ideal distribution and may be considered, in part, anomalous. Moment measure techniques, also, can be used

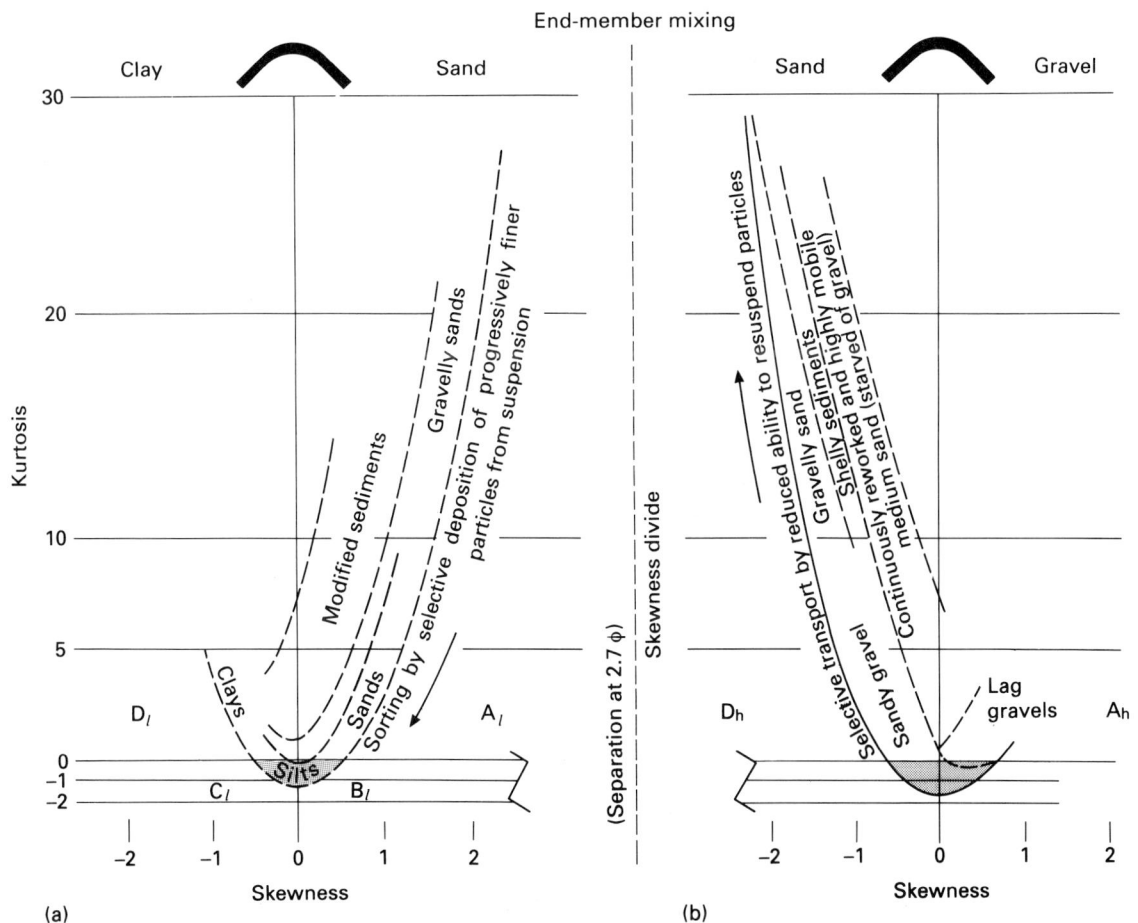

Fig. 5.6 Continuous sequence of skewness/kurtosis sectors, reflecting mixtures of (a) clay–sand and (b) sand–gravel end-members in sediments of increasingly higher energy environments. Modified after Sly *et al.* (1982).

to provide a better understanding of anomalous materials in both high and low energy regimes.

5.3.2 Sediment/energy relationships

Generally, the mean particle size of a total sediment sample is not closely correlated with formative hydraulic energy. However, there is no doubt that sedimentary deposits of decreasing particle size do reflect decreasing hydraulic energies (Sheng & Lick, 1979; Kang *et al.*, 1982). A lack of close correlation in coarse sand–cobble lag deposits can be explained by the fact that although the smallest particles may be closely associated with existing formative hydraulic shear velocities, the size of larger particles (which may predominate) is not so related. Similarly, under changing hydraulic conditions the trapping of suspended load particulates within coarse bedload materials can alter the composition of bed deposits. This obscures the relationship between dominant formative shear velocities and the mean size of bedload materials.

By restricting comparisons of shear velocity to the mean of sand-size material only, attention is drawn to the balance between erosional and depositional regimes that lie either side of the mid sand-size divide (Sly, 1989a,b). Further, since the influence of unrelated size populations is greatly reduced this comparison can be used as a more realistic approximation of *in situ* relationships between deposited materials and hydraulic velocities. However, the relationship between mean particle size and hydraulic shear is direct only when the break-point in a sample size–frequency distribution is coincident with the mean particle size. This point characterizes the intercept between traction and intermittent suspension loads (Middleton, 1976).

Owing to the effects of flocculation, there are also problems in establishing direct relationships between hydraulic shear velocities and the mean size of deposits of silty material. Flocculation generally describes the process of particle aggregation that includes complex interactions between electrolytes, particle concentrations and turbulence (Krank, 1975, 1980; Lee *et al.*, 1981; Lick, 1982; Lick & Lick, 1988). Only when suspended loads are composed of relatively low concentrations of fine silts and clays do these materials behave as discrete particles. Under these conditions, again, there is a close relationship between particle size and shear velocity.

Figure 5.6 shows the distribution of different sediment types in relation to skewness/kurtosis boundary curves. In high energy sediments, anomalous materials include lag deposits, shelly deposits (anomalous in terms of the hydraulic characteristics of this biogenic material), and materials starved of some particle size fractions (a complete range of size fractions may not always be available for inclusion in sedimentary processes in all locations). In low energy sediments, anomalous materials include ice-dropped debris, relict materials and multi-regime sediments (where sampling effectively integrates thin and multiple layers formed under conditions of changing hydraulic energy), and flocculent materials. Multi-regime sediments include turbidite deposits (Stürm, 1975; Stürm & Matter, 1978).

For convenience, formative energies of different sediment types can be displayed on a relative basis, by following the sectoral sequence defined by the changing signs of skewness and kurtosis measurements in high and low energy sediments. In Fig. 5.6, the sequence of decreasing energy is characterized by the symbols A_h–D_h in high energy sediments and by A_l–D_l in low energy sediments. The sequence is continuous. In the low energy sediments, this approach is further refined by recognizing the presence of anomalous materials, designated as E_l–H_l. If the anomalous coarse components are removed, remaining materials in the E_l–H_l samples become comparable to sediments in the sequence A_l–D_l.

5.3.3 Sediment textural data and formative origins

At this point, it should be clear that although bulk sediment textural analyses are essential as a means of describing the distributions of different sediment types, and as a basis for associated geochemical and biological studies, they are not sufficient as a means of describing the formative associations between hydraulic conditions and lake sediments. A textural analysis should include both a total sample analysis that at least quantifies the proportions of each major particle size class (cobble, gravel, sand, silt and clay), and should also provide greater detail about size/frequency composition in the sand-size class.

Two specific examples are used to illustrate the significance of detailed sediment texture data. The first focuses on the sand fraction and addresses the question of the degree to which the presence of coarse sediment implies high energy hydraulic conditions. The second seeks to establish what fundamental relationships exist in the size-composition structure of low energy sediments, and how they can be used to define low energy sedimentary processes.

An analysis of sand fraction data

Figure 5.7a presents a plot of total sample mean particle size against depth from site T in Georgian Bay, in the Laurentian Great Lakes. Superimposed on this plot are a series of curves. The solid curve represents a best-fit by eye and approximates the general textural composition of bottom sediments

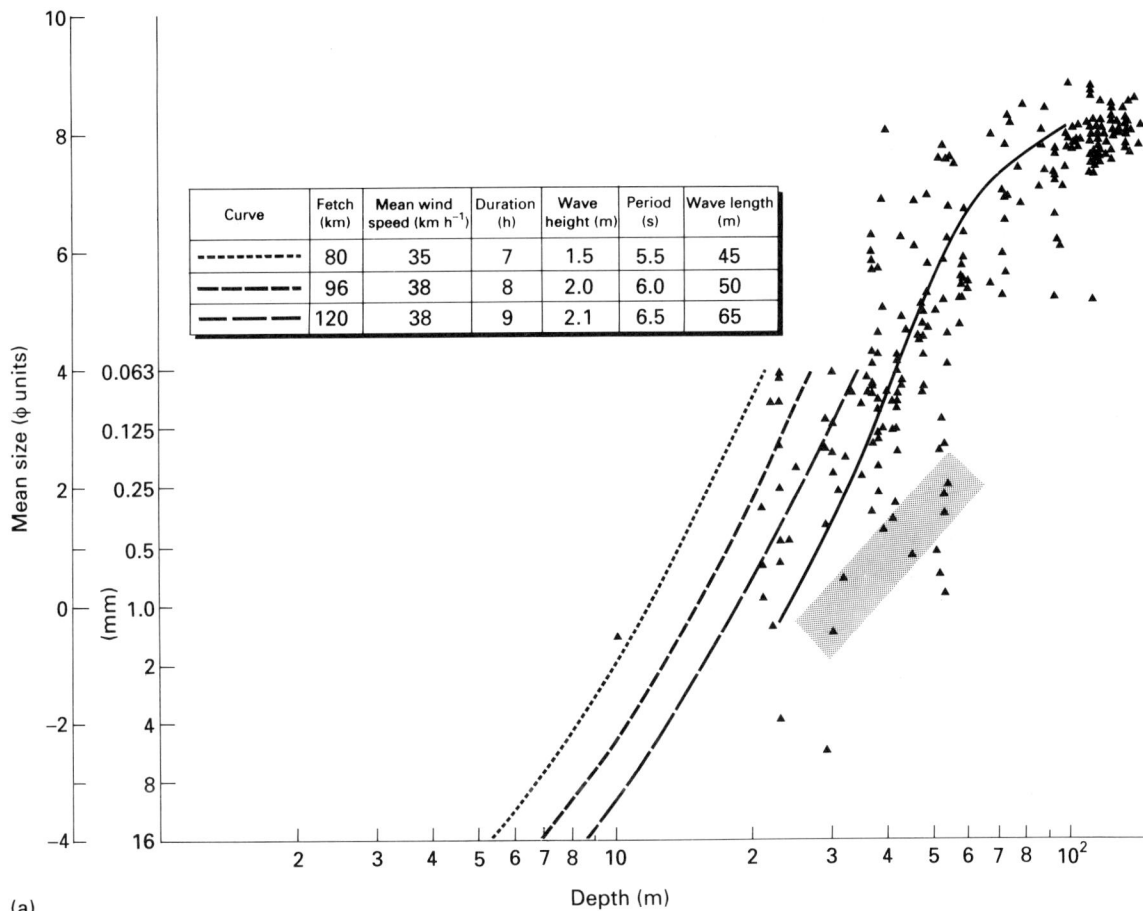

Curve	Fetch (km)	Mean wind speed (km h⁻¹)	Duration (h)	Wave height (m)	Period (s)	Wave length (m)
-------------	80	35	7	1.5	5.5	45
—— —— ——	96	38	8	2.0	6.0	50
—— — ——	120	38	9	2.1	6.5	65

(a)

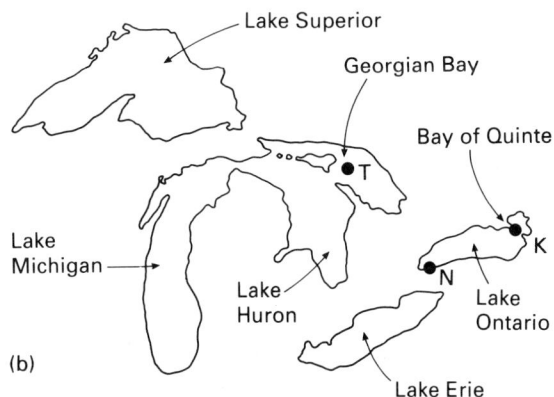

(b)

Fig. 5.7 (a) Measured and predicted relationships between mean particle size of total sediment and water depth at site T, Georgian Bay. Shaded area represents the mean size of the sand fractions only. (b) Location of sites T, N and K. Modified after Sly and Sandilands (1988).

at this site. A series of dashed lines represents estimates of mean particle size projected from climatic data (Philips & McCulloch, 1972) and wave forecast tables (US Army CERC, 1973; see also Beach Erosion Board, 1972). These estimates have been derived from the expression:

$$u_* = \pi H_s T_s \sinh (2 \pi d/L_d), \qquad (5.1)$$

(Sheng & Lick, 1979), where H_s is the significant wave height, T_s is the significant wave period, L_d is the wave length at depth (d), and the shear velocity (u_*) is expressed in cm s^{-1}. At this site, the dominant and strongest winds are from the south west but the site is largely protected from the main body of Lake Huron by the Great Barrier of shallow reefs and islands which extend north–south between Manitoulin Island and the tip of the Bruce Peninsula, forming an underwater extension of the Niagara Escarpment. The greatest fetch in Georgian Bay is about 120 km, from the south east, and therefore curve 2 is likely to be most typical of strong wave action at this site. Relationships between measured sediment particle size and the estimated size/depth curves (1–3) suggest

that either wave action at this site is very much stronger than expected or that the bottom sediments are anomalously coarse. Water levels in this region of the Laurentian Great Lakes are known to have changed considerably during the postglacial period (Sly & Sandilands, 1988; Lewis & Anderson, 1989). On this basis, it is thought that the lack of fit between measured and predicted size/depth curves may indicate that much of the bottom sediment at this site is relict material formed during earlier periods of lower lake levels. This interpretation is also supported by the presence of lag gravels, particularly at depths of 16–60 m.

The mean of the sand-size fraction has also been plotted on Fig. 5.7a and is shown by a screened band. This band lies to the extreme right of all other curves and strongly implies that sands are found at depths much greater than expected from existing hydraulic conditions. Also the slope of this band suggests that trends representing sand-only mean size and total sample mean size may closely approach or intercept in shallow water. The characteristics of sand-only mean size data, therefore, further support the sugges-

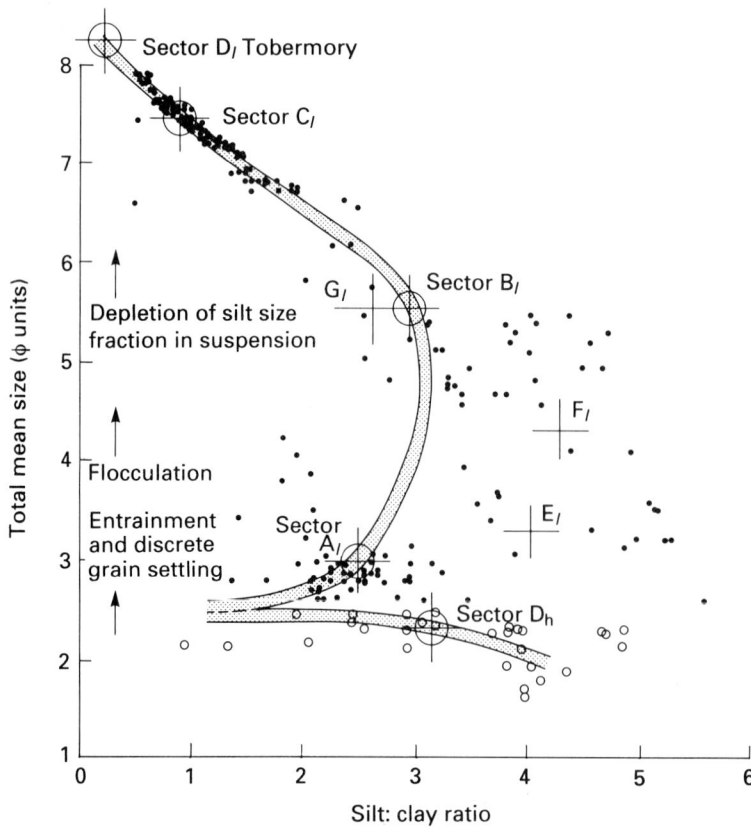

Fig. 5.8 Relationships between total mean phi size and the silt : clay ratio in sediment samples from site N (see Fig. 5.7b, p. 167), western Lake Ontario. After Sly (1989a). See text for explanation.

tion that deposits at this site include relict materials formed under previous low lake levels, and also that sediment texture may more closely correlate with modern hydraulic conditions in shallow water. Unfortunately, sediment size/depth data from less than about 20 m are not available from this site and it is not possible to present a more complete picture of the relationships that have been proposed.

Silt/clay ratios in fine sediments

Data presented in Fig. 5.8 are drawn from bottom sediment studies off the mouth of the Niagara River in western Lake Ontario (site N). The data are plotted as small open circles (high energy regime) and closed circles (low energy regime), and express the relationships between the silt/clay ratio and total sample mean size. Because sampling at the Niagara site does not extend into water depths that overlie materials of the finest particle sizes, Georgian Bay data have been used to establish the position of clay-size sample data (finer than 8 phi). It has been previously established that the use of these Georgian Bay data does not distort the trends otherwise established in the Lake Ontario data set, and they are used only to provide an overall interpretation of sediment energy relation-

ships in the following discussion (Sly, 1989a,b). Two further sets of information are presented in this figure. The mean size equivalent of each skewness/kurtosis sector (shown in Fig. 5.6) is superimposed on the sediment data. As previously noted, the sequence of skewness/kurtosis sectors can be used to express relative hydraulic energies. This sequence is shown as a screened band passing through the skewness/kurtosis sectors D_h, A_l, B_l, C_l, and D_l. The mean values of anomalous low energy sector data are plotted as points (E_l, F_l and G_l; sector H_l is not present at this site). For the most part, these anomalous data depart considerably from the screened band.

Figure 5.8 demonstrates that there is a clear trend towards decreasing low silt/clay ratios in sediments having a total sample mean particle size finer than about 6.5 phi. There is an inconsistent relationship between mean size and the silt/clay ratio in slightly coarser sediments (coarse silts–sands). The silt/clay ratios are further explored in Fig. 5.9 which plots the same sample data as per cent clay vs. per cent silt. Again, small open and closed circles are used to differentiate between high and low energy regime samples. Values of the mean phi size of total sample material are superimposed on this figure and clearly

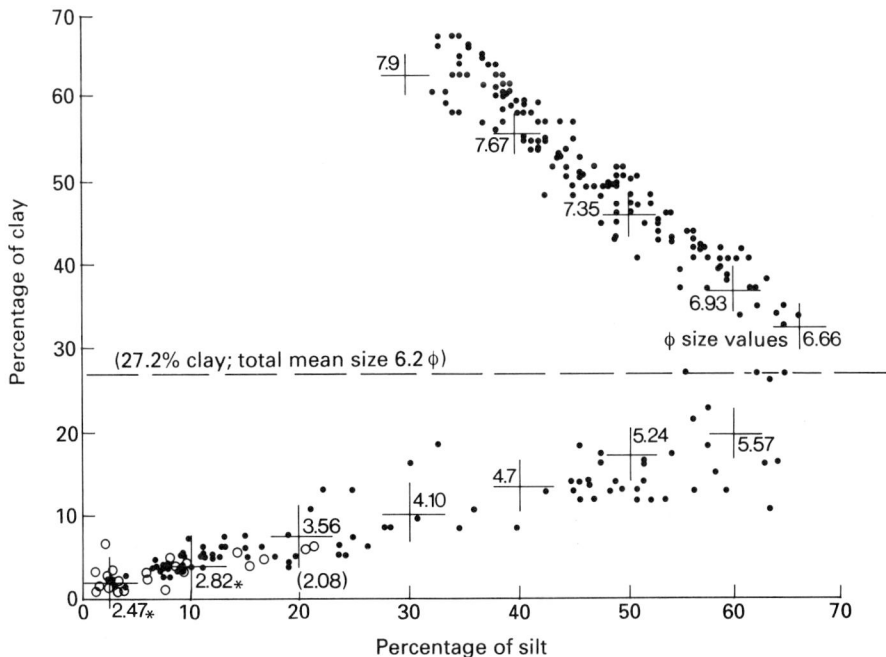

Fig. 5.9 Silt/clay size composition of site N samples (see p. 167). Crossbars relate total sample mean phi size to equivalent silt/clay values. After Sly (1989a).

establish that, as mean particle size decreases, there is first an increase and then a decrease in the silt/clay ratio, as samples fine from about 2.5 phi to > 7.9 phi. The peak in the silt/clay ratio is thought to occur when exhaustion of coarse and medium silt size material in the suspended load causes a change in particle settling behaviour, and marks a shift from flocculation to the settlement of discrete grains.

The combined evidence from Figs 5.8 and 5.9 strongly implies that highly consistent low silt/clay ratios in sediments finer than about 6.5 phi total sample mean size indicate deposition of discrete grains in response to Stokes Law. The deposition of silts and fine sands (coarser than about 6.5 phi) reflects both flocculation and an inconsistent hydraulic regime (including turbidity flows) in the discharge plume adjacent to the Niagara River mouth. Silt/clay ratios in sediments that are dominated by high energy hydraulic regimes indicate entrainment and/or entrapment of fines in materials that are by definition erosional substrates. The silt/clay ratios of fines within otherwise high energy regime coarse substrates are thought to reflect the composition of the suspended load, rather than some component selectively deposited from it.

5.4   Relationships between sediment distributions and limnological processes

The Great Lakes exhibit a wide range of sedimentary processes and associated structures and, since from a sedimentological point of view they have been studied probably more intensively than any other lake system, they are used as a basis for much of the following discussion. Where appropriate, additional information has been drawn from lake studies elsewhere.

Sedimentary processes in lakes have been defined as the response of particulate materials to a wide range of hydraulic mechanisms (Fig. 5.3). These mechanisms interact and are driven by different physical forcing functions. Suspended and bedload materials reflect immediate hydrodynamic conditions. Sediment deposits record temporal integration of particulate responses over a time-interval equivalent to the depth of material sampled. Often, it is impossible to differentiate between effects of specific hydraulic mechanisms; rather, sediment deposits reflect a net balance of formative conditions (Bennett, 1971; Boyce, 1974). Sedimentary processes differ,

not only in response to high or low energy regimes but particularly in respect to the influence of edge effect in a basin. On the basis of decreasing edge effect, the following discussion considers interface processes at the shoreline and in the nearshore zone. The influences of river inflow and lake outflow are included in both this and the following section. Processes that characterize the intermediate depth zone and the deep midlake zone are considered separately, although a number of sedimentary processes overlap this differentiation. The nearshore and intermediate zones are present in virtually all lakes, of any size. The deep midlake zone is effectively absent in many shallow lakes.

5.4.1 Shoreline and nearshore zone

This zone includes all high energy regime sedimentary environments and, also, a narrow margin of low energy regime environments, where these intercept the shoreline (such as protected areas behind barrier islands, embayments and marshes). It is dominated by the effects of changing lake level, wind-waves, blown sand, ice erosion and freeze–thaw effects (in cold climates). Shoreline and nearshore sedimentary processes include erosion, transport, deposition and non-deposition. Generally speaking, the larger the lake and the greater its shoreline exposure, the greater the vertical depth of its nearshore zone. For comparison, Sly (1993) reported that the water depth of marine sediments (Scotia Shelf, eastern Canada), at the lower limit of the high energy hydraulic regime, is nearly three times the equivalent depth in Lake Superior and about 10 times the equivalent depth in both lakes Erie and Ontario. In lakes of more moderate size (about 100 km^2), the equivalent depth zone is perhaps 25–50 times less than on exposed marine shorelines. Beach length as well as the vertical range of active sediment tends to be smaller in lacustrine than marine environments. Thus, most shoreline and nearshore sedimentary processes are shared by lakes of all sizes, and many with the marine environment (Cook, 1970; Davis et al., 1972; Komar, 1976). Proportionally, however, high energy regime processes are of less significance in lacustrine environments, particularly in small basins.

The shoreline

As noted earlier, it is only in large lakes that wave energies are sufficient to be an effective agent of bedrock erosion. However, shoreline materials frequently comprise unconsolidated sands and silty-clay materials, and these are readily eroded by wave action even in lakes only a few km² in area.

Figure 5.10 plots the change in mean particle size on a relatively high energy lake beach in eastern Lake Ontario, and demonstrates the combined effects of both seasonal water level variation and storm wave action over principal fetch distances of 50 and 280 km (Sly & Schneider, 1984). At this site, beach cobbles were actively carried nearly 2 m above mean water level, and the combination of lateral movement and roll-down effects induced cobble transport to a depth of more than 4 m below mean water level. This example is by no means extreme, and it indicates that shore processes in large lakes can be very dynamic. For comparison, Coakley and Cho (1973) reported erosion of beach sands at water depths > 5 m in western Lake Ontario, during a major storm in November 1972. Figure 5.11 shows sand laminations due to periodic deposition and reworking in the same area, and related shore erosion is shown in Figure 5.12.

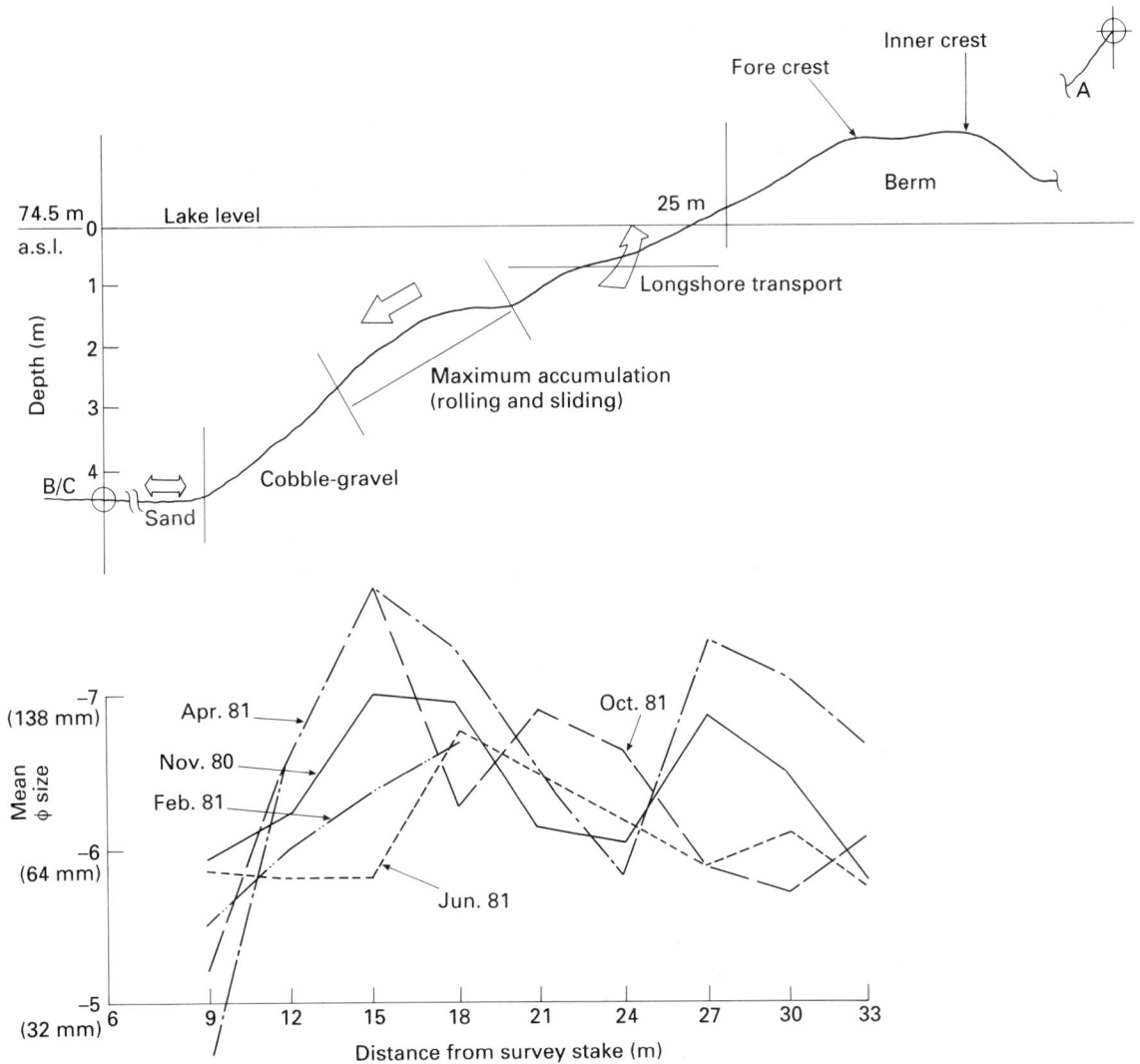

Fig. 5.10 Seasonal changes in the particle size of a cobble–gravel beach in eastern Lake Ontario, near site K (see p. 167). Modified after Sly and Schneider (1984).

Fig. 5.11 Sand laminae exposed in a cut vibro core from western Lake Ontario, water depth about 5 m (After Sly & Gardner, 1970).

Fig. 5.12 The effects of intense storm waves coupled with high lake levels: shore erosion along western Lake Ontario (photo records, Canada Centre Inland Waters, Burlington, Ontario).

Shore erosion is accentuated by the effects of terrestrial weathering and, due to freeze–thaw effects, remarkably high rates of bluff recession can occur as a result of combined slumping and wave action (Coakley & Hamblin, 1970; Gelinas & Quigley, 1973; Rukavina & Zeman, 1987). In the Laurentian Great Lakes, the recession of unprotected shorelines may exceed 2 m year^{-1} as a result of such combined effects (Environment Canada, 1975; Haras et al., 1976; IJC, 1981). Bluff slumping, often in response to arcuate slippage (and other forms of destabilization largely influenced by groundwater seepage) can affect bluff faces that exceed 40 m in height (Quigley & Tutt, 1968). Shore erosion in lakes differs very little from similar processes active along marine shore-

lines. Despite susceptibility of shore materials to terrestrial weathering, shore erosion is only significant where hydraulic energies are sufficient to remove material accumulating at the foot of above water slopes. Slumping induced by freeze–thaw effects is a particularly important shore process in boreal and sub-arctic zones and in areas of surface permafrost.

Ice push (Tsang, 1974) can accentuate shore erosion but, also, it can produce selective armouring of down-wind shores. Spring ice push frequently moves ice-bonded bed material shoreward. If bed materials are composed of coarse sediments, these are carried shoreward without size selection of the aggregate; subsequent wave action during the open water period usually removes the fine particulates and leaves a coarse lag deposit. In some areas, this may provide a protective armouring of the shoreline. Ice may also protect the shoreline from intense seasonal wave action (O'Hara & Ayers, 1972). While fixed floating winter ice rarely exceeds 0.6 m (Assel *et al.*, 1983), rafted or frozen slush-ice can bond to bed materials in water depths of 3 m or more in many areas of the Laurentian Great Lakes. Thus, ice is an extremely important factor, capable of both eroding and protecting shorelines. Ice action is obviously a characteristic of cool temperate and colder climates and, because of the salinity of oceanic water, it is a more dominant feature of freshwater lakes than marine coastal areas. Further, heat storage capacities of most lakes are very limited in comparison to oceanic waters and thus cooling takes place more rapidly in lakes than most oceanic environments.

Lake level fluctuations can have a marked effect on shoreline erosion (Haras *et al.*, 1976). For example, if beach sands dry out as a result of lowered water tables, the fine particulates are readily mobilized by aoelian processes. As a result of wind erosion, seasonal beach lowering of as much as 20 cm may occur in parts of the Laurentian Great Lakes when winter draw-down occurs before beach surfaces are adequately protected by ice and snow.

The entrapment of fines in sheltered bays and along protected shorelines may result in sediment accumulation at the shoreline, and formation of freshwater marshes and wetlands closely parallels similar features in the marine environment (IJC, 1981). Long-term fluctuations in lake levels may produce effects similar to those seen in subsiding/accreting coastal wetlands (Hatton *et al.*, 1983), resulting in periodic rejuvenation.

The nearshore

The depth of water in which silt-size particulates become a significant proportion of bottom material (>5–10%) can be taken as an approximate guide to the lower limit of the high energy hydraulic regime in lakes. Typically, this transition occurs at a depth of about 1–1.5 m in lakes with a fetch of about 1 km; the transition occurs at about 6–7 m in lakes with a fetch of about 10 km, and with a fetch of about 100 km the transition depth is about 12–15 m. For much of the Laurentian Great Lakes shoreline the transition occurs at about 20–25 m. These values are of course only a guide (Håkanson, 1977b; 1982) since not only do shoreline and nearshore morphologies and surrounding relief greatly modify wave ortho-

gonals, but wind regimes may differ greatly even within similar climatic zones.

Depending on the hydraulic energy of the nearshore zone and the supply of particulates, both in terms of quantity and range of particle size, all the normal processes associated with coastal erosion, deposition and sediment transport are equally active in lakes. The principal difference between comparable features in lakes and the marine environment is that the features are usually smaller in lakes. Also, because hydraulic energies are less and therefore movements of coarse sediments are compressed into a smaller vertical component of the water column, shore profiles in lakes with active sedimentary processes tend to have lower slope angles than marine shorelines. This characteristic is also controlled by the particle size of source materials and can often reflect a lack of available coarse particulates. Without the effects of tidal range, lake beaches are narrower than marine beaches.

Where supplies of sediment are not limited, extensive beach structures develop (Rukavina, 1976; Davidson-Arnott & Pollard, 1980). Bay-head bars reflect convergence of high energy sediment transport, headland controlled beaches form where unidirectional longshore transport is trapped by a barrier, and openshore beaches form when rates of sediment supply approximately balance their removal. The latter are particularly susceptible to seasonal variabilities in water level and wave action, and are highly dependent on material supplies from either bluff or offshore sources, or both. Offshore sources are particularly important in large lakes where water levels have risen continually over a long period of time and selective sorting has provided a mechanism for the shoreward transport of the most mobile sand-size fraction. The presence of sands in the size range of 2–2.5 phi indicates extreme mobility but because of this it is almost impossible to determine net transport directions based only on slight differences in mean particle size. Transport directions are more reliably determined from the distribution patterns of sediment accumulations (Rukavina, 1976; Davidson-Arnott & Pollard, 1980). Particle size characteristics of sediments that are much coarser or much finer than 2–2.5 phi are usually related to more stable sedimentary features of the nearshore (in that they represent areas in which vectors associated with high or low energy regimes are more consistent). Because of the temporal instabilities of most sediments in the nearshore zone, it is difficult to characterize their rates of accumulation. Under storm conditions it is possible to erode, perhaps, 0.5 m of bed material, and yet to accumulate more than this during subsequent and less dynamic seasonal conditions. Sedimentation rates are also highly variable, spatially, in the nearshore zone.

The importance of source material is borne out by the work of Fricbergs (1970), in his study of the Toronto Island area of Lake Ontario. At this site, bluff erosion supplied an estimated 3×10^5 m^3 year^{-1} of sediment but only 1.9×10^4 m^3 of this was coarse enough to be incorporated within the westward longshore drift. Deposition of this drift material has built the eastern spit of the Toronto Island complex. However, nearly 94% of the eroded bluff material was composed of fines, and these passed into deep water and were lost to the nearshore hydraulic regime.

Storms can develop intense wave action and prolonged wind–wave activities can remove large volumes of material. During 1971, for example, a major storm in Lake Huron, with winds gusting to about 100 km h^{-1}, eroded considerable stretches of the south east shoreline of the lake. Of an estimated 6×10^5 m^3 eroded from the Sarnia shoreline, only about 2.5×10^5 m^3 was later redeposited; most material was transported southward by persistent longshore currents, probably > 30 cm s^{-1} (Freeman et al., 1972). In contrast to the coarse-size fraction, concentrations of finer particulates tend to reach steady state equilibrium under erosional conditions (Lick, 1982).

Erosion and transport of coarse sediment is limited to the nearshore zone, and within this zone the process may lead to formation of anomalous lag deposits and relict features. Two different types of anomalies are particularly important and both relate to erosion of glacial deposits. Erosion of stony tills frequently results in the formation of boulder or cobble–gravel pavements within the nearshore zone (Rukavina, 1969), but at depths well beyond the capacity of hydraulic energies to move the large particles. In fact, wave action has removed only the fine materials, leaving behind a lag deposit often no more than one or two particle layers thick but extending over very large areas of lake bed. Similarly wave erosion of drift features such as drumlins, kames, moraines and eskers leave only isolated cobble–gravel and boulder patches, within an otherwise sandy substrate, marking the core of the original features.

Hydraulic energies may be quite sufficient to erode and transport sands and silts in the nearshore zone but if sediment exchange is not possible, owing to barriers between potential donor and receiving sites, both the transport zone and the receiving site remain devoid of new particulates. Examples of possible barriers include headlands that extend into deep water and channels that intercept longshore transport. Deep waters which isolate islands and reefs from nearshore sediment transport also serve as effective barriers.

Inflow structures

Although the Niagara River is a very large feature and has a discharge of about 5.4×10^3 m^3 s^{-1}, its impact on the nearshore zone of Lake Ontario shares similarities with many other rivers and streams in much smaller lakes. The nearshore transport of sediment is interrupted by the river and particulates moving along the shore are carried out to form an offshore bar as they meet the entrained flow of the river. At the present time, the bedload of the Niagara River is relatively small and almost all modern sediments in the bar are derived from shore erosion and nearshore sediment transport (Fig. 5.13). The river maintains a distinct and largely stable bathymetric channel as it crosses the nearshore zone of the lake. Channel sediments are much coarser than those in the surrounding bar, and are mostly composed of lag gravels (Sly, 1983).

Where inflowing rivers or streams contribute a greater proportion of bedload material than adjacent shorelines, their outflow structures may not differ significantly from the Niagara example cited here. The extensive tongue of sandy sediments which extends into Lac Leman (Houbolt & Jonker, 1968) marks the inflow of the River Rhône into a much smaller lake; it is also a site with a much greater sediment loading from the river than the nearshore. However, some differences occur at sites where the size composition of river-contributed material differs from that in the nearshore zone. In many of the Finger Lakes in New York State, for example, prominent delta-fans of cobble–gravel mark the entry of streams around steeply sloping shorelines. These deposits contrast strongly with the remainder of the nearshore zone which is characterized by narrow and irregular patches of sandy gravel. Inflow channels of rivers whose sediment load is dominated by materials of the most mobile sand fraction (2–2.5 phi) are characteristically shallow, and tend to be less stable than those formed with either coarser or finer material.

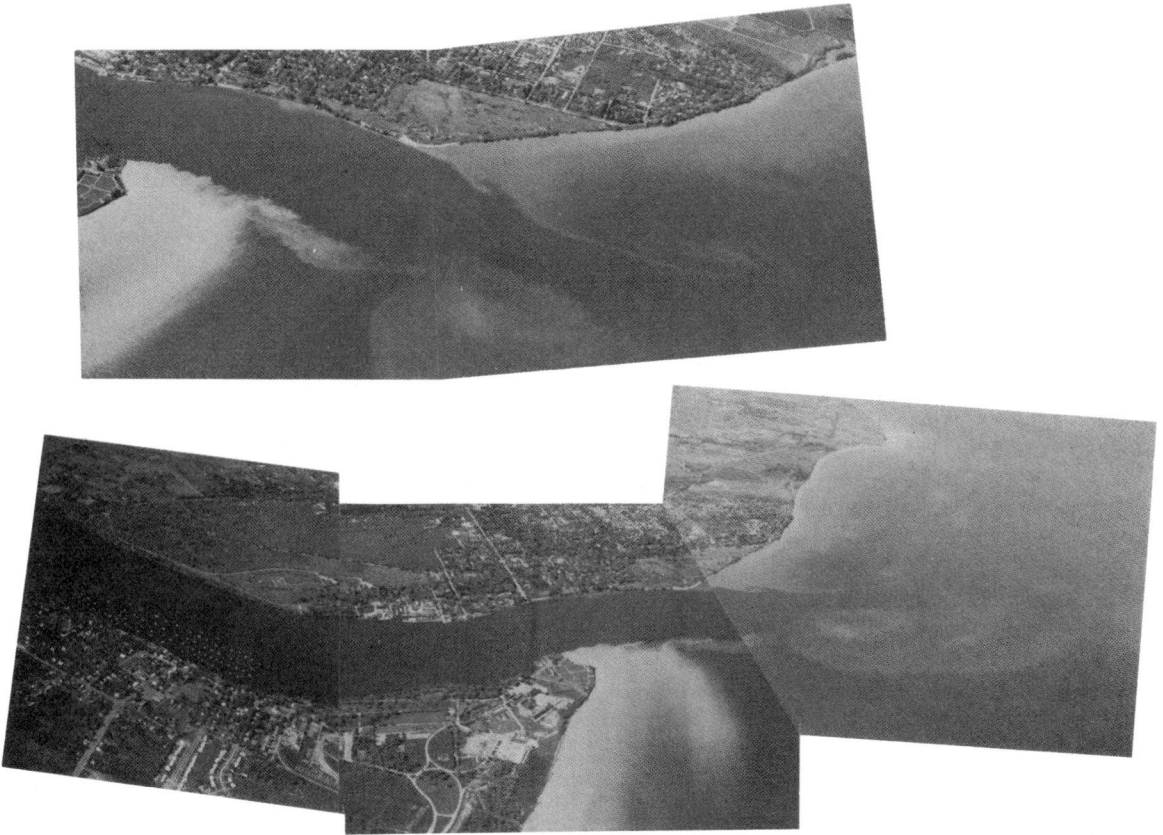

Fig. 5.13 Sheet-like structures of suspended sediments in lake water reflect thermal density stratification near site N (see p. 167), western Lake Ontario, during the spring warming period. Modified after Sly (1983).

River inflows pass into lakes as surface flows (overflows) when the density of river water is less than that of the receiving lake water (Stürm, 1975; Stürm & Matter, 1978; Carmack *et al.*, 1986). Under this condition, coarse sediments will often mark the upper part of the inflow channel but particle size will decrease rapidly with water depth. In-channel particle size will match adjacent lake sediments at or less than the maximum depth of the local in-lake high energy hydraulic regime. Where river inflows are more dense and remain entrained they plunge below surface lake waters (Gould, 1951, 1960; Lara & Saunders, 1970; Gilbert, 1975). At these sites, both channel bathymetry and particle size difference may extend below those corresponding only to in-lake processes.

Lake outflows are mostly in the form of channels that have developed over shallow outlet sills. The discharge and channel cross-section limit channel depth. Usually, such features have little influence on the lake nearshore zone except in their immediate vicinity.

5.4.2 Intermediate depth zone

Although there is no rigid boundary between any of the lake zones described in this text, the boundary between nearshore and intermediate depth zones is more easily distinguished than the lower boundary with the deep midlake zone. The lower boundary is characterized by a particle size change but it is more subtle than at the upper boundary. The lower boundary occurs where sediments have a mean particle size of about 6.5 phi and finer. In large lakes, this may be coincident with water depths well in excess of 50 m. This marks the point at which depositional material is no longer dominated by either the effects of depositional entrainment (with coarse material), or the effects of flocculation. Although sediments of the intermediate depth zone are dominated by low energy regime hydraulic conditions, they can be subject to periodic erosional reworking. The process of selective transportation also continues to influence many sediments of this zone. Sediments in this zone do not reflect the physical control of shoreline inputs to the same extent as those in the nearshore zone. Based on their geochemical characteristics, it is relatively easy to determine transport pathways within the materials of the intermediate zone (Sly, 1983).

Sediments of the intermediate depth zone are frequently composed of turbidities, flocculent materials, and precipitates. Turbidities are usually identified as an upward fining sequence within individual sedimentary layers, and are typically a few millimetres to centimetres thick. They occur as entrained density flows (Brodie & Irwin, 1970) in near surface flows, as interflows, or as bottom flows (Stürm, 1975; Stürm & Matter, 1978; Carmack et al., 1986). Entrainment may occur as a result of input loads from riverine sources, or from shore erosion. Turbidities may also form as a result of slumping (Ludam, 1974) or sliding on within-lake bottom slopes. Slumps and slides are a common feature on many steep slopes and can be induced in a number of ways, including seismic disturbance, wave action, and overloading (Matthews, 1956; Fulton & Pullen, 1969; Gilbert, 1975; Swan, 1979). Turbidities are typical of lakes with fast growing delta fronts and high input loads. They are rare in shallow lakes and they are not preserved unless below the depth of normal wave-induced sediment reworking.

Flocculation is an important mechanism that increases rates of material settlement and does not require entrainment within a water mass. It can be induced in the presence of organic matter and by salinity effects, and may be particularly important as a means of quickly resettling wave-eroded bottom materials (Lick, 1982; Lick & Lick, 1988) in shallow productive lakes.

Precipitates are frequently composed of carbonate minerals, especially calcite. In waters already highly saturated with calcium carbonate, increased alkalinities associated with primary production can induce precipitation; similarly, increased water temperatures reduce the solubility of this mineral and frequently result in seasonal production of mineral precipitates. The formation of mineral suspensions is a well-known phenomenon in many lakes (Brunskill & Ludam, 1969; Eugster & Hardie, 1978; Kelts & Hsu, 1978; Muller & Wagner, 1978; Eugster, 1980; Last, 1982; Eugster & Kelts, 1983; Robertson & Scavia, 1984). The extent to which such carbonate deposits are preserved depends, in large part, on the extent to which cold winter lake water may redissolve this freshly deposited material.

Lakes in which cyclic phenomena, such as seasonal upwelling or downwelling, or the development of thermal bars (Rodgers, 1968; Rodgers & Sato, 1970), or which retain consistent circulatory patterns over the intermediate depth zone, may develop preferential distribution patterns of silty sediments (Gould & Budinger, 1958; Lewis & Sly, 1971; Flood, 1989). These will usually occur as slight thickenings of slope sediment but, occasionally, they may occur as slightly mounded bottom features.

Flocculent materials that are deposited in this zone have very high water content and, frequently, a relatively high organic carbon content (Burns & Ross, 1972). They are more easily eroded than less flocculent low organic content sediments, or bed materials which have begun to de-water or compact, or which have become fully incorporated into biologically pelletized material (or as defined by the depth of biological reworking; Robbins, 1982). Erosion of recently deposited flocculent materials may occur at much lower shear velocities than reported in Fig. 5.5, and is largely dependent upon water content (Postma, 1967; see also Fisher et al., 1979; for a discussion of shape factors and hydraulic equivalent size of organic matter). As a result of this, seasonal reworking of recently deposited materials frequently occurs in parts of the intermediate depth zone. In Lake Ontario, there are several areas of the lake bed which accumulate flocculent sediment as a result of high summer primary production. But, in some areas, there is virtually no long-term accumulation of modern sediment (Sly, 1984). Seasonally, combined wave and current activities erode materials from the intermediate depth zone and transport them into deeper water or out of the lake basin. Figure 5.14 shows that although as much as 2 m of modern sediment has accumulated at a depth of 60–65 m in the St Lawrence Trough in eastern Lake Ontario, there are only isolated patches of this material at a depth of about 40 m. Differential erosion of material at about 40 m or less has been ascribed to the effects of hurricane storms which pass irregularly across the Laurentian Great Lakes (Sly, 1993). Modifications of ^{210}Pb profiles in Lake Michigan sediments (Edgington & Robbins, 1976) provide dramatic evidence of

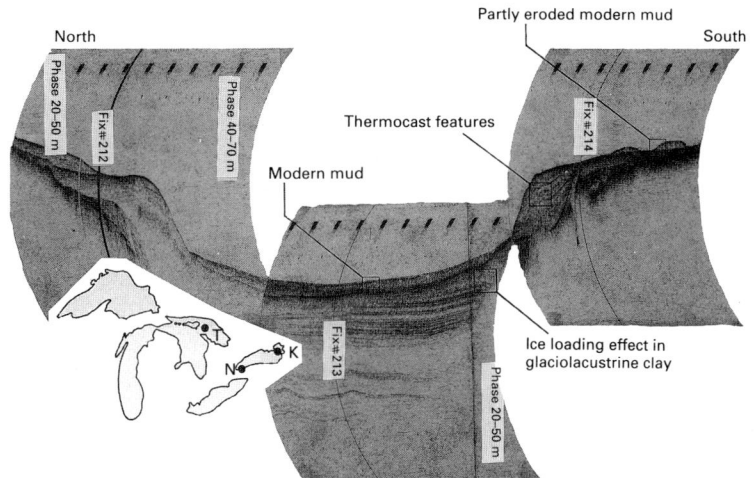

Fig. 5.14 Echo-sounding cross-section of the St Lawrence trough (site K, see p. 167) in eastern lake Ontario, showing relict structures in postglacial sediments and differential erosion of modern muds. From Sly (1983).

downslope sediment slumping, associated with major storm events.

In shallow basins, nearshore or intermediate depth zones may extend fully across a lake (Kenny, 1985). In such conditions, bottom resuspension is a rather common event and relates to the frequency with which wind velocities cause hydraulic shear to exceed critical near bed values (Sheng & Lick, 1979; Lick, 1982). Carper and Bachmann (1984) cited experiments in a small (104 ha prairie lake (maximum depth 1.7 m) where June–August winds caused resuspension of lake bed mineral sediments as much as 10% of the time over about 10% of the lake bed area. In other, larger but still shallow lakes (Lake Balaton, for example), frequent resuspension may affect almost the entire lake bed (Gyorke, 1973).

Sediment plumes

Suspended loads introduced into the intermediate depth zone from river plumes rarely conform to a consistent trajectory but, rather, vary in response to density differences between river and lake waters, discharge, and circulatory patterns within the lake.

Figure 5.15 illustrates the distribution of Hg in Lake Ontario sediments off the mouth of the Niagara River. In each of the areas A–C, Hg is associated with a distinct set of sedimentary and geochemical parameters (Sly, 1984). Area A sediments are composed of fine silts (skewness/kurtosis sector C_l), and area B and C sediments are composed of coarse silts (skewness/kurtosis sector B_l). Area A sediments accumulate from plume materials that curve west-

ward after passing through the main inflow channel. Area B sediments appear to accumulate from direct over-bar discharge, and area C accumulates from a lesser northwest flow across the bar. Regulation of these transport conditions appears to be influenced by river flow, water temperature and lake circulation.

A more dramatic example of the way in which shifts in a river plume can influence lake sediments has been given by Dominik et al. (1981). These authors presented core data from Lake Constance that show striking visual changes in fine sediments which have been related to historic changes in location of the upper Rhine inflow.

Not all suspended loads forming plume-like structures in the water column derive from river inflows. Some may be derived from shoreline and nearshore erosion of fine silts and clays that are carried lakeward as sheet-like flows, buoyed-up by thermal density differences in the lake water. This occurs in western Lake Ontario during midspring when shallow surface warming creates a strong density contrast over cold midlake water that is still close to 4°C (Fig. 5.13). Sediment plumes occurring during winter and spring often contain little organic matter and, from entrained flows, they may form turbidite deposits. At other times of the year, and when productivities and organic contents are greater, the particulates may settle as flocculent materials. Both types of deposition may occur within the same region of a lake (Fig. 5.16).

Fine particles of organic matter also accumulate in suspended layers towards the base of the water

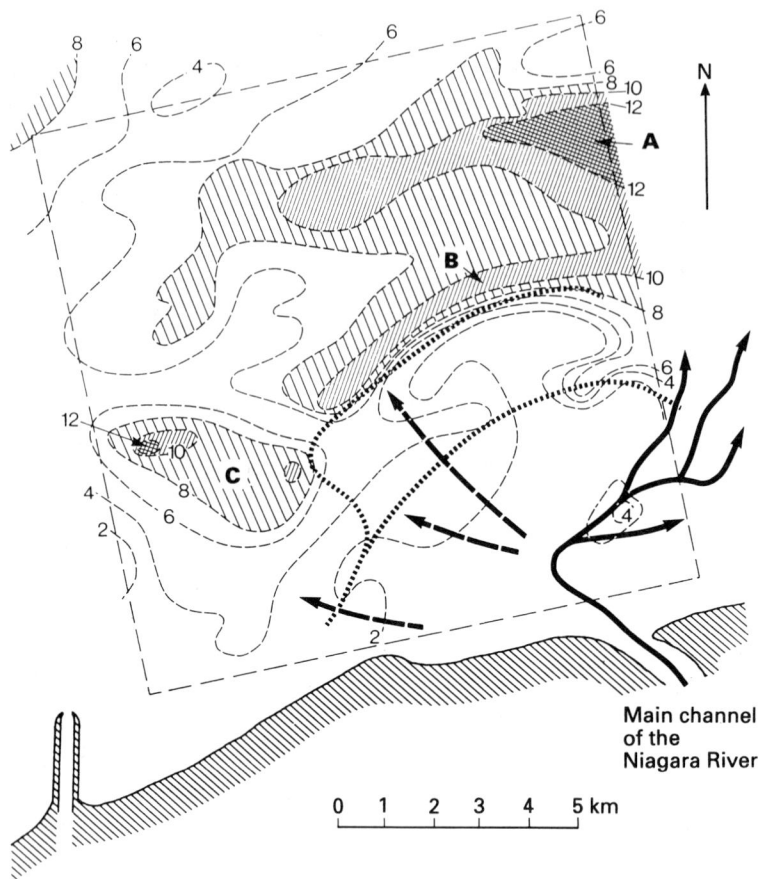

Fig. 5.15 Concentrations of quartz-corrected mercury (QC mercury) in sediments off the Niagara bar, site N (see p. 167), western Lake Ontario. Separate distributions (areas I, II and III) reflect changing patterns in the river outflow and are characterized by physical and chemical compositional differences. After Sly (1983).

column. These materials are largely derived from primary productivity in surface waters and will eventually settle to form a high water content organic bottom fluff (Burns & Ross, 1972). Where the water column is deep, the organic matter may remain in prolonged suspension and develop into a nepheloid layer. Density differences play an important part in forming and maintaining nepheloid layers.

5.4.3 Deep water midlake zone

The deep water midlake zone is characterized by low rates of sedimentation, by very fine sediments (clays and silty clays), by higher organic contents relative to other lake zones (Flannery et al., 1982; Håkanson & Jansson, 1983), and by a very substantial homogeneity, even in large lakes. The distribution of sedimentation rates in lakes is largely governed by three factors, in-lake productivity (Bloesch, 1974; Bloesch & Stürm, 1986; Fukushima et al., 1989), the quantity of material entering a lake, and the degree to which different hydraulic conditions selectively sort and transport various size fractions of material. Shallow lakes, for example, retain very little of their incoming load, and allow it to pass through virtually without any compositional change. As water depth increases, progressively more of the incoming load is retained in a lake, and size selection results in progressively lower concentrations of fine suspended sediment in the outflow. The form of a lake basin also modifies rates of sedimentation, particularly in the deep midlake zone. For any constant load of sediment, rates of very fine sediment accumulation will tend to increase as the area of this zone decreases in proportion to the nearshore and intermediate depth zones. Conversely, sedimentation rates will decrease in trough-shaped lakes where proportionally more of the lake bed is within the deep midlake zone. Thus, for example, because the size of the deepest sub-basin of Lake Erie is so much smaller than the equivalent deep water

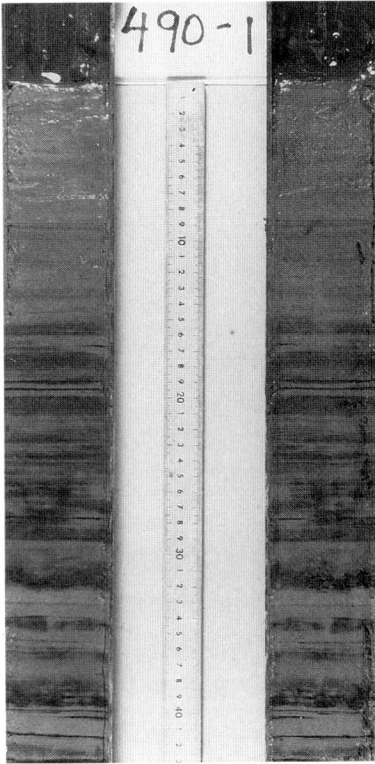

Fig. 5.16 Organic-rich layers in modern mud appear as distinct dark and sulphide-rich bands. Banding in the oxidized layer at the top of the core is less distinct. Narrow banding may be annual, wider bands obscure finer structures of annual lamination. Water depth about 50 m, northwest Georgian Bay.

areas in Lake Ontario, equivalent sedimentation rates are more than five times greater. Further, although the nearshore and intermediate depth zones in Lake Erie are also proportionally much greater than in Lake Ontario, deposition rates in these zones are also greater than in Lake Ontario because of the higher sediment loadings to Lake Erie (both from Lake Huron and as a result of shoreline erosion in Lake Erie). For comparison, in the Laurentian Great Lakes, sediment accumulation in the intermediate depth zone may range from more than 5 to a little less than 1 mm year^{-1}, and accumulation in the deep midlake zone typically ranges from about 0.2 to 2 mm year^{-1}. Kemp *et al.* (1974, 1977) and Kemp and Thomas (1976) provide detailed discussions of sedimentation rates in these lakes.

The concept of sediment focusing suggests that fine-particle-size materials follow some well-defined pathway from peripheral to midlake regions (Likens

& Davis, 1975; Davis & Ford, 1982; Hilton, 1985). However, in most lakes, the process is complex and involves selective sorting. Further, not all fine sediments go to sink in midlake environments. A high proportion of clay-size material is deposited in entrained flows as turbidities and as flocculated aggregates with very fine sand and silt (Figs 5.8 & 5.9), and is retained in the intermediate depth zone. However, the process of flocculation alters the size composition of the suspended load so that it contains progressively less silt as the host water moves away from source (Sly, 1989a).

Deep water lake sediments are largely homogeneous, both in terms of particle size and geochemical composition. This characteristic reflects sorting and the relatively long period of time required to move a particle from a shallow peripheral location to the base of the midlake water column.

Whereas entrained flow and flocculation in the intermediate depth zone may cause particles to settle to the lake bed within a few days of their introduction, in deep midlake areas most particles settle as discrete grains and may take many months to fall to the lake bed in a large lake. During that time, midlake circulatory gyres and in-lake biogenic processes effectively mix particulates from several sources peripheral to the lake, and from the atmosphere, to produce a highly integrated midlake sediment that exhibits little similarity with specific input sources. This is supported by isotope data which indicate that, in contrast to ^{137}Cs ($t_{1/2} = 30$ years), short-lived isotopes such as ^{7}Be are not concentrated in midlake sediment profiles (Eadie & Robbins, 1987).

However, Fig. 5.17 demonstrates that this is not entirely true; it also illustrates the functioning of a mechanism that effectively overrides midlake mixing. Under the influence of storms and downwelling events, it is thought that some fine sediments are removed from initial deposition locations and carried downslope into the midlake environment of Lake Ontario, probably as entrained flows, to form thin turbidities. The presence of high Mirex concentrations within the top 1 cm of bottom sediment and its consistent spread south–north across the lake argues strongly against substantial mixing of the suspended load during transport into the midlake zone.

As in the nearshore and intermediate depth zones, non-depositional areas may exist within the deep midlake zone. Non-depositional areas may occur where the water column is devoid of particulate material (perhaps because deposition has already

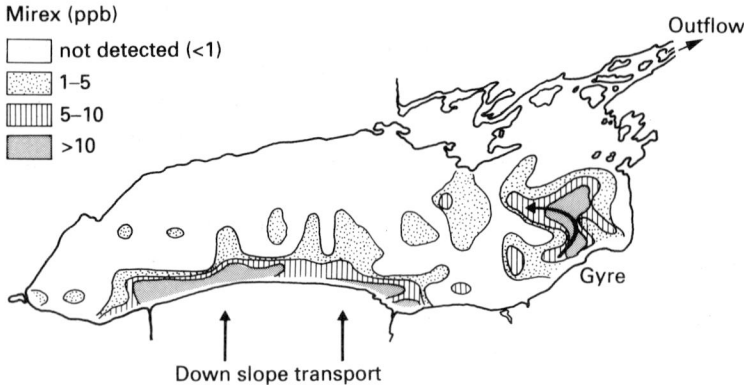

(a)

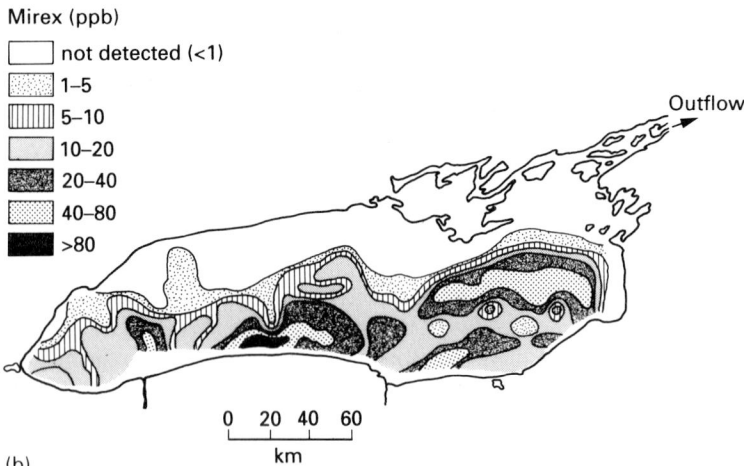

(b)

Fig. 5.17 Sediment focusing in Lake Ontario, demonstrated by the distribution contaminant mirex in (a) 1968; (b) 1977. Mechanisms involve downslope transport associated with downwelling along the south shore and transport in a circulatory gyre. Modified after Thomas *et al.* (1988).

taken place, up-drift, or because of a lack of an appropriate particle size). Since the midlake waters tend towards a homogenous composition, it seems more likely that non-deposition occurs as a result of barrier features or accelerated flow conditions, perhaps caused by bathymetric features, that are sufficient to inhibit particle settlement locally (Sly & Sandilands, 1988; Flood, 1989).

Although large lake hypolimnion currents have been reported up to 95 cm s^{-1} in Lake Erie (Burns & Ross, 1972), hypolimnion currents are generally thought to be about an order of magnitude less than epilimnion currents, and to rarely exceed 50 cm s^{-1}. Simons and Schertzer (1987, 1989), in fact, have demonstrated that near-bed current velocities in Lake Ontario are rarely more than a few centimetres per second and much below the threshold required to erode bottom sediment. Lemmin and Imboden (1987), in a much smaller lake, have also demonstrated that midlake bottom currents are too weak to cause bed erosion. Surrya (1977) suggested that resuspension of bottom sediments could occur in moderate-depth Lake Kinneret (mean depth 25 m, maximum 42 m) but some of this may have been derived from nepheloid materials; a process also implied by Santschi (1986). The presence of bottom material in transport has been noted in deep midlake areas (Eadie & Robbins, 1987) and mostly this has been ascribed to slope erosion at shallower depth, or redistribution of the nepheloid layer (Chambers & Eadie, 1981; Lemmin & Imboden, 1987). However, a recent study using rare earth tracers in Lake Superior

(Krezoski, 1989) suggests that bed material may not be completely immobile, even in very deep midlake regions (Fig. 5.18); this aspect of deep midlake sedimentology remains uncertain and requires further investigation.

Nepheloid layer

The settlement of fine particles in deep water is characterized by discrete particle fall. Midlake particulate material is derived from mineral fines from the shoreline, dust fall and biogenic material, and much of the mineral material probably falls as a continuous sediment rain (Fukushima *et al.*, 1989). However, biogenic materials behave differently. This is because, as a result of degradation of organic matter as it settles through the water column, the material loses its content of organic carbon, and structural components (such as silica tests) fracture and partly dissolve. Therefore, as these particles settle, both bulk density and particle size decrease. This has the effect of reducing the rate of particle settling. The dissolution of detrital and biogenic calcite grains in cold deep water (Robertson & Scavia, 1984; Eadie & Robbins, 1987) probably also results in a similar decrease in settling rates. In effect, as particle size decreases towards the base of the water column, the concentration of suspended matter increases, and this produces a deep water nepheloid layer. In Lake Ontario, it has been estimated that about half of the particulate material is deposited from this layer,

annually (Sly, 1993), in the deep midlake basins; the maximum residence time of particulates in this layer is about 5 years. The nepheloid layer is destroyed each year, with the seasonal overturn and upwelling (Chambers & Eadie 1981; Bell & Eadie, 1983). Comparable nepheloid layers may develop in other lake types, reflecting similar controls. Nepheloid layers may be enhanced by thermal or chemical (dissolved solids) density stratification within the water column. Nepheloid layers can form in different parts of a lake, they can be composed of virtually any form of fine particulate material (including precipitates), and their movement is governed by the host water mass which responds to the circulation of the lake. The residence tme of material in the nepheloid layer may be an important factor influencing coagulation and chemical sorption (Santschi *et al.*, 1986) before particles finally leave the water column. In the Laurentian Great Lakes, the structure and composition of nepheloid layers has been reported by Bell and Eadie (1983), Charlton (1983), Sandilands and Mudroch (1983), Eadie *et al.* (1984), Eadie and Robbins (1987) and Halfman and Johnson (1989).

The composition of suspended sediments is commonly based on settling trap data (Bloesch, 1974; Burns & Pashley, 1974; Bloesch & Burns, 1980) and both accumulation rates and composition (Bloesch & Evans, 1982; Premazzi & Marengo, 1982; Bloesch & Stürm, 1986) have been related to bottom sediments. Seston (living and dead organic matter) is recognized as an important component of bottom sediments

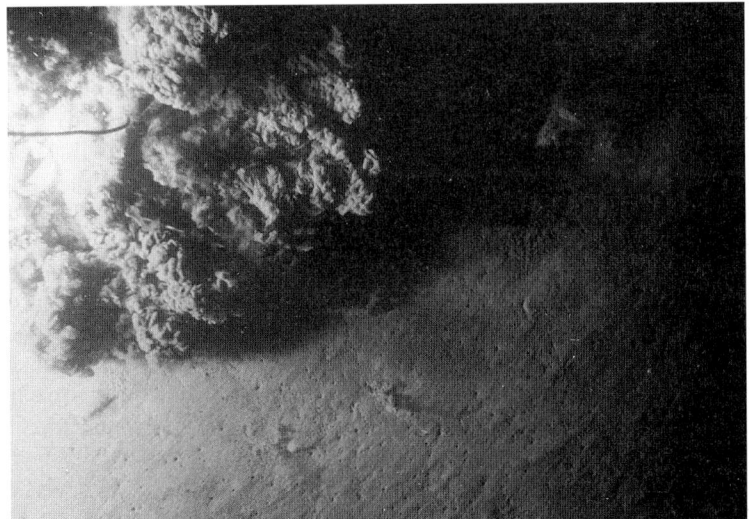

Fig. 5.18 Burrow marks in soft modern mud at a depth of 87 m in northern Georgian Bay. Fluted surface microrelief parallels bottom current direction controlled by local bathymetry (After Sly, 1970).

(Pennington, 1974) and close similarities in composition between nepheloid and bed materials have been documented by Sandilands and Mudroch (1983).

5.5 Other sedimentary processes and influences active in the Laurentian Great Lakes

5.5.1 Pro- and postglacial conditions

The previous sections have discussed modern sedimentary processes in lakes. This closing section discusses some additional interpretations that may be placed on premodern sediments, as a guide to a broader range of sedimentary processes that have occurred in some lake basins.

As previously noted, water levels throughout the Laurentian Great Lakes basin have changed considerably since the first meltwaters began to accumulate along the late Wisconsinan ice margin. The complex sequence of events that occurred as glacial ice retreated across this region is still far from completely understood but an excellent summary of more recently accepted interpretations is provided by Karrow and Calkin (1985) and Prest (1970). In very simple terms, high level lakes first formed close to the ice margin. With ice retreat, progressively lower drainage outlets opened and the lake levels dropped. Very large volumes of water were produced by rapid melting of the Wisconsin ice sheet and some of the events associated with opening of new discharge channels were also probably quite rapid. In many parts of the Great Lakes region, water levels in individual lake basins fell to extremely low elevations shortly after opening of low-point outlets into the St Lawrence and Mississippi drainages. Subsequent lake levels have risen almost continuously over the past 10 000–11 000 years, largely in response to the effects of differential isostatic adjustment. The rate of adjustment was greatest immediately following removal of overlying ice sheets. Minor variations in this trend reflect redirection of drainage within the Great Lakes system and out of it, and the effects of climatic change over this period of time.

Lake Ontario (Sly & Prior, 1984) and Georgian Bay, Lake Huron (Sly & Sandilands, 1988), may be taken as examples of the changing sedimentological conditions that followed ice retreat from these basins. Shallow seismic profiling and high resolution echo-sounding data imply the presence of moraine-like features in many parts of the Great Lakes and a late

ice-front stand in eastern Lake Superior is marked by a major feature that runs through North and South Sand Isles. Surface exposure of this feature allows confirmation of its structure. Its composition grades from fine sand to boulders several metres across. However, for the most part, morainic materials in Lake Ontario and Georgian Bay are much less prominent and remain largely covered by subsequent deposition. It is not known to what extent glacial ice was in bottom contact with bedrock in either Lake Ontario or Georgian Bay but so far there is little or no evidence of pre-Wisconsinan sediment over bedrock and away from the shoreline, within either basin. Drumlin fields are widespread across the drainage basins surrounding the Great Lakes and although drumlinized topography can be projected into the nearshore regions of the lakes, there is no evidence of its occurrence in deep midlake regions.

In deep midlake regions, bedrock appears to be overlain by a thick draping of glacio-lacustrine materials. These are largely composed of fine silts and clays, and multiple units of these deposits may be as much as 60 m thick in Georgian Bay. The glacio-lacustrine clays were deposited in early high level proglacial lakes. Figure 5.19, based on the interpretation of high-resolution seismic profiling and acoustic data, illustrates a typical cross-section of lake bed deposits in western Georgian Bay. Here, Silurian limestones and dolomites of the Niagara Escarpment are covered by a partial drape of glacial till, coarse boulder debris, and glacio-lacustrine clays that progressively overlap at higher bedrock elevations. Coarse boulder debris probably formed as a postglacial residuum following aqueous removal of fine materials in sandy till. The sandy tills are predominantly water-lain, and shoreline and on-shore exposures indicate highly variable contents of coarse and partly rounded materials. More than one sequence of tills is present and some of them have been overconsolidated by the weight of late stage ice readvance. Over-consolidated clayey tills provide a particularly resistant capping. In Lake Ontario, there is evidence that a late ice advance also caused deformation of glacio-lacustrine clays (Sly & Prior, 1984). With the loss of continuous ice cover and the exposure of steep bedrock surfaces to freeze–thaw effects, massive scree slopes formed at the foot of major scarp features where they frequently underlie subsequent deposits. Boulder debris and scree occur beneath later deposits in both Georgian Bay and Lake Ontario.

Fix positions

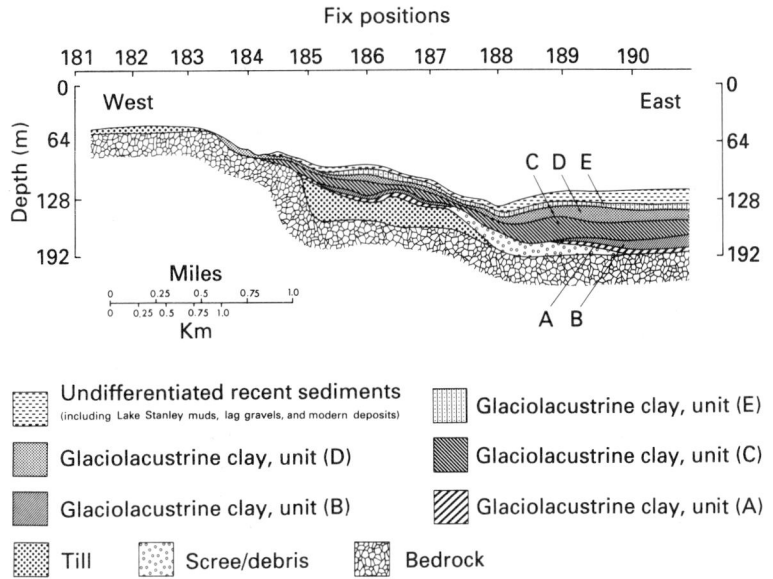

Fig. 5.19 Interpretation of a seismic profile at site T, Georgian Bay (see p. 167), showing a sequence of late glacial and postglacial deposits over bedrock. After Sly and Sandilands (1988).

The thickness of shoreline and on-shore deposits of water-lain till appears to be greater than most of the midlake basal deposits which immediately overlie bedrock, though this has not been verified. Very few midlake sediment cores have penetrated to the bedrock interface, especially through tens of metres of glacio-lacustrine clay. However, where basal units are exposed over bedrock or till, they are usually characterized by a thin layer of coarse sand or gravelly sand only a few centimetres thick. This rapidly or abruptly changes upward into laminated clays or silty clays. Based on the total thickness of these glacio-lacustrine clays in Georgian Bay and the time available for their formation (limited by the duration of various lake stages) it has been estimated that rates of glacio-lacustrine sediment accumulation may have been as high as 4–5 cm year^{-1} (Sly & Sandilands, 1988). On this basis, therefore, laminations or varve-like structures at closer intervals represent episodic phenomena at less than annual frequency. The glacio-lacustrine clays are characterized by very fine particle size, a high degree of sorting (skewness/kurtosis sector D_I, Fig. 5.6), and low organic carbon contents. Variations in the colour and composition of different units within the total sequence of glacio-lacustrine material indicate that inputs were probably derived from different sources, and at different times during the related glacial retreat. As noted by Dell (1973) and Sly and Sandilands (1988), some of the glacio-lacustrine units have relatively high carbonate con-

tent. Since deposition of these sediments occurred in deep cold waters, it may be implied that sedimentation was rapid and thus avoided the effects of high carbonate solubility in cold water. Rapid deposition was probably enhanced by density flow entrainment rather than flocculation since the particulates are both of very fine size and well sorted. At site T, there is little evidence of upward fining in many units of glacio-lacustrine clay; particle size sorting is excellent throughout and this implies that coarser fractions may not be present in areas distal to ice-front sources.

The lack of thick coarse basal deposits beneath the glacio-lucustrine clays is in contrast to glacial drift materials occurring outside existing lake basins. This suggests that there may have been a rather abrupt change between glacial conditions and the subsequent deposition of glacio-lacustrine clays. Although it is possible that little coarse material was present within ice sheets covering the lake basins this is unlikely, especially since considerable thicknesses of glacially derived material lie adjacent to the maximum extent of the Wisconsin ice margin. It is more probable that rapid melting of the ice front was sufficient to flush most coarse materials out of each basin until the early lakes were morphologically well established. In support of the concept of early flushing (Lewis & Anderson, 1989), there are examples of massive cobble-boulder pavements that formed in the spill-ways of late proglacial Lake Agassiz, north of Lake Superior (Teller & Thorliefson,

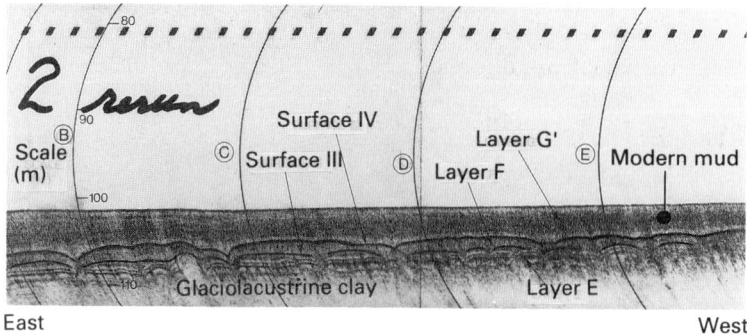

Fig. 5.20 Thermocast features (ice-wedge casts) in glacio-lacustrine clays, site N, western Lake Ontario (see p. 167). After Sly and Prior (1984).

1983). Only after ice fronts were sufficiently distal to the lakes and meltwater discharge was sufficiently reduced, did it become possible to retain sediment loads within the lake basins; at that time, most coarse materials would have been deposited beneath the melting ice and on the land surface between the ice front and lake margin. At this time, the suspended sediment load carried into the lakes was composed of large quantities of very fine material (rock-flour).

The change in deposition that occurred between the glacio-lacustrine clays and modern sediments is often marked by an erosion surface. In Georgian Bay and Lake Ontario, this is coincident with the lowest lake level stage (10 000–11 000 years BP) in each basin. In midlake areas, below the depth of this erosion surface, sedimentation has been continuous and the break between glacio-lacustrine clays and modern muds (an undefined mixture of silt and clay) is gradational. In Lake Ontario, there is evidence that periglacial conditions may have induced the formation of ice-wedge structures in glacio-lacustrine sediments exposed at or near lake level (Fig. 5.20) at that time.

Modern sediments differ from the glacio-lacustrine clays. The content of organic carbon progressively increases and particle size (often after an initial increase) fines upwards (in response to rising lake level). Most rates of modern sedimentation are significantly less than in the pro- and late glacial lakes. Rising lake levels have resulted in preservation of different types of relict material. The most common features are relict gravels that were formed during temporary lake level stands, and which were subsequently drowned by rising lake levels. In some areas, isolated ponding allowed formation of marsh and wetland deposits. Rising water levels subsequently preserved these organic-rich sediments, sometimes

with rooted vegetation in place. Crusted features are also preserved in some midlake areas now characterized by non-depositional environments. These crusts were formed (usually in clays) as a result of sub-aerial exposure, during low lake level stages (Sly & Sandilands, 1988).

Comparable conditions and sedimentary records are known from several lakes, worldwide in north-temperate latitudes, and core and sediment sampling from many lakes in central Europe have yielded interpretations, equally as fascinating as this North American record (e.g. Lister *et al.*, 1984).

5.5.2 Cultural impacts

It is important to realize that although the most dramatic changes in sedimentary processes are undoubtedly associated with major variations in water level, discharge, and climatic regime (e.g., Yamamoto, 1975; Hsu & Kelts, 1978; Stoffers & Hecky, 1978), significant changes result, also, from so-called cultural impacts (lake response to man-made changes in the surrounding catchment).

Figure 5.21, based largely on a sediment core study by Warwick (1978), illustrates subtle changes in sedimentation over nearly 3000 years in the Bay of Quinte, a long and narrow embayment on the north shore of Lake Ontario. The Erosion Index (EI) is defined as:

$$EI = (\% \text{ clay finer than 10.5 phi} \times 100)/ \newline (\% \text{ clay finer than 8.5 phi}). \quad (5.2)$$

Projected variations in water levels based on data from Lake Michigan and western Lake Ontario, and from historic records (Sly, 1986) are plotted on the right of the figure. Water levels are largely governed by precipitation and Lake Michigan and Lake

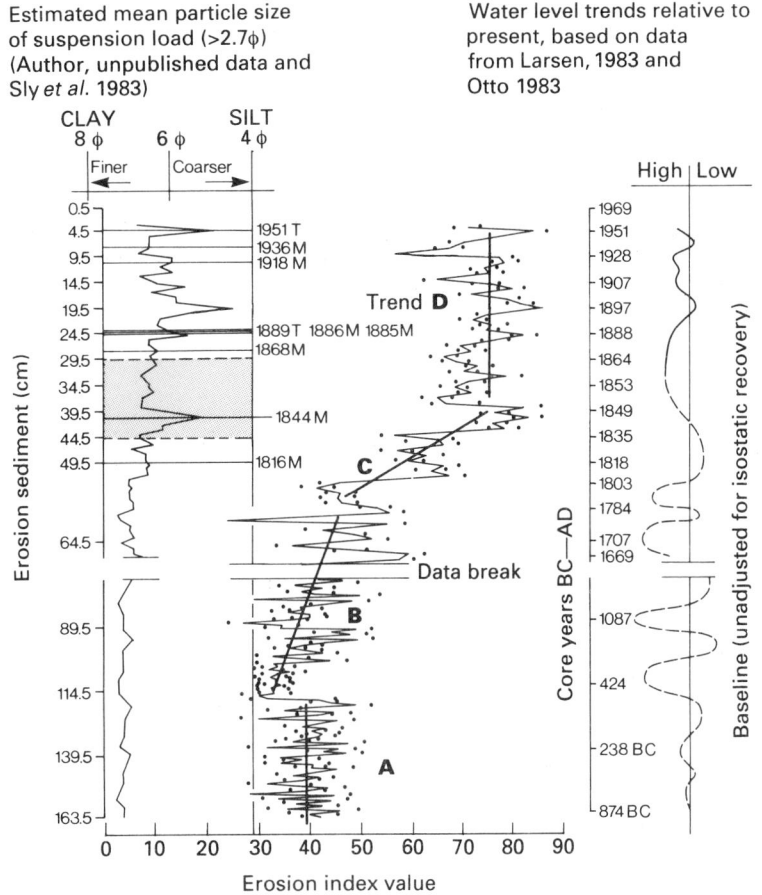

Fig. 5.21 Changes in particle size of fine sediments in the Bay of Quinte, north shore of Lake Ontario, near site K (see p. 167) (Warwick, 1978). Modified after Sly (1991).

Ontario levels are closely phased. There are four different trends apparent in the EI data.

The lowest part of the core (A), covering more than 1000 years (up to nearly AD 400) is characterized by least variability and low rates of sedimentation. This is followed by two sharp increases in long-term EI values (B and C), in which short-term EI fluctuations appear to be coincident with water-level fluctuations. Lower EI values appear to be associated with high water levels and higher EI values appear to be associated with low water levels. Finally, since about the mid 1800s, long-term EI values appear to have stabilized about a new high level (D). Trend (A) is believed to represent EI values of a naturally forested region. Trends (A and B) reflect increasing land clearance; trend (B) is in response to Indian settlement and early farming practices, and trend (C) is in response to European settlement of the area. Trend (D) reflects maintenance of minimal forest cover and

the establishment of grassland and arable farming.

It is thought that fluctuations in the water level data may reflect a balance between:
1 land erosion (fine material in increased surface runoff, due to increased precipitation), and
2 nearshore bottom sediment reworking within the bay itself, under conditions of low lake level.

Since cultural impacts are not thought to have caused a significant change in the flow regime at Warwick's study site, almost all of the effects registered in Fig. 5.21 imply changes in catchment denudation and the sediment yield to the lake. This is particularly evident during the period of maximum sediment accumulation rates (between about 1835 and 1865) when the rates of land clearance were greatest.

More recently, a number of studies have focused on the effects of landuse change and acid precipitation, particularly in eastern North America and western Europe (Sly 1987, 1991). Lake sediments and the interpretation of palaeolimnological records from core samples have provided key insights into such changes (e.g. Battarbee *et al.*, 1985, 1990), and the cultural impacts associated with extensive use of fossil fuels. Wik and Natkanski (1990), for example, have traced the movements of airborne particulates; Patrick *et al.* (1990) studied the effects of landuse practices on lake acidification; and Flower *et al.* (1990) studied the changes in water chemistry and recent palaeolimnology of upland lakes in the UK. Interpretation of many sediment records is made difficult by the need to understand conditions within lake basins, as discussed, for example, by Walling *et al.* (1979), Oldfield *et al.* (1983), Walling and Bradley (1988) and Oldfield and Richardson (1990).

5.6 References

Appleby P.G., Dearing J.A. & Oldfield F. 1985. Magnetic studies of erosion in a Scottish lake catchment: Part I, core chronology and correlation. *Limnol. Oceanogr.*, **30**, 1144–1153.

Assel R.A., Quinn F.H., Leshkevich G.A. & Bolsenga S.J. 1983. *Great Lakes Ice Atlas*. NOAA, Great Lakes Environmental Research Laboratory, Ann Arbor, Michigan.

Battarbee R.W., Appleby P.G., Odell K. & Flower R.J. 1990. ^{210}Pb dating of Scottish lake sediments, afforestation and accelerated soil erosion. *Earth Surf. Proc. Landf.*, **10**, 137–142.

Battarbee R.W., Flower R.J., Stevenson A.C. & Rippey B. 1985. Lake acidification in Galloway: a palaeoecological test of competing hypotheses. *Nature*, **314**, 350–352.

Beach Erosion Board 1972. *Waves in Inland Reservoirs. Tech. Mem.*, **132**. US Army Corps of Engineers, Washington, DC.

Bell G.L. & Eadie B.J. 1983. Variations in the distribution of suspended particulates during an upwelling event in Lake Michigan in 1980. *J. Great Lakes Res.*, **9**, 559–567.

Bennett J.R. 1971. Thermally-driven lake currents during the spring and fall transition periods. *Proc. 14th Conf. Great Lakes Res.*, pp. 535–544. International Association for Great Lakes Research.

Bloesch J. 1974. Sedimentation rates and sediment cores in two Swiss lakes of different trophic state. In: Golterman L. (ed.), *Interactions Between Sediment and Freshwater*, pp. 65–71. Junk, The Hague.

Bloesch J. & Burns N.M. 1980. A critical review of sedimentation trap technique. *Schweiz. Z. Hydrol.*, **42**, 15–55.

Bloesch J. & Evans R.D. 1982. ^{210}Lead dating of sediments compared with accumulation rates estimated by natural markers and measured with sediment traps. *Hydrobiologia*, **92**, 579–586.

Bloesch J. & Stürm M. 1986. Settling flux and sinking velocities of particulate phosphorus (PP) and particulate organic carbon (POC) in Lake Zug, Switzerland. In: Sly P.G. (ed.) *Sediments and Water Interactions*, pp. 481–490. Springer, New York.

Boyce F.M. 1974. Some aspects of Great Lakes physics of importance to biological and chemical processes. *J. Fish. Res. Bd Can.*, **31**, 689–730.

Brodie J.W. & Irwin J. 1970. Morphology and sedimentation in Lake Wakatipu, New Zealand. *New Zealand J. Mar. Freshw. Res.* **4**, 479–496.

Brunskill G.J. & Ludam S.D. 1969. Fayetteville Green Lake, New York. 1. Physical and chemical limnology. *Limnol. Oceanogr.*, **14**, 817–829.

Burns N.M. & Pashley A.E. 1974. *In situ* measurement of the settling velocity profile of particulate organic carbon in Lake Ontario. *J. Fish. Res. Bd Can.*, **31**, 291–297.

Burns N.M. and Ross C. 1972. *Project hypo*. Canada Centre for Inland Waters Paper No. 6 and US EPA Technical Report TS-05-71-208-24, Ottawa.

Carmack E.C., Wiegaud R.C., Daley R.J., Gray C.B.J., Jasper S. & Pharo C.H. 1986. Mechanisms influencing the circulation and distribution of water mass in a medium residence-time lake. *Limnol. Oceanogr.*, **31**, 249–265.

Carpart A. 1952. Le milieu geographique et geophysique. Exploration hydrobiologique du lac Tanganyika. *Inst. Roy. Soc. Nat. Belg.*, **1**, 3–27.

Carper G.L. & Bachmann R.W. 1984. Wind resuspension of sediments in a prairie lake. *Can. J. Fish. Aquat. Sci.*, **41**, 1763–1767.

Chambers R.L. & Eadie B.J. 1981. Nepheloid and suspended particulate matter in southeastern Lake Michigan. *Sedimentology*, **28**, 439–447.

Charlton M. 1983. Downflux of organic matter and phosphorus in the Niagara River area of Lake Ontario. *J. Great Lakes Res.*, **9**, 201–211.

Coakley J.P. & Cho H.K. 1973. Beach stability investigations at Van Wagner's Beach, western Lake Ontario. *Proc. 16th Conf. Great Lakes Res.*, pp. 357–376. International Association for Great Lakes Research.

Coakley J.P. and Hamblin P.F. 1970. *Investigation of Bank Erosion and Nearshore Sedimentation in Lake Diefenbaker*. Canada Centre Inland Waters Report Series, Burlington, Ontario.

Cook D.O. 1970. Models for nearshore sand transport. *Proc. 13th Conf. Great Lakes Res.*, pp. 210–216. International Association for Great Lakes Research.

Csanady G.T. 1978. Water circulation and dispersal mechanisms. In: Lerman A. (ed.) *Lakes, Chemistry, Geology, Physics*, pp. 21–64. Springer, New York.

Davidson-Arnott R.G.D. & Pollard W.H. 1980. Wave climate and potential longshore sediment transport patterns, Nottawasaga Bay, Ontario. *J. Great Lakes Res.*, **6**, 54–67.

Davis M.B. & Ford M.S. 1982. Sediment focusing in Mirror

Lake, New Hampshire. *Limnol. Oceanogr.*, **27**, 137–150.

Davis R.A., Fox W.T., Hayes M.O. & Boothroyd J. C. 1972. Comparison of ridge and runnel systems in tidal and non-tidal environments. *J. Sedim. Petrol.*, **42**, 413–421.

Dell C.I. 1973. A special mechanism for varve formation in a glacial lake. *J. Sedim. Petrol.*, **43**, 838–840.

Dominik J.A., Mangini A. & Muller G. 1981. Determination of recent deposition rates in Lake Constance with radioisotopic methods. *Sedimentology*, **28**, 653–677.

Eadie B.J. & Robbins J.A. 1987. Role of particulate matter in the movement of contaminants in the Great Lakes. In: Hites R.S. & Eisenreich S.J. (eds) *Sources and Fates of Aquatic Pollutants*, pp. 319–364. American Chemical Society, Washington DC.

Eadie B.J., Chambers R.L., Gardner W.S. & Bell G.L. 1984. Sediment trap studies in Lake Michigan: resuspension and chemical fluxes in the southern basin. *J. Great Lakes Res.*, **10**, 307–321.

Edgington D.N. & Robbins J.A. 1976. Records of lead deposition in Lake Michigan sediments since 1800. *Environ. Sci. Technol.*, **10**, 266–274.

Elder F.C. & Lane R.K. 1970. Some evidence of meteorologically related characteristics of lake surface temperature structure. *Proc. 13th Conf. Great Lakes Res.*, pp. 347–759. International Association for Great Lakes Research.

Environment Canada 1973. *Inventory of Canadian Freshwater Lakes*. Inland Waters Directorate, Ottawa.

Environment Canada 1975. *Canada–Ontario, Great Lakes Shore Damage Survey*. Technical Report, Ottawa.

Environment Canada 1976. *Canada Water Year Book*, pp. 49–55. Ottawa.

Eugster H.P. 1980. Geochemistry of evaporite lacustrine deposits. *Ann. Rev. Earth Planet. Sci.*, **8**, 35–63.

Eugster H.P. & Hardie L.A. 1978. Saline lakes. In: Lerman A. (ed.) *Lakes: Chemistry, Geology, Physics*, pp. 237–293. Springer, New York.

Eugster H.P. & Kelts K. 1983. Lacustrine chemical sediments. In Goudie A.S. & Pye K. (eds) *Chemical Sediments and Geomorphology*, pp. 321–368. Academic Press, London.

Fisher J.S., Pickral J. & Odum W.E. 1979. Organic detritus particles: initiation of motion criteria. *Limnol. Oceanogr.*, **24**, 529–532.

Flannery M.S., Snodgrass R.D. & Whitmore T.J. 1982. Deepwater sediments and trophic conditions in Florida lakes. *Hydrobiologia*, **92**, 597–602.

Flood R.D. 1989. Submersible studies of current-modified bottom topography in Lake Superior. *J. Great Lakes Res.*, **15**, 3–14.

Flower R.J., Cameron N.G., Rose N., Fritz S.C., Harriman R. & Stevenson A.C. 1990. Post-1970 water chemistry changes and palaeolimnology of several acidified upland lakes in the U.K. *Phil. Trans. Roy. Soc. Lond. B.*, **327**, 427–433.

Freeman N.G., Murthy T.S. & Haras W.S. 1972. A study of a storm surge on Lake Huron. *Collect. Abstr. 3rd Canadian Oceanogr. Symp., Burlington, Ontario*.

Fricbergs K.C. 1970. Erosion control in the Toronto area. *Proc. 13th Conf. Great Lakes Res.*, pp. 751–755. International Association for Great Lakes Research.

Fukushima T., Aizaki M. & Muraoka K. 1989. Characteristics of settling matter and its role in nutrient cycles in a deep oligotrophic lake. *Hydrobiologia*, **176/177**, 279–295.

Fulton R.J. & Pullen M.J.L. 1969. Sedimentation in Upper Arrow Lake, British Columbia. *Can. J. Earth Sci.*, **6**, 785–790.

Ganf G.G. & Viner A.B. 1973. Ecological stability in a shallow equatorial lake (Lake George, Uganda). *Proc. Roy. Soc. Lond. B.*, **184**, 321–346.

Gelinas P.J. & Quigley R.M. 1973. The influence of geology on erosion rates along the north shore of Lake Erie. *Proc. 16th Conf. Great Lakes Res.*, pp. 421–430. International Association for Great Lakes Research.

Gilbert R. 1975. Sedimentation in Lillooet Lake, British Columbia. *Can. J. Earth Sci.*, **12**, 1697–1711.

Gilbert R. & Shaw J. 1980. Sedimentation in proglacial Sunwapta Lake, Alberta. *Can. J. Earth Sci.*, **18**, 81–93.

Goldman C.R., Mason D.T. & Hobbie J.E. 1967. Two Antarctic desert lakes. *Limnol. Oceanogr.*, **12**, 295–310.

Golterman H.L. 1975. *Physiological Limnology*. Elsevier, Amsterdam.

Gould H.R. 1951. Some quantitative aspects of Lake Mead turbidity currents. *Soc. Econ. Palaeontol. Miner. Spec. Publ.* **2**, 34–52.

Gould H.R. 1960. Turbidity currents. In: Comprehensive survey of sedimentation in Lake Mead 1948–1949. *U.S. Geol. Surv. Prof. Pap.* **295**, 201–207.

Gould H.R. 1970. The Mississippi delta complex. In: Morgan J.P. (ed.) *Deltaic Sedimentation Modern and Ancient. Soc. Econ. Palaeontol. Miner. Spec. Publ.*, **15**, 3–30. Tulsa, Oklahoma.

Gould H.R. & Budinger T.F. 1958. Control of sedimentation and bottom configuration by convection currents, Lake Washington, Washington. *J. Mar. Res.*, **17**, 183–198.

Gregory K.J. & Walling D.E. 1973. *Drainage Basin Form and Process*. Arnold, London.

Gustavson T.C. 1975. Bathymetry and sediment distribution in proglacial Malaspina Lake. *J. Sedim. Petrol.* **45**, 450–461.

Gyorke O. 1973. Hydraulic model study of sediment movement and changes in bed configuration of a shallow lake. *Proc. Helsinki Symp. Hydrology of Lakes, IAHS-AISH Pub.*, **109**, 410–416.

Håkanson L. 1977a. An empirical model for physical parameters of recent sedimentary deposits of Lake Eklon and Lake Vanern. *Vatten*, **3**, 266–289.

Håkanson L. 1977b. The influence of wind, fetch and water depth on the distribution of sediments in Lake Vanern, Sweden. *Can. J. Earth Sci.*, **14**, 397–412.

Håkanson L. 1981. *A Manual of Lake Morphometry.* Springer, New York.

Håkanson L. 1982. Bottom dynamics in lakes. *Hydrobiologia*, **91**, 9–22.

Håkanson L. & Jansson M. 1983. *Principles of Lake Sedimentology.* Springer, New York.

Halfman B.M. & Johnson T.C. 1989. Surface and benthic nepheloid layers in the western arm of Lake Superior, 1983. *J. Great Lakes Res.*, **15**, 15–25.

Hamblin P. F. & Carmack E.C. 1978. River-induced currents in a fjord lake. *J. Geophys. Res.*, **83**, 885–899.

Haras W.S., Bukata R.P. & Tsui K.K. 1976. Methods for recording Great Lakes shoreline changes. *Geoscience Canada*, **3**, 174–184.

Hatton R.S., DeLaune R.D. & Patrick W.H. 1983. Sedimentation, accretion, and subsidence in marshes of the Barataria Basin, Louisiana. *Limnol. Oceanogr.*, **28**, 494–502.

Herdendorf C.E. 1982. Large lakes of the world. *J. Great Lakes Res.*, **8**, 379–412.

Hickman M. 1979. Phytoplankton of shallow lakes: seasonal succession, standing crop and the chief determinants of primary productivity, 1. Cooking Lake, Alberta, Canada, *Holarctic Ecol.*, **1**, 337–350.

Hilton J. 1985. A conceptual framework for predicting the occurrence of sediment focusing and sediment redistribution in small lakes. *Limnol. Oceanogr.*, **30**, 1131–1143.

Hjulström F. 1939. Transportation of detritus by moving water. In: Trask P.D. (ed.) *Recent Marine Sediments*, pp. 5–31. American Association for Petroleum Geologists, Tulsa, Oklahoma.

Houbolt J.J.H.C. & Jonker J.B.M. 1968. Recent sediments in the eastern part of Lake of Geneva (Lac Leman). *Geol. Mijn.*, **47**, 131–148.

Hsu K.J. & Kelts K. 1978. Late Neogene sedimentation in the Black Sea. In: Matter A. & Tucker M.E. (eds) *Modern and Ancient Lake Sediments*, pp. 129–145. Blackwell Scientific Publications, Oxford.

Hutchinson G.E. 1957. *A Treatise on Limnology; Geography, Physics and Chemistry*, Vol. 1. Wiley, New York.

Hutchinson G.E. & Loffler H. 1956. The thermal classification of lakes. *Proc. Nat. Acad. Sci.*, **42**, 84–86.

International Association of Hydrological Science (IAHS) 1977. *Erosion and Solid Matter Transport in Inland Waters. IAHS Publ.*, **122**.

International Joint Commission (IJC) 1981. *Lake Erie Water Level Study. International Lake Erie Regulations Study Board: Environmental Effects, Appendix F.* Windsor, Ontario.

Jenkins P.M. 1942. Seasonal changes in the temperature of Windermere (English Lake District). *J. Anim. Ecol.*, **11**, 248–269.

Johnson R.C. 1988. Changes in the sediment output of two upland drainage basins during forestry land use changes. In Bordas M.P. & Walling D.E. (eds) *Sediment Budgets. IAHS-AISH Publ.*, **174**, 463–471.

Kang S.W., Sheng Y.P. & Lick W. 1982. Wave action and bottom shear stress in Lake Erie. *J. Great Lakes Res.*, **8**, 482–494.

Karrow P.F. & Calkin P.E. (eds) 1985. *Quaternary Evolution of the Great Lakes. Geol. Assoc. Canada Spec. Pap.* **30**.

Kelts K. & Hsu K.J. 1978. Freshwater carbonate sedimentation. In: Lerman A. (ed.) *Lakes, Chemistry, Geology, Physics*, pp. 295–324. Springer, New York.

Kemp A.L.W. & Thomas R.L. 1976. Cultural impact on the geochemistry of the sediments of Lakes Ontario, Erie and Huron. *Geosci. Canada*, **3**, 191–207.

Kemp A.L.W., Anderson T.W., Thomas R.L. & Mudrochova A. 1974. Sedimentation rates and recent sediment history of Lakes Ontario, Erie and Huron. *J. Sedim. Petrol.*, **44**, 207–218.

Kemp A.L.W., MacInnis G.A. & Harper N.S. 1977. Sedimentation rates and a revised budget for Lake Erie. *J. Great Lakes Res.*, **3**, 221–233.

Kenny B.C. 1985. Sediment resuspension and currents in Lake Manitoba. *J. Great Lakes Res.*, **11**, 85–96.

Komar P.D. 1976. Evaluation of wave generated longshore current velocities and some transport rates on beaches. In: Davis R.A. & Ethington R.L. (eds). *Beaches and Nearshore Sedimentation, Soc. Econ. Palaeon. Min., Spec. Pub.*, **24**, 48–53. Tulsa, Oklahoma.

Krank K. 1975. Sediment deposition from flocculated suspensions. *Sedimentology*, **22**, 111–123.

Krank K. 1980. Experiments on the significance of flocculation in the settling of fine-grained sediment in still water. *Can. J. Earth Sci.*, **17**, 1517–1526.

Krezoski J.R. 1989. Sediment reworking and transport in eastern Lake Superior: in situ rare earth element tracer studies. *J. Great Lakes Res.*, **15**, 26–33.

Lara J.M. & Sanders J.I. 1970. *The 1963-4 Lake Mead Survey.* US Dept. Interior, Bur. Reclam. Report REC-OLE-70-21.

Larsen C.E. 1983. Southern Lake Michigan: evidence for Holocene lake level fluctuations. Unpublished manuscript, *US Geological Survey*, Reston, Virginia.

Last W.M. 1982. Holocene carbonate sedimentation in Lake Manitoba, Canada. *Sedimentology*, **29**, 691–704.

Lee D.-Y., Lick W. & Kang S.W. 1981. The entrainment and deposition of fine-grained sediments in Lake Erie. *J. Great Lakes Res.*, **7**, 224–233.

Lemmin U. & Imboden D. M. 1987. Dynamics of bottom currents in a small lake. *Limnol. Oceanogr.*, **32**, 62–75.

Lewis C.F.M. & Anderson T.W. 1989. Oscillations of levels and cool phases of the Laurentian Great Lakes caused by inflows from Glacial Lakes Agassiz and Barlow-Ojibway. *J. Palaeolimnol.*, **2**, 99–146.

Lewis C.F.M. & Sly P.G. 1971. Seismic profiling and geology of the Toronto waterfront area of Lake Ontario. *Proc. 14th Conf. Great Lakes Res.*, pp. 303–354. International Association for Great Lakes Research.

Lewis W.M. 1983. A revised classification of lakes based

on mixing. *Can J. Fish. Aquat. Sci.*, **40**, 1779–1787.

Lick W. 1982. Entrainment, deposition, and transportation of fine-grained sediments in lakes. *Hydrobiologia*, **91**, 31–40.

Lick W. & Lick J. 1988. Aggregation and disaggregation of finer grained lake sediments. *J. Great Lakes Res.*, **14**, 514–523.

Likens G.E. & Davis M.B. 1975. Post-glacial history of Mirror Lake and its watershed in New Hampshire, USA: An initial report. *Int. Ver. Theor. Angew. Limnol. Verh.*, **19**, 982–993.

Lister G.S., Giovanoli F., Eberli G., Finckh P., Finger W., He Q., Heim C., Hsu K.J. & Kelts K. 1984. Late Quaternary sediments in Lake Zurich, Switzerland. *Env. Geol.* **5**, 191–205.

Ludam S.D. 1974. Fayetteville Green Lake, New York. 6. The role of turbidity currents in lake sedimentation. *Limnol. Oceanogr.*, **19**, 656–664.

Miall A.D. 1976. Facies models 4. Deltas. *Geosci. Canada*, **3**, 215–227.

Mankiewicz D., Steidtmann J.R. & Borgman L.E. 1975. Clastic sedimentation in a modern alpine lake. *J. Sedim. Petrol.*, **45**, 462–468.

Martin N.V. & Olver C.H. 1976. *The Distribution and Characteristics of Ontario Lake Trout Lakes.* Ontario Ministry of Natural Resources, Toronto.

Matthews W.H. 1956. Physical limnology and sedimentation in a glacial lake. *Geol. Soc. Amer. Bull.*, **67**, 537–552.

Meybeck M. 1976. Dissolved and suspended matter carried by rivers. In: Golterman H.L. (ed.) *Interactions Between Sediment and Freshwater*, pp. 25–32. Junk, The Hague.

Middleton G.V. 1976. Hydraulic interpretation of sand size distributions. *J. Geol.*, **84**, 405–426.

Mothershill J.S. 1975. Lake Chad: Geochemistry and sedimentary aspects of a shallow polymictic lake. *J. Sedim. Petrol.*, **45**, 295–309.

Muller G. & Wagner F. 1978. Holocene carbonate evolution in Lake Balaton (Hungary): a response to climate and the impact of man. In: Matter A. & Tucker M.E. (eds) *Modern and Ancient Lake Sediments*, pp. 57–81. Blackwell Scientific Publications, Oxford.

O'Hara N.W. & Ayers J.C. 1972. Stages of shore ice development. *Proc. 15th Conf. Great Lakes Res.*, pp. 521–535. International Association for Great Lakes Research.

Oldfield F. & Richardson N. 1990. Lake sediment magnetism and atmospheric deposition. *Phil. Trans. Roy. Soc. Lond. B.*, **327**, 325–330.

Oldfield F., Barnosky C., Leopold E.B., Smith J.P., Merilainen J., Hutten P. & Battarbee R.W. 1983. Mineral magnetic studies of lake sediments: a brief review. *Hydrobiologia*, **103**, 37–44.

Otto J. 1983. Sedimentation in the 15, 16 and 20 Mile Creek Lagoons and implications for late level changes, Lake Ontario. In: Rukavina N.A. (ed.) *Proc. 3rd Work-*

shop on Great Lakes Coastal Erosion and Sedimentation, pp. 85–88. National Research Council, Ottawa, Ontario.

Patrick S.T., Timberlid J.A. & Stevenson A.C. 1990. The significance of land-use and land management change in the acidification of lakes in Scotland and Norway: an assessment utilizing documentary sources and pollen analyses. *Phil. Trans. Roy. Soc. Lond. B.*, **327**, 363–367.

Pearce A.J. 1976. Geomorphic and hydrologic consequence of vegetation destruction. *Can. J. Earth, Sci.*, **13**, 1358–1373.

Pennington W. 1974. Seston and sediment formation in five Lake District lakes. *J. Ecol.*, **62**, 215–251.

Pharo C.H. & Carmack E.C. 1979. Sedimentation processes in a short residence-time intermontane lake, Kamloops, British Columbia. *Sedimentology*, **26**, 523–541.

Philips D.W. & McCulloch J.A. 1972. *The Climate of the Great Lakes Basin. Climatological Studies*, **20**. Atmospheric Environment Service, Environment Canada, Ottawa.

Postma H. 1967. Sediment transport and sedimentation in the estuarine environment. In: Lauff G.H. (ed.) *Estuaries*, pp. 158–179. American Association for Advancement of Science., Washington DC.

Premazzi G. & Marengo G. 1982. Sedimentation rates in a Swiss–Italian lake measured with sediment traps. *Hydrobiologia*, **92**, 603–610.

Prest V.K. 1970. Quaternary geology of Canada. In: Douglas R.J. (ed.) *Geology and Economic Minerals of Canada. Geol. Surv. Canada Report*, **1**, 675–764.

Quigley R.M. & Tutt D.B. 1968. Stability of the Lake Erie north shore bluff. *Proc. 11th Conf. Great Lakes Res.*, pp. 230–238. International Association for Great Lakes Research.

Ragotzkie R.A. 1978. Heat budgets of lakes. In: Lerman A. (ed.) *Lakes, Chemistry, Geology, Physics*, pp. 1–19. Springer, New York.

Ragotzkie R.A. & Likens G.E. 1964. The heat balance of two Antarctic lakes. *Limnol. Oceanogr.*, **9**, 412–425.

Rawson D.S. 1955. Morphology as a dominant factor in the productivity of large lakes. *Verh. Internat. Verein. Limnol.*, **12**, 164–175.

Reeves C.C. 1966. Pluvial lake basins of west Texas. *J. Geol.*, **74**, 269–291.

Richardson P.J., Widmer C. & Kittle T. 1977. *The Limnology of Lake Titicaca (Peru–Bolivia), a Large, High Altitude Tropical Lake. Univ. Calif. Davis, Instit. Ecol. Publ.*, **14**.

Robbins J.A. 1982. Stratigraphic and dynamic effects of sediment reworking by Great Lakes zoobenthos. *Hydrobiologia*, **92**, 611–622.

Robertson A. & Scavia D. 1984. North American Great Lakes. In: Taub F. (ed.) *Lakes and Reservoir Ecosystems, Ecosystems of the World*, **23**, 135–176. Elsevier, Amsterdam.

Rodgers G.K. 1968. Heat advection within Lake Ontario

in spring and surface water transparency associated with the thermal bar. *Proc. 11th Conf. Great Lakes Res.*, pp. 480–486. International Association for Great Lakes Research.

Rodgers G.K. & Sato G.K. 1970. Factors affecting progress of the thermal bar of spring in Lake Ontario. *Proc. 13th Conf. Great Lakes Res.*, pp. 942–950. International Association for Great Lakes Research.

Rukavina N.A. 1969. Nearshore survey of western Lake Ontario, methods and preliminary results. *Proc. 12th Conf. Great Lakes Res.*, pp. 317–324. International Association for Great Lakes Research.

Rukavina N.A. 1976. Nearshore sediments of Lakes Ontario and Erie. *Geosci. Canada*, 3, 185–190.

Rukavina N.A. & Zeman A.J. 1987. Erosion and sedimentation along a cohesive shoreline — the north-central shore of Lake Erie. *J. Great Lakes Res.*, 13, 202–217.

Rust B.R. & Coakley J.P. 1970. Physical–chemical characteristics and post-glacial de-salination of Stanwell-Fletcher Lake, Arctic Canada. *Can. J. Earth Sci.*, 7, 900–911.

Sandilands R.G. & Mudroch A. 1983. Nepheloid layer in Lake Ontario. *J. Great Lakes Res.*, 9, 190–200.

Santschi P.H. 1986. Radionuclides as tracers for sedimentation and remobilization in the ocean and lakes. In: Sly P.G. (ed.) *Sediments and Water Interactions*, pp. 437–449. Springer, New York.

Santschi P.H., Nyffeles V.P., Li Y–H. & O'Hara P. 1986. Radionuclide cycling in natural waters: relevance of scarenging kinetics. In: Sly P.G. (ed.) *Sediments and Water Interactions*, pp. 183–192. Springer, New York.

Schindler D.W., Welch H.E., Kalff J., Brunskill G.J. & Kirtsch N. 1974. Physical and chemical limnology of Char Lake, Cornwallis Island (75 N Lat.). *J. Fish. Res. Bd Can.*, 31, 585–607.

Serrya C. 1977. Rates of sedimentation and resuspension in Lake Kinneret. In: Golterman H.L. (ed.) *Interactions Between Sediment and Freshwater*, pp. 48–56. Junk, The Hague.

Shalash S. 1982. Effects of sedimentation on the storage capacity of the High Aswan Dam reservoir. *Hydrobiologia*, 92, 623–639.

Sheng P.Y. & Lick W., 1979. The transport and resuspension of sediments in a shallow lake. *J. Geophys. Res.*, 84, (C4), 1809–1826.

Sherman I. 1953. Flocculent structure of sediment suspended in Lake Mead. *Trans. Am. Geophys. Union*, 34, 394–406.

Simons T.J. & Schertzer W.M. 1987. Stratification, currents, and upwelling in Lake Ontario, summer 1982. *Can. J. Fish. Aquat. Sci.*, 44, 2047–2058.

Simons T.J. & Schertzer W.M. 1989. *The Circulation of Lake Ontario During the Summer of 1982 and the Winter of 1982/3.* National Water Research Institute, Canada Centre Inland Waters, Burlington, Ontario.

Slaymaker O. 1982. Land use effects on sediment yield and quality. *Hydrobiologia*, 91, 93–109.

Sly P.G. 1970. Underwater photography in the Great Lakes: a report. *Proc. 13th Conf. Great Lakes Res.*, pp. 282–296. International Association for Great Lakes Research.

Sly P.G. 1978. Sedimentary processes in lakes. In: Lerman A. (ed.) *Lakes, Chemistry, Geology, Physics*, pp. 65–89. Springer, New York.

Sly P.G. 1983. Sedimentology and geochemistry of recent sediments off the mouth of the Niagara River, Lake Ontario. *J. Great Lakes Res.*, 9, 134–159.

Sly P.G. 1984. Sedimentology and geochemistry of modern sediments in the Kingston Basin of Lake Ontario. *J. Great Lakes Res.*, 10, 358–374.

Sly P.G. 1986. Review of post-glacial environmental changes and cultural impacts in the Bay of Quinte. In: Minns C.K., Hurley D.A. & Nicholls K.H. (eds) *Project Quinte: Point-source Phosphorus Control Land Ecosystem Response in the Bay of Quinte, Lake Ontario, Spec. Publ. Can. Fish. Aquat. Sci.*, 86, 7–26.

Sly P.G. 1987. Disturbance of lacustrine systems. In: Gregory K.J. & Walling D.E. (eds) *Human Activity and Environmental Processes*, pp. 145–179. Wiley, London.

Sly P.G. 1989a. Sediment dispersion: part 1, fine sediments and significance of the silt/clay ratio. *Hydrobiologia*, **176/177**, 99–110.

Sly P.G. 1989b. Sediment dispersion: part 2, characterisation by size of sand fraction and per cent mud. *Hydrobiologia*, **176/177**, 111–124.

Sly P.G. 1991. The effects of land use and cultural development on the Lake Ontario ecosystem since 1750. *Hydrobiologia*, 213: 1–75.

Sly P.G. 1993. The impact of physical processes at the sediment/water interface in large lakes. In: De Pinto J.V., Lick W. & Paul J.F. (eds) *Transport and Transformation of Contaminants near the Sediment–Water Interface*, pp. 95–113, Lewis, Chelsea, Michigan.

Sly P.G. & Gardner K. 1970. A vibro corer and portable tripod-winch assembly for through-ice sampling. *Proc. 13th Conf. Great Lakes Res.*, 297–307. International Association for Great Lakes Research.

Sly P.G. & Prior J.W. 1984. Late glacial and post-glacial geology of Lake Ontario. *Can. J. Earth Sci.*, 21, 802–821.

Sly P.G. & Sandilands R.G. 1988. Geology and environmental significance of sediment distributions in an area of the submerged Niagara Escarpment, Georgian Bay. *Hydrobiologia*, 163, 47–76.

Sly P.G. & Schneider C.P. 1984. The significance of seasonal changes on a modern cobble–gravel beach used by spawning lake trout, Lake Ontario. *J. Great Lakes Res.*, 10, 78–84.

Sly P.G., Thomas R.L. & Pelletier B.R. 1982. Comparison of sediment energy-texture relationships in marine and lacustrine environments. *Hydrobiologia*, 91/92, 71–84.

Steers J.A. 1953. *The Sea Coast.* Collins, London.

Stiller M. & Imboden D.M. 1986. ^{210}Pb in Lake Kinneret waters and sediments: residence times and fluxes. In: Sly P.G. (ed.) *Sediments and Water Interactions*, pp. 501–511. Springer, New York.

Stoffers P. & Hecky R.E. 1978. Late Pleistocene–Holocene evolution of the Kivu–Tanganyika Basin. In: Matter A. & Tucker M.E. (eds) *Modern and Ancient Lake Sediments*, pp. 43–55. Blackwell Scientific Publications, Oxford.

Stürm M. 1975. Depositional and erosional sedimentary features in a turbidity current controlled basin (Lake Brienz). *9th Internat. Congr. Sedimentol., Nice*, Theme 5, Vol. 5/2.

Stürm M. & Matter A. 1978. Turbidities and varves in Lake Brienz (Switzerland): deposition of clastic detritus by density currents. In: Matter A. & Tucker M.E. (eds) *Modern and Ancient Lake Sediments*, pp. 147–168. Blackwell Scientific Publications, Oxford.

Swan D. 1979. Large submarine landslide, Kitimat Arm, British Columbia. *1st Canadian Conf. Mar. Geotech. Engng. Calgary, Alberta*, pp. 131–139. National Research Council Canada.

Talling J.F. 1969. The incidence of vertical mixing, and some biological and chemical consequences, in tropical African lakes. *Ver. Int. Verein. Limnol.*, **17**, 998–1012.

Teller J.T. & Thorleifson L.H. 1983. The Lake Agassiz–Lake Superior connection. In: Teller J.T. & Clayton L. (eds) *Glacial Lake Agassiz. Geol. Ass. Can. Spec. Pap.*, **26**, 261–290.

Teller J. T., Thorleifson L.H., Dredge L.A., Hobbs H.C. & Schreiner B.T. 1983. Maximum extent and major features of Lake Agassiz. In: Teller J.T. & Clayton L. (eds) *Glacial Lake Agassiz. Geol. Ass. Can. Spec. Pap.*, **26**, 43–45.

Thomas R.L., Gannon J.E., Hartig J.H., Williams D.J. & Whittle D.M. 1988. Contaminants in Lake Ontario — case study. In: Schmidtke N.W. (ed.) *Toxic Contaminants in Large Lakes*, Vol. III, *Sources, Fate, and Controls of Toxic Contaminants*, pp. 327–387. Lewis, Chelsea, Michigan.

Tsang G. 1974. Ice piling on lakeshores with special reference to the occurrences on Lake Simcoe in the spring of 1973. *Inland Waters Directorate Scientific Series*, **35**.

Canada Centre Inland Waters, Burlington, Ontario.

US Army Coastal Engineering Research Centre 1973. *Shore Protection Manual, Vol. 1*. Fort Belvoir, Virginia.

Vincent W.F. 1981. Production strategies in Antarctic inland waters: phytoplankton eco-physiology in a permanently ice-covered lake. *Ecology*, **62**, 1215–1224.

Walling D.E. 1983. The sediment delivery problem. *J. Hydrol.*, **65**, 209–237.

Walling D.E. & Bradley S.B. 1988. The use of Caesium-137 measurements to investigate sediment delivery from cultivated areas in Devon, U.K. In: Bordas M.P. & Walling D.E. (eds) *Sediment Budgets. IAHS–AISH Publication*, **174**, 325–335.

Walling D.E. & Moorehead P.W. 1989. The particle size characteristics of fluvial suspended sediment: an overview. *Hydrobiologia*, **176/177**, 125–149.

Walling D.E., Peart M.R., Oldfield F. & Thompson R. 1979. Suspended sediment sources identified by magnetic measurements. *Nature*, **281**, 110–113.

Warwick W.F. 1978. *Man and the Bay of Quinte, Lake Ontario: 2800 years of cultural influence, with special reference to the Chironomidae (Diptera), sedimentation and eutrophication*. PhD Thesis, University of Manitoba, Winnipeg.

Wetzel R.G. 1975. *Limnology*. Saunders, Philadelphia.

Wik M. & Natkanski J.N. 1990. British and Scandinavian lake sediment records of carbonaceous particles from fossil-fuel combustion. *Phil. Trans. Roy. Soc. Lond. B.*, **327**, 319–322.

Wilson L. 1977. Sediment yield as a function of climate. In: *Erosion and Solid Matter Transport in Inland Waters. Int. Ass. Scient. Hydrol. Publ.*, **122**, 82–93.

Wright R.F. & Nydegger P. 1980. Sedimentation of detrital particulate matter in lakes: influence of currents produced by inflowing rivers. *Wat. Res. Res.*, **16**, 597–601.

Yamamoto A. 1975. Grain sizes of the core sediments and variations of palaeoprecipitation in Lake Biwa during the last three hundred thousand years. In: Horie S. (ed.) *Palaeolimnology of Lake Biwa and the Japanese Pleistocene*, Vol. 3, pp. 209–225. Otsu Hydrobiological Station, Kyoto University. Japan.

6 Estuarine sediment transport and deposition

K.R. DYER

6.1 Introduction

Estuaries have a global significance to continental shelf and oceanic processes because of the exchanges of water, contaminants and sediment between the estuaries and the coastal seas. They are the route by which the sediment is transported via the rivers from the interior of the land masses. During transport through the estuaries the grain-size distribution of the sediment becomes altered by continual deposition, re-erosion and transport, and certain fractions become permanently trapped, while others are transported into the sea. Consequently the estuarine processes act as a filter on the sediment input. Additionally chemical alterations can occur within the estuarine situation that can cause the mineralogical characteristics of some of the constituent particles to alter.

Sediments form a crucial link in estuarine processes. Suspended sediment concentrations are generally high, the particles are fine, cohesive, and prone to flocculate, and they are richly organic. Sediments silt up harbours and navigational channels, and influence pollutant transport. Since many of the sediments mobile in estuaries are fine-grained clay minerals, pollutants are absorbed on their surfaces and are transported with the sediment particles. Additionally they may affect the potential flocculation of the particles, their transport and deposition, as well as possibly even their mineralogical composition. Consequently the movement of contaminants can only be understood through a knowledge of the movement of particles. The suspended sediment concentrations are sufficiently high to limit light penetration and productivity. The muddy substrates can be host to a diverse and vigorous biological community, but this can be limited by the presence of layers of high concentration suspended sediment with low oxygen content intermittently present above the bed.

Estuaries are characterized by strong gradients in fluid density, in suspended sediment concentration and in chemical and biological factors. These gradients imply that the processes are changing significantly in absolute or relative magnitude, and there is close coupling between them. This suggests that the biological and chemical processes may well modulate the physical and sedimentological processes, and vice versa. Consequently in examining the transport and deposition of sediment, one needs to consider a wide range of allied processes; a task that is both difficult and fascinating.

Since estuaries in general are shallow, and sea level undergoes very drastic changes on the geological timescale, estuarine sediments may not be very widely preserved in the geological column and may be fairly thin in vertical extent. Estuaries are more likely to be numerous during transgressions than during regressions. They are ephemeral features, being fairly rapidly altered and destroyed, and having an average life of probably only thousands of years. The ephemeral nature of estuaries has been discussed by Schubel and Hirschberg (1978). It is likely that the world is unusually rich in estuaries at the present time since the Flandrian transgression has inundated the valleys cut when the rivers incised their courses to a base level, approximately 100 m below present sea level during the late Pleistocene. Much of the variation in form of the resulting estuaries depends on the volumes of sediment that the rivers have contributed to fill the valleys. Where river flow and sediment discharge was high the valleys have become completely filled and even built out into deltas. Generally deltas are best developed in areas where the tidal range is small and where currents cannot easily redistribute the sediment the rivers introduce. Deltas will have a much more marked presence in the geological column because the rate of crustal deformation due to sedimentary loading usually exceeds the rate of change of sea level, and a large thickness of sediment can accumulate.

Where sediment discharge was less, the estuaries are unfilled. These drowned river valleys, or coastal plain estuaries, still retain the main features of river valleys, having a meandering outline with frequent tributaries and a triangular cross-section.

In glaciated areas the river valleys were over-deepened by glaciers, and result in fjords. Characteristic of these is a rock bar or sill at their mouths that can be as little as a few tens of metres deep. However, inside they can be several hundred metres deep and extend hundreds of kilometres inland.

On low coastlines extensive shallow lagoons are often formed between the rivers and the sea. Within these lagoons the tidal currents are small, but they increase towards the narrow inlets which connect the lagoon to the sea. In tropical areas the lagoons can be hypersaline during the hot season, but almost entirely fresh water during the rainy season.

With such a variety of estuaries it is to be expected that there will be a diverse and complex series of processes dominating the transport and deposition of the incoming sediment.

Though the processes occurring in the smaller estuaries are likely to be insignificant geologically, they are important in the shorter term. Consequently an understanding of the dynamics of sediment movement in these estuaries and prediction of the rates of transport are of great social and environmental consequence. Therefore this chapter will concentrate on the processes of sediment transport and deposition active in coastal plain estuaries and lagoons where Man's influence is particularly important. Other reviews of note are Postma (1967), Dyer (1972), Nichols and Biggs (1985) and Dyer (1986). Dyer (1989) has reviewed the outstanding research problems concerning estuarine fine sediment.

6.2 River sediment input

It has been estimated that the worldwide total annual fluvial discharge of sediment is about 7×10^9 t

(Milliman & Meade, 1983). The majority of the sediment discharge is carried by only a small number of rivers, and these mainly occur in tropical and sub-tropical areas. The highest sediment discharge is in the east and south of Asia, where the sediment yield can exceed 1000 t Km^{-2} year^{-1}. The Asian rivers carry in excess of 75% of the world's sediment discharge, and two-thirds of that is carried by the Yellow River and the Ganges/Brahmaputra. In these rivers the river discharge is very seasonal, occurring at the times of monsoons or during snowmelt. The sediment discharge period can be even shorter. For instance, the Yellow River discharges between a quarter and a third of its annual discharge in only 2 or 3 days of flood. The sediment discharge is nearly 10×10^8 t year^{-1} which causes the coast to build outwards at a rate of about 1 km in 40 years. This river has frequently changed the position of its mouth, and in 1856 the mouth moved by about 400 km.

In most rivers there is considerable annual variation in the sediment discharge. The discharge is the product of the flow rate times the suspended sediment concentration. The relationship between these can be examined by a rating curve, and in many rivers there is considerable hysteresis in the curve throughout the year. In areas where the catchment is snow covered in winter the sediment concentration is low. However, during the spring snowmelt the material released by the frost action is quickly eroded but can become exhausted during the summer whereupon the river flow becomes less turbid. Similar hysteresis can occur within individual flood events because the availability of mobile sediment depends on a period of drying out and weathering between storms.

As the erosive power of the stream rises very rapidly with discharge, most of the sediment discharge will occur during the occasional extreme events. As an example the sediment concentration in the Susquehanna River during tropical storm Agnes was 40 times greater than any previously recorded, and the sediment discharge was equivalent to between 30 and 50 years of normal flow. Consequently in many examples over 90% of sediment discharge can occur in 5% of the total time and for 80% of the time virtually no sediment discharge occurs. This intermittency causes considerable problems for sampling as well as in estimating the effects of riverine sediments on estuarine and coastal sediment budgets.

6.3 **Types of estuary**

Within estuaries the tidal and residual water circulation patterns are important in determining the sediment transport. Fine-grained material will travel in suspension and will follow the residual or tidally averaged waterflow. The coarser-grained particles will travel along the bed and will be affected mainly by the highest velocity, moving in the direction of the maximum current. It is apparent that the patterns of sediment movement are different in the different types of estuary. Consequently a brief outline of the different estuarine types is required.

6.3.1 Highly stratified estuaries

When there is little tidal motion the river flow, being less dense than the salt water, flows over it with a marked density interface between the two water masses. Because of the stable density gradient the two water masses will not mix readily together. However, if the surface layer has sufficiently high velocity the shear can create interfacial waves on the halocline. These waves break, ejecting salt water into the fresher surface layer by a process called *entrainment*. No fresh water is mixed into the bottom layer and the mixing is entirely upwards. Thus the bottom water loses salt gradually into the surface layer and this loss is made good by a slow inflow of salt water from the sea (Fig. 6.1a). The position of the salt wedge will vary with the river flow, and the tidal range is normally microtidal, that is, a range of less than 2 m.

6.3.2 Partially mixed estuaries

With an increased tidal range, the whole water mass in the estuary moves backwards and forwards with a tidal periodicity. The friction between the water and the bed of the estuary creates considerable turbulence and this attacks the interface between the salt and the fresh water. The turbulence mixes the water column more effectively than entrainment so that the salinity difference at the interface is considerably reduced, and there is a smaller velocity shear across the interface. Turbulent mixing not only mixes the salt water into the fresher surface layer, but it also mixes the fresher water downwards. This causes a longitudinal gradient in salinity, with salinity diminishing towards the head of the estuary both in the surface and the bottom layers. There has to be a residual discharge of water towards the sea, but it now carries

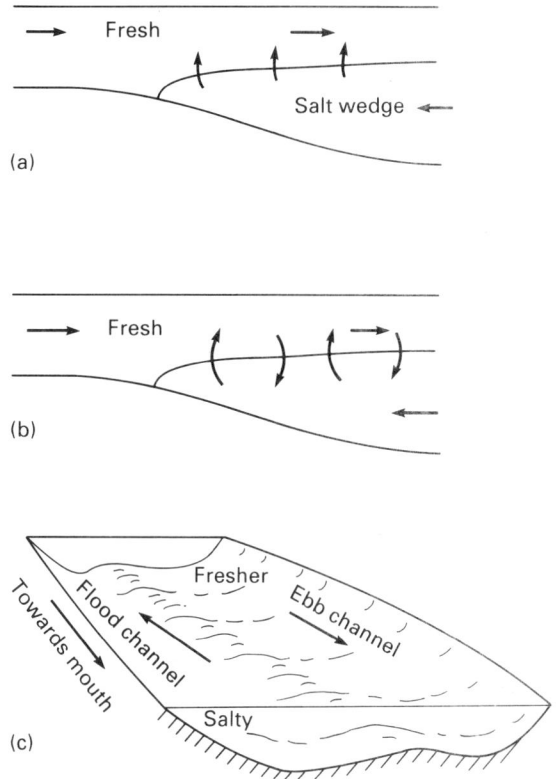

Fig. **6.1** Residual (tidally averaged) water circulation in three estuary types. (a) Salt wedge or highly stratified; (b) partially mixed; (c) well mixed.

with it the increased content of salt resulting from the enhanced mixing. The discharge from the surface layer can thus be an order of magnitude larger than the river discharge. Because of the requirements of continuity, this discharge has to be replaced by a significant landward flow within the bottom layer. Consequently at the estuary mouth a large volume of mixed water has to be discharged and the compensating inflow in the bottom layer has to be larger than in a salt wedge estuary (Fig. 6.1b).

In partially mixed estuaries the tidal range is generally mesotidal, that is, between 2 and 4 m. In this situation the tidal range can change significantly between spring and neap tides. The spring tide currents enhance the turbulent exchanges of salt and fresh water, and as a consequence the stratification can diminish considerably. This produces an increase in the mean circulation velocities and an apparent shift in the mean salinity towards the sea. At times of high river flow the partially mixed estuary will become more highly stratified, and the intensity of the mean circulation should diminish. Within partially mixed estuaries there can be considerable variation in the vertical structure along the estuary, and frequently at the head of the estuary where the water depth and the tidal range diminish, and river flow becomes comparatively more important, highly stratified characteristics can develop.

6.3.3 Well-mixed estuaries

When the tidal range is very large, that is, macrotidal conditions (>4 m), there is sufficient energy in the turbulence to breakdown the vertical salinity stratification completely so that the water column becomes effectively vertically homogeneous. In this type of estuary there can be lateral variations in salinity and in velocity, and horizontal circulation tends to develop at the expense of the vertical circulation. There can be residual flows inward on one side of the estuary and seaward on the other (Fig. 6.1c).

6.3.4 Tidal effects

In estuaries where the tidal range is large relative to water depth, considerable asymmetry can occur in the tidal curve because the effects of friction increase with decreasing water depth. Consequently the high water travels more quickly into the estuary than the low water, resulting in the tidal curve becoming sawtoothed in shape, with a quick rise at the beginning of the flood tide and a slow fall towards low water. In extreme situations the asymmetry can develop into a steep fronted bore. The tidal distortion can be represented by a combination of the main semi-diurnal M_2 component and the principal overtide M_4 with a quarter diurnal period. Depending on the phase relationship between these two constituents a variety of tidal curves can be produced, and each causes a similarly distorted tidal current pattern. For example, if the relative tidal phase is between 0° and 180° then the falling tide exceeds the rising tide in duration, and this produces a shorter enhanced flood current relative to the ebb current. This is the situation outlined above. These features have been described by Boon and Byrne (1981), Aubrey (1986), Friedrichs and Aubrey (1988), and related to the shape of the intertidal basin of the estuary. The differences in the magnitudes and durations of the ebb and flood currents can change along the estuary, with the varying frictional influence of the changing water depths, as shown by Uncles (1981) for the Severn Estuary. For passive particles in the flow these processes would cause no net movement

of sediment, unless there is a superimposed residual flow, or lags in the response between the water flow and the sediment movement.

6.4   **Turbidity maximum**

One of the most distinctive features of sediment transport in meso- and macrotidal estuaries is the presence of a turbidity maximum. This is a zone which contains suspended sediment concentrations that are higher than those in the river or further seaward in the estuary, and which is generally located at, or somewhat landward of, the head of the salt intrusion, where salinities are about 1–5‰. The energetic tidal flow is capable of maintaining high concentrations, and there are a number of processes that concentrate the suspended sediment in the zone, and that oppose the tendency for the particles to disperse.

The peak concentration of suspended sediment in the turbidity maximum varies between wide limits. Despite the differences due to sediment availability, low tidal range estuaries have maxima with concentrations of the order 100–200 p.p.m. (mg 1^{-1}), whereas high tidal range estuaries have much higher concentrations, of the order 1000–10 000 p.p.m. However, remote sensing images show that the surface suspended sediment concentration is far from smooth, but contains streaks and patches.

The turbidity maximum contains a high proportion of a narrow size range of mobile fine sediment, and plays a central role in controlling the circulation of fine sediment within the estuary, as well as the transport of sediment from the river to the sea. Since the concentrations of sediment in the turbidity maximum appear to remain almost constant when averaged over a reasonable time, the input of sediment from the river must be compensated for by a loss of material to the sea, or by accumulation.

Recent reviews of the turbidity maximum have been presented by Officer (1981), Nichols and Biggs (1985) and Dyer (1986, 1988). Descriptions of turbidity maxima have been made by, amongst others, Allen (1973), Dobereiner and McManus (1983) and Wellershaus (1981).

6.4.1   Variation with river flow

The turbidity maximum responds to changes in river flow. Increasing flow causes the maximum to move downstream, and because increased river flow normally implies a greater discharge of suspended sediment, the mass of sediment in the turbidity maximum also increases. Whether the concentration increases, as a consequence, or not, depends on the estuarine dimensions. A movement of the turbidity maximum down estuary involves expansion into an increased cross-sectional volume, and this could decrease the concentrations even though the total mass increases. In the Cumberland Basin, Amos and Tee (1989) obtained results which suggest, for that system, that an increased mass of sediment in the turbidity maximum leads to an increased maximum concentration as well as an increased longitudinal distribution.

In the Tamar estuary the turbidity maximum is stronger when it is nearer the estuary head. Uncles and Stephens (1989), in a regression analysis, found that the magnitude of the maximum against the inverse of the cross-sectional area at high water explained 60% of the variance. Additionally, a regression of distance of the peak in the maximum from the head of the estuary against run-off explained 80% of the variance in distance which varied roughly as the square root of the run-off.

The movement of the turbidity maximum in the Seine estuary is illustrated (Fig. 6.2). At low river discharge the turbidity maximum lies well within the estuary, the vertical concentration gradients are high, and the boundary with the Baie de la Seine is fairly abrupt. At high river flow the centre of mass of the maximum has been pushed downstream, and the concentration gradients are relatively less steep. There is obviously the possibility of greater loss of material from the estuary into the coastal waters for this flow situation.

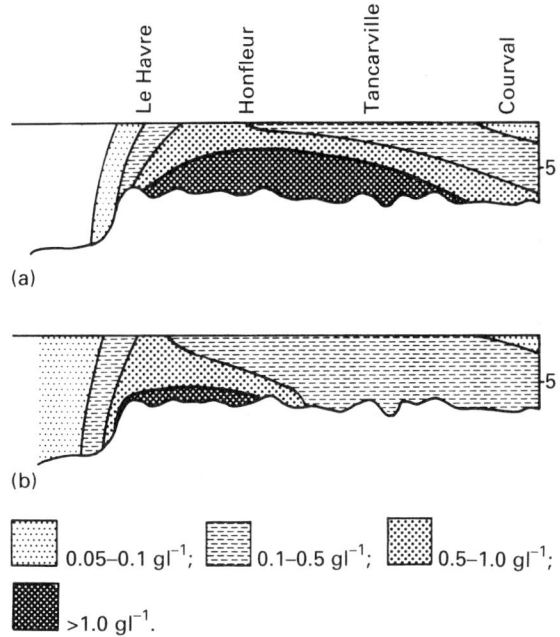

(a)

(b)

0.05–0.1 gl^{-1}; 0.1–0.5 gl^{-1}; 0.5–1.0 gl^{-1}; >1.0 gl^{-1}.

Fig. 6.2 The turbidity maximum in the Seine estuary at spring tide, for two different river discharges (R). (a) R = 200m^3 s^{-1}; (b) R = 800m^3 s^{-1}. After Avoine (1981).

An alternative way of considering the turbidity maximum movement is shown by plotting its position against river discharge (Fig. 6.3). The more or less steady position achieved at high river discharge indicates that the estuarine response tends to be

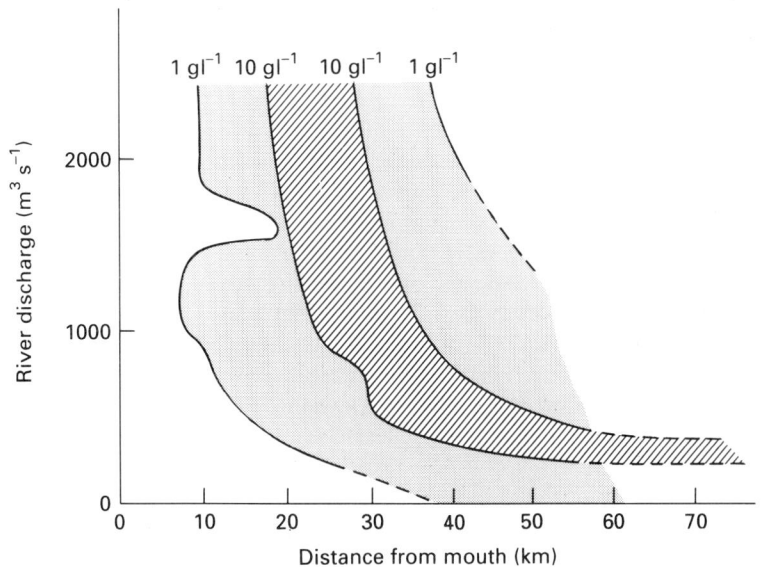

Fig. 6.3 The position of the turbidity maximum in the Gironde estuary with varying river discharge. From Allen (1973).

buffered by the stratification. Consequently, as river flow increases, the increased salinity stratification gradually decouples the upper and lower layers, so that an increasing proportion of the river-borne suspended sediment passes straight through the estuary in the upper layer. However, it may return when the stratification diminishes, and the residual landward bottom flow becomes re-established (Nichols, 1977). In the Seine estuary the turbidity maximum has migrated seaward by about 50 km due to progressive marsh reclamation, siltation, and river flow alteration (Avoine *et al.*, 1981). The turbidity maximum in the Gironde estuary varies in position by about 40 km with variation in river discharge. As this is seasonal, the sediment can accumulate in the upper reaches in spring and summer, and is redistributed down estuary in autumn and winter.

6.4.2 Variation during the tide

The ebb and flood of the tide within an estuary will cause the turbidity maximum to shift from near the head of the estuary at high tide to further downstream at low tide. Figure 6.4 shows the tidal variation within the Seine estuary for a spring tide and a river flow of 780 m^3 s^{-1}. At about high water the maximum is located well up the estuary, and concentrations are relatively low. During the ebb tide the maximum is moved seaward, and by entrainment of sediment from the bed, the concentrations increase. At low water the maximum is near the mouth, and over slack water some settling of material is apparent from the reduction of turbidity at the water surface. The reverse situation occurs during the flood tide.

It is often difficult to obtain sufficient measurements to quantify the above processes, which consequently have to be deduced from measurements taken at fixed positions within the estuary. An example of station measurements is shown in Fig. 6.5. At (A) the clearer sea water has advected up to the station. During the ebb current sediment starts being eroded from the bed (B), and is entrained through the water column to join sediment advecting down the estuary (C). At low water slack, some settling takes place at the water surface (D) in conjunction with low salinity surface water. Further erosion takes place during the flood current (E) and is entrained throughout the water column (F). This is followed by settling as the current wanes.

The structure of the turbidity maximum revealed by measurements at a single station is also shown for

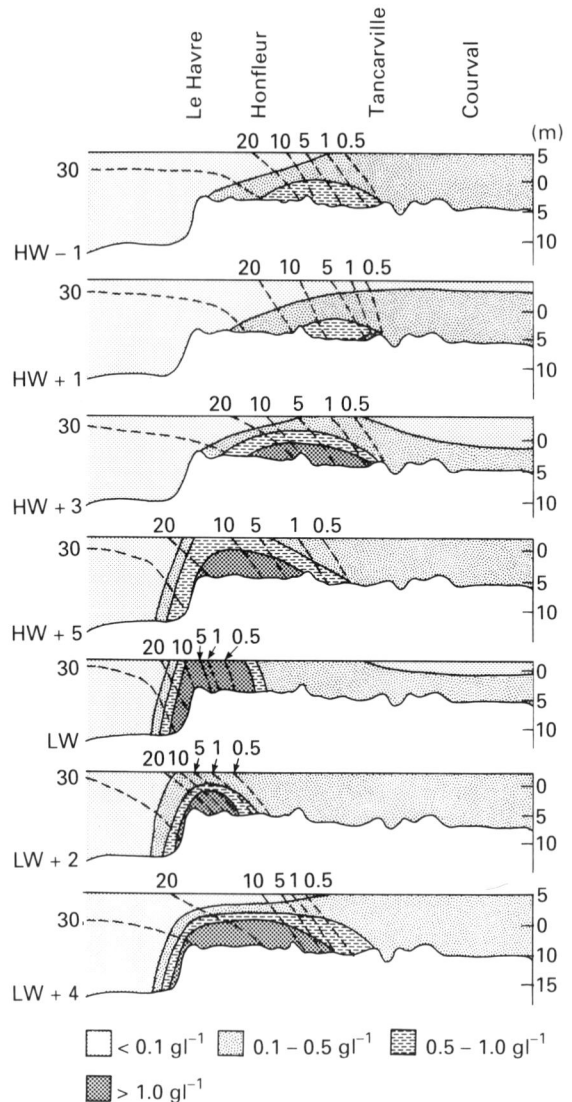

Fig. 6.4 The turbidity maximum in the Seine estuary at various intervals during a spring tidal cycle, with a river discharge of 780 m^3 s^{-1}. HW, high water; LW, low water; – – –, salinity ‰ (parts per thousand). After Avoine (1981).

a position near the head of the Tamar estuary in Fig. 6.6 (McCabe, personal communication). The depth is shown relative to the sea surface, and the measurements spanned just over 10 h of the end of the ebb tide and the beginning of the flood. The rapid deepening at about 1700 hours when the velocities

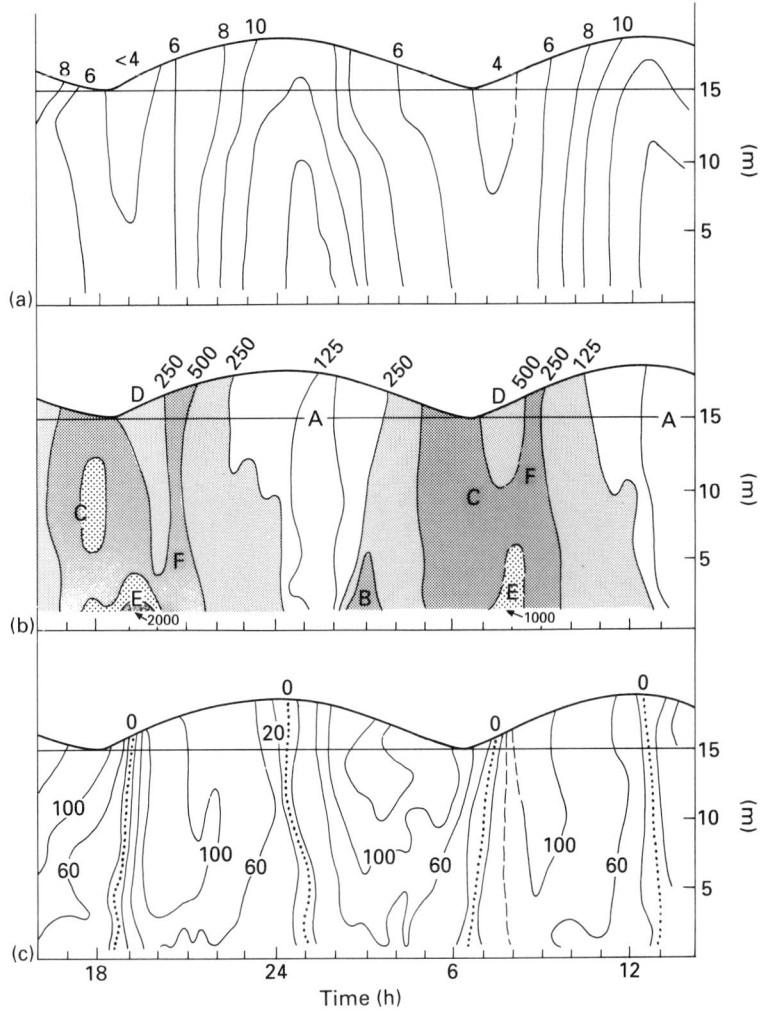

Fig. 6.5 Variation of (a) salinity (‰); (b) suspended sediment concentration p.p.m.; and (c) velocity (cm s^{-1}) for two spring tidal cycles at a station in the Weser estuary. For explanation see text. From Wellershaus (1981).

during the early flood tide reached 1 m s^{-1}, illustrates the asymmetry of the tide. The ebb tide is longer in duration, with a peak velocity of only 0.6 m s^{-1}. The turbidity maximum is more closely associated with the velocity maxima than with the salinity intrusion, though the maximum is associated with salinities of less than 2‰. This illustrates that the majority of the suspended sediment is being entrained off the bed, with a velocity of about 0.4 m s^{-1} at 0.5 m above the bed being the critical erosion threshold. Simultaneous measurement of the floc size distribution *in situ* shows that in the turbidity maximum the mean floc size is smaller than at other times. This is probably due to the combined effect of turbulent shearing and floc collisions dis-

rupting the flocs in the higher concentrations. In the saline intrusion a considerable proportion of the flocs exceed 250 µm in diameter.

A feature of macrotidal estuaries is the large difference in tidal range and in velocities between spring and neap tides. Because of the considerable variation of energy there are corresponding changes in the turbidity maximum. This effect has been examined by Allen *et al.* (1980) and Gelfenbaum (1983). During the neap–spring cycle the ratio of river flow to tidal volume changes significantly. Consequently, though the estuary may be well mixed at spring tides, at neap tides it can be partially mixed, or even well stratified. At spring tide the turbidity maximum will have its highest concentration, the currents being able

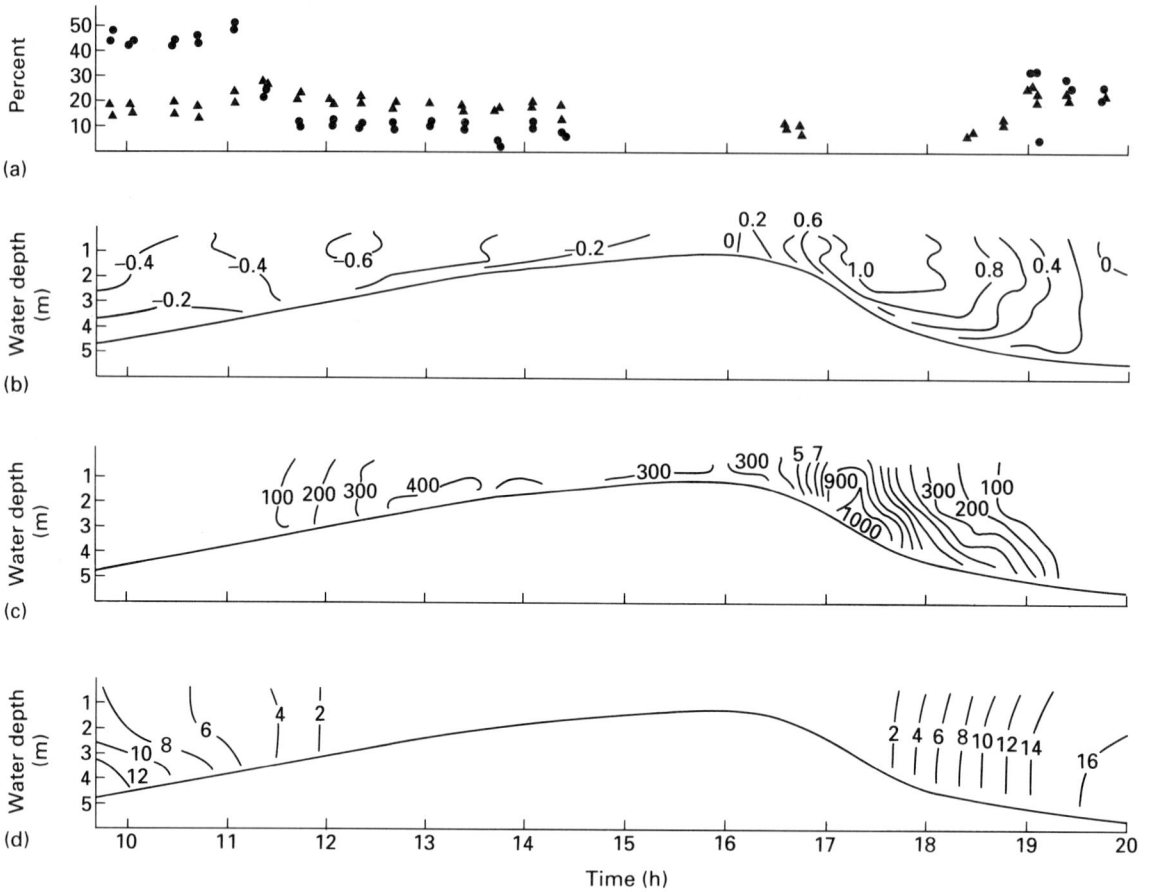

Fig. 6.6 Variation of (a) floc size, Floc size: ▲, 261.6–564 μm; ●, 160–216 μm. (%), (b) velocity (m s^{-1}), (c) concentration (mg l^{-1}) and (d) salinity (‰) during a tidal cycle in the Tamar estuary. McCabe (personal communication).

to erode and sustain more sediment in suspension, and it will be further up the estuary. This is due to the fact that there is a higher mean sea level in the upper estuary at spring than at neap tides, arising because the increased range at spring tides involves a large extra volume of water at high tide, but only a slight volume difference at low tide, relative to the neaps. During decreasing tidal amplitude towards neaps, the peak currents decrease, and less and less material is capable of being re-eroded and suspended. Additionally the durations of slack water increase, enhancing deposition. Consequently there are large differences in the position and concentration of the turbidity maximum from tide to tide. This makes it very difficult to measure the net fluxes of sediment into, or out of, the estuary, or to construct a sediment budget for the system.

The very large changes in velocity and concentration during the tide, and the dynamic movement of the water mass, makes it very difficult to separate the influences of advected structures from those produced by local erosion or deposition. This is a fundamental problem in estuarine sediment dynamics, and can only be solved by measurement campaigns at several locations simultaneously. Nevertheless, the scale of variability within the estuary means that the necessary intensity of observation is seldom achieved.

6.5   Processes forming the turbidity maximum

It should be obvious from the above general description of the turbidity maximum that it is a dynamic feature that involves interaction of the tidal flow with erosion and deposition of sediment. There are a number of processes that operate to concentrate the fine sediment at the upper end of the estuary, and to keep it there. Additionally there are estuaries where the mineralogy of the sediments indicates that the material has a marine rather than a river-borne origin. The sedimentation rates can also exceed the quantities discharged by the rivers. Consequently the processes involve the response of the whole estuary, rather than just the area of the turbidity maximum.

6.5.1 Residual circulation

In partially mixed estuaries the vertical gravitational circulation produces a residual landward bottom flow, and a seaward surface residual flow. This has long been thought to be the main mechanism for maintaining the turbidity maximum (Schubel & Carter, 1984). Suspended sediment is brought into the estuary by the river, and energetic tidal mixing maintains an even distribution of suspended sediment throughout the water depth. Because of the residual downstream flow in the river, there is a convergence in the bottom flow at a null point near the head of the salt intrusion, in salinities of about 1–5‰. Suspended sediment is brought into the estuary by the river, and in the upper estuary energetic tidal mixing transfers the sediment between the surface and lower layers (Fig. 6.7). The surface layer transports sediment downstream to the middle estuary where the particles settle into the lower layer, only to be carried headwards on the residual flow, together with particles brought in from lower down the estuary. Consequently, the maximum concentration of suspended sediment occurs at the bottom near the

null point, and the vertical gradients of suspended sediment must be related to the magnitude of the tidal mixing. This circulation process can lead to a turbidity maximum without the need for consideration of sediment properties other than settling velocity, and without any sediment erosion or deposition.

The continual circulation can be the cause of the turbidity maximum containing a high proportion of a narrow size range of fine sediment, and a change in the settling velocity leads to a variation in the suspended sediment concentration, as shown by Festa and Hansen (1978).

The vertical gravitational circulation can be perturbed along the length of the estuary by secondary circulation cells induced by bends (Dyer, 1977). Thus it is feasible that the turbidity maximum may be enhanced in areas beneath an upward flowing arm, and deposition concentrated under the downward flow. Lateral variations in concentration will then arise.

Lateral variations in residual current strength are associated with the topographically induced secondary circulation, but can also be produced by *Stokes' Drift*. Because the friction that the tidal wave experiences increases up the estuary, the tidal distortion increases towards the head of the estuary. When the tidal range becomes large compared to the water depth an effect called Stokes' Drift is produced. Near high water the unit current velocity produces a large upstream water discharge, whereas near low water the same current will only give a smaller discharge. Continuity requires the cross-sectional average of the volume discharge to be equal to the river discharge, and this causes a redistribution of the flow, so that the net discharge in the intertidal areas tends to be in the opposite direction to that in the deep channel. Consequently there will be a tendency for horizontal

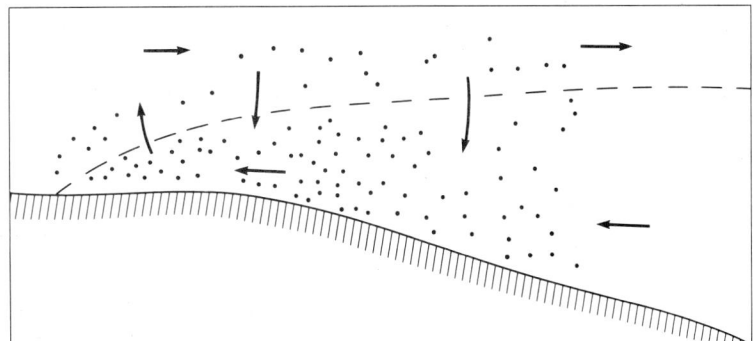

Fig. 6.7 Schematic diagram illustrating the formation of the turbidity maximum due to vertical gravitational circulation in a partially mixed estuary.

circulation cells to develop, and these can produce a circulation of sediment as well. Dalrymple *et al.* (1975) have shown inward transport in the intertidal areas of the Minas Basin, opposed by seaward movement in the channels. The opposite occurs in the Tamar estuary (Uncles *et al.*, 1986). The overall effect is likely to be sensitive to the relationship between the cross-sectional form, and the tidal water level variation.

6.5.2 Lag effects

However, as has been described above, there is considerable erosion and deposition during the tidal cycle. If the sediment particles acted as a passive tracer which responded instantaneously to the flow, the response of the sediment would not introduce further complications. However, the sediment response lags the flow. This phase difference between the suspended sediment concentration and the water velocity can produce a residual flux of sediment even when there is no residual movement of water, provided the currents are asymmetrical. Lags can be produced by a variety of causes; by interaction of the flow with the estuarine topography, as well as the erosion, entrainment, settling, and deposition of sediment.

Threshold lag

A lag can be produced by the presence of a threshold for sediment movement. When coupled to an asymmetry in the tidal currents, sediment movement can take place for a longer time on one phase of the current compared with another. As an example, in a tidal cycle with an intense short flood current and a long lower velocity ebb current, the duration of movement on the ebb tide will decrease more rapidly than that on the flood tide with an increase in the threshold of sediment movement (Fig. 6.8). In the extreme case the current on the ebb tide may not reach the threshold velocity and all of the movement then occurs on the flood tide. The asymmetry in the sediment discharge or transport rate caused by this effect will be even more marked if the transport rate has a non-linear relationship to the current velocity. In practice, measurements have shown that the sediment transport rate for bed load is normally proportional to the $\frac{3}{2}$ to $\frac{7}{2}$ power of the bed shear stress. As the asymmetry of the tide increases towards the head of the estuary the increasing magnitude of the flood current causes a transport of sediment towards the

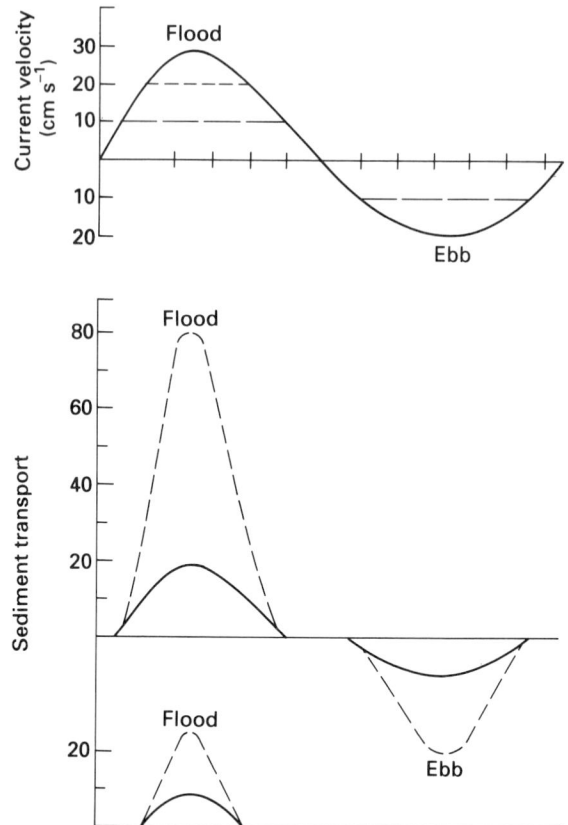

Fig. 6.8 Variation of current velocity and sediment transport (in relative units) during a tidal cycle, illustrating threshold lag. (a) Current velocity variation. (b) Sediment transport for a threshold of movement of 10 cm s^{-1}. —, Assuming linear relationship, transport proportional to $(u - u_c)$; – – –, assuming transport proportional to $(u - u_c)^{3/2}$. (c) Sediment transport for a threshold of movement of 20 cm s^{-1}. —, Sediment transport proportional to $(u - u_c)$; – – –, sediment transport proportional to $(u - u_c)^{3/2}$.

head of the estuary. This has been examined by Allen *et al.* (1980) and postulated to be a major process in macrotidal estuaries.

Erosion lag

In muddy sediments the critical erosion shear stress of the mud rises with depth into the sediment. Consequently, once the surface layers have been eroded the new surface material requires somewhat higher velocities to cause movement (Fig. 6.9). With increasing bed shear stress on the flood tide the critical erosion shear stress τ_e at the bed surface

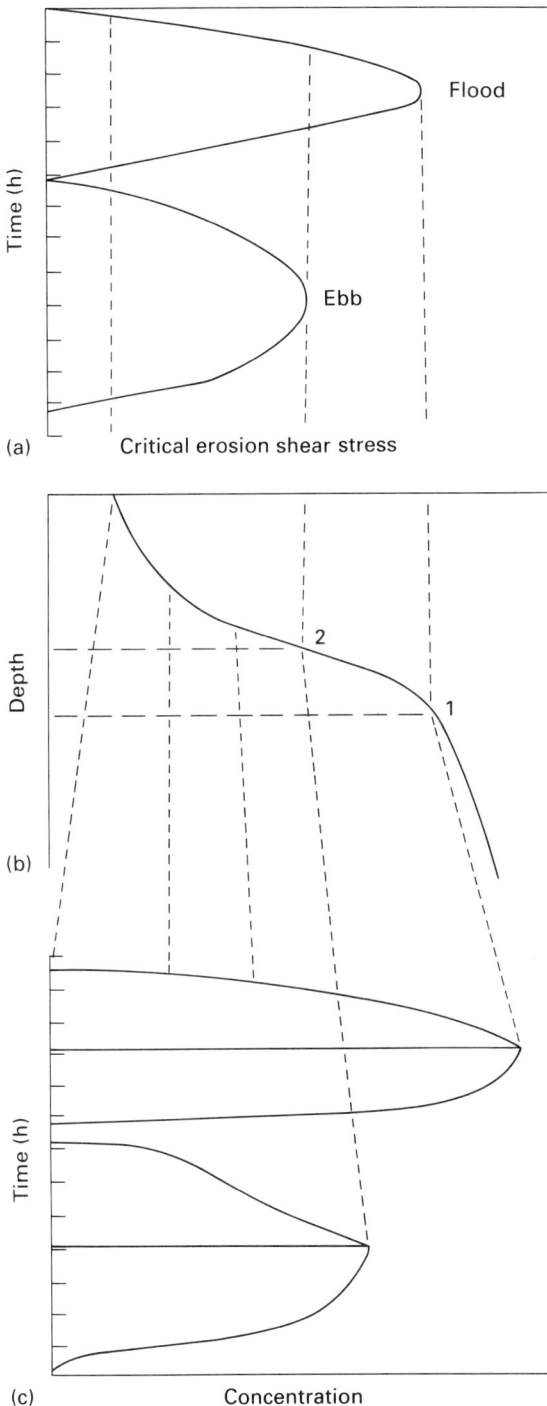

Fig. 6.9 Schematic illustration of erosion lag.
(a) Variation of bed shear stress during the tidal cycle.
(b) Variation of critical erosion shear stress with depth in
the sediment. (c) Variation of suspended sediment
concentration produced during the tidal cycle by bed
erosion.

becomes exceeded and erosion progressively pro-
ceeds into the sediment at a rate that is governed by
the variation of bed shear stress with time, as well as
the variation of erosion shear stress with depth, and
the sediment is suspended. At peak flood current
erosion has reached depth 1 where it ceases (Fig.
6.9b). During the decreasing flood phase no further
erosion can take place, and the suspended sediment is
deposited to consolidate. The depth profile of τ_e will
develop with consolidation of the mud, and since this
is a time-dependent process, we will assume that the
original profile is re- established before erosion re-
commences on the ebb tide. However, the level of
peak erosion on the ebb will be at depth 2. In many
cases the deposited mud will not have time to
consolidate fully, and remains over slack water as an
almost homogeneous turbid layer, which may be
fairly easily re-eroded. The overall effect of the
erosion lag is therefore to produce a significant phase
difference between the suspended sediment concentration and the velocity.

Scour lag

Once the sediment particles are in motion they can be
kept moving at velocities below the threshold of
initial motion. Consequently between the threshold
of erosion and the threshold of deposition material is
kept in motion, but no new erosion takes place. This
produces a *scour lag* (Postma, 1967).

Scour lag can also be defined as the time taken for
sediment, when entrained from the bed, to disperse
to higher levels in the flow (Nichols, 1986). This
means that once material is eroded from the bed it is
only gradually mixed through the water flow as it
moves downstream. Initially, the sediment will move
in the near bed layers at a velocity lower than the
depth mean current. Consequently, at higher levels in
the flow the suspended sediment concentration will
lag behind the concentrations being produced at the
sea bed. To illustrate the effect consider the simple
situation shown in Fig. 6.10. The symmetrical tide
has a decreasing maximum current towards the head
of the estuary, and the water and sediment particles
undergo a changing velocity with distance along the
channel during the tide. A particle on the bed will be
lifted into suspension as the threshold velocity is
exceeded at point 1. However, it does not achieve a
velocity equal to the depth mean until point 2, when
it will be at a level in water that started to move on
the flood at a more seaward position. It then travels
with the water trajectory until point 3, when we will

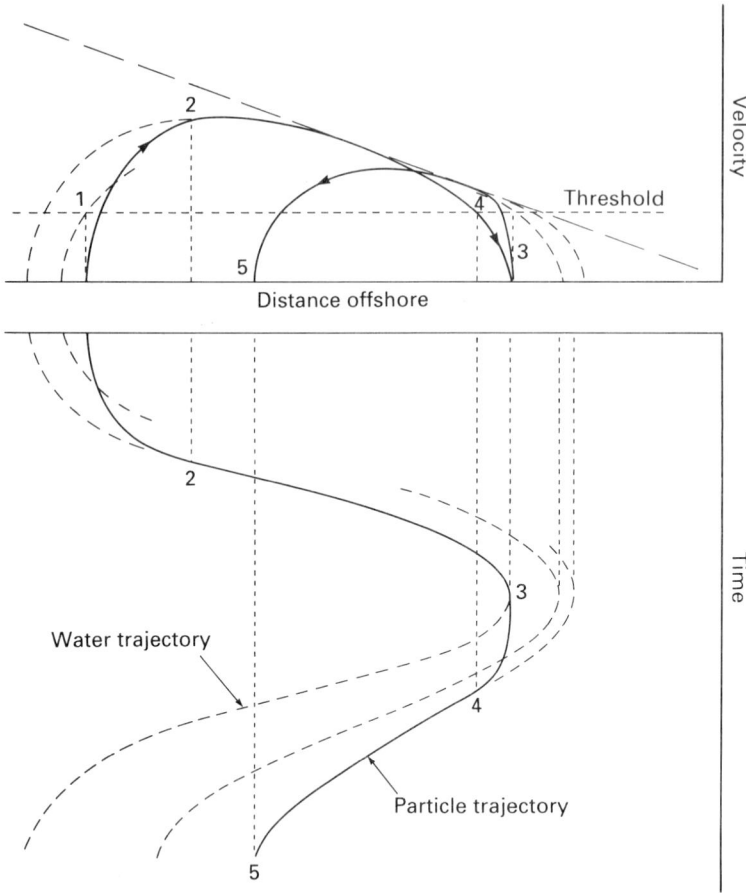

Fig. 6.10 Schematic illustration of scour lag. See text for explanation.

assume it is instantaneously deposited. On the ebb tide it will be resuspended when the threshold is exceeded, but again lags the flow until point 4 is reached. It eventually is redeposited at point 5, and considerable landward movement has been produced during the tidal cycle because of the scour lag. The magnitude of the residual movement will depend to a great extent on the relative water depths at each end of the particle trajectory, and on the turbulent mixing rates.

Settling lag

On the falling tide the particles will start to settle once the turbulence in the flow is incapable of maintaining them in suspension. As the particles settle they move along on the waning current so that they eventually reach the bed some distance from the point at which settling commenced. This effect is known as *settling*

lag, and a qualitative model describing these effects was developed by van Straaten & Kuenen (1958) and Postma (1961). Consider the same model as in Fig. 6.10, but without scour lag. At position 1 shown in Fig. 6.11, the particle is lifted off the bed and travels with the water until point 2, when it starts to settle. Because of settling lag it reaches the bed at point 3. On the following ebb tide it will not be entrained until later in the tide when the threshold velocity is reached at that position. The deposition at low water will be at position 6. Consequently, the particles gradually migrate shorewards to accumulate in the area where the maximum velocity during the tide equals the grains' threshold velocity.

This model has been quantified by Groen (1967) and Dronkers (1986). The former author showed that it is possible to develop net movement of suspended sediment in an alternating flow that has no residual component, even if the maximum flood and ebb

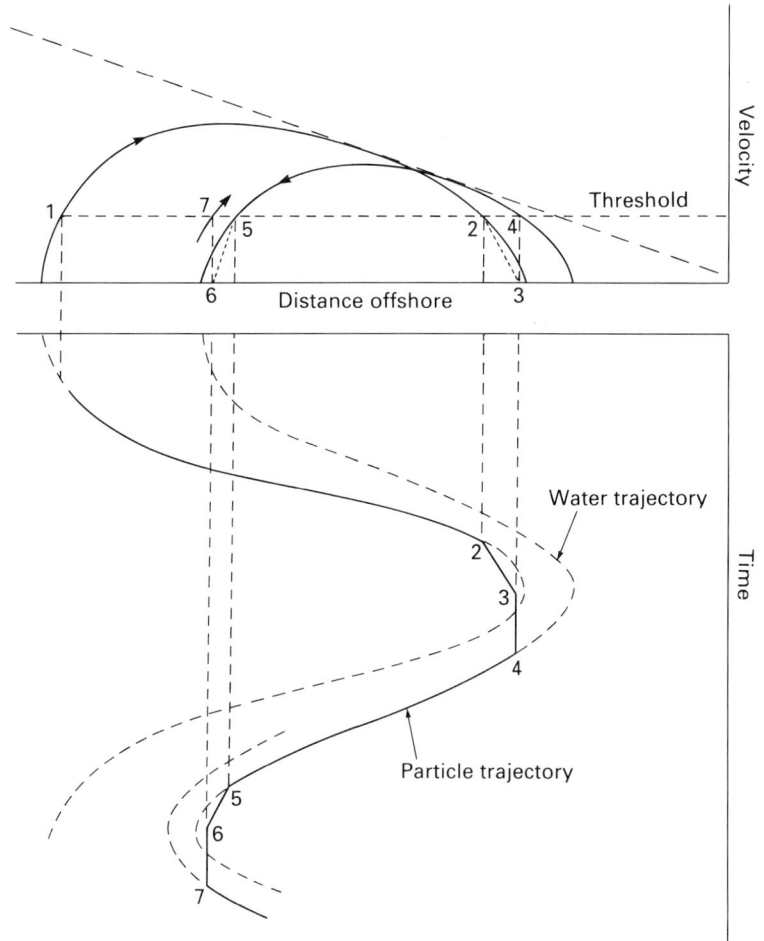

Fig. 6.11 Schematic illustration of settling lag. See text for explanation. From Dyer (1986).

currents are equal, providing the tide is asymmetrical. Dronkers (1986) emphasized the time interval during which sediment particles can settle at slack water and remain on the bottom until resuspended, and concluded that the magnitude and direction of the residual sediment flux is mainly determined by the current velocities around low water and high water slack. Depositional periods at high water generally exceed those at low water.

6.5.3 Balance of processes

Obviously many of the lag effects occur simultaneously and their effects will be difficult to separate. The relative importance of the contributions have been considered from examination of the temporal and spatial variations of velocity and suspended sediment concentrations during the tidal cycle by Dyer (1978, 1988) and Uncles *et al.* (1984, 1985a,b). The phase differences between the suspended sediment concentration and velocity produced by the lags described above create a tidal pumping of sediment. However, there are considerable problems because of the continual change in magnitude of the current velocities and the lags with time and along the estuary. Nevertheless, these studies showed the cross-sectional fluxes produced by tidal pumping were larger than those produced by residual gravitational circulation. However, the fluxes produced by the various contributions were large and of different direction, some indicating upstream fluxes, and some downstream. The overall balance was therefore the result of small differences between large opposing fluxes.

The fact that the turbidity maximum sometimes occurs landward of the salt intrusion indicates that gravitational circulation is not always dominant. Thus it seems that tidal pumping due to erosion and suspension of sediment during the tide is a major factor in generating and supporting the turbidity maximum. There is a net outflow of sediment due to the river flow. This is balanced by the tidal pumping due to resuspension on the flood tide. Asymmetry of the tidal velocities leads to lower sediment concentrations on the ebb tide, so that there is a net upstream tidal pumping. Vertical gravitational circulation is likely to be a minor contribution to the upstream transport. In the area of peak concentrations in the turbidity maximum, entrainment should be equally effective on both flood and ebb tide, with little preferential movement. At the upper end of the turbidity maximum the suspended sediment concentration reaches its peak near high slack water. This is the same as the behaviour of salinity, and leads to a downstream tidal pumping against the longitudinal suspension gradient. As tidal range increases, the entrainment of sediment will occur earlier in the tide and will be more intense. The tidal pumping should then increase, leading to alteration in the intensity and location of the turbidity maximum; it should move up the estuary towards spring tides, though there could be a lag in the response. Likewise, an increase in river flow alters the intensity of the net seaward flow, leading to a seaward shift of the turbidity maximum.

The overall effect of the gravitational circulation is probably to sharpen the peak of the turbidity maximum, rather than to create it. However, there is evidence that increased turbulence near the tip of the salt intrusion can enhance sediment resuspension (Kostaschuk & Luternauer, 1989).

When the tidal variation of concentration is plotted against velocity, a hysteresis curve is produced (Allen, 1974). In this instance we will use $u|u|$ which is proportional to the bed shear stress. In an idealized situation (Fig. 6.12a) without any sediment pick-up or settling, the advection of a suspended sediment concentration field past the measuring position would produce the same concentration on rising and falling currents. However, entrainment of sediment from the bed produces an increasing concentration on the accelerating tide, and settling lag causes the maximum concentration to lag the maximum velocity. In addition the time delay between the maximum concentration at different heights above the bed would become obvious.

The pattern of the hysteresis curve can be quite

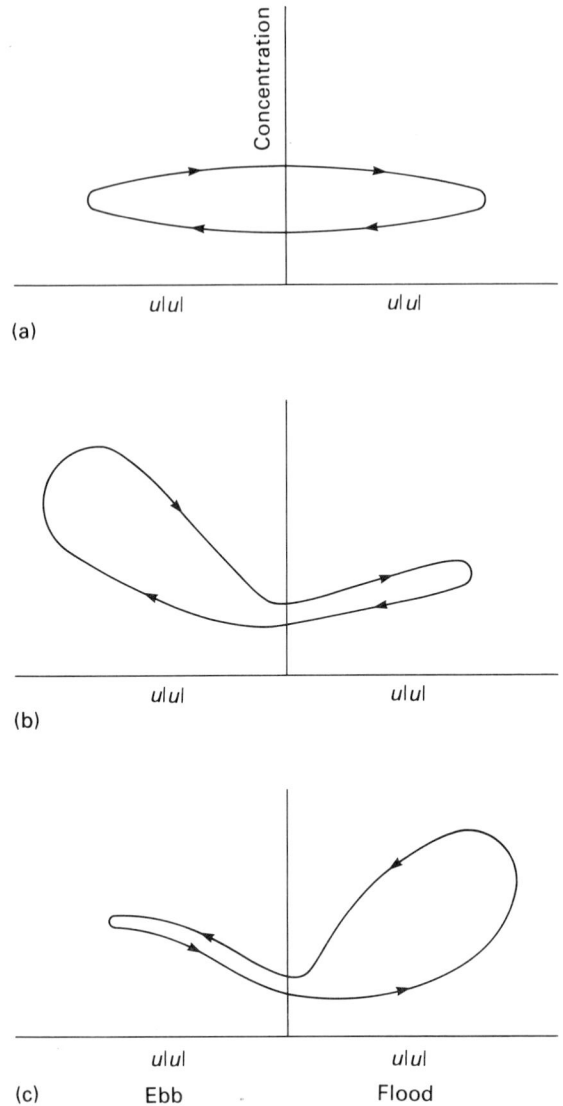

Fig. 6.12 Hysteresis curves of suspended sediment concentration against $u|u|$ (proportional to bed shear stress), for three illustrative conditions. See text for explanation.

complicated depending on the rate of change of concentration due to advection of the horizontal gradient, relative to the rate of change produced by entrainment and settling. Near the mouth on the ebb, the pick-up and the increasing concentration caused by advection work together to produce an open anticlockwise hysteresis curve (Fig. 6.12b). On the flood, however, the concentration is decreasing

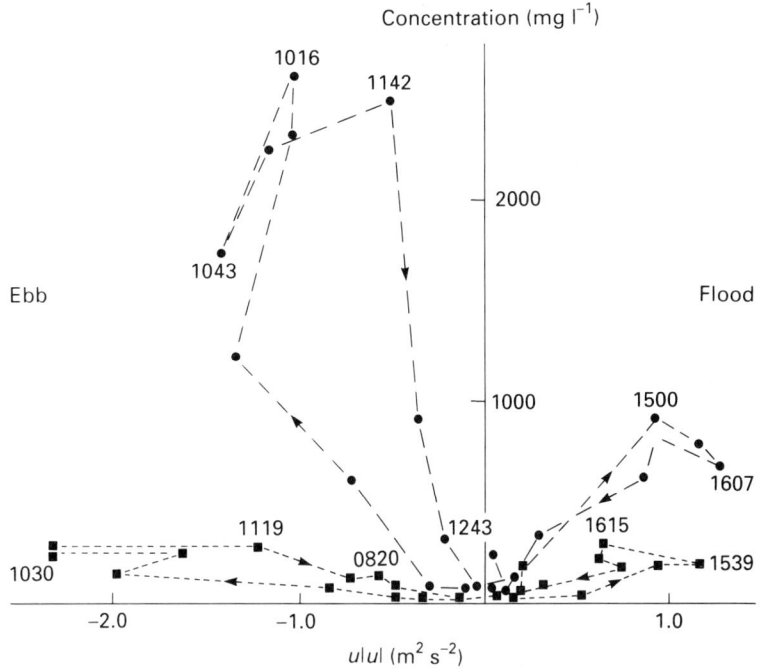

Fig. 6.13 Hysteresis curve for suspended sediment concentration during a tidal cycle for a position in the Humber estuary. Height of sampling 0.5 m above the bed. ●, Fraction finer than 40 μm; ■, fraction coarser than 40 μm.

because of advection, and increasing because of pick-up. Consequently whether the hysteresis curve is clockwise or anticlockwise depends on the relative intensities of the two processes.

On the landward side of the turbidity maximum, the opposite effect will occur (Fig. 6.12c). Examples of hysteresis are given by Nichols (1986), Grabeman and Krause (1989), and Kostaschuk *et al.* (1989). An example taken for the Humber Estuary is shown in Fig. 6.13 for two grain size fractions.

An alternative way of considering the hysteresis is to plot concentration against distance (velocity × time). This should then take into account the advective movement. Figure 6.14 shows a plot of concentration against distance for the same data shown in Fig. 6.13. The maximum velocity occurred during the period when water from between points 6 and 12 km upstream was passing the station. There was considerable entrainment of sediment during the time of maximum current, but it seems possible that an additional source of suspended sediment occurred at about 14 km upstream. By the time that water mass returned on the flood tide, at least half of the suspended sediment had settled out. Based on this type of analysis several zones of preferential entrainment have been defined for the Severn estuary by HRS (1981).

Modelling of the turbidity maximum has generally enabled the location and order of magnitude of the suspended sediment concentration to be simulated. However, the maximum is generally too diffuse, not being sufficiently limited and peaked longitudinally. This has been found by Uncles and Stephens (1989) who omitted the effects of stratification. Experiments were carried out within a simple model by Dyer and Evans (1989) which showed that the magnitude of concentrations in the turbidity maximum was sensi-

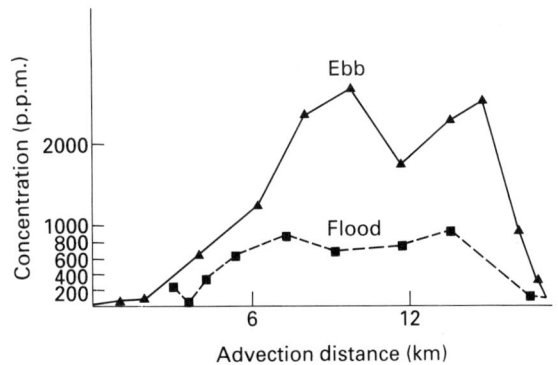

Fig. 6.14 Concentration versus advection distance for the fine fraction shown in Fig. 6.13.

tive to the settling velocity of the sediment. Settling lag contributed little to the position of the maximum, because of the short period during which settling took place, but the position was very sensitive to the erosion threshold of the bed sediment, and the degree of tidal asymmetry. Similar results have been demonstrated by Markofsky et al. (1986). Comparison between measurements and modelling point to the importance of accurately specifying the bottom shear stress and the associated deposition and resuspension (Lang et al., 1989).

It is apparent that most of the processes discussed above need to be included in mathematical models for correct simulation of the turbidity maximum.

6.6 Estuarine sedimentary cycle

The erosion, transport and deposition of muddy sediment in the turbidity maximum involves a number of further processes, which are of second order as far as the presence of the maximum may be concerned, but which are extremely important in determining the suspended sediment concentrations, and the magnitudes of lags in erosion and deposition. These processes are flocculation, settling, deposition, consolidation and re-erosion. The evolution of suspensions, and their development in response to the tidal cycle, has been described by Kirby and Parker (1983), and Parker (1986).

There are three basic forms of mud occurrence:
1 *Mobile suspensions* where the suspension is moving freely under the tidal forces or downslope under the influence of gravity. The floc sizes are likely to be in balance with the shearing stresses.
2 *Stationary suspensions* which are not moving horizontally but may be gradually settling. The floc sizes are likely to be increasing with time due to settling.
3 *Settled mud*, or mud forming part of the sea bed, resisting erosion for considerable periods and gradually consolidating. The overburden weight is predominantly supported by interparticle contacts.

During the tidal cycle, or the lunar neap–spring cycle, as the velocity of the current and consequently the shear stresses vary, the mud will preferentially appear in one or other of the above states (Fig. 6.15). A crucial factor in the processes is the development of layering within the suspension. Over a single tidal cycle, at spring tides a vertically homogeneous profile is present at maximum ebb or flood current. As the current diminishes, the load of sediment cannot be supported and that near the surface commences settling. Flocculation increases the size and settling velocity of the clusters of particles and forms a sharp step, or 'lutocline', in the profile and this gradually gets closer to the sea bed while the concentration beneath it gradually rises. All the while the suspension would be still moving along the estuary. Near to slack water the high concentration near-bed layer becomes stationary for a brief period. As the current accelerates on the opposite phase of the tide, erosion of the upper surface of the layer may occur and the gradual re-entrainment will cause the lutocline to rise through the water column towards the surface again. Alternatively, it is possible that the stationary suspension may be moved as a slug along the bottom for a while before re-entrainment.

As the tidal range gradually diminishes towards neap tides the static suspensions formed at slack water survive progressively longer into the succeeding tide, until they eventually resist erosion and

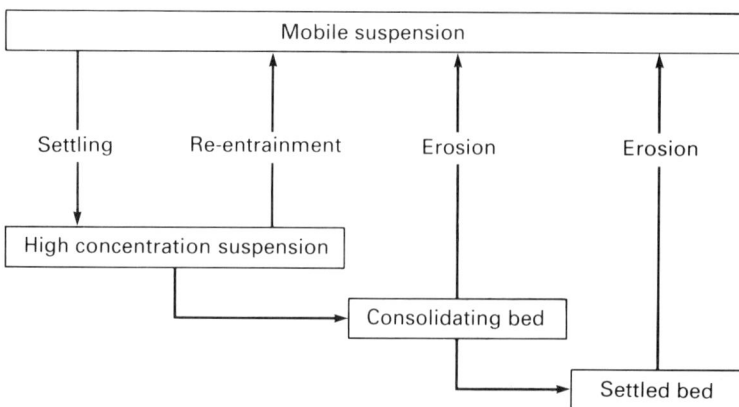

Fig. 6.15 Schematic representation of the cycles of sediment settling and erosion in an estuary. After Mehta et al. (1989).

remain throughout the complete tidal cycle. The suspension then has several tidal cycles in which to consolidate gradually. On the succeeding neap to spring cycle re-erosion of this layer may not be complete, and a small increment of deposited sediment may remain to become part of the seabed for a longer timescale. At some stage during the development of the layering the interface can be detected by echo-sounders as indistinct 'ghost echoes'. The layers are then termed *fluid mud*. Often several echoes can be distinguished within the fluid mud and comparison with the results of radioactive densimeters has shown that they correlate with density layering (Kirby & Parker, 1974). The properties of suspended mud particles and of bed sediment have been reviewed by Mehta (1986) and Mehta *et al.* (1989).

6.6.1 Flocculation

Flocculation of particles in the sea occurs as a result of the total surface charge on the particles and the encompassing electrical double layer. When the particles are in close proximity there is an attraction that leads to the formation of aggregates of particles. The collision potential of particles obviously increases with increased concentration, but is also enhanced by Brownian motion, differential settling, grain inertia and shearing (Krone, 1978). At one time it was considered that the flocculation was entirely controlled by salinity, that the riverine particles flocculated when they reached the salt water, and that deflocculation could occur when particles were recycled back into contact with the fresh water at the head of the salt intrusion. However, it is clear that organic binding is significant in most flocs, and that there is a continual break up and reforming of flocs. The suspended sediment entering the estuary from the river is already flocculated, and the particle size actually becomes finer at the salt water contact (Eisma, 1986). There are two modes that contribute to the distribution of floc size, macroflocs which reach a size of the order of millimetres and microflocs of the order of 10–20 µm. Macroflocs are about the same size as the turbulent Kolmogrov microscale, and they can be fairly readily broken down in stages by turbulent shear to form microflocs eventually (Glasgow & Luecke, 1980). Microflocs are very much more resistant to being broken up, though links between the organic content of microflocs and distance down estuary implies that these

microflocs are broken and reflocculate with new biological influence (Eisma, 1986). Microflocs interact with an increasing order of aggregation to form macroflocs.

Laboratory measurements have indicated that the modal floc size is affected by both concentration and shear stress, and the primary cause of floc disruption is by three-particle collisions (Burban *et al.*, 1989). Preliminary measurements in estuaries have shown similar results (Bale & Morris, 1987). The general form of the relationship between floc size, concentration and shear stress which explains those results and other considerations is shown in Fig. 6.16. The modal floc size scale is imprecise because of floc density/strength variations, and there is probably a time response of the flocs to the shear. Nevertheless, at low concentrations the flocs are small, and low shear increases the likelihood of floc growth due to collision. However, with increasing shear the intensity of the collisions leads to floc breaking. At higher concentrations larger size flocs are present in quiescent settling but only a small amount of shearing is necessary to disrupt them. The floc size distribution, in terms of sorting or standard deviation, is also likely to change with shear, with a narrow size distribution at low and at high shear. At moderate shear a wide range of floc sizes should be present

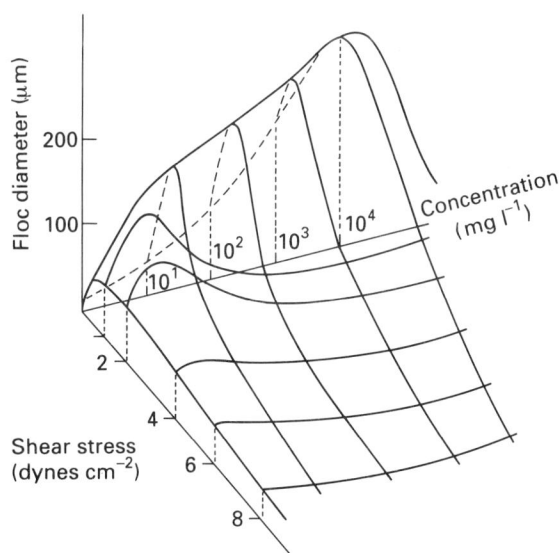

Fig. 6.16 Schematic diagram showing the relationship between floc modal diameter, concentration and shear stress. From Dyer (1989).

during floc break-up. A background of primary particles of size $< 2\,\mu m$ which are only intermittently involved in flocculation is likely.

6.6.2 Settling velocity

Settling velocity is important for modelling the mass settling to the bed. Thus the settling velocity/concentration relationship is required. There have been many laboratory studies of settling velocity, and there are discrepancies between laboratory and field results largely because of the effect of sampling in disrupting the macroflocs. There are also considerable differences in the settling velocity/concentration relationship between estuaries (Fig. 6.17), which may be the result of floc density or organic content variations, and between different tidal states. Nevertheless, at concentrations less than about

$2\text{--}5000\,\text{mg}\,\text{l}^{-1}$ the settling velocity w_s depends on concentration according to $w_s \propto c^n$, where n varies from 0.6 to 2.2. At concentrations above $10\text{--}20\,\text{g}\,\text{l}^{-1}$ the settling particles interfere with each other and 'hindered' settling occurs.

Crucial in determination of the settling velocity/size relationship of the flocs is a knowledge of floc density. Comparatively little is known about the relationship between floc size and density. This is important because the rheological properties of the suspensions are governed by volume concentration rather than by mass concentration. Figure 6.18 summarizes most of the significant results available for the density difference between the floc and the fluid against floc diameter. There is some general agreement on the trend, with flocs becoming looser with increasing size, but again the data show large scatter. Organic content of the flocs may account for a large measure of the scatter.

Even though the large flocs are of lower density, they are likely to be of most importance in determining the mass flux to the bed. However, there are many more microflocs and they have a cumulative surface area that is considerably greater than the smaller number of larger macroflocs. Therefore, though the macroflocs may dominate the sediment fluxes, the microflocs are likely to be more important from a chemical point of view.

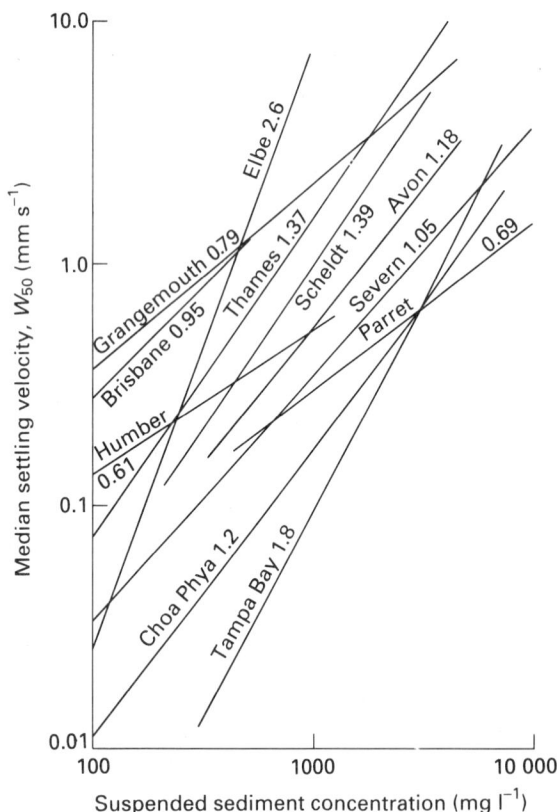

Fig. 6.17 Relationship between median settling velocity and suspended sediment concentration. From Dyer (1989).

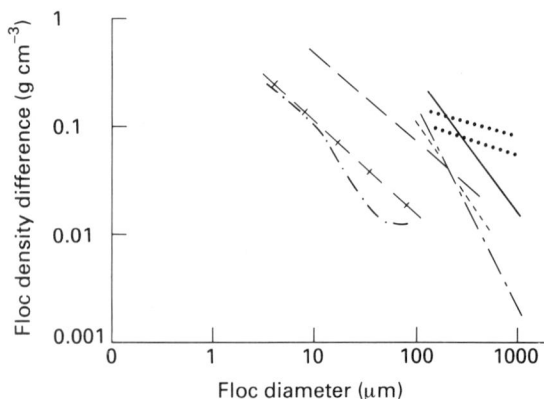

Fig. 6.18 The variation of floc density difference (floc density minus fluid density) with floc diameter. ...., Fowler & Small; ——, Joyce; – – –, Kajihara; –·–·–, Riley (all from McCave, 1973); ＋ ＋ ＋, Hawley: (1982); – – –, Gibbs (1985); – – – –, Bache & Al-ani (1988). (See Dyer, 1989 for references.)

6.6.3 Fluid mud

Because of aggregation, settling of particles towards the bed can be quite rapid towards slack water. If the flux of material is high, the water trapped between the flocs cannot readily escape and a layer of homogeneous high concentration suspension can develop overlying the bed. This is bounded on the upper surface by the lutocline and may be detectable as fluid mud. The maximum flux of particles settling towards the bed occurs in the region of about $20\,\mathrm{g\,l}^{-1}$. Judging by laboratory experiments, at somewhat higher concentration the layer has the rheological properties of a pseudo-plastic, which means that it has high viscosity at low shear rate, but with reducing viscosity at high shear rate. Consequently, though on initial formation the high concentration layer may still be actively moving along the estuary, as the current shear decreases towards slack water the layer becomes more resistant to movement, and will become static at some time. The conditions under which this happens are unclear, since the viscosity is related to the rate of shear in breaking down the floc structure. When concentrations exceed $80-220\,\mathrm{g\,l}^{-1}$ a framework structure is found in the suspension, it effectively becomes a soft bed, and has the properties of a Bingham plastic (Sills & Elder, 1986). The high concentration layers will stand on slight slopes, but a certain amount of gravitational flow must occur down slopes since ponds of high concentration suspensions are found at low points on the estuary bed. In these ponds the total mass in the vertical is considerably more than that suspended over the same area throughout the water depth at maximum current. Consequently the dynamics of the high concentration layers are not directly related to the dynamics of the tidal flow. During the accelerating flow in the next tidal phase the high concentration layer, having a high viscosity, will undergo a fluid shear at the level of the lutocline which may create some erosion of the upper surface, but failure may alternatively occur at the bottom of the layer, so that the layer moves. Once it has started moving the viscosity will decrease and the layer could very rapidly become dispersed within the body of the flow. Preliminary indications suggest that layer thickness may be crucial in separating these two modes of erosion (Odd & Rodger, 1986).

It is not certain what characteristics of the upper surface of the high concentration layer determines its detectability by echo-sounding. Both the vertical density and velocity changes are important in reflecting the sound signal, but, at high frequencies, so is the backscattering of energy by organic or other particles associated with the interface. Consequently layering may be present without necessarily being acoustically detectable, and detectability may vary with the frequency of the echo-sounder (Parker & Kirby, 1982a). Fluid muds are frequently observed in estuaries and ports.

6.6.4 Erosion properties

Quantification and prediction of the erosion characteristics of sediment are crucial to the modelling of sediment transport in estuaries. Erosion of a mud bed can be by several processes; by erosion of individual flocs, erosion of small clusters of particles, or by failure and mass erosion of a surface layer. Laboratory measurements have shown there is a correlation between density and the critical erosion shear stress for some muds, (e.g. Thorn & Parsons, 1980), but the relationship is weak, and there are many other important influencing factors such as particle mineralogy and grain size distribution, and organic content.

Because of consolidation of deposited sediment, there is a gradient of most physical and chemical properties with depth into the bed. Density and the critical erosion shear stress rise almost exponentially with depth. In many situations the mud surface comprises two layers at slack water; a thin fluid mud or loose fluffy layer, overlying a more rigid bed. The upper layer has a threshold shear stress of $0.06-0.1\,\mathrm{Nm}^{-2}$, and is fairly easily eroded by the tide. The lower layer would normally only be eroded at extreme conditions, and after weakening by biological activity.

Muddy substrates are normally ideal environments for an active infauna. Many of these are filter feeders and consequently they filter the material out of suspension, consume some of the organic content and pass it back into the water column as faeces or pseudo-faeces. It has been estimated that the total volume of water in an estuary can be filtered in a few weeks by such organisms. The packaged particles will have different densities, settling velocities and dynamic characteristics from normally agglomerated flocs (McCave, 1975). Additionally, the presence of the organisms can affect the effective roughness of the bed, and the ability of the ambient shear stress to cause erosion (Jumars & Nowell, 1984). The organic slimes and films created by the organisms will also have a binding effect on the substrate, increasing its cohesive nature. Consequently biological activity can have both a stabilizing and a destabilizing effect on muddy sediment.

Within the sediment, bioturbation by meio- and macrofauna is important in exchanging both particles and contaminants throughout the depth. Some animals feeding at depth deposit the worked sediment on the surface, whereas others filter the suspension in the overlying water and eject the material as pseudofaeces. There is normally a redox boundary which occurs at about the level of the maximum density gradient. This boundary indicates both a lack of active physical disturbance and a change in the chemical processes.

6.6.5 Sediment budgets

Within the turbidity maximum the total mass of sediment in suspension is often considerably greater than the annual input of sediment from the rivers and there is active deposition and re-erosion, and exchange of particles with the sea.

There are several potential sources that can contribute sediment to the turbidity maximum. They are rivers, streams and outfalls, the sea, erosion of estuary coastline and tidal flats, erosion of the sea bed, biological production and the atmosphere. The relative importance of these sources will vary between estuaries, and will depend to a great extent on the seasonal cycle of river discharge. The riverborne and marine sources of sediment can often be distinguished from examination of the clay mineralogy or heavy mineral content of the sediments, for example Nichols (1972).

Meade (1969) has argued strongly that the majority of sediment in estuaries of NE America is derived from the sea, and this conclusion seems to be valid for many temperate estuaries. As an example, sandy sediment has moved into the Severn estuary in the past, but the source in the Bristol Channel now appears to be exhausted (Parker & Kirby, 1982b). About 10^7 t of suspended mud is present in the water column at spring tides and this is probably two orders of magnitude greater than the annual river input. However, there are two large areas of intertidal and sub-tidal mud, one of which contains at least 1.7×10^8 t. Within this area there are three zones; one shows evidence of fairly steady deposition, one shows evidence of erosion, and between the zones is

an area of stable sea bed (Kirby & Parker, 1981). Of the 30 km^2 of sub-tidal mud about 10 km^2 is thought to be accretionary and 7.5 km^2 erosional. Presumably the mud cycles through the turbidity maximum between the zones of erosion and deposition.

This illustrates a simplified concept of the turbidity maximum as a feature through which sediment is continually cycled from one part of the estuary to another, with small amounts of material being added from the rivers and the sea, these inputs being balanced by deposition. Individual particles may spend a considerable time moving within the turbidity maximum between initial insertion and final deposition. In between, the particles may undergo many periods of temporary deposition and re-erosion.

The residence time of particles can be defined as the number of particles inside the volume divided by the number leaving per unit time (Martin et al., 1986). Some of the particles entering from the river will remain in suspension and pass through the estuary fairly quickly. However, a significant proportion will undergo many cycles of deposition on the bed and resuspension, with the deposition occurring at a number of points along the estuary which form temporary sinks for the sediment particles, operating for a variety of timescales. Consequently the mass of particles in suspension in the turbidity maximum comprises proportions of particles that may have ages (time since input) lasting from a few days to possibly years.

Obviously, extended periods of erosion or deposition in particular areas would cause an alteration in the local flow and shear stress regime leading to a gradual reduction in deposition, or a switch between erosion and deposition. Consequently within a steady state system many of the cycles of exchange are self limiting. However, there is sufficient variability in the driving forces to ensure that some of the cycles of exchange are of very long duration, and they potentially could be terminated or started by dramatic events such as storms or floods.

Consequently though large amounts of sediment are mobile within estuaries, it is difficult to determine mean fluxes and accumulation rates. Sediment sources and budgets can be established in general terms, but the short term variability is extremely large (Jouanneau & Latouche, 1982).

6.7 Intertidal areas

One of the major sources and sinks for suspended

sediments within an estuary is the intertidal area (Anderson, 1983).

The flat lying areas often have deposition rates of the order of a centimetre per year, with a gradation from low to high tide mark. The maximum rate of sedimentation occurs on the outer edge of the intertidal flats about midtide level (Dieckmann *et al.*, 1987), so that the flats build outwards and upwards, with the consequence that the active volume of the estuary gradually decreases, reducing the sedimentation rate.

Stevenson *et al.* (1986) examined 15 American estuaries and found a strong correlation between mean tidal range and accretionary balance, with high range estuaries showing accretion exceeding sea level rise. However, Nichols (1989) has examined the response of 22 American lagoons to rising sea level, and found that the majority of them had accumulation rates equivalent to the local sea level rise.

Even though an equilibrium is achieved between sedimentation and sea level rise, there is still considerable sediment movement. Signs of erosion and of deposition can often be observed in closely adjoining areas, and these form both a source and a sink for material in the turbidity maximum. The surface of the salt marshes above neap tide high water mark, and the upper part of the intertidal flats, show regular sedimentation, with layering and lamination in core samples. However, the outer edges of the salt marsh often show erosion by 'cliffing', the undercutting and erosion of small blocks of compacted salt marsh sediment. Additionally, gullies and meandering channels cross the flats and show active erosion of the banks and migration of the meanders. As the channels meander across the mudflats they transform the horizontally stratified sediments into sequences showing laminations inclined at 7–15°. These are produced by deposition on the inside of bends in the gullies while erosion occurs on the outside of the bends (Bridges & Leeder, 1976). Consequently one can envisage a continual cycle with the mudflats building up to a particular level, and then being attacked and eroded by shifts in the channels and by gullying. The eroded sediment is exchanged via the turbidity maximum to other areas of temporary deposition.

An important feature of salt marshes is the plants, which can actively trap sediment and enhance deposition (Adam, 1990; Allen & Pye, 1992). Generally there is a zonation, as one goes from neap tide to spring tide high water marks, of flora that are tolerant to decreasing frequency of immersion. In sheltered areas a height variation of only a centimetre or so is needed to distinguish clearly between two zones.

During the summer the plants are traps for settling suspended sediment, and the sediment is re-eroded during the autumn and winter. Frostick and McCave (1979) measured 5 cm of accretion on intertidal mudflats of the Deben estuary between April and September because of trapping by algae, with subsequent erosion in the winter. In contrast, Kraueter and Wetzel (1986) have shown sediment stabilization/ destabilization on a different seasonal scale. Stable conditions occurred between December and March, but increased benthic activity in the summer caused increased water content, a decrease in sediment shear strength, and increased suspended sediment concentration. Consequently there are a variety of responses depending on the dominant biological processes in the intertidal zone.

Erosion on the mudflats can be effected by ice, rain and waves (Anderson, 1983). Small amplitude waves (less than 5 cm) are capable of suspending fine-grained sediment in shallow water and can increase the suspended sediment concentration by a factor of three over calm conditions.

Laboratory measurements have shown that under wave action prolonged sub-critical movement of the water produces a fluidization of the bed that eventually leads to mobilization. Because of high turbidity, and the high effective viscosity of the suspension, the incoming waves are modified, and develop into solitary waves (Wells & Coleman, 1981). The forward velocities under the crests then cause a preferential shoreward motion. The narrow zone of high suspended sediment progresses across the intertidal area on the rising and falling tide, and can transport comparatively large quantities of sediment. During the ebb tide the zone becomes quickly concentrated into the gullies, and suspended sediment is ejected as plumes into the main channels. On the flood, the suspended sediment is transported onto the higher tidal flats where it can be trapped.

6.8 Estuary mouth

The mouths of many estuaries are restricted by the presence of spits. These arise because local coast erosion has supplied a large amount of material, and this is brought to the mouth by littoral transport within the breaker zone. The spit builds out to the stage where the constriction increases velocities to the extent that they erode as much material from the tip of the spit as is supplied. The hydrodynamic balance is illustrated by the linear relationship that has been demonstrated between the cross-sectional area of the inlet and the tidal prism (O'Brien, 1969). Topographically it is easy to define where the mouth of the estuary lies. Dynamically, however, the mouth will be somewhat to seaward of the topographic mouth, and its position will vary with river discharge, and with tidal range.

Associated with the flow through the restricted estuary mouth is the development of tidal deltas (Oertel, 1972; Hayes, 1975; Boothroyd, 1978). The interaction of the ebb and flood flow through the mouth produces a strong residual flow away from the inlet in the centre, but towards the inlet at the sides (Fig. 6.19). In the flood channels the dominant flood current carries sediment inland. Towards the head of the channel the peak velocities and the residual flow lessens as the cross-section expands. Water flowing up the flood channel on the flood tide tends to move across towards the ebb channel at high water. The ebb channel shallows seawards, and the ebb and flood channels interfinger. Between the channels there are sand banks, around which there is a circulation of sand. Because of the frequent movement the sediment becomes well sorted, the finer fraction being

winnowed away in suspension. Wave effects will modify the ebb tidal delta, and its form will be related to the coastal residual current and sediment flow (Sha, 1989). Yang (1986) has shown that the ebb tidal delta of the Oosterschelde estuary grew considerably in extent owing to erosion of sediment from within the estuary. This was due to an increase in tidal prism volume caused by closure of tidal channels, and the rise in sea level.

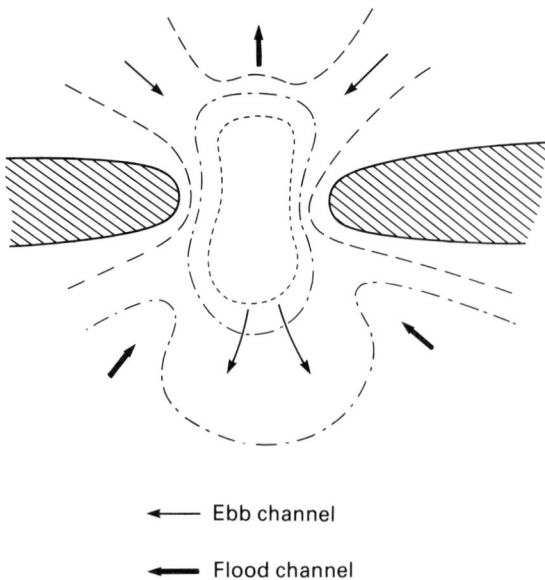

Fig. 6.19 Residual sediment transport patterns at a tidal inlet, and on the ebb and flood tidal deltas.

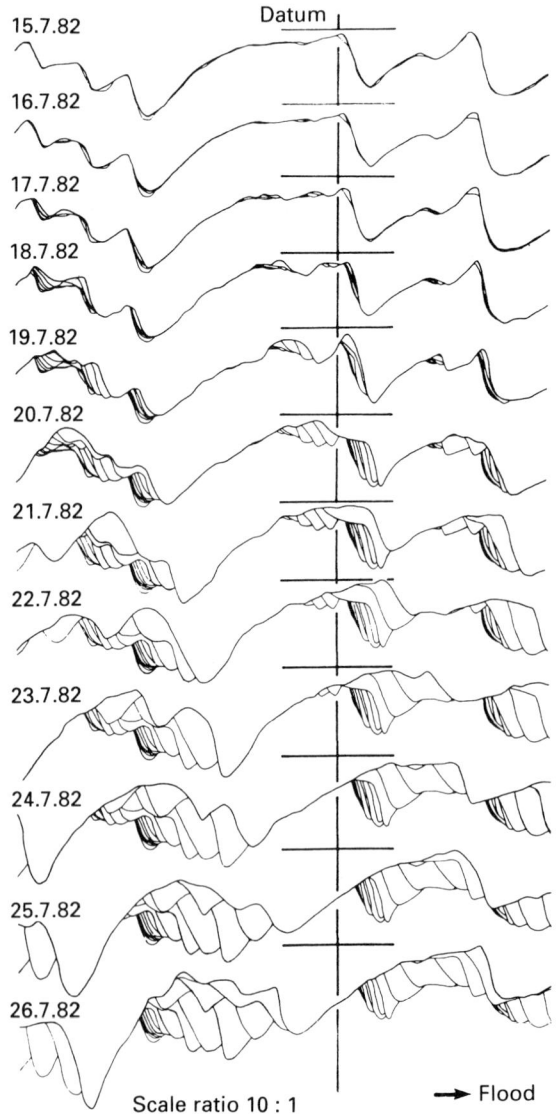

Fig. 6.20 The migration of intertidal sandwaves during a 12-day period. Illustrating the generation and destruction of internal structure. After Langhorne *et al.* (1985).

Within the estuary, the presence of ebb–flood channels, and the recirculation of sand effectively limits the penetration of sea-derived sand. Consequently there is a landward gradation of grain size, from coarse sand near the mouth, through silt to mud beneath the turbidity maximum.

Considerable interest has been shown in sand movement in the seaward parts of estuaries, and the internal structures of the banks (e.g. Nio et al., 1983). One difficulty inherent in these studies is that the bedforms can only be examined after the ebb tide. Studies in the Minas Basin (Klein, 1970) have shown that sand waves only exist when the grain size, the current velocity and the water depth exceed certain values. During the late stage of the ebb tide the sand waves are modified by flow occurring along the sand wave troughs. Pause planes should appear in the structures when the current stops, and master bedding should appear at high tide slack, when fine sediment may settle from suspension and form mud drapes. However, much of the internal structure is eroded by the progression of the trough, and interpretation of the deposited sedimentary structures becomes very difficult. This process is shown in Fig. 6.20. During 13 days of observation the bedform migrated 18 m, that is, 60% of its own length. Thus the life expectancy of the major internal structure is of the order of 1–2 spring/neap cycles. Dalrymple (1984) found no correlation of sand wave migration rates with tidal range. However, Allen and Friend (1976) have correlated migration rates with tidal range. Migration is generally less on an increasing (neap to spring) tide than on a decreasing tide.

6.9   Effects of rising sea levels

Because of global warming, it is predicted that a worldwide sea level rise of about 1 m will occur over the next century. A general rise of about 10 cm appears to have already occurred over the last 100 years. A rate of sea level rise of 1 cm per year is equivalent to the average rate during the Flandrian transgression. During the intervening period of about 5000 years of relatively stable sea level, some estuaries have been able to fill up with sediment, whereas others are still filling. Those that have filled up are in equilibrium, and though much active sediment movement takes place, as much sediment in volumetric terms is exported to the coastal zone as is imported, though the grain sizes may be different. There is obviously a considerable lag between a rise

in sea level and the estuarine equilibrium; in excess of 5000 years in many instances. Similar lags are likely with future sea level rise, and estimation of the past rate of infilling should give an estimate of the response time for individual estuaries.

The rate of estuarine infilling will depend crucially on the response of the sedimentary sources to rising sea level, and the associated effects. Rising sea level will produce considerable coast erosion or barrier beach retreat, though this may be limited by coastal defence works. Because of littoral transport much of this material will be transported to the estuary mouths. There the coarser material will become trapped in the ebb and flood tidal deltas. The finer material will be carried into the turbidity maximum by a combination of tidal pumping and gravitational circulation. There it will join material coming down the rivers.

A rise in sea level would normally be expected to reduce the rate of sediment input into the estuary, because of preferential deposition in the lower flood plains of the rivers. However, global warming is likely to increase the rainfall, and the storminess of the weather. The higher river discharge, and the increased incidence of floods are likely to largely offset the deposition, and flush sediments into the estuary.

Within the estuary, deposition will primarily occur on the outer edges of the tidal flats, so that a rise in the sea level would produce a considerable expansion of the intertidal flat levels, especially if the inner edges of the salt marshes were allowed to encroach onto the surrounding low lands. If sedimentation on the marsh surface was insufficient to keep up with sea level rise, there would be a progressive narrowing of the vegetational zones, which may lead to a further reduction in the sedimentation rate. The deeper water in the estuarine channels would lead to more active wave attack on the intertidal zone, as well as a more rapid progression of the tidal wave. The latter effect would lead to a landward movement of the locus of the turbidity maximum.

Consequently there would be an active interplay of processes, some of which would have mutually reducing effects. Quantification of these requires prediction of the total sediment budget of the estuary, and each one would be likely to have a different balance. At the moment our ability to establish a sediment budget using data on the existing system is very limited, so that prediction of future rates of infilling will require further quantitative measurement and modelling.

6.10 References

Adam P. 1990. *Saltmarsh Ecology.* Cambridge University Press, Cambridge.

Allen G.P. 1973. Etude des processus sédimentaires dans l'esturaire de la Gironde. *Mem. Instit. Geol. Bassin d'Aquitaine,* **5**.

Allen G.P., Salomon J.C., Bassoulet P., Du Penhoat Y. & De Grandpre C. 1980. Effects of tides on mixing and suspended sediment transport in macrotidal estuaries. *Sedim. Geol.,* **26**, 69–90.

Allen J.R.L. 1974. Reaction, relaxation and lag in natural sedimentary systems: general principles, examples and lessons. *Earth Sci. Rev.,* **10**, 263–342.

Allen J.R.L. & Friend P.F. 1976. Changes in intertidal dunes during two spring–neap cycles. Lifeboat Station Bank, Wells-next-the-Sea, Norfolk (England). *Sedimentology,* **23**, 329–346.

Allen J.R.L. & Pye K. (eds) 1992. *Saltmarshes: Morphodynamics, Conservation and Engineering Significance.* Cambridge University Press, Cambridge.

Amos C.L. & Tee K.T. 1989. Suspended sediment transport processes in Cumberland Basin, Bay of Fundy. *J. Geophys. Res.,* **94**, 14407–14417.

Anderson F.E. 1983. The northern muddy intertidal: A seasonally changing source of suspended sediments to estuarine waters — A review, *Can. J. Fish. Aquat. Sci.,* **40** (Suppl. 1), 143–159.

Aubrey D.G. 1986. Hydrodynamic controls on sediment transport in well-mixed bays and estuaries. In: Van de Kreeke J. (ed.) *Physics of Shallow Estuaries and Bays,* pp. 245–258. Springer, Berlin.

Avoine J. 1981. *L'estuare de la Seine: sediments et dynamique sedimentaire.* These, Docteur de Specialite. Université de Caen, France.

Avoine J., Allen G.P., Nichols M., Salomon J.C. & Larsonneur C. 1981. Suspended-sediment transport in the Seine estuary, France: effect of man-made modifications on estuary-shelf sedimentology. *Mar. Geol.,* **40**, 119–137.

Bale A.J. & Morris A.W. 1987. *In-situ* measurement of particle size in estuarine waters. *Estuar. Coast. Shelf Sci.,* **24**, 253–264.

Boon J.D. & Byrne R.J. 1981. On basin hypsometry and the morphodynamic response of coastal inlet systems. *Mar. Geol.,* **40**, 27–48.

Boothroyd J.C. 1978. Mesotidal inlets and estuaries. In: Davis R.A. (ed.) *Coastal Sedimentary Environments,* pp. 287–360. Springer, New York.

Bridges P.H. & Leeder M.R. 1976. Sedimentary model for intertidal mudflat channels with examples from the Solway Firth, Scotland. *Sedimentology,* **23**, 533–552.

Burban P.Y., Lick W. & Lick J. 1989. The flocculation of fine-grained sediment in estuarine waters. *J. Geophys. Res.,* **94**, 8323–8330.

Dalrymple R.W. 1984. Morphology and internal structure of sandwaves in the Bay of Fundy. *Sedimentology,* **31**, 365–382.

Dalrymple R.W., Knight R.J. & Middleton G.V. 1975. Intertidal sand bars in Cobequid Bay (Bay of Fundy). In: Cronin L.E. (ed.) *Estuarine Research,* Vol. 2, pp. 293–308. Academic Press, London.

Dieckmann R., Osterthun M. & Partensky H.W. 1987. Influence of water level elevation and tidal range on the sedimentation in a German tidal flat area. *Prog. Oceanog.,* **18**, 151–166.

Dobereiner C. & McManus J. 1983. Turbidity maximum migration and harbour siltation in the Tay Estuary. *Can. J. Fish. Aquat. Sci.,* **40**, 117–129.

Dronkers J. 1986. Tide-induced residual transport of fine sediment. In: Van de Kreeke J. (ed.) *Physics of Shallow Estuaries and Bays,* pp. 228–244. Springer, Berlin.

Dyer K.R. 1972. Sedimentation in estuaries. In: Barnes R.S.K. & Green J. (eds) *The Estuarine Environment.* pp. 10–32. Applied Science Publishers, London.

Dyer K.R. 1977. Lateral circulation effects in estuaries. In: *Estuaries, Geophysics and the Environment,* pp. 22–29. National Academy of Sciences, Washington, DC.

Dyer K.R. 1978. The balance of suspended sediment in the Gironde and Thames estuaries. In: Kjerfve B.J. (ed.) *Estuarine Transport Processes.* University of South Carolina Press Columbia, South Carolina.

Dyer K.R. 1986. *Coastal and Estuarine Sediment Dynamics.* Wiley, Chichester.

Dyer K.R. 1988. Fine sediment particle transport in estuaries. In: Dronkers J. & van Leussen W. (eds) *Physical Processes in Estuaries,* pp. 295–320. Springer, Berlin.

Dyer K.R. 1989. Sediment processes in estuaries: Future research requirements. *J. Geophys. Res.,* **94**, 14327–14339.

Dyer K.R. & Evans E.M. 1989. Dynamics of turbidity maximum in a homogeneous tidal channel. *J. Coastal Res. Special Issue,* **5**, 23–30.

Eisma D. 1986. Flocculation and deflocculation of suspended matter in estuaries. *Neth. J. Sea Res.,* **20**, 183–199.

Festa H.F. & Hansen D.V. 1978. Turbidity maxima in partially mixed estuaries: a two-dimensional numerical model. *Estuar. Coastal Mar. Sci.,* **7**, 347–359.

Friedrichs G.T. & Aubrey D.G. 1988. Non-linear tidal distortion in shallow well-mixed estuaries: a synthesis. *Estuar. Coastal Shelf Sci.,* **27**, 521–546.

Frostick L.E. & McCave I.N. 1979. Seasonal shifts of sediment within an estuary mediated by algal growth. *Estuar. Coastal Mar. Sci.,* **9**, 569–576.

Gelfenbaum G. 1983. Suspended-sediment response to semidiurnal and fortnightly tidal variations in a mesotidal estuary: Columbia River, USA. *Marine Geol.,* **52**, 39–57.

Glasgow L.A. & Leucke R.H. 1980. Mechanisms of deaggregation for clay–polymer flocs in turbulent systems. *Ind. Eng. Chem. Fundam.*, **19**, 148–156.

Grabemann I. & Krause G. 1989. Transport processes of suspended matter derived from time series in a tidal estuary. *J. Geophys. Res.*, **94**, 14373–14379.

Groen P. 1967. On the residual transport of suspended matter by an alternating tidal current. *Neth. J. Sea Res.*, **3**, 564–574.

Hayes M.O. 1975. Morphology of sand accumulation in estuaries: an introduction to the symposium. In: Cronin L.E. (ed.) *Estuarine Research*, Vol.II, pp. 3–22. Academic Press, New York.

HRS 1981. *The Severn estuary: silt monitoring April 1980–March 1981*. Rep. EX 995. Hydraulics Research Station, Wallingford, England.

Jouanneau J.M. & Latouche C. 1982. Estimation of fluxes to the ocean from megatidal estuaries under moderate climates and the problems they present. *Hydrobiologia*, **91**, 23–29.

Jumars F.A. & Nowell A.R.M. 1984. Effect of benthos on sediment transport: Difficulties with function grouping. *Cont. Shelf Res.*, **3**, 115–130.

Kirby R. & Parker W.R. 1974. Seabed density measurements related to echo sounder records. *Dock Harb Auth.*, **54**, 423–424.

Kirby R. & Parker W.R. 1981. *Settled mud deposits in Bridgwater Bay, Bristol Channel*. Institute of Oceanographic Sciences, Report No. 107.

Kirby R. & Parker W.R. 1983. The distribution and behaviour of fine sediment in the Severn Estuary and Inner Bristol Channel. *Can. J. Fish Aquat. Sci.*, **40** (Suppl.), 83–95.

Klein G. de V. 1970. Deposition and dispersal dynamics of intertidal sand bars. *J. Sedim. Petrol.*, **40**, 1095–1127.

Kostaschuk R.A. & Luternauer J.L. 1989. The role of the salt-wedge in sediment resuspension and deposition: Fraser River Estuary, Canada. *J. Coastal Res.*, **5**, 93–101.

Kostaschuk R.A., Luternauer J.L. & Church M.A. 1989. Suspended sediment hysteresis in a salt-wedge estuary: Fraser River, Canada. *Mar. Geol.*, **87**, 273–285.

Krauter J.N. & Wetzel R.L. 1986. Surface sediment stabilization–destabilization and suspended sediment cycles on an intertidal mudflat. In: Wolfe D.A. (ed.) *Estuarine Variability*, pp. 203–223. Academic Press, Orlando.

Krone R.B. 1978. Aggregation of suspended particles in estuaries. In Kjerfve B. (ed.), *Estuarine Transport Processes*, pp. 177–190. University of South Carolina Press, Charleston.

Lang G., Schubert R., Markofsky M., Fanger H-U, Grabemann I., Krasemann H.L., Neumann L.J.R. & Riethmuller 1989. Data interpretation and numerical modelling of the mud and suspended sediment experiment 1985. *J. Geophys. Res.*, **94**, 14381–14393.

Langhorne D.N., Malcolm J.O. & Read A.A. 1985. *Observations of the changes of intertidal bedforms over a neap–spring tidal cycle*. Institute of Oceanographic Sciences, Report 203.

Markofsky M., Lang G. & Schubert R. 1986. Suspended sediment transport in rivers and estuaries. In: van de Kreeke J. (ed.) *Physics of Shallow Estuaries and Bays*, pp. 210–217. Springer, Berlin.

Martin J.M., Mouchel J.M. & Thomas A.J. 1986. Time concepts in hydrodynamics systems with an application to [7]Be in the Gironde estuary, *Mar. Chem.*, **18**, 369–392.

McCave I.N. 1975. Vertical flux of particles in the ocean. *Deep Sea Res.*, **22**, 491–502.

Meade R.H. 1969. Landward transport of bottom sediments in estuaries of the Atlantic Coastal Plain. *J. Sedim. Petrol.*, **39**, 222–234.

Mehta A.J. 1986. Characterization of cohesive sediment properties and transport processes in estuaries. In: Mehta A.J. (ed.) *Estuarine Cohesive Sediment Dynamics*, pp. 290–325. Springer, Berlin.

Mehta A.J. Hayter E.J., Parker W.R., Krane R.B. & Teeter A.M. 1989. Cohesive sediment transport, 1. Process description. *Jour. Hydraul. Eng.*, **115**, 1076–1112.

Milliman J.D. & Meade R.H. 1983. World wide delivery of river sediment to the oceans. *J. Geol.*, **91**, 1–21.

Nichols M. 1972. Sediments of the James River Estuary, Virginia. In: Nelson B.W. (ed.) *Environmental Framework of Coastal Plain Estuaries. Geol. Soc. Amer. Mem.*, **133**, 169–212.

Nichols M.M. 1977. Response and recovery of an estuary following a river flood. *J. Sedim. Petrol.*, **47**, 1171–1186.

Nichols M.M. 1986. Effects of fine sediment resuspension in estuaries. In: Mehta A.J. (ed.) Estuarine Cohesive Sediment Dynamics, pp. 5–42. Springer, Berlin.

Nichols M.M. 1989. Sediment accumulation rates and relative sea-level rise in lagoons. *Mar. Geol.*, **88**, 201–219.

Nichols M.M. & Biggs R.B. 1985. Estuaries. In: Davis R.A. Jr (ed.) *Coastal Sedimentary Environments*, pp. 77–186. Springer, Berlin.

Nio S.D., Siegenthaler C. & Yang C-S. 1983. Megaripple cross bedding as a tool for the reconstruction of the palaeo-hydraulics in a Holocene subtidal environment, S.W. Netherlands. *Geol. Mijn.*, **62**, 499–510.

O'Brien M.P. 1969. Equilibrium flow areas of tidal inlets on sandy coasts. *J. Water. Harb. Div., Am. Soc. Civ. Eng.*, **95**, WW1, 43–52.

Odd N.V.M. & Rodger J.G. 1986. *An analysis of the behaviour of fluid mud in estuaries*. Rep. SR84. Hydraulics Research Ltd, Wallingford, England.

Oertel G.F. 1972. Sediment transport of estuary entrance shoals and the formation of swash platforms. *J. Sedim. Petrol.*, **42**, 837–863.

Officer C.B. 1981. Physical dynamics of estuarine suspended sediments. *Mar. Geol.*, **40**, 1–14.

Parker W.R. 1986. On the observation of cohesive sediment behaviour for engineering purposes. In: Mehta A.J. (ed.)

Estuarine Cohesive Sediment Dynamics, pp. 270–289. Springer, Berlin.

Parker W.R. & Kirby R. 1982a. Time dependent properties of cohesive sediment relevant to sedimentation management — European experience. In: Kennedy V. (ed.) *Estuarine Comparisons*, pp. 573–589. Academic Press, New York.

Parker W.R. & Kirby R. 1982b. Sources and transport patterns of sediment in the inner Bristol Channel and Severn Estuary. In: *Severn Barrage*, pp. 181–194. Thomas Telford, London.

Postma H. 1961. Transport and accumulation of suspended matter in the Dutch Wadden sea. *Neth. J. Sea Res.*, **1**, 148–190.

Postma H. 1967. Sediment transport and sedimentation in the estaurine environment. In: Lauff G.H. (ed.), *Estuaries, Amer. Assoc. Adv. Sci. Pub.*, **83**, 158–179.

Schubel J.R. & Carter H.H. 1984. The estuary as a filter for fine-grained suspended sediment. In: Kennedy V. (ed.) *The Estuary as a Filter*, pp. 81–105. Academic Press, New York.

Schubel J.R. & Hirschberg D.J. 1978. Estuarine graveyards, climatic change, and the importance of the estuarine environment. In: Wiley M. (ed.) *Estuarine Interactions*, pp. 285–303. Academic Press, New York.

Sha L.P. 1989. Sand transport patterns in the ebb-tidal delta off Texel inlet, Wadden Sea, the Netherlands. *Mar. Geol.*, **86**, 137–154.

Sills G.C. & Elder D.M. 1986. The transition from sediment suspension to settling bed. In: Mehta A.J. (ed.) *Estuarine Cohesive Sediment Dynamics*, pp. 192–205. Springer, New York.

Stevenson J.C., Ward L.G. & Kearney M.S. 1986. Vertical accretion in marshes with varying rates of sea level rise. In: Wolfe D.A. (ed.) *Estuarine Variability*, pp. 241–259. Academic Press, Orlando.

Thorn M.F.C. & Parsons J.G. 1980. Erosion of cohesive sediments in estuaries: an engineering guide. *Proc. 3rd Int. Symp. Dredging Tech.*, 349–358.

Uncles R.J. 1981. A note on tidal asymmetry in the Severn Estuary. *Estuar. Coast. Shelf Sci.*, **13**, 419–432.

Uncles R.J. & Stephens J.A. 1989. Distributions of suspended sediment at high water in a macrotidal estuary. *J. Geophys. Res.*, **94**, 14395–14405.

Uncles R.J., Elliott R.C.A. & Weston S.A. 1984. Lateral distributions of water, salt and sediment transport in a partly mixed estuary. *Proc. 19th Coastal Eng. Conf. Houston*, pp. 3067–3077.

Uncles R.J., Elliott R.C.A. & Weston S.A. 1985a. Observed fluxes of water, and suspended sediment in a partly mixed estuary. *Estuar. Coast. Shelf Sci.*, **20**, 147–167.

Uncles R.J., Elliott R.C.A. & Weston S.A. 1985b. Dispersion of salt and suspended sediment in a partly mixed estuary. *Estuaries*, **8**, 256–269.

Uncles R.J., Elliott R.C.A. & Weston S.A. 1986. Observed and computed lateral circulation patterns in a partly mixed estuary. *Estuar. Coast. Shelf Sci.*, **22**, 439–457.

van Straaten L.M.J.U. & Kuenen P.L.H. 1958. Tidal action as a cause of clay accumulation. *J. Sedim. Petrol.*, **28**, 406–413.

Wellershaus S. 1981. Turbidity maximum and mud shoaling in the Weser Estuary. *Arch. Hydrobiol.*, **92**, 161–198.

Wells J.T. & Coleman J.M. 1981. Physical processes and fine-grained sediment dynamics, coast of Surinam, South America. *J. Sedim. Petrol.*, **51**, 1053–1068.

Yang C-S. 1986. Estimate of sand transport in the Oosterschelde tidal basin using current-velocity measurements. *Mar. Geol.*, **72**, 143–170.

7 Beach and nearshore sediment transport

J. HARDISTY

7.1 Introduction

The marine environment can be divided into wave dominated, tidally dominated and deep water sedimentary systems, and the processes that operate in these different environments, along with the general nature of the resulting deposits, are well described in standard textbooks such as Allen (1970), Komar (1976), Leeder (1982), Pethick (1984), Dyer (1986) or Carter (1988). It is not the purpose of the present chapter to review such texts, but rather to concentrate on more recent and more advanced work in order to describe and to quantify the particular processes that operate within the wave dominated nearshore environment and to relate the processes to the sediments and sedimentary stratigraphy. It is aposite to attempt this task at the present time because the last 10 years have witnessed rapid progress in the theoretical description of wave-dominated sediment transport and the new theories have been addressed and assessed with novel instruments that have permitted, for the first time, continuous and coincident measurements to be made of the surface waves, the nearbed flow fields and, most importantly, the resulting sediment mass transport. The new insights thus gained are revealing a complex system of interactions, resonances and feedbacks whereby deposition is but one part of the beach response to changing meteorological, climatic, eustatic and tectonic conditions.

Bird (1984) estimated that there are some 500 000 km of coastline bordering the world ocean and the Shore Protection Manual (1984) suggests that 33% of the coastline of the United States (excluding Alaska) shows beach development. Assuming that the percentage applies to the totality, there are then more than 170 000 km of beaches bordering the world ocean at the present time. We begin by noting that all of these beaches exhibit a more or less smooth profile which is concave upwards and a more or less straight or gently curved plan shape. This worldwide uniformity of shape demands some universal and overriding stabilities within the complex hydrodynamic and sediment mass transport processes and some of these, at least, will become evident in the following pages.

We shall refer to these general features as *first order* beach forms, and find that the shape of the beach profile is due to *orthogonal processes* (i.e. those that operate in a vertical plane along the direction of wave advance and roughly normal to the shoreline) whilst the plan shape of the beach is due to *longshore processes* (that is those that operate in a horizontal plane in a shore parallel or coast-wise direction). First order forms often have superimposed upon them the smaller scale *second order* features (Fig. 7.1) detailed in Table 7.1 and these include features such as longshore bars or the development of beach cusps. An examination of the beach in still finer detail reveals *third order* features which range from wave-generated ripples, swash and backwash marks and drainage channels to the individual grains and internal bedding structures which constitute the very fabric of the beach deposit and may, ultimately, be preserved within the stratigraphic record.

In Section 7.2, standard hydrodynamic theory is covered and, in particular, important new long wave and infragravity theory is introduced. The sediment transport processes are described in Section 7.3 and again new results obtained with optical and acoustic backscatter devices for suspended load measurement and passive acoustic devices for bedload measure-

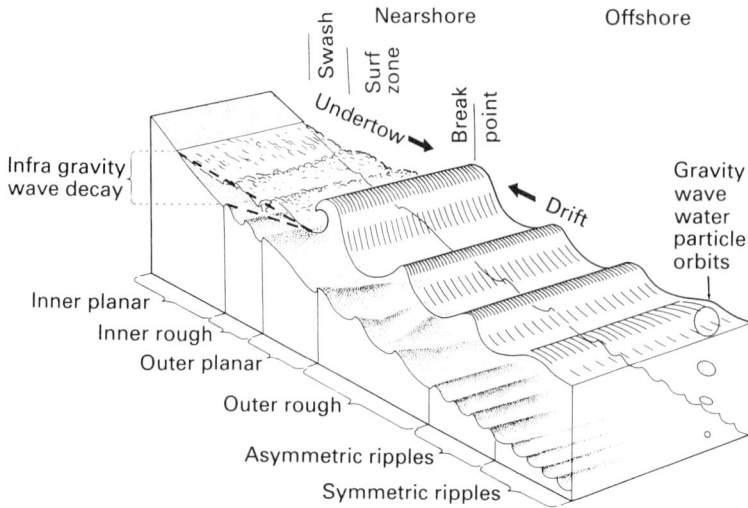

Fig. 7.1 Beach and nearshore nomenclature and sub-environments.

ments are used to exemplify the theoretical work. In Section 7.4, an attempt is made to relate the processes to the first, second and third order beach features. It will become apparent that new measurements are demonstrating the dominance of infragravity processes and yet adequate mathematical models for either the hydrodynamic or sediment dynamic processes under infragravity waves have still to be developed. Deposition, or its corollary, the palaeo-environmental interpretation of beach deposits remains difficult.

In general, this chapter will describe the processes within the text and present the corresponding equations within a series of panels. A list explaining the symbols used in the equations appears on p. 248. The objective of this device is twofold: readers who do not wish to become involved with mathematical detail should be able to appreciate the functioning of the beach system without reference to the panels and yet those who are more concerned with the erection and testing of operational beach models will be better able to work from the equations. Readers in the former group would be well advised to consider changing their allegiance because none of the mathematics presented here is very advanced and a few minutes on a personal computer will provide simple, numerical descriptions of the beach processes with an elegance that defies the most eloquent written description. The order of presentation follows a process path that starts with the independent variables (that is the incident, deep water waves), predicts the surface elevation of the waves in all water depths, and

then uses the surface elevations to predict the nearbed, oscillatory and steady currents. The currents generate the sediment mass transport rates which are finally used to appreciate the relationships between the net mass transports and the form and fabric of the beach deposits.

7.2 Waves on beaches

7.2.1 Introduction

Wind, blowing over the surface of the world ocean, transfers energy to the surface water. The energy is manifested as a series of more or less regular crests which develop in lines normal to the wind and travel in a downwind direction. These are known properly as *wind waves* or *gravity waves* (Kinsman, 1984) and should be distinguished from the much longer waves of tidal and tsumani origin. The nomenclature is confusing because the coastal ebb and flow of the periodic vertical or horizontal movement which has a coherent amplitude and phase relationship to some periodic geophysical force (Huntley, 1979) is known colloquially as a tide even though it is a manifestation of the gravitational attraction of the Sun and the Moon. The solitary and catastrophic deep ocean wave that emanates from some submarine eruption or earthquake is known colloquially as a tidal wave but correctly as a tsumani. For the present purposes we shall refer to the waves that approach a coast from offshore as the *incident waves* and use the adjectival *'gravity'* band to indicate that they have periods of

Table 7.1 Glossary of second order beach features. From Wiegel (1964), Glossary of Geology (1972) and Shepard (1973)

Backshore
The zone of the beach lying between the berm and the coastline. Shorenormal processes

Bar
(Longshore bar, ball or ridge) An elongate, slightly submerged sand ridge or ridges which may be exposed at low water. Shorenormal processes

Barrier
A sand beach (barrier beach), island (barrier island) or spit (barrier spit) that extends roughly parallel with the general coastal trend but is separated from the mainland by a relatively narrow body of water or marsh. Shoreparallel processes

Beach face
The sloping section of the beach below the berm. Shorenormal processes

Beach ridge
(Storm beach, chenier) A low lengthy ridge of beach material piled up by storm waves landward of the berm. Shorenormal processes

Berm
The nearly horizontal part of the beach landwards of the sloping foreshore. Shorenormal processes

Berm crest
The seaward limit of the berm. Shorenormal processes

Cusps
One of a series of short ridges on the foreshore extending normal to the shoreline and recurring at more or less regular intervals. Shoreparallel processes

Foreshore
The sloping part of the beach between the berm and the low tide level. Shorenormal processes

Low tide terrace
The flat portion of the beach seawards of the beach face exposed at low water. Shorenormal processes

Offshore
The zone seawards of the low tide mark. Shorenormal processes

Rip channel
Channel cut by seawards flow of rip currents which usually crosses the longshore bar. Shoreparallel and shorenormal processes

Trough
(Longshore trough, runnel or low) Elongate depression or series of depressions along the lower beach or in the offshore zone which may be exposed at low water. Shorenormal processes

less than 20 s. The surface profiles and velocity fields due to gravity waves are described in the following section. This is in order to distinguish them from the second group of waves which are generated at the shoreline and, typically, have periods of between 20 and 200 s; these we shall call *infragravity* or long waves. The surface profiles and velocity fields due to infragravity waves are described in a later section.

7.2.2 Gravity waves

Measurements of incident waves in the gravity band obtained with a pressure transducer mounted on the sea bed (Hardisty, 1988; Seymour, 1989) or a wave staff (Koonitz & Inman, 1967) or gauge (Guza & Thornton, 1989) reveal a series of more or less regular crests and troughs as shown in Fig. 7.2(a). Such observations lead to a treatment based upon properties of the circular sine or cosine functions and such an approach is detailed in the following sections.

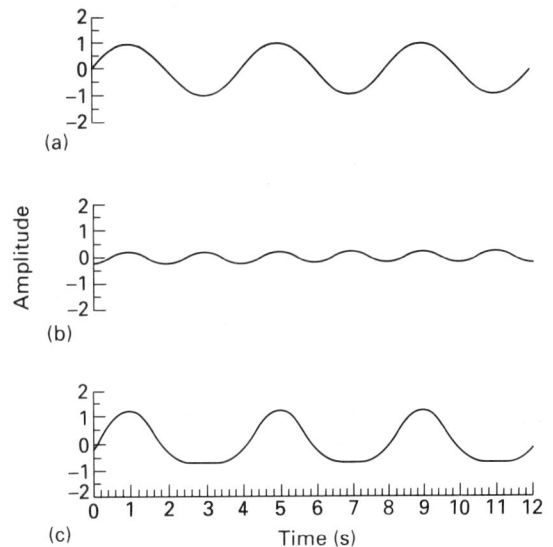

Fig. 7.2 Wave profiles and currents in the orthogonal plane: (a) the sinusoidal profile or flow field described by Airy theory, (b) the generation of the first harmonic with twice the primary frequency (amplitude 0.25), and (c) the summation of the primary and first harmonics to produce an asymmetrical surface elevation or flow field (amplitude 1).

Gravity wave period

The *wave period* is defined as the time interval between the passage of successive, corresponding points on the wave profile. It is symbolized by T and is measured in seconds. The inverse of the wave period is called the wave frequency, $f = 1/T$. The wave frequency has units of cycles per second but these are more commonly called Hertz (Hz). Since the circular sinusoidal notation is being used and there are 2π radians in each cycle, then a useful and alternative measure is the radian frequency, $\omega = 2\pi/T$, which has units of rad s^{-1}. Gravity waves with periods of up to 20 s are generated in the world ocean, and such long period gravity waves frequently occur during winter storms in, for example, the North Atlantic (Hardisty, 1990a).

Gravity wave length

The wave length is defined as the horizontal distance between corresponding points on successive wave profiles. It is symbolized by L, and has units m. It is important to recognize that the deep water wave length (which is symbolized by L_∞) decreases as the waves approach the shore and enter shallow water. In general, gravity wave theory is based upon the first wave dispersion equation (Eq. A1) which can be rearranged to predict the wave length in any water depth (Eq. A2). The equation includes a second useful abbreviation which is the *wave number*, $k = 2\pi/L$. The equation cannot, however, be used directly to predict the wave length because L occurs on both the left hand side and within the hyperbolic tangent. The equation was, therefore, rarely used and recourse was made to the approximations which are detailed by, for example, Komar (1976). These give the wave length as $L = gT^2/2\pi (\approx 1.56\,T^2)$ in water which is deeper than a quarter of a wave length (because here $\tanh(2kh) \approx 1$) and as $L = T\sqrt{(gh)}$ in water that is shallower than a twentieth of a wave length (because here $\tanh(kh) \approx kh$). These two limits have been used to define a *deep water* and a *shallow water* regime. Nowadays, however, the advent of microcomputers has greatly facilitated the use of iterative techniques to solve the dispersion equation, and the general result is expressed as the non-dimensional ratio L/L_∞ (Fig. 7.3(a)) showing that the wave length decreases from its deep water value as the shoreline is approached. For example, a 10 s wave which has a deep water length of 156 m will have a wave length of about 20 m in 3 m of water.

The *wave celerity* is defined as the horizontal distance traversed by the wave in unit time. Strictly speaking, the speed of the wave refers to the speed of the surface profile, which is called the *phase celerity* and is symbolized by C. Since the wave travels one wave length in one wave period then, by definition, $C = L/T$. In practice the phase celerity is usually faster than the speed of transmission of energy within the wave train, and the speed of transmission of energy is called the *group celerity*, C_g. The phase and group celerities are related by a third useful abbreviation, $C_g = Cn$, where the coefficient of proportionality, n, varies from 0.5 in deep water to 1 at the water's edge, as described below. The wave period remains constant during shoaling and, therefore, the wave celerity decreases with the wave length as shown in Fig. 7.3(a).

Gravity wave height

The wave height is defined as the vertical distance between the crest and the trough. It is symbolized by H, has units m, and the subscripted H_∞ refers to the deep water value. Deep water wave heights increase with wind speed, duration or fetch length (Komar, 1976) and, typically, the maximum wave height that occurs in a 50-year period beyond the edge of the continental shelf in the eastern North Atlantic is about 35 m (Draper, 1979) whilst the corresponding value in the eastern English Channel is less than 10 m (Hardisty, 1990a).

Although a certain amount of kinetic and potential energy is contained within each metre of wave crest, $(E = 1/8\rho g H^2)$ the wavelength reduces as the wave enters shallow water without energy loss and, therefore, the wave height must increase to maintain sufficient mass above the still water level to conserve the potential energy of the wave. In more detail the average rate at which wave energy is transmitted in the direction of wave propagation is called the *wave power*, P, and is equal to the product of the wave energy and the group celerity of the waves (Eq. A3, Reynolds (1877), Rayleigh (1877), Komar (1976)). The coefficient of proportionality, n, which was introduced earlier can now be seen to derive from this expression for the wave power (Eq. A4) so that $P = ECn$.

A second change in wave height is produced by the process of *wave refraction* which occurs because the wave celerity is reduced in shallow water, with the result that the crests bend to lie more parallel with the contours. Lines that run normal to the wave crest are

---

Panel A Gravity wave equations.

$$\omega^2 = gk \tanh (kh) \tag{A1}$$

$$L = \frac{g}{2\pi} T^2 \tanh \frac{2\pi h}{L} \tag{A2}$$

$$P = \frac{1}{8} \rho g H^2 \, C \frac{1}{2} \left(1 + \frac{2kh}{\sinh (2kh)}\right) \tag{A3}$$

$$n = \frac{1}{2} \left(1 + \frac{2kh}{\sinh (2kh)}\right) \tag{A4}$$

$$H = H_\infty \sqrt{\frac{C_\infty}{2Cn} \frac{s_\infty}{2_h}} \tag{A5}$$

$$\gamma_b = 0.72(1 + 6.4\tan \beta) \tag{A6}$$

$$\eta(t) = \frac{H}{2} \cos (\omega t) \tag{A7}$$

$$\eta(t) = \frac{H}{2} \cos (\omega t) + \frac{\pi H^2}{8L} \frac{\cosh (kh)[2 + \cosh (2kh)]}{\sinh^3(kh)} \cos (2\omega t) \tag{A8}$$

$$u(t) = \frac{\pi H}{T} \frac{1}{\sinh (kh)} \cos (\omega t) \tag{A9}$$

$$u(t) = \frac{\pi H}{T \sinh (kh)} \cos (\omega t) + \frac{3}{4} \frac{\pi^2 H^2}{LT} \frac{1}{\sinh^4(kh)} \cos (2\omega t) \tag{A10}$$

$$\eta_n = a_n \cos (k_n x + \omega_n + \varepsilon_n) \tag{A11}$$

$$u(t) = \sum_{n=1}^{n=\infty} \left(\frac{\omega_n}{\sinh |k_n| h}\right) \eta_n \tag{A12}$$

where the hyperbolic functions are defined as $\sinh(kh) = 0.5(e^{kh} - e^{-kh})$, $\cosh(kh) = 0.5(e^{kh} + e^{-kh})$ and $\tanh(kh) = \sinh(kh)/\cosh(kh)$.

---

called *wave orthogonals*, and wave refraction can cause wave orthogonals to converge or to diverge as the wave crosses the bathymetric contours. Therefore, when interorthogonal spacings, s, change, the amount of energy per metre of wave crest changes, and the wave height will also change. However the interorthogonal wave power must remain constant, so that deep water wave power can be related to the shallow water equivalent by the ratio $\sqrt{(s_\infty/s)}$. The ratio can then be used to calculate the local wave height in terms of the deep water value as shown by Eq. A5. The ratio is equal to unity for normal incidence but must be determined from Snells Law for simple bathymetries or from wave refraction diagrams for more complex problems. The simplest case of normal incidence can be illustrated as the ratio H/H_∞ and is shown in Fig. 7.3(b).

When waves enter water that is approximately as deep as the waves are high, they break, the crest throwing itself forward and disintegrating into bubbles and foam. Three types of breakers are commonly recognized and are described by Galvin (1972) as *spilling breakers* in which foam, bubbles and turbulent water appear at the water crest and eventually cover the front face of the wave; *plunging breakers* in which the whole front face of the wave steepens until vertical as the crest curls over the front face and falls into the base of the wave; and *surging breakers* in which the front face and crest of the wave remains relatively smooth and the wave slides up the beach with only minor production of foam and bubbles. Diagrams of the breaker types are shown in most texts, but it is important to appreciate that the classification describes parts of a continuum that

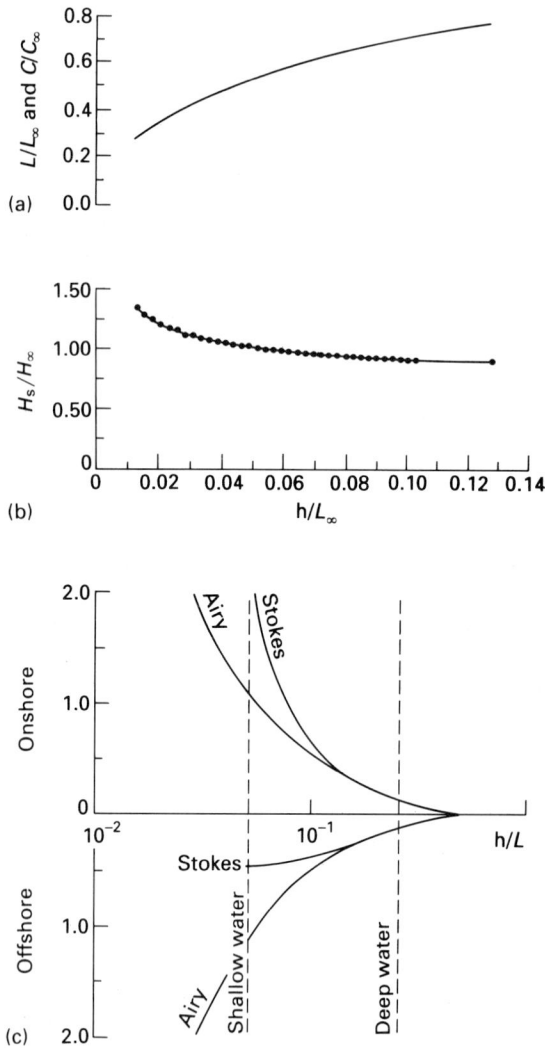

Fig. 7.3 Changes in (a) length and celerity, (b) height and (c) peak currents of incident gravity waves as they approach the shoreline. H, 0.5 m; L, 39 m; T, 5 s.

fined as the ratio of the wave uprush duration to the wave period with transitions at 1.3 and 0.7. Suhayda and Pettigrew (1977) use a *bore classification* defined as the ratio of the breaker height to the water depth. Most usefully, however, various authors attempt to link breaker and beach classifications through a *surf scaling parameter* $\varepsilon = 4\pi^2 H_b/gT^2\beta^2$ (Guza & Bowen, 1975; Guza & Inman, 1975) or its inverse the *beach reflectivity* (Huntley & Bowen, 1975; Guza & Bowen, 1977) where the former decreases during the transition from spilling to surging breakers, whilst the latter increases over the same range. In general, wide flat beaches are dissipative and have a high surf scaling parameter whereas narrow steep beaches are reflective and have a low surf scaling parameter.

Most theoretical analyses of the breaking process have been based upon the work of McCowan (1894) who demonstrated that the ratio of the wave height to the water depth at the break point, γ_b, will be 0.78, although other studies predict a range of values (cf. Hardisty & Laver, 1989). Field observations show that γ_b is also dependent on the sea bed gradient (e.g. Sverdrup & Munk, 1946; Ippen & Kulin, 1955; Kishi & Saeki, 1967; Tucker *et al.*, 1983). Galvin (1972) suggests that γ_b decreases with increasing beach slope as shown by Eq. A6. Finally, more recent work which recognizes that incident waves contain a spectrum of periods and a distribution of wave heights, suggests that waves break when the root-mean-squared wave height is 0.42 of the water depth (Thornton & Guza, 1982; Bowen & Huntley, 1984).

Gravity wave profiles

The wave equations described above were developed by 19th century mathematicians, and have been found to agree remarkably well with certain aspects of field and laboratory measurements of the phenomena. The simplest analysis is due to Airy (1845), who considers only first order powers of H. It is therefore often referred to as Airy, first-order or linear theory and predicts a wave profile that is sinusoidal (Eq. A7 and Fig. 7.2(a)). First order theory provides an adequate description of wave height and length changes during the shoaling transformations but observations of waves in the nearshore region show that the crests become more peaked whilst the troughs become longer and flatter. Second-order theory was, therefore, developed by Stokes (1847) and predicts a surface profile that is composed of two sinusoids (Eq. A8). The first sinusoid (Fig. 7.2(a)) is identical to

ranges between the extreme types. Various authors have made this point and used a range of non-dimensional numbers to attempt to quantify the breaker type. Thus, Galvin (1968, 1972) defines a *breaker coefficient* as $B_0 = H_\infty/L_\infty\beta^2$ with transitions at 10^2 and 10^{-3}. Iribarren and Nogales (1949) and Battjes (1974a,b) use *Iribarren's number* $I_r = \beta/\sqrt{H_\infty/L_\infty}$ (i.e. $B_0 = 1/I_r^2$) with transitions at 10^{-1} and 10^1. Kemp (1960, 1975) and Kemp and Plinston (1968, 1974) use the so-called *phase difference*, de-the Airy wave but the second (Fig. 7.2(b)) has a more complicated amplitude and twice the frequency of the first term (i.e. cos(2ωt)). Stokes' theory simply adds the first and second terms to generate an asymmetrical wave shape as shown in Fig. 7.2(c). This process is called the *generation of harmonics* and the first harmonic has twice the frequency of the original (or *fundamental*) wave. There are higher order and other wave theories (e.g. Lakhan, 1989) but Airy and Stokes' theory are adequate for the present purposes.

Gravity wave oscillatory currents

The surface waves generate oscillatory currents that flow shorewards beneath the wave crest and seawards beneath the wave trough (Fig. 7.1). In Airy waves a sinusoidal motion is predicted, which is circular close to the water surface but decreases in ampltitude and becomes elliptic with depth as described by Eq. A9. The same approximations that were used earlier give peak flows in *deep* water as $u_m = \pi H e^{-kh}/T$ and in *shallow* water as $u_m = H/2\sqrt{(g/h)}$ (Fig. 7.3(c)). Stokes' theory also predicts the velocity of the oscillatory currents (Eq. A10) and again the first term in these expressions is the Airy component and the second term has twice the frequency of the Airy component and is known as the first harmonic. It is positive under the crest and the trough and negative 1/4 and 3/4 wave lengths from the crest. Stokes' theory thus predicts a symmetrical orbital flow which is identical to first-order theory in deep water but which becomes increasingly asymmetric as the higher frequency term becomes larger when the wave shoals.

Gravity wave and current spectra

We have seen that a more realistic representation of natural waves is achieved by the summation of sinusoids at a fundamental and harmonic frequency, that is a graph of wave energy against frequency would show two peaks. In practice, real waves are somewhat more complex as shown in Fig. 7.4. Data were collected with pressure transducers at Spurn Head on the North Sea coast of eastern England, and a typical time series of the surface profile of the waves is shown in Fig. 7.4(a). It is now conventional to use spectral analysis to determine the graph of energy against frequency from such time series (cf. Chatfield, 1985 or Randall, 1987), and the result is called the power spectrum of the wave field (Fig. 7.4(b)). It is

apparent that there are a large number of frequencies present, and that spectral analysis offers a very powerful insight into the role of the different frequencies within the beach system. Mathematically, the surface elevation is expressed as a sum of all such sinusoids (Eq. A11) each having a particular wave number (k_n), radian frequency (ω_n) and phase relation (ε_n). It is only recently, however, that continuous time series of the surface waves along with the nearbed flows and the resulting sediment mass transport have become available, and later sections will demonstrate the use of spectral techniques to analyse the processes in more detail. In general, the incident wave field has been described by one of a number of empirical spectra such as the Pierson-Moskovitch or JONSWAP functions (e.g. Carter, 1988, p. 41).

In the presence of random waves, Guza and Thornton (1980) suggest that the equations for the horizontal velocity is given by the summation of linear theory for the n constituent sinusoids (Eq. A12). Specifically, this theory states that the spectrum of the nearbed flow is adequately described by a linear transformation of the surface elevation spectrum. In other words, if the shoaling transformations discussed in the previous chapter are correct, then the nearbed flows follow from transforming the surface elevation at each frequency using Airy theory. Comparisons between spectra of sea surface elevation and velocity, measured at the same horizontal location in relatively shallow water, have been made by various authors. Bowden and White (1966), and Simpson (1969) made observations in 4–6 m of water, Thornton and Krapohl (1974) studied long Pacific swell in 19 m of water and Cavaleri et al. (1978) took data in 16 m of water. Reporting these results, Guza and Thornton (1980) note that these measurements are of weakly non-linear waves at some distance offshore in depths where Airy theory can be expected to apply. The measured horizontal velocity spectra generally agree with spectra predicted by linear theory. Guza and Thornton (1980) extended the analysis into shallow water with measurements at Torrey Pines, San Diego, which is a gently sloping, moderately sorted, fine grained sand beach. The surface elevation was measured with dual resistance wave staffs and the orthogonal currents were measured with electromagnetic current meters. Figure 7.5 shows one of their comparisons of the predicted surface elevation from the transformed horizontal current spectrum, with the results from a wave staff. Guza and Thornton (1980) conclude that, although linear theory is, of

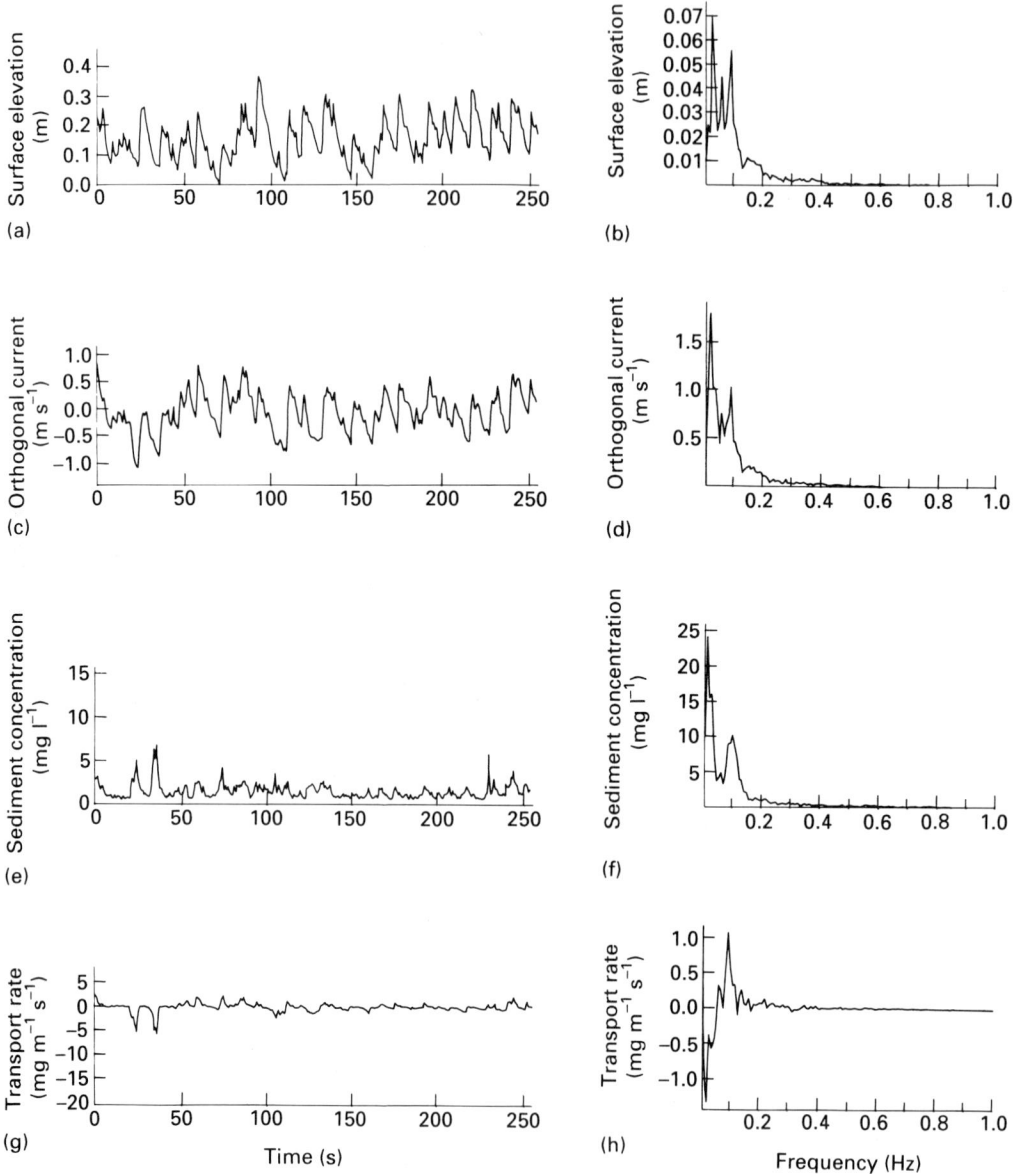

Fig. 7.4 Time series (a) and spectra (b) of surface wave elevation, corresponding orthogonal currents (c and d) and suspended load concentration (e and f) obtained with a pressure transducer, electromagnetic current meter and optical backscatter sensor at Spurn Head on the North Sea coast of eastern England. Time series (g) and cospectra (h) of u' and c' and the oscillatory transport rate $u'c'$ showing onshore transport due to the gravity wave oscillations and offshore transport due to long wave energy.

course, incapable of predicting the generation of harmonics, comparison of elevation and flow spectra show considerable agreement with the linear theory in given water depths from well outside to inside the surf zone despite the fact that non-linearities might be anticipated close to the shoreline.

7.2.3 Infragravity waves

Although the result which was introduced in the preceding section, that wave height within the surf zone is depth-limited, appears at first sight to be relatively straightforward, and almost 100 years

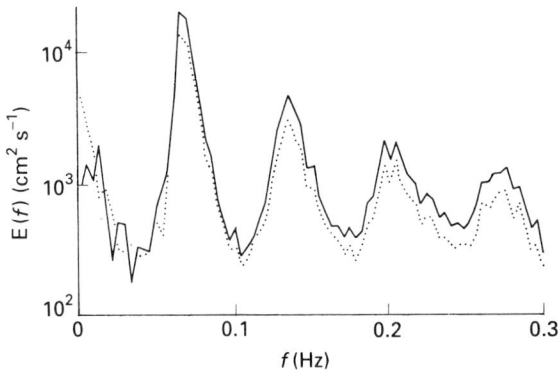

Fig. 7.5 Comparison of elevation and current spectra $E(f)$ using linear transfer functions. After Guza and Thornton (1980).

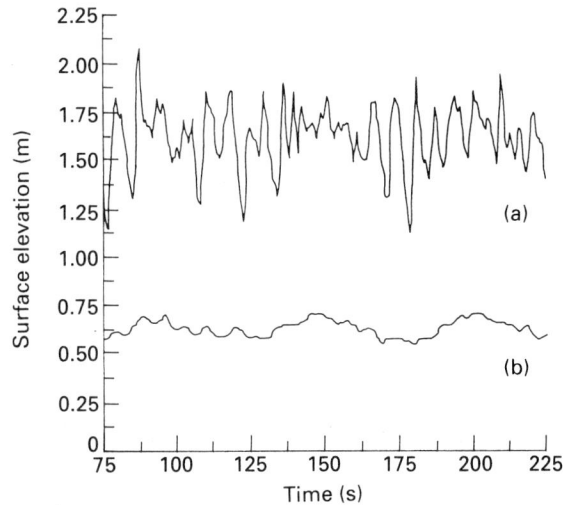

Fig. 7.6 Surface wave time series at Slapton Sands, south Devon (a) and the result of the application of low pass filtering applied to the raw data (b) offset by 1m for clarity.

have elapsed since McCowan's original work, the logical implications appear to have been missed by beach workers over the intervening century and are only now coming to light. It appears that beach research has worked on the assumption that offshore waves directly control the system. Thus storms lead to higher waves, increased offshore transport and beach draw-down and so on and so forth. However, if the surf zone waves are depth-limited, then all surf zone waves, whether storm or calm, at a particular point will be the same height, and therefore there is no mechanism by which the beach knows that an offshore storm is happening. How then does the beach respond to it?

This perplexing paradox has been addressed and, to a certain extent, has been answered, by the field deployment of a range of sophisticated monitoring devices and by the proper application of time series and spectral techniques in the last decade. Some of the developments which have become clear from this work will now be described. In general, it is suggested that the nearshore is not controlled by the incident waves, but by one or more of a family of new oscillations that are generated at the shoreline and are collectively called *infragravity* or *long* waves because they have longer periods than the gravity waves, ranging between 20 and 200 s. Presently work is continuing on still longer period motions, called far infragravity or FIG waves, but it is still too early to comment on these in detail.

Infragravity wave period

The infragravity waves become apparent if, for exam-

ple, a time series of surface elevation is taken. Figure 7.6 shows typical results from Slapton Sands in south Devon. The time series was run through a low pass filter which essentially averages the water surface elevation over (in the present case) 20-s intervals. The output shows the presence of long wave, surf beat or infragravity energy that contributes a substantial fraction of the total variance (see also Huntley & Bowen, 1973; Suhayda, 1974; Goda, 1975; Huntley, 1976; Wright *et al.*, 1979; Holman, 1981; Wright *et al.*, 1982; Guza & Thornton, 1989, and others). Guza and Thornton (1989) divide the generation and dynamics of these infragravity waves into two types. In the first, the waves are essentially two dimensional and are generated when two incident wave trains of similar but not identical frequencies interact to form groups of higher than average and lower than average waves. These groups set up a forced correction to the mean water level that travels shorewards with the group and is released as a shorewards propagating free wave when the incident waves break. It is then reflected from the beach, and the summation of the incident and reflected long waves gives a standing component to the wave pattern. These forms of long waves are known as *standing infragravity waves* (Holman, 1981). The second hypothesis recognizes that (long) wave motion travelling in an alongshore direction would be trapped by the processes of refraction described above. The hypothesis suggests that these motions could be resonant at multiples of the incident wave frequency (the sub-harmonics) or at the group frequencies and could then again establish a long wave pattern in the nearshore. These forms of long waves are called *edge waves* (Holman, 1981) and will have a three-dimensional structure. Each type is described in the following sections. The wave period of the standing long waves will be given by the beat frequency of the incident waves. For the simplest case of two sets of incident waves having frequency ω_1 and ω_2, the infragravity wave will have a frequency $\omega_{IG} = \omega_1 - \omega_2$, or, in terms of the wave periods, $T_{IG} = T_1 T_2/(T_2 - T_1)$. Thus , incident waves of periods $T_1 = 6$ s and $T_2 = 8$ s would have group periods of 24 s, whereas incident waves of periods $T_1 = 10$ s and $T_2 = 11$ s would generate group periods of 110 s. The edge waves can be resonant at this same group frequency, or alternatively they may exist as sub-harmonics of the incident waves so that the edge wave period is an integer multiple of the incident wave frequency. Huntley and Bowen (1973) report the first field measurements of edge waves, finding a peak in the cross shore current spectrum at the first sub-harmonic of the incident waves. More recent measurements appear to demonstrate that the nearshore spectrum of infragravity wave motion is broadband suggesting that various types of motion at different frequencies occur on natural beaches. In general it appears that sub-harmonic oscillations are associated with steep, reflective profiles, whilst longer periods are associated with dissipative beaches.

Infragravity wave length

The most commonly used edge wave solution (Holman, 1981) is due to Eckart (1951) and is based upon a second wave dispersion equation (compare Eq. A1 for gravity waves with Eq. B1 for edge waves) where k_e is the edge wave number (= $2\pi/L_e$) and β is the beach gradient. The mode number, n, gives the number of zero crossings in the seawards decay of the wave amplitude, and thus has values, 0, 1, 2, 3. . . . The wave length occurs only in k_e in this dispersion equation, and therefore the expression may be rearranged to determine the wavelength of the edge wave, L_e, (Eq. B2).

Infragravity wave height

We have already seen that the nearshore amplitude of

the long wave response can be broadband, and is presently difficult to predict. However, Holman (1981) and Holman and Sallenger (1984) showed from theoretical and field evidence that the non-dimensional infragravity variance parameter (equivalent to the average over frequency of the squared shoreline infragravity wave amplitude) was proportional to the significant incident wave height at breaking (H_s) as described by Eq. B3. This formula allows an estimate to be made of the amplitude, a_e, of the long wave motions at the shoreline.

Standing long wave profiles

Guza and Thornton (1989) credit the standing infragravity wave model to Munk (1949) and Tucker (1950) who suggested that incident wave groups have associated low frequency components travelling shorewards with the group and that these are released when the incident waves break, reflect off the beach or break point and then propagate seawards as a free wave. Longuet-Higgins and Stewart (1962, 1964) and Bowen *et al.* (1968) introduced the concept of radiation stress to formalize this idea. The radiation stress is defined as 'the excess flow of momentum due to the presence of the waves' (Komar, 1976). The onshore momentum flux must be balanced by an opposing force which is manifested as a water slope so that the pressure gradient of the sloping water surface balances the change (spatial gradient) in the incoming momentum. This difference is known as the set-up or set-down due to the waves. Longuet-Higgins and Stewart (1964) integrated the shorenormal radiation stresses to obtain the change in mean sea level, $\delta\eta$, given by Eq. B4. Under wave groups, as opposed to monochromatic waves, this set-up and set-down is manifested as a low frequency wave and, owing to the negative sign in Eq. B4, will have crests that correspond to minima in the combined incident wave amplitudes and troughs that correspond to the maxima as shown in Fig. 7.6. Data discussed in terms of the standing long wave theory have been published by Munk (1949), Tucker (1950) and Suhayda (1974). The offshore decay of the surface amplitude for a standing infragravity wave is given by Eq. B5 (Horikawa, 1988) where J_0 is the zeroth order, $p = 1$, Bessel function (Eq. B6) and a_{IG} is equivalent to a_e. The typical decay structure is shown in terms of the dimensionless offshore distance $\chi = \omega_{IG}^2 x/g\beta$ in Fig. 7.7.

Panel B Infragravity wave equations.

$$\omega_e^2 = g k_e (2n + 1) \tan \beta \tag{B1}$$

$$L_e = \frac{\omega_e^2}{g(2n + 1)\tan \beta} \tag{B2}$$

$$a_e = \sqrt{0.45 H_s^2 - 0.6} \tag{B3}$$

$$\delta \eta = -\frac{1}{8} \frac{H^2 k}{\sinh (2kh)} \tag{B4}$$

$$\eta(x,t) = a_{IG} J_0 \left(2\omega_{IG} \sqrt{\frac{x}{g \tan \beta}} \right) \cos (\omega_{IG} t) \tag{B5}$$

$$J_p(x) = \sum_{n=0}^{n=\infty} \frac{(-1)^n (x/2)^{2n+p}}{n!(p + n)!} \tag{B6}$$

$$u(x,t) = a \sqrt{\frac{g}{h}} J_1 \left(2\omega_{IG} \sqrt{\frac{x}{g \tan \beta}} \right) \sin (\omega_{IG} t) \tag{B7}$$

$$\eta_{IG}(x) = a_{\eta IG} \phi_n(k,x) \tag{B8}$$

0: $\quad \phi_0(k,x) = e^{-kx}$ \hfill (B9)

1: $\quad \phi_1(k,x) = (1 - 2kx) e^{-kx}$ \hfill (B10)

2: $\quad \phi_2(k,x) = (1 - 4kx + 2k^2 x^2) e^{-kx}$ \hfill (B11)

3: $\quad \phi_3(k,x) = \left(1 - 6kx + 6k^2 x^2 - \frac{4}{3} k^3 x^3 \right) e^{-kx}$ \hfill (B12)

4: $\quad \phi_4(k,x) = \left(1 - 8kx + 12k^2 x^2 - \frac{16}{3} k^3 x^3 + \frac{2}{3} k^4 x^4 \right) e^{-kx}$ \hfill (B13)

$$u_{IG}(x) = a_{uIG} \frac{\partial \phi_4}{\partial kx} \tag{B14}$$

0: $\quad \dfrac{\partial \phi_0(k,x)}{\partial (kx)} = e^{-kx}$ \hfill (B15)

1: $\quad \dfrac{\partial \phi_1(k,x)}{\partial (kx)} = \left(\frac{1}{2} - kx \right) e^{-kx}$ \hfill (B16)

2: $\quad \dfrac{\partial \phi_2(k,x)}{\partial (kx)} = (- 5 + 8kx - 2k^2 x^2) e^{-kx}$ \hfill (B17)

3: $\quad \dfrac{\partial \phi_3(k,x)}{\partial (kx)} = \left(- 7 + 18kx - 10k^2 x^2 + \frac{4}{3} k^3 x^3 \right) e^{-kx}$ \hfill (B18)

4: $\quad \dfrac{\partial \phi_4(k,x)}{\partial (kx)} = \left(- 9 + 32kx - 28k^2 x^2 + 8k^3 x^3 - \frac{2}{3} k^4 x^4 \right) e^{-kx}$ \hfill (B19)

$$k_{IG} = \frac{4\pi^2 \omega_0^2}{9g \tan \beta} \tag{B20}$$

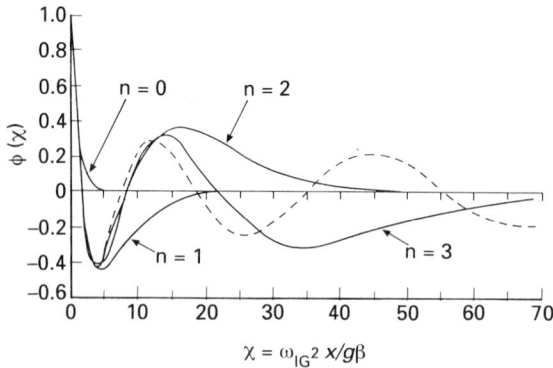

Fig. 7.7 Offshore decay of edge waves of modes $n = 0,1,2,3$ and 4 and of a standing long wave (– – –) in terms of the non-dimensional distance $\chi = \omega_{IG}^2 x/g\beta$ showing that the mode number corresponds to the number of zero crossings in the decay function.

Standing long wave currents

The offshore decay of the flow velocity amplitude beneath standing long waves is given by Eq. B7 (Horikawa, 1988) where J_1 is the first order, $p = 1$, Bessel function (Eq. B6). The result is similar in structure to, but out of phase with, the elevation equation given above.

Edge waves profiles

Concise introductions to edge waves are given by Holman (1981) and Bowen and Huntley (1984). A modified dispersion relationship was suggested by Stokes (1874), Bowen (1969b) and Bowen and Inman (1969) and has been given here as Eq. B1. If $(2n_e + 1)$ $\tan\beta > \pi/2$ there are no trapped solutions, and waves having these (ω_e, k_e) values may propagate to, or from, deep water and are consequently known as leaky modes. They include the normal incident waves. Huntley *et al.* (1981) demonstrated the applicability of Eq. B1 by measuring the energy density over a range of (ω_e, k_e) values and confirmed that maxima correspond to modal values as shown in Fig. 7.8. Sub-harmonic generation is postulated by Huntley and Bowen (1973) to explain field measurements on a steep gravel beach and by Bowen and Inman (1969) to explain laboratory observations. Wright *et al.* (1979) find that edge waves can exist on natural beaches as sub-harmonics of the incident waves, so that $m = T_{IG}/T_G$ has values of 2, 4, 6, etc. and suggest that reflective beaches coincide with low

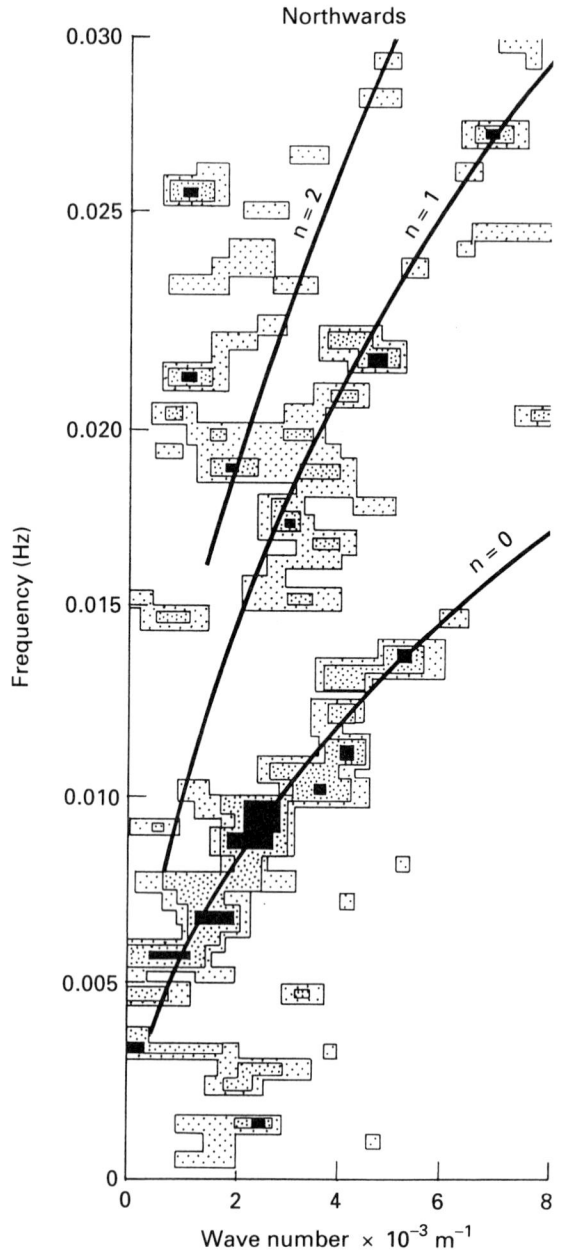

Fig. 7.8 Contoured wave number versus frequency plot showing that maximum energy lies along the integer mode number lines. After Bowen and Huntley (1984). Contour-shading increases with dot density to the solid maximum.

m values whilst dissipative beaches coincide with higher values. The offshore decay of the edge wave amplitude is described in terms of the Laguerre polynomials as shown in Eq. B8 where the corresponding values for the first four modes are given by Eqs B9–B13. The decay structure is shown in Fig. 7.7, from which it is apparent that high mode edge waves correspond closely with the standing long wave structures.

Edge wave currents

Again, in the Eckart solution, the orthogonal edge wave current is π phase shifted and its amplitude is given by Eq. B14 where the corresponding expansions are given by Eqs B15–B19.

7.2.4 Steady currents

Mass transport currents

In addition to the oscillatory, orthogonal currents described in the preceding sections, surface waves generate steady, mass transport currents in a direction normal to the wave advance. In terms of the second-order theory described above these involve an onshore drift current or mass transport in the direction of wave advance, the associated velocity, U_s, being given by Eq. C1 (after Komar, 1976, Eq. 3–33) or, at the bed (Allen, 1982, Eq. 1.54). The results show a net shoreward flow near the surface and near the bottom balanced by a return flow at mid-depth as shown in Fig. 7.9(a). Russell and Osario (1958) note that when the channel width is not constrained there is a tendency to develop a horizontal circulation and continuity need not then be confined in two dimensions. Komar (1976) notes that mass transport may be responsible for producing a net shorewards transport of sediment close to the sea bed.

Undertow

The mass transport current is generated because the orbits of second order waves are not closed. In addition, the concept of radiation stress was applied to shallow water waves by Longuett-Higgins and Stewart (1964) and Bowen *et al.* (1968). Roelvink and Stive (1989) present one of many sets of laboratory results that confirm the equations for wave set-up as shown in Fig. 7.9(b). Although there will be a horizontal balance between the radiation stress and the wave set-up, Dyher-Nielson and Sorensen (1974) pointed out that the vertical mass and momentum fluxes need not equate, leading to a seaward directed

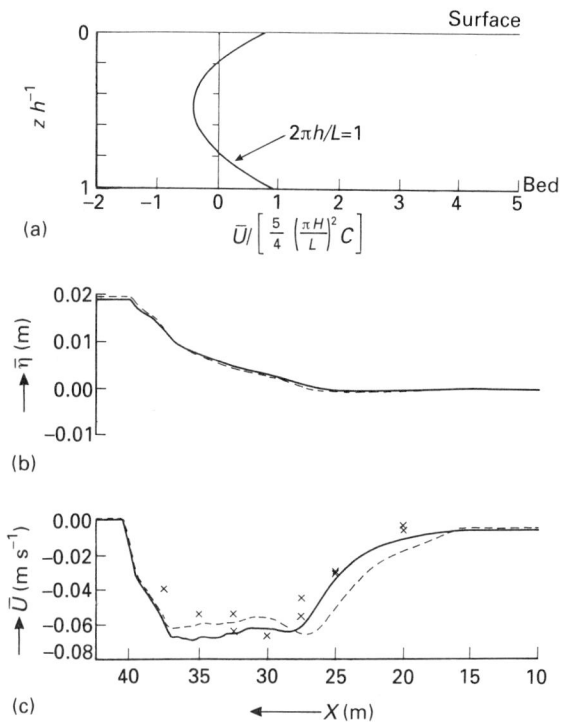

Fig. 7.9 Steady currents: (a) Mass transport currents given by Eq. C1. After Komar (1976). (b) Wave set up. After Roelvink and Stive (1989). (c) Comparison of observed (xxx) and predicted (– – –, ——) undertow. After Roelvink and Stive (1989).

offshore flow, called the *undertow*, which compensates for the shorewards mass flux above trough level. Equations for the undertow are presented by Dally and Dean (1984), Svendsen (1984), Stive and Wind (1986), Svendsen *et al.* (1987) and Roelvink and Stive (1989). One of the simpler formulations is due to Svendsen (1984) and, with the assumption that the eddy-viscosity is constant, this predicts a parabolic distribution for the undertow as in Eq. C2 (Fig. 7.9(c)), where the coefficients are given by Eqs C4, C5 and C6. Svendsen's formulae have been recast to utilize the convention that the vertical axis is measured in a positive upwards direction from the still water level, and thus, at the sea bed, the $(z + h)$ terms tend to zero and the undertow is given by α_3. A second, more realistic, model is also derived by Svendsen (1984) using an exponentially decaying vertical distribution of the eddy-viscosity, and readers are referred to Eq. 4.14 in the original paper for the full solution.

Panel C Steady current equations.

$$U_s = \frac{1}{2}\left(\frac{\pi H}{L}\right)^2 C \frac{\cosh[2k(z+h)]}{\sinh^2(kh)} \tag{C1}$$

$$U(z) = \frac{1}{2}\alpha_1(z+h)^2 + \alpha_2(z+h) + \alpha_3 \tag{C2}$$

$$\alpha_1 = c_2\left(\frac{H}{h}\right)^2 B_o\left(2\left(\frac{C_x}{C} + \frac{(H/h)_x}{H/h}\right) + \frac{B_{ox}}{B_o}\right) + g\frac{\partial\eta}{\partial x} \tag{C3}$$

$$\alpha_2 = 2\frac{U_m - U_s}{d_{tr}} - \frac{1}{3}\alpha_1 d_{tr} \tag{C4}$$

$$\alpha_3 = U_m - \frac{1}{6}\alpha_1 d_{tr}^2 - \frac{1}{2}\alpha_2 d_{tr} \tag{C5}$$

$$v = \frac{v_0(B_1 x^{P1} + Ax)}{x_b} \qquad \text{shorewards of breakers} \tag{C6}$$

$$v = \frac{v_0 B_2 x^{P2}}{x_b} \qquad \text{seawards of breakers} \tag{C7}$$

$$v_0 = \frac{5\pi}{16}\gamma b\,\zeta^2\,\frac{\tan\beta}{c_f}\,\sqrt{(gh_b)}\,\sin\alpha_b \tag{C8}$$

$$v = \frac{5\pi}{8}\frac{\tan\beta}{c_f}u_m\sin\alpha_b \tag{C9}$$

$$v = 2.7u_m\sin\alpha_b\cos\alpha_b \tag{C10}$$

$$V_c = \frac{\pi\sqrt{2}}{c_f\gamma_b^3}\left(1 + \frac{3\gamma b^2}{8} - \frac{\gamma b^2}{4}\cos 2\alpha_b\right)\frac{\partial H_b}{\partial y} \tag{C11}$$

Longshore currents

A large number of equations have been developed that attempt to predict the longshore current velocity. Putnam *et al.* (1949) utilized an energy flux model, Inman and Quinn (1952) utilized a momentum flux model, and Inman and Bagnold (1963), Eagleson (1965), and Galvin and Eagleson (1965) utilized a mass continuity model. However, the subject was placed on a sounder theoretical footing with the publication of papers by Bowen (1969a,b), Thornton (1971) and Longuet-Higgins (1970a,b) making use of the concept of radiation stress. Essentially, if waves are approaching the shore at an angle, there will be a component of the radiation stress that is directed in a longshore direction. This spatial gradient in the momentum must again be balanced by an opposing force (in the same manner as the wave set-up balances the shore normal component), and this force leads to the generation of a longshore current. Longuet-Higgins' work suggests that the cross-shore distribution in the velocity of the longshore current is given by Eq. C6 in the region shorewards of the breakers, and by Eq. C7 in the region seawards of the breakers where the reference velocity is given by Eq. C8. The ζ term is a constant factor which results from inclusion of wave set-up and the other terms are scaling and dissipation factors. Longuet-Higgins' work also suggests that a mean longshore current would be given by Eq. C9. Komar (1976) suggests that $\tan\beta/c_f$ is essentially constant for natural beaches at about 0.138, and therefore Eq. C9 becomes Eq. C10 where the cosine factor is included by Komar (1976) because of the support of laboratory data at larger angles of incidence for which $\cos\alpha_b$ is no longer equivalent to unity. Komar (1983) provides a comprehensive review of available data on the mean longshore current velocity (Fig. 7.10).

Cell circulations

The most apparent feature of cell circulation are the

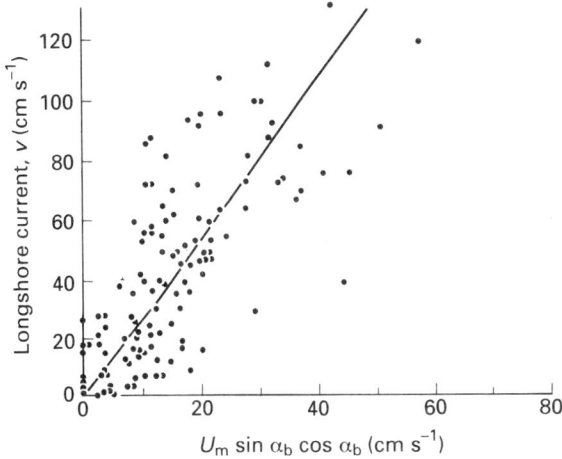

Fig. 7.10 Longshore current data compared with Eq. C9. After Komar (1976).

rip currents: strong narrow currents that flow seawards from the surf zone. These are fed by currents which run parallel to the beach and which increase in velocity from zero midway between two adjacent rips reaching a maximum just before turning seawards into the rip current itself. It is important to realize that both the cell circulation and the steady, longshore currents can both be present on the beach. The current pattern actually observed is the sum of the two processes. This is why it is often difficult to distinguish between the two types in field measurements.

There is no reason for the generation of cell circulations if the wave height at the breakers is constant along the shore. However, this is frequently not the case and then a cell circulation may develop. Komar (1976) considers the problem and finds that (based on the radiation stress approach) the velocity of the cell circulation, V_c, is given by Eq. C11 where $\delta H_b/\delta y$ is the longshore change in wave height. The cell circulation therefore depends primarily on the existence of variations in the wave height along the shore. There are two principal ways in which such variations can be produced. The more obvious is by the process of wave refraction which was discussed earlier, producing a local relative concentration in wave energy and hence locally higher waves. An example of such control is the well documented case where canyon heads cause local wave shadows and rip current development. The second cause can be the presence of the edge wave phenomena which were

discussed in a previous section and lead to a longshore variation in breaker height.

7.3 Nearshore sediment transport

7.3.1 Introduction

The preceding sections have attempted to demonstrate that fluid dynamicists, oceanographers and engineers have succeeded in erecting a general theory for waves on beaches and, on one level at least, a series of equations has been presented that permits the flow field to be described in terms of the independent, incident waves. This is but the first stage in the formulation of a quantitative understanding of beach deposits and we must now progress to examine the processes by which the wave-induced flow field erodes, transports and deposits the sedimentary particles.

The subject matter is known as the two-phase problem (Bagnold, 1966), because sediment transport involves the fluid phase (i.e. the water) in which the solid phase (i.e. the sedimentary particles which are inherently denser than the surrounding medium) tends to settle to the bed. The general physics of the two-phase problem was described in Chapter 2; here we wish to identify the fluid stresses within the beach system which overcome particle settling and transport the sediment. We wish to explain and, wherever possible, to quantify the transport rates and the net transport that results.

The basic definitions involved in the two-phase problem were given in Chapter 2. Here we shall introduce some additional definitions which are peculiar to the beach system before proceeding to a detailed description of wave-dominated sediment transport. These pecularities arise because of the oscillatory nature of the gravity and infragravity flow fields described earlier.

In particular (Fig. 7.2), it is apparent that a symmetrical sinusoidal flow would generate symmetrical onshore and offshore transports provided that the transport rate is an instantaneous and unique function of the flow, that is, that $i(t) = f(u(t))$ holds. Thus deposition, which can only result from a non-zero net transport, requires either (1) that transport is not, in fact, an instantaneous and unique function of the flow field, or (2) that the flow is asymmetric. The former condition results when the time history or acceleration of the flow is important (as has been suggested by Sleath, 1978; Hallermier, 1982a; Hanes and Huntley, 1986 and demonstrated by Hardisty, 1991a) or when bedslopes generate transport asymmetry in symmetrical flow fields (as has been suggested by Hardisty, 1986; Whitehouse, 1991 and demonstrated by King, 1991). The second condition is simpler to understand because flow asymmetry will be produced by any of the residual currents detailed earlier (i.e. second order asymmetric oscillatory flow, mass transport, undertow or long shore currents). The definitions shown in Table 7.2 are introduced in order to begin to account for the effects of such asymmetries. There are two types of *orthogonal transport formula* in use, and one of three transport rates can be predicted by each type of formula. Although each of these types of formula and transport rate have their uses, it is desirable to identify and to utilize an integrated flow formula that predicts the

Table 7.2 Orthogonal transport formulae and transport rates

Peak flow transport formulae
Relate the sediment transport rate to the amplitude of the oscillatory flow

Integrated flow transport formulae:
Relate the sediment mass transport rate to the instantaneous flow velocity $u(t)$ as given by e.g. Eq. A9

Instantaneous transport rate
The continually varying dry weight transport rate across unit width of the bed parallel to the wave crest at any instant during the passage of the wave. It is symbolized by $i(t)$, and may be subscripted for bedload, suspended load or total load: $i_b(t)$, $i_s(t)$, $i_t(t)$

Mean transport rate
The average dry weight transport rate across unit width of the sea bed parallel to the wave crest taken over a wave cycle. It is therefore the mean of the integral of the instantaneous transport rates. It is symbolized by $\bar{i}$ and may be subscripted to denote bedload, suspended load or the total load: $\bar{i}_b$, $\bar{i}_s$, or $\bar{i}_t$

Net transport
The total dry weight of sediment that is transported across unit width of the sea bed parallel to the wave crest during the passage of the wave. It is symbolized by the upper case I and is defined by the integral of the instantaneous rate through a particular time interval. Clearly the net transport is also equal to the product of the mean transport rate and the wave period, $I = \bar{i}\,T$. Again the net transport may be subscripted to represent bedload, suspended load or total load: I_b, I_s or I_t

---

net mass transport in order to examine deposition within the beach system.

7.3.2 Orthogonal sediment transport

Threshold

Particle entrainment within the oscillatory boundary layer involves explicitly unsteady forces. As well as the lift, drag and the body forces which were derived in an earlier chapter, there are others related to the horizontal pressure gradients and to the acceleration of the fluid past the bed. Although of theoretical importance, analytical criteria for plane bed thresholds in unsteady flow have met with but little success. All formulae start from a consideration of the balance of forces on a grain of sediment that rotates about a pivot point. It was shown earlier that the fluid forces tend to dislodge the grain from the bed, whereas gravity tends to return the particle to its niche position. The Shields entrainment function (θ) is a good indicator of threshold conditions, and represents a non-dimensional function of flow and sediment parameters (Eq. D1) where f_w is Jonsson's (1966) wave friction factor (which must be evaluated from, for example, the approximation of Swart (1974) as in Eq. D2), u_m is the maximum oscillatory velocity and d_o is the nearbed orbital radius (both evaluated from linear theory for monochromatic waves), k_s is the bed roughness ($k_s = 2.5D$ for flat beds), D is the sediment grain size and ρ_s and ρ are the densities of the sediment and of the water respectively. The analyses are discussed in more detail in Hardisty (1990b), but here we require functional formulae and must resort to empirical criteria. There have been many wave threshold formulations and some of these are described by Bagnold (1946), Sato and Kishi (1954), Manohar (1955), Larras (1956), Vincent (1957), Eagleson and Dean (1959), Goddet (1960), Ishihara and Sawaragi (1962), Sato *et al.* (1962), Bonnefille and Pernecker (1966), Horikawa and Watanabe (1967), Carstens *et al.* (1969), Silvester and Mogridge (1970), Chan *et al.* (1972), Komar and Miller (1973), Hallermeir (1980) and Lenhoff (1982). Examples of these various formulae are plotted in Fig. 7.11 for sand in fresh water at 20°C. Komar and Miller (1973) utilize the data of Bagnold (1946) to show that, for grain diameters less than about 0.5 mm (medium sands and finer), the threshold is best given by the equation which includes the particle volume, density, the

Panel D Sediment transport equations.

$$\theta = \frac{0.5\, f_w u_m^2}{\left(\frac{\rho_s - \rho}{\rho} - 1\right)gD} \tag{D1}$$

$$f_w = \exp\left(5.213\left(\frac{k_s}{a_s}\right)^{0.194} - 5.977\right) \tag{D2}$$

$$u_{cr} = \sqrt{0.405\,\frac{(\rho_s - \rho)gD}{\rho}\left(\frac{d_o}{D}\right)^{0.25}} \tag{D3}$$

$$u_{cr} = 33.3\,(TD)^{0.33} \tag{D4}$$

$$u_{cr} = \sqrt{0.46\,\pi\cdot\frac{(\rho_s - \rho)gD}{\rho}\left(\frac{d_o}{D}\right)^{0.25}} \tag{D5}$$

$$u_{cr} = 71.4\,(TD^3)^{0.143} \tag{D6}$$

$$M_b = \frac{\rho\, u_m^2}{(\rho_s - \rho)gD} \tag{D7}$$

$$i_b(t) = \frac{ku^3}{\tan\phi - \tan\beta} \tag{D8}$$

$$i_b(t) = \rho c_f\,\frac{e_b}{\tan\phi}\left(|u(t)|^2 u(t) - \frac{\tan\beta}{\tan\phi}|u(t)|^3\right) \tag{D9}$$

$$i_b(t) = 0.764\,(u(t)^2 - u_{cr}^2)u(t)\times 10^{-6} \tag{D10}$$

$$C_z w_s = K_s\,\frac{\partial C_z}{\partial z} \tag{D11}$$

$$C_z = C_a\exp(-z/l_s) \tag{D12}$$

$$C_a = 0.005\,\theta^3 \tag{D13}$$

$$i_s = \int_0^h C_z u_z\,dz \tag{D14}$$

$$G_\omega(f) = \sqrt{\frac{i_\omega(f)/i_{\omega 10}}{u_\omega(f)/u_{\omega 10}}} \tag{D15}$$

$$Q_L = k_L P_L^{\,f} \tag{D16}$$

$$I_L = 0.77(ECn)_b\sin\alpha_b\cos\alpha_b = 0.77\,P_L \tag{D17}$$

$$I_L = 0.28(ECn)_b\cos\alpha_b\,\frac{v}{u_m} \tag{D18}$$

density of water and the orbital diameter of the oscillatory currents (Eq. D3). Clifton and Dingler (1984) utilize appropriate densities for quartz sediment in water and find a relatively straightforward relationship between the threshold speed and the product of the wave period and grain diameter raised to the 0.33 power (Eq. D4), where T is in seconds, D is in cm and the resulting flow is in cm s^{-1}. For sediments having diameters larger than 0.5 mm (coarse sands and coarser) Komar (1976) suggests that the empirical curve of Rance and Warren (1969) is more appropriate (Eq. D5), and again Clifton and Dingler (1984) substitute for appropriate densities to find a slightly more complicated relationship with the threshold flow speed being proportional to the product of the wave period and

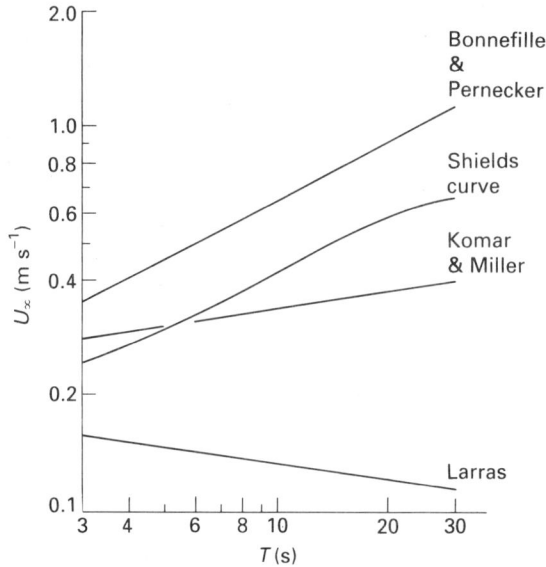

Fig. 7.11 Threshold curves for fine quartz sand in water as a function of wave period, T(s). After Sleath (1988).

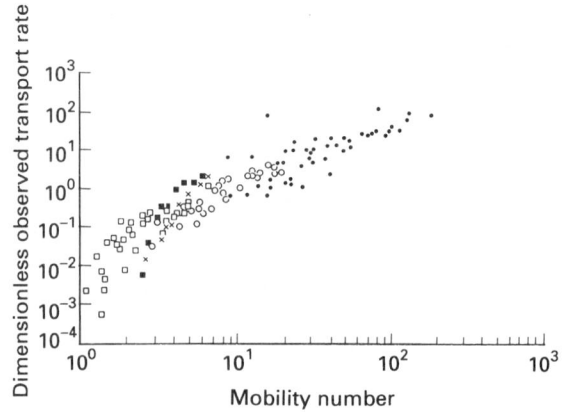

Fig. 7.12 Grain mobility number plotted against non-dimensional net semi-cycle transport rate for Woodruff (personal communication; $\times$, ■), Kalkanis (1964, ○), Abou-Seida (1965, ●) and Sleath (1978, □). After Hardisty (1991b).

the cube of the grain diameter raised to the power 0.143 (Eq. D6).

Modes and models of sediment transport

When threshold flow conditions are exceeded, sediment particles are mobilized. Initially individual grain flurries can be observed to be displaced from their niche positions and to be transported shorewards with the onshore flow beneath the wave crest. At low flow speeds, these grains settle back to a stationary position on the bed before being entrained by the returning offshore flow. Such conditions represent bedload transport, in which the grain weight is carried by intergranular forces. There have been numerous investigations of bedload movement under waves in the laboratory. The waves are almost invariably sinusoidal with a single frequency and consequently they may not be particular good representations of the natural situation. However, two of the most frequently quoted sets of laboratory experiments are those of Kalkanis (1964) and Abou-Seida (1965). In both experiments a sand bed was oscillated within a water flume, the beds containing recessed trays to collect the bedload sediment moved in each half cycle. Additional data were recently acquired with an oscillating bed in which second order flow

asymmetry was introduced and the results from these various experiments are plotted in terms of the mobility number (Eq. D7, Dyer, 1986) in Fig. 7.12. It is apparent that there is some relationship between the amplitude of the orbital velocity and the net sediment transport under these conditions. Flow conditions above threshold tend to lead to the formation of various types of ripples, and these and other bedforms are described in detail in Section 7.4. At higher flow speeds, the individual grain flurries coalesce so that the whole of the surface of the bed is mobilized. This condition is known as *sheet flow* and will tend to obliterate any bedforms developed under less severe conditions. Sheet flow conditions pertain if the shear stress expressed as a Shields parameter (defined by Eq. D1) exceeds a value of about 0.5–1 (Horikawa, 1988). The bedload transport of finer sediments can be modelled satisfactorily from the work of Bagnold and others which essentially balances the flow power with the rate of dissipation of energy within the bedload layer. Such an 'energetics' approach will be followed in later sections, and will be used to derive an integrated flow formula which predicts the net mass transport over a given time interval.

The presence of ripples, or the intensity of the flow conditions, can entrain particles into the body of the flow, and these particles are then said to undergo suspended load transport in which the submerged weight of the grain is supported by fluid-transmitted stresses. Although a significant proportion of the sediment transported about the beach system is moved in suspension, the processes are extremely complex and remain poorly understood. This is partly because, unlike the transport of bedload material, suspension transport beneath orthogonal flows can depend to an overriding extent on the vertical components of the oscillatory motion and upon the bedforms present. Inman (1957) first pointed out the importance of the suspended sand cloud over ripples and, more recently, observations of suspended sediment motion under waves have been reported by many workers (e.g. Inman & Bagnold, 1963; Sunamura *et al.* 1978). The sequence of events is shown in Fig. 7.13. The cloud of suspended sand originates in the vortex that develops in the lee of the ripples during shorewards flow. The cloud of sediment is transported upwards and backwards during the offshore flow, partially settling on the way. There are presently a number of contrasting approaches to the modelling of suspended load transport and to the prediction of the suspended load transport rate. The simplest approach is to assume that the time-averaged suspended sediment concentration profile can be represented by a diffusion equation with the absolute values referenced to a nearbed concentration related to the bedload rate. The resulting concentration profile is then multiplied by the mean flow velocity profile to determine the suspended load rate and the net transport. This approach is detailed in the following sections, but we shall see that it describes inadequately some of the aspects of the process which are being revealed by recent field studies. In particular, spectral and co-spectral analyses are revealing that the long wave motions described earlier may be responsible for the majority of nearshore sediment transport, and it is not presently clear how these transport loads should be modelled.

Orthogonal bedload transport and transport rates

The majority of predictive equations for orthogonal bedload transport assume that the sediment responds in a quasi-steady manner (Roelvink & Stive, 1989) to the instantaneous, time-varying flow and to the downslope gravity force. A large number of empirical equations are presently available, and we shall see that many can be cast in a similar form to the results developed by Bagnold (1954, 1963, 1966). The details are covered in other chapters in this book and by Hardisty (1990b) and will be summarized here. Bagnold (1954) reported the gravity-free dispersion of shearing grains submerged within the annular space between two rotating cylinders. The inner cylinder consisted of a flexible membrane and was used to measure the dispersive pressure generated by the grain shear and to confirm that the dynamic Coulomb yield criterion (Coulomb, 1773) was applicable to the system. Using this criterion, Bagnold was able to equate the available fluid power with the rate of dissipation of energy within the granular-shear layer and hence derive a transport formula relating flow parameters to the bedload rate. Details of the mechanisms have also been investigated by, for example, Bailard and Inman (1979), Hanes and Bowen (1985) and Hanes (1986) and recently a number of individual grain models have been demonstrated (e.g. Haff, 1991). Energy dissipation within the vertically integrated bedload layer is equated by Bagnold to a flow power taken to be the product of the bed stress and a nearbed velocity (cf. Hardisty, 1983). Since the stress is related by the quadratic law to the square of the flow velocity, the result is equivalent to a cubic relationship as given in Eq. D8 where the denominator accounts for the effect of bedslope on the bedload rate.

Bailard (1981, 1982) extends the work of Bagnold (1966), Bowen (1980) and Bailard and Inman (1981)

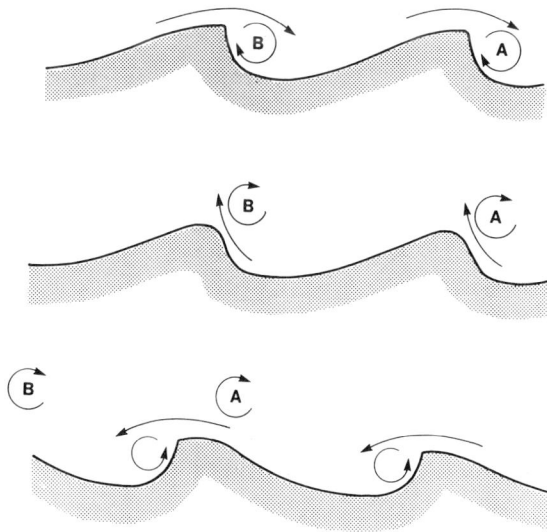

Fig. 7.13 Schematic representation of the suspended transport process, showing upwards and onshore movement of vortex clouds A and B in an asymmetric flow field.

to produce an analogous equation which is here shown as Eq. D7. Bailard (1981, 1982) also follows Bagnold (1966) to produce a similar expression for the depth integrated, suspended load transport but, for the reasons detailed below, that route is not followed here. Similar bedload expressions have been developed from laboratory and field data by other groups. Madsen and Grant (1976) use the Abou-Seida and Kalkanis data to demonstrate that the average bedload transport rate, $\bar{i}_b$, is proportional to θ^3 (i.e. to u^6 or τ^3). Hallermeier (1982a) uses these and other data sets to demonstrate that the mean transport rate is proportional to u_{max}^3/T. Vincent et al. (1981) relate the bedload concentration at any instant to $(\theta - \theta_{cr})$, that is to $(u^2 - u_{cr}^2)$, and then multiply this concentration by the instantaneous flow speed and integrate over the wave semi-cycle to evaluate the average transport rate (i.e. the result is proportional to $\int(u^2 - u_{cr}^2)u\,du$ for each semi-cycle). Sleath (1978, 1982) suggests that the maximum bedload rate is 8/3 times the average value and also relates the rate to a threshold inclusive term $(\theta - \theta_{cr})^{3/2}$. Shibayama and Horikawa (1982) report a net rate which scales with the sixth power of the maximum flow velocity. It is clear that most of these equations have the same general structure but it is also clear that none of them properly accounts for the full complexities of the orthogonal transport problem. The issues are discussed further in Hardisty (1990b) where the bedload function of Bagnold (1963) was examined with the laboratory data of Kalkanis (1964) and set against Watanabe's (1982) formula. The result is shown as Eq. D10 although it should be noted that bedslope and acceleration effects are explicitly excluded and the calibration applies to medium sand. In general the expression predicts a bedload transport rate that is small at low flow speeds (i.e. $u(t) \approx u_{cr}$) and that increases to become proportional to the cube of the flow speed at high rates (i.e. $u(t) \gg u_{cr}$).

Suspended load concentration profile

The concentration of suspended sediment close to the sea bed varies continuously through the wave cycle, as has been demonstrated by Homma and Horikawa (1963), Homma et al. (1965), Horikawa and Watanabe (1967), MacDonald (1973) and Sleath (1982). However, of more practical interest is the mean concentration and concentration profile over the wave cycle because the orthogonal suspended load

rate is the product of the velocity and concentration profiles. In direct analogy with the unidirectional case, the problem divides into determining the reference concentration, determining the concentration and velocity profiles and then determining the transport rate as the vector product of these two quantities. It is for this reason that a depth integrated suspended load transport formula cannot be accepted given that, for example, the mean flow profile depends upon the presence and nature of any or all of the steady current terms discussed above (see also Roelvink & Stive, 1989, p. 4796).

The diffusion approach (Dyer, 1986; Hill et al., 1988) assumes steady state conditions in which the downwards flux of settling particles (Cw_s where C is the sediment concentration at a height z and w_s is the sediment fall velocity) is equal to the upwards flux given by the product of the concentration gradient ($\delta C/\delta z$) and an eddy diffusion coefficient (K_s) as shown in Eq. D11. In the simplest case, when turbulence and mixing are presumed vertically invariant (e.g. Black & Rosenberg, 1991), K_s is independent of z so that the diffusion equation integrates to give the concentration at any height, z, as a simple function of a reference concentration (Eq. D12) where l_s is a mixing length, $l_s = w_s/K_s$. Field work (e.g. Kennedy & Locher, 1972; Black & Rosenberg, 1991) shows that the use of a more sophisticated form for K_s does not significantly improve the fit of Eqs D11 and D12 to experimental observations. Black and Rosenberg (1991) utilized optical backscatter devices and acoustic current meters at Apollo Bay in southern Australia and found good support for theory, obtaining values of l_s ranging from 0.066 to 0.071 beyond the breakers, rising to 0.055–0.058 near the breakpoint and to 0.252–0.386 inside the surf zone. Black and Rosen-

Table 7.3 Suspension coefficients for unidirectional flows. Full references in Hill et al. (1988)

$\gamma_s = \dfrac{Ca\tau}{\tau - \tau_{cr}}$	Source
1.30×10^{-4}	Hill et al. (1988)
1.95×10^{-3}	Smith and McClean (1977a)
2.40×10^{-3}	Smith and McClean (1977b)
5–30×10^{-4}	Glenn (1983)
1.6×10^{-5}	Wiberg and Smith (1983)
1.5×10^{-2}	Kachel and Smith (1986)
5–20×10^{-5}	Drake and Cacchione (1988)
0.78×10^{-4}	Dyer (1980)
0.84×10^{-4}	Dyer (1986)

berg comment that the higher values reflect the greater time-averaged turbulent intensity beneath broken waves, and that the offshore values are within the empirical range suggested by Deigaard *et al.* (1986).

Suspended load reference concentration

It is apparent that the diffusion equation offers a time-averaged approach to the determination of the suspended load concentration profile provided that a suitable value can be chosen for the corresponding reference concentration. Considering, firstly, the case of an unsteady unidirectional flow, such as occurs in laboratory flumes, rivers or within tidal channels, Hill *et al.* (1988) relate the reference concentration to a non-dimensional shear stress where the coefficient of proportionality, γ_s, is called the suspension coefficient and quote the various values shown in Table 7.3. Although the principle of this approach is the same, a more circuitous route must be followed in oscillatory flow because it is rarely possible to measure τ beneath waves. Instead the shear stress is determined from the quadratic stress law and incorporated into a Shields parameter, θ, the cube of which is related to the reference concentration (Nielsen, 1986 and Eqs D1, D2 & D13). Support for such determinations of the reference concentration and concentration profile is reported by Black and Rosenberg (1991) when applied to the time series averaged (i.e. over about 17 min) concentration measurements obtained in Apollo Bay and for the 92nd percentile of the velocity distribution in random waves for u_m whilst Wright *et al.* (1982) recommend a smaller coefficient in Eq. D13 and a larger multiplier in the roughness length calculation to account for the presence of ripples.

Total load rates and net transport

Qualitatively, the total load sediment transport rate is equal to the sum of the bedload and the suspended load rates. The bedload rate is taken to be an instantaneous response to the flow field and thus, over a time interval, t_1 to t_2, may be modelled using Eq. D8, D9 or D10 with the instantaneous velocity being given by the sum of the gravity and infragravity oscillatory currents and the mass transport, undertow (and longshore) steady currents. Eq. D10. The suspended load is taken to be given as the vertical integral of the product of the time-averaged suspension concentration and residual current profiles. The suspension concentration profile is based upon the diffusion approach and is tied to a mean reference concentration given by the cube of the wave Shields parameter based on, perhaps, the 92% of the orthogonal flow distribution. The corresponding steady current profile is due to the mass transport and undertow currents plus any asymmetry due to, and depending upon, the phase relationships between the gravity and infragravity oscillatory currents. The vector product must be integrated vertically from the sea bed to the water surface and then both the bedload and suspended load rates must be integrated over the period of interest (Fig. 7.14). The result is expressed mathematically as Eq. D14.

Spectral and cospectral results

We have already seen that the analysis of beach

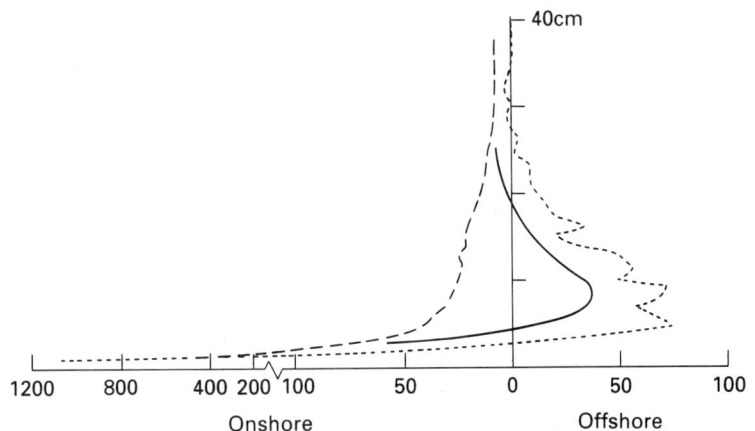

Fig. 7.14 The total suspended sediment flux (g cm^2 s^{-1}) and its two components. – – –, $\overline{U}$, $\overline{C}$; · · ·, $\overline{U}_w \overline{C}_w$; ——, total. After Green and Vincent (1991).

processes in the frequency rather than the time domain has provided a useful insight into the role of periodicities within the beach system. Thus, a random incident wave field can be described by its surface elevation spectrum and the nearbed oscillatory current spectrum can, it appears, be predicted from a linear transform of that surface elevation spectrum. Again, the leaking of energy across frequencies describes the growth of harmonics in a non-linear wave field or the transfer of energy to infragravity long wave motions which may be present and resonant in the nearshore. However, sediment dynamicists were denied this powerful approach because, until recently, it has not been possible to make continuous measurements of the sediment transport rate. The last decade has, however, seen the development of optical (e.g. Huntley & Hanes, 1987) and acoustic (e.g. Vincent & Green, 1990) backscatter devices which have made available this new data. Unfortunately, the instantaneous and continuous measurement of bedload processes is still difficult, although the development and deployment of passive acoustic devices that 'listen' to the noise of intergranular collisions and hence estimate the bedload transport rate is progressing for beaches (e.g. Hardisty, 1991a; Jagger & Hardisty, 1991) and has already provided a powerful insight into the role of unsteady turbulent events in gravel-bedded tidal channels (e.g. Heathershaw & Thorne, 1985).

Results which are typical of those emerging with the new techniques were shown in Figure 7.4 from a recent deployment at Spurn Head on the North Sea coast of eastern England and the following description is based largely on a recent paper by Beach and Sternberg (1991). In particular, the assumption is made that the time-averaged suspended sediment transport rate $<uc>$ can be decomposed into its steady component $<u><c>$ and its fluctuating component $<u'c'>$ where u is the flow velocity, c is the (scalar) sediment concentration, the primes represent the fluctuating components about mean values, and $<\ >$ represents a time average. Thus, the surface elevation and orthogonal flow time series (Fig. 7.4(a) & (c)) drive an instantaneous suspended load concentration (Fig. 7.4(e)) which leads to a sediment transport rate (Fig. 7.4(g)). The demeaned values of u and c are then subjected to spectral analysis (Fig. 7.4(d) & (f)) which reveals that, although considerable suspension is associated with gravity wave frequencies, the transport is dominated by processes in the long wave band. This type of analysis was introduced by Huntley

and Hanes (1987) who worked with a miniature optical backscatter device and electromagnetic flow meters at Pte Sapin in New Brunswick, Canada and introduced the co-spectrum of the fluctuating components of cross-shore velocity and suspended sediment concentration. The co-spectrum (Fig. 7.4(h)) is particularly useful because it indicates the magnitude and direction of sediment transport as a function of frequency. The net cross-shore sediment flux is the integral of the co-spectrum over all frequencies. The results of Huntley and Hanes' (1987) co-spectral analysis indicated onshore transport due to the incident waves with offshore transport due to long period motions near the sea bed. These type of results are confirmed by the Spurn Head data set (Fig. 7.4) showing a positive (i.e. onshore) co-spectrum in the gravity band and offshore (negative co-spectrum) at long wave frequencies. Co-spectral analyses are also reported by Beach and Sternberg (1988, 1991), Russell et al. (1991) and Greenwood et al. (1991). In general it appears that:

1 Under low energy conditions, incident wave processes dominate, cross-shore sediment flux is onshore and the fluctuating component is of the same order as mean components.

2 Under storm conditions, infragravity motions dominate, transporting sediment in either the onshore or offshore direction depending upon the horizontal and vertical phase relationships between sediment resuspension and flow fields. The fluctuating component can be an order of magnitude larger than the mean component.

Extending this type of analysis towards a predictive capability, Hardisty (1991b) defined a spectral gain function (cf. Jenkins & Watts, 1968), to relate the spectral amplitude of the fluctuating components of the sediment transport rate to the corresponding flow values. It appears that the gain function is remarkably flat over the gravity and infragravity frequencies in this example, although Hardisty (1991a) used passive acoustic sensors at Slapton Sands in Devon, England to demonstrate that the transport of coarser sediment attenuates with frequency and postulated that this could be attributable to acceleration effects through the history integral and added mass terms (cf. also Madsen, 1991). Nevertheless, we might define the spectral gain function by Eq. D15 and it is possible that this approach could be equally applicable to the turbulent description of the sediment mass transport process in fluvial channels or in aeolian environments.

7.3.3 Longshore sediment transport

In comparison with the shorenormal transport of sediment discussed in the preceding chapter, shoreparallel sediment transport has received far more widespread attention, largely because it manifests itself wherever the movement is prevented by the construction of jetties, breakwaters and groynes. Such structures act as dams to the shoreparallel transport, causing a build-up of the beach on the updrift side and a corresponding erosion in the down drift direction. Shoreparallel transport has been termed longshore transport, littoral drift or littoral transport. In the 19th century it was commonly believed that tidal currents and coastal ocean currents were primarily responsible for longshore transport. We now know that the wave-induced longshore currents, which were detailed earlier, are the chief cause of the sediment movement; the other currents are only effective under exceptional circumstances. Although only effective close to, or within, the breaker zone the background theory of longshore sediment transport is briefly described in this section. Fuller reviews can be found in Komar (1976, 1983), Savage (1962), Das (1971), King (1972), Greer and Madsen (1978), Hallermeier (1982b), Seymour (1989), Bodge (1989) and Bodge and Kraus (1991).

The empirical formulae rely on a presumed correlation between the longshore transport rate and a measure of the longshore component of the incident wave energy. In general, the result is expressed as a volumetric longshore transport rate equal to the product of a calibration coefficient and the longshore wave power raised to an exponent (Eq. D16). Results reported by Watts (1953), Caldwell (1956), Savage (1959), Ijima *et al.* (1960, 1964), Komar and Inman (1970), Das (1971) and the Shore Protection Manual (1984) are summarized by Hardisty (1990b, chapter 11) and show values for the coefficient in the range 0.06–6.2 with the majority of values being less than 0.5 and values for the exponent in the range 0.54–1.00 with the majority of values being given as unity. Recent work (reviewed by Bodge, 1989) emphasizes the relationship between the coefficient and the type of breakers. Specifically, it appears that the proportionality coefficient is more or less linearly related to the surf similarity parameter and that, in general, plunging or collapsing breakers lead to greater total longshore transport for a given level of longshore wave energy flux.

A series of papers (Bagnold, 1963; Inman & Bag-

nold, 1963; Komar & Inman, 1970) provides a firmer theoretical basis for the prediction of the longshore sediment transport rate by equating the immersed weight sediment transport rate to the available fluid power as described earlier. This approach is detailed by Komar (1976, 1983) and, since the flow power is again taken to be the product of the bed stress and the (longshore) flow velocity and the bed stress is in turn equal to the energy flux divided by the (oscillatory) flow velocity, then the resulting transport is proportional to the wave power and to the ratio of the longshore to the orbital current velocities. Komar and Inman (1970) use the results to deduce the equation for the longshore current given earlier (Eq. C9). For practical purposes, Komar (1976) recommends that the immersed weight transport rate be calculated from his best fit line through the empirical data (Eq. D17) or from the theoretical equation (Eq. D18) with the appropriate steady current.

7.4 **Beach deposits**

7.4.1 Introduction

Komar (1976) describes an idealized transect from offshore towards the beach. In the deepest water (out to the shelf edge) the ripples would be rounded by bioturbation, being activated only during storm events and having only relatively short crest lengths. Moving shorewards, ripple wave lengths become shorter and crests become longer. Approaching the breaker zone, the flow asymmetries due to the generation of the higher harmonics described above appear to produce a corresponding asymmetry in the bedform profiles, with steeper shorewards faces which the longer but lower velocity offshore flow cannot remodify. Within the surf zone, Komar (1976) notes that wave formed ripples may co-exist with current ripples formed by longshore or cell circulation flows. The sedimentary characteristics and bedform regimes of each sub-environment are described in the following section, and a complete description is given in Carter (1988, chapter 3). However, it will be apparent that whereas the hydrodynamic and, to a lesser extent, the sediment dynamic theories described above are well advanced, the application of the theories to explain and to predict the sedimentological characteristics of wave-dominated environments remains a largely qualitative or descriptive procedure. Much progress will doubtless be made in the near future; here we must rely upon a little algebra to suggest sedimentological boundaries that may direct the field work.

7.4.2 Modern beach sediments

Grain size characteristics

Numerous studies from Wiegel (1964) onwards report a loose correlation between beach gradient and grain size and, for gravel beaches at least,it is argued that infiltration losses into the beach are responsible for the steeper gradients (cf. Bagnold, 1963) at the shoreward limit of the orthogonal profile. However, it is usually assumed that the grain size and grain size distribution of nearshore deposits as a whole bear some relationship to the maximum flow parameters, and particularly to the theoretical threshold conditions described above. However, there have been few practical assessments of the theory (Horn & Hardisty, 1990; Horn, 1991) and the work carried out by Jago and Barusseau (1981) remains one of the most comprehensive attempts at testing the theory against field observations. Jago and Barusseau (1981) used a large Van Veen grab to obtain closely spaced sea bed sediment samples in water depths between 5 and 40 m along shorenormal transects in the microtidal Golfe de Lion in the western Mediterranean. They found that the sands of the inner shelf were well graded and displayed a seaward-fining textural gradient (cf. also Passega *et al.*, 1967) with mean diameters decreasing from some 250 μm in the nearshore to less than 100 μm in the offshore. Jago and Barusseau (1981) utilized a complete 1-year wave record from a nearby Datawell wave-rider buoy to compare the observed grain size characteristics with the predictions of the Komar and Miller (1975a,b) threshold equation described above, a theoretical null point equation (Johnsson & Eagleson, 1966) and an incipient motion equation (Johnsson & Eagleson, 1966). They found, firstly, that although the null point and threshold approaches gave comparable predictions and thus appeared to be describing the same state, the incipient motion equation consistently predicted coarser grain size characteristics. Secondly, the predicted grain sizes were consistently finer than the observations in deeper water and coarser than the observations in shallow water, that is, the spatial gradient of the grain size parameters was correctly modelled, but the derivative of the gradient was too large in the theoretical analyses. Finally, and perhaps most importantly, Jago and Barrusseau (1981) found

that, towards the shoreward end of the transect, threshold conditions were exceeded for more than 40% of the year and they thus concluded that the grain size of the shallow water sands does not reflect the maximum energy of the environment.

Symmetrical wave ripples

The profile form of ripples is defined in terms of the wave length, λ, and the ripple crest height, H_r, giving a derived parameter which Allen (1982) calls the *ripple vertical form index*, λ/H_r. Inman (1957) remains perhaps the most comprehensive study of the bedforms, describing wavelengths between the surf zone and depths of about 50 m off the Californian coast. In general, the results demonstrated a relatively simple relationship between ripple wavelength and the orbital diameter, $\lambda = C_1 d_o$ where d_o is the nearbed orbital diameter under linear wave theory ($d_o = H/\sinh(kh)$) and C_1 is a constant between 0.65 (Miller & Komar, 1980) and 0.8 (Komar, 1976). However, wave channel studies (Hunt, 1882; Evans, 1942; Scott, 1954; Kennedy & Falcon, 1965; Mogridge & Kamphuis, 1973), work in pulsating water tunnels (Carstens *et al.*, 1969) and observations using a section of bed oscillating through still water (Bagnold, 1946; Manohar, 1955) seem to suggest a more complicated relationship with ripples reaching a maximum wave length, which depends upon their grain size, and then developing a smaller wave length with increasing orbital diameter as shown in Fig. 7.15. Clifton (1976) suggests classifying the linear section as *orbital ripples* to distinguish them from the *anorbital ripples* which exist in the upper part of the graph and have a wavelength that is independent of d_o. However, this problem may have more to do with the use of linear theory in defining d_o than any physical process. Nevertheless, Clifton and Dingler (1984) state that the maximum orbital diameter (which defines the upper limit of orbital ripples) is given by $\lambda = 14700\, D^{1.68}$ where both the grain diameter D and the ripple wave length are in centimetres (Miller & Komar, 1980). Dingler and Inman (1977) stated that the ratio of the ripple height to spacing for orbital ripples was a constant such that $H_r/\lambda = C_2$ where $C_2 \approx 0.16$. A distinction must be drawn between the *rolling-grain ripples* which are generated by the to-and-fro movement of sediment on the sea bed and the *vortex ripples* which are distinguished by flow separation enabling the maintenance of ripple form through constant agitation of

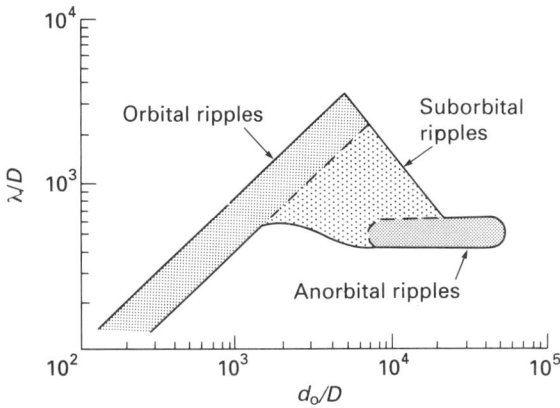

Fig. 7.15 General non-dimensional diagram showing that the ripple wave length increases with orbital diameter but that the relationship becomes more complex in the 'sub-orbital' and 'anorbital' region. After Clifton and Dingler (1984).

sediment from crest to trough and back again as was depicted in Fig. 7.13.

Asymmetrical wave ripples

The symmetrical wave ripples described above can take an asymmetric form due to the co-existence of oscillatory motion and some residual flow field. Allen (1982) refers to these forms as *wave-current ripples* and notes that asymmetry depends upon a net sediment transport on a spatial scale larger than the forms themselves. Defining the horizontal distance from the ripple crest to the trough across the 'upstream' stoss slope as a_r and the corresponding distance across the steeper 'downstream' lee slope as b_r (so that $a_r = b_r$ in symmetrical ripples and $a_r > b_r$ as asymmetry develops) a *ripple asymmetry index* is defined as $A_r = (a_r/b_r - 1)$. Allen (1982) reports A_r from 0.04 using Inman's (1957) data to more than 2 for Tanner (1971). Allen (1982) uses a range of data to show that the ripple asymmetry index is about four times as large as the ratio of the steady or mass transport current (Eq. C1) to the peak oscillatory current calculated from linear theory (Eq. A9), and this tentative relationship is here included as Eq. E1. The shoaling tranformations and the generation of harmonics and mass transport currents described above all appear to suggest that wave formed ripples will become more asymmetric as they are traced from deep into shallow water. There has, as yet, been no discussion of the effect of undertow or long wave induced flow asymmetries on ripple generation or ripple asymmetry.

Bars and other periodic orthogonal features

On natural beaches, longshore bars are a common feature, but their appearances may vary considerably. Single or multiple formations, with longshore parallel or slightly oblique crests or crescentic features may occur both inside and outside the direct influence of the surf zone. There are, possibly, four classes of models that have been suggested to explain the development of these formations:

1 *Reflected incident wave field longshore bars.* Carter *et al.* (1973) and Lau and Travis (1973) introduced a class of model that relates bar formation to reflection of the incident wave field, although gravity wave reflection produces bar spacings that are shorter than field observations. However, the idea was extended by, for example, Short (1975) and Bowen (1980) who showed that field observations of bar spacings were in reasonable agreement with the reflected long waves described above. A further extension was made by Symonds *et al.* (1982) who suggested that variations in the position of the breakers could generate leaky long waves which could, in turn, be responsible for bar formation.

2 *Resonant interaction longshore bars.* A second class of models is based upon a resonant interaction between the incident wave field and the submerged features themselves. Examples are given by Mei (1985) who describes resonant reflection of the waves by periodic bar systems. Similarly, Boczar-Karakiewicz and Davidson-Arnott (1987) describe bar formation by the non-linear interaction of the bed features and higher order harmonics in the wave train.

3 *Edge-wave generated longshore bars.* A series of models, discussed in more detail below, attribute the length scales of various sea bed features, including longshore bars, to resonant frequencies within a standing or progressive edge wave field (e.g. Bowen & Inman, 1971; Guza & Inman, 1975; Dolan *et al.*, 1979). The theoretical exercise by Holman and Bowen (1982) convincingly shows how complex, rhythmic topographies may be explained as a result of the combination of edge wave modes.

4 *Surf zone steady current longshore bars.* Unlike the preceding mechanisms, the fourth class of model explicitly includes surf zone currents in the processes

Panel E Sedimentological equations.

$$\left(\frac{a_r}{b_r} - 1\right) = 5 \frac{\pi H}{L} \frac{1}{\sinh (kh)} \tag{E1}$$

$$\lambda_c = (g/2\pi)T^2 \sin(2n + 1)\beta \tag{E2}$$

Seaward boundary of inner rough zone:

$$\frac{\pi H}{T w_s} \frac{1}{\sinh (kh)} = \frac{252 D^{0.5}}{w_s} \tag{E3}$$

Seaward boundary of the outer planar zone:

$$\frac{\pi H}{T w_s} \frac{1}{\sinh (kh)} = \frac{0.4 L}{T w_s} \tag{E4}$$

Seaward boundary of inner offshore zone:

$$\frac{\pi H}{T w_s} \frac{1}{\sinh (kh)} = 156 T \tag{E5}$$

Seaward boundary of outer offshore zone:

$$\frac{\pi H}{T} \frac{1}{\sinh (kh)} = 33.3 \, (TD)^{0.33} \tag{E6}$$

$$\frac{dz}{dt} = -\frac{1}{\rho_s} \left(\frac{\partial i_t}{\partial x} + \frac{1}{u_s} \frac{\partial i}{\partial t}\right) \tag{E7}$$

related to bar formation. In particular, Dyher-Nielson and Sorensen (1974) qualitatively relate longshore bar formation to gradients in the radiation stress and therefore to the undertow discussed earlier. Dally and Dean (1984) develop the approach for periodic waves, whilst randon waves are included in models proposed by Stive and Battjes (1984), Stive (1986) and, with convincing support from flume experiments, in the model developed by Roelvink and Stive (1989).

Cusps and other periodic longshore features

Beach cusps are amongst the oldest known and most familiar of foreshore structures, being commonest and best developed around high tide mark. They are a series of regularly spaced, shallow embayments that face and plunge seaward and join at sharp, seaward pointing horns (Allen, 1982). Typically, cusp spacings are from as little as 0.1m (on the shores of small lakes) to as much as 50–75 m (on open ocean coasts) although they may form part of a hierarchical group

including much longer features. The literature on beach cusps is very extensive and a decent review is given by Allen (1982, pp. 455–461). Here we are particularly concerned with the proposed relationships between beach cusp wavelengths and the formation or preservation of the nearshore long wave motions described above. The idea appears to be attributable to Longuet-Higgins and Parkin (1962) and Allen (1982) compares data from Longuet-Higgins and Parker (1962) and Komar (1973) to one half of the edge wave wave length given by Eq. E2 which is the Ursell as opposed to the Eckart solution. The edge wave period is taken as an integer multiple of the incident wave period, usually a sub-harmonic at twice the incident wave period and the zeroth ($n = 0$) mode appears most likely to be excited.

Shore to shelf: a modern sedimentological transect

The last two decades have seen an increasing number of detailed field studies of nearshore sedimentology which have led to a general conceptual model for wave-dominated environments. It is clear that the process work is now approaching the sedimentological work as a series of quantifiable steps emerges from the conceptual model. The work is reviewed by, for example, Clifton and Dingler (1984), Jago and Hardisty (1984) and Shipp (1984). Initially, effort was concentrated on the definition of sub-environments by examining the texture, sedimentary structures and biogenic features of the nearshore (Howard & Reineck, 1972a,b). Later, these initial ideas were developed further by relating the near-shore sub-environments to 'an oscillatory flow regime concept' (Clifton *et al.*, 1971; Davidson-Arnott & Greenwood, 1974; Clifton, 1976; Clifton & Dingler, 1984) which essentially recognizes the increasing energy in the shorewards direction and a change from symmetrical, Airy flow offshore to increasingly asymmetrical higher order flow in the nearshore as detailed above. The sequence was depicted in Fig. 7.1 and illustrated in Fig. 7.16 and involves a shorewards hierarchy from inactivity, through symmetrical and asymmetrical ripples, cross ripples and megaripples to flat beds. Clifton *et al.* (1971) described the high energy shoreface of southern Oregon on the northwest coast of the USA, and their classification has since been extended by the other papers detailed above. Typically, five zones are recognized and the following description is based upon Reineck and Singh (1973). The sixth category has been included here to allow the transition from symmetrical to asymmetrical ripples to be discussed in detail. However, it should be noted that not all of the sub-environments appear to be present at any one location; in particular the shoremost sub-environments appear only to be well developed on a high energy coast. In general, the purpose of the following sections is to propose quantitative definitions for the stratigraphic boundaries which mark the transitions between the different sub-environments.

Inner planar zone. This zone is characterized by a planar surface, and essentially comprises the swash zone. The main bedding developed is the beach or swash laminations (i.e. evenly laminated sand). Hydrodynamically this zone is controlled by the swash processes. The seaward limit of the zone can be taken as the maximum egress of the waters' edge, and will thus be some vertical distance below the mean sea level due to both the nearshore wave characteristics, and to the long wave and set-up effects described above.

Inner rough zone. The inner planar zone is bordered by a zone showing bed irregularity. On steeper beaches this zone consists of a series of 3–10 symmetrical ripples. Such ripples are steep sided, 15–20 cm in height, 30–60 cm in length, and are composed of coarse sand or fine gravel. The crests of these ripples are rather continuous and can be followed over long distances, running parallel to the shore. The first ripple seaward of the planar zone is generally made up of the coarsest sediment. In the seaward direction the grain size of sediment composing the ripples decreases continuously, together with the ripple height. The internal structure of the ripples is complex; the foresets may dip either landward or seaward.

On gently sloping beaches the inner rough zone shows sets of depressions (troughs) which are 1–2 m across and 10–15 cm deep. The longer axis of the trough runs parallel to the shoreline. The troughs are separated by broad flat ridges. The seaward side of the troughs are steeper and coarser grained, and they slowly migrate seaward. Thus, in such regions, major bedding of the inner rough zone is trough cross-bedding, with sets varying in thickness from 4 to 100 cm, and dipping seawards. Longshore currents show maximum velocity over the inner rough zone and may modify the shape of the troughs and the direction of their migration. The steeper side shifts toward the upcurrent side of the trough and cross-bedding shows a pronounced longshore component. Under very strong longshore currents the bedding may take the form of small scale dunes with crests in a shorenormal direction. The inner rough zone is 3–10 m wide, although this zone may be completely absent on some beaches. Hydrodynamically, this zone is controlled by the interactions of the surf and the swash.

Outer planar zone. Seaward of the inner rough zone follows the outer planar zone. This occupies the outer part of the surf zone and conditions similar to upper flow regime are achieved with the sand transport being dominated by sheet flow over an essentially planar surface. Bedding in this zone is therefore horizontally laminated sand. We might, tentatively, suggest that the landward boundary of the outer planar zone corresponds to a transition from sheet flow to large-scale bedforms and thus to the limiting conditions for sheet flow described above. We now introduce a non-dimensional facies number, $F = \pi H / T w_s \sinh(kh)$, which recurs in the following sections

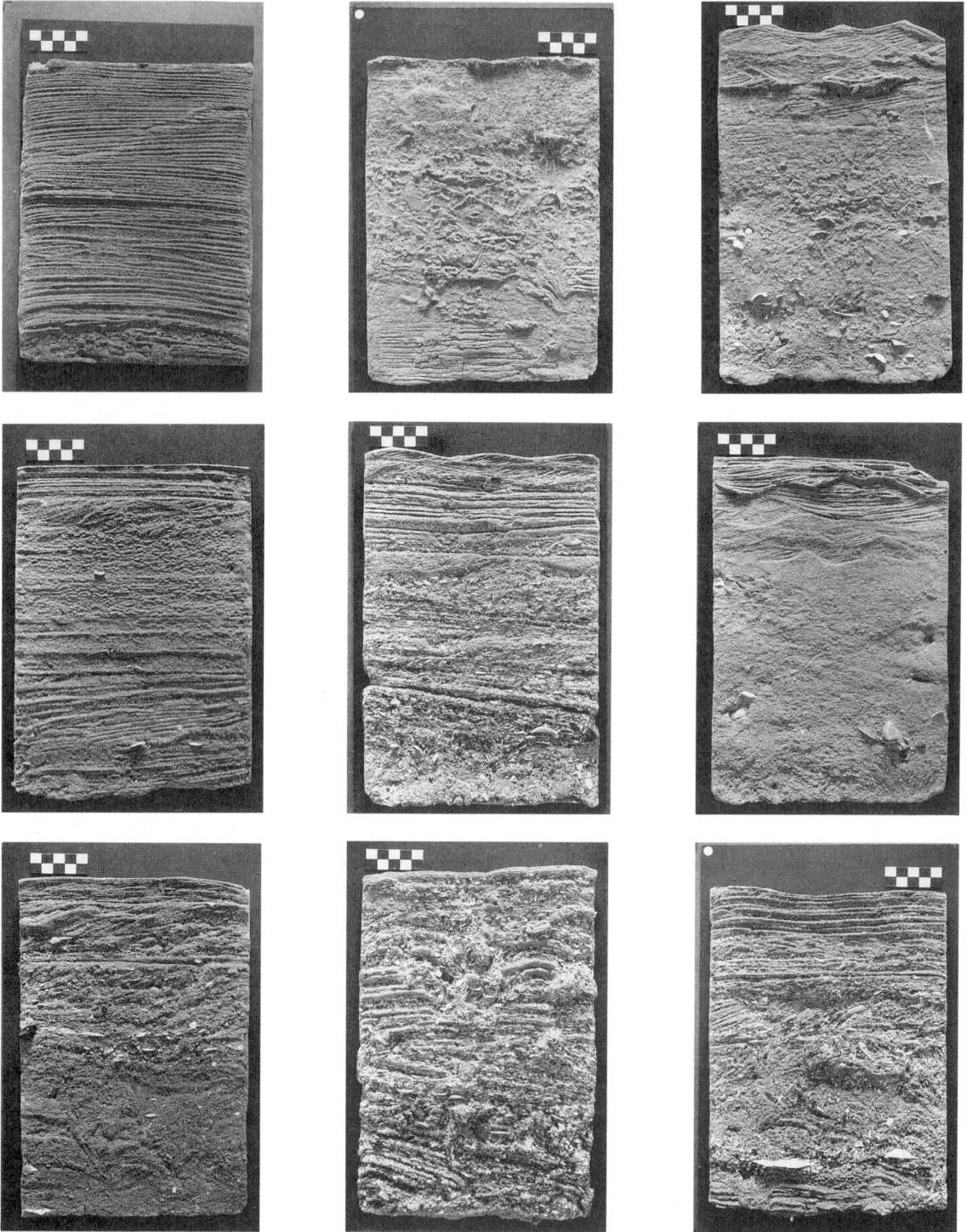

Fig. 7.16 Box cores of beach deposits, bedforms and grain size variations from the upper foreshore (left hand column) mid-foreshore (centre column) and lower foreshore (right hand column) from the macrotidal profile at Pendine Sands, SW Wales. Photographs provided by C.F. Jago.

and appears indicative of the facies boundaries. Taking $\theta \approx 1$ in Eq. D1, and substituting for the specific gravity of quartz sediment, for the gravitational acceleration, for u_m from linear theory and a wave friction factor of 0.02 for shoaling waves over sandy bottoms (Carter, 1988), then the shoreward limit of the outer planar zone can be defined in terms of the facies number by Eq. E3. Furthermore, the seaward boundary appears to correspond with the breaker zone, and therefore the simple, monochromatic breaker criterion $\gamma_b = H_b/h_b \approx 0.8$ may be utilized to define the boundary as given by Eq. E4.

Outer rough zone. The surface of this zone is covered with large-scale bedforms. For example, Clifton *et al.* (1971) reported landward facing lunate megaripples, each with a landward dipping lee side and with amplitudes of between 10 and 30 cm. They report a rapid landward migration of the megaripples, but this was clearly indicative of net beach accretion at the time of their observations rather than being representative of any quasi-steady-state condition. Given such migrations, bedding in this zone is cross-bedding with steeply dipping foresets towards land. It appears that, hydrodynamically, this zone is representative of shallow water, highly asymmetric wave action with strong oscillatory and drift current fields.

Inner offshore zone. This zone is characterized by asymmetrical wave ripples, which are 2–5 cm in height and 10–20 cm in length. Occasionally, sets of interference ripples are observed. The bedding of this zone is therefore small-scale cross-bedding. The seaward boundary of asymmetrical ripples is difficult to define since, again, it represent a point in a continuum, but in order to make progress here we take a 5% asymmetry as recognizable and defined by $(a_r - b_r) = 0.05\, b_r$, which corresponds to $A_r = 0.05$. The boundary can be determined in terms of the facies number by combining the drift and oscillatory current equations used to derive Eq. E1 with this 5% value and substituting $L = CT$, and the deep water approximation, $C = 1.56T$, into Eq. E1 to derive the result is shown as Eq. E5.

Outer offshore zone. This zone lies between the seaward limit of the asymmetrical ripple fields and wave base at which wave-induced currents become too small to transport material. Hydrodynamically, the zone is controlled by first order, essentially symmetrical oscillatory currents. The seaward

boundary of grain motion can be equated with the presence of symmetrical ripple marks, and can be determined by combining the maximum value of the linear, oscillatory current (i.e. Eq. A9 with $\cos(\omega t) = 1$) with the threshold equation for fine, clastic sediment (Eq. D4) to derive Eq. E6.

7.4.3 Ancient beach sediments

Deposition

The axiomatic, two-dimensional sediment deposition equation is (Exner, 1920; Allen, 1970, 1982) Eq. E9, where z is the elevation of the sedimentary surface, ρ_s is the sediment bulk density, i_t is the total transport rate and x and t are the spatial and temporal scales. Hardisty (1986, 1990b) refers to the left hand side, dz/dt, as the sediment response function and, in investigating geomorphological equilibrium, (i.e. $dz/dt = 0$) identifies stable and unstable states represented by positive and negative values of d^2z/dt^2, showing that the beach system, in the short time scale, exists in a state of negative feedback and therefore geomorphological stability. Sedimentologists are, however, concerned with non-zero and specifically with positive values of dz/dt, so that deposition is occurring and continues to occur over a longer period of time leading to the preservation of wave-dominated sediments in the stratigraphic record. Equation E7 shows that there must then be negative values of $\delta i_t/\delta x$ (that is a negative spatial gradient in the transport rates) or of $\delta i_t/\delta t$ (that is a negative temporal gradient in the transport rate). In general, an inspection of the various panels reveals that there are only three possibilities that lead to this result. The total, local transport rate decreases with a decrease in incident wave height, an increase in the wave radian frequency (i.e. a decrease in incident wave period) or an increase in $\sinh(kh)$ which, in turn, separates into the decrease in wave length and hence period that has already been noted or an increase in the local water depth. Essentially, deposition occurs only if wave conditions ameliorate or water level rises.

Stratigraphical examples

The beach is a key feature in stratigraphic analysis (Pettijohn *et al.*, 1987) because it is a sand generator and separates the terrestrial environments from the finer sediments offshore. Clifton *et al.* (1971) point to several examples of ancient shoreface sediments, including the Pleistocene terraces on the coast of southern Oregon and central California and the marginal marine Branch Canyon Sandstone (Hill *et al.*, 1958) in the southeastern Caliente Range of southern California. Homewood and Allen (1981) and Allen (1984) describe the Tertiary clastic sediments of the Upper Marine Molasse of western Switzerland, noting that wave ripple marks were very common in the sub-tidal shoals of the nearshore facies belt and in the sandy and pebbly sediments of the offshore facies belt.

It is therefore possible to work in a reverse direction through the various processes detailed above in order to reconstruct a palaeo-environment from measurements of sediment texture and of bedforms preserved in the stratigraphic record. Some of these steps are illustrated by Dott (1974), Clifton (1981), Homewood and Allen (1981), Allen (1981, 1984) and Clifton and Dingler (1984). For example, Clifton (1981, pp. 176–177) worked on a Miocene shoreline sandstone from California and inferred sheet flow deposition in about 8 m of water with an open ocean wave period of 8–12 s. In Devonian lacustrine sandstones of the Shetland Islands, Allen (1981) estimated water depths to be mostly about 5 m and, from a short wave period of about 3–4 s, inferred a lake about 20 km wide.

7.5   List of symbols used in the equations

a_{IG}	infragravity wave shoreline amplitude
a_r	stoss slope ripple crest to trough distance
a_s	roughness length
a_{uIG}	infragravity wave shoreline velocity amplitude
A_r	ripple asymmetry index ($= a_r/b_r - 1$)
b_r	lee slope ripple crest to trough distance
B_o	coefficient in the undertow formulae (≈ 0.07) or breaker coefficient
B_1, B_2	coefficients in the longshore current equation
c_f	drag coefficient in the longshore or bedload equations
C	wave phase celerity
C_1	ripple coefficient ($= \lambda/d_o$)
C_2	ripple coefficient ($= H_r/\lambda$)
C_a	suspended sediment reference concentration
C_g	wave group celerity
C_z	sediment concentration at height z
d_o	near-bed orbital diameter
d_{tr}	trough depth in the undertow formulae
D	grain diameter
e_b	bedload transport efficiency
E	wave energy
f	longshore transport exponent
f_w	wave friction factor
g	gravitational acceleration
$G_\omega(f)$	transport gain function
h	water depth
H	water height
H_r	ripple wave height
i_b	bedload transport rate
i_s	suspended load transport rate
i_t	total load transport rate
$i_\omega(f)$	amplitude of transport spectrum
$i_\omega 0.1$	amplitude of transport spectrum at 0.1 Hz
I_L	immersed weight longshore transport rate
I_r	Iribarrens number
J_p	Bessel function
k	wave number ($= 2\pi/L$)
k_e	edge wave number
k_L	longshore transport coefficient
k_s	roughness length
K_s	eddy diffusivity coefficient
l_s	mixing length
L	wave length
M_b	sediment mobility number
n	parameter in $C = C_g n$
p	order of the Bessel function
p_1, p_2	coefficients in the longshore current equation
P	wave power
P_L	longshore power gradient
Q_L	volumetric longshore transport rate
s	interorthogonal spacing
t	time
T	wave period
U_s	drift current velocity
$u, u(t)$	oscillatory current velocity
u_{cr}	threshold velocity
u_m	maximum oscillatory current
u_s	mean sediment velocity
u_z	orthogonal mean current speed at height z
$u_\omega(f)$	amplitude of flow spectrum
$u_\omega 10$	amplitude of flow spectrum at 10 Hz
U_m	depth mean $U(z)$ in the undertow formulae Svendsen (1984) quotes $U_m^2 \approx 0.03\, u_w^2$ where u_w is the mean oscillatory current.
$U(z)$	undertow current
v	longshore current velocity
v_o	longshore current reference velocity

V_c	combined longshore current
w_s	sediment fall velocity
x	horizontal axis
z	vertical axis
α	incident gravity wave direction
α_{1-3}	coefficients in the undertow equation
β	beach gradient
$\delta\eta$	wave set up
ε	surf scaling parameter
ϕ	Laguerre polynomial or angle of internal friction
γ_b	breaker parameter ($= H_b/h_b$)
η	surface elevation
λ	ripple wavelength
λ_c	cusp wavelength
θ	Shields parameter
ρ	density of water
ρ_s	density of sediment
τ	bed shear stress
ω	wave angular frequency ($= 2\pi/T$)
ω_o	central frequency
χ	dimensionless offshore distance

Subscripts

b	values at the breakpoint
∞	values in deep water
e	edge wave values
h	local values in water depth h
IG	infragravity wave values
n	values in frequency steps
s	significant values (e.g. the significant wave height, H_s)
cr	threshold value

7.6 References

Abou-Seida M.M. 1965. *Bed-load function due to wave action.* University of California Hydraulic Engineering Laboratory, HEL-2-11.

Airy G.B. 1845. Tides and waves. *Encyc. Metrop.,* **192**, 241–396.

Allen J.R.L. 1970. *Physical Processes of Sedimentation.* Allen and Unwin, London.

Allen J.R.L 1982. *Sedimentary Structures: their Character and Physical Basis*, Vols I & II. Elsevier, Amsterdam.

Allen P.A. 1981. Wave-generated structures in the Devonian lacustrine sediments of southeastern Shetland and ancient wave conditions. *Sedimentology,* **28**, 369–379.

Allen P.A. 1984. Miocene waves and tides in the Swiss Molasse. *Mar. Geol.,* **60**, 455–473.

Bagnold R.A. 1946. Motion of waves in shallow water.

Interaction between waves and sand bottoms. *Proc. R. Soc. Lond., A,* **187**, 1–18.

Bagnold R.A. 1954. Experiments on the gravity free dispersion of large solid spheres in a Newtonian fluid under shear. *Proc. R. Soc. Lond., A,* **225**, 49–63.

Bagnold R.A. 1963. Beach and nearshore processes, Part 1. Mechanics of marine sedimentation. In: M.N. Hill (ed.) *The Sea*, Vol. 3, pp. 507–528. Wiley-Interscience, New York.

Bagnold R.A. 1966. An approach to the sediment transport problem from general physics. *US Geol. Surv., Prof. Pap.,* **422-I**.

Bailard J.A. 1981. An energetics total load model for a plane beach. *J. Geophys. Res.,* **C86**, 10938–10954.

Bailard J.A., 1982. Modeling on–offshore sediment transport in the surf zone. *Proc. 18th Int. Conf. Coastal Eng.,* pp. 1419–1438. American Society of Civil Engineers, New York.

Bailard J.A. & Inman D.L. 1979. A re-examination of Bagnold's granular-fluid model and bedload transport equation. *J. Geophys. Res.,* **84**, 7827–7833.

Bailard J.A. & Inman D.L. 1981. An energetics bedload model for a plane sloping beach: local transport. *J. Geophys. Res.,* **86**, 2035–2043.

Battjes J.A. 1974a. *Computation of set-up, longshore currents, run-up and overtopping due to wind generated waves.* Comm. on Hydraulic., Delft University of Technology, Report 74-2.

Battjes J.A. 1974b. Surf similarity. *Proc. 14th Int. Cont. Coastal Eng.,* pp. 446–480. American Society of Civil Engineers, New York.

Beach R.A. & Sternberg R.W. 1988. Suspended sediment transport in the surf zone: response to cross-shore infragravity motion. *Mar. Geol.,* **80**, 61–79.

Beach R.A. & Sternberg R.W. 1991. Infragravity driven suspended sediment transport in the swash, inner and outer-surf zones. In: Kraus N.C., Gingerich K.J. & Kriebel D.L. (eds) *Coastal Sediments '91,* pp. 114–128. American Society of Civil Engineers, New York.

Bird E.C.F. 1984. *Coasts: an Introduction to Coastal Geomorphology.* Blackwell Scientific Publications, Oxford.

Black K.P. & Rosenberg M.A. 1991. Suspended sediment load at three time scales. In: Kraus N.C. Gingerich K.J. & Kriebel D.J. (eds) *Coastal Sediments '91,* pp. 313–327. American Society of Civil Engineers, New York.

Boczar-Karakiewicz B. & Davidson-Arnott R.G.D. 1987. Nearshore bar formation by nonlinear wave processes — A comparison of model results and field data. *Mar. Geol.,* **77**, 287–304.

Bodge K.R. 1989. A literature review of the distribution of longshore sediment transport across the surf zone. *J. Coastal Res.,* **5**, 307–328.

Bodge K.R. & Kraus N.C. 1991. Critical examination of longshore transport rate magnitude. In: Kraus N.C., Gingerich K.J. and Kriebel D.L. (eds) *Coastal Sediments '91,* pp. 139–155. American Society of Civil Engineers, New York.

Bonnefille R. & Pernecker L. 1966. Le debut d'entrainment des sediments sous action de la houle. *Bull. C.R.E.C.*, **15**, 27–32.

Bowden K.F. & White R.A. 1966. Measurements of the orbital velocities of sea waves and their use in determining the directional spectrum. *Geophys. J. Roy. Astron. Soc.*, **12**, 33–54.

Bowen A.J. 1969a. The generation of longshore currents on a plane beach. *J. Mar. Res.*, **27**, 206–215.

Bowen A.J. 1969b. Rip currents, 1: theoretical investigations. *J. Geophys. Res.*, **74**, 5467–5478.

Bowen A.J. 1980. Simple models of nearshore sedimentation; beach profiles and longshore bars. In: McCann S.B. (ed.) *The Coastline of Canada. Geol. Surv. Can. Pap.*, **80–10**, 1–11.

Bowen, A.J. & Huntley D.A. 1984. Waves, longwaves and nearshore morphology. *Mar. Geol.*, **60**, 1–13.

Bowen A.J. & Inman D.L. 1969. Rip currents, 2: laboratory and field observations. *J. Geophys. Res.*, **74**, 5479–5490.

Bowen A.J. & Inman D.L. 1971. Edge waves and cerscentic bars. *J. Geophys Res.*, **76**, 8662–8671.

Bowen A.J., Inman D.L. & Simmons V.P. 1968. Wave set-up and set-down. *J. Geophys. Res.*, **73**, 2569–2577.

Caldwell J. M. 1956. Wave action and sand movement near Anaheim Bay, California. *U.S. Army Corps. Eng., Beach Erosion Board, Tech. Memo*, **44**.

Carstens M.R., Nielson F.M. & Altinbilek H.D. 1969. Bed forms generated in the laboratory under an oscillatory flow: analytical and experimental study. *US Army Corps Eng., Beach Erosion Board, Tech. Memo*, **28**.

Carter R.W.G. 1988. *Coastal Environments*. Academic Press, London.

Carter T.G., Liu P.L.F. & Mei C.C. 1973. Mass transport by waves and offshore sand bedforms. *J. Waterways, Harbors, Coastal Eng. Div. A.S.C.E.*, **99**, 165–183.

Cavaleri L., Ewing J.A. & Smith N.D. 1978. Measurement of the pressure and velocity field below surface waves. In: *Turbulent Fluxes Through the Sea Surface: Wave Dynamics and Predictions. NATO Conf. Serv.* **V**, pp. 257–272. Plenum Press, New York.

Chan K.W., Baird M.H.I. & Round G.F. 1972. Behaviour of dense particles in a horizontally oscillating liquid. *Proc. R. Soc. Lond., A330*, 537–559.

Chatfield C. 1985. *The Analysis of Time Series*, 3rd edn. Chapman & Hall, London.

Clifton H.E. 1976. Wave-generated structures — a conceptual model. In: Davies R.A. & Ethington R.L. (eds) *Beach and Nearshore Processes. Soc. Econ. Pal. Min., Special Publication*, **24**, 126–148.

Clifton H.E. 1981. Progradational sequences in Miocene shoreline deposits, southeast Caliente Range, California. *J. Sedim. Petrol.*, **51**, 165–184.

Clifton H.E. & Dingler J.R. 1984. Wave formed structures and palaeoenvironmental reconstruction. *Mar. Geol.*, **60**, 165–198.

Clifton H.E., Hunter R.E. & Phillips R.L. 1971. Depositional structures and processes in the high-energy non-barred nearshore. *J. Sedim. Petrol.*, **41**, 651–670.

Coulomb C.A. 1773. L'Essai sur une application des regles de maximis et minimis a quelque problemes de statique relatifs a l'architecture. *Acad. R. Sci. Mem. Math. Phys.*, **7**, 343–382.

Dally W.R. & Dean R.G. 1984. Suspended sediment transport and beach profile evolution. *A.S.C.E. J. Waterways, Port, Coastal and Ocean Eng. Div.*, **110**, 15–33.

Das M.M. 1971. Longshore sediment transport rates: a compilation of data. *US Army Corps Engnrs, C.E.R.C. Misc. Pap.*, **1–71**. Vicksburg MS 39180–0631.

Davidson-Arnott R.G.D. & Greenwood B. 1974. Bedforms and structures associated with bar topography in the shallow-water wave environment. *J. Sedim. Petrol.*, **44**, 698–704.

Deigaard R., Fredsoe J. & Hedegaard I.B. 1986. Suspended sediment in the surf zone. *J. Waterway, Port, Coastal and Ocean Eng.*, **112**, 115–128.

Dingler J.R. & Clifton H.E. 1984. Tidal-cycle changes in oscillation ripples on the inner part of an estuarine sand flat. *Mar. Geol.*, **60**, 219–233.

Dingler J.R. & Inman D.L. 1977. Wave-formed ripples in nearshore sands. *Proc. 15th Int. Conf. Coastal Eng.*, pp. 2109–2126. American Society of Civil Engineers, New York.

Dolan R., Hayden B. & Felder W. 1979. Shoreline periodicities and edge waves. *J. Geol.*, **87**, 175–185.

Dott R.H. 1974. Cambrian tropical storm waves in Winsconsin. *Geology*, **2**, 243–246.

Draper L. 1979. Wave climatology of the UK continental shelf. In: Banner F.T., Collins M.B. & Massie K.S. (eds) *Northwest European Shelf Seas: The Sea Bed and Sea in Motion*, Vol. I, *Geology and Sedimentology*, pp. 353–368. Elsevier, Amsterdam.

Dyer K.D. 1986. *Coastal and Estuarine Sediment Dynamics*. Wiley, Chichester.

Dyher-Nielsen M. & Sorensen T. 1974. Sand transport phemonema on coasts with bars. *Proc. 12th Int. Conf. Coastal Eng.*, pp. 855–866. American Society of Civil Engineers, New York.

Eagleson P.S. 1965. Theoretical study of longshore currents on a plane beach. *MIT Hydrodynamic Lab., Tech. Rpt.*, **82**. Cambridge, Mass.

Eagleson P.S. & Dean R.G. 1959. Wave-induced motion of bottom sediment particles. *Proc. A.S.C.E. J. Hydraulic. Div.*, **85**, 53–79.

Eckart C. 1951. Surface waves on water of variable depth. *Wave Rpt.*, **100**. Scripps Institute of Oceanography, University of California.

Evans O.F. 1942. The relationship between the size of wave-formed ripple marks, depth of water, and the size of the generating waves. *J. Sedim. Petrol.*, **12**, 31–35.

Exner F.M. 1920. Sitzbungsber. *Akad. Wiss. Wein., Math-Naturw. Kl., Abt. 2a*, **129**, 929–952.

Galvin C.J. 1968. Breaker type classification on three laboratory beaches. *J. Geophys. Res.*, **73**, 3651–3659.

Galvin C.J. 1972. Wave breaking in shallow water. In: Meyer R.E. (ed.) *Waves On Beaches*, pp. 413–456. Academic Press, New York.

Galvin C.J. & Eagleson P.S. 1965 Experimental study of longshore currents on a plane beach. *U.S. Army C.E.R.C. Tech. Memo.*, **10**.

Glossary of Geology and Related Sciences 1972. American Geological Institute, Washington.

Goda Y. 1975. Irregular wave deformation in the surf zone. *Coast. Eng. Japan*, **18**, 13–27.

Goddet J. 1960. Etude du debut d'entrainment des materiaux mobiles sous l'action de la houle. *La Houille Blanche*, **15**, 122–135.

Green M.O. & Vincent C.E. 1991. Field measurements of time-averaged suspended-sediment profiles in a combined wave and current flow. In: Soulsby R. & Bettess R. (eds) *Sand Transport in Rivers, Estuaries and the Sea*, pp. 25–30. Balkema, Rotterdam.

Greenwood B., Osborne P.D. & Bowen A.J. 1991. Measurements of suspended sediment transport: prototype shorefaces. In: Kraus N.C., Gingerich K.J. & Kriebel D.L. (eds) *Coastal Sediments '91*, pp. 284–299. American Society for Civil Engineers, New York.

Greer M.N. & Madsen O.S. 1978. Longshore sediment transport data: a review. *Proc. 16th Int. Conf. Coastal Eng.*, pp. 1563–1576. American Society of Civil Engineers, New York.

Guza R.T. & Bowen A.J. 1975. The resonant instabilities of long waves obliquely incident on a beach. *J. Geophys. Res.*, **80**, 4529–4534.

Guza R.T. & Bowen A.J. 1977. Resonant interactions from waves breaking on a beach. *Proc. 15th Coastal Eng. Conf.*, pp. 560–579. American Society of Civil Engineers, New York.

Guza R.T. & Inman D.L. 1975. Edge waves and beach cusps. *J. Geophys. Res.*, **80**. 2997–3012.

Guza R.T. & Thornton E.B. 1980. Local and shoaled comparisons of sea surface elevations, pressures and velocities. *J. Geophys. Res.*, **85**, 1524–1530.

Guza R.T. & Thornton E.B. 1989. General measurements. In: Seymour R.J. (ed.) *Nearshore Sediment Dynamics*, pp. 51–60. Plenum Press, New York.

Haff P.K. 1991. Basic physical models in sediment transport. In: Kraus N.C., Gingerich K.J. and Kriebel D.L. (eds) *Coastal Sediments '91*, pp. 1–14. American Society of Civil Engineers, New York.

Hallermeier R.J. 1980. Sand motion initiation by water waves: two asymptotes. *Proc. A.S.C.E., J. Water. Port. Coastal. Div.*, **106**, 299–318.

Hallermeier R. 1982a. Oscillatory bedload transport: data review and simple formulation. *Cont. Shelf Res.*, **1**, 159–190.

Hallermeier R. 1982b. Bedload and wave thrust computations of alongshore sand transport. *J. Geophys. Res.*, **87**, 5741–5751.

Hanes D.M. 1986. Grain flows and bed-load sediment transport: review and extension. *Acta Mech.*, **63**, 131–142.

Hanes D.M. & Bowen A.J. 1985. A granular fluid model for steady intense bed-load transport. *J. Geophys. Res.*, **90**, 9149–9158.

Hanes D.M. & Huntley D.A. 1986. Continuous measurements of suspended sand concentration in a wave dominated nearshore environment. *Cont. Shelf Res.*, **6**, 585–596.

Hanes, D.M. & Inman D.L. 1985. Observations of rapidly flowing granular-fluid materials. *J. Fluid Mech.*, **150**, 357–380.

Hardisty J. 1983. An assessment and calibration of formulations for Bagnold's bedload equation. *J. Sedim. Petrol.*, **53**, 1007–1010.

Hardisty J. 1986. A morphodynamic model for beach gradients. *Earth Surf. Proc. Landf.*, **11**, 327–333.

Hardisty J. 1988. Measurement of shallow water wave direction for longshore sediment transport. *Geo-Marine Lett.*, **8**, 35–39.

Hardisty J. 1990a. *British Seas: an Introduction to the Oceanography and Resources of the North-west European Continental Shelf*. Routledge, London.

Hardisty J. 1990b. *Beaches: Form and Process*. Unwin-Hyman, London.

Hardisty J. 1991a. Bedload transport under low frequency waves. In: Kraus N.C., Gingerich K.J. & Kriebel D.L. (eds) *Coastal Sediments '91*, pp. 726–733. American Society of Civil Engineers, New York.

Hardisty J. 1991b. Bedload transport on a sloping surface in asymmetrically oscillating flow. In: Soulsby R. & Bettess R. (eds) *Sand Transport in Rivers, Estuaries and the Sea*, pp. 189–196. Balkema, Rotterdam.

Hardisty J. & Laver A.J. 1989. Breaking waves on a macrotidal barred beach: a test of McCowan's criteria. *J. Coastal Res.*, **5**, 79–82.

Heathershaw A.D. & Thorne P.D. 1985. Sea-bed noises reveal role of turbulent bursting phenomenon in sediment transport by tidal streams. *Nature*, **316**, 339–342.

Hill M.L., Carlson S.A. & Dribblee T.W. 1958. Stratigraphy of Cuyama Valley–Caliente Range area, California. *Bull. Am. Ass. Petrol. Geol.*, **42**, 2973–3000.

Hill P.S., Nowell A.R.M. & Jumars P.A. 1988. Flume evaluation of the relationship between suspended sediment concentration and excess boundary shear stress. *J. Geophys. Res.*, **93**, 12499–12509.

Holman R.A. 1981. Infragravity energy in the surf zone. *J. Geophys. Res.*, **86**, 6442–6450.

Holman R.A. & Bowen A.J. 1982. Bars, bumps and holes: Models for the generation of complex beach topography.

J. Geophys. Res., **87,** 457–468.

Holman R.A. & Sallenger, 1984. Longshore variability of wave run-up on natural beaches. *Proc. 22nd Int. Conf. Coastal Eng.,* pp. 1896–1912. American Society of Civil Engineers, New York.

Homewood P. & Allen P.A. 1981. Wave-, tide and current-controlled sand bodies of Miocene Molasse, western Switzerland. *Am. Ass. Petrol. Geol.,* **65,** 2534–2545.

Homma M. & Horikawa K. 1963. A laboratory study on suspended sediment due to wave action. *Proc. 10th Cong. I.A.H.R., Lond.,* 213–220.

Homma M., Horikawa K. & Kajima R. 1965. A study on suspended sediment due to wave action. *Coastal Eng. Japan,* **8,** 85–103.

Horikawa K. (ed.) 1988. *Nearshore Dynamics and Coastal Processes.* University of Tokyo Press, Tokyo.

Horikawa K. & Watanabe A. 1967. A study on sand movement due to wave action. *Coastal Eng. Japan,* **10,** 39–57.

Horn D.P. 1991. Computer simulation of shore-normal variations in sediment size. In: Kraus N.C., Gingerich K.J. & Kriebel D.L. (eds) *Coastal Sediments '91,* pp. 875–889. American Society of Civil Engineers, New York.

Horn D.P. & Hardisty J. 1990. The application of Stokes' wave theory under changing sea levels in the Irish Sea. *Mar. Geol.,* **94,** 341–351.

Howard J.D. & Reineck H.E. 1972a. Georgia coastal region, Sapelo Island U.S.A.: Sedimentology and Biology. IV. Physical and biogenic sedimentary structures of the nearshore shelf. *Senckenbergiana Marit,* **4,** 81–123.

Howard J.D. & Reineck H.E. 1972b. Georgia coastal region, Sapelo Island, U.S.A.: Sedimentology and Biology. VIII. Conclusions. *Sencken. Marit.,* **4,** 217–222.

Howard J.D. & Reineck H.E. 1981. Depositional facies of high-energy beach-to-offshore sequence: Comparison with low-energy sequence. *Am. Ass. Petrol. Geol. Bull.,* **65,** 807–829.

Hunt A.R. 1882. On the formation of ripple-mark. *Proc. R. Soc. Lond., A,* **34,** 1–19.

Huntley D.A. 1976. Long period waves on a natural beach. *J. Geophys. Res.,* **81,** 6441–6449.

Huntley D.A. 1979. Tides on the North West European continental shelf. In: Banner F.T., Collins M.B. & Massie K.S. (eds) *The North West European Shelf Seas: the Sea Bed and the Sea in Motion,* Vol. II *Physical and Chemical Oceanography, and Physical Resources,* pp. 301–352. Elsevier, Amsterdam.

Huntley D.A. & Bowen A.J. 1973. Field observations of edge waves. *Nature,* **243,** 160–162.

Huntley D.A. & Bowen A.J. 1975. Field observations of edge waves and their effect on beach material. *J. Geol. Soc. Lond.,* **131,** 68–81.

Huntley D.A. & Hanes D.M. 1987. Direct measurements of suspended sediment transport. *Proc. Coastal Sediments '87,* pp. 723–737. American Society of Civil Engineers, New York.

Huntley D.A., Guza R.T. & Thornton E.B. 1981. Field observations of surf beat, 1, progressive edge waves. *J. Geophys. Res.,* **83,** 1913–1920.

Ijima T., Sato S., Aono H. & Ishii K. 1960. Wave and coastal sediment characteristics at Fukue coast, Atsumi Bay. *Proc. 7th Int. Conf. Coastal. Eng.,* pp. 69–79. American Society of Civil Engineers, New York.

Ijima T., Sato S. & Tanaka N. 1964. On the coastal sediment of Kashima Harbour coast. *Proc. 11th Conf. Coastal. Eng.,* pp. 175–180. American Society of Civil Engineers, New York.

Inman D.L. 1957. Wave generated ripples in nearshore sands. *US Army Corps Eng. Beach Erosion Board, Tech. Memo.,* **100.**

Inman D.L. & Bagnold R.A. 1963. Littoral processes. In: M.N. Hill (ed.) *The Sea.* Vol 3, pp. 529–554. Interscience. Wiley, New York.

Inman D.L. & Quinn W.H. 1952. Currents in the surf zone. *Proc. 2nd Int. Conf. Coastal Eng.,* pp. 24–36. American Society of Civil Engineers, New York.

Ippen A.T. & Kulin G. 1955. The shoaling and breaking of the solitary wave. *Proc. 5th Int. Conf. Coastal Eng.,* pp. 27–49. American Society of Civil Engineers, New York.

Iribarren R. & Nogales C. 1949. Protection des ports. *17th Int. Naval Cong., Lisbon,* Section II–4, 31–82.

Ishihara T. & Sawaragi T. 1962. Laboratory studies on sand drift, the critical velocity and the critical water depth for sand movement and the rate of transport under wave action. *Coastal Eng. Japan,* **5,** 59–65.

Jagger K.A. & Hardisty J. 1991. Higher frequency acoustic measurements of course bedload transport. In: Kraus N.C., Gingerich K.J. & Kriebel D.L. (eds) *Coastal Sediments '91,* pp. 2187–2198. American Society of Civil Engineers, New York.

Jago C.F. & Barusseau J.P. 1981. Sediment entrainment on a wave-graded shelf, Roussillon, France. *Mar. Geol.,* **42,** 279–299.

Jago C.F. & Hardisty J. 1984. Sedimentology and morphodynamics of a macrotidal beach, Pendine Sands, SW Wales. *Mar. Geol.,* **60,** 123–154.

Jenkins G.M. & Watts D.G. 1968. *Spectral Analysis and its Applications.* Holden-Day, San Francisco.

Johnsson I.G. 1966. Wave boundary layers and friction factors. *Proc. 10th Int. Conf. Coastal. Eng.,* 127–148.

Johnsson J.W. & Eagleson P.S. 1966. Coastal processes. In: Ippen A.T. (ed.) *Estuary and Coastline Hydrodynamics,* pp. 405–493. McGraw Hill, New York.

Kalkanis G. 1964. Transportation of bed material due to wave action. *US Army C.E.R.C., Tech. Memo.,* **2.**

Kemp P.H. 1960. The relation between wave action and beach profile characteristics. *Proc. 7th Int. Coastal Eng. Conf.,* pp. 262–276. American Society of Civil Engineers, New York.

Kemp P.H. 1975. Wave assymmetry in the nearshore zone and breaker area. In: Hails J.R. and Carr A. (eds)

Nearshore Sediment Dynamics and Sedimentation, pp. 47–67. Wiley, London.

Kemp P.H. & Plinston D.T. 1968. Beaches produced by waves of low phase difference. *J. Hyd. Div. Am. Soc. Civ. Eng.*, **94**, 1183–1195.

Kemp P.H. & Plinston D.T. 1974. Internal velocities in the uprush and backwash zone. *Proc. 14th Int. Coastal Eng. Conf.*, pp. 575–585. American Society of Civil Engineers, New York.

Kennedy J.F. & Falcon M. 1965. Wave-generated sediment ripples. *MIT Hydrodynamics Laboratory Rpt.*, **86**.

Kennedy J.F. & Locher F.A. 1972. Sediment suspension by water waves. In: Meyer R.E. (ed.) *Waves on Beaches and Resulting Sediment Transport*, pp. 249–295. Academic Press, New York.

King C.A.M. 1972. *Beaches and Coasts*, 2nd edn. Arnold, London.

King D.B. 1991. The effect of beach slope on oscillatory flow bedload transport. In: Kraus N.C., Gingerich K.J. & Kriebel D.L. (eds) *Coastal Sediments '91*, pp. 734–744. American Society of Civil Engineers, New York.

Kinsman B. 1984. *Wind Waves*. Dover, New York.

Kishi T. & Saeki H. 1967. The shoaling breaking of the solitary wave on impermeable rough slopes. *Proc. 10th Int. Conf. Coastal Eng*, pp. 27–49. American Society of Civil Engineers, New York.

Komar P.D. 1973. Computer models of delta growth due to sediment input from rivers and longshore transport. *Geol. Soc. Am. Bull.*, **84**, 2217–2226.

Komar P.D. 1975. Nearshore currents: generation by obliquely incident waves and longshore variations in breaker heights. In: Hails J.R. & Carr A. (eds) *Nearshore Sediment Dynamics and Sedimentation*, pp. 17–45. Wiley, London.

Komar P.D. 1976. *Beach Processes and Sedimentation*. Prentice-Hall, New Jersey.

Komar P.D. 1983. Nearshore currents and sand transport on beaches. In: John B. (ed.) *Physical Oceanography of Coastal and Shelf Seas*, pp. 67–109. Elsevier, Amsterdam.

Komar P.D. & Inman D.L. 1970. Longshore sand transport on beaches. *J. Geophys. Res.*, **76**, 713–721.

Komar P.D. & Miller M.C. 1973. The threshold of sediment movement under oscillatory water waves. *J. Sedim. Petrol.*, **43**, 1101–1110.

Komar P.D. & Miller M.C. 1975a. Sediment threshold under oscillatory waves. *Proc. 14th Int. Conf. Coastal Eng.*, 756–775. American Society of Civil Engineers, New York.

Komar P.D. & Miller M.C. 1975b. On the comparison of the threshold of sediment motion under waves and unidirectional currents with a discussion of the practical evaluation of the threshold. *J. Sedim. Petrol.*, **45**, 362–367.

Koonitz W.A. & Inman, D.L. 1967. A multipurpose data acquisition system for field and laboratory instrumentation of the nearshore environment. *US Army Coastal Engineering Research Centre, Tech. Memo.*, **21**.

Lakhan V.C. 1989. Computer simulation of the characteristics of shoreward propagating deep and shallow water waves. In: Lakhan V.C. & Trenhaile A.S. (eds) *Applications in Coastal Modelling*, pp. 107–158. Elsevier, Amsterdam.

Larras J. 1956. Effets de la houle et du clapotis sur les fonds de sable. *IV Journ. Hydraulic Rep.*, **9**. Paris.

Lau J. & Travis B. 1973. Slowly varying Stokes waves and submarine longshore bars. *J. Geophys. Res.*, **78**, 4489–4497.

Leeder M.R. 1982. *Sedimentology: Process and Product*. George Allen and Unwin, London.

Lenhoff L. 1982. Incipient motion of sediment particles. *Proc. 18th Int. Conf. Coastal Eng.*, American Society of Civil Engineers, New York.

Longuet-Higgins M.S. 1970a. Longshore currents generated by obliquely incident sea waves, 1. *J. Geophys. Res.*, **75**, 6778–6789.

Longuet-Higgins M.S. 1970b. Longshore currents generated by obliquely incident sea waves, 2. *J. Geophys. Res.*, **75**, 6790–6801.

Longuet-Higgins M.S. & Parkin D.W. 1962. Sea waves and beach cusps. *Geog. J.*, **128**, 194–201.

Longuet-Higgins M.S. & Stewart R.W. 1962. Radiation stress and mass transport in gravity waves, with application to 'surf-beats'. *J. Fluid Mech.*, **13**, 481–504.

Longuet-Higgins M.S. & Stewart R.W. 1964. Radiation stress in water waves, a physical discussion with applications. *Deep-Sea Res.*, **75**, 6790–6801.

Losada M.A. & Gimenez-Curto L.A. 1981. Flow characteristics on rough permeable slopes under wave action. *Coast Eng.*, **4**, 187–206.

MacDonald T.C. 1973. Sediment transport due to oscillatory waves. *Univ. Calif. Hyd. Lab. Tech. Rep.*, **HEL–2–39.**

Madsen O.S. 1991. Mechanics of cohesionless sediment transport in coastal waters. In: Kraus N.C., Gingerich K.J. & Kriebel D.L. (eds) *Coastal Sediments '91*, pp. 15–27. American Society of Civil Engineers, New York.

Madsen O.S. & Grant W.D. 1976. Sediment Transport in the Coastal Environment. *M.I.T. Ralph M. Parsons Lab. Report*, **209**.

Manohar M. 1955. Mechanics of bottom sediment movement due to wave action. *US Army Corps Eng., Beach Erosion Board, Tech. Memo.*, **75**.

McCowan J. 1894. On the highest wave of permanent type. *Phil. Mag.*, **38**, 351–357.

Mei C.C. 1985. Resonant reflection of surface waves by periodic sand bars. *J. Fluid Mech.*, **152**, 315–335.

Miller M.C. & Komar P.D. 1980. Oscillation ripples generated by laboratory apparatus. *J. Sedim. Petrol.*, **50**, 173–182.

Mogridge G.R. & Kamphuis J.W. 1973. Experiments on bed form generation by wave action. *Proc. 13th Int. Conf. Coastal Eng.*, 1123–1142.

Munk W. 1949. Surf beats. *EOS Transcripts*, **30**, 849–854.

Nielsen P. 1986. Suspended sediment concentration under waves. *Coastal Eng.*, **10**, 1–21.

Passega R., Rizzini A. & Borghetti G. 1967. Transport of sediment by waves, Adriatic coastal shelf, Italy. *Bull. Am. Assoc. Pet. Geol.*, **51**, 1304–1319.

Pethick J.S. 1984. *An Introduction to Coastal Geomorphology*. Edward Arnold, London.

Pettijohn F.J., Potter P.E. & Siever R. 1987. *Sand and Sandstone*, 2nd edn. Springer, New York.

Putnam J.A., Munk W.H. & Traylor M.A. 1949. The prediction of longshore currents. *Trans. Am. Geophys. Uni.*, **30**, 337–345.

Rance P.J. & Warren N.F. 1969. The threshold movement of coarse material in oscillatory flow. *Proc. 11th Int. Conf. Coastal Eng.*, 487–491.

Randall R.B. 1987. *Frequency Analysis*. Bruel and Kjaer, Naerum, Denmark.

Rayleigh L. 1877. On progressive waves. *Proc. Lond. Math. Soc.*, **9**, 21–26.

Reineck H.-E & Singh I.B. 1973. *Depositional Sedimentary Environments*. Springer, New York.

Reynolds O. 1877. On the rate of progression of groups of waves and the rate at which energy is transmitted by waves. *Nature*, **36**, 343–344.

Roelvink J.A. & Stive M.J.F. 1989. Bar-generating cross-shore flow mechanisms on a beach. *J. Geophys. Res.*, **94**, 4785–4800.

Russell P., Davidson M., Huntley D.A., Cramp A.D., Hardisty J. & Lloyd G. 1991. The British beach and nearshore dynamics programme. In: Kraus N.C., Gingerich K.J. & Kriebel D.L. (eds) *Coastal Sediments '91*, pp. 371–384. American Society of Civil Engineers, New York.

Russell R.C.H. & Osario J.D.C. 1958. An experimental investigation of drift profiles in a closed channel. *Proc. 6th Int. Conf. Coastal Eng.*, pp. 171–183. American Society of Civil Engineers, New York.

Sato S. & Kishi T. 1954. Shearing force on sea bed and movement of bed material due to wave motion. *J. Res. Public Works Res. Inst.*, **1**, 1–11.

Sato S., Ijima T. & Tanaka N. 1962. A study of the critical depth and mode of sand movement using radioactive glass sand. *Proc. 8th Int. Conf. Coastal Eng.*, pp. 304–323. American Society of Civil Engineers, New York.

Savage R.P. 1959. Laboratory study of the effect of groynes in the rate of littoral transport. *US Army Corps. Eng., Beach Erosion Board, Tech. Memo.*, **75**.

Savage R.P. 1962. Laboratory determination of littoral transport rates. *J. Water. Harb. Div. A.S.C.E.*, **88**, 69–92.

Scott T. 1954. Sand movement by waves. *US Army Corps Eng., Beach Erosion Board, Tech. Memo.*, **48**.

Seymour R.J. (ed.) 1989. *Nearshore Sediment Transport*. Plenum Press, New York.

Shepard F.P. 1973. *Submarine Geology*, 3rd edn. Harper and Row, San Francisco.

Shibayama T. & Horikawa K. 1982. Sediment transport and beach transformation. *Proc. 18th Int. Coastal Eng. Conf.*,

pp. 1439–1458. American Society of Civil Engineers, New York.

Shipp R.C. 1984. Bedforms and depositional sedimentary structures of a barred nearshore system, eastern Long Island, New York. *Mar. Geol.*, **60**, 235–259.

Shore Protection Manual 1984. Coastal Engineering Research Center, Dept of the Army, Vicksburg, Mississippi. Vols. I & II.

Short A.D. 1975. Multiple offshore bars and standing waves. *J. Geophys. Res.*, **80**, 3838–3840.

Silvester R. & Mogridge G.R. 1970. Reach of waves to the bed of the continental shelf. *Proc. 12th Int. Conf. Coastal Eng.*, pp. 487–491. American Society of Civil Engineers, New York.

Simpson J.H. 1969. Observations on the directional characteristics of waves. *Geophys. J. R. Astron. Soc.*, **17**, 93–120.

Sleath J.F.A. 1978. Measurement of bedload in oscillatory flow. *J. Water. Harb. Coastal Ocean Div. A.S.C.E.*, **104**, 291–307.

Sleath J.F.A. 1982. The suspension of sand by waves. *J. Hydraulic Res.*, **20**, 439–452.

Sleath J.F.A. 1984. Sea Bad Mechanics. Wiley, New York. 355 pp.

Stive M.J.F. 1986. A model for cross shore sediment transport. *Proc. 20th Int. Conf. Coastal Eng.*, pp. 1550–1564.

Stive M.J.F. & Battjes J.A. 1984. A model for offshore sediment transport. *Proc. 19th Int. Conf. Coastal Eng.*, pp. 1420–1436. American Society of Civil Engineers, New York.

Stive M.J.F. & Wind H.G. 1986. Cross-shore mean flow in the surf zone. *Coastal Eng.*, **10**, 325–340.

Stokes G.G., 1874. On the theory of oscillatory waves. *Trans. Camb. Phil. Soc.*, **8**.

Suhayda J.N. 1974. Standing waves on beaches. *J. Geophys. Res.*, **79**, 3065–3071.

Suhayda J.N. & Pettigrew N.R. 1977. Observations of wave height and celerity in the surf zone. *J. Geophys. Res.*, **82**, 1419–1429.

Sunamura T., Bando K. & Horikawa K. 1978. An experimental study of sand transport mechanisms and rate over assymmetrical ripples. *Proc. 25th Japanese Conf. Coast. Eng.*, pp. 250–254. Japanese Society of Civil Engineers.

Svendsen I.A. 1984. Mass flux and undertow in a surf zone. *Coastal Eng.*, **8**, 347–365.

Svendsen I.A., Schaffer H.A. & Hansen J.B. 1987. The interaction between the undertow and the boundary layer flow on a beach. *J. Geophys. Res.*, **92**, 11845–11856.

Sverdrup H.U. & Munk W.H. 1946. Theoretical and empirical relations in forecasting breakers and surf. *Trans. Am. Geophys. Union*, **27**, 828–836.

Swart D.H. 1974. A schematization of onshore–offshore transport. *Proc. 14th Int. Conf. Coastal Eng.*, pp. 884–900. American Society of Civil Engineers, New York.

Symonds G., Huntley D.A. & Bowen A.J. 1982. Two-dimensional surf beat: Long wave generation by a time-varying breakpoint. *J. Geophys. Res.*, **87**, 492–498.

Tanner W.F. 1971. Numerical estimates of ancient waves, water depth and fetch. *Sedimentology*, **16**, 71–88.

Thornton E.B. 1971. Variation of longshore current across the surf zone. *Proc. 12th Int. Conf. Coastal Eng.* pp. 291–308. American Society of Civil Engineers, New York.

Thornton E.B. & Guza R.T. 1982. Energy saturation and phase speeds measured on a natural beach. *J. Geophys. Res.*, **87**, 9499–9508.

Thornton E.B. & Krapohl R.F. 1974. Water particle velocities measured under ocean waves. *J. Geophys. Res.*, **79**, 847–852.

Tucker M.J. 1950. Surf beats: sea waves of 1 to 5 min period. *Proc. R. Soc. Lond., A*, **202**, 565–573.

Tucker M.J., Carr A.P. & Pitt E.G. 1983. The effect of an offshore bank in attenuating waves. *Coastal Engng*, 7, 133–144.

Vincent C.E. 1957. Contribution to the study of sediment transport on a horizontal bed due to wave action. *Proc. 6th Int. Conf. Coastal Eng.*, pp. 326–355. American Society of Civil Engineers, New York.

Vincent C.E. & M.O. Green 1990. Field measurements of the suspended sand concentration profiles and fluxes and of the resuspension coefficient γ_0 over a rippled bed. *J. Geophys. Res.*, **95**, 11591–11601.

Vincent C.E., Young R.A. & Swift D.J.P. 1981. Bedload transport under waves and currents. *Mar. Geol.*, **39**, 71–80.

Watanabe A. 1982. Numerical models of nearshore currents and beach deformation. *Coastal Engng. in Japan*, **25**, 147–161.

Watts G.M. 1953. A study of sand movement at South Lake Worth Inlet, Florida. *US Army Corps Eng., Beach Erosion Board, Tech. Memo.*, **42**.

Whitehouse R.S.J. 1991. Slope-inclusive bedload transport: experimental assessment and implications for models of bedform development. In: Soulsby R. & Bettess R. (eds) *Sand Transport in Rivers, Estuaries and the Sea*, pp. 215–222. Balkema, Rotterdam.

Wiegel R.L. 1964. *Oceanographical Engineering*. Prentice-Hall, New Jersey.

Wright L.D., Chappel J., Thom B.G., Bradshaw M.P. & Cowell P. 1979a. Morphodynamics of reflective and dissipative beach and inshore systems: southeastern Australia. *Mar. Geol.*, **32**, 105–140.

Wright L.D., Nielsen P., Short A.D. & Green M.O. 1982. Morphodynamics of a macrotidal beach. *Mar. Geol.*, **50**, 97–128.

8 Deep sea processes of sediment transport and deposition

D.A.V. STOW

8.1 Introduction

8.1.1 Definitions

For the purposes of this chapter, we can define all sedimentary environments that occur below storm wave base (approximately 200 m water depth) as deep sea. This provides us, therefore, with a very large playground in which to operate, including all areas of oceans and marginal seas beyond the shelf-slope break, and hence beyond the reach of most processes discussed in other chapters in this book. The shelf edge typically occurs at depths of about 100–200 m, although locally it may be as shallow as 20 m and so well within the influence of waves and tidal currents. Enclosed and semi-enclosed basins, on broad shelf areas or in the central parts of fjords, may also provide sites of sedimentation below wave base.

Collectively, these areas present a remarkable variability in: (1) *physiography* — including narrow deep-sea trenches, broad flat abyssal plains, gentle gradients of open slopes, steep margins off reef complexes and volcanic seamounts, and irregular topography associated with mid-ocean ridge zones; and (2) *tectonic setting* — including active convergent and oblique slip margins such as those surrounding much of the Pacific Ocean, mature divergent margins that typify both the west and east Atlantic, arc-related and more complex marginal basins such as the Mediterranean Sea, and young rift basins such as the Red Sea. Consequently, there is an equal range of bathymetry and scale, as well as variety of processes that operate and factors that control them.

For clastic sedimentary particles to accumulate in the deep sea they must be eroded from land or from the sea floor, be transported beyond the shelf-slope break, and then deposited. Much biogenic material is synthesized directly in the oceans and then subjected to similar transport–depositional processes, although biogenic detritus may also be derived by erosion. Authigenic minerals are precipitated directly from sea water at or near the sediment–water interface, and they too may be subsequently reworked.

In this chapter, we first of all consider the origin and supply of sedimentary materials to the deep-sea realm, the fluids and forces that act within this milieu, and then define the range of processes that operate to redistribute the materials supplied. We then discuss each of the main processes in turn, the typical bedforms they generate and characteristic properties of the sediments deposited. A final section outlines the main allocyclic and autocyclic controls that influence these processes.

The data presented draw heavily on previous syntheses of deep-water processes (e.g. Stow, 1985, 1986, 1992; Pickering *et al.*, 1989) as well as the most recent literature on the subject.

8.1.2 Sediment supply

The primary source of terrigenous material in deep-sea deposits is the physical and chemical weathering and erosion of pre-existing rocks. The nature and rate of supply will be affected principally by tectonic, climatic and sea-level controls in the source and transitional regions. Collectively, the world's rivers transport by far the largest volume of suspended and dissolved loads to the sea, most of the former being deposited initially in paralic and shallow shelf environments. At high latitudes, melting and calving from glaciers and floating ice adds directly to both shelf and deep-sea deposition. Eolian processes can

Table 8.1 Sediment flux to the ocean basins

Source	Supply $\times 10^9$ t year^{-1}
Rivers (suspended load)	18.3
Rivers (solution load)	4.2
Groundwater (solution load)	0.48
Ice (ice shelves and bergs)	3.0
Coastal erosion	0.25
Wind-blown dust	0.6
Volcanic ejecta	0.15
Biogenic carbonate	1.4
Biogenic silica	0.49

Data from Goldberg (1974) and Open University (1984).

Table 8.2 Principal properties of sea-water

Salinity:	
General average	35‰
Surface waters	32.4–39.8‰
Deep water (>1000 m)	34.5–35‰
Temperature	
General average	3.52°C
Surface range	−1.87°–30.0°C
Deep water (>1000 m)	−3.56°–8.9°C
Density (expressed as specific gravity)	1.024–1.029
pH	
General average	8.0

contribute large volumes of finer grained material derived from low latitude arid and semi-arid regions as well as from explosive volcanic eruptions.

The relative flux from these different sources is poorly constrained at present (Table 8.1) and will have varied markedly in the past.

Biogenic material, including calcareous and siliceous tests as well as soft organic tissue, is supplied from primary productivity in surface waters of the oceans and seas, from organic growth in shoal areas followed by erosion and resedimentation, and as runoff from biogenic-rich areas of the continents and coastline (e.g. swamps and rainforests, river biota, shoreface molluscs, etc.). Calculating the flux of this material is even more complex than for terrigenous flux, as so much is dissolved and recycled, often many times over, before reaching the deep-sea floor.

The oceans are truly a chemical soup with an ionic composition built up from the dissolved loads of fluid input, dissolution of biogenic material following death of the organism, progressive corrosion of unstable clastic components such as ferromagnesian minerals, feldspars and amorphous compounds, and primary volcanic/hydrothermal emissions from mid-ocean ridges. There appears to be a crude steady state chemistry within the oceans, involving large material flux in and out of solution.

8.1.3 Fluids and forces

Sea-water properties

The ambient fluid in the oceans is sea water, having physical and chemical characteristics that lie within certain well-defined limits (Table 8.2).

Although it is not possible to isolate the deep sea as a separate fluid, when the average mixing time of the oceans is in the order of 1500 years, the oceans can be considered in terms of distinct water masses that have recognizable and measurable parameters and that show progressive mixing across their boundaries.

Physical forces

The physical forces that have most affect on deep-sea water, both within and between the separate water masses, include: (1) temperature and salinity differences that drive the thermohaline circulation; (2) internal tides and waves; (3) the Coriolis force that acts on moving water masses; (4) sea surface topography coupled with major storm events that transmit surface potential energy downwards through the water column; and (5) sediment suspensions held in semi-permanent nepheloid layers and in episodic sediment gravity flows that move by virtue of the downward acting gravitational force.

The sea-water/sediment interface acts as a critical transition zone in the deep sea, and is very variable from one area to another. In some mud-rich areas it has a soft spongy texture with 70–80% water content and full interchange with the superjacent fluid; in others, it has been swept and winnowed repeatedly by strong bottom currents leaving a coarse sand or gravel lag substrate. Incipient cementation may produce hard grounds on carbonate bottoms, whereas precipitation of ferromanganese compounds may result in nodules and pavements.

The sediments below the sea floor act as an enormous reservoir for porewaters (approximately equal to 20% of the ocean volume), and show continuous interchange with the ocean reservoir across the water/sediment boundary.

The initiation of movement in various particulate materials on the sea floor, both from the margins into the deep sea and within the ocean basins, can be considered in terms of (1) the mechanics of sediment failure on a slope; or (2) the critical shear velocity required to erode and move sedimentary materials on a plane bed.

In the first case, sediment deposited on a slope will only begin to move downslope when the shear stress exerted by the force of gravity exceeds the shear strength of the sediment (Watkins & Kraft, 1978; Karlsrud & Edgers, 1982) along a slip plane within the sediment column (Fig. 8.1(a)). The shear strength is a function of the cohesion between the grains plus the intergranular friction. Sediment failure therefore results either from an increase in shear stress, due to a steepening of the slope or thickening of the sedi-

ment pile, or from a decrease in shear strength due to the sudden shock of earthquakes, storms, etc., causing fluidization or thixotrophy in the sediment. The weight of rapidly deposited sediments may exert a similar strain effect.

In the second case, sediment lying on a plane bed will begin to move as the fluid shear stress is increased and the critical threshold for grain movement is reached. Each sediment grain will experience a drag force due to the fluid shear velocity at the bed and a lift force due to the Bernoulli effect. When these combined fluid forces exceed the normal weight force due to gravity the grain will begin to move (Fig. 8.1(b)). Storms, internal waves, normal bottom currents and turbidity currents can all initiate sediment movement in this way.

Various experimental investigators have attempted to determine the threshold of motion for different grain types and sizes. The much used Hjulstrom (1939) diagram relating erosion of a particular grain size to current velocity (Fig. 8.1(c)) is poorly established and incorrect for grain sizes finer than

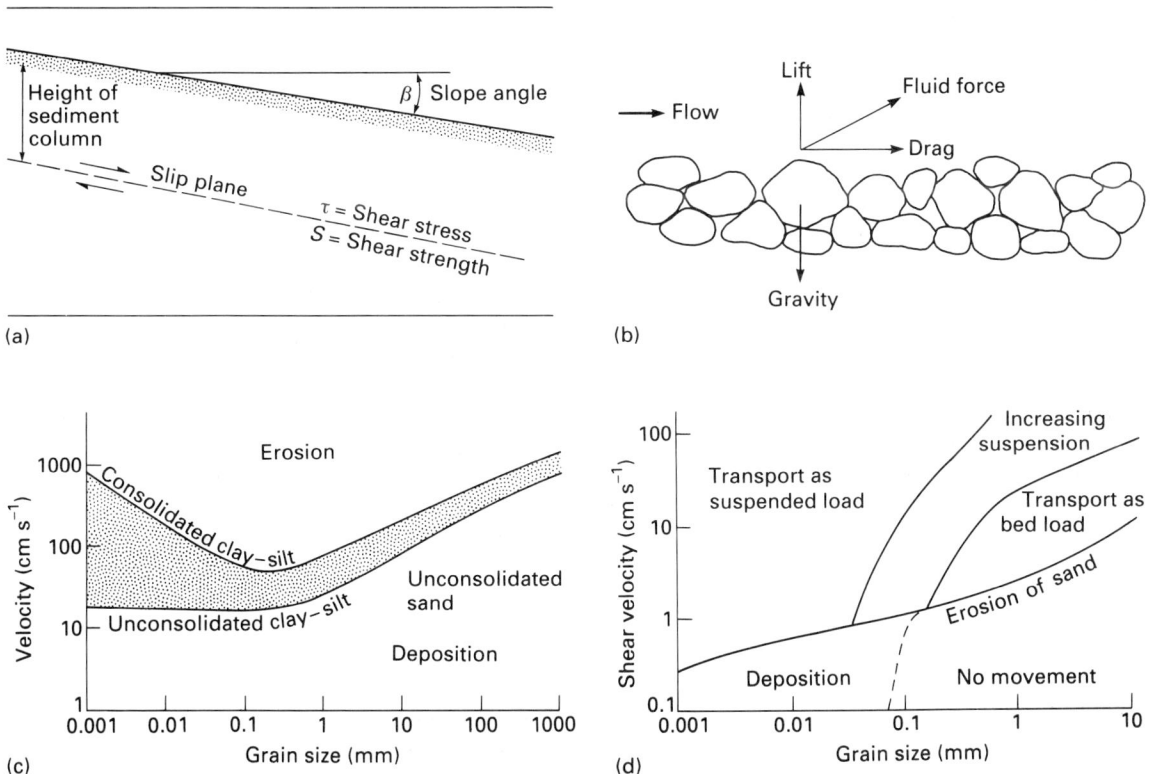

Fig. 8.1 Erosion, transport and deposition of sedimentary materials. (Stow, 1986).

sands. Shields (1936) related the grain Reynolds number to a dimensionless shear stress, but also had few data for the finer grain sizes, whereas Miller *et al.* (1977) present much better data in this size range. A more recent synthesis (McCave, 1984) plots grain size against shear velocity showing transport–deposition fields for fine sediment and transport–erosion fields for coarser materials (Fig. 8.1(d)). McCave argues that erosion of fine cohesive sediment is not simply a function of grain size and velocity and cannot therefore be plotted on the same diagram.

Once in the water column and on the move, then a range of different forces and processes acts to further transport and eventually deposit sedimentary material. These processes are discussed in detail in the following sections.

Chemical forces

Various chemical forces must also be taken into account as they can very significantly affect certain types of deep-sea sediment. Below the carbonate compensation depth (CCD), bottom waters are highly aggressive (i.e. having low pH values) towards both biogenic and inorganic carbonate material such that, unless deposition and burial is extremely rapid, all $CaCO_3$ will be dissolved either in the water column and/or on the sea floor. Only an insoluble silicate residue will be preserved as sediment.

The depth of the CCD in modern oceans is a function of biogenic productivity, oceanic circulation and the geochemical cycle of carbon. It averages 5.3 km in the North Atlantic, decreasing to 4.4 km in the North Pacific, and has varied markedly in the past. The aragonite compensation depth (ACD) is generally about 1 km shallower.

Biogenic opaline silica is chemically unstable throughout the water column but both rapid sedimentation and protection by a thin organic film surrounding the siliceous tests allows its accumulation in certain parts of the deep sea. Most organic matter is rapidly oxidized and recycled during its descent through the water column but preservation in the sediment can result from very high rates of production and/or the development of zones with anoxic (or very low oxic) bottom waters. This generally acts to suppress biological degradation of organic material.

Different chemical environments in the deep sea can also lead to authigenic precipitation of certain minerals including clays, zeolites, ferromanganese crusts, phosphates and carbonate hardgrounds.

Biological forces

Biological forces that influence deep-water sediments either prior to and/or post-deposition include: (1) primary productivity, which has a direct affect on sediment supply and on chemical partitioning in the oceans; (2) bacterial decay processes, which serve to alter the organic component of sediment once deposited and hence also affect the pore water chemistry; and (3) the activity of burrowers and borers on and within sediments, which mix and homogenize, stir and suspend or break down and erode almost any substrate that is left uncovered for even relatively short periods of time.

The range of processes

Within the deep sea, three main groups of processes are capable of eroding, transporting and depositing terrigenous, biogenic, volcanigenic and other particulate materials (Fig. 8.2): *resedimentation processes, bottom currents* and *surface currents with pelagic settling.* For completeness we have added *authigenic processes* on Fig. 8.2, recognizing the *in situ* origin of some deep-sea materials. Several attempts have been made to classify these processes, so that a plethora of terminology and confusing (partial) synonyms exists (see review by Nardin *et al.*, 1979). The classification in Table 8.3 is based on the mechanical behaviour of the flow, the transport mechanisms and sediment support system (Dott, 1963; Middleton & Hampton, 1976; Moore, 1977; Lowe, 1979; Nardin *et al.*, 1979; Pickering *et al.*, 1989).

The 15 conceptually distinct processes listed (not including authigenic processes), are in fact part of a continuum of mechanical behaviour, ranging from elastic through plastic to viscous fluid and viscous settling (Fig. 8.3). The transition from slides to sediment gravity flows involves a change in the physical state of the sediment mass towards greater internal disaggregation by breakdown of the metastable grain packing and incorporation of more fluid. The transition from debris flow to liquefied or fluidized flows and turbidity currents involves further remoulding and dilution of the flow. During any single event of transport and deposition (Fig. 8.3) these various processes may operate at the same time or in temporal sequence, as demonstrated experimentally (Middleton, 1967) and from field evidence (Hein, 1982).

The extreme end member of sediment gravity flows, a very low-concentration, low-velocity turbidity current will be deflected by the Coriolis force from its downslope path to a direction along the slope. At this point it may grade imperceptibly into a normal bottom current, also known as a contour current, which is driven by the deep thermohaline circulation in the oceans (Stow & Lovell, 1979) rather than by the gravitational effect of its sediment load. Other bottom currents (Fig. 8.2) are also caused by normal oceanic circulation, and all behave as viscous fluids. When there is no horizontal advection and dilution is extreme, simple vertical settling of particles occurs. The influence of sediment-free or non-depositing bottom currents may be important in supplying dissolved chemical species for authigenic mineral growth in some instances.

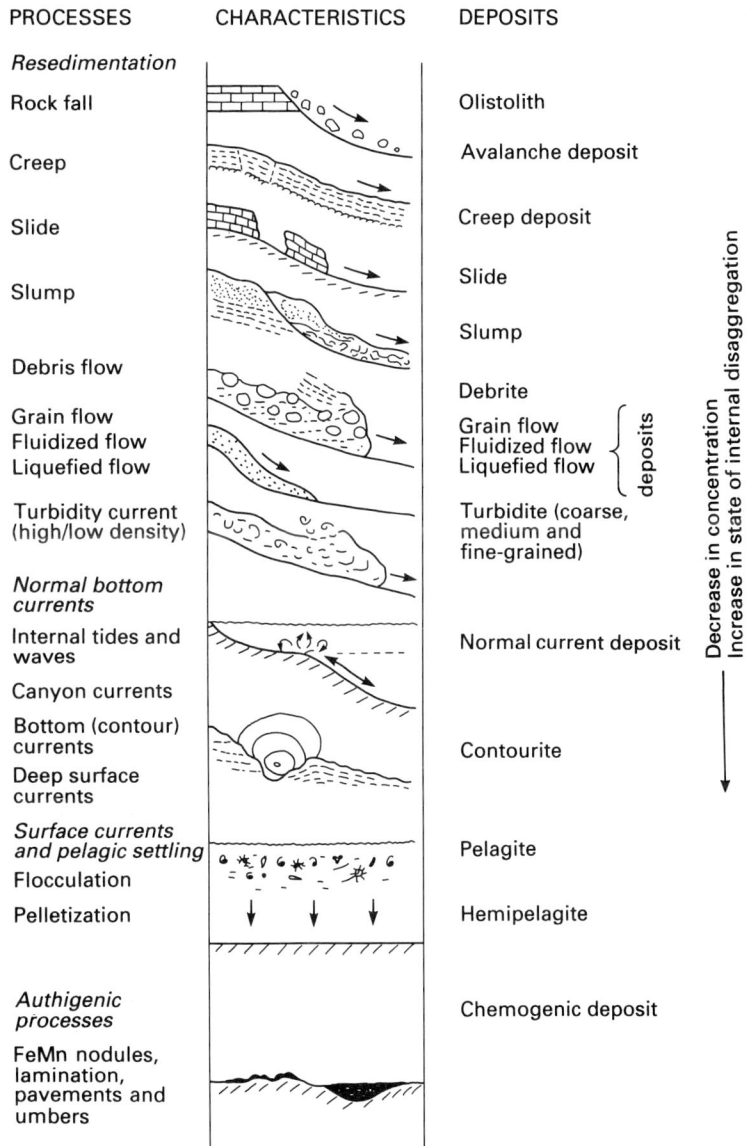

Fig. 8.2 The range of processes that operate in the deep sea. (Stow, 1986.)

PROCESSES

Resedimentation

Rock fall

Creep

Slide

Slump

Debris flow

Grain flow
Fluidized flow
Liquefied flow

Turbidity current
(high/low density)

Normal bottom currents

Internal tides and waves

Canyon currents

Bottom (contour) currents

Deep surface currents

Surface currents and pelagic settling

Flocculation

Pelletization

Authigenic processes

FeMn nodules, lamination, pavements and umbers

CHARACTERISTICS

DEPOSITS

Olistolith

Avalanche deposit

Creep deposit

Slide

Slump

Debrite

Grain flow
Fluidized flow } deposits
Liquefied flow

Turbidite (coarse, medium and fine-grained)

Normal current deposit

Contourite

Pelagite

Hemipelagite

Chemogenic deposit

Decrease in concentration
Increase in state of internal disaggregation

Table 8.3 Definitions of depositional processes in the deep sea (modified from Nardin et al., 1979) and estimates of their chief physical characteristics. Reprinted from Stow (1986)

Depositional process	Transport and sediment support mechanisms	Slope	Dimensions	Concentration	Velocity (cm s⁻¹)	Duration	Transport distance (km)	Average sedimentation rate
RESEDIMENTATION†								
Rock fall	*Elastic** Freefall and rolling of blocks and clasts, no internal deformation of clasts	Very steep	Clasts can be >10 m	Solid	Freefall	?min to h	<0.5	High
Sediment creep	Slow strain and downslope movement along decollement zone due to load-induced stress with little internal deformation	Gentle	20–80 m thick	'Solid'	V. slow (imperceptible)	Semi-continuous	?<0.5	As for background
Slide (glide)	Shear failure along discrete shear planes with little internal deformation	>About 1°	Max. 300 km³, 500 m thick (+ complete range)	Almost 'solid'	?	?h	0.001–?100	High
Slump	Shear failure accompanied by rotation along discrete shear surfaces	>About 1°	As above	Almost 'solid'	?	?h	0.001–?100	High
Debris flow (mudflow)	*Plastic** Shear distributed throughout sediment mass, slow plastic flow, clast buoyancy and matrix strength support mechanisms	>About 1°	Up to few 10s of m thick	Dense slurry	?1–20	?h	?Max 350	Moderate to high
Grain flow	*Viscous fluid (flow)** Quasi-visco-plastic flows of cohesionless grains, dispersive pressure support mechanism, localized, small-scale events	>18°	Up to few cm thick	Few data available	Few data available	?min to h	?<0.1	Do not usually operate as separate processes
Fluidized flow	High-viscosity, short-lived flow of cohesionless grains, supported by upward-moving pore waters	>3°	<10 cm thick	Few data available	Few data available	?min to h	?<0.1	Do not usually operate as separate processes
Liquefied flow	Cohesionless sediment supported by upward-escape of pore waters as flow collapses and freezes, very short-lived	>About 0.5°	Basal few 10 cm of flow	Few data available	Few data available	?min to h	?<0.05	Do not usually operate as separate processes

	Mechanical behaviour	Slope	Dimensions	Concentration	Velocity	Duration	Horizontal transport	Sedimentation rate
Turbidity current (high density)	Low-viscosity flow of mixed grains supported by fluid-turbulence (autosuspension)	>About 0.5°	Length and width up to 10s of km, thickness up to 100 s of m	50–250 g l⁻¹	Max. 250	?h to about 1 day	Up to about 1000	<5 cm to >5 m per 1000 years
Turbidity current (low density)	Very low-viscosity flow of mixed grains supported by fluid turbulence (autosuspension)	Almost no slope	As above	0.025–3 g l⁻¹	Average 10–50	?h to few days	Up to several 1000s	<5 cm to >5m per 1000 years
NORMAL BOTTOM CURRENTS‡								
Internal tides and waves	Medium to large-scale oscillations at density discontinuities within upper few hundred metres of water column, can suspend sediment by fluid turbulence	No slope	Up to few 10s of m amplitude	?	5–300	Semi-continuous currents often with marked periodicities	?	V. low
Normal canyon currents	Essentially 'clear-water' flows, up and down slope canyons and channels, tidal or higher periodicity, minor sediment suspension by fluid turbulence	Up and down slope <few°	Up to few 10s of m thick	?<0.3 mg l⁻¹	0–30	Semi-continuous currents often with marked periodicities	Up to several 100s	Low
Bottom (contour) currents	Deep, slow, essentially 'clear-water' flows driven by thermohaline circulation, can be associated with bottom nepheloid suspensions (fluid turbulence)	No slope or gentle slopes	Width up to few 10s of km, thickness up to 100s of m	0.025–0.25 mg l⁻¹	Max. 200 Mean 10	Semi-continuous currents often with marked periodicities	Up to several 1000s	<10cm per 1000 years
Deep surface currents	Deep, slow, essentially 'clear-water' flows that are deep parts of surface wind-driven ocean currents	No slope or gentle slope	As above	?As above	?As above	Semi-continuous currents often with marked periodicities		?As above
PELAGIC SETTLING Pelagic settling	*Viscous fluid** Vertical settling of individual grains, flocs and pellets through water column (viscous fluid)	Ubiquitous	Settling through 100s to 1000s of m water column	Extremely low	0.002–0.005 settling rate (or more if flocs)	Semi-continuous	No horizontal transport	Mean <1 cm per 1000 years

* Mechanical behaviour; † resedimentation (= mass gravity transport); ‡ normal bottom currents (= semi-permanent bottom currents).

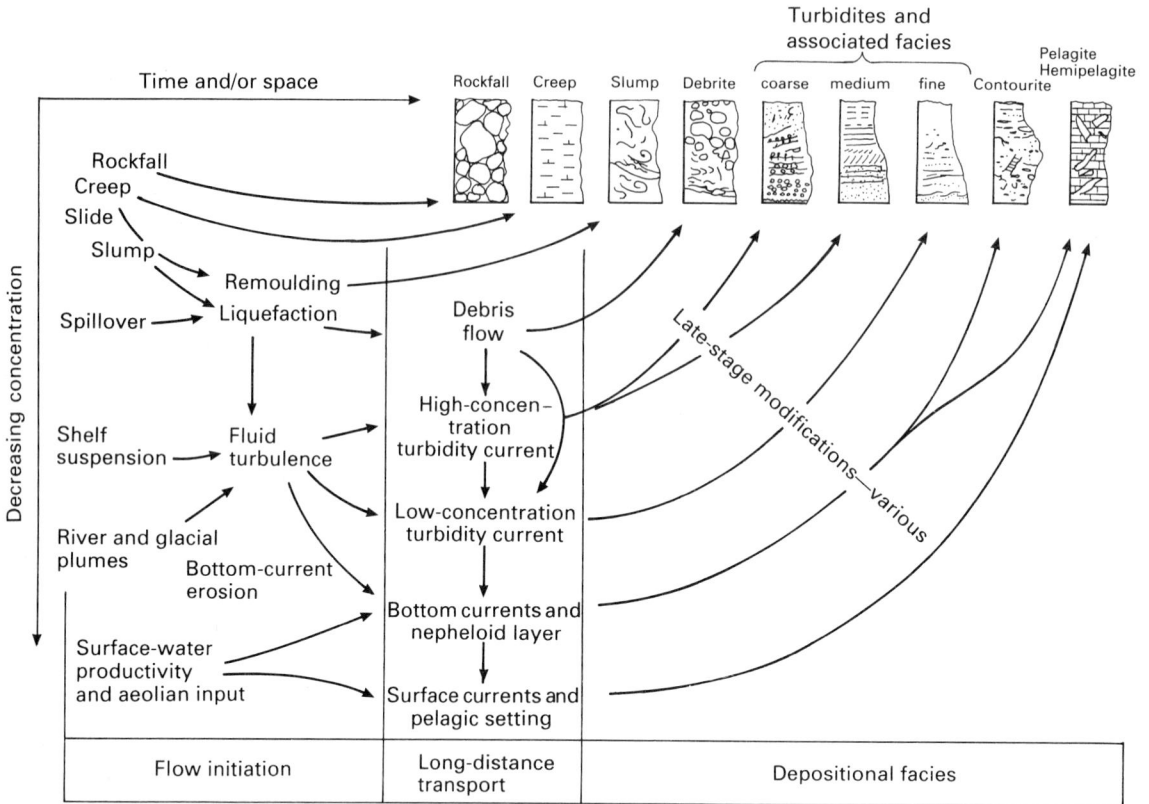

Fig. 8.3 Deep-sea process interaction and continuum in relation to main sedimentary deposits. Framework is one of time and/or space, and concentration of flows. Idealized facies models that result from deposition by the different processes are also shown. Post-depositional modification can involve current reworking, liquefaction and bioturbation. (From Pickering, Stow *et al.*, 1986; modified after Walker, 1978.)

8.2 Resedimentation processes

8.2.1 Introduction

Resedimentation processes (including both *mass gravity transport* and *sediment gravity flow*) are all those processes that move sediment downslope over the sea floor from shallower to deeper water and are driven by gravitational forces (Fig. 8.2, Table 8.3). Many of these processes have close analogues with downslope processes on land and with pyroclastic processes associated with explosive volcanic eruptions, both of which have been the subject of intensive study by geomorphologists, engineers and volcanologists, as well as by sedimentologists. Our understanding of the deep-sea equivalents is being slowly pieced together through laboratory experiments, a growing body of direct observational data and by careful scrutiny of the resultant deposits.

It is clear that many of the resedimentation processes are extremely complex in nature and do not lend themselves readily to mathematical modelling, although we have tried to present the current state-of-the-art in the following pages. Three important properties we should mention at the outset are mechanical behaviour, cohesiveness of the debris involved and sediment-support mechanisms.

Mass movements such as rockfall, creep and slides, are characterized by generally elastic behaviour in which strain is directly proportional to stress and material displacement occurs without significant internal deformation. Debris flows behave plastically, with shear deformation being distributed throughout the moving wet sediment mass. Sediment gravity flows, including grain flow, fluidized flow, liquefied flow and turbidity currents, exhibit viscous fluid behaviour, ranging from very high viscosity (quasi-viscous) to very low viscosity flows.

In nature, the material involved may be entirely granular or cohesionless (e.g. sand, gravel) or dominantly cohesive (e.g. mud, clay), and this will influence the style of behaviour and movement. Muddy sediment is more likely to creep or slide downslope than move as a rockfall. Sandy debris flows, exhibiting mainly frictional strength and grain collision support mechanisms, will flow less far than muddy (cohesive) debris flows. Of course, many natural flows display a mixed granular–cohesive behaviour.

Downslope movement or flow of sediment–water mixes will continue provided that the shear stress generated by movement exceeds frictional resistance to flow, and that the grains or clasts are inhibited from settling by one of several support mechanisms (Pickering et al., 1989). These mechanisms include: (1) turbulence of the fluid, at high Reynold's numbers and Froude numbers greater than unity; (2) buoyancy provided by a high density flow matrix; (3) dispersive pressure generated from grain collisions; (4) trapped or escaping pore fluids creating an upward fluid drag; and (5) internal strength, either frictional or cohesive, of the moving sediment mass. Natural resedimentation events commonly display an interaction of support mechanisms (Table 8.3).

8.2.2 Rock falls and avalanches

Rock falls are sudden, rapid, freefall events that are common in mountainous areas on land or along sea cliffs but are relatively rare at sea because the slopes are mostly too gentle. They occur only on steep slopes of faulted or carbonate margins or in the heads of deeply incised submarine canyons, and are initiated by undercutting and erosion, and by earthquake shocks. Single displaced clasts (olistoliths) may be very large (> 10 m) and bounce or roll downslope for several tens or hundreds of metres before coming to rest (Johns, 1978). Talus slopes are common off reef mounds and poorly sorted, loose carbonate debris is well known from many ancient and modern carbonate margins where the slope angle is around 30° or more (Kenter, 1990). A combination of rock fall, rock avalanche, sliding and grain flow is believed responsible for such deposits.

Shor et al. (1990) have described gravel/boulder scree deposits from submersible dives along a steep, eroded, thalweg margin within the main Laurentian Fan channel. These were derived from loose gravel wave deposits on the main floor of the incised channel.

8.2.3 Sediment creep

Sediment creep is a process of slow strain due to load-induced stress (or the downslope weight of sediment), that may extend over periods ranging from hours to thousands of years (Watkins & Kraft, 1978). It is most commonly observed as *soil creep* down hillslopes on land, for which Carson and Kirkby (1972) recognize two classes of mechanism.

Continuous or rheological creep occurs as a result of the constant breaking and reforming of the tiny electrochemical bonds between individual particles. The cumulative effect of this is very slow downslope creep. *Intermittent* creep (usually diurnal or seasonal) is the result of physical or biological displacement of particles, at first normal to the slope and then with a downslope readjustment. Heating and cooling, wetting and drying, freezing and thawing are all mechanisms that cause soil creep, but seem unlikely to act on deep-water sediments. Burrowing organisms, coupled with internal slippage, suspension transport within the soil of the finest particles, and rheological creep most probably dominate on submarine slopes.

With so many mechanisms combining to determine the final sediment movement, it is difficult to derive a mathematical model for creep. However, Allen (1985) derived a model for soil creep in which the creep rate, U, at depth, y, is given by:

$$U = U_o - \frac{(\gamma_g \sin \beta)}{2\eta_a} y^2, \qquad (8.1)$$

where U_o is the surface creep rate, γ is the soil bulk density, β is the slope angle, η_a is the apparent viscosity of soil (held constant).

Allen (1985) found that the velocity profile given by Eq. 8.1 was not especially like observed profiles. He therefore introduced an exponential increase in the apparent viscosity with depth and derived a model giving more realistic results, noting that some movement was predicted at all depths.

Interestingly, Hill et al. (1982) interpreted seismic records of a compressionally folded, surface unit on the Canadian Beaufort Sea slope as indicating downslope creep of the upper few tens of metres with displacement at depth being taken up by movement along an internal decollement zone (Fig. 8.4).

Few other direct observations of submarine sediment creep have been made so that the processes, rates of movement, thickness of section involved, and the conditions that most favour creep are little

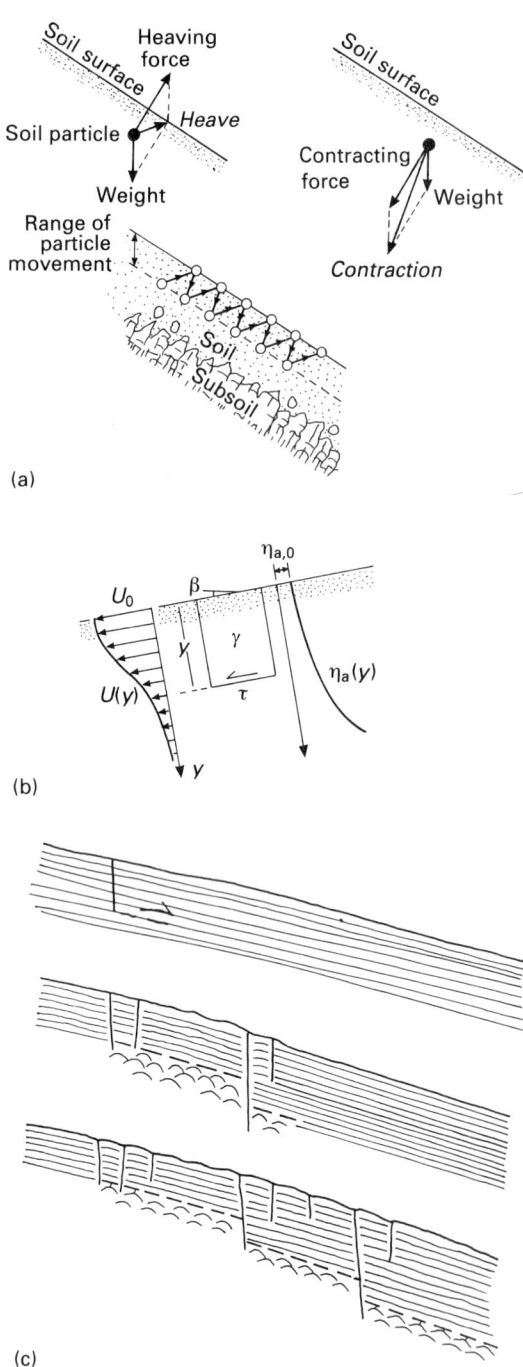

(a)

(b)

(c)

Fig. 8.4 Models for sediment creep on a gentle submarine slope. (a) Movement of soil particles during soil creep; (b) definition of terms for soil creep (see text for explanation); (c) progressive development of sediment creep along a 30–50 m subsurface decollement zone. (After Allen, 1985 and Stow, 1986.)

known. However, it is believed to be a widespread and very significant process in the deep sea.

At high ratios of shear stress to shear strength, creep deformation may accelerate rapidly to creep rupture and may thus act as a precursor to slide or slump failure.

8.2.4 Sliding and slumping

Sliding and slumping are more or less synonomous terms that describe downslope displacement of a semi-consolidated sediment mass along a basal shear plane while retaining some internal (bedding) coherence. Sliding emphasizes the lateral displacement along either simple *translational* or slightly *rotational* shear planes with little internal disturbance, whereas slumping emphasizes the internal disturbance and folded shear planes. These processes are very widespread on slopes of all gradients greater than about $0.5°$ and range in volume from less than $1 m^3$ to over $100 km^3$ and can be several hundreds of metres thick (Morgenstern, 1967; Saxov & Nieuwenhius, 1982). Embley (1980) claims that at least 40% of the continental rise off eastern North America is covered by a veneer of mass-flow deposits (slides and debrites).

Sediment instability on slopes is affected by numerous interacting variables (Schwarz, 1982; Lee, 1989) including: (1) the slope angle; (2) high rates of sedimentation, leading to high water content and low shear strength; (3) repeated cyclic stress, commonly caused by seismicity but also influenced by oceanographic factors such as currents and internal waves; (4) high primary productivity and/or bottom water anoxicity, both leading to high organic carbon content in the sediment; and (5) generation of gas in the sediment due to clathrate decomposition and organic matter decay.

Assessing sediment stability on slopes has generally used a static infinite slope model (Moore, 1961; Morgenstern, 1967) in which the lateral extent of the slope is much greater than the thickness of the sediment. This model expresses a safety factor, *SF*, as the ratio of the resisting force to the shear force, given by:

$$SF = \left[\frac{1 - \Psi}{\rho_s g' Z \cos^2\beta} \right] \left[\frac{\tan \phi}{\tan \beta} \right], \qquad (8.2)$$

where Ψ is excess pore fluid pressure, Z is depth below the surface, ρ_s is sediment density, ρ is water density, g' is $g\int(\rho_s - \rho)/\rho_s$ in reduced gravitational constant, β is the slope angle, ϕ is the angle of internal friction.

It has been shown that the probability of failure is

low for $SF > 1.3$, and failure is almost certain for $SF < 0.9$ (Athanasiou-Grivas, 1978). Hein and Gorsline (1981) concluded from their study of geotechnical characteristics of slope sediments in the Californian Continental Borderland that sedimentation rates in excess of 30 mg cm^{-2} year^{-1} must be attained before slope failures become common.

A large slide on a gentle slope typically has the morphology shown in Fig. 8.5 (Lewis, 1971; Dingle, 1977). The head is characterized by tensional structures such as faults, slump scars and bed deficiency. Above the head area retrogressive slumping may have occurred, involving successive sediment failure and the upslope progradation of unstable slump scar surfaces. The main body of the slide mass can be relatively undisturbed or contain several distinct slump blocks, whereas the toe area displays compressional structures such as thrusting and overriding of beds.

The complexity of slide features on modern slopes has been amply demonstrated by recent studies (Lee,

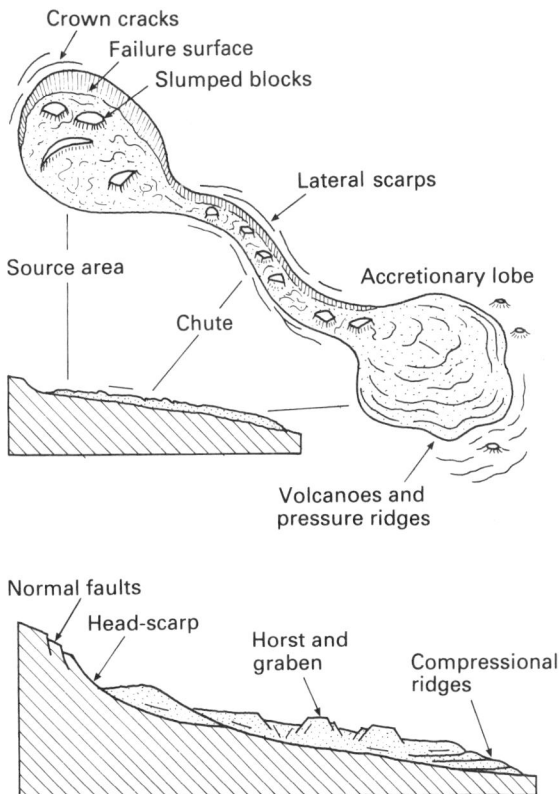

Fig. 8.5 Characteristics of sediment slides. (After Allen, 1985).

1989; Barnes & Lewis, 1991; Simm *et al*, 1991), which clearly show the gradation that exists in nature between slide, debris flow and turbidity current processes and products.

8.2.5 Debris flows

Debris flows are highly concentrated, highly viscous, sediment dispersions that possess a yield strength and display plastic flow behaviour (Johnson, 1970; Hampton, 1972). They are slurry-like or glacier-like, slow laminar flows that advance down slopes in excess of only 0.5°, either continuously or intermittently. They are best known from modern sub-aerial settings (Pierson, 1981; Takahashi, 1981; Johnson, 1984), where they are capable of carrying boulders up to 2.7×10^6 kg, may move at speeds of 20 m s^{-1} and have bulk densities in the range of $2–2.5 \times 10^3$ kg m^{-3}. Only a small amount (about 5%) of interstitial matrix (i.e. mud and water) is required to allow flow over even gentle slopes (Hampton, 1975, 1979).

Numerous recent surveys of marine slopes have shown the widespread importance of debris flows in moving large volumes of material, often many tens or hundreds of kilometres downslope. Catastrophic debris flows may transport enormous slabs up to about 2.3×10^9 kg (immersed weight, Marjanac, 1985).

Debris flows move when the critical yield strength is exceeded and deformation (flow) begins in a basal zone of highest shear stress. Here there is a complex region of sliding, rolling, bouncing and laminar flow. Higher in the flow, where shear stresses are less, the material may be rafted along as a semi-rigid plug, Deposition occurs by downward thickening of this plug until the entire mass freezes. Flow margin freezing may also occur and result in the construction of debris levees.

The motion of debris flows is perhaps even more difficult to model than that of slides. Lowe (1982) advocates a strictly cohesive model based on Bingham plastic behaviour, Johnson (1970) favoured either a Bingham-plastic or a Coulomb-viscous rheological model, whereas Takahashi (1981) proposed a 'dilatant-fluid' rheological model based entirely on dispersive pressure. Allen (1985) has used the Bingham-plastic model to derive a mathematical model in which flow velocity, U, is given by:

$$U = \frac{1}{\eta_a} \left[\frac{(\gamma - e)g \sin \beta}{4} (R^2 - r^2) - \tau_{yd}(R - r) \right], \quad (8.3)$$

where η_a is the apparent viscosity (Bingham viscosity), γ is the bulk density of the debris flow, e is the density of the fluid medium, β is the slope angle, R is the channel radius in which the debris flow moves, r is the radius of the rigid plug of debris flow, τ_{yd} is the plastic yield strength.

However, it appears most likely that natural debris flows display a wide range of rheological properties, being more or less cohesive-plastic, viscous-fluid and granular-collisional in behaviour (Shultz, 1984; Pickering *et al.*, 1989). The clast support mechanisms would also vary, including a combination of buoyancy, frictional strength, matrix strength, elevated pore pressures of the matrix, and dispersive pressure. Pickering *et al.* (1989) also discuss the possibility of some very large flows being turbulent in part.

The principal features of submarine debris flows include: (1) a head region or gathering area; (2) a channelized portion that may develop debris levees and raft large blocks in a central plug zone, and (3) a lobe/sheet depositional area. Commonly, the front of the flow forms a steep scarp up to 30 m or more in height, but on steeper slopes the flow is thinner, more rapid and has a lower elevation to the mud nose (Fig. 8.6). As debris flows advance downslope they load the underlying deposits and may induce secondary failure on the sea bed. They can also give rise to slumping where either the nose of the flow or the slope of the sea bed becomes oversteepened.

8.2.6 Grain flow, liquefied flow, fluidized flow

In an early classification scheme for sediment gravity flows (Middleton & Hampton, 1973; Middleton & Southard, 1984), four process end members were recognized: debris flows, liquefied flows (originally called fluidized flows), grain flows and turbidity currents. These were recognized as idealized, and it has become clear that the only sediment–gravity processes capable of long-distance transport over relatively gentle slopes are debris flows and turbidity currents. However, both these processes may involve or pass through stages of grain flow and liquefied/fluidized flow.

True grain flows are quasi-visco-elastic flows characterized by grain-to-grain collisions that result in a dispersive pressure support mechanism (Bagnold, 1954). They require slopes in excess of about 18° and so are probably a very localized process in the deep sea, perhaps occurring as small-scale sand avalanches in the heads of submarine canyons (Shepard & Dill, 1966). From an analysis of the mechanisms of grain flow, Lowe (1976) demonstrated a near-parabolic

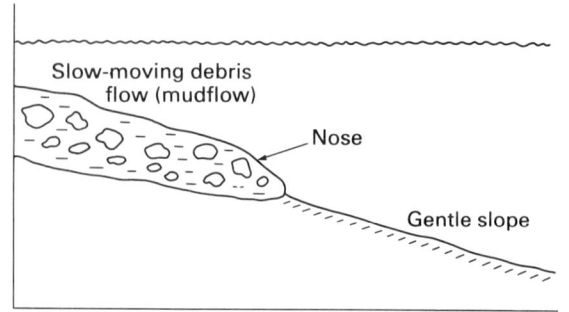

(a)

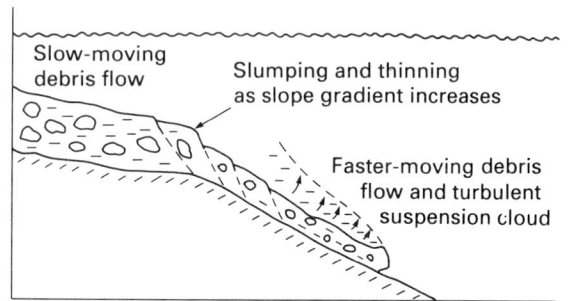

(b)

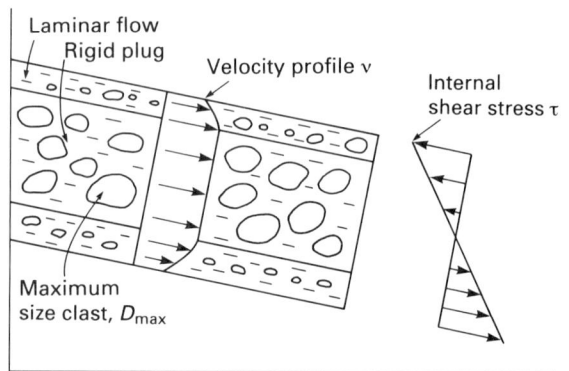

(c)

Fig. 8.6 Model and characteristics of sub-aqueous debris flows. Shear strength, $S = c + \sigma \tan \phi$; flow initiation, $\tau \geqslant S$. Steady flow, $\tau = S + \eta$; competence, $D_{max} = 8.8c/g(\rho_s - \rho_f)$. (After Stow, 1986.)

velocity profile with a thin surficial plug of non-shearing grains moving passively above an active shear plane. Lowe further concluded that sandy grain flows cannot be thicker than a few centimetres and so cannot be solely responsible for the deposition of thick massive sand beds.

Liquefied and fluidized flows are related processes that involve the collapse of a metastable fabric and partial or full grain support by upward-moving pore fluids. The grains become suspended and the sediment strength is reduced to zero. Loosely packed silt and sand are especially susceptible to fluidization, whereas gravel is usually too porous and in muds the cohesive forces resist fluidization (Lowe, 1975, 1979; Middleton & Hampton, 1976). Fluidized sand behaves like a fluid of high viscosity and can flow rapidly down slopes in excess of 2–3°. The excess pore fluid pressures dissipate quickly, from minutes to a few hours, depending on flow thickness and grain size. Deposition occurs through a short period of liquefied flow in which the grains settle rapidly and the flow freezes bottom to top.

8.2.7 Turbidity currents

Definitions and flow initiation

Turbidity currents are a type of density current or gravity current (Simpson, 1982) in which the denser fluid is a relatively dilute suspension of sediment. Many other types of density current exist in nature, including *overflows* such as plumes of fresh mud-laden water at river mouths, *underflows* in the atmosphere (sea-breeze fronts and dust storms), powder-snow avalanches, pyroclastic base-surges and *nuées ardentes* emanating from volcanic eruptions, and *vertical density flows* caused by plumes of volcanic ash falling through the atmosphere and oceans (Allen, 1985).

The turbidity current is the main type of sediment-driven underflow in the oceans, seas and lakes, the most important and perhaps best known of the resedimentation processes. Although much studied in the laboratory and theoretically, a full-sized prototype has never yet been observed in nature. The evidence for their occurrence comes from observations of density underflows in lakes, incipient, small, dilute turbidity currents in the heads of canyons, and the sudden, unexpected but sequential breaking of submarine telegraph cables (Heezen *et al.*, 1954; Krause *et al.*, 1970). From the almost ubiquitous

occurrence of their characteristic deposits, turbidites, they are known to occur widely throughout the deep sea.

Depending on the manner in which the flow is initiated and on subsequent sediment supply, two main types of turbidity current can be identified: relatively short-lived *surge-type* flows, and relatively long-lived *steady-* or *uniform-type* flows. Flow initiation occurs in one of four main ways: (1) from the transformation of slumps or debris flows by mixing with seawater; (2) from sand-spillover, grain flows or rip-currents feeding sediments into the heads of submarine canyons; (3) by storm stirring of unconsolidated bottom sediments and the build-up of a concentrated shelf nepheloid layer; and (4) directly from suspended sediments delivered to the sea by rivers in flood or by glacial meltwaters.

Mathematical models

In reality, turbidity currents are non-uniform, unsteady, non-linear, free-boundary flows, that move by virtue of dispersed sediment, which may both be deposited during flow and eroded from the substrate over which the flow passes. With these characteristics, Allen (1985) has argued, it is not yet possible to construct a comprehensive theoretical model, although attempts have been made to explore certain aspects of turbidity flow under severe constraints. For 'uniform' flows (Fig. 8.7), Middleton (1966a) proposed a Chezy-type equation, in which the velocity of the body of the turbidity current, U_B, is given by:

$$U_B^2 = \left[\frac{8g}{(f_o+f_i)}\right]\left[\frac{\Delta\rho}{(\rho+\Delta\rho)}\right]d_B\tan\beta, \quad (8.4)$$

where $\Delta\rho$ is the density difference between the flow and seawater, ρ is the density of seawater, d_B is the thickness of body, β is the slope angle, f_o is the dimensionless friction coefficient at the base of the flow, f_i is the dimensionless friction coefficient at the top of the flow.

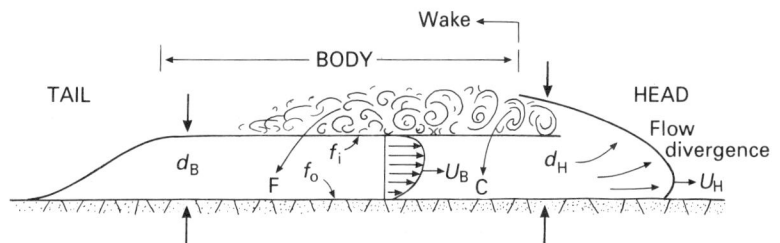

Fig. 8.7 Model and characteristics of uniform turbidity current. (After Pickering *et al.*, 1989.)

For the velocity of the head of the current, U_H, Middleton (1966b) gives:

$$U_H^2 = 0.56 g d_H \left[\frac{\Delta\rho}{(\rho + \Delta\rho)} \right]. \qquad (8.5)$$

One problem with using Eq. 8.4 is lack of knowledge of the friction coefficients. Middleton and Southard (1984) suggest that for large flows ($f_o + f_i$) is likely to be about 0.01, and that the difference between f_o and f_i will vary considerably depending on whether the flow is super-critical or sub-critical (i.e. respectively with or without intense mixing at the upper interface).

Equation 8.5 indicates that the velocity of the head is independent of the slope angle and this is believed to be true for relatively gentle slopes (i.e. < 1.24°). For steeper slopes, from perhaps 2 to 10°, Hay (1983) has proposed a modified flow equation that includes a dependence on bottom slope. The ratio of head velocity to body velocity, U_H/U_B, is found from experimental work to be approximately unity for gentle slopes but less than 1.0 on steeper slopes. As coarse-grained material from the body enters the head there is a corresponding loss of fine suspension from intense mixing in the neck region and settling back into the top of the body flow. Thus an equilibrium is maintained and a lateral grading developed from a coarser suspension in the head to a finer suspension in the tail.

The slope also has a key effect on whether turbidity currents are sub-critical (Froude number < 1.0) or super-critical (Froude number > 1.0). For a reasonable friction factor, $f = 0.02$, Komar (1971) demonstrated that turbidity currents would be super-critical on slopes > 0.5°, which would commonly be the case on many basin slopes and upper fan settings. Where the slope gradient decreases in the base-of-slope region then transition of sub-critical flow occurs, probably involving a hydraulic jump with intense turbulence and flow homogenization.

Autosuspension and ignition

In all turbidity currents the sediment-support mechanism which keeps the sediment particles in suspension is provided primarily by the upward component of fluid turbulence, which is mainly sustained by friction at the boundary between the flow and both the floor and the ambient fluid. It has been argued that turbidity flow can be sustained in the form of autosuspension (Bagnold, 1962). Autosuspension is a process of flow self-maintenance (Southard & Mackintosh, 1981) whereby a state of dynamic equilibrium

is achieved in which (1) the excess density of the suspended sediment propels the flow; (2) the flow generates friction and fluid turbulence; and (3) the turbulence keeps the sediment particles in suspension, and so on (Fig. 8.8). All that is needed to keep the loop intact is that the loss of energy by friction be compensated for by a gain in gravitational energy as the flow travels downslope. In this theoretical model it is possible for a turbidity current to travel over long distances without appreciable erosion or deposition as long as the slope remains constant.

Later work (Southard & Mackintosh, 1981; Middleton & Southard, 1984) has shown that true autosuspension is not always achieved and that as energy is lost from the system so some of the suspended load settles through the flow and is deposited. Pantin (1979) introduced an efficiency factor, e, into the autosuspension model, such that:

$$e\beta U_s > w, \qquad (8.6)$$

where β is the slope angle, U_s is the transport velocity of suspended sediment, w is the grain settling velocity. Pantin (1979) showed that flow density is the main determinant on whether autosuspension will occur. Below a critical density a turbidity current will subside and deposit, whereas above this value the

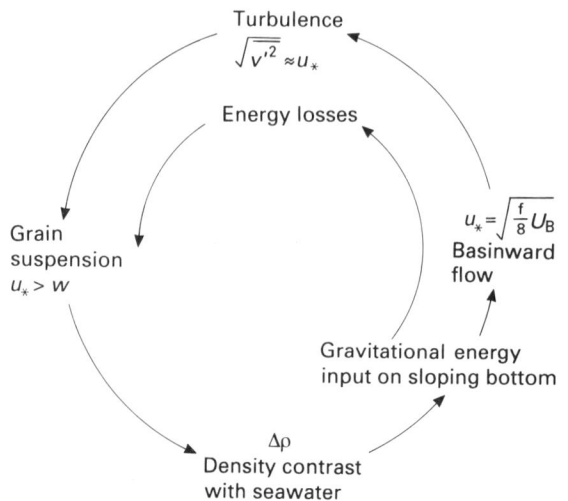

Fig. 8.8 Conceptual diagram to explain autosuspension. If gravitational energy input = energy losses, the flow will be self maintaining, and grains with settling velocity w will be maintained in suspension by vertical velocity fluctuations of average strength $\sqrt{\overline{v'^2}}$ approximated by u_*. (Middleton 1976). u_* is related to body velocity. U_B, through friction factor, f. See text for explanation. (After Pickering *et al.*, 1989.)

flow will 'explode', increasing in density and velocity and achieve autosuspension. This is known as flow *ignition* (Fukushima *et al.*, 1985).

Flow characteristics

Observations of atmospheric and pyroclastic density flows together with experiments on sub-aqueous density suspensions have shown that turbidity currents develop a characteristic longitudinal anatomy of head, neck, body and tail (Fig. 8.7) (Middleton, 1966a; Middleton & Hampton, 1976; Allen, 1985). The head of a turbidity current has a characteristic shape and flow pattern. In plan view, the head appears lobate with local divergencies of flow direction and a spanwise arrangement of regularly spaced lobes and clefts (Simpson, 1969, 1972; Allen, 1971). Inside the head a forward and upward sweeping, circulatory flow pattern exists. The coarsest grains tend to become concentrated in the head. The body is the part behind the head where the flow is almost uniform in thickness. Deposition may take place from the body while the head still erodes. The tail is the part where the flow thins rapidly and becomes very dilute. Mixing between the flow and the ambient fluid produces a dilute entrained layer. On slopes greater than 1.24° the head is thicker than the body, whereas on lesser slopes the body is thicker than the head (Komar, 1971). This is important for the type of sediment overflow in channelized environments. Mixing of the flow with water, loss of sediment by deposition and by flow separation in the neck will slacken and eventually stop the turbidity current. In an average turbidity current most coarse sediment will be deposited in a timespan of hours, though complete settling of the fine-grained tail may take a week (Kuenen, 1967).

Turbidity currents in the oceans are several orders of magnitude larger than those produced in laboratory flumes, so that the extent to which experimental results can be applied to turbidity currents in nature is somewhat problematic. The closest we have come to high density currents in nature is noting the occurrence of sequential breakages of submarine cables. The classic example is the Grand Banks earthquake of 1929 that triggered an enormous slump and an ensuing turbidity current that travelled downslope for hundreds of kilometres on to the Sohm Abyssal Plain (Heezen & Ewing, 1952; Piper & Normark, 1982). The maximum velocity attained by this current was some $70 \ \mathrm{km \ h^{-1}}$ $(25 \ \mathrm{m \ s^{-1}})$ (Menard, 1964). Other well-documented examples have occurred off the coast of Algeria, from the canyon systems off the

mouths of the Congo and Magdalena rivers and in the western New Britain Trench (see summary by Heezen & Hollister, 1971). Estimated velocities were again of the order of tens of kilometres per hour.

Some idea of the width and thickness of turbidity currents and the distances they travel can be deduced from the resulting depositional topography. The natural levees of submarine channels are believed to be produced from the overflow of channelized turbidity currents. Such currents must therefore be up to several kilometres wide and several hundreds of metres thick (Komar, 1969; Nelson & Kulm, 1973; Stow & Bowen, 1980). The length of deep-sea channels and of the flat expanses of abyssal plains both indicate that turbidity currents can travel as far as 4000–5000 km (Curray & Moore, 1971; Chough & Hesse, 1976; Piper *et al.* 1984; Stow *et al.*, 1990).

Various attempts have been made to estimate the physical features of low-density turbidity currents (Shepard *et al.*, 1977, 1979; Stow & Bowen, 1980; Bowen *et al.*, 1984) (Table 8.3). They vary in thickness from a few metres to channel-full flows over 800 m thick and have velocities in the range 10–50 cm s^{-1}.

Flow lofting

Turbidity currents generally carry warm, less dense water (normal marine or brackish) into denser deep-sea medium by virtue of an excess density due to suspended sediment. However, after sufficient particles have been deposited, the current will become progressively more buoyant, cease its lateral motion and ascend to form a plume in a process known as *lift-off* or *flow lofting* (Sparks *et al.*, 1993).

Sparks *et al.* (1992) discuss the concept of such sediment-laden gravity currents that come to display a reversing buoyancy from theoretical and experimental evidence. They postulate a vertical mixing of the dilute suspension and warmer water into the overlying weakly stratified ocean. They also suggest that the fine sediment from the lofted fluid may become much more widely dispersed than the main turbidity current deposit.

Arguing from observations made on sediments recovered from the distal Bengal Fan in the central Indian Ocean, Stow and Wetzel (1990) propose a very similar process, which they believe is commonplace in distal turbidite settings. They suggest that the 'dying' turbidity current discharges its suspension into the water column, up to hundreds or more than a thousand metres above the ocean floor. Further

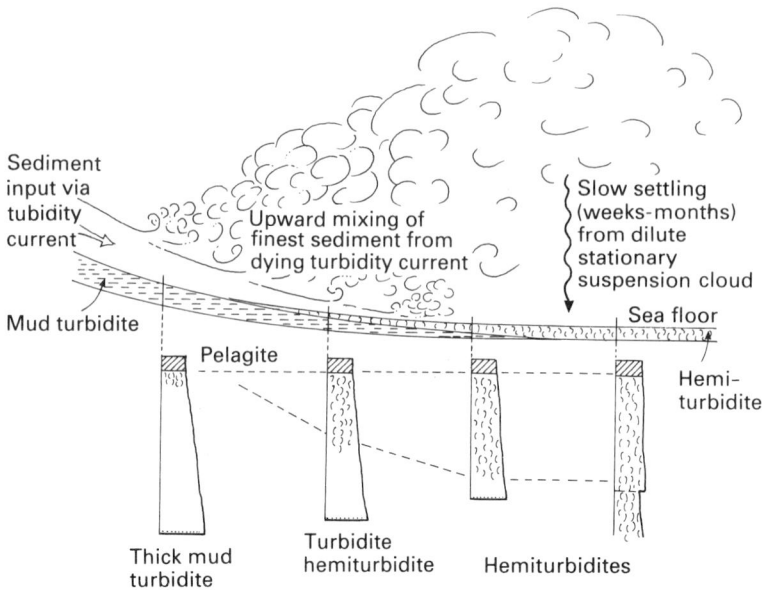

Fig. 8.9 Model for flow lofting and hemiturbidite deposition. (After Stow & Wetzel, 1990.)

material is added to this suspension cloud as the tail of the turbidity current arrives in the area over a period of, perhaps, a few days to a week or more. Material from this cloud then settles very slowly to the sea floor both above and beyond the distal feather edge of the muddy turbidite deposited by the original current. On the basis of continuous bioturbidation throughout, Stow and Wetzel (1990) propose a time scale of a few weeks to many months for deposition of a 1-m thick unit. They call these deposits *hemiturbidites* (Fig. 8.9).

8.3 Bottom current processes

8.3.1 Introduction

The second main group of processes that operate in the deep sea, actively eroding, transporting and depositing sediment, are collectively known as bottom currents (or 'normal bottom currents' — Stow, 1986). These include all types of deep current that are *not* normally driven by sediment suspensions but that result from normal oceanographic forces such as winds, tides, waves and thermohaline circulation. The main types and characteristics of these currents are shown in Table 8.3.

The ocean waters, shallow and deep, can be compartmentalized into different water masses, each having distinctive temperature/salinity (T–S) characteristics. A profile across the North Atlantic Ocean

(Fig. 8.10) shows the layered distribution of these water masses. More than 75% of the world ocean and almost all of the water below 1000 m has T–S characteristics within the relatively narrow limits of $-1°C–5°C$ and 34.4–35%. Only the near-surface water can be significantly warmer and marginal seas (e.g. Red Sea, Mediterranean Sea) more saline.

In this section, we are mainly concerned with the origin, nature and movement of these deep water masses and with how they become intensified into discrete currents.

8.3.2 Bottom (contour) currents

Origin and intensification

Deep ocean bottom water is formed by the cooling and sinking of surface water at high latitudes (Gill, 1973; Killworth, 1973) and the deep, slow thermohaline circulation of these polar water masses throughout the world's oceans (Neuman, 1968). Highly saline but warm water also flows out of the Mediterranean Sea as an intermediate-level water mass.

Antarctic Bottom Water (AABW), the densest and deepest water in the oceans, forms in the region of the coast-hugging westward-directed surface Polar Current, with localized areas of major generation, such as the Weddell Sea, where the formation of sea ice, at about 1.9°C increases the salinity of the water remaining. This water then runs down the continental

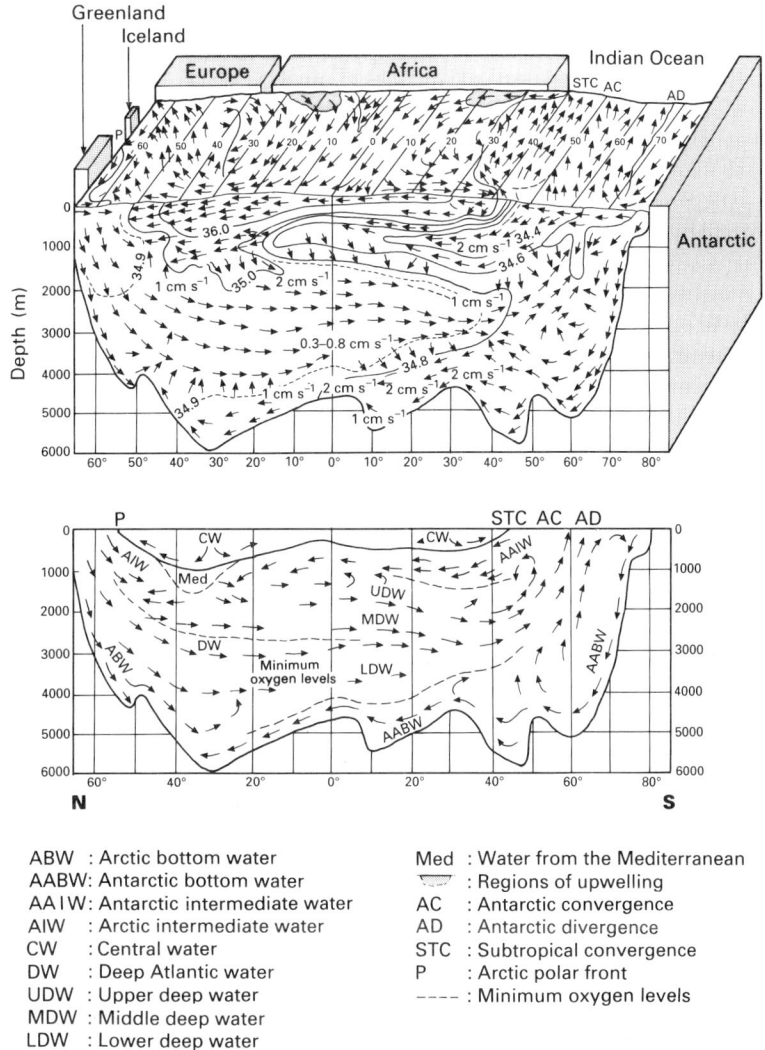

Fig. 8.10 Bottom water masses in the North Atlantic Ocean.

ABW   : Arctic bottom water
AABW: Antarctic bottom water
AAIW: Antarctic intermediate water
AIW   : Arctic intermediate water
CW   : Central water
DW   : Deep Atlantic water
UDW   : Upper deep water
MDW   : Middle deep water
LDW   : Lower deep water

Med   : Water from the Mediterranean
⌣   : Regions of upwelling
AC   : Antarctic convergence
AD   : Antarctic divergence
STC   : Subtropical convergence
P   : Arctic polar front
---- : Minimum oxygen levels

slope, circulates eastwards around the Antarctic continent, perhaps several times, and then flows northwards into the Atlantic, Indian and Pacific Oceans. It can still be recognized, although considerably warmer, in the central and northern Atlantic Ocean where it mixes with and contributes to the North Atlantic Deep Water (NADW). AABW is also the main source of Pacific and Indian Ocean Common Water (PIOCW).

A major source of NADW appears to be the sub-polar gyre in the Norwegian and Greenland Seas, although it remains partly trapped by an irregular topographic barrier known as the Scotland–Iceland–Greenland Ridge. Once the Norwegian–Greenland Seas basin is filled with this cold, dense water mass

then intermittent overflow to the south occurs through narrow channels cutting across the Ridge. Overlying water is entrained by turbulent shearing and helps impart distinctive characteristics to a NE Atlantic Deep Water (NEADW) and a NW Atlantic Deep Water (NWADW) mass. NWADW mixes with Labrador Sea Water (LSW), which is formed by surface cooling in the Labrador Sea. NEADW in part moves south within the eastern Atlantic basin and part crosses the Mid-Atlantic Ridge to mix with and partly overlie NWADW.

These slowly moving water masses are affected by the Coriolis Force, due to the Earth's spin, which deflects moving bodies to the right in the northern hemisphere, to the left in the southern hemisphere and is zero at the equator. The result is that deep-water flows are banked up against the continental slopes on the western margins of the ocean basins, are unable to move upslope against gravity and so become restricted and intensified, forming Western Boundary Undercurrents (WBUC).

Deep water in each ocean spreads slowly eastwards from these WBUCs and wells up through the main thermocline. Mostly, the upwelling is a slow, uniform upward diffusion, but in some coastal and equatorial regions it is enhanced. On reaching the upper layers of the oceans, the water becomes part of the normal wind-driven circulation and eventually finds its way back to the polar regions. The generalized pattern of deep circulation is shown in Fig. 8.11 (Stow & Lovell, 1979), which agrees well with that modelled by Stommel (1957) and predicted by Stommel and Aarons (1960b).

Bottom currents are also locally intensified by flow restriction through narrow passages in the deep sea, for example in flowing through Fracture Zone gaps in the Mid-Ocean Ridge system.

Mathematical models

The complexity of models needed to describe deep-sea circulation and its coupling with the general oceanic circulation is beyond the scope of this chapter, and better left to more specific texts on dynamic oceanography (e.g. Tolmazin, 1985). Nowell and Hollister (1985) outline the essence of an Ekman layer–sediment transport model as a prelude to the High Energy Benthic Boundary Layer Experi-ment (HEBBLE, see later), and McLean (1985) anal-yses some of the preliminary results.

Flow characteristics

Whereas much of the deep-sea floor is swept by very slow currents (< 2 cm s^{-1}), the western boundary currents commonly attain velocities of 10–20 cm s^{-1} and these may be greater than 100 cm s^{-1} where the flow is particularly restricted (Stow & Lovell, 1979; McCave & Tucholke, 1986).

Although these bottom currents are more or less continuous and sufficiently competent in parts of the ocean to erode, transport and deposit sediment, they are also highly variable in both velocity and direction (Luyten, 1977; Richardson et al., 1981; Nowell & Hollister, 1985). Large-scale eddies peel off and move at right angles to the main flow, and the average velocity decreases from the core to the margins of the current. Both seasonal (Shor et al., 1980) and tidal (McCave et al., 1980) periodicities have been re-corded, and current reversals are common. The currents vary from a few kilometres to tens of kilometres in width and can flow at different levels within the water column depending on the relative densities of adjacent water masses.

Well-developed nepheloid layers (Fig. 8.12), or tur-bid bottom waters with marked concentrations of suspended matter, are commonly associated with the higher velocity bottom currents in many parts of the ocean basins (Eittreim et al., 1976; Biscaye & Eit-treim, 1977). These currents appear to maintain fine (average size 12 µm) particles in suspension by

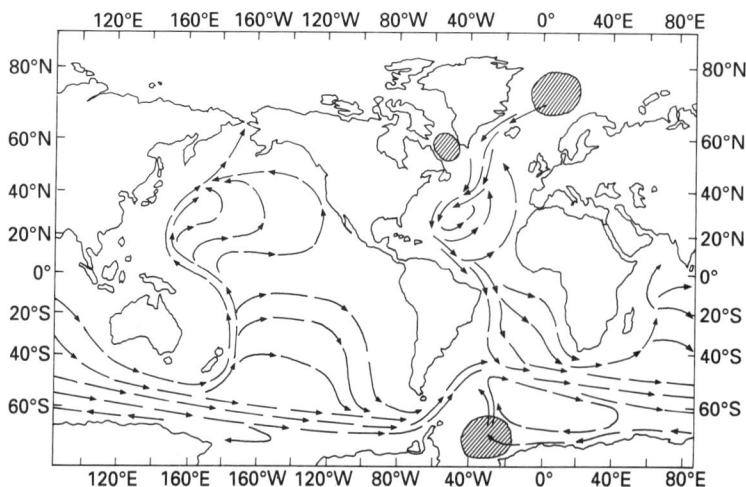

Fig. 8.11 Global pattern of abyssal circulation. (After Pickering et al., 1989).

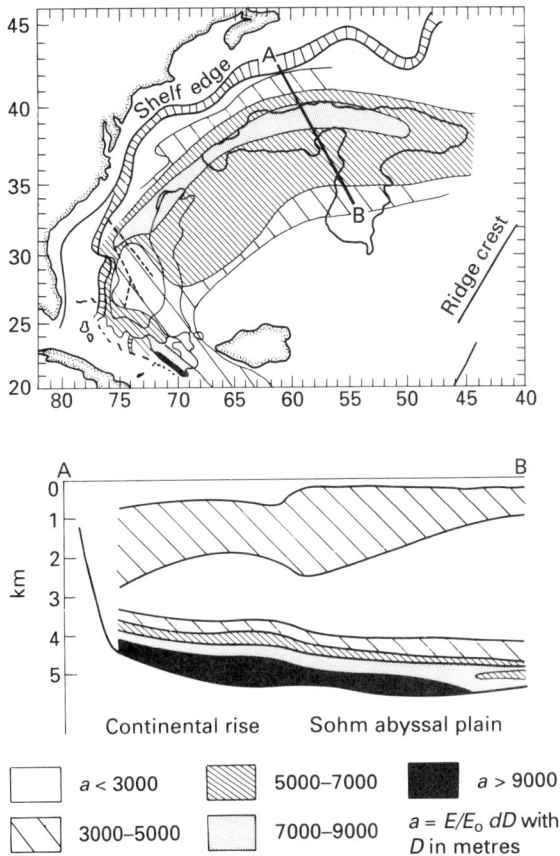

Fig. 8.12 Nepheloid layers in the NW Atlantic Ocean. (After Stow, 1986.)

Legend for the figure:

$a < 3000$	5000–7000	$a > 9000$
3000–5000	7000–9000	$a = E/E_0 \, dD$ with D in metres

silts and sands by bottom currents compared with transport and deposition. However the result of the HEBBLE programme, on the lower continental rise off Nova Scotia (Hollister & McCave, 1984), and of several detailed studies of the nepheloid layer (McCave, 1986) have, more recently, shed much light on this whole problem.

We know that the hydrodynamic processes acting on the deep-sea floor are the result of an interplay of deep-sea circulation and surface current activity, which is mainly controlled by atmospheric conditions. What the HEBBLE work clearly demonstrated is that temporary very high surface energy conditions may propagate downward and induce high energy over the deep-sea floor. Very high surface kinetic energy in the oceans, as shown by maximum variability in the level of the sea surface (Richardson, 1983; Cheney et al., 1983), is only observed in relatively few areas, such as in the NW Atlantic along the eastern US margin, or in the SW Atlantic along the Argentine margin. These are also areas of significant contourite sedimentation.

The variation of bottom kinetic energy conditions in the HEBBLE area results in an alternation of short (days to weeks) episodes of erosion associated with high velocity currents, and longer periods (weeks to months) of deposition associated with lower velocity. Episodes of high current velocity, called 'benthic', 'abyssal' or 'deep-sea' storms (Gardner & Sullivan, 1981; Hollister & McCave, 1984), correspond to high surface kinetic energy, due to local very low atmospheric pressure. During these episodes, a large volume of material is resuspended and thus contributes to a very high density nepheloid layer. This material may then be transported over long distances in the Western Boundary Undercurrent before eventual deposition. Between abyssal storm events, the current velocities are much lower and very high sedimentation rates occur (up to 1.4 cm month^{-1}), over these short time periods. However, the net sedimentation rate at the scale of Holocene deposits on this part of the continental rise, is very low (5.5 cm per 10^3 years) as a result of the very frequent and active erosional episodes (estimated annual deposition/preservation ratio of 3100). At a greater geological scale (Neogene), similar rates of deposition (2–10 cm per 10^3 years) are observed on the giant contourite drifts in the North Atlantic Ocean.

Resuspension of sea-floor sediments is not the only process feeding the nepheloid layer and hence bottom-current transport. Several other processes

turbulent eddy diffusion for a residence period of about 1 year (Eittreim & Ewing, 1972). Concentrations of deep-sea nepheloid layers are extremely low (0.01–0.3 mg l^{-1}, McCave & Swift, 1976) and their thicknesses vary from less than 100 m to over 1000 m.

Although bottom currents are clearly efficient agents of particle transport and redistribution, both terrigenous and biogenic, and play a major role in shaping deep-sea morphology (see later), for many years we have had no very good understanding of the exact nature of erosional and depositional processes by bottom currents. Whereas, the velocities commonly recorded by current-metre measurements were adequate to transport fine-grained material, they were insufficient to effect marked erosion of very cohesive deep-sea muds. There has also been a debate centred around the role and extent of reworking of may be involved including: (1) inputs of terrigenous particles from the adjacent margins by turbidity currents; (2) advection along isopycnal surfaces via suspension cascading or some other hemipelagic process; (3) direct settling of biogenic pelagic particles and pellets; (4) resuspension of particles by the burrowing activity of benthic organisms. Whatever the origin of the particles, it appears that there is a close relationship been high turbidity nepheloid layers and active bottom-current transport, including both erosion and deposition.

8.3.3 Major surface currents

The very large wind-driven current systems developed in the surface layers of the oceans can also have a direct effect even at very great depths (several kilometres). This is true of the deep Gulf Stream gyres of the North Atlantic (McCave & Tucholke, 1986), the deep Kuroshio Current off Japan and the Antarctic Circumpolar Current driven by the West Wind Drift.

8.3.4 Internal waves and tides

Surface waves and tides are some of the most important physical processes affecting sediments and biota in shallow water. As the sea is clearly a heterogeneous body, undulation swells or internal waves can also form between sub-surface water layers of varying density in the upper few hundreds of metres, most notably at the thermocline (Lafond, 1962). Such internal waves are very widespread and vary considerably in amplitude and periodicity. They may exceed surface waves in amplitude, although their speed of progression is usually slow (5–300 cm s^{-1}). Similar large-scale oscillations at density discontinuities have been shown to have a tidal period and are known as internal tides (Rattay, 1960).

The breaking and turbulent eddies caused by internal waves and the velocities attained by both internal waves and tides probably cause significant sediment stirring and erosion at the shelf break, on the tops of seamounts or in relatively shallow slope and shelf-basins (Shepard, 1973b). They are also thought to contribute to up-and-down canyon currents (Shepard et al., 1979).

8.3.5 Canyon currents

Whenever current metres have been placed in deep-sea channels or canyons, originally with the intention

of measuring turbidity current flow, it has been found that semi-permanent currents are everywhere present, even at depths in excess of 4 km (Shepard et al., 1979). Generally, however, these currents alternate directions, flowing both up and down canyon with periodicities ranging from 15 min to 24 h and with velocities commonly up to 30 cm s^{-1} (Fig. 8.13). A tidal periodicity seems most usual in the deeper parts, but a higher frequency of flow reversal normally occurs in the head region. Other flow periods and directions have also been recorded, probably related to internal waves, surface currents, storm surges or cold-water cascading currents.

In some cases, following major storm periods

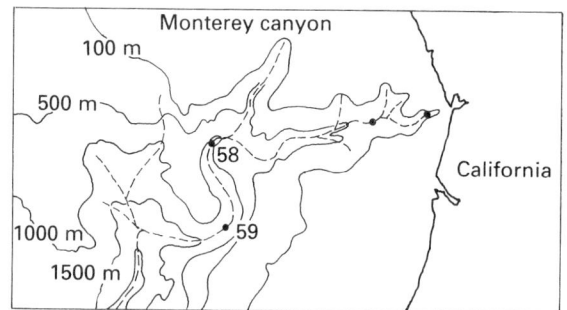

(a)

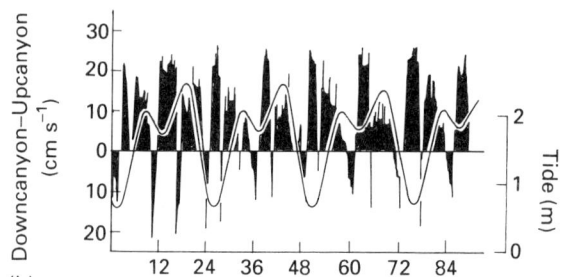

(b)

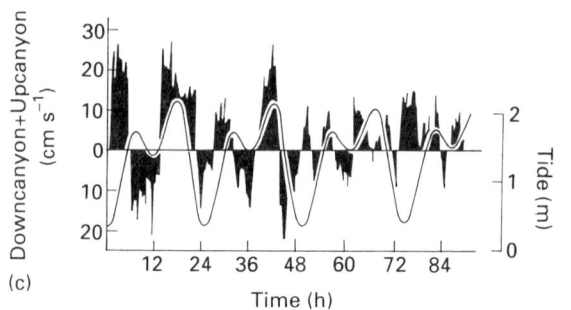

(c)

Fig. 8.13 Canyon currents within Monterey Canyon, off Western California. (a), Location. (b) Station 58, depth 1061 m. (c) Station 59, depth 1445 m. (After Stow, 1986.)

leading to the build up of water in the shallow shelf regions near canyon heads, prolonged downcanyon flows can develop attaining speeds of 50–100 cm s^{-1} lasting for several days. These are probably true low-density turbidity currents and not equivalent to the more common non-turbid canyon currents.

From the measured frequency and velocity of canyon currents it is clear that they have considerable effect on the winnowing, entrainment and movement of sediment, the moulding of canyon and channel morphologies and on keeping canyon floors generally free from pelagic or hemipelagic sedimentation.

8.4   Pelagic and hemipelagic processes

8.4.1 Introduction

Slow vertical settling of microscopic biogenic and non-biogenic particles through the water column, in the absence of significant effects of other processes at any depth within the open oceans, is generally referred to as *pelagic settling*. It is less important for clastic than for biogenic sediments as the materials involved are largely the tests of calcareous and siliceous planktonic organisms and their associated organic matter which has been biosynthesized in the surface layers of the oceans. These form the pelagic deposits of the deep sea (Jenkyns, 1986).

In many areas of the deep sea, particularly on slopes and in basins close to land, terrigenous elements (clays, quartz, feldspar, volcanic dust and other minerals) with a high proportion of silt-sized grains can form a significant part of the settling material and hence of the resulting *hemipelagic* deposit. Such materials are transported by surface currents, winds and floating ice and mix with pelagic biogenic components during settling (Stow, 1985).

8.4.2 Vertical settling of pelagic material

A simple way to model pelagic settling is to consider the free fall velocity, V_g, of a perfect sphere through the water column:

$$V_g = \frac{gd^2(\sigma - \rho)}{18\mu}, \qquad (8.7)$$

where g is the acceleration due to gravity, d is the particle diameter, σ is the particle density, ρ is the fluid density, μ is the fluid viscosity.

This is known as Stokes' Law and yields fall velocities of the order of 10^{-4}–10^{-6} m s^{-1} for the finest silt- and clay-sized particles.

However, in nature, a number of factors mitigate against this ideal situation (Allen, 1985). Firstly, most particles are not spherical and many, including platy clays and elaborate siliceous or calcareous tests, are very far indeed from being spherical. Experimental results show a combination of tumbles, steady fall and oscillations for the settling behaviour of solitary discs in a stagnant fluid (Allen, 1985). Other work has measured differential fall velocities for a range of planktonic microfossils. Secondly, numerous observations of suspended sediment in the open oceans have revealed that particles very rarely settle in isolation. Instead, physical mechanisms of flocculation and biogenic processes of pelletization ensure that most material settles much more rapidly (10^{-2}–10^{-3} m s^{-1}) as flocs or pellets. Thirdly, falling particles are subjected to progressive dissolution of calcareous and siliceous tests (particularly below the Carbonate Compensation Depth, CCD), oxidation of organic matter and lateral transport by bottom currents, turbidity currents or other slow advection processes.

8.4.3 Settling and horizontal advection

Somewhere between vertical pelagic settling and low-density turbidity currents is a range of overlapping mechanisms that can conveniently be called *hemipelagic processes*. The materials involved are an admixture of terrigenous and primary biogenic particles (Jenkyns, 1986). Deposition is by slow settling, typically coupled with a component of current-induced lateral advection of suspended sediment in mid- or bottom-water nepheloid layers. This may involve suspension cascading, lutite flows, up and down canyon currents and very thin turbid-layer flows (McCave, 1972; Drake *et al.*, 1978; Stow, 1985) (Fig. 8.14).

8.5.   Sediment nature and distribution

8.5.1 Introduction

Our main discussion thus far has concentrated on the nature of long-distance transport mechanisms into and within the deep sea. However, the character of the sediments deposited depends partly on these processes but partly also on the final mode of deposition and early postdepositional deformation.

In this section, then, we focus on the nature of deep-sea sediments, both modern and ancient, and what we can thereby learn of the final depositional

Inputs
Aeolian
Fluvial
Glacial
Coastal

Aeolian transport

Surface current transport

Cross-shelf transport

Primary biological productivity

Pelagic processes

{Settling}

Mid-water suspen.

INITIATION AND RECHARGING OF RESEDIMENTATION PROCESSES
Slides, resuspension, slope erosion, direct sediment discharge

Suspension

Cascading

Canyon currents and bottom nepheloid suspension

(high/low density thin/thick overflow)

turbidity currents

{Pelletization flocculaton and settling}

HEMIPELAGIC PROCESSES

Bottom currents, eddies

Bottom currents and nepheloid layers

Bottom current processes

TURBIDITY CURRENT PROCESSES

Dissolution and Authigenesis

Chemogenic processes

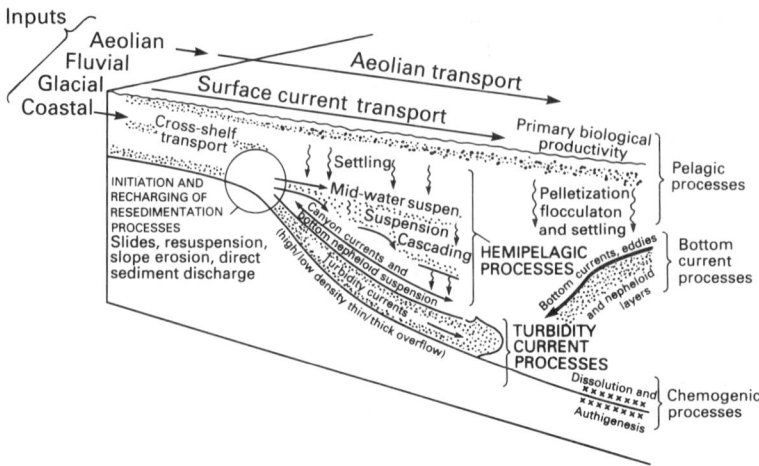

Fig. 8.14 Processes of fine-grained sedimentation in the deep sea indicating the range of processes contributing to hemipelagic deposition. (After Stow, 1985.)

processes involved. There have been many hundreds of research papers (and books) describing deep-sea sediments of every age and depositional setting from around the world. A synthesis of these data into a series of facies models seems the most useful approach to take here.

The analysis of sedimentary structures in both core and outcrop sections allows inference about the processes that caused them. However, only recently have we been able to observe directly and systematically the present day deep-sea sediment surface and bedforms from submersible dives and deep-tow near-bottom surveys. We begin our discussion with some of these observations, before considering how they may be preserved in ancient successions.

8.5.2 Bedforms and other sea-floor features

Mass wasting on slopes

All slopes surrounding deeper water basins, whether they are shelf, slope or oceanic in nature, are subject to mass wasting and show, to a greater or lesser degree, surface features that reflect these processes. The main characteristic of such slopes, particularly those that are relatively steep (> 5°), tectonically active and/or with a thick unstable sediment cover, is a generally irregular surface topography (Prior & Suhayda, 1979a; Hill, 1984). This is made up of: (1) erosional features such as slide, slump and debris flow scars, grooves and depressions formed by falling and passing olistoliths, and scour marks caused by high energy sediment gravity flows; and (2) depositional masses from rockfalls, slides, slumps and

debris flows, chaotic rafted blocks, or slow sediment creep rucking the surface into semi-regular wrinkles or 'waves'. None of these is a true bedform shaped by fluid flow, but they are nevertheless very characteristic sediment surface forms caused by mass wasting (Fig. 8.15).

Turbidity current bedforms

A broad range of bedforms may be constructed on the sea floor during the passage of turbidity currents. In deep-sea channels covered with gravels and pebbly sands, deposited and moved by high concentration turbidity currents and associated sediment gravity flows, large-scale asymmetric gravel waves (wave-length 30–70 m, amplitude 5–10 m) and symmetrical macrodunes (wave length c. 300 m, amplitude 2–5 m) have recently been described (Hughes et al., 1990). Earlier observations have rarely reported such regular gravel bedforms but have recorded more irregular or bar-like gravel lenses (Malinverno et al., 1988). These have been inferred previously from studies of ancient series (Hein, 1982).

A new generation of deep-tow instrument packages is now being used to make very detailed observations of sea-floor features from, for example, channels on the Monterey Fan in the east Pacific. Towed Ocean Bottom Instrument (TOBI) data have revealed lenticular to irregularly-shaped channel bars, wave and dune fields, erosional steps transverse to the channel axis, and trains of roughly circular scours or pock marks oriented at an acute angle to the channel in overbank deposits (Fig. 8.16) (D. Masson personal communication).

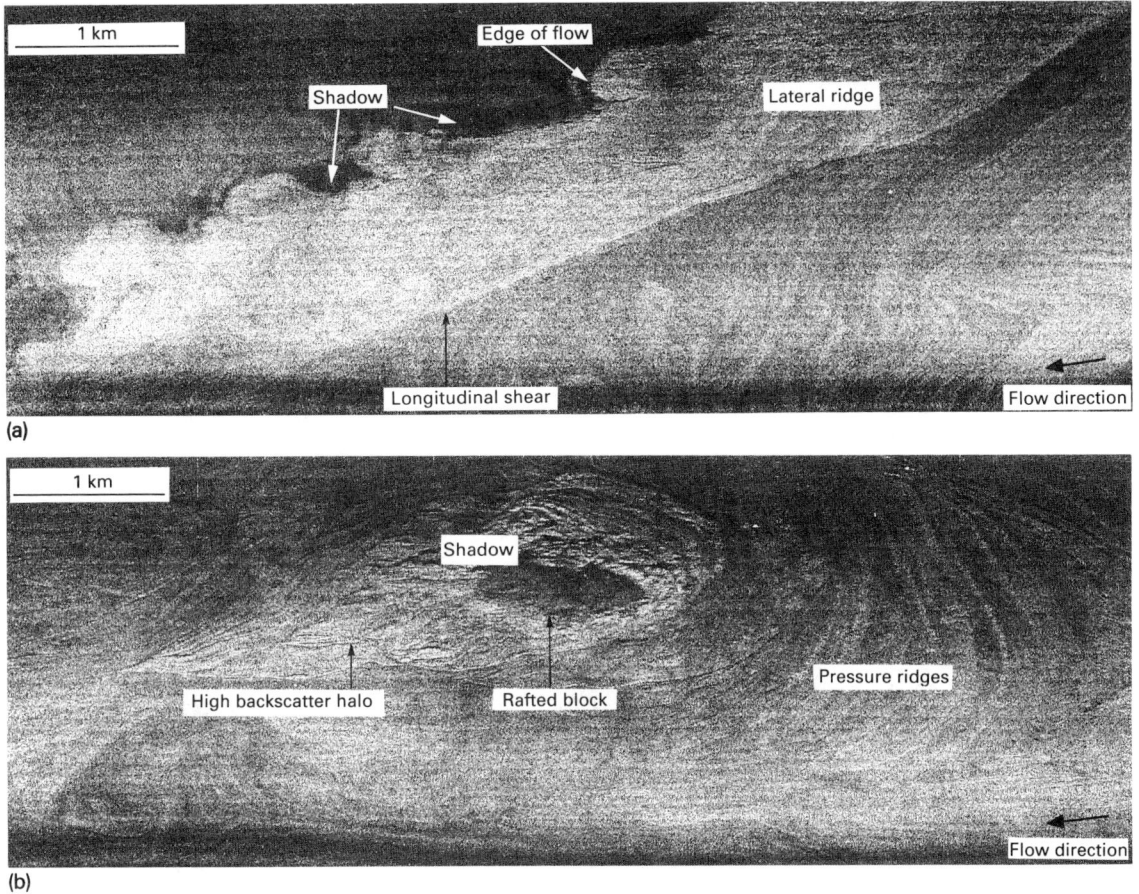

Fig. 8.15 Deep-towed 30 kHz sidescan sonograph (TOBI image) from the Saharan Debris Flow deposit in the northeast Atlantic. (a) Edge of debris flow deposit (lobate); (b) rafted block within debris flow deposit. (From Masson *et al.*, 1993.)

In some areas, channel levees are covered by regular large-scale sediment waves that typically show an upslope (up-current) migration when examined with high-resolution seismic profiling (Damuth, 1975; Normark *et al.*, 1980). These sediment waves typically show wave lengths of 0.5–2 km, amplitudes of 10–40 m, and are constructed in fine-grained muds and silty muds. Normark *et al.* (1980) considered them to be some kind of stationary wave or giant antidune bedform beneath a large low-concentration turbidity current that overtopped the confining bank on the outside of a channel meander. They noted a general decrease in amplitude and increase in wave length away from the channel.

Smaller-scale bedforms, such as dunes, ripples, lineated and smoothed surfaces, have been variously noted from beneath and presumed turbidity current pathways mainly from sea-bottom photographic sur-

veys (e.g. Heezen & Hollister, 1971).

The erosional behaviour of turbidity currents is also well-known from longitudinal furrows and grooves as well as horse-show shaped flute marks scoured into soft substrates. Large and very large scale erosive features, megaflutes and giant flutes, have been described by Normark *et al.* (1979) from the Navy Fan and by Shor *et al.* (1990) from the Laurentian Fan (Fig. 8.16). Megaflutes may be 5–30 m across and 1–2 m deep, whereas the giant scour recently observed in the latter example is over 1 km long and 1 km wide at the flared end and up to 100 m deep near the apex.

Shor *et al.* (1990) also described smaller closed depressions that are 10–20 m deep and occur with regular spacing of 1.5–3 km along valley floor channels. These they believe are analogous to 'pools' developed on bedrock river floors.

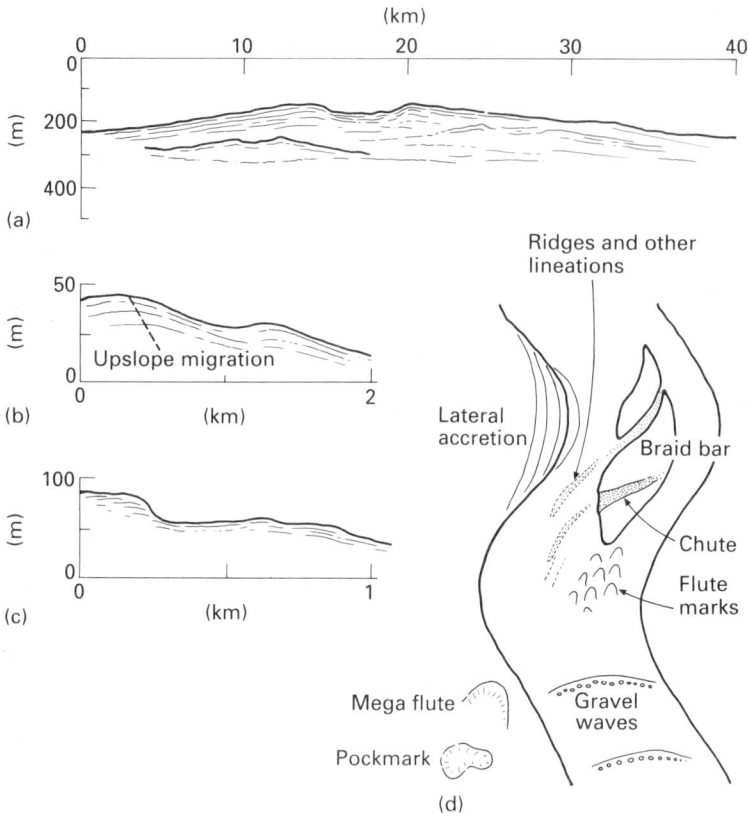

Fig. 8.16 Turbidity-current bedforms related to channel and non-channel/levee areas. Line drawings from seismic reflection profiles showing : (a) mid-fan channel-levee complex; (b) upslope migration of sediment waves on back side of levee; (c) longitudinal section of large-scale scour or megaflute; (d) schematic plan of portion of sinuous channel (approx. 1 km wide).

Bottom current bedforms

Bedforms controlled by bottom (contour) currents have been described in detail by many authors (Heezen & Hollister, 1971; Allen, 1982; Stow, 1982; Nowell & Hollister, 1985; McCave & Tucholke, 1986 among others; see also Stow & Faugeres, 1993). They occur at a range of scales from large sediment waves and erosional furrows to smaller scale dunes, ripples, lineation, scour and tail marks, etc. (Fig. 8.17). Bedform assemblages have been related to bottom-current intensity (Hollister & McCave, 1984), although the exact processes of formation still remain questionable for some of them, especially for giant sediment waves.

According to different authors working in different areas, sediment waves (Fig. 8.18) may be parallel, perpendicular or at an angle to current flow; they may propagate downstream or upstream, and if they are formed on a gentle slope, they can migrate downslope or upslope (Hollister *et al.*, 1974; Asquith, 1979; Lonsdale & Hollister, 1979; Embley *et al.*, 1980; Kolla *et al.*, 1980; Flood & Shor, 1988). Furthermore,

these sediment waves are similar in dimensions and morphology to the sediment waves formed by turbidity currents outlined in the previous section (Normark *et al.*, 1980). The problem of a contour current or turbidity current origin for sediment waves is highlighted by examples from the Tyrrhenian Sea cited by Mariani *et al.* (1993).

Other sea-floor features

The deep-sea floor is moulded and modified by many other processes, which we do not have space to discuss further in this chapter. These include: (1) various types of wet-sediment injections (mud diapirs, volcanoes and ridges) that are particularly characteristic of active margin slopes and slope basins (Jones & Preston, 1987; Pickering *et al.*, 1989); (2) pock marks and other soft-sediment disturbances caused by shallow gas release, common beneath areas of active upwelling and high biogenic productivity (Summerhayes, 1992); (2) bottom relief and sediment drape caused by neotectonic fault activity (e.g. Pickering, *et al.* 1989; Redbourn *et al.*, 1993); and (4) iceberg scour marks and dropstone depressions on high lati-tude slope systems (Dowdeswell & Scourse, 1990).

8.5.3 Facies models and their hydrodynamic interpretation

Introduction

Each of the various processes discussed in Sections 8.2, 8.3 and 8.4 above gives rise to a characteristic deposit. These can be described in terms of a series of facies models (Figs. 8.19–21) which show the stan-dard sequence of structures and other sedimentary features attributed to a particular process.

Slumps, slides and debrites

Slumps and slides occur at all scales and can involve any lithology. Internally, the beds are mainly coher-ent in slides, apart from a shear zone along the base, compressional features in the toe region and ten-sional structures at the head. There is a more perva-sive bed disruption in slumps including several types of folds, thrusts, balls, overfolds, bed rotation and scar surfaces, although no standard sequence of these structures has yet been identified. The direction of slumping, and hence of the slope, is generally taken as perpendicular to the mean of azimuths of the slump fold axes, and is determined from the sense of overturning (Woodcock, 1976).

Debrites (debris flow deposits, olistostromes) are very variable in nature, ranging from mud dominated with only scattered pebbles or boulders to clast dominated with less than 5% muddy matrix. Bed thickness also varies up to several tens of metres, and appears to be positively correlated with maximum clast size. Debrites may be quite structureless or show part or all of the standard sequence illustrated in the model: (I) sheared fissile-lensoid lamination; (II) fault-, slump- and convolute deformation; and (III) matrix-supported clast-rich zone, with or without water-escape features.

These divisions reflect the complex dynamics of debris flows. In part there may be a sliding mecha-nism involved resulting in basal shear (I). Continued slow forward advance during top-down freezing of the flow results in fault-fold disruption of the layer immediately overlying the shear zone (II). The upper division (III) represents the semi-rigid plug that was being rafted downslope; where this was more fluid and faster moving, then rapid deposition causes upward escape of pore fluids and the development of sub-vertical water-escape pipes. A number of debrites

have been described that are clearly coupled, both vertically and/or laterally, with a capping turbidite.

Turbidites

Three different *turbidite models* have been recog-nized, each with its own standard sequence of struc-tures through a single depositional unit or bed, and each resulting from turbidity current deposition, in some cases with late-stage flow modification. The individual structural features or divisions for the coarse-grained, medium-grained and fine-grained turbidites are best described by reference to Fig. 8.19. Complete sequences are rarely encountered, with par-tial sequences (top-absent, mid-absent, base-absent) being the rule. Normal grading is common, together with reverse grading particularly at the base of thick coarse-grained beds, and sequential grading through a series of silt laminae in the finer-grained beds.

Bouma (1962) first presented what has now become the classical model for medium-grained sand–mud turbidites. This was soon interpreted as the result of flow deceleration of relatively low-concentration turbidity currents accompanied by rapid deposition (Division A), followed by a progression of bedforms (Divisions B, C, D) representing fallout during high to low flow regime, and finally deposition from the turbidity current tail together with pelagic/hemipelagic settling (Harms & Fahnstock, 1965; Walker, 1965). The absence of a megaripple division between B and C has been discussed by various authors (see summary in Pickering *et al.*, 1989).

High-concentration flows carrying gravel and coarse sand leave deposits quite different from those of classical turbidites, and are the result of a some-what different set of depositional processes (Picker-ing *et al.*, 1989). These include: (1) rapid *en masse* deposition due to increased intergranular friction; (2) inverse grading at base due to grain collisions/dispersive pressure; (3) development of traction car-pets, driven by shear from the overriding flow, and sequential deposition by 'freezing'; (4) grain-by-grain deposition from suspension with little subsequent traction transport; (5) formation of irregular stratifi-cation by alternate deposition of bedload and sus-pended load; and (6) syn- and/or post-depositional escape of pore-fluids forming dishes, pipes and other escape features.

Several authors have proposed structural se-quences of coarse-grained turbidites (e.g. Hiscott & Middleton, 1979; Hein, 1982; Lowe, 1982) and we have summarized Lowe's (1982) scheme in Fig. 8.19.

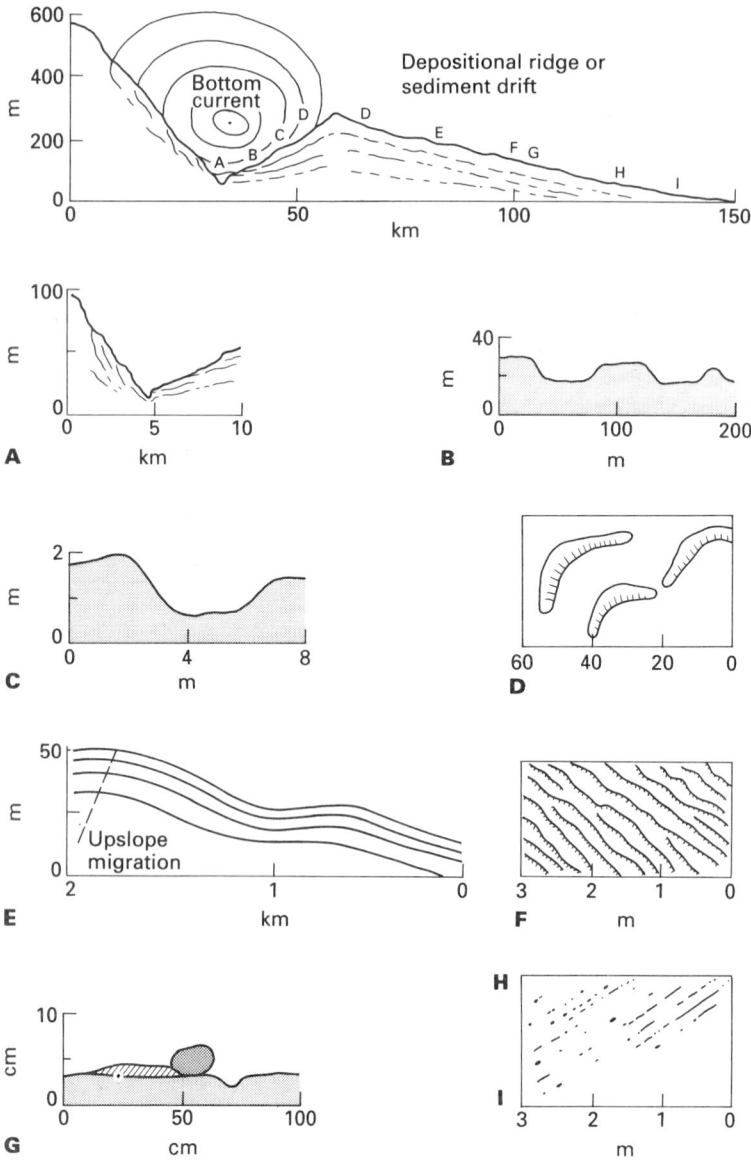

Fig. 8.17 Bottom-current bedforms. (a) Composite diagram. A, moat; B, large furrow; C, small furrow; D, dunes and waves; E, giant sediment waves; F, ripples; G, scour and tail; H, lineation and coarse lag; I, smoothed sediment surface. (Stow, 1981); (b) bottom photograph of valley floor adjacent to Faro Drift, Gulf of Cadiz, showing zones of linguoid ripples, 1, and lineation, 2, (from Faugeres et al., 1985).

(a)

There has been some dispute as to whether or not super-critical flow of turbidity currents could produce antidune bedforms at the base (or within) such a sequence (see Hand et al., 1972; Skipper & Bhatta-charjee, 1978). However, the beds that were origi-nally described as containing antidunes have subsequently been reinterpreted by Pickering and Hiscott (1985) as flow reversal by reflection and deflection in a constricted basin. Such flow reversal has gained general acceptance from experimental observations (Pantin & Leeder, 1987).

Low-concentration flows, transporting mainly silt- and clay-sized material, may also deposit a regular sequence of structures as a result of waning flow velocity (Piper, 1978; Stow, 1979; Stow & Bowen, 1980; Stow & Shanmugam, 1980). The idealized model (Fig. 8.19) is believed to result from suspen-sion fallout and traction (T_0–T_2), shear-sorting of silt grains and clay flocs in the bottom boundary layer (T_3–T_5), and suspension fallout (T_6–T_8). Many fine-grained turbidites develop only partial sequences (Piper & Stow, 1991), and excessively thick mud

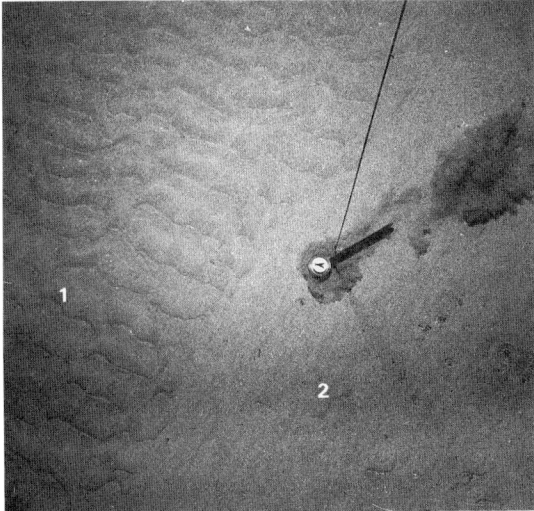

(b)

Fig. 8.17 *Continued*

turbidites appear to show a more complex sequence including multiple repetitions of certain divisions (Porebski *et al.*, 1991).

When a large dilute, mud-charged turbidity current nears the end of its run, perhaps having entered a flat basin plain or having encountered some topographic barrier, two process transitions appear to be possible. The first is for the incoming tail of the turbidity current to provide more and more suspended sediment to the slowing current, so that the concentration builds up and the flow transforms into a high-concentration, slow-moving mud layer (almost a debris flow), that eventually settles out rapidly (or 'freezes') to form a thick structureless mud unit (McCave & Jones, 1988). The second is for flow dilution to continue, and for the finest grained suspension to be thrown up and beyond where the true turbidity current deposits its final thin bed. This material mixes with the basin waters and with the falling pelagic sediment, depositing sufficiently slowly that bioturbation continues but still retaining some of the original characteristics of the parent flow, the 'hemiturbidites' of Stow and Wetzel (1990).

Contourites

The problem of diagnostic lithological features of both modern and ancient *contourites* has long been addressed (Hollister & Heezen, 1972; Stow & Lovell, 1979; Lovell & Stow, 1981; Gonthier *et al.*, 1984;

Stow & Faugeres, 1993). It still remains a partly unresolved problem as contourites may mimic sediments deposited by other processes, and bottom-currents may rework, slightly or significantly, other types of deposits.

Depending on the composition of the sediment supply, contourites may be siliciclastic, volcaniclastic or biogenic. Depending on the grain size that the bottom-current is able to transport, contourites range from muddy to sandy, with a range of transitional facies composed of admixtures of sand, silt and clay typically displaying a mottled appearance. Very strong currents may scour the sea floor and result in gravel-lag contourites.

The majority of these deposits are strongly bioturbated, and any primary current structures (lamination, ripples, erosional surfaces) are consequently not well preserved. Better preservation of primary features may occur where there has been an interaction of different processes (e.g. bottom currents and turbidity currents), or more rapid deposition from stronger flows in the case of shallower-water bottom-current deposits.

A standard sequence or contourite model (Fig. 8.20) has been recognized (Faugeres *et al.*, 1984; Gonthier *et al.*, 1984), in which the grain size varies from fine mud, through mottled silty mud to silt and fine sand grade. Where this variation is mirrored by subtle changes in sedimentary structures and concomitant changes in the biogenic/terrigenous composition, then it is possible to interpret the sequence in terms of long-term variation in mean current velocity (Stow *et al.*, 1986). Unlike the standard Bouma (1962) sequence in turbidites, which is the result of instantaneous deposition, the contourite sequence is built up gradually over longer periods, in the order of tens of thousands years.

However, it is not always possible to use grain-size parameters alone as tracers of bottom-current intensity, because various other factors such as variations in terrigenous sediment supply or biogenic productivity may also influence grain size. The interplay of different controlling factors may explain the more irregular 'sequences' or lack of sequences recorded in some drift deposits (Stow & Faugeres, 1993).

8.6 Controls on process and deposit

8.6.1 Introduction

There are many different but overlapping variables that control processes, style and rates of sedimenta-

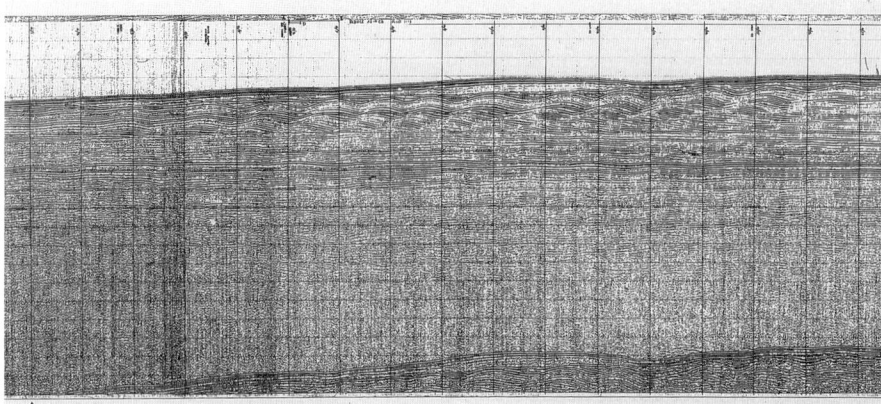

(a)

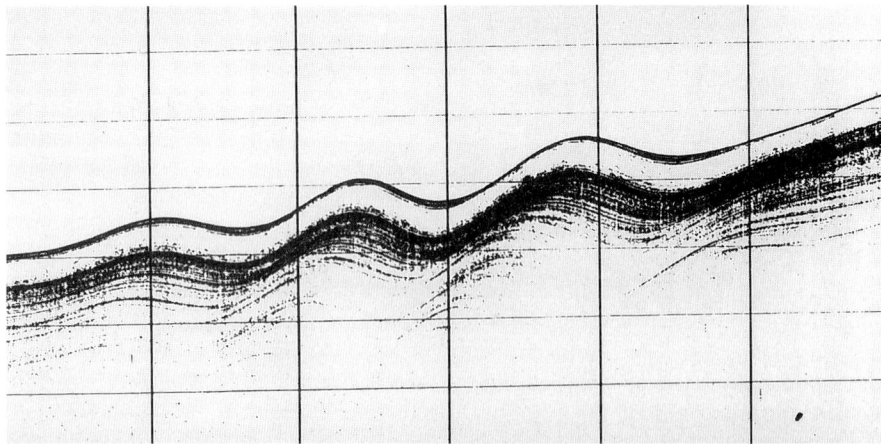

(b)

Fig. 8.18 Sediment waves formed under bottom current and turbidity current flow. (a) Buried sediment wave field, northwest UK Continental Margin (from Richards *et al.*, 1987); (b) detail of sediment waves on backside of small drift in the northeastern area of the Rockall Trough. (From Howe *et al.*, 1993.)

tion in the deep sea. These variables also influence processes in other environments but their effects are necessarily different. In this final section, we briefly review the most important controls including: (1) sediment type and supply; (2) eustatic and local sea-level changes; and (3) tectonic setting and activity. Other factors that have been dealt with in previous sections are: (4) oceanic circulation; (5) sea-water chemistry; and (6) basin size and shape. More wide-ranging controls that we will not discuss specifically here include the relative rates of generation and destruction of oceanic crust, the disposition of the continents and global climate. Good general reviews of controls on deep-sea sedimentation can be

found in Stow (1985, 1986), Stow *et al.* (1985), Bouma *et al.* (1985) Jenkyns (1986), and Pickering *et al.* (1989).

8.6.2 Sediment type and supply

Various types of sediment are available for deposition in the deep sea. Terrigenous material is the most abundant worldwide, with muds being between two and ten times as important volumetrically as sands and gravels. Biogenic debris from carbonate reefs and platforms is common at low latitudes, and calcareous and siliceous oozes may drape basins and slopes worldwide and be locally redeposited from areas of

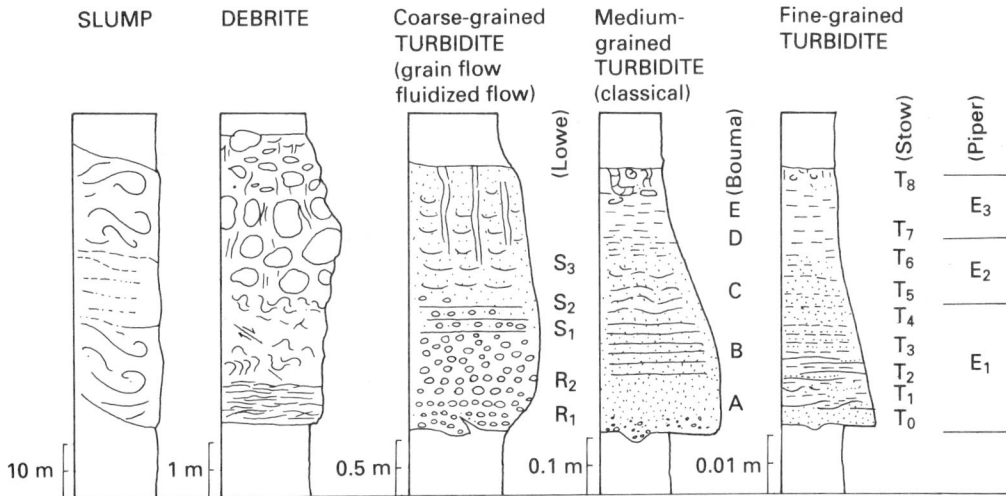

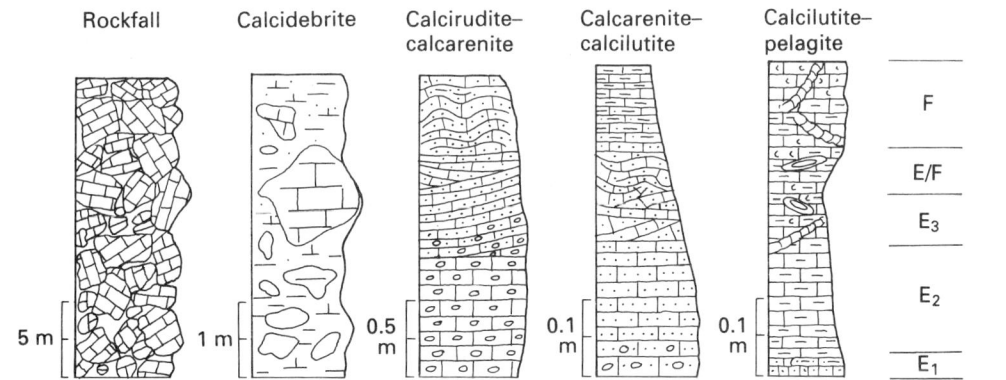

Fig. 8.19 Facies models for resedimented deposits. (a) Resedimented clastic facies models for slumps, debrites and turbidites, showing the idealized structural sequences. The scale bars give an indication only of typical unit thickness, which may vary widely in practice. Grain-size increases to the right for each column. (b) Resedimented carbonate facies models for rock falls, debrites and turbidites. Scale bars give an indication only of typical unit thickness, which may vary widely in practice. Grain size increases to the right for each column. (After Stow, 1986.)

high pelagic accumulation. Evaporites, volcaniclastics and organic-carbon-rich sediments can all occur as turbidites and associated facies. The sediment grain size affects the process and distance of transport and hence the geometry of the deposit. Biogenic particles behave differently from terrigenous grains during transport, so that carbonate facies differ from their clastic counterparts.

The volume and rate at which sediments are supplied to an area, and therefore made available for redeposition, are other important variables. Major river-delta systems, such as the Ganges, Indus, and

Mississippi, can provide a large and rapid supply of sediment to the shelf, although the availability of this material for downslope resedimentation will depend on sea level and shelf width. Wave-stirred, canyon-indented shelves will generally provide less sediment to the outer margin. In high latitudes, glaciers and floating ice shelves may greatly increase the supply of terrigenous material to the shelf margin. Low-latitude carbonate platforms and topographic highs covered with pelagic material commonly provide lower rates of sediment supply.

The number and spacing of input points along a

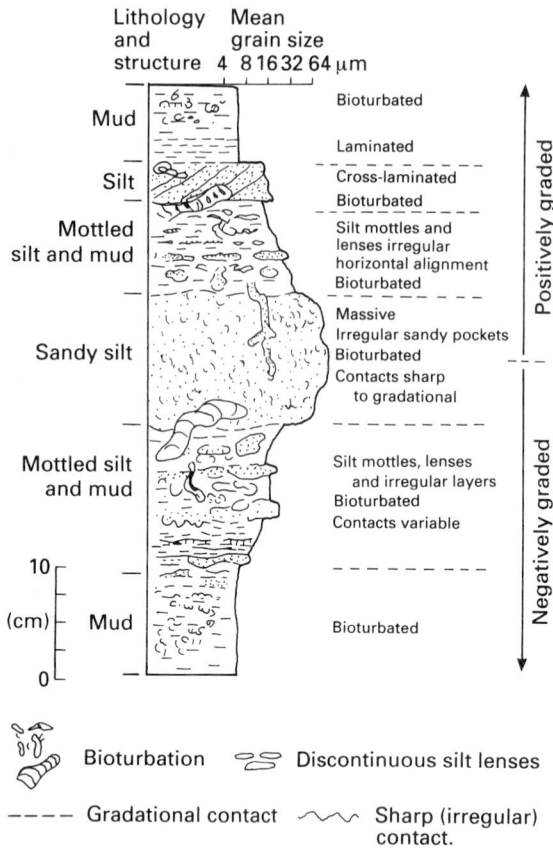

Fig. 8.20 Facies models for bottom current deposits. (After Gonthier *et al.*, 1984.)

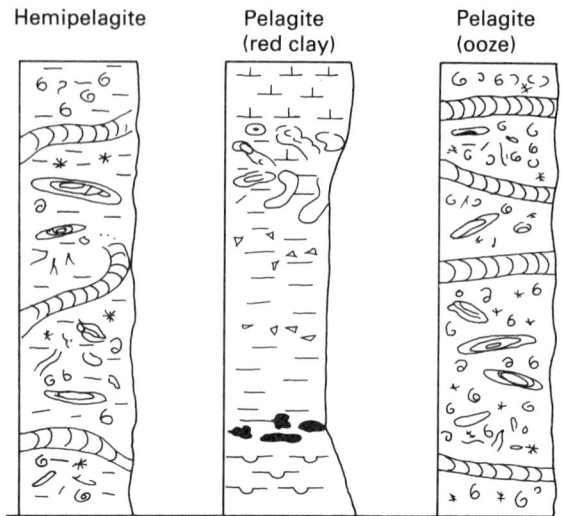

Fig. 8.21 Facies models for pelagites and hemipelagites. (After Stow, 1986.)

given margin will determine whether single or isolated fans are developed, whether an overlapping-fan/slope-apron system is produced, or whether a basin is filled completely or partially.

8.6.3 Eustatic and local sea-level changes

Fluctuation in sea level not only affects the nearshore realm of sedimentation, but also profoundly influences deep-sea depositional and resedimentation patterns. Shoreline sources such as rivers or littoral drift cells either may have direct access to basin slopes during periods of low sea level or indirect access through paralic and continental shelf environments during periods of high sea level.

Sea-level changes may be global (eustatic) or regional in nature. Eustatic fluctuations occur as a result of a change in the total volume of ocean basins or a change in the volume of sea water. Various attempts have been made to chronicle the eustatic fluctuations and construct an average worldwide curve of sea-level variation (e.g. Haq *et al.*, 1987). Regional fluctuations in sea level that result from local tectonic and isostatic factors can be much greater, and hence locally more significant, than eustatic changes.

Highstands of sea level are associated with an amelioration in climate, increased biological diversity, reduced oxygenation of sea water, shallowing of the CCD, condensed pelagic successions and mud-drape abandonment facies over terrigenous systems. Lowstands, by contrast, are associated with global cooling, widespread unconformities or hiatuses on the shelves and slopes, and increased clastic input into the deep sea.

8.6.4 Tectonic setting and activity

The main tectonic settings in which deep-sea sedimentation occurs include mature passive margins, active rifting margins, convergent margins with or without arc/trench systems, transform margins, marginal seas and back-arc basins, oceanic basins and intracratonic (shelf or mid-continent) basins.

These tectonic settings exert a first-order control on the style and nature of sedimentation by affecting the rates of uplift and denudation, drainage patterns, coastal plain and shelf widths, continental margin gradients, gross sediment budgets, the morphology of receiving basins, and local sea-level changes. The specific tectonic parameters that are particularly important are the size and internal geometry of the basin, including gradients of the base margin and floor. The style and frequency of seismic activity and faulting, both in the original and transitional source areas, are also of primary significance since these factors influence the frequency and volume of sediment gravity flows feeding the basin, for example mature passive margins experience infrequent, but commonly large, earthquakes, which may trigger very large slumps that develop into debris flows and turbidity currents. Frequent earthquakes along active margins do not permit a large build-up of sediment in transitional settings.

8.7 References

Allen J.R.L. 1971. Mixing at turbidity current heads, and its geological implications. *J. Sedim. Petrol.*, **41**, 97–113.

Allen J.R.L. 1982. *Sedimentary Structures: Their Characters and Physical Basis. Developments in Sedimentology*, **30**, I & II, Elsevier, Amsterdam.

Allen J.R.L. 1985. *Principles of Physical Sedimentology*. Allen & Unwin, London.

Asquith J.M. 1979. Nature and origin of the lower continental rise hills, off the east coast of the United States. *Mar. Geol.*, **32**, 165–190.

Athanasiou-Grivas D. 1978. Reliability analysis of earth slopes. *Proc. Soc Engineering Sci. 15th Annual Meeting, Gainesville, University of Florida,*, pp. 453–458.

Bagnold R.A. 1954. Experiments on a gravity-free dispersion of large solid spheres in a Newtonian fluid under shear. *Proc. R. Soc. Lond.*, *A*, **225**, 49–63.

Bagnold R.A. 1962. Auto-suspension of transported sediment; turbidity currents. *Proc. R. Soc. Lond.*, *A*, **265**, 315–319.

Barnes P.M. & Lewis K.B. 1991. Sheet slides and rotational failures on a convergent margin: the Kidnappers Slide, New Zealand. *Sedimentology*, **38**, 205–221.

Biscaye P.E. & Eittreim S.L. 1977. Suspended particulate loads and transports in the nepheloid layer of the abyssal Atlantic Ocean. *Mar. Geol.*, **23**, 155–72.

Bouma A.H. 1962. *Sedimentology of Some Flysch Deposits*. Elsevier, Amsterdam.

Bouma A.H., Barnes N.E. & Normark W.R. (eds). 1985. *Submarine fans and related turbidite systems*. Springer, New York.

Bowen A.J., Normark W.R. & Piper D.J.W. 1984. Modelling of turbidity currents on Navy Submarine Fan, California Continental Borderland. *Sedimentology*, **31**, 169–85.

Carson M.A. & Kirkby M.J. 1972. *Hillslope Form and Process*. Cambridge University Press, Cambridge.

Cheney R.E., Marsh J.G. & Beckley B.D. 1983. Global mesoscale variability from collinear tracks of SEASAT altimeter data. *J. Geophys. Res.* **88**, 4343–4354.

Chough S. & Hesse R. 1976. Submarine meandering talweg and turbidity currents flowing for 4000 km in the Northwest Atlantic Mid-ocean Channel, Labrador Sea. *Geology*, **4**, 529–533.

Curray J.R. & Moore D.G. 1971. Growth of the Bengal deep-sea fan and denudation in the Himalayas. *Bull. Geol. Soc. Am.*, **82**, 563–572.

Curray J.R. & Moore D.G. 1974. Sedimentary and tectonic processes in the Bengal deep-sea fan and geosyncline. In: Burk C.A. & Drake C.L. (eds) *The Geology of Continental Margins* pp. 617–627. Springer, New York.

Damuth J.E. 1975. Echo-character of the western equational Atlantic floor and its relationship to the dispersal and distribution of terrigenous sediments. *Mar. Geol.*, **18**, 17–45.

Dingle R.V. 1977. The anatomy of a large submarine slump on a sheared continental margin (SE Africa). *J. Geol. Soc. Lond.*, **134**, 293–310.

Dott R.H. Jr 1963. Dynamics of subaqueous gravity depositional processes. *Bull. Am. Ass. Petrol. Geol.*, **47**, 104–128.

Dowdeswell J.A. & Scourse J.D. (eds). 1990. *Glacimarine Environments: Processes and Sediments. Geol. Soc. Spec. Publ.*, **53**. Geological Society Publishing House, Bath.

Drake D.E., Hatcher P.G. & Keller G.H. 1978. Suspended particulate matter and mud deposition in Upper Hudson submarine canyon. In: Stanley D.J. & Kelling G. (eds) *Sedimentation in submarine canyons, fans, and trenches*, pp. 33–41. Dowden, Hutchinson & Ross, Stroudsberg, Philadelphia.

Eittreim S. & Ewing M. 1972. Suspended particulate matter in the deep waters of the North American Basin. In: Gordon A.L. (ed.) *Studies in Physical Oceanography*, **2**, 123–168. Gordon & Breach, New York.

Eittreim S., Thorndike E.M. & Sullivan L. 1976. Turbidity distribution in the Atlantic Ocean. *Deep-Sea Res.*, **23**, 1115–1128.

Embley R.W. 1980. The role of mass transport in the distribution and character of deep-ocean sediments with special reference to the North Atlantic. *Mar. Geol.*, **38**, 23–50.

Embley R.W., Hooje P.J., Lonsdale P., Mayer L. & Tucholke B.E. 1980. Furrowed mud-waves on the western Bermuda rise. *Bull Geol. Soc. Am.*, **91**, 731–740.

Faugeres J.C., Gonthier E. & Stow D.A.V. 1984. Contourite drift molded by deep Mediterranean outflow. *Geology*, **12**, (5) 296–300.

Flood R.D. & Shor A.N. 1988. Mud waves in the Argentine basin and their relationship to regional bottom circulation patterns. *Deep-Sea Res.*, **35**, 943–972.

Fukushima Y., Parker G. & Pantin H.M 1985. Prediction of

ignitive turbidity currents in Scripps Submarine Canyon. *Mar. Geol.*, **67**, 55–81.

Gardner W.D. & Sullivan L.G. 1981. Benthic storms: temporal variability in a deep ocean nepheloid layer. *Science*, **213**, 329–331.

Gill A.E. 1973. Circulation and bottom water production in the Weddell Sea. *Deep-Sea Res.*, **20**, 111–140.

Goldberg E.D. (ed.) 1974. *The Sea, Vol 5 (Marine Chemistry)*. Wiley-Interscience, New York.

Gonthier E.G., Faugeres J.C. & Stow D.A.V. 1984. Contourite facies of the Faro Drift, Gulf of Cadiz. In: Stow D.A.V. and Piper D.J.W. (eds) *Fine-grained Sediments: Deep-water Processes and Facies. Geol. Soc. (Lond.) Spec. Publ.*, **15**, 275–292. Blackwell Scientific Publications, Oxford.

Hampton M.A. 1972. The role of subaqueous debris flow in generating turbidity currents. *J. Sedim. Petrol.*, **42**, 775–793.

Hampton M.A. 1975. Competence of fine-grained debris flows. *J. Sedim. Petrol.*, **45**, 834–844.

Hampton M.A. 1979. Buoyancy in debris flows. *J. Sedim. Petrol.*, **49**, 753–758.

Hand B.M., Middleton G.V. & Skipper K. 1972. Antidune cross-stratification in a turbidite sequence, Cloridorme Formation, Gaspe, Quebec. *Sedimentology*, **18**, 135–138.

Haq B.U., Hardenbol J. & Vail P.R. 1987. Chronology of fluctuating sea levels since the Triassic. *Science*, **235**, 1156–1167.

Harms J.C. & Fahnestock R.K. 1965. Stratification, bed forms and flow phenomena (with an example from the Rio Grande). In: Middleton G.V. (ed.) *Primary Sedimentary Structures and their Hydrodynamic Interpretation.* Spec. Publ. Soc. Econ. Paleont. Mineral, **12**, 84–115. SEPM, Tulsa.

Hay A.E. 1983. On the frontal speeds of internal gravity surges on sloping boundaries. *J. Geophys. Res.*, **88**, 751–754.

Heezen B.C. & Ewing M. 1952. Turbidity currents and submarine slumps, and the 1929 Grand Banks earthquake. *Am. J. Sci.*, **250**, 849–873.

Heezen B.C. & Hollister C.D. 1971. *The Face of the Deep.* Oxford University Press, New York.

Heezen B.C., Ericson D.B. & Ewing M. 1954. Further evidence for a turbidity current following the 1929 Grand Banks earthquake. *Deep-Sea Res.*, **1**, 193–202.

Heezen B.C., Hollister C.D. & Ruddiman W.F. 1966. Shaping of the continental rise by deep geostrophic contour currents. *Science*, **152**, 502–508.

Hein F.J. 1982. Depositional mechanisms of deep-sea coarse clastic sediments, Cap Enrage Formation, Quebec. *Can J. Earth Sci.*, **19**, 267–287.

Hein F.J. & Gorsline D.S. 1981. Geotechnical aspects of fine-grained mass flow deposits: California Continental Borderland. *Geo-Marine Lett.*, **1**, 1–5.

Hill P.R. 1984. Sedimentary facies of the Nova Scotian upper and middle continental slope, offshore eastern Canada. *Sedimentology*, **31**, 293–309.

Hill P.R., Moran K.M. & Blasco S.M. 1982. Creep deformation of slope sediments in the Canadian Beaufort Sea. *Geo-Mar. Lett.*, **2**, 153–70.

Hiscott R.N. & Middleton G.V. 1979. Depositional mechanics of thick-bedded sandstones at the base of a submarine slope, Tourelle Formation (Lower Ordovician), Quebec, Canada. In: Doyle L.J. & Pilkey O.H. (eds) *Geology of Continental Slopes. Spec. Publ. Soc. Econ. Paleont. Mineral*, **27**, 307–326. SEPM, Tulsa.

Hjulstrom F. 1939. Transportation of detritus by moving water. In: Trask P.D. (ed.) *Recent Marine Sediments: A Symposium. Spec. Publ. Soc. Econ. Paleont. Mineral*, **4**, 5–31. SEPM, Tulsa.

Hollister C.D. & Heezen B.C. 1972. Geological effects of bottom currents: western North Atlantic. In: Gordon A.L. (ed.) *Studies in Physical Oceanography*, **2**, pp. 37–66. Gordon and Breach, New York.

Hollister C.D. & McCave I.N. 1984. Sedimentation under deep-sea storms. *Nature*, **309**, 220–225.

Hollister C.D., Flood R.D., Johnson D.A., Lonsdale P. & Southard J.B. 1974. Abyssal furrows and hyperbolic echo-traces on the Bahama Outer Ridge. *Geology*, **2**, 395–400.

Hughes Clarke J.E., Shor A.N., Piper D.J.W. & Mayer L.A. 1990. Large-scale current-induced erosion and deposition in the path of the 1929 Grand Banks turbidity current. *Sedimentology*, **37**, 613–629.

Jenkyns H.C. 1986. Pelagic environments. In: Reading H.G. (ed.) *Sedimentary Environments and Facies*, 2nd edn, pp. 343–397. Blackwell Scientific Publications, Oxford.

Johns D.R. 1978. Mesozoic carbonate rudites, megabreccias and associated deposits from central Greece. *Sedimentology*, **25**, 561–573.

Johnson A.M. 1970. *Physical Processes in Geology*. Freeman, Cooper, San Francisco.

Johnson A.M. 1984. (with contributions by Rodine J.R.) Debris flow. In: Brunsden D. & Prior D.B. (eds) *Slope Instability*, pp. 257–362. Wiley, New York.

Jones M.E. & Preston R.M.F. (eds) 1987. *Deformation of Sediments and Sedimentary Rocks. Geol. Soc. (Lond.) Spec. Publ.*, **29**. Blackwell Scientific Publications, Oxford.

Karlsrud K. & Edgers L. 1982. Some aspects of submarine slope stability. In: Saxov S. & Nieuwenhuis J.K. (eds) *Marine Slides and Other Mass Movements*, pp. 63–81. Plenum Press, New York.

Kenter J.A.M. 1990. Carbonate platform flanks: slope angle and sediment fabric. *Sedimentology*, **37**, 777–794.

Killworth P.D. 1973. A two dimensional model for the formation of Antarctic bottom water. *Deep-Sea Res.*, **20**, 941–971.

Kolla V., Eittreim S., Sullivan L., Kostecki J.A. & Burckle L.H. 1980. Current-controlled abyssal microtopography and sedimentation in Mozambique basin, Southwest

Indian Ocean. *Mar. Geol.*, **34**, 171–206.

Komar P.D. 1969. The channelized flow of turbidity currents with application to Monterey deep-sea fan channel. *J. Geophys. Res.*, **74**, 4544–4558.

Komar P.D. 1971. Hydraulic jumps in turbidity currents. *Bull. Geol. Soc. Am.*, **82**, 1477–1488.

Krause D.C., White W.C., Piper D.J.W. & Heezen B.C. 1970. Turbidity currents and cable breaks in the western New Britain Trench. *Bull. Geol. Soc. Am.*, **81**, 2153–2160.

Kuenen Ph.H. (1967) Emplacement of flysch-type sand beds. *Sedimentology*, **9**, 203–243.

Lafond E.C. 1962. Internal waves. In: Hill M.N. (ed.) *The Sea*, Vol. 1, pp. 731–751. Wiley Interscience, London.

Lee H.J. 1989. Undersea landslides: extent and significance in the Pacific Ocean. In: Brabb E.E. & Harrod B.L. (eds) *Landslides: Extent and Economic Significance*, pp. 367–380. Proc. 28th Int. Geol. Cong., Symposium on Landslides.

Lewis K.B. 1971. Slumping on a continental slope inclined at 1°–4°. *Sedimentology*, **16**, 97–110.

Lonsdale P. & Hollister C.D. 1979. A near-bottom traverse of Rockall Trough: hydrographic and geologic inferences. *Oceanologica Acta*, **31**, 91–105.

Lovell J.P.B. & Stow D.A.V. 1981. Identification of ancient sandy contourites. *Geology*, **9**, 347–349.

Lowe D.R. 1975. Water escape structures in coarse-grained sediments. *Sedimentology*, **22**, 157–204.

Lowe D.R. 1976. Subaqueous liquified and fluidized sediment flows and their deposits. *Sedimentology*, **23**, 285–308.

Lowe D.R. 1979. Sediment gravity flows: their classification and some problems of application to natural flows and deposits. In: Doyle L.J. & Pilkey O.H. (eds) *Geology of Continental Slopes. Spec. Publ. Soc. Econ. Paleont. Mineral*, **27**, 75–82. SEPM, Tulsa.

Lowe D.R. 1982. Sediment gravity flows: II. Depositional models with special reference to the deposits of high-density turbidity currents. *J. Sedim. Petrol.*, **52**, 279–297.

Luyten J.R. (1977) Scales of motion in the deep Gulf Stream and across the Continental Rise. *J. Mar. Res.*, **35**, 49–74.

Malinverno A., Ryan W.B.E., Auffret G.I. & Pautot G. 1988. Sonar images of the path of recent failure events on the continental margin off Nice, France. *Spec. Pap. Geol. Soc. Am.*, **229**, 59–75.

Mariani M., Argnani A., Roveri M. & Trincardi F. 1993. Sediment drifts and erosional surfaces in the central Mediterranean: seismic evidence of bottom current activity. *Sedim. Geol.*, **82**, 207–220.

Marjanac T. 1985. Composition and origin of the megabed containing huge clasts, flysch formation, middle Dalmatia, Yugoslavia. In: *Abstracts and Poster Abstracts*, pp. 270–273. *6th European Regional Meeting, Lleida, Spain, Int. Assoc. Sedimentologists*.

McCave I.N. 1972. Transport and escape of fine-grained sediment from shelf areas. In: Swift D.J.P., Duane D.B. & Pilkey O.H. (eds) *Shelf Sediment Transport: Process and Pattern*. pp. 225–248. Hutchinson & Ross, Stroudsburg PA.

McCave I.N. 1984. Erosion, transport and deposition of fine-grained marine sediments. In: Stow D.A.V. & Piper D.J.W. (eds) *Fine Grained Sediments: Deep-Water Processes and Facies. Spec. Publ. Geol. Soc. Lond.*, **15**, 35–69. Blackwell Scientific Publications, Oxford.

McCave I.N. & Jones P.N. 1988. Deposition of ungraded muds from high-density non-turbulent turbidity currents. *Nature*, **333**, 250–252.

McCave, I.N. & Swift S.A. 1976. A physical model for the rate of deposition of fine-grained sediments in the deep sea. *Bull. Geol. Soc. Am.*, **87**, 541–546.

McCave I.N. & Tucholke B.E. 1986. Deep current-controlled sedimentation in the western North Atlantic. In: Vogt P.R. & Tucholke B.E. (eds) *The Geology of North America, Vol. M: The Western North Atlantic Region, Decade of North American Geology*, Geology Society of America, pp. 451–468. Boulder, Colorado.

McCave I.N., Londsdale P.F., Hollister C.D. & Gardner W.D. 1980. Sediment transport over the Hatton and Gardar contourite drifts. *J. Sedim. Petrol.*, **50**, 1049–1062.

McLean S.R. 1985. Theoretical modelling of deep-ocean sediment transport. *Mar. Geol.*, **66**, 243–265.

Menard H.W. 1964. *Marine Geology of the Pacific*. McGraw Hill, New York.

Middleton G.V. 1966a. Experiments on density and turbidity currents: I. Motion of the head. *Can. J. Earth Sci.*, **3**, 523–546.

Middleton G.V. 1966b. Experiments on density and turbidity currents: II. Uniform flow of density currents. *Can. J. Earth Sci.*, **3**, 627–637.

Middleton G.V. 1966c. Small scale models of turbidity currents and the criterion for auto-suspension. *J. Sedim. Petrol.*, **36**, 202–208.

Middleton G.V. 1967. Experiments on density and turbidity currents: III. Deposition of sediment. *Can. J. Earth Sci.*, **4**, 475–505.

Middleton G.V. & Hampton M.A. 1973. Sediment gravity flows: mechanics of flow and deposition. In: Middleton G.V. & Bouma A.H. (eds) *Turbidites and Deep Water Sedimentation*, pp. 1–38. *Short Course Notes, Pacific Sect. Soc. Econ. Paleont. Mineral.*

Middleton G.V. & Hampton M.A. 1976. Subaqueous sediment transport and deposition by sediment gravity flows. In: Stanley D.J. & Swift D.J.P. (eds) *Marine Sediment Transport and Environmental Management*, pp. 197–218. John Wiley, New York.

Middleton G.V. & Southard J.B. 1984. *Mechanics of sediment transport*. 2nd edn. Soc. Econ. Paleont. Mineral. Eastern Section Short Course No. 3, Providence.

Middleton G.V. & Southard J.B. 1978. *Mechanics of Sedi-*

ment Movement. *SEPM Short Course*, **3**. Tulsa.

Miller M.A., McCave I.N. & Komar P.D. 1977. Threshold of sediment motion under unidirectional currents. *Sedimentology*, **24**, 507–528.

Moore D.G. 1961. Submarine slumps. *J. Sedim. Petrol.*, **31**, 343–57.

Moore D.G. 1977. Submarine slides. In: Voight B. (ed.) *Rockslides and Avalanches*, Vol. 1. *Developments in Geotechnical Engineering*, **14A**, 563–604.

Morgenstern N. 1967. Submarine slumping and the initiation of turbidity currents. In: Richards A.F. (ed.) *Marine Geotechnique*, pp. 189–220. University of Illinois Press, Urbana.

Nardin T.R., Hein F.J., Gorsline D.S. & Edwards B.D. 1979. A review of mass movement processes, sediment and acoustic characteristics and contrasts in slope and base-of-slope systems versus canyon–fan–basin floor systems. In: Doyle L.J. & Pilkey O.H. Jr. (eds) *Geology of Continental Slopes. Spec. Publ. Soc. Econ. Paleont. Mineral.*, **27**, 61–73. SEPM, Tulsa.

Nelson C.H. & Kulm L.D. 1973. Submarine fans and channels. In: *Turbidites and Deep Water Sedimentation*, pp. 39–78. *Soc. Econ. Paleont. Mineral, Pacific Section, Short Course*. Anaheim.

Neuman G. 1968. *Ocean Currents*. Elsevier, Amsterdam.

Normark W.R., Hess G.R., Stow D.A.V. & Bowen A.J. 1980. Sediment waves on the Monterey Fan levees: a preliminary physical interpretation. *Mar. Geol.*, **37**, 1–8.

Normark W.R., Piper D.J.W. & Hess G.R. 1979. Distributary channels, sand lobes, and mesotopography of Navy Submarine Fan, California Borderland, with applications to ancient fan sediments. *Sedimentology*, **26**, 749–774.

Nowell A.R.M. & Hollister C.D. (eds) 1985. Deep Ocean sediment transport. *Mar. Geol.*, **66**, 1–4.

Open University. 1984. *Oceanography, Units 11 and 12, Sediments*. The Open University Press, Milton-Keynes.

Pantin H.M. 1979. Interaction between velocity and effective density in turbidity flow: phase plane analysis, with criteria for autosuspension. *Mar. Geol.*, **31**, 59–99.

Pantin H.M. & Leeder M.R. 1987. Reverse flow in turbidity currents: the role of internal solitions. *Sedimentology*, **34**, 1143–1155.

Pickering K.T. & Hiscott R.N. 1985. Contained (reflected) turbidity currents from the Middle Ordovician Cloridorme Formation, Quebec, Canada: an alternative to the antidune hypothesis. *Sedimentology*, **32**, 373–394.

Pickering K.T., Hiscott R.N. & Hein F.J. 1989. *Deep Marine Environments: Clastic Sedimentation and Tectonics*. Unwin Hyman, London, 416 pp.

Pierson T.C. 1981. Dominant particle support mechanisms in debris flows at Mt Thomas, New Zealand, and implications for flow mobility. *Sedimentology*, **28**, 49–60.

Piper D.J.W. 1978. Turbidite muds and silts on deep sea fans and abyssal plains. In: Stanley D.J. & Kelling G. *Sedimentation in Submarine Canyons, Fans and Trenches*, pp. 163–175. Dowden, Hutchinson & Ross, Stroudsburg, PA.

Piper D.J.W. & Normark W.R. 1982. Effects of the 1929 Grand Banks earthquake on the continental slope off eastern Canada. In: *Current Research, Part B. Geol. Surv. Can. Pap.*, **82–1B.**

Piper D.J.W. & Stow D.A.V. 1991. Fine-grained turbidites. In: Einsele G. & Seilacher A. (eds) Cycles and Events in Stratigraphy, pp. 360–376, 2nd edn. Springer-Verlag, Berlin.

Piper D.J.W., Normark W.R. & Stow D.A.V. 1984. The Laurentian Fan–Sohm Abyssal Plain. *Geol. Mar. Lett.*, **3**, 141–146.

Porebski S.J., Meischner D. & Gorlich K. 1991. Quaternary mud turbidites from the South Shetland Trench (West Antarctica): recognition and implications for turbidite facies modelling. *Sedimentology*, **38**, 691–715.

Prior D.B. & Suhayda J.N. 1979a. Submarine landslide morphology and development mechanisms, Mississippi Delta. *Proc. 11th Conf. Ocean Technol., Paper OTC 3482*, 1055–1058.

Prior D.B. & Suhayda J.M. 1979b. Application of infinite slope analysis to subaqueous sediment instability, Mississippi Delta. *Engng. Geol.*, **14**, 1–10.

Rattay M. 1960. On the coastal generations of internal tides. *Tellus*, **12**, 54.

Redbourn L., Bull J., Scrutton R.A. & Stow D.A.V. 1993. Channels, echocharacter mapping and tectonics from 3.5 kltz profiles, distal Bengal Fan. *Mar. Geol.* (in press).

Richardson M.J., Wimbush M. & Mayer L. 1981. Exceptionally strong near-bottom flows on the continental rise off Nova Scotia. *Science*, **213**, 887–888.

Richardson P.L. 1983. Eddy kinetic energy in the North Atlantic from surface drifters. *J. Geophys. Res.*, **88**, 4355–4367.

Saxov S. & Nieuwenhuis J.K. (eds) 1982. *Marine Slides and Other Mass Movements*. Plenum Press, New York.

Schwarz H.-U. 1982. Subaqueous slope failures — experiments and modern occurrences. *Contrib. Sedimentol.*, **11**.

Shepard F.P. 1973. *Submarine Geology*, 3rd edn. Harper and Row, New York.

Shepard F.P. & Dill R.F. 1966. *Submarine Canyons and Other Sea Valleys*. Rand McNally, Chicago.

Shepard F.P., Marshall N.F., McLoughlin P.A. & Sullivan G.G. 1979. Currents in submarine canyons and other seavalleys. *Stud. Geol. Am. Ass. Petrol. Geol.*, **8**.

Shepard F.P., McLoughlin P.A., Marshall N.F. & Sullivan G.G. 1977. Current-meter recordings of low-speed turbidity currents. *Geology*, **5**, 297–301.

Shields A. 1936. Mitt. Preuss. Vers. Anst. Wasserb. u. Schiffb., Berlin, Heft 26.

Shor A., Lonsdale P., Hollister C.D. & Spencer D. 1980. Charlie–Gibbs fracture zone: bottom-water transport and its geologic effects. *Deep-Sea Res.*, **27A**, 325–345.

Shor A.N., Piper D.J.W., Hughes Clarke J.E. & Mayer L.A.

1990. Giant flute-like scour and other erosional features formed by the 1929 Grand Banks turbidity current. *Sedimentology*, **37**, 631–645.

Shultz A.W. 1984. Subaerial debris-flow deposition in the Upper Paleozoic Cutter Formation. *J. Sedim. Petrol.*, **54**, 759–72.

Simm R.W., Weaver P.P.E., Kidd R.B. & Jones E.J.W. 1991. Quaternary mass movement on the lower continental rise and abyssal plain off western Sahara. *Sedimentology*, **38**, 27–40.

Simpson J.E. 1969. A comparison between laboratory and atmospheric density currents. *Q.J.R. Met. Soc.*, **95**, 758–765.

Simpson J.E. 1972. Effects of the lower boundary on the head of a gravity current. *J. Fluid Mech.*, **53**, 759–768.

Simpson J.E. 1982. Gravity currents in the laboratory atmosphere and ocean. *Ann. Rev. Fluid Mech.*, **14**, 213–234.

Skipper K. & Bhattacharjee S.B. 1978. Backset bedding in turbidites: a further example from the Cloridorme Formation (Middle Ordovician), Gaspe, Quebec. *J. Sedim. Petrol.*, **48**, 193–202.

Southard J.B. & Mackintosh M.E. 1981. Experimental test of autosuspension. *Earth Surf. Proc. Landf.*, **6**, 103–111.

Sparks R.S.J., Bonnecaze R.T., Huppert H.E., Lister J.R., Hallworth M.A., Mader H & Phillips J. 1993. Sedimen-laden gravity currents with reversing buoyancy. *Earth Planet. Sci. Lett.*, **114**, 243–257.

Stommel H. 1957. A survey of ocean currents theory. *Deep-Sea Res.*, **4**, 149–184.

Stommel H. 1958. The abyssal circulation. *Deep-Sea Res.*, **5**, 80–82.

Stommel H. & Aarons A.B. 1960a. On the abyssal circulation of the World ocean — I. Stationary planetary flow patterns on a sphere. *Deep-Sea Res.*, **6**, 140–154.

Stommel H. & Aarons A.B. 1960b. On the abyssal circulation of the World ocean — II. An idealized model of the circulation pattern and amplitude in oceanic basins. *Deep-Sea Res.*, **6**, 217–233.

Stow D.A.V. 1979. Distinguishing between fine-grained turbidites and contourites on the deep-water margin off Nova Scotia. *Sedimentology*, **26**, 371–387.

Stow D.A.V. 1982. Bottom currents and contourites in the North Atlantic. *Bull. Inst. Geol. Bassin d'Aquitaine*, **31**, 151–166.

Stow D.A.V. 1985. Deep-sea clastics: where are we and where are we going? In: Brenchley P.J. & Williams B.J.P. (eds) *Sedimentology: Recent Developments and Applied Aspects. Geol. Soc. Lond. Spec. Publ.*, **18**, 67–93.

Stow D.A.V. 1986. Deep clastic seas. In: Reading H.G. (ed.) *Sedimentary Environments and Facies*, pp. 398–444,

Blackwell Scientific Publications, Oxford.

Stow D.A.V. 1992. *Deep-Water Turbidite Systems.* International Association of Sedimentologists, No. 3. Blackwell Scientific Publications, Oxford.

Stow D.A.V. & Bowen A.J. 1980. A physical model for the transport and sorting of fine-grained sediments by turbid-ity currents. *Sedimentology*, **27**, 31–46.

Stow D.A.V. & Faugeres J.C. (eds) 1993. Contourites and bottom currents. *Sedim. Geol.*, **82**. Special Issue.

Stow D.A.V. & Lovell J.P.B. 1979. Contourites: their recognition in modern and ancient sediments. *Earth-Sci. Rev.*, **14**, 251–291.

Stow D.A.V. & Shanmugam G. 1980. Sequence of structures in fine-grained turbidites: comparison of recent deep-sea and ancient flysch sediments. *Sedim. Geol.*, **25**, 23–42.

Stow D.A.V. & Wetzel A. 1990. Hemiturbidite: a new type of deep-water sediment. In: Cochran J.R., Stow D.A.V. *et al.* (eds) *Proc. Ocean Drilling Programme Sci. Results*, **116**, 25–34.

Stow D.A.V., Faugeres J.C. & Gonthier E.G. 1986. Facies distribution and drift growth during the late Quaternary, Faro Drift, Gulf of Cadiz. *Mar. Geol.*, **72**, 71–100.

Stow D.A.V., Howell D.G. & Nelson C.H. 1985. Sedimentary, tectonic, and sea-level controls. In: Bouma, A.H., Normark W.R. & Barnes N.E. (eds) *Submarine Fans and Related Turbidite Systems*, pp. 15–22. Springer, New York.

Stow D.A.V., Amano K., Bulson P.S. *et al.* 1990. Sediment facies and processes on the distal Bengal Fan, Leg 116. In: Cochran J.R., Stow D.A.V. *et al.* (eds) *Proc. Ocean Drilling Programme Sci. Results*, **116**, 377–396.

Summerhayes C.P. (ed.) 1992. Upwelling systems: evolution since the early Miocene. *Geol. Soc. (Lond.) Spec. Pub.*, **64**. Geological Society Publishing House, Bath.

Takahashi T. 1981. Debris flow. *Ann. Rev. Fluid Mech.*, **13**, 57–77.

Tolmazin D. 1985. *Elements of Dynamic Oceanography.* Allen and Unwin, London.

Walker R.G. 1965. The origin and significance of the internal sedimentary structures of turbidites. *Proc. Yorks. Geol. Soc.*, **35**, 1–32.

Watkins D.J. & Kraft L.M. 1978. Stability of continental shelf and slope off Louisiana and Texas: geotechnical aspects. In: Bouma A.H., Moore G.T. & Coleman J.M. (eds) *Framework Facies and Oil-Trapping Characteristics of the Upper Continental Margin. Stud. Geol. Am. Ass. Petrol. Geol.*, **7**, 267–286. Tulsa.

Woodcock N.H. 1976. Structural style in slump sheets: Ludlow series, Powys, Wales. *J. Geol. Soc. Lond.*, **132**, 399–415.

9 Aeolian sediment transport and deposition

W.G. NICKLING

9.1 Introduction

The entrainment, transport and deposition of sediment by wind are important geological processes which modify the Earth's surface. However, these processes exhibit great temporal and spatial variability on both geological and human timescales. Moreover, short-term changes in climate and landuse patterns have dramatic effects on both the frequency and magnitude of wind erosion. This was dramatically demonstrated during the 1930s in the Great Plains of the United States and the prairies of western Canada when decreased precipitation coupled with intensive agriculture caused a significant increase in wind erosion giving rise to the term 'dirty thirties'. More recently, drought conditions in the Sahel region of Africa, combined with overgrazing, extensive cutting of vegetation for firewood and other landuse pressures resulting from overpopulation in some areas, resulted in major increases in wind erosion on both agricultural and undisturbed desert soils during the 1980s. Although this increase in wind erosion is significant, and has caused famine and terrible hardships for the people living in the Sahel, it is not an isolated event in recent or geologic time (Morales, 1979).

Considerable insight has been gained into aeolian processes and landforms since Bagnold published his classic work *The Physics of Blown Sand and Desert Dunes* in 1941, setting the foundation for almost all subsequent aeolian research. However, it is only through continued research that we are able to understand more fully those factors affecting the frequency and magnitude and the spatial and temporal variation of these processes. With this knowledge we can develop control measures and strategies to reduce wind erosion in affected areas. Furthermore, through the interpretation of aeolian sediments and bedforms incorporated into the ancient rock record and deep ocean sediments we can improve our understanding of the Earth's geological and climatological history.

9.2 Global distribution of aeolian processes and deposits

In general, wind erosion is not as widespread or severe as water erosion. However, it is of major importance in environments where soil moisture deficits are frequent. Such areas are commonly thought of as deserts but may also include both arid and semi-arid regions.

Although many definitions for the term 'desert' exist in the literature, the most generally accepted definitions are based on climatic data that are used to develop specific moisture indices. The distribution of the world's drylands based on Meigs' (1953) classification scheme, is shown in Fig. 9.1. Most of the world's drylands are concentrated on the major continents near the tropics in the intertropical convergence zone. Overall, drylands comprise approximately one third of the Earth's land surface area with 4% classified as extremely arid, 15% as arid and 14.6% as semi-arid (Cooke & Warren, 1973).

Despite the frequent association of wind erosion with arid environments, even in more humid temperate regions a considerable amount of soil can be lost by wind each year. This may occur in localized areas, during periods of high wind velocities and moisture deficits, and where landuse activities increase the erosion potential of the surface. For example, it has been estimated that the total annual soil loss by wind

293

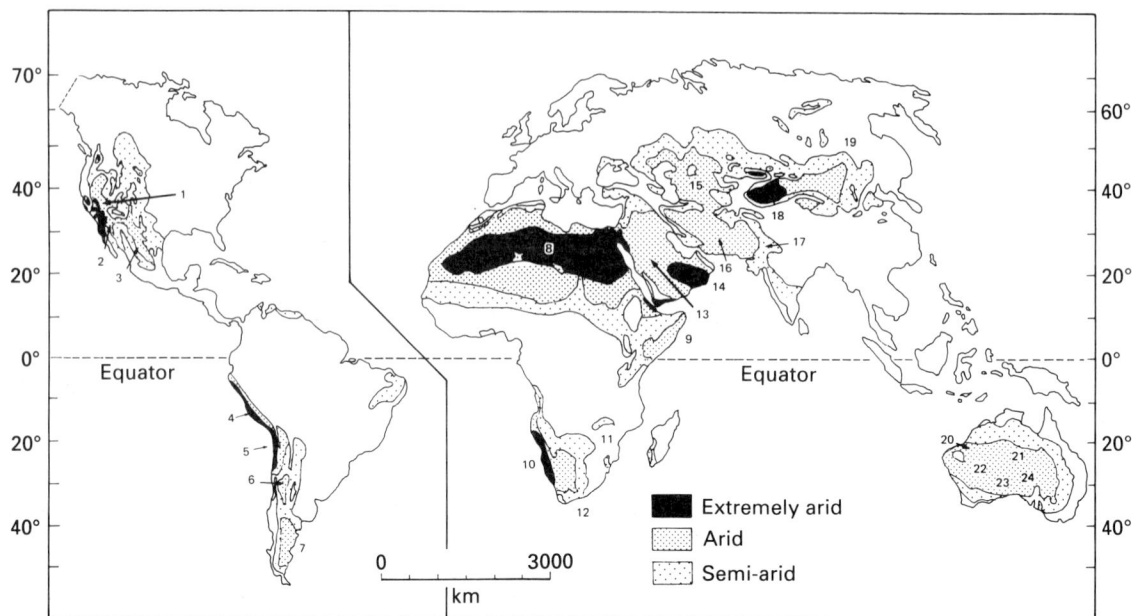

Fig. 9.1 Distribution of world drylands. Major deserts incude: (1) Great Basin, (2) Sonoran, (3) Chihuahuan, (4) Peruvian, (5) Atacama, (6) Monte, (7) Patagonian, (8) Sahara, (9) Somali-Chabli, (10) Namib, (11) Kalahari, (12) Karroo, (13) Arabian, (14) Rub al Khali, (15) Turkestan, (15) Iranian, (17) Thar, (18) Taklimakan, (19) Gobi, (20) Great Sandy, (21) Simpson, (22) Gibson, (23) Great Victoria, and (24) Sturt. Modified from Cooke and Warren (1973) and Greeley and Iversen (1985).

on the prairies of western Canada is approximately 160×10^6 t compared with an annual soil loss of 117×10^6 t by water (Coote *et al.*, 1981). Similar soil losses by wind have also been reported for the Great Plains of the USA (Lyles, 1975).

Although wind erosion of cultivated lands is a serious global problem, the deflation of 'natural' surfaces in arid and semi-arid environments has also been shown to have important and sometimes devastating environmental and geological impacts. For example, Peterson and Junge (1971) have estimated that more than 500×10^6 t of dust (diameter $\leqslant 20\,\mu\text{m}$) are transported annually by wind. Prospero and Nees (1977) have suggested that high dust concentrations in the atmosphere from deflation may be partially responsible for recent droughts in the Sahel region of Africa. Other studies have emphasized the environmental effects of high dust concentrations on human health hazards and transport of pathogens (Bar-Ziv & Goldberg, 1974; Leathers, 1981), degradation of ambient air quality (Hagen & Woodruff, 1973; Nickling & Gillies, 1989), localized changes in global energy balance (Idso & Brazel, 1977) and

decreased visibility leading to traffic accidents and airport closures (Patterson & Gillette, 1977; Buritt & Hyers, 1981).

The transport of sand-size material (diameter 62.5–$200\,\mu\text{m}$) can also have significant economic impacts. Drifting sand in arid and semi-arid environments frequently covers roadways and runways causing disruption of road and air traffic. Sand-size material deposited on coastal beaches is frequently blown inland to form dune ridges which may pile up and around buildings (Fig. 9.2) and other structures (Nordstrom & McCluskey, 1984). In agricultural areas, drifting sand can bury cultivated lands, resulting in decreased soil fertility and crop productivity. Moreover, agricultural crops as well as other plants can be severely damaged or killed by the abrasion of blowing sand.

The geological record clearly shows both the continued importance as well as the temporal and spatial variability of aeolian processes in the development of the Earth's surface. Extensive ancient and recent aeolian deposits are found in many parts of the world. For example, the Coconino Sandstone of

Fig. 9.2 Encroachment of a blowout dune over a family cottage, Prince Edward County, north shore Lake Ontario, Canada.

Permian Age, ranges from tens to hundreds of metres thick and covers thousands of square kilometres of the Colorado Plateau in the southwestern USA. The distinctive large cross-beds and other related structures are indicative of the aeolian origin of this sandstone (Greeley & Iversen, 1985). Similarly, vast areas of the Earth's surface are covered with comparatively recent wind-blown silt deposits termed loess, that vary in thickness from less than 1 m to over 300 m. Extensive, thick loess deposits occur in

China, Soviet Central Asia, the Ukraine, Central and Western Europe, Argentina, and the Great Plains of the USA (Fig. 9.3) (Pye, 1987). Pecsi (1968) suggests that loess and loess-like deposits cover 10% of the Earth's land surface, although Pye (1987) states that primary aeolian loess probably covers only about 5% of the surface area.

Recent research has also shown that aeolian dust is a major source of sediment in the deep-ocean basins, in some locations comprising up to 80% of the total

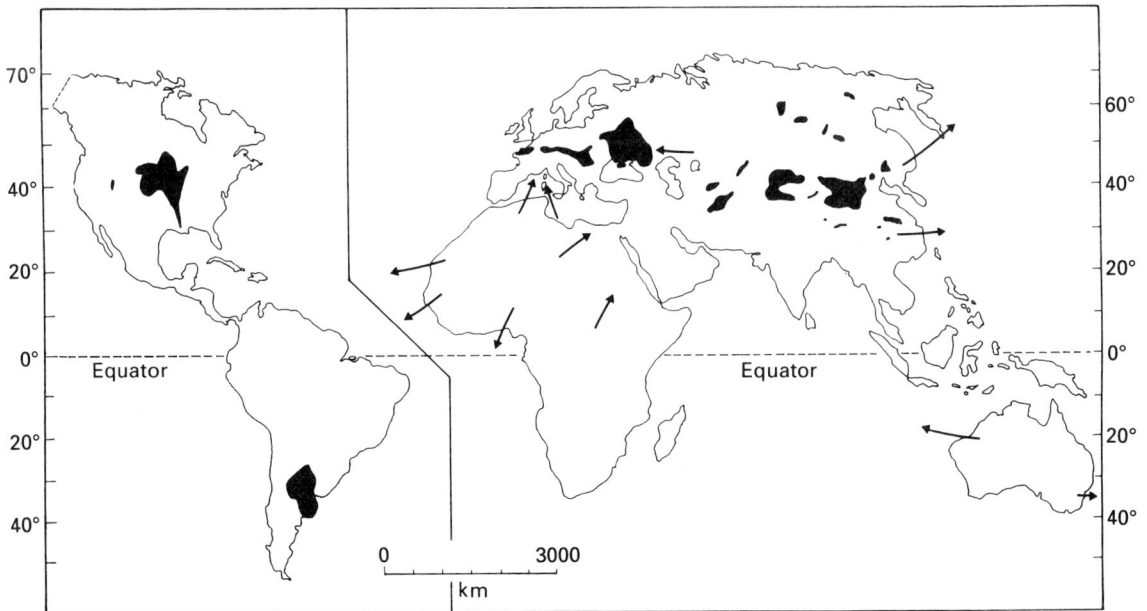

Fig. 9.3 Distribution of major loess deposits showing dominant dust transporting winds (arrowed). Modified from Pye (1987).

accumulated material (Pye, 1987). Sedimentary deposits derived from aeolian dust fallout are most extensive in the eastern Atlantic and northwest Pacific but are also important in the northwest Indian Ocean, southwest Pacific and eastern equatorial Pacific. Janecek and Rea (1983), on the basis of a deep-sea core taken in the Pacific at 21°N, 174°E, have concluded that the average dust deposition over the past 700 000 years is approximately 0.15 t km^{-2} year^{-1}. Somewhat higher rates of deposition of 1.29 t km^{-2} year^{-1} have been recorded in surface sediments from a core taken in the Pacific at 9°N, 104°W (Rea *et al.*, 1985).

Dust accumulation observed in deep-sea cores have provided useful information on climatic change and shifts in global wind patterns during the Quaternary. The oceanic sedimentary record is relatively complete compared to the continental record, with relatively good stratigraphic and age control provided by the biogenic components in the sediments (Pye, 1987).

Vast sheets of aeolian sand, frequently called ergs or sand seas, are also found in many parts of the world, usually in regions where precipitation is less than 15 cm year^{-1} (Wilson, 1971) (Fig. 9.4). Ergs vary greatly in size, from approximately 100 km^2 to over 500 000 km^2 with a modal size of approximately 188 000 km^2 (Wilson, 1971). The largest erg is the Rub al Khali in Saudi Arabia with an area greater than 560 000 km^2, approximately 2.2 times the area of Great Britain. Wilson (1972a) has estimated that 99.8% of aeolian sand is found in ergs larger than 125 km^2 and 85% is found in ergs larger than 32 000 km^2.

Despite the extent of present and past aeolian activity on Earth, recent research has indicated that wind erosion processes are also important on other planets, particularly Mars, Venus and Titan (Greeley & Iversen, 1985). Detailed investigations into aeolian processes and landforms on Mars and other planets (e.g. Breed *et al.*, 1979; Greeley *et al.*, 1980, 1982; Iversen & White, 1982) through theoretical modelling and wind tunnel simulation has provided new insight and understanding of aeolian processes on these planets as well as on Earth.

Wind erosion, like many other geological and climatological processes, has both positive and negative attributes in relation to human activity. The erosion, transport and deposition of sediment by wind can cause significant environmental problems including deterioration of agricultural soils, disruption of road and air traffic, as well as increased health

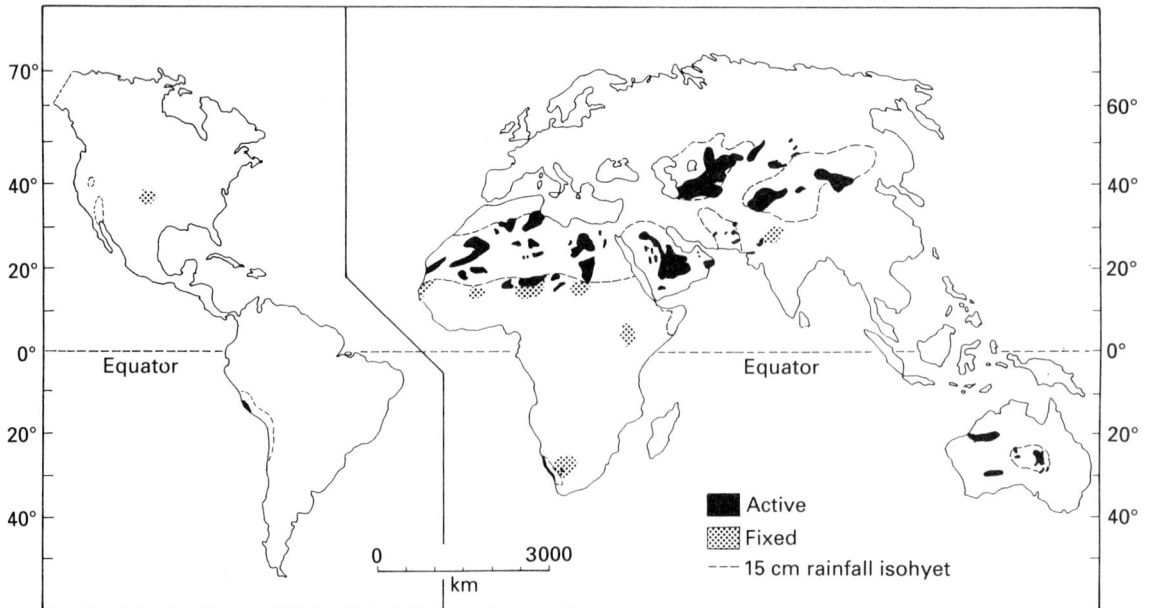

Fig. 9.4 World distribution of major ergs in relation to the 15 cm rainfall isohyet. Modified from Cooke and Warren, (1973).

hazards to humans and animals (Fig. 9.5). Nevertheless, wind-deposited sediments are of extreme importance to human existence. The naturally fertile loess soils are major areas of crop production throughout the world. Moreover, dust deposition has been shown to increase soil fertility by the addition of airborne nutrients (Yaalon & Ganor, 1973; Beavington & Cawse, 1979). In some locations wind-blown sediments, and in particular ancient sand seas, form important reservoirs for water and petroleum (Greeley & Iversen, 1985; Pye & Tsoar, 1990).

The study and understanding of aeolian processes are important for many different reasons. From an academic point of view, the study of aeolian processes and deposits allows us to gain further insight into the development of the Earth's surface and the nature of climatic change during geological time. This may be accomplished by comparing active aeolian processes and deposits to sediments in the rock record and deep-sea cores. In a more applied context, understanding of aeolian processes aids in the development of more suitable controls, measures and strategies to reduce wind erosion on susceptible soils. Ultimately, predictive models may be developed for planning and development purposes.

9.3 Factors affecting the entrainment and transport of sediment by wind

9.3.1 The wind profile

When air blows across a fixed bed at relatively high velocities the flow is usually turbulent and is characterized by eddies of varying size moving with different speeds and directions. As a result of viscous frictional effects, the wind speed near the bed is retarded (Fig. 9.6). If the surface is comprised of very fine particles ($< 80\,\mu m$) the surface is said to be aerodynamically smooth ($Re_p \lesssim 5$). A very thin laminar flow layer, usually less than 1-mm thick, develops adjacent to the bed even for flows in which most of the boundary layer is turbulent. In the laminar

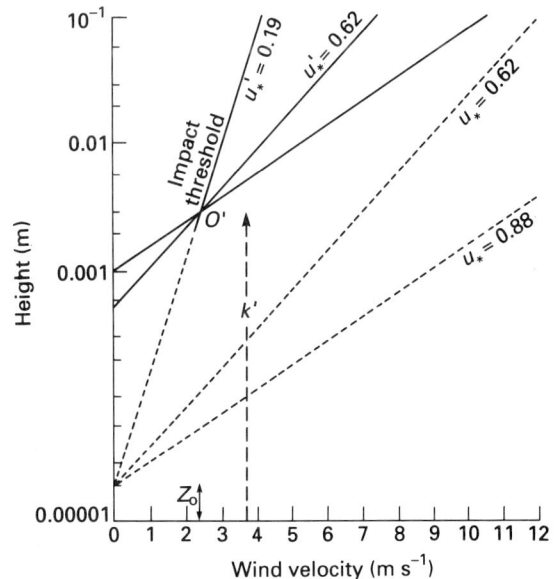

Fig. 9.5 Recent drought conditions in the Sahel region of Africa has resulted in significant increases of wind erosion causing severe hardships for local inhabitants. The degree of wind erosion is dramatically evident by the deflation of soil that has exposed the roots of a tree in central Mali.

Fig. 9.6 Wind velocity profiles over a stationary and mobile sand bed showing the relative shear velocities (u_*) as defined in Eq. 9.1.

sub-layer the velocity increases approximately in a linear manner with distance above the surface.

In contrast, when surface particles or other roughness elements are relatively large the surface is aerodynamically rough ($Re_p \geq 70$). The laminar sub-layer ceases to exist and is replaced by a viscous sub-layer for which the velocity profile is not well understood (Middleton & Southard, 1984). Under neutral stability conditions, the velocity profile above the viscous sub-layer for aerodynamically rough surfaces can be characterized by the Prandtl–von Kármán equation:

$$\frac{U}{u_*} = \frac{1}{k} \ln \frac{z}{z_0}, \qquad (9.1)$$

where U is the velocity at height z, z_0 is the characteristic roughness, length or height of the surface, u_* is the shear velocity and k is von Kármán's constant ($\simeq 0.4$). Typical values of roughness heights for various surfaces are shown in Table 9.1.

Over a stationary bed the velocity distribution with height plots as a straight line on semi-logarithmic graph with the intercept representing the roughness height (z_0). Bagnold (1941) and numerous other investigators have noted that no matter how strongly the wind blows (i.e. increases in u_*) all the velocity profiles tend to converge to the same intercept or roughness height (Fig. 9.6).

For sand surfaces, z_0 has been found by numerous investigators to be approximately equal to 1/30 the mean particle diameter, but also varies with the shape and average distance between individual particles or other roughness elements as shown in Fig. 9.7. In situations where the surface is covered by tall vegetation or high densities of other large roughness elements, the wind velocity profile is displaced upwards from the surface to a new reference plane which is a function of the height, density, porosity and flexibility of the roughness elements (Oke, 1978). The upward displacement is termed the zero plane displacement height (d). For these situations the wind profile equation becomes:

$$\frac{U}{u_*} = \frac{1}{k} \frac{(z-d)}{z_0}. \qquad (9.2)$$

The shear velocity (u_*) is proportional to the slope of the wind velocity profile when plotted with a logarithmic height scale (Fig. 9.6) and is related to the shear stress (τ_0) at the bed and the air density (ρ_a) by:

$$u_* = \sqrt{\frac{\tau_0}{\rho_a}}. \qquad (9.3)$$

When the wind velocity is great enough to move sand particles the wind velocity profile is altered because of momentum losses from the air to the sand grains. The velocity distributions with height remain as straight lines as shown by Bagnold (1941) but tend to cross at a particular point (Fig. 9.6). The height of the focus, k', appears to be associated with the height of ripples that form on the surface and is analogous to the relationship between the roughness height, z_0, and the grain diameter.

Table 9.1 Aerodynamic properties of natural surfaces. Modified from Oke (1978)

Surface	Remarks	z_0 Roughness length	Roughness displacement (m)	d Zero plane (m)
Water*	Still—open sea		0.1–10.0×10^{-5}	–
Ice	Smooth		0.1×10^{-4}	–
Snow			0.5–10.0×10^{-4}	–
Sand, desert			0.0003	–
Soils			0.001–0.01	–
Grass*		0.02–0.1 m	0.003–0.01	≤ 0.07
		0.25–1.0 m	0.04–0.20	≤ 0.66
Agricultural crops*			0.04–0.20	≤ 3.0
Orchards*			0.5–1.0	≤ 4.0
Forests*	Deciduous		1.0–6.0	≤ 20.0
	Coniferous		1.0–6.0	≤ 30.0

* z_0 depends on wind speed.

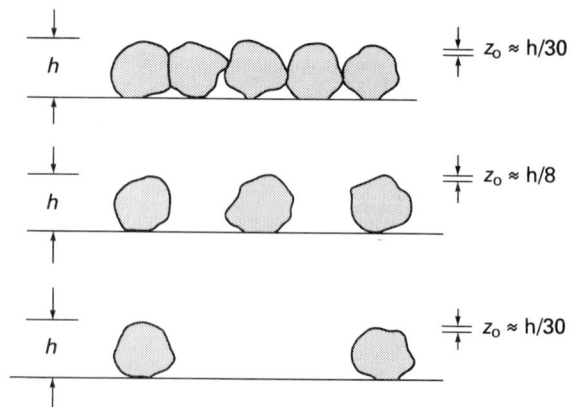

Fig. 9.7 Effect of particle spacing on the roughness length z_0. For closely spaced particles such as a sand surface the roughness length is approximately 1/30 the mean grain diameter (h). The maximum roughness length of 1/8 the grain diameter occurs when the centre to centre distance is about twice the diameter. From Greeley and Iversen (1985).

9.3.2 Particle forces

Consider a sand-size particle sitting at rest on the surface of a loose granular bed over which a fully turbulent air stream is moving. In this situation several forces act on the particle (Fig. 9.8). For an idealized case with spherical grains the forces acting on the exposed grain include fluid drag (D_F), lift (L_F), moment (M), weight (W) and interparticle cohesion (I_p). The drag and lift forces as well as the resultant moment are caused by the fluid flow around and over the exposed particles. The lift force results from the decreased fluid static pressure at the top of the grain as well as the steep pressure gradient near the grain surface (Bernoulli effect). The weight and cohesive forces are related to physical properties of the surface particles including size, density, mineralogy, shape, packing arrangement, moisture content and the presence or absence of bonding agents such as soluble salts.

At the moment of entrainment the combined retarding effect of the weight and cohesive forces will be just overcome by the resultant drag and lift forces. The particle will tend to pivot about point P in a downstream direction. The balance of forces just prior to entrainment can be obtained by a summation of moments about the pivot point P. Thus:

$$D_F a + L_F b + M = Wb + I_p c, \qquad (9.4)$$

where a, b and c are the moment arms (Fig. 9.8).

Despite numerous studies of the entrainment process, these forces have not been determined with accuracy because of the difficulty of measurement. Moreover, the complex interaction between these forces, and cohesion is presently not well understood.

9.3.3 Theoretical and empirical expressions for particle threshold

General considerations regarding threshold forces

Theoretical values of the drag force and moment for a sphere resting on a plane bed in uniform shear for $Re < 0.45$ have been derived by Goldman *et al.* (1967). Saffman (1965, 1968) has derived related expressions for the lift force in similar flow conditions.

Direct measurement of the lift force on spheres have been made by Einstein and El-Samni (1949) and Chepil (1958) and for the drag force by Coleman (1972) and Coleman and Ellis (1976a,b). Although there is some discrepancy in the experimental results, drag and lift forces appear to be of the same order of magnitude for relatively high Reynolds numbers.

Greeley and Iversen (1985) present general expressions for the drag, lift and moment forces as:

$$D = K_D \rho_p u_*^2 D_p^2$$
$$L = K_L \rho_p u_*^2 D_p^2 \qquad (9.5)$$
$$M = K_M \rho_p u_*^2 D_p^3,$$

where K_D, K_L and K_M are drag, lift and moment coefficients and ρ_p is the grain density. Typical values for these coefficients are presented in Table 9.2.

The moment arm lengths are proportional to the grain diameter and thus $a = a_1 D_p$, $b = b_1 D_p$ and $c = c_1 D_p$. Substituting Eq. 9.5 into Eq. 9.4 gives:

$$K_D \rho_p u_*^2 D_p^2 (a_1 D_p) + K_L \rho_p u_*^2 D_p^2 (b_1 D_p) + K_M \rho u_*^2 D_p^3$$
$$= W_t (b_1 D_p) + I_p (c_1 D_p), \qquad (9.6)$$

which can be simplified and written in the dimensionless form:

$$A^2 = \frac{(\pi b_1/b^6)\,[1 + (6c_1/\pi b_1)\,I_p/\rho_p\,gD_p^3]}{K_D a_1 + K_M + b_1 K_L}, \qquad (9.7)$$

where A is termed the dimensionless threshold friction speed and g the acceleration due to gravity.

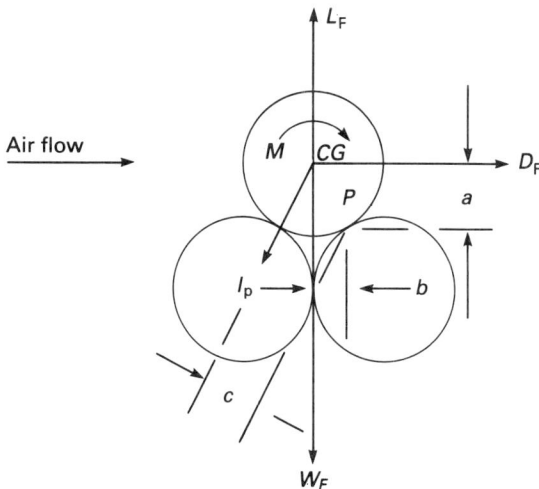

Fig. 9.8 Forces acting on a stationary grain, where D_F, I_p, L_F, M and W_F are the drag, inter-particle lift, moment and weight forces, respectively, on the grain; CG and P are the cente of gravity and pivot point, respectively; and a, b and c are the moment arm lengths.

Table 9.2 Experimental and theoretical values of small particle shear flow force coefficients. After Greeley and Iversen (1985)

Drag coefficient (K_D)	Lift coefficient (K_L)	Moment coefficient (K_M)	Reynolds number (Re_p)	Fluid medium
8.01		0.74	<0.45	Theoretical
	0.808 Re_p		<0.45	Theoretical
5.44			0.95	Hydroxyethyl-cellulose solution
15.42			130–13 200	Water
	2.42		3600	Water
3–4.7	2.2–5		1000–1400	Air

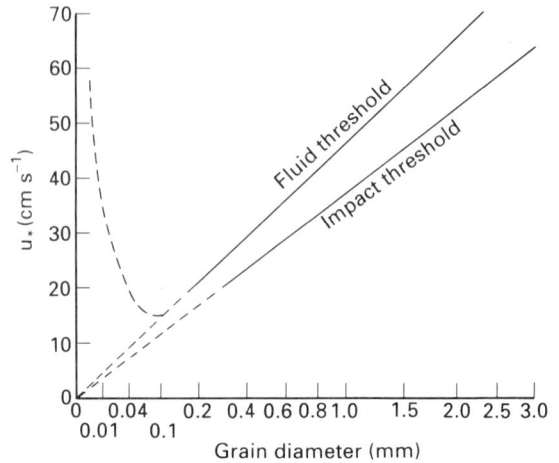

Fig. 9.9 Relationship of threshold shear velocity (u_*) to grain size. Modified from Bagnold (1941).

The Bagnold threshold equation

One of the first researchers to investigate the entrainment of sediment by wind was Bagnold (1941). On the basis of theoretical considerations and empirical observations, Bagnold (1941) developed a relatively simple expression to define the critical shear velocity (u_{*t}) at which the movement of loose particles begins:

$$u_{*t} = A \sqrt{\left(\frac{\rho_p - \rho_a}{\rho_a} gd\right)}, \qquad (9.8)$$

where A is an empirical coefficient dependent on the surface grain characteristics and approximately equal to the dimensionless friction speed of Eq. 9.7. Bagnold found from empirical observation that A had a value of approximately 0.1 for particle friction Reynolds numbers (Re_p) greater than 3.5.

The relationship between threshold shear velocity and particle diameter based on the above expression is shown in Fig. 9.9. For grains larger than approximately 80 μm the threshold shear velocity increases linearly with the square root of grain diameter. In this situation the relatively large grains protrude into the air stream and carry a greater proportion of the fluid drag than the general surface. In contrast, for Re_p <3.5 the particles lie below the viscous sub-layer and the drag, instead of being carried by a few isolated grains, is distributed more uniformly over the entire surface. In such cases the surface is said to be aerodynamically smooth and the value of the coefficient A rises rapidly. Consequently, u_{*t} is no longer proportional to the square root of grain diameter but becomes dependent on A.

Bagnold (1941) suggested that once the critical threshold shear velocity was reached, stationary surface grains begin to roll or slide along the surface by the direct pressure of the wind. Once particles begin to gain speed they start to bounce off the surface into the air stream in a process termed saltation. More recent work by Chepil (1959), Bisal and Nielsen (1962) and Iversen and White (1982) has indicated that particles moving into saltation do not usually roll or slide along the surface prior to upward movement into the air stream. On the basis of wind tunnel observations they suggested that the intitial movement into saltation is caused by instantaneous air pressure differences near the surface which act as impulse (i.e. lift) forces. They observed that as wind velocity was increased across a sand surface, particles began to vibrate with increasing intensity and, at some critical point, left the surface instantaneously as if ejected. Few particles were seen to roll along the surface and bounce into the airstream as described by Bagnold (1941). Lyles and Krauss (1971) have also reported similar observations but suggest that the vibrations were seldom steady. They found that particle vibration would occur in flurries of 3–5 vibrations (1.8 ± 0.3 Hz for 0.59–0.84 mm sand grains) before the particles instantaneously left the bed.

During the downwind saltation of grains, the velocity, and hence momentum, is increased before they fall back to the surface. On striking the surface, the moving particles may ricochet off other grains and become re-entrained into the airstream or embedded in the surface. In both cases, momentum is transferred to the surface in the disturbance of one or more stationary grains. As a result of the impact of saltating grains, particles are ejected into the airstream at shear velocities lower than that required to move a stationary grain by direct fluid pressure. This new, lower threshold required to move stationary grains after the initial movement of a few particles has been termed the *dynamic* or *impact threshold* (Bagnold, 1941). Wind tunnel experiments by Bagnold (1941) indicate that the dynamic impact threshold for a given sediment follows the same square root function as the fluid threshold (Eq. 9.8) but with a lower coefficient A of 0.08 instead of 0.1 (Fig. 9.9).

Other approaches to particle threshold

More recent investigations by Iversen *et al.* (1976a, b), White *et al.* (1976), Greeley *et al.* (1977) and Iversen and White (1982) have reconsidered Bagnold's (1941) model and, in particular, the relative importance of lift and cohesive forces as well as the theoretical basis of Bagnold's A parameter. These authors have questioned the assumption that the A coefficient is a unique function of the particle friction Reynolds number (Re_p). They suggest that for very small grains interparticle forces due to electrostatic charges, moisture films and other cohesive forces are very important and independent of particle density (Greeley & Leach, 1978). Very small particles, even when thoroughly dry, cohere on contact with other particles or solid surfaces. Iversen *et al.* (1976a) and Greeley and Leach (1978) also argue that interparticle forces become much more important when gravitational forces are reduced on other planets such as Mars.

Iversen *et al.* (1976a) and Greeley *et al.* (1977) contend that, provided particle size and density are known, u_{*t} can be predicted from Bagnold's equation (Eq. 9.8) if the function $A(Re_p)$ is known. In order to investigate the form of this function these authors have carried out detailed theoretical and experimental investigations over a number of years. In contrast to most previous aeolian research, these scientists have investigated aeolian processes in a planetary context by the extrapolation of results from Earth-based research to Mars, Venus and Titan. Experimental testing was carried out in specially designed wind tunnels that have the capability of duplicating atmospheric conditions on these planets as well as on Earth. The following presents a brief summary of their approach and results.

These researchers suggest that Bagnold's (1941) threshold equation can be rewritten in the dimensionless form:

$$A = u_{*t} \, (\rho_p / \rho_a \, g \, D_p)^{1/2}, \qquad (9.9)$$

where A is the dimensionless threshold friction speed and is related to the A coefficient in Eq. 9.8. In Eq. 9.9 the air density term in the numerator of Eq. 9.8 is eliminated since $\rho_p \ggg \rho_a$.

Iversen *et al.* (1976a,b) carried out a large number of wind tunnel tests at one atmosphere pressure on a wide range of particles with different mean sizes and densities. Results of these tests are illustrated in Fig. 9.10 and show the measured threshold friction speed plotted against a threshold parameter, $(\rho_p \, g \, D_p / \rho_a)^{1/2}$. The figure clearly indicates that for relatively large particles the threshold friction parameter, A, (i.e. the slope of the curve) is essentially constant. If the parameter, A, was constant for all grain sizes the data would fall on a straight line passing through the origin. However, as can be clearly seen in Fig. 9.10, there is a minimum below which there is an upswing and increased scatter in the data. This suggests that A is not solely a function of particle diameter for small particles less than approximately 80 μm.

Iversen *et al.* (1976a) have also plotted their threshold results in dimensionless form and compared them with threshold curves derived by other investigators (Fig. 9.11). For $Re_p \geqslant 3$ the results are comparable but for smaller Reynolds numbers they are somewhat disparate. The authors suggest that the scatter may have resulted from differences in particle size distribution of the test materials, the inherent problems in direct measurement of particle threshold but most importantly, the effects of interparticle cohesion that cause A not to be a sole function of Re_p for small particles.

Subsequent research by Greeley *et al.* (1980) and Iversen and White (1982) was carried out to investigate the combined effects of atmospheric pressure and particle density using specially designed wind tunnels. Results of these tests confirmed the importance of interparticle forces for particles < 80 μm.

On the basis of theoretical considerations and wind tunnel tests, Iversen and White (1982) and Greeley and Iversen (1985) suggest that the general form of the threshold equation can be defined by:

$$A = A_1 f \, (Re_p) \, (1 + K_1/\rho_p \, g \, D_p^n)^{1/2}, \quad (9.10)$$

where K_1 and n are the interparticle force coefficient

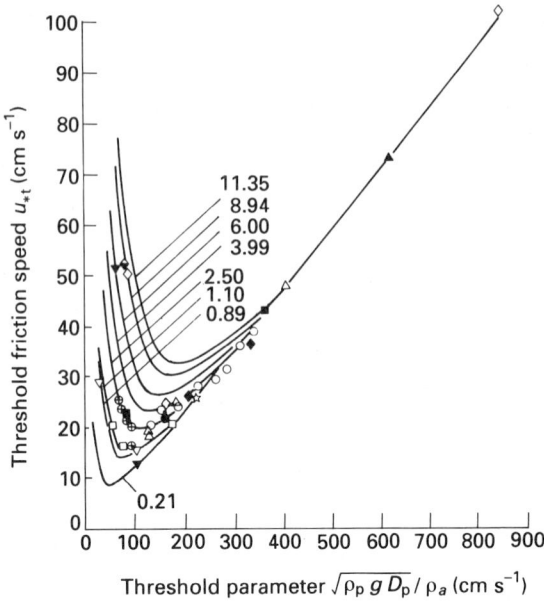

Fig. 9.10 Threshold friction speeds at one atmosphere pressure for particles of different sizes and densities. From Iversen *et al.* (1976a).

Material		Density (g cm^{-1})	Diameter (μm)
▼	Instant tea	0.21	719
▽	Silica gel	0.89	17;169
□	Nut shell	1.10	40 to 359
■	Clover seed	1.30	1290
☆	Sugar	1.59	393
⊕	Glass	2.42	31 to 48
○	Glass	2.50	38 to 586
◆	Sand	2.65	526
⬦	Aluminium	2.70	36 to 204
△	Glass	3.99	55 to 519
◔	Copper oxide	6.00	10
▲	Bronze	7.80	616
◇	Copper	8.94	12,37
◆	Lead	11.35	8,720

$$A = 0.2 \, (1 + 0.006/\rho_p \, g \, D_p^{2.5})^{1/2}/(1 + 2.5 \, Re_p)^{1/2},$$

for $0.03 \leqslant Re_p \leqslant 0.3$; (9.11)

$$A = 0.129 \, (1 + 0.006/\rho_p \, g \, D_p^{2.5})^{1/2}/$$
$$(1.928 \, Re_p^{0.092} - 1)^{1/2},$$

for $0.3 \leqslant Re_p \leqslant 10$; (9.12)

$$A = 0.120 \, (1 + 0.06/\rho_p \, g \, D\rho^{2.5})_p^{1/2}$$
$$\{1 - 0.0858 \exp \left[- 0.0617 \, (Re_p - 10) \right]\},$$

for $Re_p \geqslant 10$. (9.13)

9.3.4 Complicating effects of textural and surficial conditions on entrainment

Textural conditions

Nickling (1988) has argued that despite the fact that fluid threshold can be closely defined for a uniform sediment size greater than approximately 0.1 mm, it cannot be defined for most natural sediments because of several complicating factors. Natural sediments, no matter how well sorted, usually contain a range of grain sizes and shapes as well as variations in grain density and packing that can cause variation in both fluid and dynamic threshold. As a result, fluid and dynamic threshold cannot be defined by finite values and should be viewed as threshold ranges that are a function of the size, shape, sorting and packing of the surface sediments. Nickling (1988) used a sensitive laser monitoring system (Fig. 9.12) in conjunction with a high speed counter to detect initial grain motions and count individual grain movements in wind tunnel tests using natural sands and glass beads. These results indicated that, as velocity was slowly increased over the sediment surface, the smaller or more exposed grains were first entrained by fluid drag and lift forces in either surface creep or saltation. As the velocity continued to rise, larger or less exposed grains were also moved by fluid forces. On striking the surface, saltating grains imparted momentum to other stationary grains. Impacting grains were found to rebound from the surface as well as to eject one or more stationary grains at shear velocities lower than that required to entrain the stationary particles by direct fluid pressures and lift forces. The newly ejected particles moved downwind and impacted the surface, displacing an even larger number of stationary grains. As a result, there is a cascade effect with a few grains of varying sizes and shapes being entrained primarily in drag and lift

and exponent, respectively and A_1 is the cohesionless threshold coefficient. This expression is a generalization of Eq. 9.7 in terms of the particle friction Reynolds number (Re_p) and assumes the interparticle force (I_p) is proportional to particle diameter to the exponent ($3-n$).

Using linear regression on the wind tunnel data, Iversen and White (1982) have estimated values for A_1, K_1 and n for a wide range of particle friction Reynolds numbers and have derived the following threshold equations:

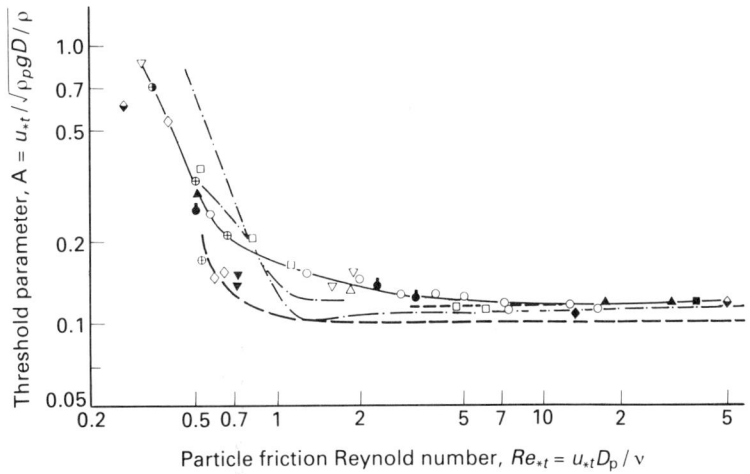

Fig. 9.11 Threshold friction speed parameter versus particle friction Reynolds number comparing the data curves of Bagnold (1941), Chepil (1945b, 1959) and Zingg (1953). From Greeley and Iversen (1985).

Material		Density (g cm^3)	Diameter (μm)
▼	Instant tea	0.21	719
▽	Silica gel	0.89	17;169
□	Nut shell	1.10	40 to 359
■	Clover seed	1.30	1290
☆	Sugar	1.59	393
⊕	Glass	2.42	31 to 48
○	Glass	2.50	38 to 586
◆	Sand	2.65	526
◖	Aluminium	2.70	36 to 204
△	Glass	3.99	55 to 519
◑	Copper oxide	6.00	10
▲	Bronze	7.80	616
◇	Copper	8.94	12,37
◈	Lead	11.35	8,720

——— , Bagnold (1941); —·—· , Chepil (1945, 1959); ———— , Zingg (1953).

forces. These grains, in turn, set in motion a rapidly increasing number of grains primarily by saltation impact (impact threshold range). This progression from fluid to dynamic threshold, based on number of grain movements is characterized by plots of the number of grain movements against increasing shear velocity measured in the wind tunnel tests (Fig. 9.13).

In each case the grain count data near threshold can be fitted by a hyperbolic function having the general form:

$$C = \frac{a}{(u_* - b)^2},\qquad (9.14)$$

where C is the number of grain movements and a and b are empirical coefficients for $0 < u_* < b$.

The general shape and, in particular, the radius of curvature in the transition section as well as the relative position of the curves shown in Fig. 9.13 are a function of the a and b coefficients, respectively. It is evident that the asymptote and general position of the transition section of the curves tend to shift along the shear velocity axis as mean grain size increases. There is also a general trend for the radius of curvature in the transition section to decrease as the sorting of the sample improves. This is clearly demonstrated in Figs 9.14 and 9.15 which show the changes in the a and b coefficients with mean size and sorting of the test materials.

Nickling (1988) has also shown that the shear velocity associated with the minimum radius of curvature on the grain count plots (Fig. 9.13) closely approximates the fluid threshold determined from Bagnold's (1941) threshold equation.

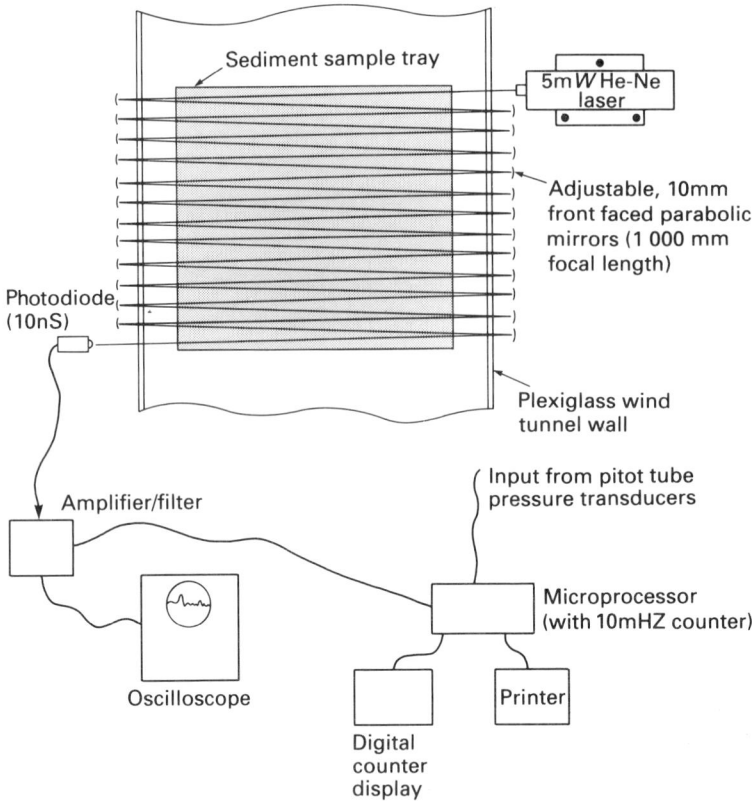

Fig. 9.12 Laser particle counting system used for the determination of particle threshold. From Nickling (1988).

Surface moisture

Field observations and wind tunnel studies have shown that surface moisture content is an extremely important variable controlling both the entrainment and transport of sediment by wind. From wind tunnel experiments, Belly (1964) has shown that gravimetric moisture contents of approximately 0.6% can more than double the threshold velocity of medium-sized sands. Above approximately 5% gravimetric moisture content, sand-sized material is inherently resistant to entrainment by most natural winds.

Belly (1964) modified Bagnold's (1941) threshold equation to account for moisture content:

$$U = A \left(\frac{\rho_p - \rho_a}{\rho_a} \right)^{0.5} (1.8 + 0.6 \log_{10} M), \quad (9.15)$$

where M is moisture content expressed as per cent dry weight.

In contrast, Azizov's (1977) wind tunnel experiments on agricultural soils showed an exponential relationship between the increase in moisture content

and threshold velocity which differs significantly from the results of Belly (1964) and Bisal and Hsieh (1966). The empirical relationships presented by these and other investigators (e.g. Chepil, 1956) most likely result from differences in experimental procedures and the textural characteristics of the sediments tested which affect the interparticle forces between grains.

In an attempt to develop a more generally applicable relationship, McKenna-Neuman and Nickling (1989) have derived and tested a theoretical moisture model based on capillary force equations derived by Haines (1925), Fisher (1926) and Allberry (1950). The model is based on the capillary forces developed at interparticle contacts surrounded by isolated wedges of water. The capillary forces (F_c) are inversely proportional to moisture tension and directly proportional to the geometric properties of the contacts. The capillary force at the contact can be defined by:

$$F_c = \frac{\pi T^2}{P} G \quad (9.16)$$

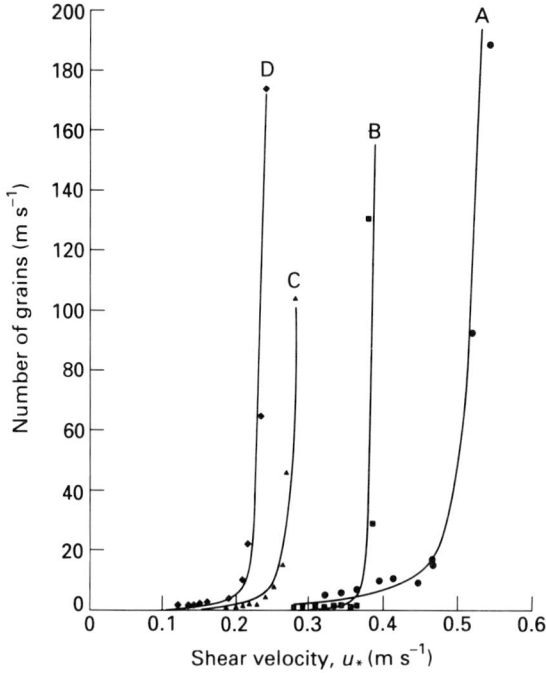

Fig. 9.13 Increase in number of grain entrainments with increasing shear velocity for four sands measured with a laser particle counting system. Sands have different mean sizes and sorting characteristics that are reflected in the shape and relative positions of the grain count curves. Sand A (●) Mean size = 0.37 φ (0.77 mm); sorting = 0.39φ; $c = 0.14/(u_* - 0.549)^2$. Sand B (■) Mean size = 0.96φ (0.51 mm); sorting = 0.15φ; $c = 7.0 \times 10^{-4}/(u_* - 0.383)^2$. Sand C (▲) Mean size = 1.90φ (0.27 mm); sorting = 0.34φ; $c = 1.14 \times 10^{-2}/(u_* - 0.285)^2$. Sand D (◆) Mean size = 2.43φ (0.19 mm); sorting = 0.29φ; $c = 8.6 \times 10^{-3}/(u_* - 0.240)^2$. After Nickling (1988).

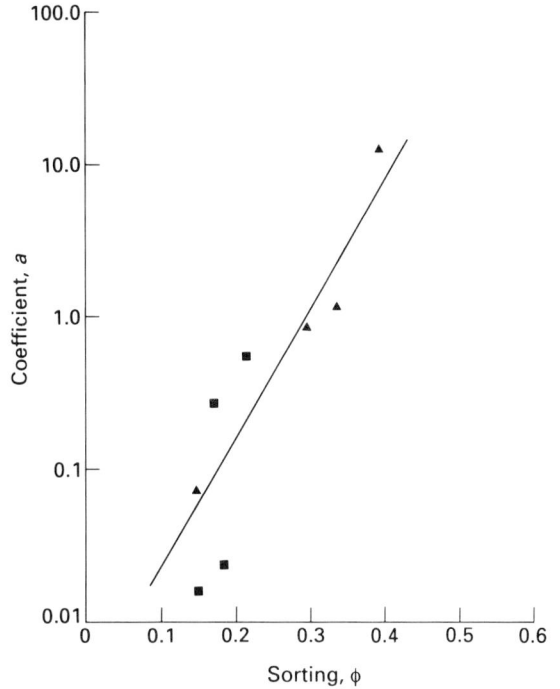

Fig. 9.15 Change in a, coefficient of the grain count curves, with sorting of the test sands. ▲, Sands; ■, unimodal glass beads; $\log a = 8.55\phi - 2.48$; $r^2 = 0.74$. Nickling (1988).

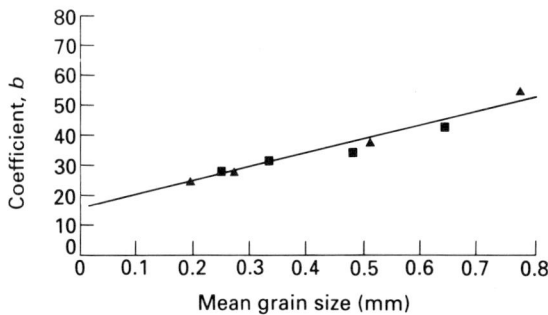

Fig. 9.14 Change in b, coefficient of the grain count curves with increasing grain size. ▲, Sands; ■ unimodal glass beads; $b = 16.22 + 47.01 \bar{x}$. $r = 0.94$.

where T is the surface tension of water, G is a dimensionless geometric coefficient describing the shape of the contacts between grains and P is the moisture tension developed in the water wedge at the grain contact.

Equation 9.16 can be expressed as a capillary force moment for grains with either open or closed packing arrangements. The capillary force moment has been incorporated into Bagnold's (1941) threshold equation providing a general expression for the threshold velocity of a moist surface:

$$u_{*\mathrm{tw}} = A\left[\frac{\rho_g - \rho_a}{\rho_a}\right]^{0.5}\left[\frac{6 \sin 2\beta}{\pi D^3 (\rho_p - \rho_a) g \sin \beta} F_c + 1\right]^{0.5}$$

(9.17)

where β is the particle resting angle. A comparison of predicted and observed results from wind tunnel tests is shown in Fig. 9.16.

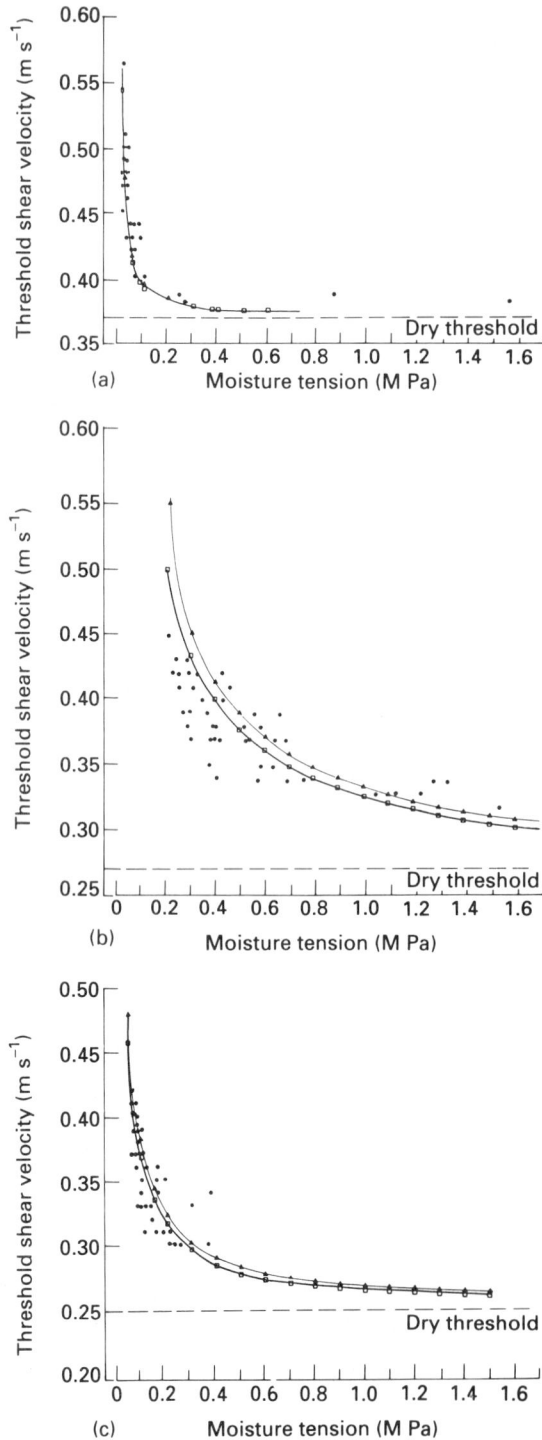

Fig. 9.16 (*see opposite column for legend*)

Bonding agents

In nature, particles at the surface are not always found in a dry, loose state. Rather they are frequently bound together by various agents to produce erosion-resistant structural units or more continuous surface crusts. These bonding agents include silt and clay, organic matter and precipitated soluble salts.

Chepil and Woodruff (1963) observed that the effectiveness of silt and clay as bonding agents depends on their relative proportion in relation to the quantity of sand-sized material. They suggested that soils having 20–30% clay, 40–50% silt and 20–40% sand produce the greatest number of non-erodible clods with the highest degree of mechanical stability and are least affected by abrasion.

Work by Chepil (1951) on agricultural soils showed that the presence of organic matter increases the ability of particles to form aggregates that are less susceptible to entrainment by wind than the individual grains. Very high organic contents in dry soils, however, increase erosion potential because of the very loose soil structure (Hopkins *et al.*, 1946).

Nickling (1978, 1984) and Nickling and Ecclestone (1981) have shown that even low concentrations of soluble salts can significantly increase threshold velocity by the formation of cement-like bonds between individual particles. Lyles and Schrandt (1972) noted that sodium chloride was more effective than magnesium and calcium chloride in reducing soil movement because sodium tends to produce a surface crust that protects the underlying soils.

Surface crusts

Surface crusts formed by various processes can also inhibit wind erosion by increasing the surface thresh-

Fig. 9.16 (*opposite*) Comparison of theoretical and observed threshold velocities with increases in soil moisture tension (i.e. decrease in gravimetric moisture content). Theoretical curves have been computed for open and closed packing arrangements using Eq. 9.17.
(a) 0.51-mm sand. Predicted values from model: ▲, close packing; □, open packing; ●, measured values; G = 0.425; $u*_{td} = 0.37$ m s^{-1}. (b) 0.27-mm sand. Predicted values from model: □, open packing; ▲, close packing; ●, measured values; G = 2.6; $u*_{td} = 0.27$ m s^{-1}. (c) 0.19-mm sand. Predicted values from model: □, open packing; ▲, close packing; ●, measured values; G = 0.225; $u*_{td} = 0.25$ m s^{-1}. From McKenna-Neuman and Nickling (1989).

old velocity and decreasing the supply of grains to the air stream. Clay-rich crusts can be formed by raindrop impact on bare soils (Chen *et al.*, 1980). The presence of algae and fungi has also been shown to contribute to crust formation (Foster & Nicolson, 1980; Van den Ancker *et al.*, 1985) as have soluble salts (Nickling, 1978; Pye, 1980; Gillette *et al.*, 1980, 1982).

The effect of crustal strength on threshold velocity was measured for a wide range of soils using portable field wind tunnels in the Mojave Desert by Gillette *et al.* (1980, 1982) and in the Sonoran Desert by Nickling and Gillies (1989). Gillette *et al.* (1980, 1982) found that in the case of undisturbed soils, even weak crusts (modulus of rupture < 0.07 MPa) will protect the soil from wind erosion. Nickling and Gillies (1989) found significantly higher threshold velocities on undisturbed crusted desert surfaces compared to similar surfaces disrupted by off-road vehicles and cattle trampling.

Surface roughness

Surface characteristics such as the presence of vegetation, gravel lag deposits (Fig. 9.17) or other large roughness elements are also important to the entrainment process. Lyles *et al.* (1974) and Lyles (1977) showed that, as surface roughness increases, a greater proportion of the wind shear stress is taken up by the larger non-erodible roughness elements leaving less shear stress to entrain the finer more erodible particles. More recent wind tunnel experiments by Logie (1982) demonstrated the importance of roughness

Table 9.3 The minimum and maximum measured value of the inversion point for various sizes of spheres and gravel. From Logie (1982)

Roughness elements	Diameter (mm)	Inversion point (%)
Gravel	2–3	3.0–6.3
	7	7.4–10.5
	8.7	7.0–10.2
	15	14.0–22.0
	17.5	10.5–24.5
Spheres	13.4	11.0–13.9
	16.9	15.0–20.4
	25	18.8–27.0

element spacing on threshold velocity and the nature of particle entrainment. Low densities of roughness elements (glass spheres and gravel particles) tended to reduce threshold velocity and cause increased erosion around the roughness elements because of the development of turbulent eddies. By contrast, higher densities of roughness elements increased the threshold and limited erosion. Logie showed that for any size of roughness element a certain critical cover density exists, termed the inversion point, where the influence changes from protection to activation of erosion (Table 9.3). It was also found that the irregular-shaped gravel produced more turbulence than the glass spheres and, consequently, a higher density of roughness elements was required in the former case to stop deflation.

Fig. 9.17 Lateritic gravel lag deposit caused by deflation of fine-grained material from fluvial sediments in central Mali. Photo courtesy of B. Tegler.

Fig. 9.18 Raising of dust during tillage operations in southwestern Ontario. Photo courtesy of C. Baldwin.

Anthropogenic effects

Human activity can have a significant effect on the emission of fine-grained particulates. The ejection of dust during tillage (Fig. 9.18) on dry soils is common in agricultural areas. Similarly, large quantities of dust can be raised during construction activities, by military manoeuvres in desert areas (Marston, 1986) and by vehicles on dirt roads (Wilshire, 1980; Fig. 9.19). Hall (1981) has estimated that a four-wheel drive vehicle travelling at 60 km h^{-1} along a dirt road containing 12% silt will raise 3.7 kg km^{-1} of dust. The amount of dust emitted increases as vehicle speed and silt content of the soil increases. Bagnold (1960) and Nickling and Gillies (1989) have ob-

served that cattle, sheep and goats moving over even heavily crusted soils can cause large quantities of dust to be raised.

9.3.5 Entrainment of fine grained sediments

As previously discussed (Section 9.3.3), particles smaller than approximately 80 μm are inherently resistant to entrainment by wind. Nevertheless, major dust storms are common occurrences in many parts of the world (Goudie, 1983; Middleton *et al.*, 1986), even though shear velocities of > 1.0 m s^{-1} are relatively uncommon. This is explained by the fact that silt- and clay-size particles are rarely entrained by direct fluid pressures and lift forces

Fig. 9.19 Large quantities of dust being ejected from a road track by a passing vehicle in central Mali.

(Chepil, 1945b; Gillette, 1974; Nickling & Gillies, 1989). In most cases, small particles are ejected by the impact of saltating sand grains.

The inherent resistance of silt-size material to entrainment in nature is further demonstrated if one considers the extensive loess deposits found in many parts of the world (Fig. 9.3). These wind-blown silts, once deposited, tend to resist re-entrainment by wind unless acted upon by some additional external force including saltating sand grains, or disturbance by animal or vehicular traffic (Fig. 9.19). In the case of crusted soils, abrasion by impacting particles moving in saltation has been observed to be important in breaking up of surface crusts and aggregates thereby releasing fine particulates into the air stream (Chepil, 1945c; Gillette *et al*, 1974; Nickling, 1978; Hagen, 1984).

Chepil and Woodruff (1963) argued that wind erosion will usually only occur when soil grains capable of moving in saltation are present in the soil and that dust clouds are initiated by the particles moving in saltation. Gillette (1977), in detailed field studies, found that the vertical emission of small particulates, expressed as a ratio of the amount of soil moved in saltation and creep, is approximately proportional to u_*^4. Similar observations have been made on loam textured soils in Arizona by Nickling and Gillies (1989). Other researchers have also found that sand-size aggregates of silt and clay may be transported in saltation but tend to break up during transport, releasing fines which are carried away in suspension (Gillette & Goodwin, 1974; Gillette, 1977; Nickling, 1978; Hagen, 1984).

9.4 Transport of sediment by wind

9.4.1 Modes of particle transport by wind

Bagnold (1941) suggested that three distinct modes of transport can take place depending primarily on grain size. Very small particles ($< 60-70 \, \mu m$) are transported in *suspension* and kept aloft for relatively long distances by the turbulent eddies in the wind. Relatively large particles (approximately 60–1000 μm) move downwind in a series of bounces or jumps, a process termed *saltation*. In true saltation, particles are ejected into the air stream at fairly steep angles. After reaching some maximum height, they are carried by the wind in relatively smooth trajectories back to the surface where they impact and bounce back into the air or become embedded (Fig. 9.20). Larger or less exposed particles ($> 500 \, \mu m$) are pushed or rolled along the surface primarily by the impact of saltating grains in *surface creep*.

True suspension occurs when the particle settling velocity (i.e. velocity a particle of a given diameter will fall in a still fluid, U_f) is very small in relation to the shear velocity of the wind. Pure saltation occurs when the turbulent vertical component of velocity has no significant effect on the particle trajectories. No sharp distinction exists between true suspension and saltation. Rather, there is a transition where both inertia and settling velocity influence the particle motion. This transition has been termed *modified saltation* (Hunt & Nalpanis, 1985; Nalpanis, 1985) and is characterized by a more random particle trajectory. The transition between true saltation and

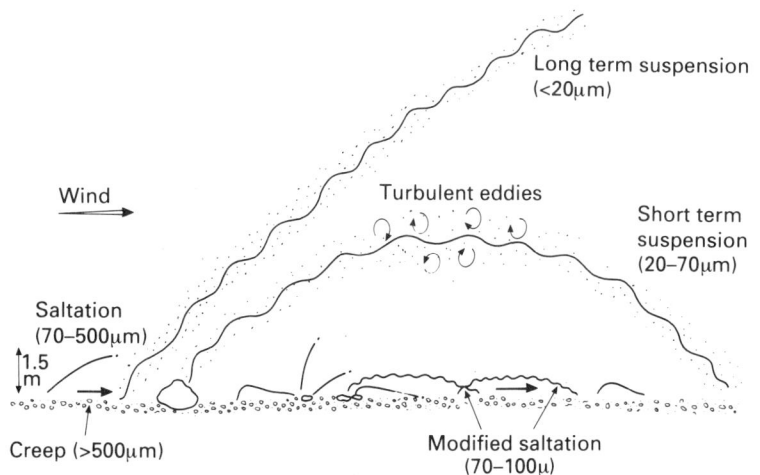

Fig. 9.20 Modes of particle transport by wind. Typical particle size ranges transported during moderate wind storms are shown. Modified from Pye (1987).

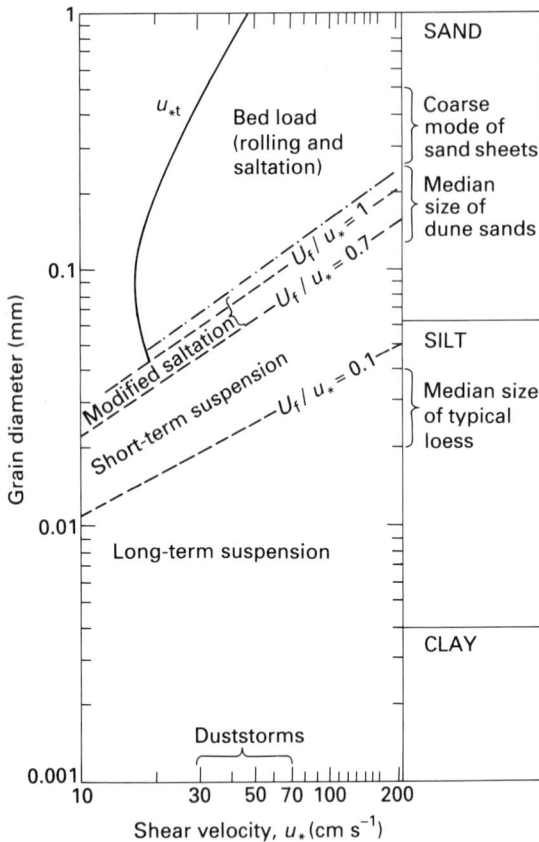

Fig. 9.21 Modes of transport for quartz spheres of different diameters at different shear velocities. From Tsoar and Pye (1987).

suspension as a function of grain size has been determined theoretically by Nalpanis (1985) and is shown in Fig. 9.21.

Tsoar and Pye (1987) have also made a distinction between suspended grains that remain aloft for long periods of time (*long-term suspension*) and those that settle back to the surface relatively quickly (*short-term suspension*), as shown in Fig. 9.20. More detailed consideration of suspension transport will be given in Section 9.5.

9.4.2 Saltation trajectories

Numerous investigators have noted that particles moving in saltation are characterized by trajectories with an initial steep vertical ascent followed by a much more horizontal movement eventually return-

ing to the bed with relatively small impact angles (Fig. 9.20). Bagnold (1941) and Chepil (1945a) measured ascent angles of 75°–90°. White and Shultz (1977) determined the average ejection angle to be approximately 50° for particles 586 μm in diameter and u_* of 40 cm s^{-1}. The discrepancy between the findings of White and Shultz (1977) and earlier work may in part be due to the differing methods of observation. White and Shultz (1977) were able to record trajectories as low as 0.5 cm in height whereas Bagnold (1941) and Chepil (1945a) primarily recorded the angles of particles with much higher trajectories, implying that particles with higher trajectories have higher ejection angles.

White and Shultz (1977) found that actual filmed particle trajectories were higher than those predicted by theoretical equations of motion. They suggested that this occurs because of a lift force that is generated by the spinning of the grain as it moves through the air. The vertical lift force on a particle that is aloft has been termed the *Magnus effect* and is related to the rate of spinning and the steep velocity gradient adjacent to the particle's surface. White (1982) observed mean particle spin rates of 350–400 rps, similar to those of 200–1000 rps recorded by Chepil (1945a). White and Shultz (1977) found that when the Magnus effect was included in the equations of motion, theoretical trajectories were in much better agreement with the observed trajectories in their wind tunnel experiments.

Owen (1964) calculated a trajectory height of 0.81 u_*^2 g^{-1} and length of 10.3 u_*^2 g^{-1} for an idealized case assuming that the initial take-off speed is of the same order as the shear velocity (u_*). These theoretical predictions compare favourably with actual measurements of trajectory heights and lengths made by White and Shultz (1977).

Generally, the higher the saltating particle's apogee the greater the velocity at which the particle will be carried in the air stream. This results from the steep velocity gradient above the surface. Additionally, higher apogees give rise to longer saltation path lengths. Once particles have attained their apogee they descend approximately linearly, impacting with the surface at incidence angles of 10–16° (Bagnold, 1941), or 4–28° with an average angle 13.9° (White & Schulz, 1977). The impact angle of a saltating particle is dependent on the particle size, and height of the trajectory and the fluid velocity, decreasing with increasing wind velocity and decreasing with increasing particle size (Jensen & Sorensen, 1986).

9.4.3 Transport equations

Bagnold (1941) was one of the first investigators to develop a mathematical expression to account for the quantity of sand transported as a function of the shear stress exerted by wind. The importance of this early work rests in the fact that it formed the basic theoretical background and supporting empirical evidence for almost all future research.

Bagnold (1941), from theoretical considerations in conjunction with field and wind tunnel observations, found that the sediment transport mass flux in saltation and creep, q could be defined by:

$$q = C\left(\frac{d}{D}\right)^{0.5}\frac{\rho}{g}\,u_*^{3}, \qquad (9.18)$$

where (d/D) is the ratio of the mean size of a given sand to that of a 'standard' 0.25 mm sand. The coefficient C is a sorting coefficient and has the following values: 1.5 for nearly uniform sand; 1.8 for naturally graded sand; 2.8 for poorly sorted sand; and 3.5 for a pebbly surface.

Thus, holding other factors constant, the sediment flux, q, increases from a minimum over a nearly uniform sand bed to somewhat higher values for more poorly sorted sands and attains a maximum value over a pebble surface. This is because coarse sand grains saltate more effectively and rebound is more perfect from a hard pebbly surface.

Subsequent to Bagnold's (1941) work, several other investigations have developed both theoretical and empirical equations to describe the transport of sediment by wind. These equations, although frequently derived from different points of view, are similar in general form to that initially proposed by Bagnold (1941). Based on wind tunnel tests, Zingg (1953) modified the Bagnold equation to:

$$q = C\left(\frac{d}{D}\right)^{0.75}\frac{\rho_a}{g}\,u_*^{3}. \qquad (9.19)$$

A fundamental problem with both the Bagnold and Zingg equations is that they do not include a threshold term and thus predict sediment transport at shear velocities below that required to initiate particle movement. This problem was addressed by Kawamura (1951) who proposed a somewhat different equation that included a threshold shear velocity (u_{*t}) term:

$$q = K\frac{\rho}{g}(u_* - u_{*t}(u_* + u_{*t})^2, \qquad (9.20)$$

where K is an empirical coefficient which is a function of the textural characteristics of the sediment and

must be determined by experiment. Kawamura obtained $K = 2.78$ for moderately well sorted sand with a mean diameter of 0.25 mm. A threshold shear velocity term was also included in the transport equation derived by Lettau and Lettau (1978):

$$q = C\left(\frac{d}{D}\right)^{n}\rho\,u_*^{2}\,(u_* - u_{*t}), \qquad (9.21)$$

where the exponent n has values ranging from 0.5 to 0.75.

A graphic comparison of sediment flux with increasing shear velocity for similar particle sizes predicted by several different transport equations is shown in Fig. 9.22. It is apparent that, despite the rather similar form of the equations and similar mean particle diameters used in the calculations, the equations give significantly different results.

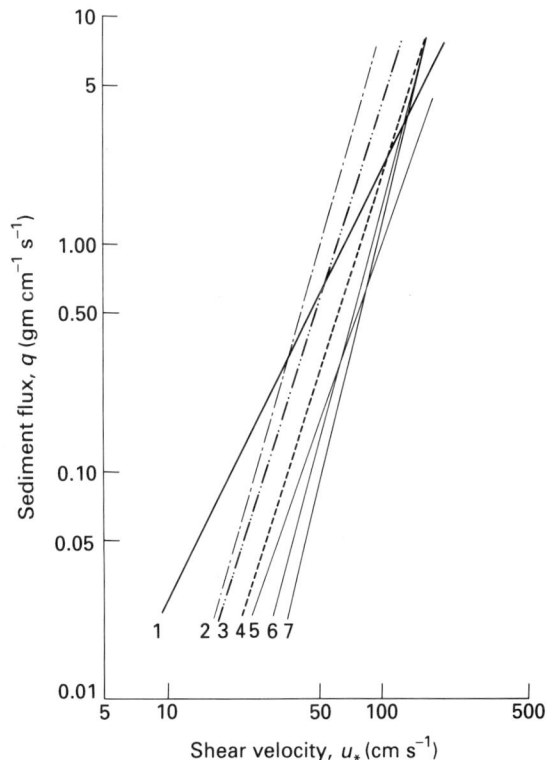

Fig. 9.22 Comparison of sediment transport equations. 1, Belly (1964) 0.3 mm; 2, O'Brien & Rindlaub (1936) 0.02 mm; 3, Kawamura (1951) 0.25 mm; 4, Bagnold (1941) 0–25 mm; 5–7, Williams (1964): 5, spheres; 6, sand; 7, crushed quartzite.

9.4.4 Textural controls on aeolian transport

Most sediment transport equations contain various empirically derived coefficients related to the grain size and sorting characteristics of the sediments considered. Differences in the predicted sediment flux of these equations for increasing shear velocity (Fig. 9.22) have been attributed to differences in these empirical coefficients. However, Williams (1964), and Willetts et al. (1982) argue that at least some of the differences may well be related to undocumented textural differences and, in particular, to the grain shape of the test sands used by the different investigators in determining these empirical coefficients.

Wind tunnel tests by several authors have clearly indicated the importance of particle shape and surface roughness on the sediment transport rate. Williams (1964) carried out wind tunnel experiments on three different materials (glass spheres, natural quartz sand and crushed quartzite) with different shape characteristics. He found that sediment flux (q) varied as a power function of shear velocity (u_*), but with exponents considerably different from the value of three suggested by Bagnold (1941) and others. He observed that, in general, the value of the exponent increased as the sphericity of the sediment increased. Exponent values ranged from 2.76 for crushed quartzite to 3.42 for the natural sand and 4.10 for spheres. Similar observations have been reported by Willetts et al. (1982), Willetts (1983) and Finlan (1987).

Williams' (1964) results also showed that at low shear velocities ($u_* < 75$ cm s^{-1}) there is a tendency for sediment transport rate to increase for a given shear velocity as particle shape becomes more irregular. At higher shear velocities this pattern is reversed and sediment transport rate increases with increasing sphericity. He also observed that at low wind velocities the initial movement of spherical particles began much further downwind than did the more angular crushed quartzite and natural sands. From this observation he concluded that particle shape has a significant influence on threshold velocity, which has been confirmed by Nickling (1988) and Willetts (1983). Willetts and Rice (1986) also noted that grain shape has a pronounced effect on the kinetics of the saltation path and on the nature of the bed collisions that eject particles in suspension and move particles forward in surface creep. In wind tunnel tests using two sands with similar mean size but different particle shapes (compact and platy), they found that collision was a more efficient mechanism for maintaining saltation in the case of the compact quartz particles than for the platy sands. They observed that the platy sand lost less forward momentum on impact than did the more compact quartz grains and underwent a much smaller change of vertical momentum. They concluded that collision impacts maintain a smaller load of platy particles than compact particles.

9.5   Transport of dust by wind

The transport of suspended sediment during dust storms is a common occurrence in many parts of the world, but particularly in arid and semi-arid environments. During such events, thick dust clouds, often reaching 3000 m in height, are transported hundreds of kilometres and cover areas greater than 500 000 km^2 (Goudie, 1983). Sediment transported during these dust storms can be derived from a wide range of surface types (Table 9.4).

Dust storms are internationally defined as meteorological events in which visibility at eye level is reduced to < 1000 m by dust raised from the surface. Despite this convention, several authors have adopted other visibility criteria that were thought to be more appropriate for their particular study (e.g. Oliver, 1945: < 700 m; Péwé et al., 1981: < 800 m; Nickling & Brazel, 1984: < 1600 m). A dust storm day is defined as a period of 24 h in which the visibility is reduced below 1000 m for all or part of

Table 9.4 Major dust-producing terrain types. After Pye (1987)

Glacial outwash plains and braided fluvioglacial channels
Dry wadi beds
Dry lake beds
The surfaces of coastal sabkhas
Alluvial fans
Stormy deserts with high weathering rates
Areas of exposed argillaceous bedrock
Areas of loess where vegetation cover has been reduced by climatic change and/or cultivation
Areas of deeply weathered regolith where vegetation has been reduced by climatic change and/or human activities
Alluvial floodplain sediments, particularly those that have been cultivated
Areas of formerly stabilized dunes that have been reactivated

Table 9.5 Dust storm frequencies for selected locations. After Goudie (1983)

Location	Frequency per year of dust storms creating visibilities of less than 1000 m
Abadan, Iran	13
Amritsar	10
Allahabad, India	8
New Delhi, India	8
Ganganagar, India	17
Pathankot, India	8
Paotou, China	19
Paoting, China	18.5
Baghdad, Iraq	21.5
Basra, Iraq	14.7
Aqaba, Jordan	10.7
H4, Jordan	16.3
Al Hummar, Jordan	7.0
Kano, Nigeria	23.0
Maidiguri, Nigeria	23.0
Nguru, Nigeria	26.0
Mexico City, Mexico	68
Kazakhstan, USSR	60
Kantse, China	35
Hami, China	33
Beersheva, Israel	27
Shaibah, Iraq	37.6
Hinaidi, Iraq	33.2
Diwaniyah, Iraq	35.0
Kuwait Airport, Kuwait	27.1
Abu Kamal, Syria	9.7
Mersa Matruh, Egypt	9.6
Riyadh, Saudi Arabia	12.7
Dharan, Saudi Arabia	11.3
Tabouk, Saudi Arabia	10.8
Khartoum, Sudan	24

Table 9.6 Dustfall deposition rates. After Goudie (1983)

Location	Rate ($t\ km^{-2}\ year^{-1}$)
Kuwait	55
Bulgaria	44.5
Arizona	54
Negev (Israel)	50–200
Mexico City	< 10 – > 50
Caspian Sea	39.5
Kansas, USA	56
Northern Nigeria	137–181

high, characteristically in the order of $50\ t\ km^{-2}$ $year^{-1}$ (Table 9.6).

9.5.1 The transport process

Despite the wealth of information existing on the horizontal flux of sand-size material in saltation and creep, relatively few detailed field measurements have been carried out on the vertical and horizontal flux of fine-grained particulates in suspension from natural surfaces. Chepil and Woodruff (1957) provided the first detailed data on the variation of sediment concentration with height during dust storms in Colorado and Kansas. More recent studies have been undertaken in the High Plains of Texas by Gillette (1974, 1977), Gillette and Walker (1977) and Gillette et al., (1972, 1974), in Canada by Nickling (1978, 1983) and in Arizona by Nickling and Gillies (1989).

Direct field observations by several investigators have shown that, over eroding surfaces, the concentration of suspended sediment decreases as a power function of height (Chepil & Woodruff, 1957; Gillette et al., 1972; Gillette, 1977; Nickling, 1978).

Small particulates (< 20 μm diameter) are usually transported by wind in suspension. Although these particles can be entrained into the air stream by direct fluid pressure and lift forces, they are frequently ejected by the impact of saltating sand grains (Section 9.4). In other cases they are released by abrasion, either during transport in saltation and creep of sand-sized aggregates of silt and clay or by the bombardment of larger stationary aggregates by saltating grains. Once entrained, the fine particles are carried by turbulent eddies in the air stream and have the potential to be transported to great heights and over great distances.

the period. The highest reported frequency of dust storms is in the Seistan Basin of Iran where dust storms are encountered on average 80.7 days per year (Middleton, 1986b). Dust storm frequencies at selected stations are shown in Table 9.5.

Although the frequency and magnitude of dust storms is spatially variable, at the global scale they represent an important geomorphic process. Dust storms are associated with high erosion rates and in downwind areas, high deposition rates. It has been estimated that dust from soil erosion contributes approximately 500×10^6 t of particulates to the atmosphere each year (Peterson & Junge, 1971). Rates of dust deposition in many areas of the world are also

The long-range transport of small grains in suspension is a function of many interrelated factors, including:
1 the nature of the eroding surface (textural conditions, presence or absence of crusts, vegetation cover, moisture content, etc.),
2 ejection rate of grains into the air stream,

3 particle size distribution and shape characteristics of the ejected particles, and
4 the turbulent structure of the wind which is affected by the stability conditions of the atmosphere.

Suspended sediment concentrations measured by Nickling (1978) during 15 dust storms generated over glacio-fluvial outwash sediments in the Yukon Terri-

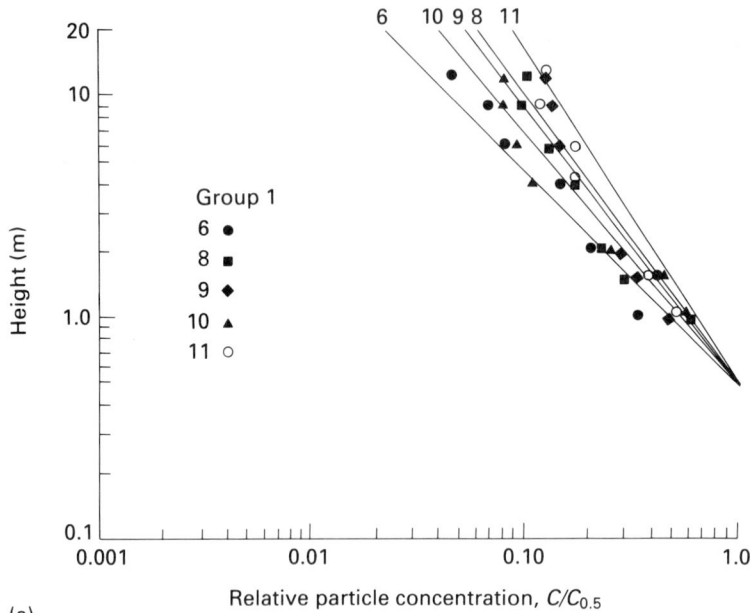

(a)

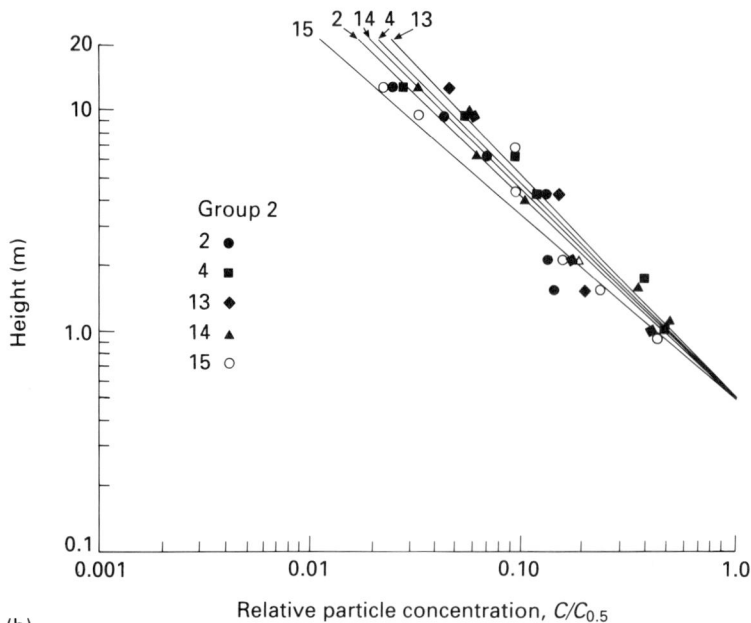

(b)

Fig. 9.23 Variation of relative particle concentration with height measured during dust storms in the Slims River Valley, Yukon. Dust was generated from fluvial–glacial outwash sediments. Symbols represent individual dust storms. From Nickling (1978).

tory, Canada, are shown in Fig. 9.23. A consistent exponential decrease in concentration with height was observed for all dust storms but there was considerable variation in the slope of the curves from one storm to another. This was attributed, in part, to the degree of turbulent exchange in the air stream as indicated by a modified form of the Richardson number.

These field observations, and those made by Chepil and Woodruff (1957) and Gillette (1977), conform to theoretical concentration models such as that proposed by Rouse (1937). He showed that the relative concentration of suspended particulates of a particular size at a given height is given by:

$$\frac{C}{C_a} = \frac{h-y}{y} \cdot \left(\frac{a}{h-a}\right)^z, \tag{9.22}$$

where C/C_a is the relative concentration (number of particles per cm^3) at any level, y, referenced to an arbitary elevation, a, above the surface (where $a < y$), and, h, is the elevation at which the particle concentration is zero. For the lower atmosphere, $h \gg y$ and Eq. 9.22 can be simplified:

$$\frac{C}{C_a} = \left(\frac{a}{y}\right)^z. \tag{9.23}$$

The exponent $z = U_f/ku_* = 2.5 (U_f/u_*)$, where U_f is the particle settling velocity which can be defined as (Allen, 1985):

$$U_f = KD^2, \tag{9.24}$$

where D is the particle diameter and $K = \rho_g\, g/18\,\mu$ for which ρ_g is the particle density, g the acceleration due to gravity and μ is the dynamic viscosity of the air.

In the case of small particles that obey Stokes' Law, Eq. 9.23 can be modified by solving for the particle settling velocity (U_f) which gives:

$$\frac{C}{C_a} = \exp 2.5 \frac{KD^2}{u_*} \ln\left(\frac{a}{y}\right). \tag{9.25}$$

Tsoar and Pye (1987) have used Eq. 9.25 to compute the relative concentration above the surface for spherical quartz particles of different sizes, assuming a shear velocity of $70\ cm\ s^{-1}$ (Fig. 9.24). The calculations indicate that particles $> 20\,\mu m$ are carried only a few metres above the ground, while particles $< 20\,\mu m$ are distributed almost evenly with height up to at least 100 m. The relative concentration of grains $> 30\,\mu m$ at an elevation of 100 m was found to be 26% of the concentration at 0.5 m. On the basis of their calculations, they conclude that coarser silt particles, such as those typically associated with loess deposits, travel closer to the ground and are much more likely to be trapped by vegetation and other surface roughness elements.

Nickling (1983) found that, on average, there was

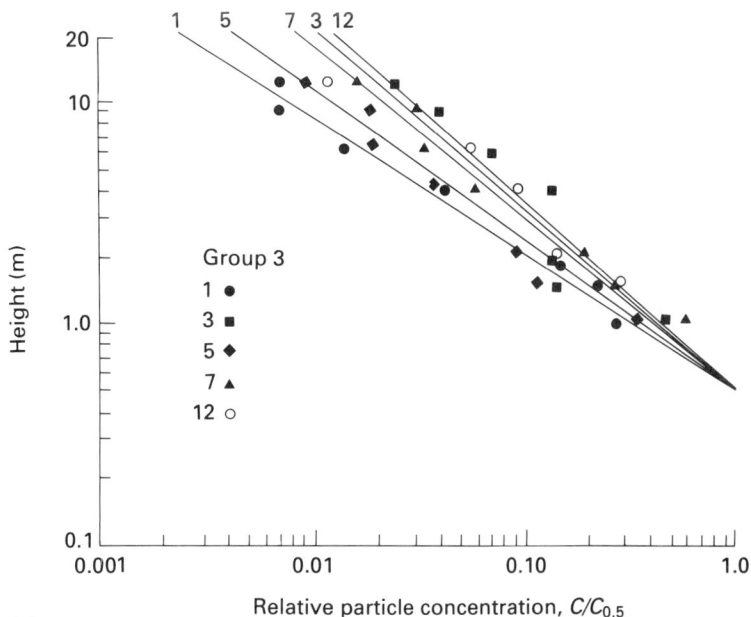

Fig. 23 *Continued*   (c)

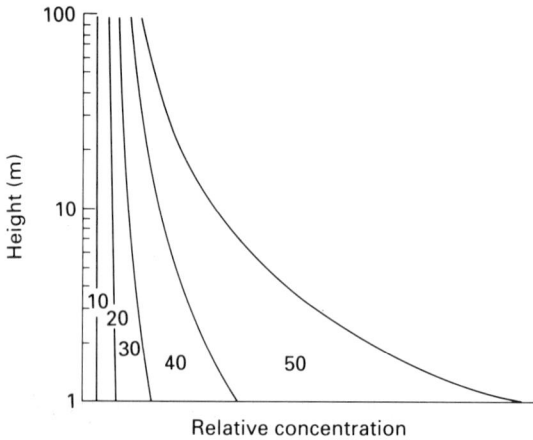

Fig. 9.24 Predicted vertical changes in relative concentration of 10, 20, 30, 40 and 50 µm size particles (quartz spheres) under severe wind storm conditions. From Tsoar and Pye (1987).

an exponential decrease in mean grain size with height during the 15 dust storms (Fig. 9.25) which is similar to observations made by Sundborg (1955), Chepil and Woodruff (1957), Gillette et al. (1974) and Goossens (1985). However, Nickling (1983)

noted that there was considerable storm-to-storm variation in the decrease of mean size with height and it could be somewhat irregular because of turbulent mixing and the range of grain sizes supplied to the air stream. The sorting of the suspended sediment with height during each dust storm was also found to be rather irregular but, in general, tended to improve from the surface up to 1.5 m, above which it was increasingly more poorly sorted. Nickling (1983) argued that the improvement in sorting up to 1.5 m results from the selective removal of relatively coarse grains in modified saltation as height increases. Above 1.5 m, all particles are relatively small and have a tendency to become more uniformly mixed by the turbulent eddies.

Goossens (1985), in a study of dust generated from a road construction site, also found suspended sediment to be granulometrically stratified, with coarse silt moving primarily near the bottom of the dust cloud while finer silt was transported at both top and bottom.

Vertical particle flux

The quantity of dust transported in suspension is

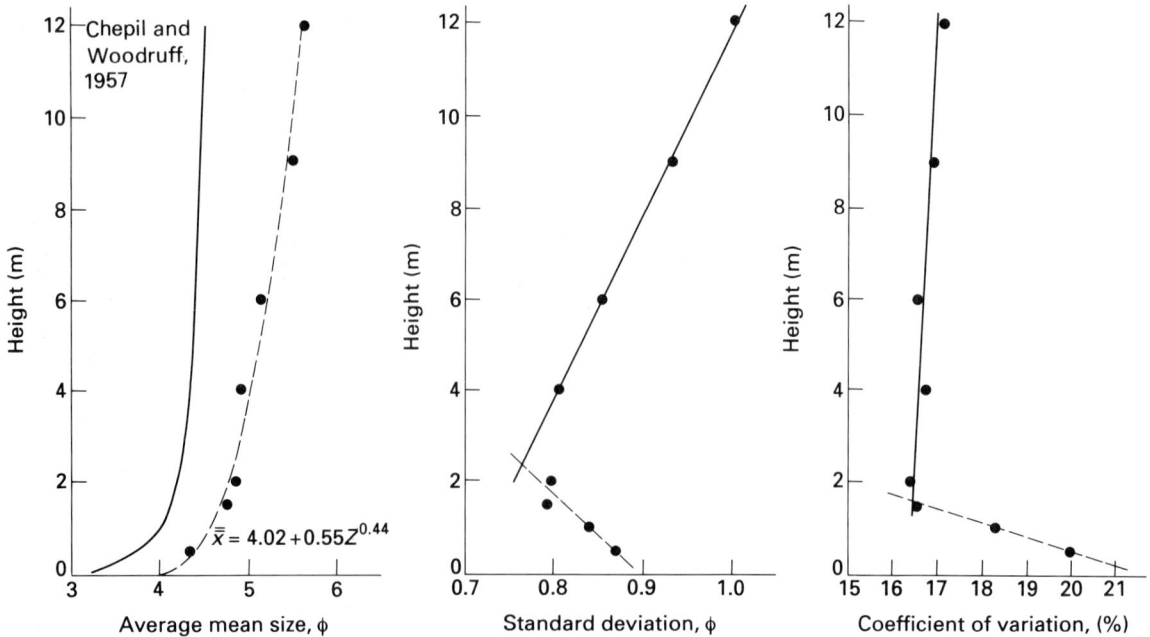

Fig. 9.25 Changes in mean size and sorting of suspended sediment with height measured during dust storms in the Yukon, Canada. Modified from Nickling (1983).

Fig. 9.26 Meteorological and sampling towers used to measure suspended sediment during dust storms in Mali, west Africa. Towers are 10 m in height.

directly related to the rate at which grains are ejected into the air stream from the surface. Gillette *et al.* (1972) suggested that the vertical particle flux (F) can be determined from the particle concentration profile by:

$$F = \varepsilon \frac{dC}{dz}, \qquad (9.26)$$

where F is the vertical particle flux (g cm^{-2} s^{-1}, ε is the eddy diffusion coefficient and C is the particle concentration at height z. Under neutral stability conditions:

$$\varepsilon = u_* kz. \qquad (9.27)$$

The most detailed vertical particle flux data-set available for natural surfaces is that presented by Gillette (1977). In this study, tests were conducted over a 3-year period on relatively flat fields consisting of erodible soils with uniform textures. Aerosols were collected using specially designed membrane filter samples (Gillette *et al.*, 1974) placed at heights of 1.5 and 6.0 m, or 1.0 and 6.8 m. Mean wind velocity was measured at the same heights using cup anemometers (Figs. 9.26, 9.27 & 9.28). The observed particle fluxes (F) as a function of shear velocity for the nine sites are shown in Fig. 9.29. The sandy soils show a fairly

Fig. 9.27 Isokinetic suspended sediment samplers used on the towers. The samplers have a 1.3-cm inlet orifice and face automatically into the wind. Air flow through the nozzle is controlled by a needle valve and flowmeter incorporated into the vacuum line. Suspended sediment is collected on 0.1 μm membrane filters.

Fig. 9.28 Field sediment trap used to collect sand-sized material in saltation and creep. The trap automatically aligns itself into the wind and is 1.0 m in height with a 1.0-cm wide sampling orifice. Sediment entering the sampler is funnelled down the back of the trap and collected in a container located below the surface.

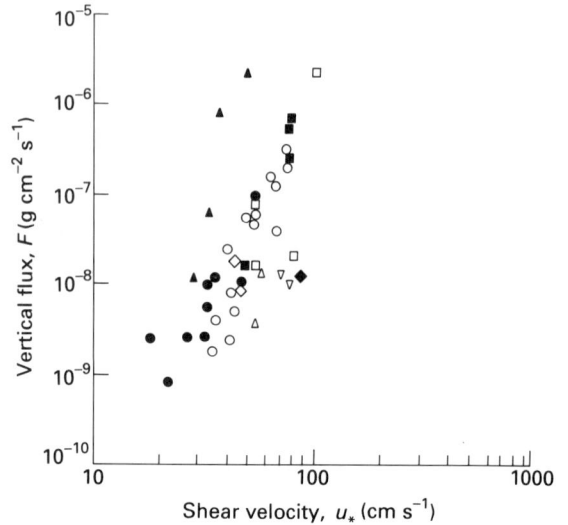

•, Sand; ■, sand; ▲, loamy sand; ○, sand; □, sand; △, sandy clay loam; ◇, sand; ◆, loamy sand; ▽, clay loam.

Fig. 9.29 Vertical particle flux versus shear velocity for different soils observed over different soil types in Texas. From Gillette, (1977).

uniform trend of increasing vertical particle flux with shear velocity. By contrast, the loamy soils show a greater scatter in particle flux, most probably because of their widely different dry aggregate structures. The relatively low values of vertical particle flux associated with the clay soil result from a high threshold shear velocity and the resistance to break-up of aggregates containing montmorillonite clay.

A portable field wind tunnel (Fig. 9.30) was used by Nickling and Gillies (1989) to investigate the relative vertical particle fluxes from different surface types in southern Arizona. The surface types included active agricultural lands, abandoned farm fields, undisturbed and disturbed desert surfaces, construction sites and mine tailings. They found a general increase in vertical particle flux with increasing shear velocity (Fig. 9.31) which compared favourably with the results of Gillette (1977). They also showed that the strength of the relationship was improved if the data set was partitioned on the basis

of textural and surface characteristics. Strong linear relationships were found between F and u_* for loamy soils (silt and clay contents > 25%) and soils with sand contents > 85%. On average, for a given shear velocity, particle emission from the surface was greater for the loam textured soils in comparison to the sandy soils. Highest emission rates were associated with loamy textured soils that had been heavily disturbed by vehicular traffic such as that encountered at construction sites and on off-road vehicle tracks. Their observations also indicated that surface roughness can have a significant effect on the vertical particle flux. The presence of roughness elements, such as standing vegetation, clods or pebbles, tended to decrease the emission of particulates on similar textured soils at approximately the same shear velocities because of a shielding effect.

Long-range dust transport

When dust particles are ejected into the air they are kept aloft by the turbulent eddies in the wind. Tsoar and Pye (1987) suggest that, at a given instant, a turbulent wind can be viewed as having a horizontal velocity component (u) and a vertical velocity com-

Fig. 9.30 Portable field wind tunnel used to measure soil losses. The tunnel is 12.5 m in length with a 0.75 m × 1.0 m cross-section and has no floor in the working section. Air flow is generated by a 0.38-m centrifugal fan driven by a 35 h.p. diesel engine. The wind tunnel, engine and fan are carried on a 7-m heavy duty trailer that can be pulled by a four-wheel drive vehicle.

ponent (w). Over a given time interval the mean values for the velocity components are designated $\overline{u}$ and $\overline{w}$. The fluctuations in the vertical component is defined by w' and is equal to $(w - \overline{w})$. Under neutral stability conditions near the surface, but above the saltation layer, the fluctuating vertical velocities have normal distribution with a mean of zero. The standard deviation of the vertical fluctuating velocity distribution is equal to the root mean square for the values of w' ($\sigma = \sqrt{\overline{w'^2}}$) and represents the force opposing the tendency for small particles to settle out of the air stream. The rate at which small particles tend to settle out of the air stream is dependent on the settling velocity of the particles, U_f (Eq. 9.24). Consequently, a small particle should remain in suspension if $\sqrt{\overline{w'^2}}$ is greater than U_f.

The standard deviation of the fluctuating vertical velocities can be related to the shear velocity by:

$$\sqrt{\overline{w'^2}} = Au_*, \qquad (9.28)$$

where A is a constant with a range of approximately 0.7–1.4 with and a mean of about 1.0 (Lumley & Panofsky, 1964). As a result, there will be a tendency for grains to remain in suspension when the ratio U_f/u_* is < 1.0. However, Gillette (1974, 1977) and Tsoar and Pye (1987) argued that the difference between saltation and suspension is not distinct as the turbulent structure of the wind and the textural characteristics of the sediment influence the settling velocity. They suggested an arbitrary upper limit for pure suspension of $U_f/u_* = 0.7$.

In order for particles to remain suspended for long periods and over large distances, particles must be kept aloft by the upward velocity fluctuations of the wind. Gillette (1974, 1977) calculated the ratio of upward to downward motions of particles in air

having a normal vertical velocity distribution (Fig. 9.32). He found that particles with $U_f/u_* = 0.4$ have a ratio of upward to downward movements of 0.5. Since, in this case, there are two downward movements for each upward movement it is unlikely that a particle of this size would be carried to any great height or for any great distance. Particles < 20 μm are sufficiently small that their sedimentation velocities are usually < 0.1 u_* for almost all eroding winds and, consequently, will remain in suspension to great heights or over long distances.

Tsoar and Pye (1987) used the ratio $U_f/u_* = 0.1$ to differentiate between long-term suspension and short-term suspension. They showed that under typical wind storm conditions ($u_* = 0.2$–0.6 m s^{-1}), medium and coarse silt grains, which make up most loess deposits, are transported primarily in short-term suspension and modified saltation (Fig. 9.21).

The vertical and horizontal transport of small particulates in suspension has been shown to be related to the coefficient of turbulent exchange (eddy diffusion coefficient) in the atmosphere (Taylor, 1915). More recently, Tsoar and Pye (1987) used this approach to calculate the height and distance that small particulates can be transported in suspension.

Based in part on early equations by Taylor (1915), Tsoar and Pye (1987) also determined that the maximum likely distance and time travelled by a dust particle that obeys Stokes' Law, is inversely proportional to the fourth power of the particle diameter and can be estimated by:

$$L = \overline{U}^2\, \varepsilon/K^2 D^4, \qquad (9.29)$$

where L is the distance travelled by the suspended particle, $\overline{U}$ is the mean wind velocity and $K = \rho_p\, g/18\, \mu$. Using Eq. 9.29 they calculated the distance different

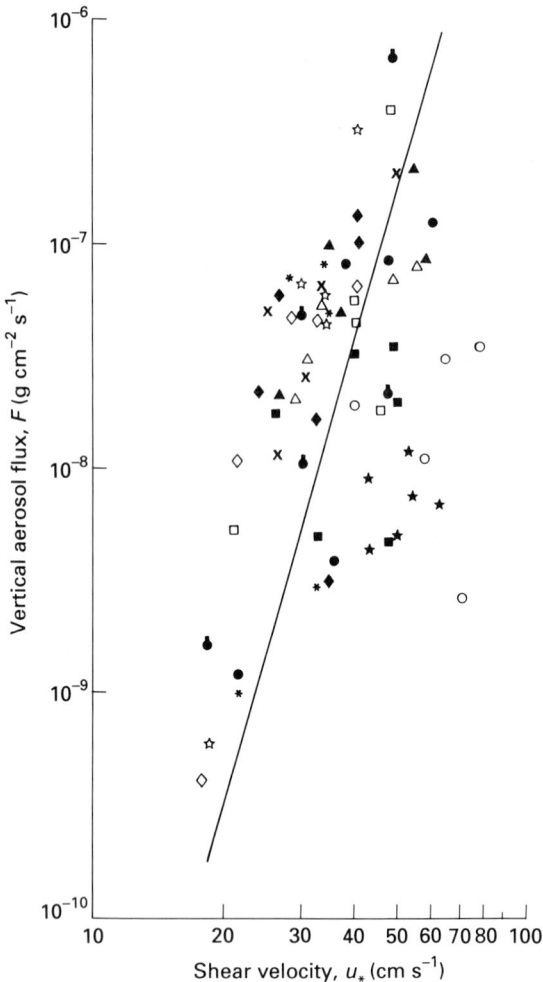

Location:

★, Mesa Ag. Site; ●, Glendale Const.; ○, Maricopa Ag.;
▲, Yuma Dist. Desert; △, Yuma Ag; ■, Algodones;
□, Yuma Scrub; ☆, Santa Cruz River; ◢, Tucson Const.;
✕, Ajo; ◆, Hayden; ◇, Salt River; ✳, Casa Grande.

Fig. 9.31 Vertical particle fluxes versus shear velocity measured over different soils in Arizona using a portable field wind tunnel. $F = 2.33 \times 10^{-11} u_*$: 1.889; $r = 0.42$; $n = 67$. From Nickling and Gillies (1989).

sizes of particles will be transported in suspension with a mean wind velocity of 15 m s^{-1} and values of ε ranging from 10^3 to 10^7 cm^2 s^{-1} (Fig. 9.33).

Their calculations indicate that, during moderate dust storms, particles > 20 μm are unlikely to travel more than about 30 km from the source, while

particles < 10 μm may be dispersed over a distance of approximately 500 km when $\varepsilon = 10^4$ m s^{-1}. For values of $\varepsilon = 10^6$ m s^{-1} particles 20–30 μm in diameter could be transported up to 3000 km. The calculations of Tsoar and Pye (1987) provide important numerical evidence to help explain the large areal extent and global distribution of loess deposits on the Earth's surface.

Factors influencing dust deposition

As a dust cloud moves away from a source area there is a constant tendency for particles to be deposited. The rate of deposition is a complex function of the turbulent wind structure, the textural characteristics of the particles which affect their fall velocity, and the nature of the surface over which the dust cloud travels. This is particularly true of particles transported in short-term suspension.

Smooth, dry surfaces such as bedrock or crusted playas can be viewed as 'reflective' surfaces (Pye, 1987). Particles settling on these surfaces can be easily re-suspended if the shear velocities are sufficiently high to overcome threshold. Sand grains that move easily over these surfaces at lower threshold velocities may also eject particles on impact. However, the hard surfaces themselves provide only a limited input of new particles and, consequently, the particle concentration of the dust cloud maintains a state of equilibrium (Pye, 1987).

By contrast, a large proportion of the settling grains in a dust cloud passing over open water bodies, marshes, or wet soil surfaces is deposited in a relatively short distance (Foda, 1983; Foda et al., 1985). These surfaces are very retentive to dust particles because of cohesive effects and do not supply additional grains to the air stream. As a result, the dust concentration and dust sedimentation rate decreases rapidly downwind.

Large changes in surface roughness also lead to a rapid increase in sedimentation rates near the roughness boundary. If dust-laden air passes from an unvegetated to vegetated surface there is a reduction of the velocity gradient above the canopy and the shear velocity may decrease below the critical value required for the re-suspension of the settled grains. In general, forests are more effective in trapping dust than steppe or tundra vegetation because of their greater roughness (Tsoar & Pye, 1987).

In deserts, dry surfaces and sparse vegetation results in dust not being permanently trapped. Settled

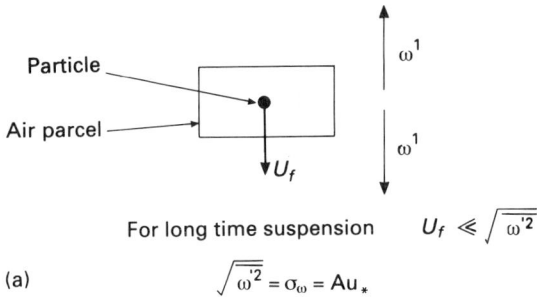

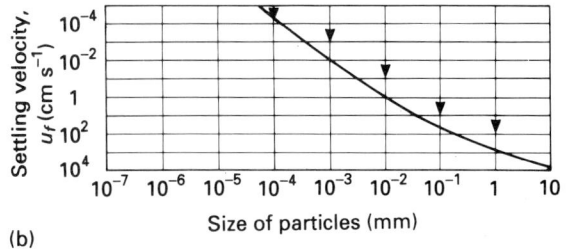

Fig. 9.32 (a) Schematic representation of the balance of forces necessary for long-term suspension; (b) settling velocities for different sized particles; (c) distribution of upward to downward motions for a particle having a settling velocity (U_f) in air with a mean of zero and a standard deviation equal to u_*. Modified from Gillette (1977).

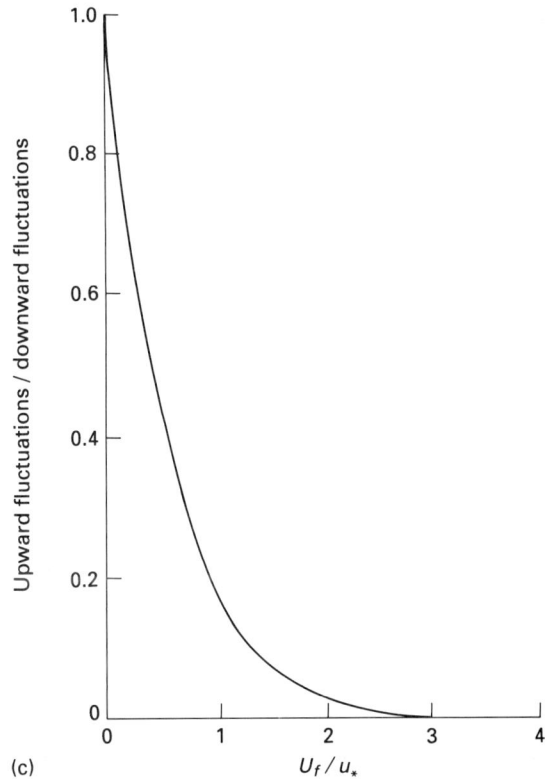

dust can be re-suspended during successive wind storms if the shear velocity is sufficiently high to entrain the particles directly. Alternatively they may be ejected at a lower shear velocity from impact by saltating sand grains. The constant re-suspension of dust in deserts in part accounts for the lack of loess deposits in these environments. Re-suspended dust in deserts is transported to the semi-arid margins where it is frequently trapped by vegetation. Thick loess accumulation is more likely if the more vegetated semi-arid area is close to the dust source (Tsoar & Pye, 1987).

Loess deposits typically show an exponential decrease in thickness with distance from the source (Fig. 9.34). The rate of thinning is related to the nature of the vegetation over which the dust clouds moved. Rapid rates of thinning from the source have been associated with loess deposits in forested environments in the USA (Snowden & Priddy, 1968; Frazee et al., 1970). Loess accumulations in central Europe and Russia (Zolotun, 1974; Fink & Kukla, 1977) tend to thin more gradually with distance from the source because of the less rough, steppe vegetation that dominated this area during the Pleistocene.

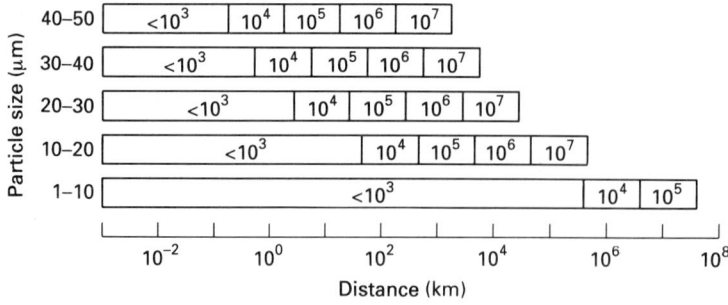

Fig. 9.33 Maximum distances likely to be travelled by quartz spheres of different sizes for a mean wind speed of 15 m s^{-1} and coefficient of turbulent exchange (ε) ranging from 10^3 to 10^7 cm^2 s^{-1}. From Tsoar and Pye (1987).

9.5.2 Dust-transporting wind systems

A fundamental control on the distribution of dust storms is the meteorological conditions that produce winds of sufficient velocity to entrain and transport dust. Middleton *et al.* (1986) identify four major types of meteorological conditions that produce major dust storms: (1) low pressure fronts with intense baroclinal pressure gradients; (2) the convergence zone between cold air masses in regions of monsoonal air flow; (3) downdraughts associated with intense convectional activity; and (4) katabatic drainage. On a more localized scale, intense dust devils caused by strong heating of the surface can also eject large quantities of dust into the air.

The major dust-generating wind conditions listed above have many localized variations. Reviews of meteorological conditions associated with dust storm activity at the global and more localized scales have been presented by Goudie (1978, 1983), Middleton (1984, 1986a,b), Middleton *et al* (1986) and Pye (1987).

The most important dust-transporting winds on a global scale are associated with the passage of low pressure fronts across the southern Sahara and the Sahel region of Africa. These *Harmattan* winds have been estimated to carry approximately half of the world's mineral aerosols (Junge, 1979). In winter the Harmattan winds affect a large part of west Africa south of latitude 15°N. During this season, dust is derived from the northeastern Sahara in the alluvial plain of Bilma (18°N, 12°E) in Niger and Faya Largeau (18°N, 12°E) in Chad (Goudie, 1983). The dust is transported in a southwesterly direction towards the Gulf of Guinea reaching northern Nigeria in 24–48 h.

During the northern hemisphere summer, dust storms are frequent in the southern Sahara and Sahel. At this time the Inter Tropical Convergence Zone (ITCZ), which separates hot dry Saharan air from moister tropical air to the south, lies across the northern Sahel. Squall lines, with strong dust-raising winds develop several degress of latitude south of the monsoon front. They usually occur between 10°N

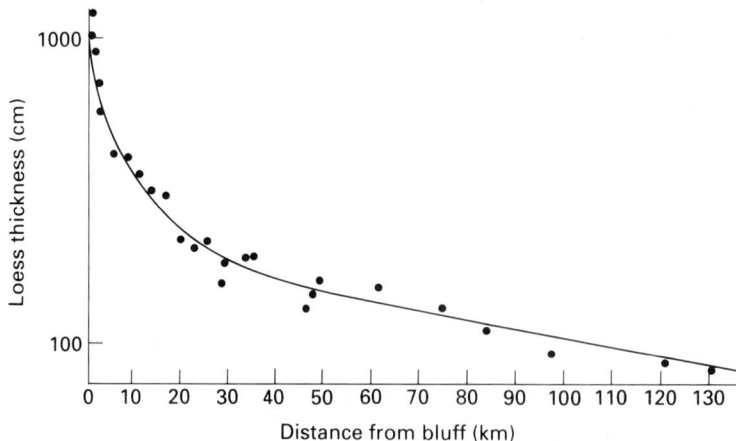

Fig. 9.34 Changes in loess thickness with distance from the sediment source. Thickness: ●, measured; —, predicted. (From Frazee *et al.*, 1970.)

and 15°N and move from east to west. The squalls are usually short lived, with only a few reaching the west African coast. However, during summer, Saharan dust is transported westward over the Atlantic in association with waves developed in the mid-tropospheric easterly flow. Dust associated with these outbreaks has been traced as far west as the Caribbean, Florida and eastern South America (Prospero *et al.*, 1970, 1981).

In eastern Africa, the convergence associated with the summer monsoon channels dust from Ethiopia, Somalia and northern Kenya across the Arabian Sea to the region north of Karachi, Pakistan. A similar convergence further north in Sudan generates dust storms that cross the Red Sea into Saudi Arabia.

In the Middle East, major depressions moving eastward from the Mediterranean across Turkey and northern Iraq are the principal cause of dust storms in this region. The *Shamal* winds, that transport dust from Iran and adjacent regions, are frequently associated with low pressure systems anchored over southern Iran. These systems form strong baroclinal gradients with a semi-permanent anticyclone over northern Saudi Arabia (Middleton *et al.*, 1986). The convergence zone between the two pressure systems generates strong winds that entrain and transport sediments from the Tigris–Euphrates floodplain.

Dust storms associated with the passage of strong low pressure fronts are frequent in many parts of the world. Low pressure fronts moving in an easterly direction across Eurasia to the arid Soviet steppes and interior deserts are responsible for the high

frequency of dust storms in this region. Similar easterly moving frontal conditions have also been associated with moderately intense dust storms of relatively long duration in the Sonoran Desert region of Arizona. These meteorological conditions and associated dust storms occur primarily in late autumn, winter and spring (Idso & Brazel, 1977; Brazel & Nickling, 1986). In Australia, dust storms frequently follow the passage of strong low pressure fronts moving eastward across the southern portion of the continent (Loewe, 1943). During the summer monsoon, the convergence between high and low pressure systems may also channel dust from the interior of the Simpson Desert westward and out over the Indian Ocean. This type of convergence may occur simultaneously with the movement of low pressure fronts across southeastern Australia.

In the central region of Sudan, most major dust storms are generated by the intense downdraught and outflow of cool air from cumulonimbus clouds that develop as a result of convective activity during strong surface heating. These storms have been termed haboobs, from Arabic meaning violent wind. It should be noted that in Sudan the term haboob is used to describe any dust storm despite its origin. Pye (1987) suggested that the term *downdraught haboob* be applied specifically to those dust storms generated by downdraughts from intense convective activity.

The leading edge of these storms appears as a solid wall of dust with height ranging from 300 to 3000 m (Fig. 9.35) and have the form characteristic of a density current head (Idso, 1967). In the Khartoum

Fig. 9.35 A major downdraught haboob passing through Tempe, Arizona.

region of Sudan these storms occur about 24 times each year usually during summer when warm monsoonal air and cooler northerly air converge. They travel at an average forward speed of 50 km h^{-1} and last between 30 min and 1 h (Freeman, 1952).

Dust storms associated with intense downdraughts are frequent in many other parts of the world but usually with much lower frequency. For example, in the Phoenix area of Arizona, 50% of all dust storms can be classified as downdraught haboobs (Idso *et al.*, 1972; Brazel & Nickling, 1986). The remainder are associated with the passage of fronts or as rare tropical disturbances and upper level cut-off lows (Brazel & Hsu, 1981; Brazel & Nickling, 1986). Downdraught haboobs are important dust-raising systems in the Gobi desert (Middleton *et al.*, 1986) and in the Thar Desert of India and Pakistan where they are known locally as *andhi* (Joseph *et al.*, 1980).

In mountainous regions, foehn and valley winds (katabatic winds) are frequently responsible for the generation of dust storms. Foehn winds, which are termed *chinooks* and *Santa Ana* winds in North America, develop on the lee side of mountain ranges when air is forced to flow over them by the regional pressure gradient. Air ascending over the mountains cools and frequently loses moisture due to precipitation. On its descent, the drier air warms rapidly and is accelerated due to adiabatic compression. Dust storms generated by these types of winds are frequent in areas such as Turkmenstan in winter and spring when large depressions over the Near East and Saudia Arabia cause easterly airflow over the mountains of northern Iran and down the north facing slopes of Kopet-Dag (Petrov, 1968). Chinook winds flowing over the Rocky Mountains have resulted in severe wind erosion and dust-storm generation in the Canadian prairies usually during the late winter and early spring.

The downslope drainage of cool air from steep mountain slopes is termed katabatic or valley winds. Middleton (1986b) reported that winds of this type draining down valleys in the Hindu Kush and Karakoram Ranges deflate the plains of the Indus and its tributaries as well as the Quaternary lake beds and alluvial fans of Afghanistan and eastern Iran. Point-source dust storms in these areas are characterized by high velocity surface winds and dense dust clouds. Down-valley glacier winds have been found to generate localized dust storms in the Slims River valley located in the Yukon Territory, Canada (Nickling, 1978) and in the Mantanuska and Kwik River valleys

of Alaska (Trainer, 1961). In both these cases major dust storms are caused or intensified by topographic funnelling of pressure gradient winds.

9.6 Aeolian deposits and bedforms

To many individuals, sand dunes and extensive sand deposits known as ergs or sand seas are synonymous with deserts. Although sand deposits occupy up to 25% of the world's deserts, most desert surfaces are frequently dominated by erosional and depositional forms associated with fluvial processes and various weathering phenomena. Moreover, major aeolian deposits are not confined to deserts. Extensive loess deposits are found at desert margins and in present humid temperate environments (Section 9.2). Thick wind-derived sediments in the ocean basins and large dune complexes in coastal areas are further indications of aeolian processes beyond the desert environment. In all, sand dunes and associated aeolian deposits represent major geomorphological forms and processes which have been investigated extensively.

Loose sand deposits are often characterized by the development of bedforms with varying size, form and pattern. Bedforms tend to develop as regularly repeating patterns resulting from dynamic equilibrium between the shearing force of the wind and the sediment on the surface (Wilson, 1972a). Many researchers have argued that aeolian bedforms develop as a hierarchal arrangement of superimposed forms (e.g. Allen, 1968; Wilson, 1972a,b). Cooke and Warren (1973) suggest that ripples are almost always found on the surface of dunes, with dunes commonly being superimposed on larger forms termed *draa* in North Africa.

Several classifications of aeolian bedforms and related sand deposits are cited in the literature. Although these classifications vary, most are similar to those developed by Bagnold (1941) and Hack (1941). A generalized classification scheme modified from McKee (1979) and Breed and Grow (1979) is shown in Table 9.7. A schematic representation of the dunes is also shown in Fig. 9.36.

Wilson (1972a) outlined six main factors that he considered fundamental to the development and maintenance of aeolian bedforms:
1 Bedforms require an initial irregularity on the bed to act as a nucleus for initiation of the form. The nucleus can take several forms such as a chance irregularity at the bed due to differential erosion or

Table 9.7 Basic dune types. Modified from McKee (1979) and Greeley and Iversen (1985)

Name	Form	Type*	Slipface(s)	Wind†
Transverse				
Barchan	Crescent in plan-view	S,C	1	Transverse
Barchanoid ridge	Rows of connected crescents in plan-view	S,C,CX	1	Transverse
Transverse ridge	Asymmetric ridge in cross-section	S,C,CX	1	Transverse
Longitudinal	Symmetric ridge in cross-section	S,C,CX	2	Parallel
Parabolic	U-shaped in plan-view	S,C	1 or more	Parallel
Dome	Circular or elliptical mound	C,CX	none or poorly defined	—
Star	Central peak with three or more arms	S,C,CX	3 or more	Multiple

* S, simple; C, compound; CX, complex.
† Orientation of dune axis with respect to wind direction or the vector of more than one wind direction.

the presence of an obstruction to the airflow (e.g. pebble, standing vegetation or rock outcrop).

2 The bedform pattern on the surface develops spontaneously and does not depend on pre-existing forms on or within the bed.

3 The development of bedforms and their pattern results from individual grain movements rather than dry internal deformation of the bed.

4 There is a two-way interaction between the developing forms and the airflow over them.

5 Bedforms and their pattern migrate downwind as a result of net accumulation on the lee slopes and erosion of the stoss slopes.

6 If flow conditions and local topography remain constant, an equilibrium will be established and maintained between the bedform shape and flow pattern over the form.

Wilson (1972a,b) observed that the wave length of aeolian bedforms in the North African Sahara appeared to be related to the grain size of the 20th percentile (P_{20}; coarse tail) obtained from sand at the crest of these deposits (Fig. 9.37). Although the wave lengths of the groups overlap, the data plot in three distinct groupings associated with ripples, dunes and draas. He concluded that the distinct lack of intermediate forms indicated that smaller structures could not grow into larger ones.

Within each group there was a distinct tendency for the wave length of the bedform to increase with grain size, implying that larger features were formed under greater wind speeds (Fig. 9.37). He argued that the wave length of impact ripples (ripples formed by the impact of saltating grains) was controlled by the mean saltation path length of the sand grains, as

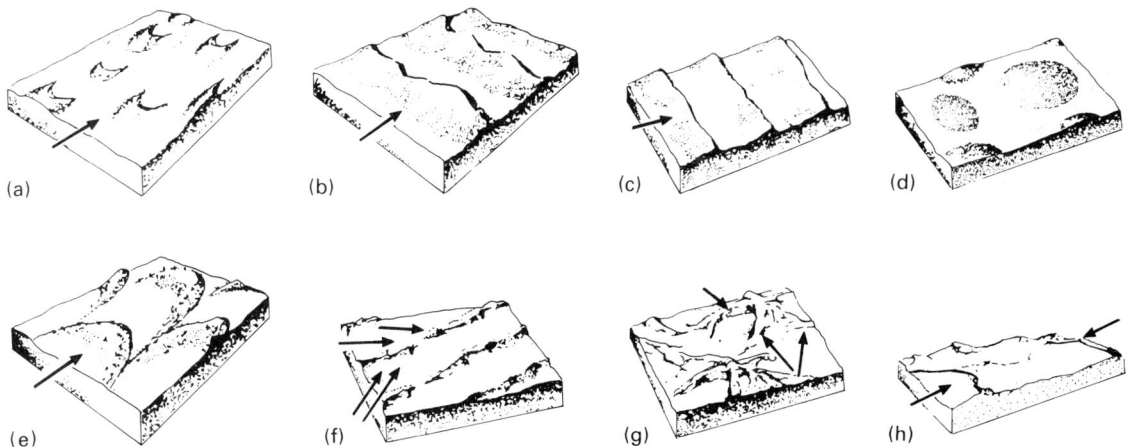

Fig. 9.36 Schematic representation of common dune forms. After McKee (1979). (a) Barchan dunes; (b) barchanoid ridge; (c) transverse ridge; (d) dome dunes; (e) parabolic dunes; (f) longitudinal dunes; (g) star dunes; (h) reversing dunes.

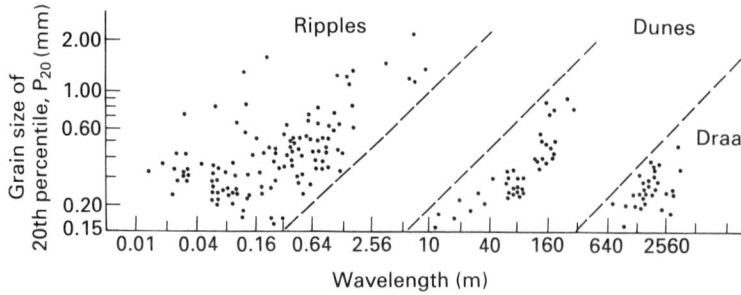

Fig. 9.37 Relationship between aeolian bedform wavelength and the grain size of the 20th percentile (P_{20}) of the crestal sands. From Wilson (1972a).

suggested by Bagnold (1941). He further suggested that the wave lengths of dunes and draas were related to the size of secondary atmospheric flow elements. On the basis of his observations he concluded that the relationship between wave length and grain size was universal in nature.

However, the universality of this relationship has been contested by Wasson and Hyde (1983) and by Lancaster (1988a). Data on the spacing of dunes from four dune fields in Australia (Fig. 9.38) in relation to the grain size of crestal sands presented by Wasson and Hyde (1983) did not conform to the 'universal' relationship presented by Wilson (1972b). They found that ripple spacing was within the range shown by Wilson (1972b) but that the distance between

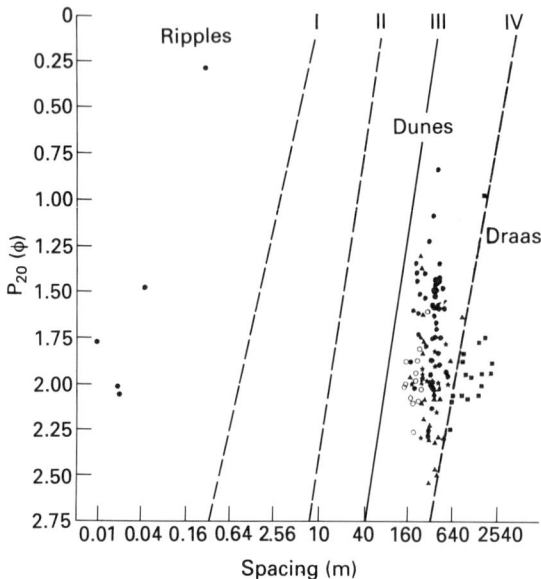

Fig. 9.38 Bedform wave length and 20th percentile (P_{20}) grain size relationships found in selected deserts of Australia. From Wasson and Hyde (1983).

dune crests in the Simpson, Strzelecki, Kulwin and Innesowen dune fields fell largely between the wave lengths indicated for dunes and draas of Wilson's diagram (Fig. 9.37). Moreover, most of the widely spaced dunes found in the Simpson–Strzelecki area fall within the draa category of Wilson. These observations confirmed the view that the North African separation of dunes from draas does not apply to Australian dunefields.

Wasson and Hyde (1983) did not negate the granulometric control on dune spacing found by Wilson (1972b) in the Sahara but doubted its universal applicability. The wider range of particle sizes, and relatively coarse P_{20} values associated with the crestal sands of the Saharan dunes, may accentuate the effect of grain size on dune spacing when compared to the finer sands of the Australian dunes. The lack of coarse sand grains in Australian dune crests was attributed to the fact that these dunes are generally derived from fine-grained alluvium.

Complementary data on dune spacing from dune fields and sand seas in Namibia, the southwestern Kalahari and the Gran Desierto of Mexico were presented by Lancaster (1988a). His data for crescentic, linear and star dunes showed no general relationship between dune spacing and grain size. In a given sand sea, the spacing of some dune types appeared to be correlated with the P_{20} but other types showed no significant relation. He also indicated that the spacing of complex and star dunes in Namibia and the Gran Desierto is unrelated to P_{20} but that spacing tended to increase with a coarsening of P_{20} for simple and compound crescentic dunes in these regions. He further suggested that differences in sand transport rates at the base and crest of stoss slopes of these dunes caused an increase in dune length (measured parallel to the wind) and crest spacing with increasing grain size (Lancaster, 1985; Tsoar, 1989). Lancaster (1988a) concluded that variations in winds and transport rates at differing temporal and spatial scales are the most important controls on dune size and spacing.

The height and spacing of individual and superimposed dunes in active sand seas tend toward an equilibrium in relation to contemporary sand transport rates and directions. In contrast, the size and spacing of draas are more closely related to long-term patterns of sand accumulation in certain areas of the sand sea that are determined by regional scale wind patterns and sand transport rates. Although superimposed dunes and draas are formed by two different scales of air flow, they can co-exist in the same flow in a manner similar to that found for sub-aqueous bedforms (Lancaster, 1988b).

9.6.1 Ripples

Ripples are the smallest and most common of all aeolian bedforms. Three basic ripple types have been identified in the literature. These are: (1) impact or ballistic ripples, also termed sand ripples and normal ripples; (2) fluid drag or aerodynamic ripples; and (3) megaripples also known as granule ripples, erosion ripples, sand ridges and giant ripples. Ripples generally develop transverse to the wind direction with crests that are straight or slightly sinuous. Except for very large ripples, the crests extend for only a few metres before they bifurcate or terminate (Greeley & Iversen, 1985; Pye & Tsoar, 1990).

Ripple wave lengths range from approximately 0.5 cm to > 25 m, with heights from 0.5 cm to over 25 cm. They most frequently occur in sands with mean grain sizes of approximately 0.3–2.5 mm, although ripples with low density particles of up to 1 cm in diameter have been observed in megaripples (Greeley & Peterfreund, 1981).

Impact ripples

Impact ripples develop by the impact of saltating grains (Bagnold, 1941; Sharp, 1963; Wilson 1972a; Ellwood *et al.*, 1975; Seppala & Lindé, 1978). Sharp (1963) found that impact ripples in the Kelso dunes of California have typical wave lengths of 7–14 cm and heights of 0.5–1.0 cm for sands 0.30–0.35 mm in diameter. He also observed that an increase in grain diameter or wind velocity resulted in ripples with longer wave lengths. These ripples were observed to have asymmetrical cross profiles with upwind slopes of 8–10° and lee slopes of 20–30°. Lee slopes were generally composed of two distinct parts; a short

slope with a steep angle of repose (avalanche or talus-like), and a somewhat longer and gentler concave slope that merged with the trough. Ripples in coarser sand tended to be more asymmetric in cross profile with gentler windward slopes and steeper leeward slopes. Similar observations on ripple morphology in the Kelso Dune area have also been made by Werner *et al.* (1986).

Sharp (1963) investigated the internal structure of ripples by impregnating the bedforms in the field with a quick drying plastic. Thin sections of the casts were made in the laboratory and examined under a petrographic microscope. He observed that almost all the ripples examined seemed to constitute 'a pile of relatively coarse sand grains resting on a firm, generally smooth base of thinly bedded finer sand' (Sharp, 1963, p. 621). The ripples contained low concentrations of fine sand that tended to be evenly mixed throughout the coarser material (Fig. 9.39). Most of the ripples had no discernible internal structure although a few displayed faint foreset beds. Occasionally, distinct foreset beds of predominantly fine sand were also observed, apparently deposited during periods of low wind velocity when movement of coarser grains was minimal.

Sharp (1963) also noted a concentration of coarse grains at the crest. Similar observations have been made by other investigators. These large grains move up the backslope of the ripple in surface creep by the impact of smaller saltating grains. On reaching the crest they tumble over the brink and come to rest on the lee slope where they are eventually buried and re-incorporated into the ripple.

Sharp (1963) thought that the fine base over which the coarser main body of the ripples moved was developed by the trapping of fine material moving in

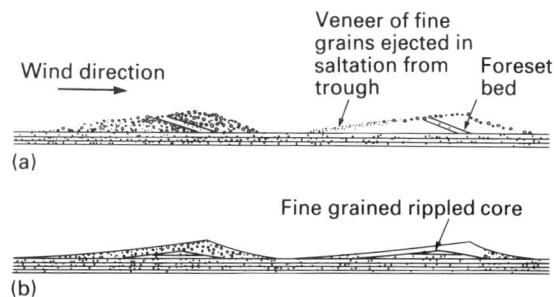

Fig. 9.39 Typical ripple cross-sections produced from plastic-impregnated field samples. From Sharp (1963).

saltation within the spaces between the coarser grains. As the ripple migrated forward, the fine grains settled downward through the coarser grains, eventually being added to the basal layer. He suggested that the fine grained core found in some ripples (Fig. 9.39) may be indicative of this sorting process. Hunter (1977) and Kocurek (1981a) have also attributed size grading in various sedimentary features to interstitial settling of finer grains.

Impact ripple formation

Many hypotheses have been put forward for the initiation and migration of ripples. Early models proposed that direct fluid forces on the sand surface or in the saltation layer could generate wave-like motions which produce ripple forms on the bed (e.g. Cornish, 1914). For the most part, these notions have been discredited. Later investigators formulated a variety of models for ripple formation based on the movement of surface grains by saltation impacts (Bagnold, 1941; Sharp, 1963; Anderson, 1987).

Through wind tunnel and field observations, Bagnold (1941), ascribed the development of ripples to the characteristic path length and impact angles of saltating grains (Fig. 9.40). He argued that for well-sorted sands, once movement was initiated in saltation, the grains would move with a relatively narrow range of trajectory path lengths and impact angles (α). On a perfectly flat smooth bed, the rate of bombardment would be fairly uniform across the surface. However, the presence of chance irregularities (Fig. 9.40) causes more grains to strike section AB than BC, which is sheltered by the rise at B. As a result, more grains are ejected from the bed by saltation impact in AB, forming a zone of net erosion. The sheltered zone, BC, is a zone of net accumulation as relatively few grains leave the bed. Grains derived from AB will move downwind a distance equal to the

characteristic path length. These grains propagate the erosion zone downwind and move the entire system forward with a ripple spacing approximately equal to the saltation path length. For a given wind speed and grain size, a maximum ripple height develops. Any further increase in amplitude would raise the crest into a zone of high velocity, where crest grains would be quickly removed.

Bagnold (1941) noted that this model holds only for sands with relatively uniform grain size distributions. In the case of more poorly sorted sands, the coarse grains move along the surface in creep caused by the impact of saltating grains. When these coarser grains reach the crest their potential rate of movement decreases as a result of the low rate of impacts in the shadow zone (BC in Fig. 9.40). Numerous investigators confirm the concentration of coarse grains at the crest (e.g. Sharp, 1963). In general, steep tall ripples with more asymmetric profiles are formed in sands containing a large proportion of coarse grains. More uniform fine-textured sands form low amplitude ripples.

Sharp (1963) disputed Bagnold's (1941) argument that ripple wave length is determined by saltation path length. He contended that ripple height has a more direct control on wave length. He further suggested that the coarse grains found in the main body of most ripples (Fig. 9.40) are moved in surface creep by saltation impact. Where surface irregularities exist, these grains pile up until an equilibrium height is attained that is controlled by the angle of incidence of saltating grains. The increase in height of the ripple brings the crest into a zone of higher wind velocity. Since the saltation path length is a function of wind speed, saltating grains entrained from higher crests will move further downwind, producing ripples with longer wave lengths.

Anderson (1988) presented a theoretical model for aeolian impact ripples that gave careful consideration to experimental work on saltation grain impacts at the bed by Rumpel (1985), Willetts and Rice (1986) and Ungar and Haff (1987). He suggested that high energy impacts from saltating grains gave rise to two distinct populations of subsequent trajectories: (1) a single high energy rebound, where the impacting grain usually had a long trajectory and constant impact angle in successive saltation as described by Rumpel (1985); and (2) a short trajectory, low impact energy saltation which he termed *reptation* (from the Latin *to crawl*) and which included material moving in surface creep (Fig. 9.41). The high energy re-

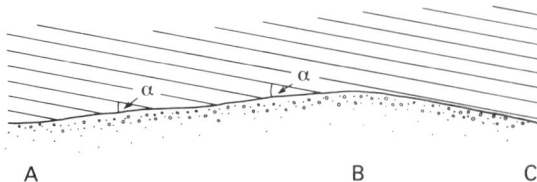

Fig. 9.40 Pattern of saltating grains striking a somewhat irregular surface resulting in impact ripple formation. A, B, and C represent zones of relative erosion and deposition. Modified from Bagnold (1941).

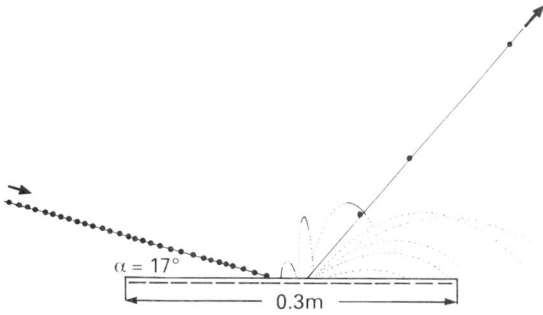

Fig. 9.41 Pattern of particle ejections when a 4-mm steel pellet impacts a bed of identical pellets recorded at successive instants by a high speed strobe. Two distinct groups of trajectories are identified: (1) a single high energy rebound, usually the impacting grain itself; and (2) approximately 10 low energy ejecta having ejection speeds an order of magnitude lower than the impacting grain. Modified from Anderson (1987).

bounds represented an average of about 60% of the impact energy. Low energy ejecta (up to 10 grains per high energy impact) emerged shortly after impact from a region approximately 10 grain diameters across and centred slightly upwind of the impact site. Mean ejection speeds for these grains were approximately 3% of the impact speed and, in total, accounted for about 1% of the impact energy. The remaining large proportion of impact energy induced local transient dilation of the bed, inelastic deformation of the bed grains and frictional rotation of bed grains.

Anderson (1987) used a form of the sediment continuity equation to define mass flux, in conjunction with an expression for the spatial variation in ejection rates for reptating grains. He considered a sinusoidally perturbed bed and a probability distribution of reptation path lengths to predict the initial growth rates and translation speeds of incipient ripples. Shortly after the initiation of simulated movement, a dominant wave length emerged which is scaled by grain transport distance. The relevant length was found to be the mean reptation length and not the characteristic path length as suggested by Bagnold (1941). Anderson found that the fastest growing ripple wave length was approximately six times the mean reptation distance and corresponded to the wave lengths of incipient ripples in wind tunnel experiments. The model, although promising, applies only to incipient ripple development. As the ampli-

tudes increase other physical effects are introduced that complicate the analysis (Anderson, 1987).

Aerodynamic ripples

In cases where fine, well sorted sand is transported by high winds, long low-amplitude forms termed fluid drag ripples (Bagnold, 1941) or aerodynamic ripples (Wilson, 1972a) may develop. These ripples are similar to some sub-aqueous ripples described by Allen (1968) and are thought to result from secondary flow phenomena.

If wind velocities are sufficiently high, small particles may be transported in either suspension or modified saltation (Section 9.4). The trajectories of these particles are long, and somewhat irregular, as they are affected by air turbulence. As a result, the geometry of these ripples is primarily controlled by local variations in surface shear stress and sediment transport rate, rather than by the length of the saltation trajectories as with impact ripples (Greeley & Iversen, 1985). Aerodynamic ripples have wave lengths ranging from 0.2 to 2 m and heights from 0.2 to 5 cm and are characterized by more sinuous crests than impact ripples (Bagnold, 1941; Wilson, 1972a). Wilson (1972a) suggests that the sinuosity in these ripples may result from longitudinal flow elements caused by Taylor–Görtler type vortices in a manner similar to that suggested by Allen (1968) for the development of longitudinal bedform elements in water.

Megaripples

Several researchers have observed large, rather symmetrical, ripple forms with wave lengths in excess of 20 m. These have been termed granule ripples (Sharp, 1963) and megaripples (Greeley & Peterfreund, 1981; Greeley & Iversen, 1985). Megaripples are usually formed in sediments with distinct bimodal grain size distributions with a large proportion of coarse grains. These ripples are typically veneered with a lag of coarse grains on their surface that is most concentrated at the crest.

Bagnold (1941) and Sharp (1963) have suggested that this type of ripple is primarily formed by the movement of the larger grains in surface creep. The rate of surface creep is controlled by the impact of saltating grains derived either upwind or from the surface layers. Since the finer, saltating grains are transported more rapidly and have a lower threshold velocity than the coarser grains, lower wind velocities and a lag of coarse grains develops on the ripple surface with a maximum concentration of the coarsest grains at the crest. The form develops an equilibrium height that is determined by the average wind velocity, the diameter of the crest grains and the (relative) impact angle of the finer saltating grains.

9.6.2 Dunes

Sand dunes represent one of the most characteristic and aesthetic forms found in desert environments. However, dunes are not confined to deserts and are common to coastal areas (e.g. Lancaster, 1982a; Pye, 1983; Pye & Tsoar, 1990) as well as more humid temperate environments. Dunes and associated sand deposits have also been identified on other planets such as Mars (Greeley & Iversen, 1985).

Dunes have been defined by Bagnold (1941) as mounds or ridges of sand that exist independently of surrounding topography. Dunes have also been considered as deformable obstructions to air flow that are free to move and are not dependent on fixed obstructions for their maintenance (Bloom, 1978). These somewhat stringent conditions have been relaxed by other authors to describe dune or dune-like features formed by the trapping or fixing of sand by vegetation such as coppice and shadow dunes (e.g. Hesp, 1981; Fig. 9.42; McKee, 1982) or by deposition behind topographic features as illustrated by echo and climbing dunes (e.g. Tsoar, 1983a). Comprehensive discussion and reviews of various dune types have been presented by Warren (1969), Lancaster (1982b, 1989a,b), Pye (1983), Tsoar (1989) and Pye and Tsoar (1990).

The fascination with sand dunes has resulted in an enormous literature on the development, morphology and structure of these forms by a large number of researchers. Coincident with this research has been the introduction of a bewildering number of terms into the literature to name and describe dunes of differing size, shape and possible mode of origin. These terms are frequently derived from local names for these features or have been applied to denote the general morphology or formation process. In an attempt to reconcile the terminology for similar forms in different locations Breed and Grow (1979) have compiled a detailed list of references comparing local terms for various dune types.

Many classification schemes for dunes have been suggested in the literature (e.g. Melton, 1940; Hack, 1941; McKee, 1979). Most of these classifications are based on local descriptions, possible mode of origin or morphological characteristics. McKee (1979) and his colleagues carried out an exhaustive study of sand deposits and dunes in the world's sand seas using Landsat imagery. This work compared sand deposits in a global context and resulted in a classification scheme based on dune shape and the arrangement and number of slip faces (if present), as well as the orientation of the dominant wind or the vector sum of more than one wind (Table 9.7).

Dune types can be grouped, in a general manner, on the orientation of their primary elements in

Fig. 9.42 Coppice dunes located in the coastal region of the Namib Desert. Sand, saltating across the surface, is trapped by vegetation forming elongated forms 2–3 m in length and up to 1.5 m in height.

Fig. 9.43 Small barchan dune approximately 1.5 m in height formed by limited sand supply from fluvio-glacial outwash in the Slims River Valley, Yukon.

relation to the dominant wind direction. Barchans, barchanoid ridges and transverse ridges tend to be aligned perpendicular to the dominant wind. Barchans are isolated crescentic forms (Fig. 9.43) that may coalesce to form barchanoid ridges. Transverse dunes have a less sinuous and more sharply defined crest than barchanoid ridges. Despite their general transverse form, barchanoid ridges have distinct transverse and longitudinal components termed barchanoid and linguoid elements (Cooke & Warren, 1973; Fig. 9.44).

Simple longitudinal dunes form parallel to the wind and have two basic sub-types; sinuous sharp crested forms termed seifs and more rounded, long, relatively straight forms commonly found in the Simpson and Kalahari deserts called simple linear dunes (Lancaster, 1982b). Star dunes are characterized by their large size and radiating sinuous arms and are associated with multidirectional or complex wind regimes. The major arms in star dunes tend to be aligned transverse or slightly oblique to the dominant sand-transporting directions, with the minor arms aligned parallel to the dominant wind and transverse to the subordinate winds (Lancaster, 1989b).

Dunes have been further classified on the basis of their degree of complexity (Breed & Grow, 1979). Simple dunes are mounds or ridges of sand that exist as discrete forms. Compound dunes are sand forms on which smaller dunes of similar type and slipface orientation are superimposed. Complex dunes are combinations of two or more dune types (Table 9.7).

The initiation of dunes is not well understood but

is thought to result from the complex interaction of several factors, including a source of suitably sized material, an appropriate wind regime and some mechanism that promotes deposition. Deposition will occur when the transporting ability of the wind is reduced to the point that it can no longer carry sand in saltation. This may occur at gentle dips in the surface, behind chance irregularities or at locations of convergent secondary flow.

Bagnold (1941) has argued that, in order for true

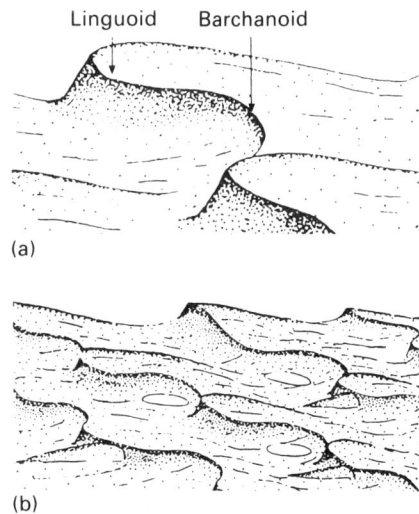

Fig. 9.44 Schematic representation of a barchanoid ridge pattern (aklé) showing (a) the distinct linguoidal and barchanoidal elements and (b) aklé dunes. Modified from Cooke and Warren (1973).

dunes to develop, a sand patch of a critical size (4–6 m) must first be established that can effectively trap saltating sand from upwind. Saltating sand moving over a hard or pebble-covered surface is transported at a greater rate, and with longer trajectories, than over a sand surface as indicated by the coefficient, C, in Eq. 9.18.

As the wind passes from a rough pebble surface to a smoother sand patch, there will be an initial increase in erosion and sand transport rate because of the relatively high shear velocity established over the rougher surface. However, the increased saltation load causes a retardation of the wind through the loss of momentum from the air to the saltating grains which eventually leads to a reduced equilibrium sand transport rate over the patch. If the sand patch is small (i.e. < 4–6 m long) the wind will not be sufficiently retarded by the saltation cloud so that sand, originating upwind, will pass over the sand patch. Conversely, if the sand patch is relatively large there will be a net accumulation and the sand patch will grow in size until some equilibrium height, which is a function of the average wind speed and mean grain size, is reached.

At some point in time the sand mound may become sufficiently steep on its downwind side that the sand will begin to avalanche. This process does not usually occur across the entire face of the dune but as continuous isolated, small failures or tongue-shaped grain flows. Downwind migration of the dune form results from deposition of these avalanche deposits in conjunction with the deposition of saltating grains that pass over the dune crest (Hunter, 1977; Fryberger & Schenk, 1981; Anderson, 1988).

Barchan dunes

The simplest dune forms are barchans and generally form in areas where: (1) the sand supply is limited; (2) the wind tends to be unidirectional; (3) the regional surfaces are relatively hard (e.g. desert pavement); and (4) the vegetation cover is limited.

Barchans are typically relatively small with a rather symmetrical form having a single slip face between two distinctive arms or horns (Figs 9.36 & 9.43). Barchans are usually between 0.3 and 10 m high, although heights of over 50 m have been observed. Widths (measured from arm to arm) also vary greatly but are often about 10 times the dune height and tend to be somewhat narrow if developed under relatively strong winds (Allen, 1968). Windward slopes range

from 5 to 15°, with slip face slopes of 30 to 40° depending on the grain size characteristics and wind speed.

Hastenrath (1967), from measurements of barchans in Peru, found widths ranging from approximately 20 to 50 m with associated heights of 1–6 m. Similar observations have been made in Peru by Finkel (1959). Lancaster (1982a) found most barchans to be 2–5 m high, with some larger forms 25–30 m high, in the Skeleton coast dunefield of Namibia. The rather consistent shapes and heights of barchans in a given area is indicative of the tendency of these dunes to develop and equilibrium form that is a function of the wind regime, grain size characteristics and sand supply in the region (Howard et al., 1978; Rubin & Hunter, 1982).

In general, barchans tend to maintain their form even if the topography over which they are moving becomes somewhat altered. However, in some cases, one of the horns may become extended if the wind pattern or sand supply is altered. Bagnold (1941) has suggested that the over extension of one arm may initiate the development of longitudinal (seif) dunes.

The rate of barchan advance is quite variable and has been related to wind speed, grain characteristics of the surface over which they move and, most importantly, to dune height (Bagnold, 1941; Finkel, 1959; Long & Sharp, 1964; Hastenrath, 1967; Tsoar, 1984). Movement rates of 17–47 m year^{-1} have been measured in Peru (Hastenrath, 1967) and 7.5 m year^{-1} in the eastern Sahara (Haynes, 1989). More recently, numerical models have been developed by Howard et al (1978) and by Wipperman and Gross (1986) to simulate the development and migration of barchans based on established sand transport equations. Model outputs were found to be in relatively good agreement with observed morphology and migration rates.

Barchanoid ridges and transverse dunes

Barchanoid ridges are formed by the coalescence of simple barchan dunes that form a parallel wavy ridge perpendicular to the dominant wind (Fig. 9.45). They appear to develop under the same unidirectional wind conditions as simple barchans but where sand supply is greater (McKee, 1982). Barchanoid ridges are common in many areas including the Rub al' Khali, Saudi Arabia; Takla Makan and Ala Shan deserts, China; Thar Desert, Pakistan; the Gran Desierto, Mexico; the Namib Desert, and the Alg-

Fig. 9.45 Well-developed barchanoid ridges at Great Sand Dunes National Monument, Colorado.

odones Dunes and Nebraska Sand Hills of the USA (Breed & Grow, 1979; McKee, 1982; Lancaster, 1982a; Sweet *et al.*, 1988).

Transverse dunes are related to barchanoid ridges but form under unidirectional winds where sand supply is even greater. In contrast to barchanoid ridges, these forms have a relatively straight crest with slip faces only on their downwind side. Transverse dunes, although similar in general form, differ in internal structure. McKee (1966) found that the most diagnostic feature of transverse dunes in the White Sands National Monument, New Mexico, was the great extent of horizontal, or nearly horizontal, parallel laminae which were generally absent in barchanoid forms.

Like other dune forms, barchanoid ridges and transverse dunes tend to develop an equilibrium form and pattern that is related to sand supply, sediment size, and average wind velocity. Lancaster (1982a) and Lancaster *et al.* (1987) found good correlations between dune height and spacing for barchanoid and transverse dunes in the Gran Desierto and Namib deserts. Complementary data have also been presented by Breed and Grow (1979) for larger forms identified from Landsat imagery and air photographs in several sand seas.

Compound barchanoid ridges and transverse dunes have also been identified in many of the world's deserts as having complex forms (Fig. 9.46). For example, Lancaster *et al.* (1987) found crescentic ridges, 500 m wide and up to 20–30 m high with crest-to-crest spacing up to 1500 m, in the Gran Desierto upon which were superimposed smaller

ridges 2–5 m high with spacings of 100–200 m. Similar features have been identified in the Great Sand Dunes, Colorado (McKee, 1979), the Liwa Oasis, Saudi Arabia (Bagnold, 1941) and in the coastal areas of the Namib Desert (McKee, 1982). The sinuous crests of barchanoid ridges have distinct alternating transverse and longitudinal elements, although the general form is perpendicular to the dominant wind (Fig. 9.44). In early literature, it was thought that these two elements represented fast and slow lanes of secondary flow parallel to the main flow with fast flow moving the linguoidal elements forward more quickly. However, this simple flow model cannot account for the common pattern found in areas having repeating barchanoid ridges, termed *aklé* patterns (Cooke & Warren, 1973). In these cases, the barchanoid element of one ridge is followed downwind by a linguoid element in the next ridge (Figs 9.44 & 9.47). Wilson (1970), quoted by Cooke and Warren (1973), suggested that this feature can be explained by considering the combined transverse and longitudinal secondary flow patterns including return flow from the lee eddy (Fig. 9.47). Consideration of this flow pattern produced dune ridge elements that are oblique to the flow, and which account for the half wave length shift of the downwind barchanoid element. The angle of the longitudinal element to the main flow (ϕ) for an idealized case can be defined by $\tan \phi = 1/2\,\tau$, where τ is the transverse wave length. This compares favourably with actual field measurements, the discrepancies being attributed to differences in the longitudinal and transverse element wave lengths associated with the measured dune ridges; Cooke and

Fig. 9.46 Compound transverse ridges in the Namib Sand Sea with superimposed barchans and barchanoid ridges. The main ridges are spaced approximately 800 m apart. Photo courtesy of N. Lancaster.

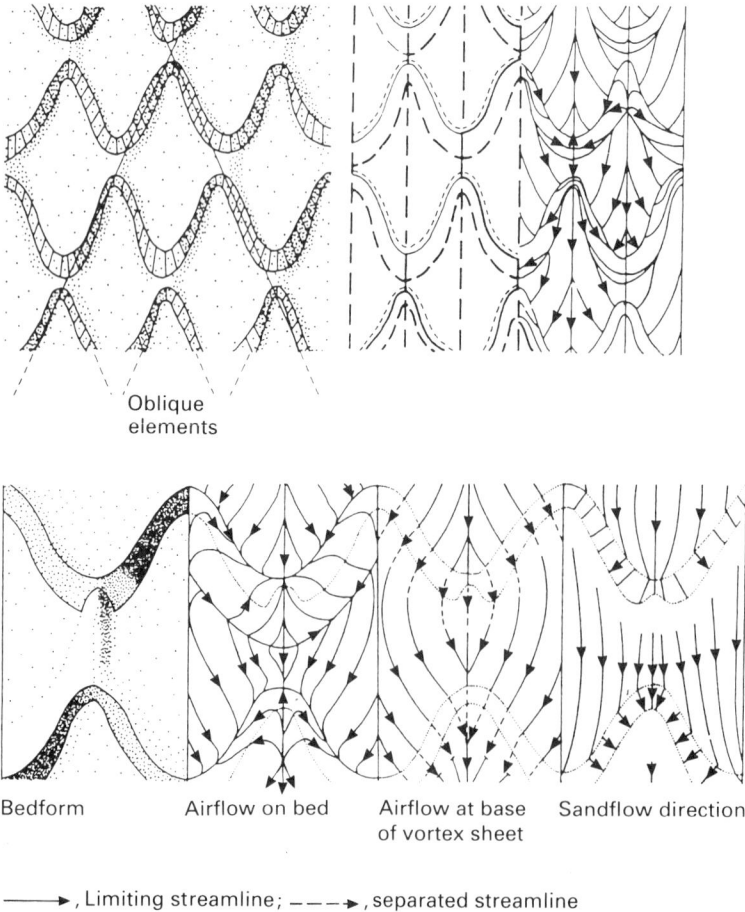

Oblique elements

Bedform Airflow on bed Airflow at base of vortex sheet Sandflow direction

⟶ , Limiting streamline; – – –▶ , separated streamline

Fig. 9.47 Barchanoid ridge patterns (aklé) resulting from three-dimensional flow patterns with laterally displaced vortices. From Wilson (1970) as modified by Cooke and Warren (1973).

Warren (1973) concluded that these bedform patterns are preserved because of the development of a spiral lee eddy at the base of the slipface.

Linear dunes

Linear dunes are long, relatively low ridges of sand that tend to be aligned with the dominant wind direction or to the resultant vector of multiple wind directions. These dunes most frequently occur as multiple parallel ridges separated by wide sandy, gravelly or rocky areas; the interdune corridors are frequently vegetated (Fig. 9.48). Under some conditions linear dunes can occasionally be found as a solitary ridge. Lancaster (1982b) suggested that linear dunes are the most common of all dune forms and cover approximately one half to two-thirds of the global sand seas. They are the most dominant dune form in the Simpson desert of Australia, as well as the southwestern Kalahari, the Namib and in the southern and southwestern ergs of the Sahara. They are also a common feature in the Great Sand Sea of Libya, the northern Mauritania sand seas, the ergs of Bilma and Ténéré and the southern and western Rub al' Khali in Saudi Arabia (Breed & Grow, 1979; Lancaster, 1982b; McKee, 1982).

Linear dunes can be simple, compound or complex with the larger forms being of draa size. Breed and Grow (1979) have identified solitary simple linear dunes in excess of 100 km in length with the largest dunes, having parallel ridge systems, found in the Qa'miyat region of the southwestern Rub al' Khali

where individual ridges 0.3–2.7 km wide can be traced for up to 190 km.

Simple linear dunes are characterized by rather symmetrical or somewhat asymmetrical profiles having slopes of 10–20° with slip faces formed on alternating sides. They can have straight or sinuous crests, which may be either rounded, as found in the relatively straight Australian linear dunes (e.g. Mabbutt, 1968; Folk, 1971a) or sharp crested, as observed in the Sinai (e.g. Tsoar, 1983b), Libyan (Bagnold, 1941) and Namib deserts (Lancaster, 1980; McKee, 1982). The Australian linear dunes commonly merge to form Y-junctions with angles of 30–50° opening into the wind (Folk, 1971a), whereas the Libyan forms display distinct zig-zag patterns. Simple linear dunes are typically 0.2 km wide, 5–20 km high and spaced 0.7–0.9 km apart, but vary greatly in both form and size.

Compound and complex linear dunes are usually wider and higher than simple forms and have linear dunes, or other dunes, superimposed on their flanks. Compound forms frequently consist of two or more closely spaced parallel or converging ridges on the crest of wider linear dunes, and may have an anastomosing pattern. Complex dunes are large features in excess of 150 m high and frequently have star-shaped forms regularly spaced along their crests, with smaller secondary dunes being observed transverse or oblique to the main dune trend (Lancaster, 1982b).

At least two types of simple linear dunes are identified in the literature; long, straight dunes having rounded crests and frequently vegetated plinths

Fig. 9.48 Complex linear dunes located in the Namib Sand Sea. Spacing is approximately 2.0–2.5 km with heights ranging from 100 to 170 m. Photo courtesy of N. Lancaster.

Fig. 9.49 Simple linear dune located in the Simpson Desert, Australia. The dune is approximately 10 m in height and is characteristic of the straight, round crested, vegetated linear dunes of Australia.

(lower flanks of the dunes), such as those found in the Simpson and Kalahari deserts (Fig. 9.49), and the distinctive sinuous, sharp crested forms with numerous slip faces termed seifs (Fig. 9.50), which are common in the Libyan desert (Bagnold, 1941) and in the Sinai (Tsoar, 1983a,b). However, Tsoar (1989) suggested that simple linear dunes should be subdivided into three main types: (1) vegetated linear dunes with low rounded profiles typified by the convergence of adjacent dunes in Y-junctions; (2) unvegetated sharp-crested self dunes formed under bidirectional wind regimes; and (3) lee dunes that form downwind of obstructions such as cliffs, rock outcrops or vegetation, with their size being proportional to that of the obstacle.

Despite their common occurrence and numerous detailed studies, the origin and development of linear dunes remains a perplexing problem. This situation arises primarily from the difficulty in making simultaneous observations of the three-dimensional air flow patterns over the form (both short and long term) in relation to the sediment transport and internal structure of the dune. Many different hypotheses have been put forward to explain their origin, and have been summarized by Cooke and Warren (1973), Lancaster (1982b), Tsoar (1989), Pye and Tsoar (1990), Thomas and Tsoar (1990) and Livingstone and Thomas (1993).

Early hypotheses argued that linear dunes were remnant forms, eroded from fluvial or lacustrine

Fig. 9.50 Seif-type dune with a typical sharp wavy crest located in southeastern California. Photo courtesy of G. Kocurek.

sediments (e.g. King, 1960). Critical evidence against such notions has come from more recent studies of the internal structure of linear dunes where typical aeolian cross-bedded strata have been found (e.g. McKee & Tibbits, 1964; Tsoar, 1974).

Another rather popular explanation for the origin of linear dunes advocates the presence of helical or roller vortices that develop in the atmosphere with horizontal axes parallel to the wind direction. Bagnold (1953) was one of the first proponents of this model and proposed that differential heating of a desert surface would lead to convective cells. He argued that the regional wind organized these cells into pairs of roller vortices, spaced at a distance three times the height of the convectional layer. The opposing horizontal spirals were thought to transport sand obliquely across the desert surface into long parallel ridges spaced equally to the size of the vortices (Fig. 9.51). Support for the helical flow theory has been given by Hanna (1969), Folk (1971b, 1976, 1977), and Warren (1979). However, it has been severely criticized by Leeder (1977), Lancaster (1982b) and Tsoar (1983b) for two basic reasons. Firstly, if linear dunes are created by helical roller vortices, the resultant dunes should be spaced 4–12 km apart, which does not conform to the observed maximum spacing of 3–3.5 km (Breed & Grow, 1979) or the relatively narrow spacings of < 1 km commonly found in the Simpson and Kalahari deserts. Moreover, the vortices would have to be positioned at exactly the same place during each wind episode in order for the dune to grow. However, as

Greeley and Iversen (1985, p. 169) point out, once the dune forms become of a significant size and extent they will exhibit 'feedback' to the winds and affect the boundary-layer wind pattern. Secondly, no field evidence has been provided to support the argument that sand is swept from the interdune areas to the dune (Lancaster, 1982b).

Some linear dunes appear to develop by the extension of other dune types as a result of a change in wind direction or sand supply (e.g. Warren, 1976; Lancaster, 1980; Tsoar, 1984; Hastenrath, 1987). Bagnold (1941) argued that seif dunes could form by the extension of one arm of a barchan in a bidirectional wind regime. He suggested that strong storm winds blowing obliquely to the barchan would add sand preferentially to one arm, which would subsequently be extended by the dominant wind. The repetition of this cycle would result in a linear form with a sharp sinuous crest and repeating summits.

Dunes of linear type are also associated with the trailing arms of parabolic or 'blow-out dunes' (e.g. Verstappen, 1968; McKee, 1982). In some cases the horns of barchan-type dunes may become stabilized because of vegetation growth during relatively wet periods. During drier, windy conditions sand is deflated from the higher central portion of the dune between the arms leaving long, parallel or subparallel trailing arms open into the wind with a slip face on the opposite side of the arms. In some cases, passage of parabolic dunes may leave long parallel sand ridges with little indication of the original dune from which they initiated (Greeley & Iversen, 1985).

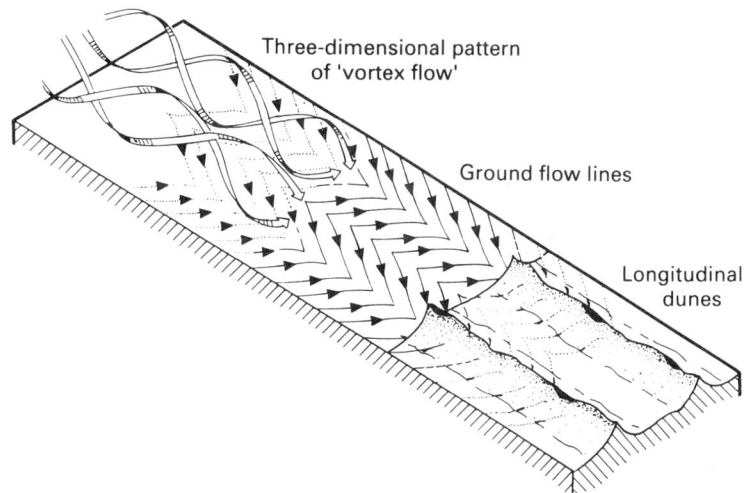

Fig. 9.51 Three-dimensional helical flow model showing the relationship between the resultant ground flow lines and the longitudinal dunes. Modified from Cooke and Warren (1973).

Linear ridges up to 12 km in length associated with parabolic dunes have been observed from Landsat imagery by Breed and Grow (1979).

Tsoar's (1983b) detailed measurements of wind-flow patterns and sand movements over linear dunes have provided important insight into the development of these forms. He found that when winds strike the crest of a linear dune at an oblique angle, it is deflected parallel to the dune with flow separation occurring in the lee of the crest (Fig. 9.52). He argued that because the lee eddy has two components (parallel and perpendicular to the dune) it has a helical form. In the re-attachment zone, the perpendicular component is small relative to the parallel component, and the resulting wind direction on the lee side will be dominantly parallel to the dune. The velocity of the parallel wind is dependent on the dune shape, as measured by the ratio of dune height to the horizontal length of the windward slope, and the angle between the wind approach direction and the dune crest. Maximum lee slope windspeeds occur when this angle is between 35 and 50° (Tsoar *et al.*, 1985). At these critical angles, sand transported over the crest is moved along the slope by deflected wind with sand eroded from the lee slope itself. When approach angles are 50–90°, the deflected wind speed is decreased, resulting in the possible deposition of sand carried over the crest.

Thus, on a sinuous crest, there will be alternating points of erosion and deposition, giving rise to the characteristic 'peaks and saddles' of some linear dunes. Tsoar (1983b) argued that the development of linear dunes requires: (1) bidirectional wind regimes

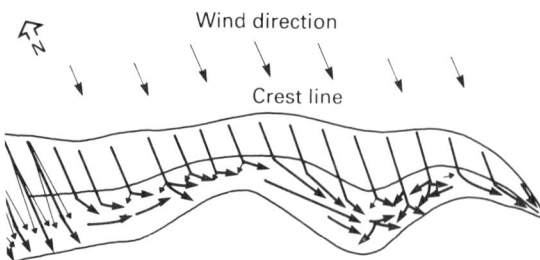

Fig. 9.52 Model for the development of seif-type linear dunes. Winds blowing obliquely to the dune separate at the sharp crest and re-attach downslope on the lee side. Airflow is deflected as a vector down the dune resulting in sand transport in the same direction. Secondary flow transports some sand in other directions. Modified from Tsoar (1983b).

where the angle between the dominant winds is not greater than 150–180° to the winds that blow obliquely across the dune; and (2) a sharp crest that causes flow separation to occur on the lee side.

Although Tsoar's (1983b) model can explain the development and maintenance of linear dunes, it cannot account for their initial formation. However, Tsoar did provide evidence that linear dunes may be secondary forms developed from barchans, transverse or zibar dunes (low transverse sand ridges lacking slip faces covered with well-packed coarse sand).

Star dunes

Star dunes are the largest aeolian bedform found in many sand seas, reaching heights in excess of 300 m and widths up to 3000 m. These dunes, also known as *rhourds* and *oghurds*, are characterized by a distinct pyramidal shape with three or four arms radiating from a central peak. The arms vary in length but have sinuous sharp crests with multiple slip faces and deep hollows often present between the arms (Fig. 9.53).

Star dunes appear to develop in areas with either multidirectional or complex wind regimes, especially during the months when most sand transport occurs (Lancaster, 1989a,b). As a result of the complex multidirectional wind patterns, net sand transport on these dunes is relatively low, suggesting that these forms migrate very slowly or remain relatively fixed in place.

Breed and Grow (1979) have identified three basic star dune forms:

1 sharp, pointed, radiating pyramidal dunes typical of the Gran Desierto, Mexico; Ala Shan Desert, China; part of the southern Sahara in Niger and the Grand Erg Oriental in Algeria.

2 Rounded compact star dunes with short arms such as those found at the southern margin of the Grand Erg Oriental.

3 Sharp-crested forms with greatly elongated arms in a preferred direction as in the Namib desert.

The general form and orientation of the arms are thought to reflect differences in wind regimes and sand supply. Lancaster (1989a,b) noted that the major arms of star dunes in the Namib and Gran Desierto sand seas tend to be aligned transverse or slightly oblique to the major sand-transporting directions. The minor arms are generally parallel to these directions and transverse to the subordinate wind direction. Similar observations have been made for

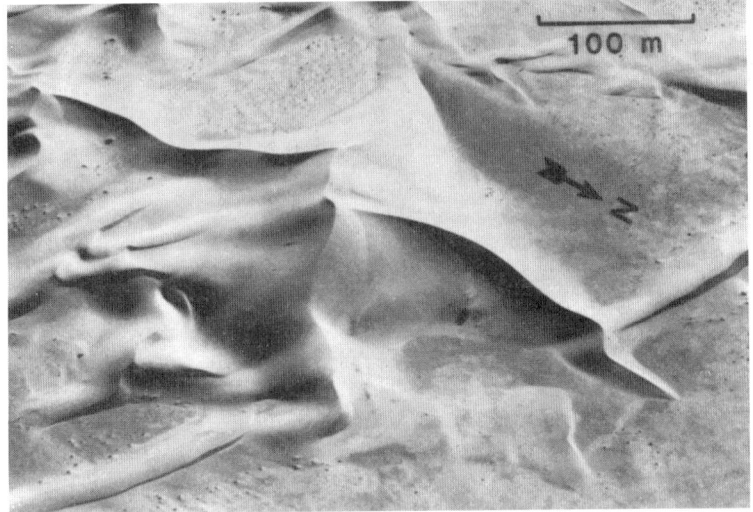

Fig. 9.53 Mature star dune located in the Algodones dune field, California. Photo courtesy of G. Kocurek.

star dunes observed at Dumont, California by Nielson and Kocurek (1987). By contrast, the arms of star dunes in the Grand Erg Oriental appear to have a complex pattern of alignments.

Star dunes occur singly, in aligned chains (Fig. 9.54), or as clusters that frequently have a distinct orientation and can be located within or at the margins of sand seas (Greeley & Iversen, 1985). A strong association has also been noted between the

occurrence of star dunes and topographic barriers. Escarpments and mountain massifs modify regional winds and increase their variability, enhancing star dune development.

While the general characteristics of star dunes are fairly well documented in the literature, little is known about the effects of multidirectional winds on the morphology and pattern of sand deposition on these forms. Even less is known about their origin

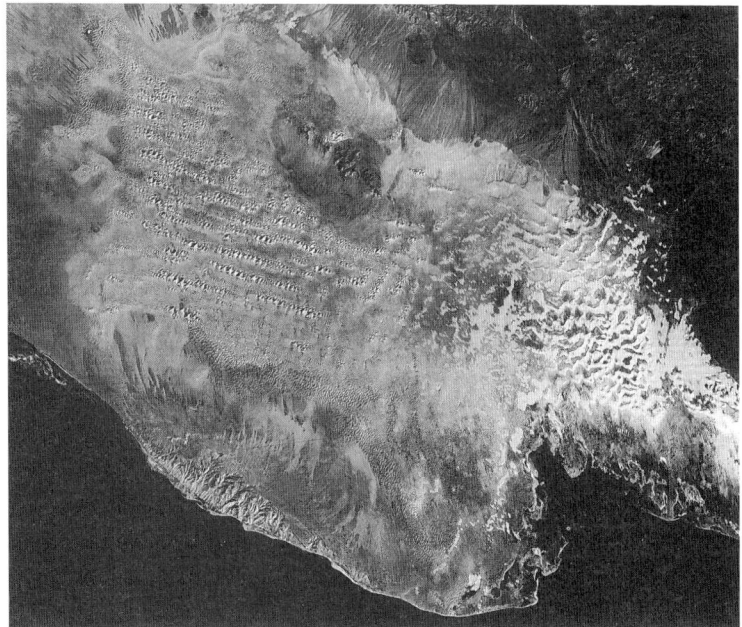

Fig. 9.54 Chains of star dunes in the Gran Desierto, Mexico. Photo courtesy of N. Lancaster.

and possible migration rates. The most detailed information available has been presented by Nielson and Kocurek (1987) and Lancaster (1989b).

Lancaster (1989b) has carried out a detailed study of erosion and deposition in relation to surface wind velocities on a relatively small star dune in the Gran Desierto sand sea. His observations indicate that the interactions between dune form and seasonally changing air flow patterns and the existence of major lee-side secondary circulation play a major role in the development of this dune. These complex interactions lead to sand deposition in the central parts of the dune giving rise to a pyramidal shape and some extension of the arms.

Lancaster's (1989b) observations led him to conclude that star dune development in the Gran Desierto follows a recognizable sequence, and may be associated with the modification of pre-existing dunes that migrate into a region having opposing multidirectional wind regimes (Fig. 9.55).

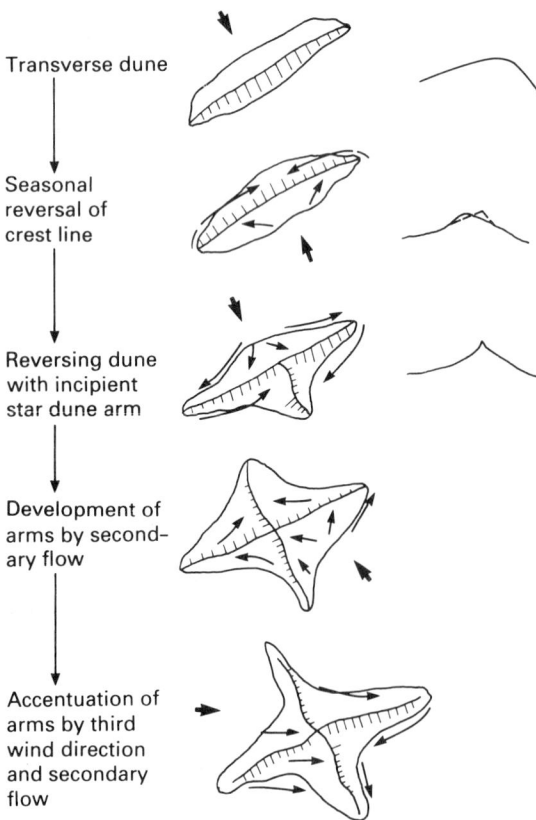

Transverse dune

Seasonal reversal of crest line

Reversing dune with incipient star dune arm

Development of arms by second-ary flow

Accentuation of arms by third wind direction and secondary flow

Fig. 9.55 A model for star dune development in the Gran Desierto, Mexico. From Lancaster (1989a).

He suggested that star dune development can begin with a simple transverse dune of limited width that is subjected to winds from two directions opposed by approximately 180°. Reversed air-flow over the crest results in separation on the lee side, giving rise to inflow of air around the edge of the ridge. This initiates a secondary flow cell with two opposing elements that converge opposite the centre of the dune ridge. Deposition at the convergence results in the formation of an arm approximately parallel to the mainflow. With continued seasonal reversal of flow direction, a further arm may develop on the opposite side of the dune. A third wind direction, as observed in the Gran Desierto, may be transverse to the minor arms of the dune and will tend to build these up by lee-side deposition. Over time, the dune will tend to increase in height by deposition in the central part. The increased height will increase the flow acceleration and intensity of separation over the crest. This will give rise to a stronger lee-side circulation, reinforcing the tendency for dune growth.

Other dune types

Numerous other types of dunes have been described in the literature. Frequently these forms represent local variations of the dunes discussed in Section 9.6.2 or are sand deposits with dune-like characteristics formed under rather specific wind conditions or in relation to some obstruction. *Lee dunes*, which are similar in form to many linear dunes, develop in the lee of major topographic barriers such as bedrock outcrops, volcanic cones or craters. Obstructions of this type can cause a horseshoe vortex to develop around the obstruction with parallel inward spiralling vortices extending downwind (Greeley & Iversen, 1985; Pye & Tsoar, 1990). Sand transported around the obstruction may be deposited in the shadow zone in the lee of the obstruction between the vortices, either as a single or two parallel ridges.

Smaller, somewhat related forms, termed *coppice dunes* result from the trapping of sand in front of, and in the lee of, vegetation. As the vegetation grows and traps more sand the form increases in size, often with an elongated downwind orientation. Coppice dunes may grow to over 2 m in height.

Low, rounded mounds of sand lacking slipfaces ranging from a few hundred metres to over a kilometre in diameter are termed *dome dunes* (McKee, 1966; Breed & Grow, 1979). These features are often

Fig. 9.56 A climbing dune that has accumulated against the Bandiagara Escarpment in central Mali. The escarpment is approximately 100 m in height.

complex in form having other superimposed dunes migrating across them.

Climbing dunes, echo dunes and *falling dunes* are frequently found in association with large topographic barriers such as escarpments and hills. Dunes migrating towards the cliff or hill may accumulate at the base and eventually migrate up and over the crest. Such forms have been termed climbing dunes (Fig. 9.56). If the hill or escarpment is relatively high, air flow towards the barrier may be impeded, resulting in flow separation and reverse flow on the upwind side. Sand being transported towards the barrier may encounter the reverse flow and be deposited at some distance from the hill or cliff face forming an echo dune. Some large dunes formed in this way run parallel to scarps for many kilometres in the Sahara (Cooke & Warren, 1973). Dunes that migrate across a plateau or hill may cascade over the crest to form falling dunes.

A variety of dune forms and associated sand deposits are also associated with coastal environments (Pye, 1983; Pye & Tsoar, 1990; Nordstrom *et al.*, 1990). Many dunes found in coastal environments are similar, but frequently smaller than those found in desert areas and include barchans, transverse ridges, barchanoid ridges and parabolic (blowout) dunes. However, in coastal environments vegetation tends to play a more important role in dune initiation, development and stabilization. Some forms, such as *retention ridges* (or *precipitation ridges*), owe their existence to vegetation. These

features form long narrow dunes parallel to the shore on the landward margins of beach and foredune deposits. They develop where dune sand accumulates against and within vegetation barriers, forming a stable dune that may eventually be colonized by trees and other plants. This type of dune is common along the Great Lakes of Canada and the United States and the west coast of the United States.

9.6.3 Internal structure of sand dunes and associated deposits

Our understanding of the internal structure of dunes is somewhat limited and based primarily on relatively few detailed studies. The most comprehensive research has been conducted by McKee and his colleagues (e.g. McKee & Tibbits, 1964; McKee, 1966, 1982; McKee & Douglas, 1979; McKee & Moiola, 1975) and is summarized in McKee (1979). Additional information on bedform patterns and stratification in recent and ancient aeolian deposits has been provided by Brookfield (1977), Hunter (1977), Kocurek (1981b, 1988), Rubin and Hunter (1982, 1985) and Tsoar (1982).

Most of the work cited above has been carried out in the southwestern United States through the trenching of active aeolian dunes and associated deposits. Although this work has led to a much greater understanding of the internal structure of dunes, some caution must be taken in universally extrapolating these results as the dunes of the southwestern United

States may not be typical of those found in larger, usually drier, sand seas in other parts of the world (Brookfield, 1984).

The internal structures of aeolian dunes have been found to vary considerably because of the many factors affecting both dune form and type. These include: (1) wind direction, speed and variability; (2) sand supply; and to a lesser extent (3) the textural characteristics of the sand.

McKee (1979) has identified three distinct structures associated with aeolian dunes. Firstly, dunes frequently display distinct cross-strata with foreset beds dipping downwind at angles of 30–34°. These beds are usually associated with dune slipfaces and reflect the angle of repose of the sands. Cross-beds of this type can be observed in aeolian sandstone beds where they occur in great thickness with few intervening beds containing different lithologies or other sedimentary structures (Brookfield, 1984). The second characteristic feature is the common occurrence of frequently massive, tabular–planar cross-strata that often thin from the base upward. Thirdly, *bounding surfaces* between sets of cross-strata are also common to many active swollen dunes as well as

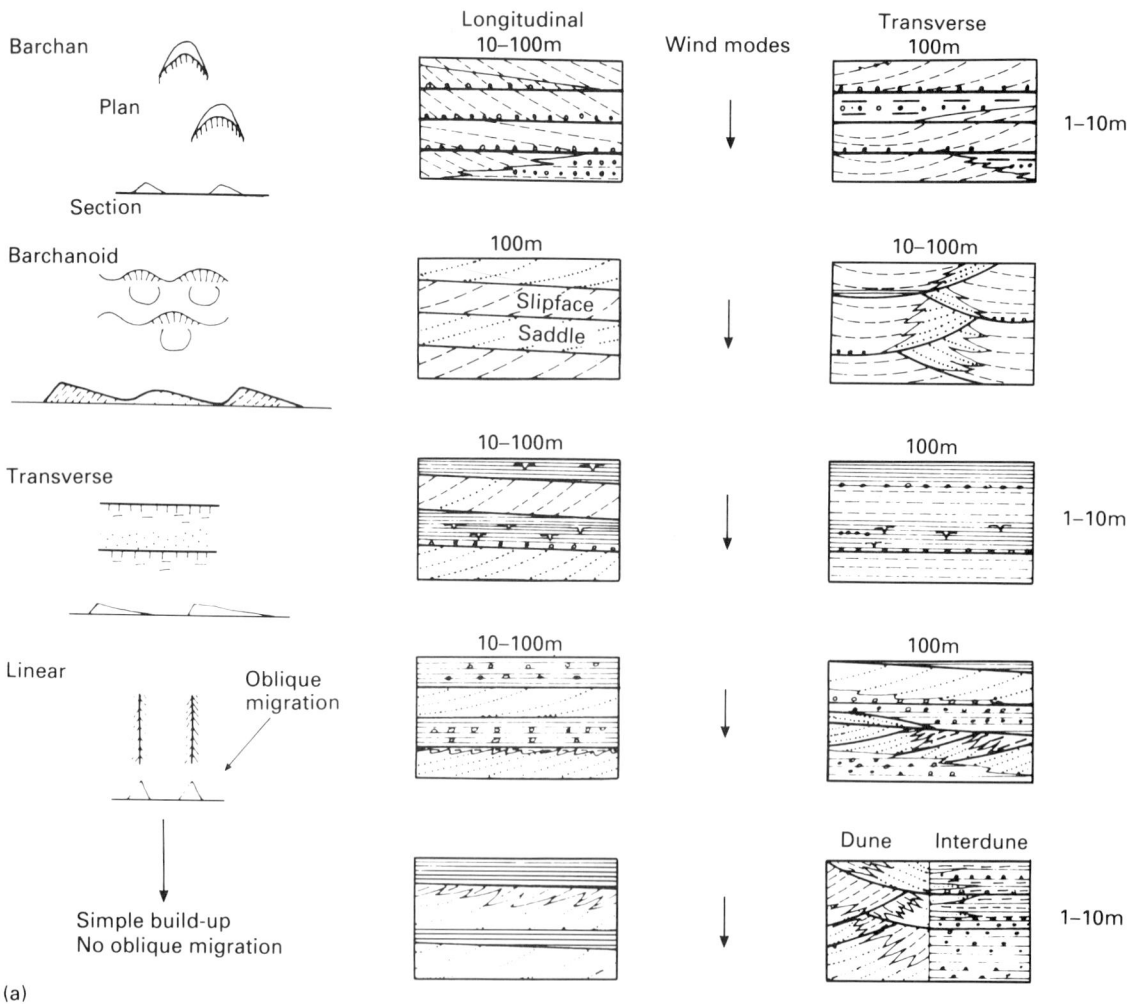

Fig. 9.57 Stratification models for different dune types: (a) simple dune forms; (b) compound and complex dune forms. Transverse sections are perpendicular to the resultant wind direction; longitudinal sections are parallel to the resultant wind. From Brookfield (1984).

sandstones of aeolian origin. Typical structures associated with aeolian dunes are schematically represented in Fig. 9.57.

Brookfield (1977) suggested that bounding surfaces result from the migration of dunes or dune–draa combinations over interdune areas and are best observed in aeolian sandstone deposits associated with sand seas. Three types of bounding surfaces have been identified. *First order* surfaces are flat-lying bedding planes cutting across all other aeolian structures and have been attributed to the migration of draas. *Second order* bounding surfaces are found between first order surfaces, usually dipping downwind, and are associated with dunes migrating down the lee slopes of draas or with the lateral migration of longitudinal dunes across the lee slopes of draas. *Third order* surfaces are located above and below cosets of cross-laminae and are indicative of erosion followed by renewed deposition due to local fluctuations in wind speed and direction and may be viewed

as reactivation surfaces. Simple dune systems frequently lack second order bounding surfaces but in cases where smaller dunes overtake larger ones, or in reversing dune systems, they may be present.

Despite the generalizations on the internal structure of dunes, characteristic features are associated with specific dune types. Barchans and simple transverse ridges that form perpendicular to the dominant wind have distinct high angle cross-beds inclined downwind with a narrow range of orientations. The cross-beds are frequently truncated by numerous low angle erosion surfaces that also dip gently downwind.

Thick sets of tabular cross-beds have been associated with transverse ridges at White Sands National Monument, New Mexico (McKee, 1966). He found tabular–planar beds with closely spaced erosion surfaces near the top of the ridges but with more widely spaced erosion surfaces dipping downwind at moderate angles in the lower part of the dune. Barchanoid

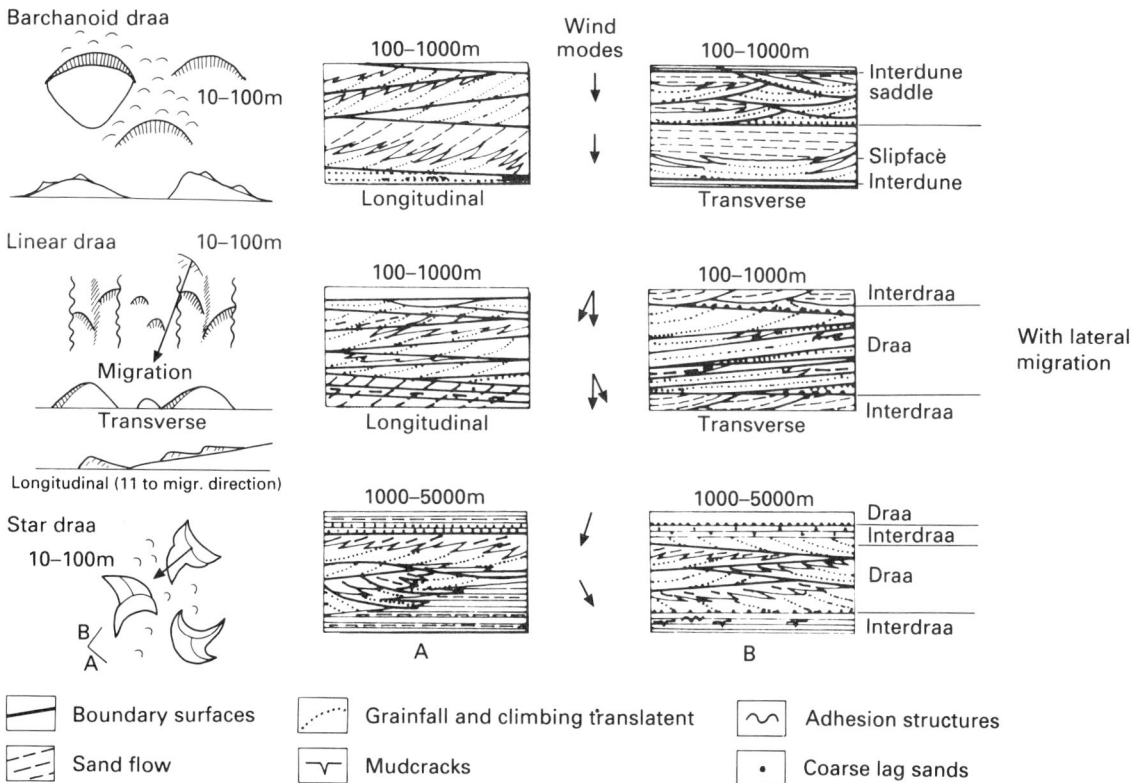

(b)

Fig. 9.57 (*continued*)

ridges that have a more sinuous crest frequently display wedge-shaped sets of dipping beds also observed in parabolic dunes. Concave downward foresets beds have also been identified in parabolic dunes and are attributed to cross winds that undercut the base of the foreset beds as the dune migrates forward (McKee, 1966).

Longitudinal dunes, when viewed in cross-section perpendicular to the wind, are characterized by sets of foreset beds dipping in almost opposite directions away from the crest (Fig. 9.57). However, seif-type longitudinal dunes may have a more complex bedding pattern because of their sharp sinuous crests.

Linear dunes have rarely been identified in the rock record despite the fact that they are the most common dune form in many deserts. Rubin and Hunter (1985) suggest one reason for this may be that longitudinal dunes can migrate laterally when they are not exactly parallel with the long-term sand transport direction. As a result, these dunes may deposit cross-strata having unimodal dip directions resembling transverse dune deposits.

The structures formed in large complex star dunes are almost completely unknown from direct field observation. It is generally assumed that because of their complex morphology, the dip direction of the cross-bedding within these forms is also likely to be rather variable and complex (Brookfield, 1984).

The occurrence of large-scale cross-bedding has frequently been associated with aeolian dunes and draas and used to aid the identification of aeolian sandstones. However, similar cross-bedding patterns may also occur in some submarine bedforms and this has led to controversy over the origin of some sandstones. However, Hunter (1977) states that the determination of aeolian origin for sandstones is frequently based on limited and questionable criteria such as large-scale cross-stratification, excellence of sorting, or the occurrence of rare features such as raindrop imprints, vertebrate tracks and ventifacts. He argues further that detailed examination of individual laminae within the deposit may provide more reliable criteria on which to distinguish aeolian from submarine sands.

From his study of small coastal dunes, Hunter (1977) identified four main types of laminae which are characteristic of aeolian sand deposits. These are planebed laminae; climbing ripple laminae; and sandflow cross-strata. Planebed laminae are produced by wind velocities too high for ripple forma-

tion and are analogous to upper planebed laminae in aqueous deposits.

Climbing ripple laminae closely resemble similar aqueous forms and are divided into two sub-types. *Translatent strata* occur when only the bounding surface between ripples are visible; *ripple form strata* occur where the ripple foresets can be identified. Both these sub-types are inversely graded and closely packed. *Grainfall* laminae are produced in zones of flow separation by deposition from suspension frequently in the lee of a dune. Packing in these laminae is intermediate between the closely packed traction deposits associated with climbing ripples and more loosely packed sandflow deposits. *Sandflow* cross-stratifications are formed by the avalanching of non-cohesive sand on dune slip faces and merge with grainfall laminae at their base.

9.7   Acknowledgements

I express my sincere appreciation to B. Morrison for the time and effort that went into the typing of this manuscript and to M. Puddister for her careful and always cheerful drafting of the figures. Thanks are also given to S. Wolfe, J. Gillies and Ken Pye for their critical comments on the content of the manuscript.

9.8   References

Allberry E.C. 1950. On the capillary forces in an idealized soil. *J. Agric. Sci.*, **40**, 134–141.

Allen J.R.L. 1968. *Current Ripples*. North-Holland, Amsterdam.

Allen J.R.L. 1985. *Principles of Physical Sedimentology*. Allen and Unwin, London.

Anderson R.S. 1987. A theoretical model for aeolian impact ripples. *Sedimentology*, **34**, 943–956.

Anderson R.S. 1988. The pattern of grain fall deposition in the lee of aeolian dunes. *Sedimentology*, **35**, 175–188.

Azizov A. 1977. Influence of soil moisture on the resistance of soil to wind erosion. *Soviet Soil Sci.*, **9**, 105–108.

Bagnold R.A. 1941. *The Physics of Blown Sand and Desert Dunes*. Methuen, London.

Bagnold R.A. 1953. The surface movement of blown sand in relation to meteorology. In: *Desert Research, Proceedings International Symposium, Jerusalem. Res. Coun. Isr. Spec. Pub.*, **2**, 89–93.

Bagnold R.A. 1960. The re-entrainment of settled dusts. *Int. J. Air Poll.* **2**, 89–93.

Bar-Ziv J. & Goldberg G.M. 1974. Simple siliceous pneumoconiosis in Negev Bedouins. *Arch. Env. Health*, **29**, 121–126.

Beavington F. & Cawse P.A. 1979. The deposition of trace

elements and major nutrients in the dust and rainwater in northern Nigeria. *Sci. Total Env.*, **13**, 263–274.

Belly P.Y. 1964. Sand movement by wind. *Tech. Mem.*, **1**, US Army Corps of Engineers, Coastal Engineering Research Centre.

Bisal F. & Hsieh J. 1966. Influence of moisture on erodibility of soil by wind. *Soil Sci.*, **102**, 143–146.

Bisal F. & Nielson K.F. 1962. Movement of soil particles in saltation. *Can. J. Soil Sci.*, **42**, 81–86.

Bloom A.L. 1978. *Geomorphology: A Systematic Analysis of Late Cenozoic Landforms.* Prentice-Hall, Englewood Cliffs, NJ.

Brazel A.J. & Hsu S. 1981. The climatology of hazardous Arizona dust storms. *Geol. Soc. Am. Spec. Pap.* **1120A**, 29–36.

Brazel A.J. & Nickling W.G. 1986. The relationship of weather types to dust storm generation in Arizona (1965–1980). *J. Climatol,* **6**, 255–275.

Breed C.S. & Grow T. 1979. Morphology and distribution of dunes in sand seas observed by remote sensing. In: McKee E.D. (ed) A Study of Global Sand Seas. *US Geol. Surv. Prof. Pap.,* **1052**, 253–308.

Breed C.S., Grolier M.J. & McCauley J.F. 1979. Morphology and distribution of common 'sand' dunes on Mars. Comparison with the Earth. *J. Geophys. Res.*, **84**, 8183–8204.

Brookfield M. 1977. The origin of bounding surfaces in ancient aeolian sandstones. *Sedimentology,* **24**, 303–332.

Brookfield M. 1984. Eolian sands. In: Walker R.G. (ed.) *Facies Models. Geoscience Canada Reprint Series,* **1**, 91–103.

Buritt B. & Hyers A.D. 1981. Evaluation of Arizona's highway dust warning system. *Geol. Soc. Am. Spec. Pap.,* **186**, 429–431.

Chen Y., Tarchitzky J., Brouwer J., Morin J. & Banin A. 1980. Scanning electron microscope observations on soil crusts and their formation. *Soil Sci.*, **130**, 49–55.

Chepil W.S. 1945a. Dynamics of wind erosion: I. Nature of movement of soil by wind. *Soil Sci.*, **60**, 305–320.

Chepil W.S. 1945b. Dynamics of wind erosion: II. Initiation of soil movement. *Soil Sci.*, **60**, 397–411.

Chepil W.S. 1945c. Dynamics of wind erosion: IV. The translocating and abrasive action of the wind. *Soil Sci.*, **61**, 169–177.

Chepil W.S. 1951. Properties of soil which influence wind erosion. V. Mechanical stability of surface. *Soil Sci.*, **52**, 411–478.

Chepil W.S. 1956. Influence of moisture on erodibility of soil by wind. *Soil Sci. Soc. Am. Proc.,* **20**, 288–292.

Chepil W.S. 1958. The use of equally spaced hemispheres to evaluate aerodynamic forces on a soil surface. *EOS,* **39**, 397–403.

Chepil W.S. 1959. Equilibrium of soil grains at the threshold of movement by wind. *Soil Sci. Soc. Am. Proc.,* **23**, 422–428.

Chepil W.S. & Woodruff N.P. 1957. Sedimentary characteristics of dust storms. II. Visibility and dust concentration. *Am. J. Sci.*, **255**, 104–114.

Chepil W.S. & Woodruff N.P. 1963. The physics of wind erosion and its control. *Adv. Agron.,* **15**, 211–302.

Coleman N.L. 1967. A theoretical and experimental study of drag and lift forces acting on a sphere resting on a hypothetical streambed. *Proc. 12th Cong. Int. Ass. Hydrol. Res. Ft. Collins, Colorado,* **3**, 185–192.

Coleman N.L. 1972. The drag coefficient of a stationary sphere on a boundary of similar spheres. *La Houille Blanche,* **1**, 17–21.

Coleman N.L. & Ellis W.M. 1976a. Model study of the drag coefficient of a streambed particle. *Proc. 3rd Fed. Interag. Sediment. Conf. Denver, Colorado,* 4–12.

Coleman N.L. & Ellis W.M. 1976b. A streambed-particle model-study facility using hydroxythylcellulose solutions as a fluid. *US Dept. Agr. Res. Ser.,* **ARS-S-147.**

Cooke R.U. & Warren A. 1973. *Geomorphology in Deserts.* University of California Press, Berkeley.

Coote D.R., Dumanski J. & Ramsey J.F. 1981. An assessment of the degradation of agricultural lands in Canada. *Agricultural Land Resource Research Institute, Contribution,* **118**.

Cornish V. 1914. *Waves of Sand and Snow.* Fisher-Unwin, London.

Einstein H.A. & El-Samni E. 1949. Hydrodynamic forces on a rough wall. *Rev. Mod. Phys.,* **21**, 520–524.

Ellwood J.M., Evans P.D. & Wilson I.G. 1975. Small scale aeolian bedforms. *J. Sedim. Petrol.,* **45**, 554–561.

Fink J. & Kukla G.J. 1977. Pleistocene research in the northeastern foothills of the Alps and in the Vienna Basin. *Acta Geol. Acad. Scient. Hung.,* **22**, 111–124.

Finkel J.H. 1959. The barchans of southern Peru. *J. Geol.,* **67**, 614–647.

Finlan C.A. 1987. *The effects of particle shape and size on the entrainment and transport of sediment by wind.* Unpublished MSc Thesis, University of Guelph, Canada.

Fisher R.A. 1926. On the capillary forces in an idealized soil: Correction of formulae given by W.B. Haines. *J. Agric. Sci.,* **141**, 178–184.

Foda M.A. 1983. Dry fall of fine dust on sea. *J. Geophys. Res.,* **88**, 6021–6026.

Foda M.A., Khalaf F.I. & Al-Kadi A.S. 1985. Estimation of dust fallou rates in the northern Arabian Gulf. *Sedimentology,* **32**, 595–603.

Folk R.L. 1971a. Longitudinal dunes of the northwestern edge of the Simpson Desert, Northern Territory, Australia: I. Geomorphology and grain size relationships. *Sedimentology,* **16**, 5–54.

Folk R.L. 1971b. Genesis of longitudinal and oghurd dunes elucidated by rolling upon grease. *Bull. Geol. Soc. Am.,* **82**, 3461–3468.

Folk R.L. 1976. Rollers and ripples in sand, streams and sky: rhythmic alteration of transverse and longitudinal vortices in three orders. *Sedimentology*, **23**, 649–669.

Folk R.L. 1977. Longitudinal ridges with tuning fork junction in the laminated interval of flysch beds: evidence for low order helicoidal flow in turbidities. *Sed. Geol.*, **19**, 1–6.

Foster S.M. & Nicolson T.H. 1980. Microbial aggregation of sand in a maritime dune succession. *Soil Biol. Biochem.*, **13**, 205–208.

Frazee C.J., Fehrenbacher J.B. & Krumbein W.C. 1970. Loess distribution from a source. *Proc. Soil Sci. Soc. Am.*, **34**, 296–301.

Freeman M.H. 1952. Dust storms of the Anglo-Egyptian Sudan. *Met. Rep.*, **11**, HMSO, London.

Fryberger S.G. & Schenk C. 1981. Wind sedimentation tunnel experiments on the origins of aeolian strata. *Sedimentology*, **28**, 805–821.

Gillette D.A. 1974. On the production of soil wind erosion aerosols having the potential for long-term transport. *J. Réch. Atmos.*, **8**, 735–744.

Gillette D.A. 1977. Fine particle emissions due to wind erosion. *Trans. Am. Soc. Agr. Eng.*, **20**, 169–179.

Gillette D.A., Adams J., Endo A. & Smith D. 1980. Threshold velocities for input of soil particles into the air by desert soils. *J., & Geophys. Res.*, **85**, 5621–5630.

Gillette D.A. Adams J., Muhs D. & Kihl R. 1982. Threshold friction velocities and rupture moduli for crusted desert soil for the input of soil particles into the air. *J. Geophys. Res.*, **87**, 9003–9015.

Gillette D.A. & Goodwin P.A. 1974. Microscale transport of sand-sized soil aggregates eroded by wind. *J. Geophys. Res.*, **79**, 4080–4084.

Gillette D.A. & Walker T.R. 1977. Characteristics of airborne particles produced by wind erosion of sandy soil, high plains of west Texas. *Soil Sci.*, **123**, 97–110.

Gillette D.A., Blifford I.H. & Fenster C.R. 1972. Measurements of aerosol-size distributions and vertical fluxes of aerosols on land subject of wind erosion. *J. App. Meteorol.*, **11**, 977–987.

Gillette D.A., Blifford I.H. & Fryrear D.W. 1974. The influence of wind velocity on the size distributions of aerosols generated by the wind erosion of soils. *J. Geophys. Res.*, **79**, 4068–4075.

Goldman A.R., Cox R.G. & Brenner H. 1967. Slow viscous motion of a sphere parallel to a plane wall. *Chem. Engng Res.*, **22**, 890–897.

Goossens D. 1985. The granulometric characteristics of a slowly-moving dust cloud. *Earth Surf. Proc. Landf.*, **10**, 353–362.

Goudie A.S. 1978. Dust storms and their geomorphological implications. *J. Arid Env.*, **1**, 291–311.

Goudie A.S. 1983. Dust storms in space and time. *Prog. Phys. Geog.*, **7**, 502–529.

Greeley A.S. & Iversen J.D. 1985. *Wind as a Geological Process*. Cambridge University Press, Cambridge.

Greeley R. & Leach R. 1978. A preliminary assessment of the effects of electrostatics on aeolian processes. Rep. Planet. Geol. Prog. 1977–78. *NASA Tech. Mem.*, **79729**, 236–237.

Greeley R. & Peterfreund A.R. 1981. Aeolian 'megaripples': examples from Mono Craters, California and northern Iceland. Abs. with *Prog. Geol. Soc. Am.*, **13**, 463.

Greeley R., Leach R., White B.R., Iversen J.D. & Pollack J.B. 1980. Threshold windspeeds for sand on Mars: wind tunnel simulations. *Geophys. Res. Lett.*, **7**, 121–124.

Greeley R., Leach R., Williams S.H., White B.R., Pollack J.B., Krinsley D.H. & Marshall J.R. 1982. Rates of wind abrasion on Mars. *J. Geophys. Res.*, **87**, 10009–10024.

Greeley R., White B.R., Pollack J.B., Iversen J.D. & Leach R.N 1977. Dust storms on Mars: considerations and simulations. *NASA Tech. Mem.*, **78423.**

Hack J.T. 1941. Dunes of the western Navajo country. *Geog. Rev.*, **31**, 240–263.

Hagen, L. 1984. Soil aggregate abrasion by impacting sand and soil particles. *J. Am. Soc. Agr. Eng.* **27**, 805–808.

Hagen L. & Woodruff N.O. 1973. Air pollution in the Great Plains. *Atmos. Env.*, **7**, 323–332.

Haines W.B. 1925. Studies on the physical properties of soils: II. A note on the cohesion developed by capillary forces in an idealized soil. *J. Agr. Sci.*, **15**, 529–535.

Hall F.F. 1981. Visibility reductions from soil dust in the western United States. *Atmos. Env.*, **15**, 1929–1933.

Hanna S.R. 1969. The formation of longitudinal sand dunes by large helical eddies in the atmosphere. *J. App. Meteorol.*, **8**, 874–883.

Hastenrath S.L. 1967. The barchans of Arequipa Region, southern Peru. *Zeit Geomorph.*, **NF 11**, 300–311.

Hastenrath S. 1987. The barchan dunes of southern Peru revisited. *Zeit Geomorph.*, **NF 31**, 167–178.

Haynes C.V., Jr 1989. Bagnold's barchan: a 57-yr record of dune movement in the eastern Sahara and implications for dune origin and paleoclimate since Neolithic times. *Quat. Res.*, **32**, 153–167.

Hesp P.A. 1981. The formation of shadow dunes. *J. Sedim. Petrol.*, **51**, 101–112.

Hopkins E.S., Palmer A.E. & Chepil W.S. 1946. Soil drifting control in the Prairie Provinces. *Pub. Can. Dept. Agr.*, **568** (Farmer's Bulletin 32).

Howard A.D., Morton J.B. Gad-el-Hak M. & Pierce D. 1978. Sand transport model of barchan dune equilibrium. *Sedimentology*, **25**, 307–338.

Hunt J.C.R. & Nalpanis P. 1985. Saltating and suspended particles over flat and sloping surfaces. I. Modelling concepts. In: Barnddorff-Nielson O.E., Moller J.T., Romer-Rasmussen K., & Willets B.B. (eds) *Proceedings of International Workshop on the Physics of Blown Sand, Aarhus, 28–31 May 1985. Dept. Theoretical Statistics, Institute of Mathematics, University of Aarhus Mem.*, **8**, (I), 9–36.

Hunter R.E. 1977. Basic types of stratification in small eolian dunes. *Sedimentology*, **24**, 361–387.

Idso S.B. 1967. Dust storms. *Sci. Am.*, **235**, 108–114.

Idso S.B. & Brazel A.J. 1977. Planetary radiation balance as a function of atmospheric dust: climatological consequences. *Science*, **198**, 731–773.

Idso S.B., Ingram R.S. & Pritchard J.M. 1972. An American haboob. *Bull. Am. Met. Soc.*, **53**, 930–955.

Iversen J.D. & White B.R. 1982. Saltation threshold on Earth, Mars and Venus. *Sedimentology*, **29**, 111–119.

Iversen J.D., Greeley R. & Pollack J.B. 1976b. Windblown dust on Earth, Mars and Venus. *J. Atmos. Sci.*, **33**, 2425–2429.

Iversen J.D., Pollack J.B., Greeley R. & White B.R. 1976a. Saltation threshold on Mars: the effect of interparticle force, surface roughness and low atmospheric density. *Icarus*, **29**, 381–393.

Janecek T.R. & Rea D.K. 1983. Eolian deposition in the northwest Pacific Ocean: Cenozoic history of atmospheric circulation. *Bull. Geol. Soc. Am.*, **94**, 730–738.

Jensen J.L. & Sorensen M. 1986. Estimation of some aeolian saltation transport parameters: A re-analysis of William's data. *Sedimentology*, **33**, 547–555.

Joseph P.W., Raipal D.K. & Deka S.N. 1980. 'Andhi': the convective dust storm of northwest India. *Mausam*, **31**, 431–432.

Junge C.E. 1979. The importance of mineral dust as an atmospheric constituent in the atmosphere. In: Morales C. (ed.) *Saharan Dust*, pp. 49–60. Wiley, New York.

Kawamura R. 1951. Study of sand movement by wind. *Institute of Science and Technology, Tokyo, Report*, **5**, (3–4), 95–112. Tokyo, Japan.

King D. 1960. The sand ridge deserts of South Australia and related aeolian landforms. *Trans. R. Soc. S. Austr.*, **83**, 99–108.

Kocurek G. 1981a. Significance of interdune deposits and bounding surfaces in aeolian dune sands. *Sedimentology*, **28**, 753–780.

Kocurek G. 1981b. Erg reconstruction: the Entrada Sandstone (Jurassic) of Northern Utah and Colorado. *Paleogeogr., Paleoclimatol., Paleoecol.*, **36**, 125–153.

Kocurek G. 1988. First-order and super bounding surfaces in eolian sequences – bounding surfaces revisited. *Sed. Geol.*, **56**, 193–206.

Lancaster N. 1980. The formation of seif dunes from barchans – supporting evidence of Bagnold's model from the Namib desert. *Zeit. Geomorph.*, **NF 24**, 160–167.

Lancaster N. 1982a. Dunes on the Skeleton Coast, Namibia: geomorphology and grain size relationships. *Earth Surf. Proc.*, **7**, 575–587.

Lancaster N. 1982b. Linear dunes. *Prog. Phys. Geog.*, **6**, 475–504.

Lancaster N. 1985. Variations in wind velocity and sand transport on the windward flank of desert sand dunes. *Sedimentology*, **32**, 581–593.

Lancaster N. 1988a. Controls of eolian dune size and spacing. *Geology*, **16**, 972–975.

Lancaster N. 1988b. The development of large aeolian bedforms. *Sed. Geol. Petrol.*, **55**, 69–89.

Lancaster N. 1989a. The dynamics of star dunes: an example from the Gran Desierto, Mexico. *Sedimentology*, **36**, 273–289.

Lancaster N. 1989b. Star dunes. *Prog. Phys. Geog.*, **13**, 67–91.

Lancaster N. Greeley R. & Christensen P.R. 1987. Dunes of the Gran Desierto sand–sea, Sonora, Mexico. *Earth Surf. Proc. Landf.*, **12**, 277–288.

Leathers C.R. 1981. Plant components in desert dust in Arizona and their significance for man. *Geol. Soc. Am. Spec. Pap.*, **186**, 191–206.

Leeder M.R. 1977. Folk's bedform theory. *Sedimentology*, **24**, 863–864.

Lettau K. & Lettau H.H. 1978. Experimental and micrometeorological field studies on dune migration. In: Lettau H.H. & Lettau K. (eds) *Exploring the World's Driest Climate*. University of Wisconsin-Madison, Institute for Environmental Studies, IES Report, **101**, 110–147.

Livingstone I. & Thomas D.S.G. 1993. Modes of linear dune activity and their palaeoenvironmental significance: an evaluation with reference to Southern African examples. In: Pye K. (ed.) *The Dynamics and Environmental Context of Aeolian Sedimentary Systems. Geol. Soc. Spec. Publ.*, **72**, 91–101, Geological Society Publishing House, Bath.

Loewe F. 1943. Dust storms in Australia. *Comm. Bur. Met. Bull.*, **28**, 124–134.

Logie M. 1982. Influence of roughness elements and soil moisture of sand to wind erosion. *Catena*, **1**, (Suppl.) 161–173.

Long J.T. & Sharp R.P. 1964. Barchan dune movement in the Imperial Valley, California. *Bull. Geol. Soc. Am.*, **75**, 147–156.

Lyles L. 1975. Possible effects of wind erosion on soil productivity. *J. Soil and Water Conserv.*, **30**, 279–283.

Lyles L. 1977. Wind erosion: processes and effect on soil productivity. *Trans. Am. Soc. Agr. Eng.*, **20**, 880–884.

Lyles L. & Krauss R.K. 1971. Threshold velocities and initial particle motion as influenced by air turbulence. *Trans. Am. Soc. Agr. Engnrs*, **14**, 563–566.

Lyles L. & Schrandt R.L. 1972. Wind erodibility as influenced by rainfall and soil salinity. *Soil Sci.*, **114**, 367–372.

Lyles L., Schrandt R.L. & Schneidler N.F. 1974. How aerodynamic roughness elements control sand movement. *Trans. Am. Soc. Agr. Engnrs*, **17**, 134–139.

Mabbutt, J.A. 1968. Aeolian landforms in central Australia. *Austr. Geog. Stud.*, **6**, 139–150.

Marston R.A. 1986. Maneuver-caused wind erosion impacts, South Central New Mexico. In: Nickling W.G. (ed.) *Aeolian Geomorphology*, pp. 273–290. Allen and Unwin, Boston.

McKee E.D. 1966. Structures of dunes at White Sands National Monument, New Mexico (and a comparison with structure of dunes from other selected areas). *Sedimentology*, **7**, 1–69.

McKee E.D. (ed.) 1979. *A Study of Global Sand Seas. U.S. Geol. Surv. Prof. Pap.* **1052**.

McKee E.D. 1982. Sedimentary structures in dunes of the Namib Desert, Southwest Africa. *Geol. Soc. Am. Spec. Pap.*, **188**.

McKee E.D. & Douglas S.R. 1979. Growth and movement of dunes at White Sands National Monument, New Mexico. *U.S. Geol. Surv. Prof. Pap.*, **750-D**, 108–114.

McKee E.D. & Moiola R.J. 1975. Geometry and growth of the White Sands dune field, *J. Res. U.S. Geol. Surv.*, **3**, 59–66.

McKee E.D. & Tibbits G.C. 1964. Primary structures of a seif dune and associated deposits in Libya. *J. Sedim. Petrol.*, **34**, 5–17.

McKenna-Neuman C. & Nickling W.G. 1989. A theoretical and wind tunnel investigation of the effect of capillary water on the entrainment of sediment by wind. *Can. J. Soil Sci.*, **69**, 79–96.

Meigs P. 1953. World distribution of arid and semi-arid homoclimates. In: *Reviews of Research on Arid Zone Hydrology*, pp. 203–209. UNESCO, Paris.

Melton F.A. 1940. A tentative classification of sand dunes. *J. Geol.*, **48**, 113–173.

Middleton N.J. 1984. Dust storms in Australia: frequency, distribution and seasonality. *Search*, **15**, 46–47.

Middleton N.J. 1986a. Dust storms in the Middle East. *J. Arid Env.*, **10**, 83–96.

Middleton N.J. 1986b. A geography of dust storms in southwest Asia. *J. Climatol.*, **6**, 183–196.

Middleton G.V. & Southard J.B. 1984. *Mechanics of Sediment Movement, 2nd edn. S.E.P.M. Short Course*, **3**. Eastern Section of the Society of Economic Paleontologists and Mineralogists, Providence, RI.

Middleton N.J., Goudie, A.S. & Wells, G.L. 1986. The frequency and source areas of dust storms. In: Nickling, W.G. (ed.) *Aeolian Geomorphology*, pp. 237–259. Allen and Unwin, Boston.

Morales C. (ed.) 1979. *Saharan Dust — Mobilization, Transport, Deposition*. Wiley, New York.

Nalpanis P. 1985. Saltating and suspended particles over flat and sloping surfaces. II. Experiments and numerical simulations. In: Barndorff-Nielson O.E., Moller J.T., Romer-Rasmussen K. & Willets B.B. (eds.) *Proc. Inter. Workshop on the Physics of Wind Blown Sand, Aarhus, 28–31 May 1985, Department of Theoretical Statistics, Institute of Mathematics, University of Aarhus Mem.*, **8**, (I), 37–66.

Nickling W.G. 1978. Eolian sediment transport during dust storms: Slims River Valley, Yukon Territory. *Can. J. Earth Sci.*, **15**, 1069–1084.

Nickling W.G. 1983. Grain-size characteristics of sediments transported during dust storms. *J. Sedim. Petrol.*, **53**, 1011–1024.

Nickling W.G. 1984. The stabilizing role of bonding agents on the entrainment of sediment by wind. *Sedimentology*, **31**, 111–117.

Nickling W.G. 1988. The initiation of particle movement by wind. *Sedimentology*, **35**, 499–511.

Nickling W.G. & Brazel A.J. 1984. Temporal and spatial characteristics of Arizona dust storms (1965–80). *J. Climatol.*, **4**, 645–660.

Nickling W.G. & Ecclestone M. 1981. The effects of soluble salts on the threshold shear velocity of fine sand. *Sedimentology*, **28**, 505–510.

Nickling W.G. & Gillies J.A. 1989. Emission of fine-grained particulates from desert soils. In: Leinen M. & Sarnthein M. (eds) *Paleoclimatology and Paleometeorology: Modern and Past Patterns of Global Atmospheric Transport*, pp. 133–165, Kluwer, Dordrecht.

Nielson J. & Kocurek G. 1987. Development, processes, migration and deposits of star dunes, Dumont dunefield, California. *Bull. Geol. Soc. Am.*, **99**, 177–186.

Nordstrom K.F. & McCluskey J.M. 1984. Considerations for the control of house construction in coastal dunes. *Coastal Zone Management J.*, **12**, 385–402.

Nordstrom K.F., Psuty N.P. & Carter R.W.G. (eds) 1990. *Coastal Dunes: Form and Process*. Wiley, New York.

O' Brien M.P. & Rindlaub B.D. 1936. The transportation of sand by wind. *Civil Engng.*, **6**, 325–327.

Oke T.R. 1978. *Boundary Layer Climates*. Methuen, New York.

Oliver F.W. 1945. Dust storms in Egypt and their relation to the war period, as noted in Maryut, 1939–45. *Geog. J.*, **106**, 26–49.

Owen P.R. 1964. Saltation of uniform grains in air. *J. Fluid Mech.*, **20**, 225–242.

Patterson E.M. & Gillette D.A. 1977. Measurements of visibility vs. concentration for airborne soil particles. *Atmos. Env.*, **11**, 193–196.

Pecsi M. 1968. The main genetic types of Hungarian loess and loess-like sediments. In: Schultz C.B. & Frye J.C. (eds) *Loess and Related Eolian Deposits of the World*, pp. 317–320. University of Nebraska Press, Lincoln NB.

Peterson S.T. & Junge C.E. 1971. Sources of particulate matter in the atmosphere. In: Kellogg W.W. & Robinson G.D. (eds) *Man's Impact on the Climate*. pp. 310–320. MIT Press, Cambridge, MA.

Petrov M.P. 1968. Composition of eolian dust in southern Turkmenia, as in dust storm of January 1968. *Int. Geol. Rev.*, **13**, 1178–1182.

Péwé T.L., Péwé A.E., Péwé R.H. Journaux A. & Slatt R.M. 1981. Desert dust: characteristics and rates of deposition in central Arizona, USA. *Geol. Soc. Am. Spec. Pap.*, **186**, 169–190.

Prospero J.M. & Nees R.T. 1977. Dust concentration in the atmosphere of the equatorial North Atlantic; possible relationship to Sahelian drought. *Science*, **196**, 1196–1198.

Prospero J.M., Bonatti E., Schubert C. & Carlson T.N. 1970. Dust in the Caribbean atmosphere traced to an African dust storm. *Earth Planet. Sci. Lett.*, **9**, 287–293.

Prospero J.M., Glaccum R.A, & Nees R.T. 1981. Atmospheric transport of soil dust from Africa to South America. *Nature*, **289**, 570–572.

Pye K. 1980. Beach salcrete and eolian sand transport: evidence from North Queensland. *J. Sedim. Petrol.*, **50**, 257–261.

Pye K. 1983. Coastal dunes. *Prog. Phys. Geog.*, **7**, 531–557.

Pye K. 1987. *Aeolian Dust and Deposits.* Academic Press, London.

Pye K. & Tsoar H. 1990. *Aeolian Sand and Sand Dunes.* Unwin Hyman, London.

Rouse H. 1937. Modern conceptions of the mechanics of fluid turbulence. *Trans. Am. Soc. Civil Eng.*, **102**, 403–453.

Rea D.K., Leinen M. & Janecek T.R. 1985. Geologic approach to the long-term history of atmospheric circulation. *Science*, **227**, 721–725.

Rubin D.M. & Hunter R.E. 1982. Bedform climbing in theory and nature. *Sedimentology*, **29**, 121–138.

Rubin D.M. & Hunter R.E. 1985. Why deposits of longitudinal dunes are rarely recognized in the geologic record. *Sedimentology*, **32**, 147–157.

Rumpel, D.A. 1985. Successive aeolian saltation: studies of idealized collisions. *Sedimentology*, **32**, 267–280.

Saffman P.G. 1965. The lift on a small sphere in a slow shear flow. *J. Fluid Mech.*, **22**, 385–400.

Saffman P.G. 1968. Corrigendum. *J. Fluid Mech.*, **31**, 624.

Seppala M. & Linde K. 1978. Wind tunnel studies of ripple formation. *Geog. Ann.*, **60**, 29–40.

Sharp R.P. 1963. Wind ripples. *J. Geol.*, **71**, 617–636.

Snowden J.O. & Priddy R.R. 1968. Geology of Mississippi loess. *Bull. Miss. Geol. Surv.*, **111**, 13–203.

Sundborg A. 1955. Meteorological and climatological conditions for the genesis of aeolian sediments. *Geog. Ann.*, **37**, 94–111.

Sweet M.L., Nielson J., Havholm K. & Farrelley J. 1988. Algodones dune field of southeastern California: case history of a migrating modern dune field. *Sedimentology*, **35**, 939–952.

Taylor G.I. 1915. Eddy motion in the atmosphere. *Phil. Trans. R. Soc. Lond. A*, **215**, 1–26.

Thomas D.S.G. & Tsoar H. 1990. The geomorphological role of vegetation in desert dune systems. In: Thornes J. (ed.) *Vegetation and Erosion*, pp. 471–489. Wiley, Chichester.

Trainer F.W. 1961. Eolian deposits of the Matanuska Valley agricultural area, Alaska. *Bull. U.S. Geol. Surv.*, **1121C**, C1–C35.

Tsoar H. 1974. Desert dunes morphology and dynamics, El-Arish, *Zeit Geomorph., Supp. Bd.*, **20**, 41–61.

Tsoar H. 1982. Internal structure and surface geometry of longitudinal (seif) dunes. *J. Sedim. Petrol.*, **52**, 823–831.

Tsoar H. 1983a. Wind tunnel modeling of echo and climbing dunes. In: Brookfield M.E. & Ahlbrandt T.S. (eds) *Eolian Sediments* and Processes, pp. 247–259. Elsevier, Amsterdam.

Tsoar H. 1983b. Dynamic processes acting on a longitudinal (seif) dune. *Sedimentology*, **30**, 567–578.

Tsoar H. 1984. The formation of seif dunes from barchans—a discussion. *Zeit Geomorph.*, **NF 28**, 99–103.

Tsoar H. 1989. Linear dunes — forms and formation. *Prog. Phys. Geog.*, **13**, 507–528.

Tsoar H. & Pye K. 1987. Dust transport and the question of desert loess formation. *Sedimentology*, **34**, 139–153.

Tsoar H., Rasmussen K.R. Sørensen M. & Willets B.B. 1985. Laboratory studies of flow over dunes. In: Barndorff-Nielsen O.E., Moller J.T., Romer-Rasmussen K. & Willets B.B. (eds) *Proceedings of International Workshop on the Physics of Blown Sand, 28–31 May, 1985. Department of Theoretical Statistics, University of Aarhus, Memoir* **8**, 327–349.

Ungar J.E. & Haff P.K. 1987. Steady-state saltation in air. *Sedimentology*, **34**, 289–299.

Van den Anker J.A.M., Jungerius P.D. & Muir L.R. 1985. The role of algae in the stabilization of coastal dune blowouts. *Earth Surf. Proc. Landf.*, **10**, 189–192.

Verstappen H.T. 1968. On the origin of longitudinal (seif) dunes. *Zeit. Geomorph.*, **NF 12**, 200–212.

Warren A. 1969. A bibliography of desert dunes and associated phenomena. In: McGinnies W.G. & Goldman B.J. (eds) *Arid Land in Perspective*, pp. 75–99. University of Arizona Press, Tucson.

Warren A. 1976. Morphology and sediments of the Nebraska sand hills in relation to Pleistocene winds and the development of aeolian bedforms. *J. Geol.*, **84**, 685–700.

Warren A. 1979. Aeolian processes. In: Embleton C. and Thornes J. (eds) *Process in Geomorphology*, pp. 325–351. Edward Arnold, London.

Wasson R.J. & Hyde R. 1983. A test of granulometric control of desert dune geometry. *Earth Surf. Proc.*, **8**, 301–312.

Werner B.T., Haff P.K., Livi R.P. & Anderson R.S. 1986. Measurement of eolian sand ripple cross-sectional shapes. *Geology*, **14**, 743–745.

White B.R. 1982. Two-phase measurements of saltating turbulent boundary layer flow. *Inter. J. Multiphase Flow*, **8**, 459–473.

White B.R. & Shultz J.C. 1977. Magnus effect in saltation. *J. Fluid Mech.*, **81**, 497–512.

White B.R., Iversen J.D. & Pollack J.B. 1976. Estimated grain saltation in a Martian atmosphere. *J. Geophys. Res.*, **81**, 5643–5650.

Willetts B.B. 1983. Transport by wind of granular materials of different grain shapes and densities. *Sedimentology*, **30**, 669–679.

Willetts B.B. & Rice M.A. 1986. Collision in aeolian

transport: the saltation/creep link. In: Nickling W.G. (ed.) *Aeolian Geomorphology*, pp. 1–17. Allen and Unwin, Boston.

Willetts B.B., Rice M.A. & Swaine S.E. 1982. Shape effects in aeolian grain transport. *Sedimentology*, **29**, 409–417.

Williams G. 1964. Some aspects of the aeolian saltation load. *Sedimentology*, **3**, 257–287.

Wilshire H.G. 1980. Human causes of accelerated wind erosion in California's desert. In: Coastes D.R. & Vitek J.D. (eds) *Thresholds in Geomorphology*, pp. 415–434. Allen and Unwin, London.

Wilson I.G. 1970. *The External Morphology of Wind-laid Sand Bodies.* Unpublished PhD Thesis, University of Reading.

Wilson I.G. 1971. Desert sandflow basins and a model for the development of ergs. *Geophys. J.*, **137**, 180–199.

Wilson I.G. 1972a. Aeolian bedforms — their development and origins. *Sedimentology*, **19**, 173–210.

Wilson I.G. 1972b. Universal discontinuities in bedforms produced by the wind. *J. Sedim. Petrol.*, **42**, 667–669.

Wipperman F.K. & Gross G. 1986. The wind-induced shaping and migration of an isolated dune: a numerical experiment. *Boundary-Layer Meteorol.* **36**, 319–334.

Yaalon D.H. & Ganor E. 1973. The influence of dust on soils during the Quaternary. *Soil Sci.*, **116**, 146–155.

Zingg A.W. 1953. Wind tunnel studies of the movement of sedimentary material. In: *Proc. Firth Hydraul. Conf. Stud. in Engineering, Bull.* **34**, 111–135. University of Iowa.

Zolotun V.P. 1974. Origin of loess deposits in the southern part of the Ukraine. *Soviet Soil Sci.*, **1**, 1–12.

10 Volcaniclastic sediment transport and deposition

R.V. FISHER & H.-U. SCHMINCKE

10.1 Introduction

In the last two decades, the importance of volcaniclastic sedimentology has been firmly established, following recognition that volcanism, which produces large volumes of volcaniclastic materials, is closely associated with consuming plate boundaries, accreting plate boundaries and intraplate tectonic environments (Fig. 10.1). The record of tectonism is commonly preserved only within fillings of sedimentary basins, therefore knowledge of volcaniclastic transport and deposition as well as their tectonically significant compositional characteristics can help determine past tectonic activity and environments. Research on modern and ancient volcaniclastic fragments and deposits has also added to our knowledge about the transport and depositional processes of sediment gravity flows such as nuée ardentes which can be considered as volcanically generated analogues of turbidity currents. Moreover, volcanic ash beds are

very helpful in event stratigraphy and stratigraphic correlation (tephrochronology), and therefore an excellent research tool of correlation for stratigraphic facies extending into non-volcanic regions. Tephra layers are the only type of geological bed extending across both marine and terrestrial environments (Fig. 10.2). Thus, a background knowledge of the products of eruptions and transport and depositional processes of volcaniclastic materials is necessary for many different kinds of stratigraphic and tectonic studies.

Throughout geologic time, volcaniclastic sediments have been an important part of the Earth's sedimentary budget. Volcaniclastic sediments, for example, were inevitably a major part of the sedimentary system of the Earth's early volcanic crust, and being derived from the early crust, all ancestral sediments were initially volcanic. Volcaniclastic materials were the original filter through which elements of the lithosphere were initially passed by chemical weathering on the Earth's surface.

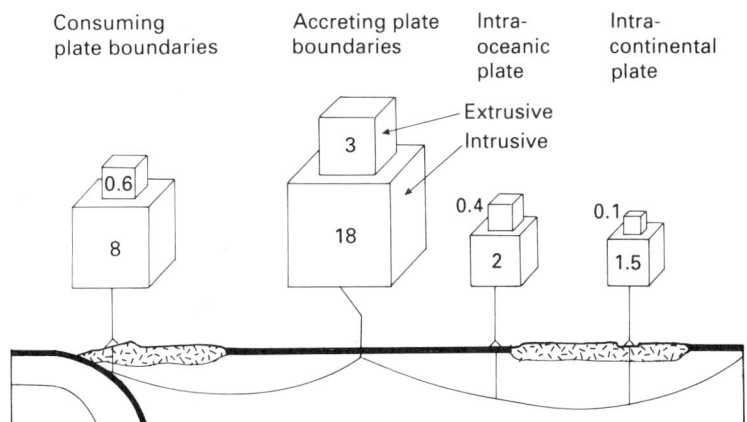

Fig. 10.1 Global volumes of igneous rocks intruded and extruded annually. From Schmincke (1982).

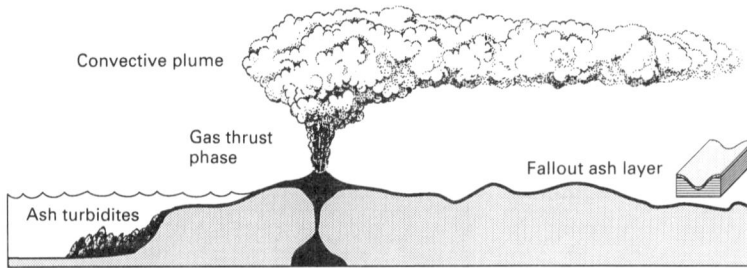

(a)

(b)

(c)

Fig. 10.2 Common types of tephra eruptions, modes of transport and depositional environments. From Schmincke and Bogaard (1991).

As much as 3–4 km³ of volcaniclastic material may be generated yearly by volcanism (Fisher & Schmincke, 1984). During an eruption, much coarse-grained material is directly deposited at or near a volcanic centre, and modern research has shown that if the eruption is sub-aerial, an equal or greater volume of fine-grained ash from the same eruption can be distributed thousands of kilometres from the source. Such fine-grained volcaniclastics containing abundant metastable glass that rapidly alters to clays and zeolites generally goes unrecognized within the

stratigraphic record. Also, vast quantities of erosionally and diagenetically derived volcanic clay, silt and sand are being shed into the oceans from rivers draining volcanic areas of the Pacific Rim, Central Europe, Mediterranean and East African regions. Moreover, large volumes of clastic material are formed completely submerged at spreading centres in the oceans, and especially on seamounts and around oceanic island volcanoes. Over geologic time, a large amount of these submarine volcaniclastics is either accreted against overriding plates in regions of sub-

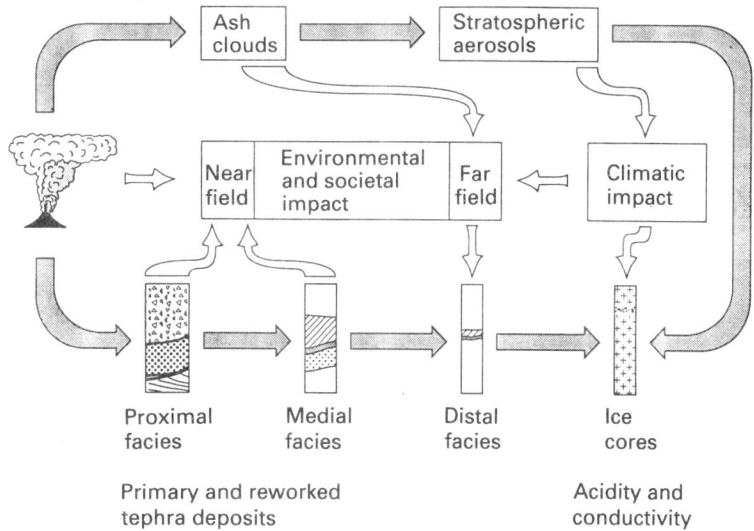

Fig. 10.3 Near and far field effects of tephra eruptions. From Schmincke and Bogaard (1991).

duction as they and penecontemporaneous sediments are 'scraped off' the descending plates (Hamilton, 1988) or is subducted along with the descending plate. Volcaniclastic rocks are estimated to form 25% of ancient geosynclinal (i.e. convergent plate margins) assemblages (Ronov, 1968). A conservative estimate of from 5 to 10% by volume of the yearly clastic input from volcanic sources to the world's sedimentary budget is not unreasonable.

Volcanism can also affect the Earth's climate, and therefore sedimentation procesess without contributing solid particles to the sedimentary budget (Fig. 10.3). Volcanic aerosols, largely droplets of sulphuric acid, may remain suspended in the atmosphere for years, resulting in absorption and backscattering of the Sun's rays into the stratosphere causing cooling (Pollack *et al.*, 1976); Devine *et al.*, 1984) that may increase storm frequency and severity hence increased erosion and sedimentation. Solid particles, though they may be fine enough to circle the Earth as dust, have a relatively short residence time in the atmosphere and do not perturb the environment atmosphere very long (Carey & Sigurdsson, 1982).

Research on volcaniclastic rocks was long neglected, largely because of the 19th century agreement to sub-divide rocks into igneous, sedimentary and metamorphic categories. This placed volcaniclastic materials within a hybrid class by default, and because of the accident of classification, sedimentary rocks composed of volcanic particles were essentially ignored. They were left unstudied for many years by

igneous petrologists because they are sediments. Sedimentary petrographers tended to ignore these rocks because they were perceived as belonging to the domain of igneous rocks, a situation that illustrates the profound influence that classification and nomenclature can have upon thought processes and directions of research.

10.2 Volcaniclastic particles

A sediment may be defined as a mass of particles that has been deposited or is being transported on the Earth's surface from one place to another and deposited by flow or fallout processes, a combination of these, or by chemical precipitation. By this definition, volcanic particles are deposited as sediments, the principal differences with non-volcanic sediments being in some of the physical processes by which the particles are formed. Some volcanic particles are generated by weathering and erosion (epiclastic, discussed below) and therefore differ only in composition from non-volcanic clasts. Other volcanic particles are formed instantly by explosive processes and are propelled at high velocities (>100 m s^{-1}) along the surface of the Earth or high into the atmosphere (>40 km above the Earth).

10.2.1 Particle types

Volcaniclastic particles are defined to include all clasts with a volcanic composition irrespective of

Table 10.1 Classification of pyroclastic, hydroclastic and epiclastic fragments and deposits. After Fisher (1961)

Clast size	Pyroclast or hydroclast	Pyroclastic or hydroclastic deposit (primary or reworked)		Epiclastic deposit		Non-genetic terms (includes pyroclastic and hydroclastic)	
		Unconsolidated tephra	Consolidated rocks	Unconsolidated	Consolidated	Unconsolidated	Consolidated
64 mm	Bomb, block	Agglomerate, bed of blocks or bombs or bomb/block tephra	Agglomerate, pyroclastic or hydroclastic breccia	Epiclastic volcanic, boulder, block, or cobble bed or layer	Epiclastic volcanic breccia or conglomerate	Volcanic boulder, block or cobble	Volcanic breccia or volcanic conglomerate
	Lapillus	Layer/bed of lapilli or lapilli tephra	Lapillistone	Epiclastic volcanic, pebble bed or layer		Volcanic pebble	
2 mm	Coarse ash grain	Coarse ash	Coarse (ash) tuff	Epiclastic volcanic sand	Epiclastic volcanic sandstone	Volcanic sand	Volcanic sandstone
1/16 mm	Fine ash grain (dust grain)	Fine ash (dust)	Fine (ash) tuff (dust tuff)	Epiclastic volcanic silt (or clay)	Epiclastic volcanic siltstone (or claystone)	Volcanic silt (clay)	Volcanic siltstone, (claystone)

origin (Fisher, 1961). They form in several ways, the main processes being pyroclastic, hydroclastic and epiclastic (Table 10.1). *Lithoclasts* (Fig. 10.4) composed of pre-existing rocks may also originate by explosive processes.

Pyroclastic fragments (*pyroclasts*) are formed by expansion of gases within the magma during decompression. Magma disintegrates when vesicles explosively expand to create a glass foam (pumice) that breaks into smaller lapilli and shards consisting of walls of broken vesicles (bubble wall shards) and micropumice clasts (Fig. 10.5). During explosive

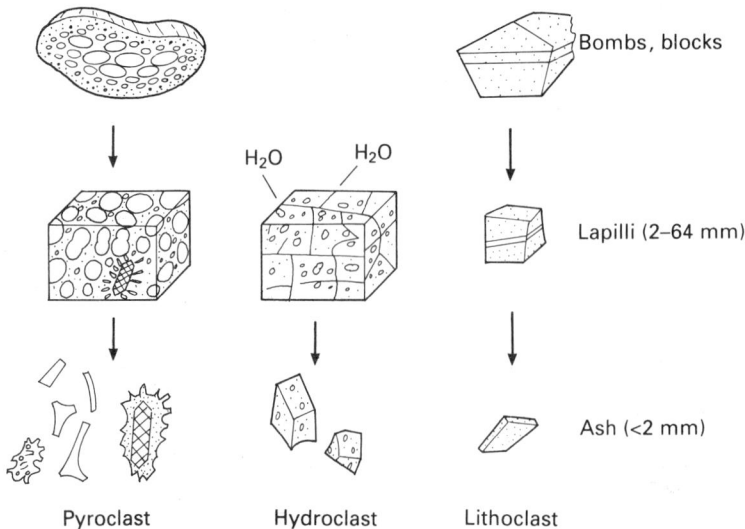

Fig. 10.4 Main types of particles in pyroclastic rocks. From Schmincke (1988).

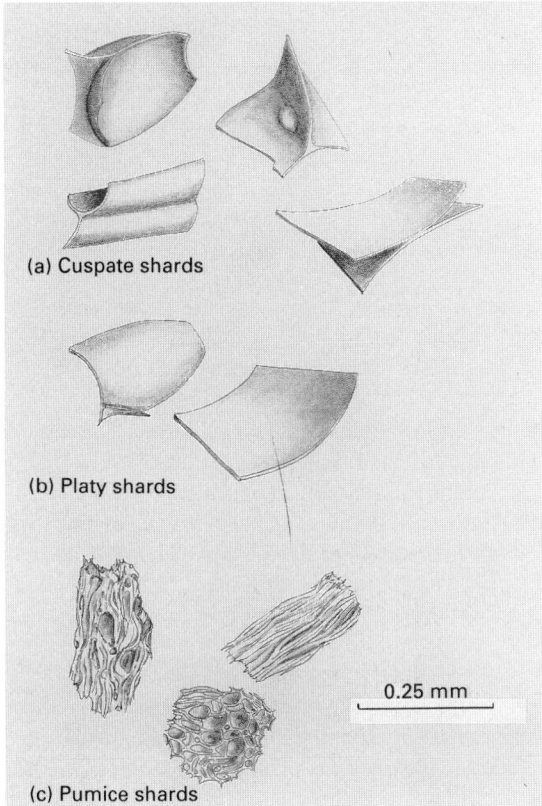

(a) Cuspate shards

(b) Platy shards

0.25 mm

(c) Pumice shards

Fig. 10.5 Common types of glass shards. From Fisher and Schmincke (1984).

eruptions, lithic fragments are broken from the walls of the magma chamber, vent and crater by explosive expansion, friction, abrasion, and collapse of conduit walls and roofs of evacuated magma reservoirs. Xenoliths may be picked up in the magma before fragmentation.

Hydroclastic fragments (*hydroclasts*) result from magma–water interactions during which abundant chilled glass particles are formed (Fig. 10.6). There are two major hydroclastic processes, one explosive and one non-explosive. Abundant basaltic glass (sideromelane) fragments can be generated from hot lava in contact with water. Cooling of basaltic lava blebs in air commonly produces glass rendered nearly opaque by microcrystalline iron/titanium oxides (tachylite). Thus, hydroclastites commonly contain sideromelane shards. Non-explosive breakage is by thermal contraction following quenching. This results in angular, chunky, poorly vesiculated shards in shallow or deep water, fresh or marine. Vesicular glassy hydroclasts are produced by the contact of vesiculating basaltic magma and water, giving fragments with some of the vesicles broken by thermal shock and others broken by internally expanding gas. Explosively produced hydroclasts cannot form in deep water because pressures inhibit steam expansion.

The word 'pyroclastic' is generally used for volcanic particles ejected through volcanic conduits, but

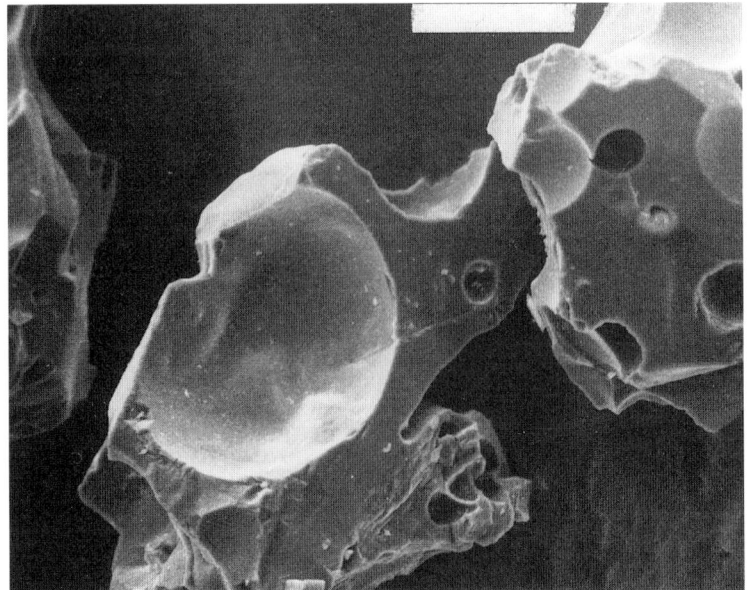

Fig. 10.6 SEM photomicrograph of hydroclastic shards. Bar scale = 100 microns. From Heiken (1974, plate 24A).

over the past decade it has been realized that pyro-clastic and explosive hydroclastic processes are equally important in propelling fragments from a vent (Fisher & Schmincke, 1984). Accordingly we propose expanding the definition of *tephra* to include pyroclastic and hydroclastic fragments that move along the surface as pyroclastic flows and surges, or through the atmosphere. An inclusive term is necessary because pyro- and hydroclastic explosive mechanisms are different. Also, some of the physical characteristics of pyro- and hydroclastic fragments differ according to environment of fragmentation, transportation and deposition, as revealed by hand lens, microscope and SEM analyses (Heiken & Wohletz, 1985; Marshall, 1987). Three main varieties of pyroclastic and hydroclastic ejecta according to origin are: (1) essential (or juvenile); (2) cognate (or accessory); and (3) accidental. Essential pyroclasts are derived directly from the erupting magma and consist of non-vesicular or inflated particles of chilled magma, or crystals that were in the magma prior to eruption (phenocrysts and xenocrysts). Cognate particles are fragmented comagmatic volcanic rocks from previous eruptions of the same volcano. Accidental fragments are derived from the sub-volcanic basement and therefore may be of any composition.

There are other ways to form volcaniclastic particles. Weathering of volcanic rocks results in *epiclastic* particles which are commonly lithic clasts and minerals. Because tephra particles cannot always be distinguished from epiclastic particles, we recommend that 'volcaniclastic' include all fragmental material of volcanic composition as originally defined (Fisher, 1961). Other types of volcaniclastic particles are *autoclastic* formed by mechanical friction or gravity crumbling of spines, domes or lava flows, and *alloclastic*, formed by disruption of pre-existing volcanic rocks by igneous processes beneath the Earth's surface, with or without intrusion of fresh magma.

It is important to distinguish between particle origins for interpreting tectonic, physiographic and sedimentary environmental settings, and to determine whether volcanic activity was penecontemporaneous with deposition, or recycled from older rocks by weathering and transport. Pyroclasts and hydroclasts can be distinguished from epiclasts by physical characteristics. Interpreting penecontemporaneity is not straightforward, however, because newly formed pyroclastic particles, and particles derived from weathering of volcanic rocks (lava flows or tuff and volcanic breccia) are frequently mixed during trans-port. In the interest of clarity, strict separation of the processes that create the particles (e.g. pyroclastic, hydroclastic, epiclastic) from the processes that transport the particles (e.g., wind, running water, ice, volcanic explosion, and gravity transfer by avalanche) is required. For example, pyroclastic particles carried by rivers, wind, etc., are 'reworked pyroclastic' particles, not 'epiclastic' which infers a different particle origin.

10.2.2 Volcaniclastic sediments

Classification

Volcaniclastic sediments are classified according to grain size into ash (<2 mm), lapilli (2–64 mm), and bombs or blocks (>64 mm) (Fisher, 1961; Schmid, 1981) (Table 10.1). Other criteria include composition, origin and vesicularity. Rock names for mixtures of pyroclastic and epiclastic fragments are given in Table 10.2).

Accretionary lapilli

Accretionary lapilli are spherical- to spheroidal-shape aggregates of ash that form in wet eruption clouds (pyroclastic and hydroclastic), pyroclastic flows and surges (Schumacher & Schmincke, 1991) (Fig. 10.7). Armoured lapilli (Waters & Fisher, 1971) form when wet ash becomes plastered around a solid nucleus such as crystal, pumice or lithic fragments during a hydroclastic eruption. They commonly occur as discrete 'pisolites' enclosed in an ash matrix but in places are so abundant that they form individual layers. Accretionary lapilli occur in sub-aerial as well as sub-aqueous tephra deposits.

Two structural types of accretionary lapilli have been identified (Schumacher & Schmincke, 1991): (1) the R-type consists of a coarse-grained core surrounded by a fine-grained rim or multiple rims; and (2) the C-type has no rim, consisting essentially only of a core. Rims, single or multiple, are graded from coarse to fine outward toward the outer surface of the lapilli. R-type accretionary lapilli are restricted to proximal deposits within a few kilometres from source. C-type accretionary lapilli occur in distal deposits or in vesiculated tuff layers deposited from ash clouds that entrap large amounts of external water. Deposits from ground-hugging ash cloud surges (Section 10.3.1) concentrate accretionary lapilli in their upper parts, whereas in fallout ashes

Table 10.2 Classification of pyroclastic and epiclastic mixtures. After Fisher and Schmincke (1984).

Pyroclastic*		Tuffites (mixed pyroclastic-epiclastic)	Epiclastic (volcanic and/or non-volcanic)	Average clast size (mm)
Agglomerate, agglutinate pyroclastic breccia Lapillistone		Tuffaceous conglomerate, tuffaceous breccia	Conglomerate, breccia	64
(Ash) tuff	coarse	Tuffaceous sandstone	Sandstone	2
	fine	Tuffaceous siltstone	Siltstone	1/16
		Tuffaceous mudstone, shale	Mudstone, shale	1/256
100%	75%	25% (increase)	0% by volume	

← —————————————————————— Pyroclasts ——————————————————————
(increase)

————————————————— Volcanic + non-volcanic epiclasts (+ minor amounts —————————————→
of biogenic, chemical sedimentary and authigenic
constituents)

* Terms according to table 5–1, Fisher and Schmincke (1984).

(Section 10.3.2) aggregates are enriched at the base of the deposit.

Pumice, scoria and cinders

Three kinds of particles, distinguished in part by their degree of vesicularity and without reference to size, are scoria, cinders and pumice. Pumice is a white or pale grey to brown, highly vesicular glass foam, generally of evolved and more rarely of basaltic composition with a density of $<1\,\mathrm{g\,cm^{-3}}$; bubble walls are thin and composed of translucent glass (Fig. 10.8). Scoria (also called cinders) are usually mafic particles less inflated than pumice and with a density greater than water with vesicles more widely spaced and thicker-walled than pumice. Scoria is generally composed of tachylite.

10.2.3 Rock classification

Ash-, lapilli-, bomb- or block-size fragments can be of juvenile, cognate or accidental origin in any combi-

Fig. 10.7 Accretionary lapilli in rhyolitic fallout tephra layers deposited >50 km from their source. Miocene Ellensburg Formation (Washington, USA). From Schmincke (1967).

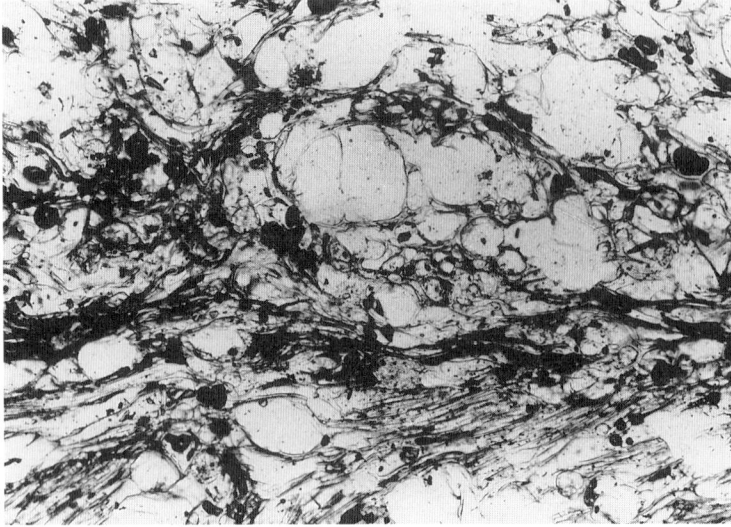

Fig. 10.8 Photomicrograph of pumice structure. Spherical, ovoid and highly stretched vesicles typical of silicic pumice. From pumice block of pyroclastic flow, Mount Mazama, Oregon (USA). Scale, 1 cm = 0.8 mm.

nation and may be composed of crystals, glass shards, pumice and lithic fragments in any proportion. Tuff is the consolidated equivalent of ash, and lapillistone is the consolidated equivalent of lapilli. The term 'volcanic breccia' is broader in scope, applying to all volcaniclastic rocks composed predominantly of angular volcanic particles greater than 2 mm in size (Fisher & Schmincke, 1984); thus it parallels the term 'breccia' widely used for non-volcanic sedimentary rocks. Further classification is made according to environment of deposition (lacustrine, submarine, sub-aerial) or manner of transport (fallout, pyroclastic flow, pyroclastic surge).

10.3 Pyroclastites

The name *pyroclastite* (used by Guerrera & Veneri, 1989) is proposed for pyroclastic deposits of which there are two main categories: (1) pyroclastic sediment gravity flow deposits that form by deposition from currents that move along the surface of the Earth; and (2) fallout deposits composed of tephra that is ejected into the atmosphere or water, or rises from hot pyroclastic flows, and falls back to the ground surface. Fallout and flow processes commonly start from the same eruptive event but also occur independently of one another. Each year, large volumes of pyroclastites are added to sedimentary basins from tephra fallout and, more locally, by pyroclastic surges and flows, many of which become

mixed with non-volcanic sediments during normal gravity-driven transport.

10.3.1 Pyroclastic sediment gravity flow deposits

Pyroclastic sediment gravity flows are hot, gas-particle, density currents that originate from disruption of magma, and enter the atmosphere by eruption through vents to become part of the sedimentological domain of the Earth's surface. Clastic components are crystals, glass shards and pumice, and lithic fragments in highly variable proportion depending upon (1) the composition of the magma; (2) the country rock through which the material rises; and (3) the ability of the currents to erode the surface over which they flow. Xenoliths may form the bulk of the lithic fragments.

There are two end-member kinds of deposits: (1) *pyroclastic flow deposits* that are relatively thick, poorly sorted, commonly but not invariably containing abundant fine-grained ash in the matrix (<63 μm) and with crude or no internal bedding; and (2) *pyroclastic surge deposits* that are relatively thin, better sorted than flow deposits, with or without abundant matrix fines, and well-bedded to cross-bedded. Surge deposits may occur beneath or on top of pyroclastic flow deposits, or by themselves.

Pyroclastic flow deposits rich in pumice and glass shards are known as ignimbrite. Depending upon emplacement temperature, ignimbrites range from unconsolidated, to cemented by vapour phase minerals, to welded ignimbrites (welded tuff). Pyroclastic flow deposits composed of mixtures of non-vesicular, to partially or wholly vesicular, fine- to coarse-grained juvenile lithic particles, are known as block-and-ash flow deposits (Table 10.3). Sub-aqueous pyroclastic flow deposits are discussed in this section. Hydroclastic sediment gravity flow deposits are discussed in Section 10.4.

Pyroclastic sediment gravity flows can move rapidly for long distances, their deposits generally being much thicker in valleys than on ridges. Deposits from single flows range in volume from less than 0.1 km^3 to over 3000 km^3. Some pyroclastic flows of large volume are erupted at such high temperatures that they become welded. Structures caused by high temperature are discussed by Smith (1960), Ross and Smith (1961) and Fisher and Schmincke (1984).

Ideas about the transport of particles and emplacement of deposits are based on observations of small-volume eruptions such as 1902 Mt Pelée (Lacroix, 1904) and 1980 Mount St Helens (Lipman

Table 10.3 Classification of pyroclastic flows and surges. After Fisher and Schmincke (1984; table 8–2, p. 221).

Essential fragment	Eruptive mechanism	Pyroclastic flow	Deposit	Comments
Vesiculated				
	Eruption column collapse or 'boiling-over'	Pumice or ash flow	Ignimbrite; pumice and ash flow deposit	Intermediate-volume deposits formed by continuous collapse of a Plinian eruption column as envisaged by Sparks *et al.* (1978). Silicic in composition
				Small-volume deposits probably formed by interrupted column collapse. Intermediate to silicic in composition
		Scoria flow	Scoria and ash deposit	Small-volume deposits probably formed by interrupted eruption column collapse produced by short explosions (see Nairn & Self, 1978). Basalt to andesite in composition. Some deposits contain large unvesiculated blocks (e.g. those of Ngauruhoe, 1975)
Decreasing average density of juvenile clasts				
	Lava or dome collapse, or lateral projection	Explosive nuée ardente—Lava debris flow	Block and ash deposits	Small-volume deposits, usually andesitic or dacitic in composition
		Gravitational nuée ardente—Lava debris flow	Block and ash deposits	Small-volume deposits usually andesitic or dacitic in composition. These are the hot avalanche deposits of Francis *et al.* (1974)
Nonvesiculated				

& Mullineaux, 1981; Hoblitt, 1986), and deductions made from physical volcanological studies of ancient deposits, and from experimental, computer modelling and theoretical studies.

Fragmental components and grain size

Components of pyroclastic flow and surge deposits are dominantly juvenile pumice, bubble wall shards, phenocrysts and lithics, mixed with various amounts of accessory and accidental lithics and crystals. Large to intermediate volume flows are commonly composed of ash-sized glass shards, phenocrysts and lithic particles that enclose variable amounts of lapilli and blocks of pumice and juvenile to accidental lithic fragments. Small-volume pyroclastic sediment gravity flow deposits may be pumice-rich (e.g. afternoon eruption, 18 May 1980 Mount St Helens; Wilson & Head, 1981), or primarily lithic if they are derived from domes (Perret, 1973; Mellors *et al.*, 1988) or the collapse of fronts of dacitic lava flows (Rose *et al.*, 1977).

Phenocrysts in pyroclastic sediment gravity flow deposits are generally more broken than those in lavas. Crystal abundance, ranging from near 0 to about 50%, is greater in the matrix than in enclosed pumice fragments. This, together with rounding of pumice and abundance of fines (particles $< 63\,\mu m$; $4\,\phi$) is evidence that pumice is abraded during flowage. Large volumes of fine-grained ash are elutriated from such flows to form ash shards whose deposits may outdistance those of the flow by > 1000 km. Most large-volume pyroclastic flow deposits are

calc-alkaline dacite to rhyolite, thus phenocrysts include euhedral, doubly terminated quartz, sanidine and plagioclase with minor amphibole, pyroxene, biotite, iron-titanium oxide cystals and minor zircon and sphene. In more alkaline rocks, anorthoclase is the dominant feldspar.

Lithic fragments rarely exceed 5% by volume of intermediate- to large-volume and some small-volume pumiceous pyroclastic flow deposits.

Inman sorting values (σ_ϕ) of pyroclastic flow deposits are usually greater than 2.0ϕ, but tend to decrease with length of transport. Many pyroclastic surge deposits are finer grained than pyroclastic flow and fallout deposits and have sorting values intermediate between the two, but there is wide overlap (Fig. 10.9).

Pyroclastic flow deposits

Poor sorting, subtle grading or its absence, and poor or no bedding, characterize pyroclastic flow deposits (Fig. 10.10). Although pyroclastic deposits tend to be massive, graded basal zones, discontinuous trains of large fragments, alternating coarse- to fine-grained layers, orientation of elongate or platy clasts, and colour or composition changes may produce a crude layering. Slight differences in size of fragments in different layers give an irregular and indistinct stratification to some deposits. Flat fragments within pyroclastic flow deposits near their basal parts are commonly strongly oriented parallel to depositional surfaces or are imbricated, dipping up-flow (Schmincke *et al.*, 1973; Mimura, 1984).

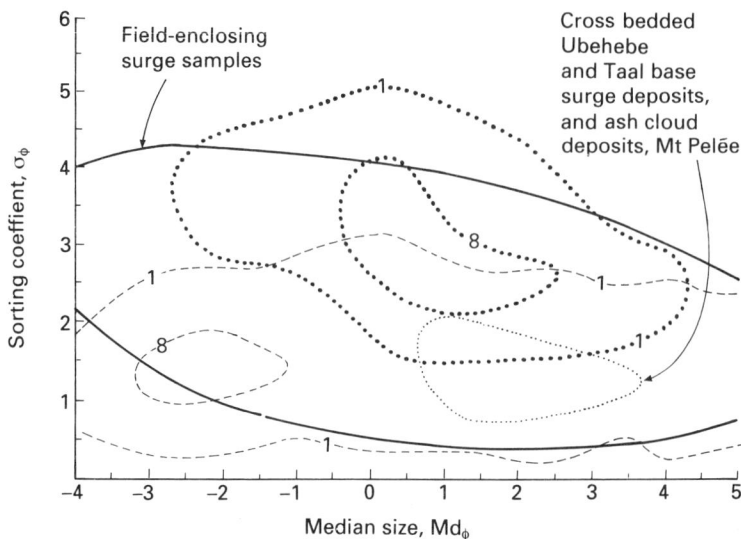

Fig. 10.9 Relationship between median size and sorting coefficients of fallout (dashed lines), pyroclastic flows (large dots), base surges (small dots) and all pyroclastic surges (solid line). Selected contours 1 and 8 enclose percentages of samples. From Fisher and Schmincke (1984).

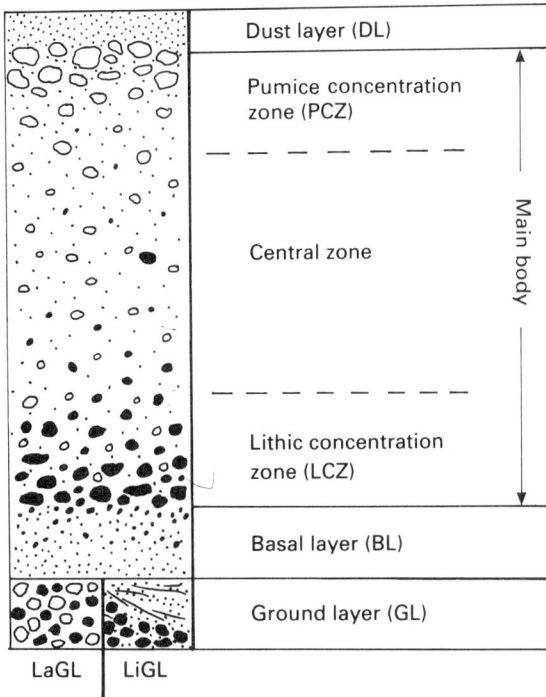

Fig. 10.10 Diagram of sub-division of flow units of pyroclastic flow deposits. Crowe and Fisher (1973), Schmincke *et al.* (1973), Sparks (1976), Wright (1979), Tokunaga and Yokoyama (1979), Self *et al.* (1980), Fisher and Heiken (1982). LaGL = lapilli-rich ground layer. LiGL = lithic-rich ground layer. From Freundt and Schmincke (1986).

Pumice fragments tend to show inverse grading, the largest occurring at the top of a flow. Denser lithic fragments concentrate toward the base. Inverse grading of pumice is caused by buoyant rise during flow. Inverse grading of lithic fragments is likely due to shear effects at the lower boundary (Schmincke, 1967; Fisher & Mattinson, 1968; Sparks, 1976) (Fig. 10.11). Maximum sizes of lithic and pumice fragments decrease with distance from source (Kuno *et al.*, 1964; Fisher, 1966).

Multiple grading (and bedding) may develop from separate flows of the same composition repeated at relatively short time intervals (Sparks, 1976), by topographic splitting of the flow front which advances around obstacles and reunites on their opposite side (Fisher, 1990), and mechanical segregation of different fragment sizes due to shearing within a high concentration flow (Fisher & Schmincke, 1984).

Pyroclastic flow deposits are derived from high concentration sediment gravity flows which are non-turbulent and partially fluidized (Sparks, 1976). The fluidized state is defined as a system consisting of a mixture of particles suspended by an upward escaping gas where the upward drag force exerted by the fluid is equal to the weight of the particles (Wilson, 1980, 1984). Particles are suspended in moving flows by upward moving gases, by grain-to-grain contact and also by basal boundary-to-grain contact where momentum is exchanged, and by the mass property called strength. The concentration of particles is high with relatively little pore space, thus dilation must occur for the particles to move past one another and

Fig. 10.11 Photograph of inverse grading of pumice at top and lithics at base of flow unit. Knife for scale. Ignimbrites of Mt St Helens eruption of May 18, 1980, Washington (USA).

the mass to flow. In such a fluid, the higher the velocity, the lower the strength. Conversely, the strength rises as velocity decreases rise until it reaches a threshold value, and movement of the mass as a whole ceases. In such a fluid, it is difficult for large and small particles to become separated (i.e. sorted). During motion, shear forces between particles can cause them to be redistributed according to size differences. Oriented fragments are common and generally occur in the 'least energy' position relative to flow direction, and can result in crude layering. Compacted pumice has the same elongate appearance as particles oriented by flow processes; it may be difficult to tell the difference.

Pyroclastic surge deposits

Compositionally, pyroclastic surge deposits are much like pyroclastic flow deposits, the main difference being that surges are commonly richer in crystals and lithics than flows and are better sorted (Fig. 10.12). Lithic percentage in pyroclastic surge deposits derived from ignimbrite or deposited during the same ignimbrite-forming eruption are low in lithic fragments (< 5%), but those derived from domes may contain over 90% lithics (e.g. blast deposit of 18 May 1980 eruption of Mount St Helens). Juvenile lithics are usually accompanied in the matrix by broken crystals with adhering matrix derived from the explosive disruption of the neck and dome of the volcano (Fisher & Heiken, 1982).

Pyroclastic surge deposits are thinly to thickly laminated and many have planar but slightly wavy bedded structures. Their most characteristic feature is wavy-, lenticular- or low angle cross-bedding (Fisher & Schmincke, 1984), but in many instances they are thin and massive and some closely resemble dunes of medium-sized sand (Sigurdsson *et al.*, 1987). Many contain lenses of well-sorted and well-rounded pumice lapilli. In cases where deposits are planar and poor in fines similar to fallout deposits, they are distinguished from fallout because large fragments move into place during flow rather than impact from fall and thus do not form bedding sags.

Pyroclastic surge deposits are better sorted and more enriched in crystal and lithic fragments than pyroclastic flow deposits. Median diameters range from -3.0 to $+4.0\phi$ but most occur within the $1.0–2.0\phi$ range, and Inman sorting values range from about 1.0 to 4.0ϕ with most at about 1.5ϕ (Walker, 1984; Sigurdsson *et al.*, 1987). Thus, surge deposits have a more restricted grain size than pyroclastic flow deposits, lacking both the very fine and the coarse fractions. Sorting values of surges are intermediate between flows and fallout, but there is wide overlap.

Pyroclastic surges are relatively low-concentration (i.e. highly expanded) density currents with particles supported mainly by turbulence (Sparks, 1976) (rather than gas fluidization) where particle fall velocity is small relative to scale of turbulence. The surge is driven by the momentum of expanding gas, by momentum of the particles, and by gravity, depending upon slope. Particle concentration increases within the lower part of the surge (i.e. the surge becomes density stratified, Valentine, 1987) and sedimentation takes place in a bedload region where

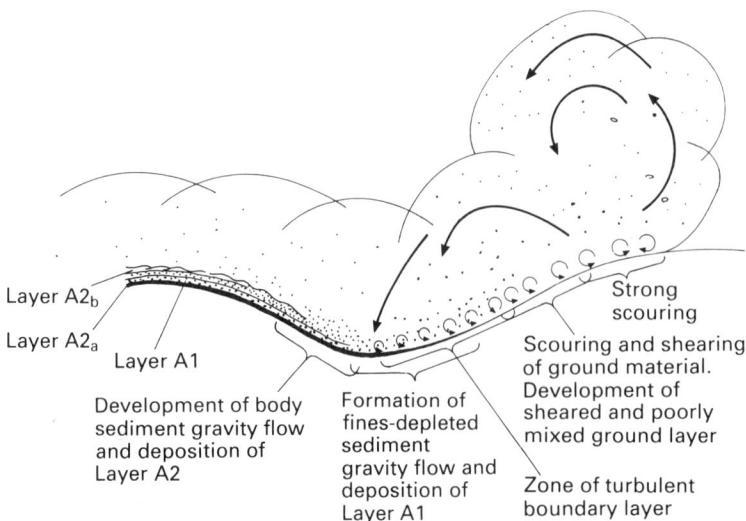

Layer A2$_b$

Layer A2$_a$

Layer A1

Development of body sediment gravity flow and deposition of Layer A2

Formation of fines-depleted sediment gravity flow and deposition of Layer A1

Strong scouring

Scouring and shearing of ground material. Development of sheared and poorly mixed ground layer

Zone of turbulent boundary layer

Fig. 10.12 Diagram of flowing pyroclastic surge showing domains of scouring, shearing and sedimentation which result in a stratigraphic sequence that includes the ground layer and layers A1, A2a and A2b. Not shown is fallout layer from ash elutriated into atmosphere from flow system. From Fisher (1990).

particles may move by saltation and traction or by mass movement as high concentration sediment gravity flows. The bedload region of the surge may become detached from the surge and move independently according to slope direction and, if concentration is high enough, it may transform to a pyroclastic flow that moves down valleys and ponds in low areas (Fig. 10.13) (Fisher, 1990).

Pyroclastic sediment gravity flow facies

The interaction of pyroclastic sediment gravity flows with topographic irregularities (Freundt & Schmincke, 1985, 1986) are two important factors in the development of pyroclastic facies. Pyroclastic surges can override the sides of a valley and their deposits may mantle topography similar to fallout tephra, but unlike fallout tephra they are traceable into thicker pyroclastic flow deposits in valleys. This relationship can be interpreted in two different ways, both of which probably operate. One interpretation is that they are overbank deposits from pyroclastic flows that moved downvalley, overtopping their sides (Schumacher & Schmincke, 1990), and another interpretation is that they developed from surges or flows that moved across the landscape leaving thin beds on the uplands, draining and coalescing in low places (Crowe & Fisher, 1973; Fisher, 1977, 1990; Hoblitt et al., 1981; Wilson, 1985; Fisher et al., 1987).

Topography profoundly affected distribution patterns, and thickness and grain sizes changes of the 18 May 1980 Mount St Helens blast surge deposits (Fisher, 1990), indicated by the fact the blast surge travelled twice as far to the west where ridges parallel flow direction than to the east where ridges are at right angles to the flow direction. High ridges perpendicular to the surge were maximum roughness elements that greatly increased frictional drag. Greater velocity on ridge tops and lowered velocity on the lee sides resulted in large differences in thickness patterns. This same effect on a smaller scale occurred with small obstacles such as tree trunks and small gulleys.

To the west, north and northeast at Mount St Helens, the turbulent top of the blast surge was many times higher than the mountain ridges that it crossed. Where the depositing volcaniclastic sediment was blocked or drained into separate valleys, 'rootless' pyroclastic flows formed in valley bottoms and ponded within depressions. If the volume had been many times greater, the deposits in separate valleys would resemble a once-continuous sheet.

The mechanism of crossing mountainous barriers at Mount St Helens may be applied to pumice-rich pyroclastic flows that are known to have crossed topographic barriers of considerable height. For example, the 22 000 years BP Ito pyroclastic flow (Japan) travelled 70 km over topographic barriers as high as 600 m (Yokoyama, 1974). The Taupo Ignimbrite, although only ∼ 30 km³ in volume, is spread out over a ∼ 20 000 km³ area and mantles mountains up to 1500 m higher than the inferred vent as

Fig. 10.13 Photograph of transition from surge (left) into massive flow deposits in the proximal facies. Quaternary Laacher See tephra deposits, Eifel, Germany. Photograph by Schmincke.

far as 45 km from the source (Wilson, 1985). The 33.5 ka Campanian ignimbrite, Phlegrean Field (Italy) (Paterne *et al.*, 1988) travelled farther than 100 km from its inferred source near Pozzuoli, covered an area of about 7000 km^2 and is 30–40 m thick in lowland areas (Barberi *et al.*, 1978). It moved over many limestone peaks of 1000 m and greater elevation, and in one place crops out at 950 m, ~ 50 km from its inferred source. It is suggested that these pyroclastic flows were transported as expanded sediment gravity flows many times thicker than the highest peaks, and that topographic blocking and runoff resulted in pyroclastic flow deposits in valleys.

Calculations by Sparks (1978) suggest that ground-hugging pyroclastic flows with velocities of 100 m s^{-1} are capable of overcoming barriers several hundred metres high. According to Sheridan (1979), the slope of the 'energy line' traces the potential flow head from the top of the gas-thrust region of an eruption column to the distal toe of a flow along the line of transport. Theoretically, a pyroclastic flow could surpass all topographic barriers that do not extend above the line. Although calculations permit the speculation that pyroclastic flows can move along the ground over high mountains as continuous, fast-moving, flowing pyroclastic sheets, observations at Mount St Helens suggest that isolated pockets of pyroclastic flows found in separate basins across mountain ranges may also be emplaced from density stratified turbulent currents.

Origins of pyroclastic sediment gravity flows

Pyroclastic flows and surges can form in several ways: (1) gravitational collapse of a vertical eruption column (Sparks *et al.*, 1978); (2) the 'boiling over' of a highly gas-charged magma from a crater (Taylor, 1958); (3) inclined blasts from the base of an emerging spine or dome (Lacroix, 1904); (4) lateral blasts following release of pressure caused by collapse of part of a volcano edifice (Bogoyavlenskaya *et al.*, 1985; Siebert *et al.*, 1987); (5) collapse of a growing dome (Mellors *et al.*, 1988); (6) ash fountaining (Hoblitt, 1986); and (7) explosive disruption from the front of a lava flow (Rose *et al.*, 1977) (Fig. 10.14).

The collapse of vertical eruption columns to form pyroclastic flows was recognized at the 1929 eruption of Komagatake, Japan (Kozu, 1934) and postulated from sedimentological data at St Vincent, BWI by Hay (1959). Using development of a base surge from

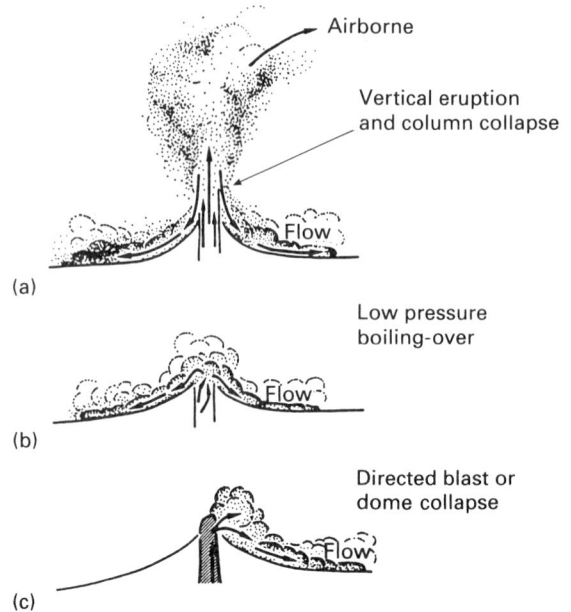

Fig. 10.14 Common origins of pyroclastic flows and pyroclastic surges. (a) Mayon Volcano, Philippines; Soufriere, St. Vincent. (b) Mt Lamington, Papua. (c) Mt St Helen's, Washington, USA. From Fisher and Schmincke (1984).

a 1947 nuclear explosion at Bikini Atoll (south Pacific) as a model, column collapse ('bulk subsidence') was suggested as a cause of pyroclastic flows leading to development of ignimbrites by Fisher (1966), and the process of column collapse was described from a series of photographs showing the development of a surge at Capelinhos (Azores) (Waters & Fisher, 1971). The connection between column collapse and the origin of pyroclastic flow and surge deposits was quantitatively established by Sparks and Wilson (1976) and Sparks *et al.* (1978). However, some pyroclastic flows and surges originate without accompanying vertical eruption columns. For example, at Mount St Helens on 22 July and 7 August, 1980 (Hoblitt, 1986), pyroclastic density currents began as fountains of gases and pyroclasts around the vent prior to development of a vertical eruption column.

Sub-aqueous volcaniclastic flows

Volcaniclastic mass flow deposits interbedded with

rocks known to be sub-aqueous because of the presence of fossils and evidence such as pillow lavas, are called sub-aqueous volcaniclastic flow deposits. Most known examples are marine (Fisher & Schmincke, 1984). Internal textures and structures are used to interpret their origins as pyroclastic flows, but in ancient deposits, sedimentological or paleontological evidence from associated sediments is used to determine unequivocal sub-aqueous deposition.

The most convincing case for marine deposition is from the Wadaira Tuff Member of the Tokiwa Formation, Japan, which is a conformable, richly fossiliferous mudstone sequence containing five main, separate, graded sub-aqueous pyroclastic flow sequences (Fiske & Matsuda, 1964). Foraminifera within the mudstone above the pyroclastic flow deposits indicate open sea deposition at depths of 150–500 m. Depositional units consist of a Lower Massive Division consisting of a thick, massive and structureless lower part, overlain by the Upper Bedded Division, with an aggregate thickness of about one-third that of the lower division. Each bed in the upper division is graded from coarse to fine, and upward, each bed becomes thinner and finer grained, a type of grading called *double grading* (Fiske & Matsuda, 1964). The entire sequence is interpreted to indicate derivation from an initial large submarine explosion (Lower Massive Division) followed by smaller, waning eruptions (Bedded Upper Division) (Fig. 10.15).

Pyroclastic flows are known to have entered the sea following sub-aerial eruptions (e.g. 1815 Tambora, Sigurdsson & Carey, 1989; 1902 Mt Pelée, Lacroix, 1904; 1976 Augustine Volcano, Fisher & Schmincke, 1984). It is probable that pyroclastic flows generated during the 1883 Krakatau (Java) eruption entered the sea and caused destructive tsunamis (Self & Rampino, 1981). Remobilized deposits from areas of high deposition rates of tephra on the slopes of shoreline volcanoes and volcanic islands may form debris flows with structures and textures similar to pyroclastic flows, and may further give rise to turbidity current deposits (ash turbidites) (Wright & Mutti, 1981). There is no convincing evidence that tephra in the form of hot pyroclastic flows can originate from a completely underwater eruption, although it is theoretically possible (Sparks *et al.*, 1980) However, it is not yet established, whether or not a hot density current from land can flow into water and maintain its integrity as a hot pyroclastic flow. The Roseau Ash in the Caribbean was originally interpreted to be the

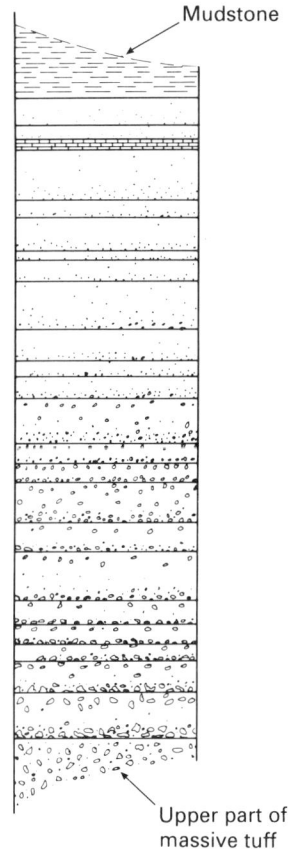

Fig. 10.15 Diagram of double grading of sub-aqueous pyroclastic flow deposits. From Fisher and Schmincke (1984).

continuation of a hot pyroclastic flow that moved from land into water (Carey & Sigurdsson, 1980), but it has been since re-interpreted as the deposits from numerous small sediment gravity flows (Whitham, 1989).

10.3.2 Fallout deposits

Fallout deposits are derived from explosive eruptions that eject ballistic fragments and/or eruption plumes into the atmosphere. *Sub-aerial fallout tephra* is deposited on land from eruption columns or from atmospheric winds that attenuate the eruption columns and carry the tephra. *Sub-aqueous fallout tephra* (marine or lacustrine tephra) is composed mostly of pyroclasts derived from eruptions on land that fall on water and sink through the water column to the bottom. Dispersal patterns of marine fallout layers are modified by marine currents and by bioturbation or can be destroyed by postdepositional redistribution by sediment gravity flows (debris flows, turbidity currents). Products of sub-aqueous eruptions that occur wholly underwater are discussed in the hydroclastite section.

Eruption columns

Sustained explosive volcanic eruptions into the atmosphere commonly produce eruption columns ranging from a few hundred metres height (basaltic Hawaiian lava fountains) up to 50 km (felsic Plinian eruption columns). Column height and wind vectors determine extent and geometry of particle distributions from volcanic eruptions, thus the most extensive tephra is felsic. If the density of the column becomes greater than the atmosphere, the column can collapse, or if the column is denser than air upon emergence from the vent, pyroclastic density currents may form (Sparks & Wilson, 1976; Sparks *et al.*, 1978; Wilson *et al.*, 1978, 1980; Wilson & Walker, 1987; Valentine & Wohletz, 1989). Most Plinian eruption columns develop from explosions of highly evolved rhyolitic to dacitic, trachytic and phonolitic magmas.

At temperature inversions, especially at the troposphere–stratosphere boundary (tropopause), and in regions where eruption columns become neutrally buoyant, convective plumes mushroom to form umbrella clouds (Sparks, 1986; Sparks *et al.*, 1986). Lateral displacement is pronounced at the tropopause which is characterized by strong winds (jet stream). The lateral spread and speed of many plumes has been traced by satellite photography or remote sensing techniques such as radar (e.g. Harris *et al.*, 1981). If mass discharge rates are sufficiently high, as is the case in many Plinian eruptions, plumes will penetrate this level and ascend well into the stratosphere.

Distribution and thickness

Sub-aerial tephra is carried upward by explosive expansion within turbulent eruption plumes and returned to Earth by (1) ballistic trajectory; or is carried for long distances by (2) turbulent suspension with energy supplied initially from explosive expansion from the eruption; and (3) later by atmospheric wind. These three main threshold settling velocity

values for any given set of conditions (wind factors, eruption factors) affect the distribution, textures and structures of fall deposits (Fisher & Schmincke, 1984).

Explosive energy can project greater volumes of particles of larger size to higher elevations than wind is able to pick up. Thus, tephra can travel much longer distances than fragments that are distributed by atmospheric wind. Another difference between sub-aerial eruption products and non-volcanic wind-blown sediments is that volcanic ash can enter the atmosphere in wet as well as dry environments. The distribution patterns of sub-aerial fallout tephra deposits depend upon wind strength and commonly form fan-shaped sheets.

Medial to distal thin tephra layers on land or in the sea form excellent stratigraphic markers because they are widespread, are laid down simultaneously in many sedimentary basins and environments, and can be dated radiometrically to infer the age of the eruption and deposition alike (Bogaard *et al.*, 1989).

Sub-aqueous fallout tephra is less easily eroded than sub-aerial tephra. Marine fallout layers may survive for hundreds of millions of years, commonly diagenetically altered to layer silicates (bentonites). They are globally useful for assessing paleovulcanicity, and for determining the evolution of volcanically active continental margins, oceanic islands and especially island arcs. The spacing of marine ash layers within stratigraphic successions can give information about the cyclicity of volcanism, volcanic production rates and volumes, and the influence of large explosive eruptions on climate (Kennett, 1983; Bitschene & Schmincke, 1990; Schmincke and Bogaard, 1990). Fall deposits in sub-aerial environments are more apt to be redistributed by surface erosional agents, but can be useful for correlation purposes if they are deposited and preserved in lakes or swamps (Bogaard & Schmincke, 1985).

Fragments with small fall velocities compared to wind strength may reside in the atmosphere for days before settling as a thin film of dust on land and beneath the sea. Very small particles commonly stick together by small electric forces or by surface films of water on particles to form snow-flake-like ash clusters or concentric accretionary lapilli that fall faster than the individual particles comprising the lapilli. Thus, large volumes of very small particles may fall out in proximal areas along with coarse particles. The size and density sorting of fall deposits is better than other kinds of pyroclastic deposits, but particle aggregation can cause poor sorting. Poor sorting may also be caused when fine particles fall prematurely in rain drops. Even so, maximum particle sizes and sorting coefficients decrease exponentially as distance from the source increases (Pyle, 1989). Directions and dispersal area of airborne tephra depend upon wind vectors at different altitudes.

Accretionary lapilli commonly survive in sub-aerial conditions but in some cases may break down *in situ* (e.g. 1980 Mount St Helens eruption, Sisson, 1982), and they even survive within sub-aqueous depositional environments. Because the outer skins of many accretionary lapilli are composed of very fine-grained (hence cohesive) ash, they can remain intact while sinking through water.

Thickness and distribution patterns of tephra layers are best shown by isopach maps. Isopachs form base maps upon which other data (e.g. median diameter, sorting) can be plotted directly to show their relationships to one another. This gives an added dimension for interpreting wind strength and direction, mean eruption rates and other parameters.

Postdepositional compaction can substantially alter thickness, and therefore change isopach patterns. Compaction of deep-sea tephra, for example, can reduce its thickness to one-half the thickness of equivalent dry tephra on land and may be greatly affected by upward mixing by bioturbation. It is difficult and expensive to measure enough points of tephra thickness on the sea bottom. Thickness plotted against distance for deep-sea Toba tephra resulted in an estimated volume of 1000 km^3 (Ninkovich *et al.*, 1978). The tephra was derived from an eruption 70 000 years BP that produced many ignimbrites and Toba caldera, Sumatra.

Structures

Sedimentary structures of tephra are the result of transport modes and depositional environments: (1) the nature of basal and upper contacts, either sharp or diffuse; (2) vertical grading relative to composition and density of pyrogenic crystals; (3) the relative amounts and kinds of pumice with respect to position in the beds; (4) the kind and amount of admixed non-volcanic debris; (5) the type of internal lamination; and (6) the morphology of the grains.

The types of eruptions (e.g. Hawaiian, Plinian), proximity to source, and depositional environment (land or water) are important factors that determine bedding features. Fallout tephra tends to be coarser grained and thicker near the source and exponentially becomes finer grained and thinner away from the source. Thickness and size maxima may occur some distance from the source. Fallout deposits are better sorted than flow deposits, and normal grading is more prevalent in fallout than density current deposits, with sorting and normal grading better developed in sub-aqueous than in sub-aerial varieties.

Near-source sub-aerial fall deposits are commonly poorly sorted, rich in xenoliths and are alternately layered coarse-grained to finer-grained layers without sharp bedding planes (Fig. 10.16). A continuous, voluminous and irregularly pulsating eruption of high intensity (Plinian) lasting several hours or days may deposit thick alternating and gradational coarse-to fine-grained beds (Fig. 10.17). Thick homogeneous beds may also build up over millions of years in areas receiving periodic light ash falls from distant volcanoes (Fisher & Rensberger, 1973).

Many features of sub-aqueous fallout resemble sub-aerial deposits, although there are some differences. For example, the lower contacts of submarine fallout ash layers are generally quite sharp as in sub-aerial tephra, but upper boundaries are commonly diffuse. In sub-aqueous tephra, density grading and sorting are very well developed with phenocrystic feldspar,

Fig. 10.16 Photograph of coarse-grained proximal fallout deposits rich in xenoliths, about 1.5 km from source. Quaternary phonolitic Laacher See tephra deposits, Eifel, Germany. Scale is 1 m long.

Fig. 10.17 Photograph of trachytic (white) and basaltic (dark) compositionally zoned Plinian fallout deposit, rim of Canadas caldera, Pico de Teide, Canary Islands.

quartz, biotite and pyroxene concentrated at the base, and glass shards prevalent at the top.

The abundance of shards relative to non-volcanic components decreases either gradually or irregularly upward, a major cause being biologic mixing (bioturbation). Significant amounts of ash can be biologically mixed through a thickness of sediment ranging up to 40 cm (Ruddiman & Glover, 1972). Even though thin layers may be completely dispersed, they are equal to discrete ash layers in importance for evaluating volcanic activity. If dispersed pyroclastic material forms more than 15% of a deposit, it is considered to represent significant explosive volcanic activity.

Mantle Bedding. Fallout tephra characteristically mantles topographic surfaces < 25°, but gravitational modification causes fall deposits to be thinner on ridges than in valleys or smaller topographic features depending upon slope rate of burial (Fig. 10.18). Under sub-aerial conditions, sheet flooding or rill development is inhibited in freshly deposited, coarse-

Fig. 10.18 Photograph by Fisher of mantle bedding, Oshima, Japan.

grained permeable tephra because the water quickly soaks into the ground, but in fresh, fine-grained ash, light rain causes crusts which reduce permeability and accelerate erosional processes (Segerstrom, 1950). Thick ash may disrupt or destroy original drainage pattern, killing vegetation and causing accelerated erosion during which new streams quickly cut to levels below old stream-bed levels.

Graded bedding. Fall deposits from the atmosphere and through water are graded according to fall velocity of particles, usually with the largest and heaviest particles at the base (Bramlette & Bradley, 1942). However, normal grading under sub-aerial conditions is rarely perfect because of fluctuations in eruption energy, wind direction and strength, turbulence in the eruption column, and rain flushing. At long distances from the volcano, however, transport is controlled by wind factors rather than explosive expansion, so that settling velocity becomes the dominant depositional control and normal grading develops.

Size parameters

The size characteristics and consequently sedimentary structures of fallout layers are governed initially by both the atmospheric and/or sub-aqueous media through which they fall. Fall deposits commonly have sorting values between 1.0ϕ and 2.0ϕ and characteristically are better sorted than pyroclastic flow and lahar deposits. In many fallout deposits, sorting becomes better with distance. At short distances from the vent (about 30 km, depending on grain size), sorting values may increase and then decrease, with much of this variation at a few kilometres from the vent caused by pronounced differences in density between pumice, crystals and lithics. At Laacher See, Germany, there is a change from unimodal, coarse-grained and positively skewed tephra to unimodal and fine-grained non-skewed tephra in many pumice fall deposits (Bogaard & Schmincke, 1985). The coarse mode is dominated by pumice and lithic fragments, while that of the fine-grained mode (Md < 2 mm) is dominated by glass shards.

Grain size patterns of submarine tephra have been used to estimate eruption column heights, magnitude and energy of eruptions. Ninkovich *et al.* (1978) compared the relationship between median diameter and distance to source for samples from the basal parts of three deep-sea tephra layers and sub-aerial

samples of the 1947 Hekla ash. However, they point out that the coarsest one percentile ($\sigma 1$), which is a sympathetic measure of maximum grain size (Suzuki *et al.*, 1973), is a more easily measured and useful grain-size parameter than $Md\phi$. Ninkovich *et al.* (1978) estimated the eruption column height from the Toba eruption to have been 50 km using size data from the Toba deep-sea ash layer compared with the known eruption column height, wind velocity and size data from the 1947 Hekla eruption.

Maximum size of components. Inferences about vent locations, relative volcanic energy, inclination of eruption column, and wind directions are enhanced using averages of maximum diameters (using 3, 5 or 10 fragments) of lithic and pumice fragments (Fisher & Schmincke, 1984). Maximum size is also used to determine eruption column heights and maximum wind speeds during eruption. Maximum diameters of pumice and lithics exponentially decrease away from the source with lithic maxima commonly decreasing sharply in the first 2 km or so and then more gradually away from the source. This can be attributed to ballistic transport of the largest lithic fragments near the source. Lithic maxima are commonly about half that of pumice in the same outcrop depending on their densities. Pumice is brittle and may decrease in size more rapidly than lithic fragments because of breakage upon impact with the ground. Distribution patterns of isograd maps that differ from isopach maps of the same deposit near the source may be attributed to directional wind patterns at different altitudes or to inclined eruption columns.

Carey and Sigurdsson (1986) and Carey and Sparks (1986) analysed *en masse* cross-wind effects on maximum particle transport distance. The geometry of lithic isopleths as determined by measurements of lithoclasts within Plinian pumice fall deposits can be used to determine quantitatively the maximum eruption column height and average wind speed for individual fallout layers. The half-width of an isopleth measured perpendicular to the main dispersal axis is mainly a function of the eruption column height. The maximum downwind range along the axis is controlled by the column height and the average wind speed (Carey & Sparks, 1986)

Median diameter. Median particle diameters of fallout layers exponentially decrease away from the source (Pyle, 1989), with the decrease commonly dispersed around a mean. Median diameters may provide a more sensitive indicator of inclined eruptions or wind variations during an eruption than do isopach maps alone but require more laboratory time to determine than field measurement of maximum diameters. Also, sampling from thick tephra layers with large vertical variations in size parameters can introduce errors. Because tephra deposits are polycomponent, the type of fragment, for example pumice, accidental lithic or crystal, must be recorded.

10.4 Hydroclastites

Hydroclastites are deposits composed of volcanic particles formed by explosive or non-explosive interaction of water and magma. The name *hydroclastic* applies specifically to the origin of fragments from magma–water interactions. Historically, the name *hyaloclastic* was given to rocks produced by non-explosive spalling and granulation of rinds of pillow lavas (Rittmann, 1958) (Fig. 10.19).

Explosive magma–water eruptions include: *phreatic eruptions*, caused by conversion of ground water to steam resulting in explosive expulsion of pre-existing rock fragments but no juvenile ejecta; *phreatomagmatic eruption*, caused by mixing of ascending magma with ground or surface water resulting in expulsion of juvenile as well as pre-existing lithic material; *sub-aqueous explosions* resulting from magma–water reactions in shallow water bodies or beneath ice, a variant of phreatomagmatic eruptions; and *littoral explosions* that occur where sub-aerial lava flows or hot pyroclastic flows intersect with water along shorelines. Littoral cones are formed at the edge of the sea from lava entering the water. Various kinds of explosive hydroclastic processes generate maar volcanoes, low-rimmed craters.

10.4.1 Deposits

Hydroclastites are commonly well bedded and finer grained than pyroclastic deposits and contain features such as cross-bedding and U-shaped channels that simulate non-volcanic water-deposited sediments.

Deposits of hydroclastites are composed of vitric shards, crystals and lithic fragments. *Vitric shards* from hydroclastic eruptions are characteristically blocky, with equant shapes and fracture-bounded surfaces transecting sparse vesicles. Blocky mafic shards (sideromelane) form a large percentage of deposits of maar volcanoes and littoral cones. Vesicular shards are also present where vesiculating magma was quenched by water or steam. These deposits may contain clasts intermediate between blocky fragments and those that are slightly vesicular with scalloped edges to highly vesicular clasts.

Surfaces of many *lapilli and bombs* from hydroclastic eruptions have an irregular, intricately cracked structure variously described as cauliflower, crackled or bread crust. This texture resembles that of bread crust bombs, but forms by thermal contraction rather than from internal gas expansion. Cauliflower bombs are commonly dense (Fig. 10.20).

Accretionary lapilli are common in many hydroclastic layers owing to abundant water and steam in

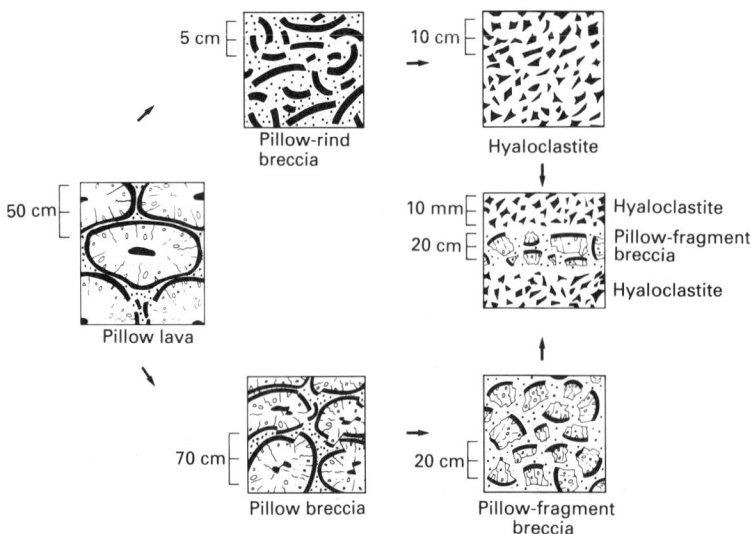

Fig. 10.19 Schematic diagram of formation of hyaloclastite and pillow fragment breccias. From Schmincke (1988).

Fig. 10.20 Photograph of cauliflower bombs in debris flow deposits of 1877 eruption of Cotopaxi volcano, Ecuador.

the eruption column occurring in conjunction with abundant fine-grained tephra.

Structures

Hydroclastic beds are from a few millimetres to several tens of centimetres thick and commonly extend outward over the crater rims (Fig. 10.21). They form during (1) numerous short eruptive pulses characteristic of hydroclastic eruptions; or (2) are

sedimentation units consisting of several beds derived from the passage of a single base surge. Structures include parallel-bedded to cross-bedded units, mudcracks, small rills, channels and convolute bedding. Transport directions given by imbrication of platy fragments and cross-bedding geometry show a radial current configuration directed outward from a crater.

'Convolute laminations' are a common type of soft sediment deformation consisting of folded beds

Fig. 10.21 Photograph of rim beds of Cerro Colorado maar volcano, northern Mexico. Crater to the right. Antiform axis of rim migrated outward during upward growth of rim beds.

between undeformed layers that may be several centimetres thick and may extend laterally for several metres (Fig. 10.22). These form by gravity sliding of sloping water-saturated tephra or from shear-deformation caused by an overriding pyroclastic surge.

Vesicles, common in base surge beds (Lorenz, 1974), are generally sub-spherical, with smooth interior walls coated by very fine-grained ash. Large voids (up to 1 cm) commonly have irregular shapes and consist of several coalesced bubbles. Vesicles are most common in beds with soft sediment deformation structures, but also occur in lahars, in tuff beds with mud cracks, and even in tuff plastered on vertical surfaces. Many tuff beds with vesicles contain accretionary lapilli which themselves may contain small vesicles. Gases responsible for the vesicles may be: (1) from the gaseous phase of the depositing system; (2) given off by hot pyro- and hydroclasts; (3) air rising from the underlying ground; or (4) from water evaporating from snow or water-soaked soil beneath a layer of hot ash. Still another source may be (5) rain that turns to steam as it percolates downward into hot ash.

Sandwave beds

Sandwave beds (Sheridan & Updike, 1975) are wavy bedded or undulating bedding planes, or laminae inclined to a depositional surface that form dunes, antidunes, ripples and internal cross laminations within those structures (Schmincke *et al.*, 1973). Wave lengths near the explosion centre at Taal volcano attained 19 m, and decreased systematically to about 4 m at 2.5 km from the crater centre (Waters & Fisher, 1971). The dunes migrated away from explosion centre as indicated by orientation of internal laminations and shape of dune crests (Moore, 1967). Low dips of foreset (lee-side) laminations (10–15°) indicate that dunes advanced by flow entrainment of particles rather than gravitational rolling of loose debris. Wave lengths of the Taal dunes vary directly with total thickness of the deposit, bedding thickness, size parameters of ash and the distance from the source, suggesting that the carrying capacity decreased progressively with distance as velocity slowed.

Cross-beds in base surge deposits change laterally from very large dune-like structures (Fig. 10.23) near the source to smaller more subdued dunes. Still farther, dunes grade into plane parallel beds (Fig. 10.24). These downcurrent changes suggest that the bed load was dropped rapidly near the source and the energy of transport or capacity to carry a load then decreased more slowly (Schmincke *et al.*, 1973).

Plane parallel beds

Plane parallel conformable sequences closely

Fig. 10.22 Photograph of convolute lamination in Quaternary phreatomagmatic maar deposits. Kilbourne Hole, New Mexico (USA).

Fig. 10.23 Photograph of proximal surge dunes and chute-and-pool structures, 1.5 km from source. Phreatomagmatic Upper Laacher See Tephra deposits, Eifel, Germany. Scale at base is 1 m long.

resemble fallout layers, but often grade laterally into zones of cross-bedding and dune shapes. Platy clasts may be imbricated. Internal laminae commonly show subtle cross-bedding or are lenticular over short distances.

Reverse grading of plane parallel beds suggests lateral mass transport and deposition by non-turbulent flow as is the case for lahars. This is especially evident in beds where large blocks rest on lower contacts without significant deformation indi-cating emplacement by flow. They are deposited from rather sluggish, cool pyroclastic flows.

Massive beds

Thick beds tend to be poorly sorted and internally massive, but may show pebble trains or vague inter-nal textural variations giving a crude internal strati-fication that is either planar or wave-like. Such beds commonly have reversely graded basal zones.

Fig. 10.24 Plane parallel beds of the distal facies of phreatomagmatic surge deposits (2.5 km from source). Phreatomagmatic Upper Laacher See Tephra deposits, Eifel, Germany. Scale is 2 m long.

Fig. 10.25 U-shaped channel cut and filled by base surges. Koko Crater, Hawaii (USA). From Fisher (1977).

U-shaped channels

U-shaped channels in base surge deposits have broad curving bottoms and clearly cut underlying layers (Fig. 10.25), and in cross-section commonly measure from 0.3 m width and 0.2 m depth. Infilling beds reflect the shape of the channels, but the curvature of individual beds decreases upward (beds thicken toward centres of channels). Thus, the upper beds in the channels extend uniformly across the channel conformably with the sequence above and outside the channel. The U-shape of channels is believed to be caused by lobate geometry of advancing density currents (Fisher, 1977). The fronts of density currents consist of lobes separated by clefts that spread outward from the source. Moving down a widening slope of a volcano, individual lobes diverge to follow independent paths and carve diverging furrows straight down the slope. The concentration of particles within the lobes is probably greatest along their central axes, where boundary effects are least and velocity effects are greatest.

Bedding sags

Bedding sags form by the impact of ballistically ejected bombs, blocks and lapilli upon beds capable of being plastically deformed. They are common in hydroclastic deposits of many maar volcanoes, tuff rings and tuff cones (Fig. 10.26). Beds beneath the fragments may be completely penetrated, dragged

down and thinned, folded, or show microfaulting (Heiken, 1971). Deformation is commonly asymmetrical, showing the angle and direction of impact if three-dimensional exposures are available. These differ from dropstones in glacial environments in that dropstones fall perpendicular to the bottom, symmetrically indenting bedding and rarely, if ever, penetrating.

Fig. 10.26 Photograph of ballistically emplaced block. Zuni Salt Lake maar volcano, New Mexico (USA).

10.4.2 Origins

Important physical parameters of hydroclastic deposits are: (1) grain-size distributions; (2) particle vesicularity, crystallinity and shape; (3) tephra particle abundances; (4) deposit thickness and areal extent; (5) bedform types; and (6) angle of depositional surfaces (cohesivity) (Wohletz, 1986). These parameters reflect aspects of eruption mechanisms such as flux of magma and water within the volcano, the depth of interaction, and the geometry of the vent. Thermodynamics of hydroclastic systems are very complex, but varying the mass ratios of water and magma produces a systematic effect. Mass ratios less than about 0.3–0.4 can be expected to be in the dominantly superheated state, depending upon magma composition. For ratios greater than about 0.3–0.4, expansion will produce saturated steam (Wohletz, 1983).

10.4.3 Base surges

Base surges are hydroclastic sediment gravity flows that form by collapse of steam-saturated eruption columns and travel outward along the ground surface. The name comes from a surge that developed from the base of an explosive column from an underwater nuclear explosion in 1947 at Bikini Atoll (South Pacific). A volcanically produced base surge was recognized during the 1965 phreatomagmatic eruption of Taal Volcano, Philippines (Moore, 1967). Base surges are composed of turbulent mixtures of water vapour or condensed droplets and solid particles. The surge may initially be dry if the water is vaporized, but during transport if the water condenses, the mixture will become wet with water droplets mixed with particles coated with thin water films. Upon deposition, surface tension of water cohesively bonds the particles and causes the deposit to behave plastically.

Base surge deposits are moderately to poorly sorted, decreasing in grain size and thickness logarithmically away from the source, with local thickness controlled by topography. The deposits commonly display sedimentary structures resembling non-volcanic structures but are unique to volcanic deposits. Base surge deposits commonly extend <3 km from their source.

10.5 **Lahars**

A lahar is a debris flow composed of a significant component of volcanic materials (> 25%) (Fisher & Schmincke, 1984), a descriptive definition that can be applied in the field from observations of deposits without requiring a judgement about synchroneity of volcanism.

Lahars, many of which are highly destructive, occur worldwide (Fig. 10.27). One of the most common places of origin is on andesite stratovolcanoes where they occur in abundance on the cone and dominate the surrounding sedimentological domain, forming low-sloping, relatively flat-surfaced and widespread ring plains (Hackett & Houghton, 1989). Thus, stratovolcanoes are sediment reservoirs

Fig. 10.27 Photograph of lahar deposit generated at Nevado de Ruiz volcano, Colombia on 5 November, 1985, covering the destroyed village of Armero.

holding vast volumes of volcanic detritus easily reworked into contiguous marine and non-marine sedimentary basins. The sediments stratigraphically record the time and place of volcanism that produced them. Because most stratovolcanoes occur in magmatic arcs, lahars and their derivative sediments are important in reconstructing paleovolcanitectonic events.

10.5.1 Deposits

The composition, grain size and thickness of lahar deposits are quite variable. They may be monolithologic or heterolithologic, and composed of coarse to fine ash or contain abundant angular to rounded blocks. Common characteristic features are poor to no internal bedding, matrix supported large fragments, and reverse-to-normal grading. Thicknesses range from a few centimetres up to 200 m (where ponded). In eroded ancient volcanic terranes, many thick laharic breccia deposits appear to be structureless and non-graded at close range, although close examination usually shows subtle inhomogeneities such as changes in clast concentration, slight erosional breaks along minute textural changes that are not traceable for long distances. Colour changes and textural breaks become more obvious away from the source because thick-appearing beds may be compound lahars composed of innumerable relatively thin beds, 1–20 m thick. Exposures of thick compound lahar deposits are less common in the modern record because lahars rapidly deposit material in basins and plains adjacent to growing volcanoes thereby covering the older materials.

Components of lahar deposits

Lahar deposits from active volcanoes characteristically contain dense angular to sub-angular andesitic to dacitic rocks mixed with ash-sized minerals and lithic particles. Rounded particles may be inherited from pre-existing alluvial debris incorporated in a lahar by bulking, a process by which sediment is incorporated into a flow by erosion at the flow boundary (Scott, 1988). Monolithologic lahars are likely to be derived directly from an eruption, whereas heterolithologic lahars are derived from the collapse of crater walls, scavenging of debris of different sequences from steep volcanic slopes, or epiclastic materials from old volcanic terranes.

Pumice- or juvenile lithic-rich lahar deposits re-

semble deposits of hot, dry pyroclastic flows (block-and-ash flows), but in some cases, the cool lahars can be distinguished from deposits of hot flows by thermoremanent magnetic temperature determinations (Hoblitt & Kellogg, 1979). Charred wood within lahars suggests that hot pyroclastic flows were transformed into lahars as they entered a river system such as at Mount St Helens in 1980 (Scott, 1988).

Grain size distribution

Lahar deposits are typically coarse grained and poorly sorted. Poor sorting results from the inability of fragments of differing fall velocities to separate effectively from one another during flow. Lahar deposits commonly appear to be coarser grained than shown by granulometric analysis because matrix supported boulders and cobbles, if present, are visually more impressive than the smaller particles, giving a false impression of true size values. A characteristic feature of some lahar deposits is the presence of large, matrix-supported clasts up to boulders exceeding 1 m in diameter.

Lahar deposits are not completely characterized by standard size analyses because boulders cannot be included. Thus, Fisher and Schmincke (1984) compare matrix phases (sand/silt/clay recalculated to 100%) of different debris flows. Lahars tend to have less clay-size material than non-volcanic debris flows, one possible reason being that many lahars contain more explosively comminuted debris and less fine-grained materials derived by weathering. The abundance of clay within the matrix of some lahars indicates source areas rich in hydrothermally altered rocks (Crandell, 1971).

Grading

There are two kinds of size grading in lahars — *reverse-to-normal* and *bottom-to-top* grading. Reverse-to-normal grading is most common and is recognized where there is an upward increase in fragment size from the base of a layer, attaining a maximum within the lower part of the bed and then progressively decreasing upward toward the top of the bed (Schmincke, 1967). Such grading also occurs in many pyroclastic flow deposits. Less common is bottom-to-top grading (Fig. 10.28) where larger fragments progressively increase in size to the very top of a deposit.

Reverse grading develops from dispersive forces

Fig. 10.28 Photograph of bottom-to-top grading. Block-and-ash flow, 1929 eruption of Mount Pelée, Martinique.

that act normal to flow boundaries during movement of concentrated dispersions (Bagnold, 1954, 1955). Dispersive forces are derived from the bouncing of fragments off the rigid lower boundary. This transfers momentum from grain to grain (or close grain encounters) thereby supporting individual grains upward throughout the flowing bed. Reverse grading from dispersive pressure is limited to the basal part of a debris flow deposit and results in reverse-to-normal grading. Bottom-to-top reverse grading appears to have a different origin than reverse-to-normal grading. Middleton's (1970) 'kinetic sieve' mechanism, whereby clasts fall between the larger fragments which in turn progressively work themselves upward, best explains this kind of inverse grading. Bottom-to-top reverse grading may be restricted to debris-flow deposits with high clay contents (Smith, 1987a), but there are block-and-ash flow deposits that lack clay showing that it is not a requirement for its development.

Structures

Debris flow deposits are rich in internal structural features, manifested as crude bedding, include oriented flat particles, twigs and tree trunks oriented sub-parallel to the depositional surface, imbricated pebbles, reverse-to-normal grading, smaller fragments grouped against larger fragments suggesting differential movement, and discontinuous particle trains aligned parallel to the depositional surface (Fisher & Schmincke, 1984). Development of these features is governed by mechanisms of high concentration plastic flow. Matrix strength in debris flows may produce a rigid plug where shear stress is below the yield threshold throughout (Johnson, 1970), and this plug rides on a zone of non-turbulent flow where shear stress is greater than the yield threshold. Debris flows stop (i.e. are deposited) when the plug expands to the base of the flow at the expense of the zone of non-turbulent flow. Thus, evidence of shearing adjacent to the base of flow is frozen in place during the last stages of flow and preserves the clast orientations, textures and structures of the debris flow.

10.5.2 Fluid properties

Debris flows are granular fluids with high bulk density, and exhibit the property of strength resulting from particle interactions due to high concentration of particles, resulting in poor sorting, characteristic grading and other textural and structural features of the deposits. At volume concentrations of less than about 20–30%, particle support in a solids–water mixture is mostly by turbulence, but above that concentration (up to about 60% volume %), particle interactions greatly modify flow behaviour with particles being supported by a combination of turbulence and particle interactions. At still higher concentrations, particle support is primarily by particle interactions and the fluid may be described as plastic (Costa, 1984).

Internal resistance to flow results from: (1) electrostatic forces causing *cohesive resistance to flow* arising from clay–water mixtures; or (2) from friction caused by inertial interaction of large fragments (greater than medium-silt size) causing *inertial resistance to flow* or frictional resistance (cohesionless or density-modified grain flows; Lowe, 1982). The relative importance of these two causes of flow resistance depends upon the amount of admixed clay–size grains, a small amount ($\sim 5\%$) of which can cause major changes in flow behaviour. Grains are supported in debris flows by mass effects caused by high concentration (e.g. cohesive strength, frictional strength, viscous resistance and dispersive pressure), and by turbulence and pore-fluid expulsion.

Debris flows consist of: (1) a continuous phase (matrix phase or fluid phase) composed of water mixed with particles <2 mm, and (2) a coarse-grained phase of large particles >2 mm (Fisher, 1971; Scott, 1988). Therefore, even with a continuum of grain sizes from clay to boulders, it is possible to consider mass properties of high concentration flows conceptually (e.g. viscosity, density and strength) without knowing the individual properties of the particles: the matrix phase can be considered to be the fluid that transports the large fragments, even though the large particles are part of the fluid. A coarse-grained debris flow can be most easily characterized by grain-size parameters of the matrix phase, and matrix competence can be characterized by a measure of the largest fragments (e.g. the average of the five largest particles within a specified area, e.g. 1 m^2). Complete analyses of coarse-grained lahar deposits containing boulders, cobbles and pebbles entail various methods of size determinations including wet or dry sieving of the sand-size and coarser parts of a sample (up to about 16 mm) and analysis of the silt- and clay-size fractions.

During downstream movement in water-rich environments, lahars may progressively mix with water and transform to hyperconcentrated flood flows. Hyperconcentrated flood flows lack the strength and cohesion of lahars, but can carry high sediment loads with fragments supported both by turbulence and particle interactions (Pierson & Scott, 1985; Smith, 1986; Scott 1988). Such floods can move as far as 250 km or more down valleys, and because of their high loads, will impact an entire river system such as at Mount St Helens in 1980.

10.5.3 Origin of lahars

Three major categories of lahars by origin are:

1 Those formed by the direct and immediate result of eruptions through crater lakes, snow or ice, and heavy rains falling during or immediately after an eruption on abundant unstable loose material. They may also form by dewatering of debris avalanches (Pierson & Scott, 1985; Scott, 1988) and by pyroclastic surges flowing over and melting snow and ice (Lowe *et al.*, 1986; Major & Newhall, 1989).

2 Lahars may form indirectly from eruptions such as commonly occur shortly after eruptions by triggering of lahars from earthquakes or rapid drainage of lakes dammed by erupted products (Glicken *et al.*, 1989).

3 Many lahars are unrelated to contemporaneous volcanic activity, occurring by mobilization of loose tephra by heavy rain, or meltwater on steep slopes of volcanoes by rain, or meltwater from snow seeping into loose debris at any stage of volcano cone-building or cone degradation. Moreover, volcanic terrain in areas of extinct volcanic activity and without a volcanic edifice can give rise to debris flows composed mainly of epiclastic volcanic fragments.

Common non-volcanic processes by which lahars and other debris flows form are by heavy rains falling upon loose debris or by loose debris becoming saturated with water from melting snow, glaciers or heavy rains (Osterkamp *et al.*, 1986). Water-saturated material can move downhill like wet concrete when its internal strength is exceeded.

10.6 Volcaniclastic sedimentation and facies

10.6.1 Introduction

The interaction between volcanism and sedimentation, and development of concurrent facies are largely governed by two important factors. These are that: (1) active volcanism produces abundant sediment that is rapidly delivered to sites of deposition; and (2) lateral changes are the result of flow transformations (Section 10.6.4). During eruptions, large volumes of pyroclastic and hydroclastic sediment are released far more rapidly than any process of production of epiclastic particles (Kuenzi *et al.*, 1979; Walton, 1979; Vessel & Davies 1981; Ballance, 1988; Houghton & Landis, 1989). The episodic nature of eruptions may profoundly disrupt sedimentary environments and processes resulting in rapid changes in depositional systems through time. Removal and transfer of these materials from active volcanoes occur through flow transformations as material is carried into contiguous basins of deposition. Sediment is carried from the volcano to the sea to be stored for a time in sub-aqueous borderland environments, and then remobilized and carried into deep marine basins (Fisher, 1984). During times of quiescent volcanism, smaller volumes of pyroclastic, hydroclastic and volcanic epiclastic sediment are remobilized by similar flow transformations (Walton, 1979).

Volcaniclastic facies depend ultimately upon magma composition which governs eruptive rates, types of particles, manner of emplacement, total volume and therefore type of volcano. In subduction settings, andesite to dacite suite magmas construct high-standing stratovolcanoes with large volumes and great heights, and therefore large reservoirs of sediment (Hackett & Houghton, 1989). They erode rapidly, providing large volumes of reworked pyroclastic and hydroclastic particles together with epiclastic volcanic debris that are deposited into surrounding basins. Large calderas, commonly built in extentional back-arc regions, are as voluminous as stratovolcanoes, but they are low-standing volcanoes. Very large craters of calderas are initially closed sedimentary basins in which lacustrine sediments and slump blocks from crater walls are deposited. Differences between volcanoes require that different facies aspects be considered in order to reconstruct volcanic areas. These facies aspects are: (1) distance-related facies; (2) the type of source volcano; and (3) whether vents were single, multiple, central or flank.

The presence of vitric fragments (shards, pumice) within sedimentary sequence indicates a pyroclastic or hydroclastic origin. Moreover, glass is metastable and readily alters to clays and zeolites, and therefore does not appear as an epiclastic fragment.

10.6.2 Facies

Volcaniclastic facies are defined by distance from source, type of transporting agent, environment of deposition, and in some cases, by composition. First-order volcaniclastic facies are generally defined by position of the rock body relative to source within non-marine or marine environments, for example *proximal, medial and distal facies*. These designations are generalized and depend upon the size and volume of deposits. For example, at Mount St Helens, the 18 May 1980 blast surge went no farther

than 24 km from source, therefore all of the proximal, medial and distal facies occur within that limit (Fisher, 1990). However, at Aso caldera Japan, one pyroclastic flow deposit, which travelled at least 155 km from source, is considered to be proximal out to 45 km (Suzuki-Kamata, 1988).

The proximal facies may include the source volcano (Vessel & Davies, 1981), but where the source is not exposed, proximal facies rocks can be defined by type of transport such as lava flows (short travel distance), lahars, and fallout layers (most far-travelled) and, in the case of reworked pyroclastics or volcanic epiclastic materials, on their coarsest and thickest parts (Smith, 1988a,b). Pyroclastic facies may be divided into different sub-facies, such as lahar or pyroclastic flow and pyroclastic surge sub-facies (mechanisms of transport), lacustrine, submarine fan or alluvial sub-facies (environment of deposition), etc. These criteria are the foundation for defining larger-scale facies environments such as source volcanoes and their surroundings (Fisher & Schmincke, 1984; Hackett & Houghton, 1989).

10.6.3 Depositional units and multiple beds

Individual transport events result in the deposition of single layers or several layers that change in aspect such as thickness, texture or composition away from source. Changes in textures and structures of layers result from changing physical behaviour within a single transportational event such as a block-and-ash flow deposit laid down by a single nuée ardente or, in marine regions, turbidity currents that deposit Bouma sequences. Unlike non-volcanic events, changes in grain size and transportation characteristics can also be ascribed to changing variables of a volcanic eruption at the source as, for example, the widening of a vent during a Plinian eruption leading to eruption column collapse resulting in development of a pyroclastic flow on top of earlier fallout deposits (Sparks, 1976; Sparks *et al.*, 1978).

One example of a multilayered deposit from a single event comes from the 18 May 1980, 0832 h eruption of Mount St Helens, Washington (USA). Five different layered units were formed from a single blast erupted laterally from the north side of the volcano. The layers include a ground layer containing a poor mixture of material from the original ground surface with some juvenile lithics from the eruption, overlain by blast surge deposits, capped by an accretionary lapilli fallout layer (Fisher, 1990).

The concept of *tephra event unit* has been developed which includes proximal to distal facies, all produced in a geologically very brief time interval (Schmincke & Bogaard, 1991).

10.6.4 Rock sequences and volcanoes

Rock sequences deposited within marine or nonmarine basins that are derived from depositional events originating from many eruptions can be divided into large-order facies groups that reflect the history and dynamics of volcanism through sedimentary analysis (Busby-Spera, 1988b). Such basinal sequences occur adjacent to volcanic fields, including magmatic arcs with large stratovolcanoes close to marine basins and island volcanoes.

The growth rate of andesitic stratovolcanoes, with consequent influence upon depositional environments, is geologically extremely rapid, being of the order of a few hundred to a few thousand years. These large constructional landforms are composed of great volumes of easily remobilized fragmental material. Their growth is therefore reflected almost instantly in the sedimentary record of the surrounding region (Kuenzi *et al.*, 1979; Vessel & Davies, 1981) by direct deposition from airborne tephra, by deposition of ground-hugging pyroclastic flows, or from eruption-related debris avalanches, lahars and fluvial materials. Rapid construction of a volcano results in an increase in rate of erosion as slopes steepen and local climates are altered. Large volcanoes create climatic barriers, where rainfall and consequent erosion can be dramatically high on the windward side.

Andesitic volcanism produces three conditions necessary to lahar development, that is, steep slopes, relatively high rainfall and abundant loose fragmental material. With high enough elevations glaciers may also form, including their consequent abundant outwash. Rapid growth of volcanoes can result in oversteepened unstable slopes leading to collapse of sections of the mountain and development of debris avalanches (Glicken, 1986; Siebert *et al.*, 1987) (Fig. 10.29).

Rapid growth rates of volcanoes profoundly influence the progressive facies changes associated with an entire volcano system. For example, later products from a stratovolcano at its maximum height and volume can be carried farther than its earlier products when it was smaller. In sub-aqueous environments, as a volcano grows from deep through shallow water to sub-aerial environments, explosivity increases which leads to greater production of particles and their more efficient dispersal (Schmincke, 1982; Staudigel & Schmincke, 1984) (Fig. 10.30). Thus, coarsening upward, progradational sequences in adjacent marine basins, as demonstrated by Busby-Spera (1988a), result from actively growing stratovolcanoes. Incision occurs during inactive periods with reworked primary pyroclastic and epiclastic volcanic debris being carried away by fluvial systems leaving little or no record of sedimentation

Fig. 10.29 Photograph of surface of 1980 Mount St Helens debris avalanche.

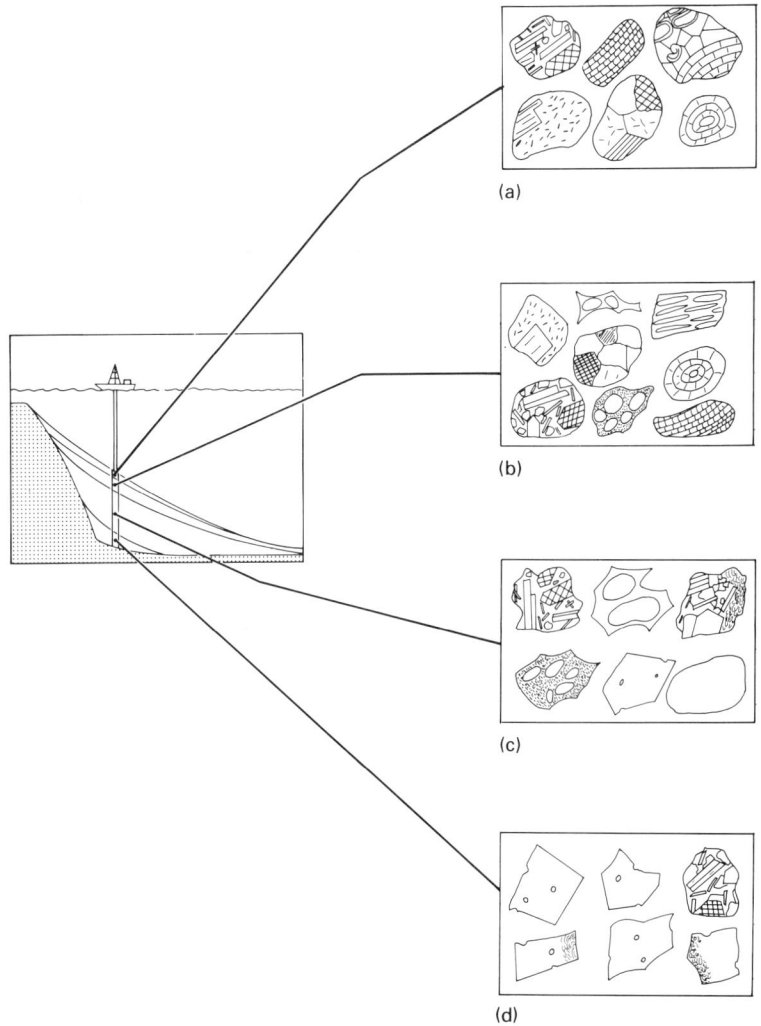

Fig. 10.30 Evolution of volcanic island as reflected in volcanic and non-volcanic particles in deposits from volcaniclastic apron.
(a) Retrograde erosional stage;
(b) volcanic island and erosional stage; (c) shallow water and shield stage; (d) DWS, deep water stage. From Schmincke (1988).

near the source (Smith, 1987a; Smith *et al.*, 1988). Fining upward sequences develop in sedimentary basins as a volcano lowers by erosion, with products being dominantly of epiclastic and reworked pyroclastic origin as shown by the Great Valley sequence of California (USA) (Ingersoll, 1978).

Flow transformation and facies lineage

All of the volcaniclastic sediments discussed in previous sections of this chapter can be accommodated within a stratovolcano facies framework that is linked by flow transformations (Fisher, 1983). A flow transformation, which occurs within single-event sediment gravity flows, can be defined as the change from laminar to turbulent behaviour (or vice versa) involving: (1) separations caused by gravity (gravity transformations); (2) a change without much variation in water or gas content (body transformation) as when slope changes; and (3) separations caused by turbulent mixing with ambient fluid above a flow surface (surface transformation) (Fig. 10.31). Freundt and Schmincke (1986) show that pyroclastic flows may transform from surge on the higher slopes of a volcano to plug flow on the plains via a hydraulic jump.

Scott (1988) extends the concept of flow (laminar-turbulent) transformation to include changes in transport agents whereby lahars are transformed to lahar-runout flows (hyperconcentrated flows), or hot

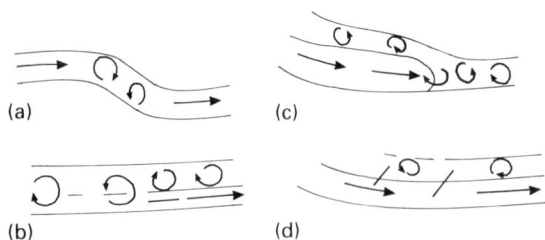

Fig. 10.31 Diagram showing types of flow transformations: (a) Body transformation; (b) gravity transformation; (c) surface transformation; and (d) elutriation transformation. From Fisher (1983).

pyroclastic flows are transformed to lahars. Weirich (1989) demonstrates that sub-aqueous debris flows transform to turbidity currents by hydraulic jumps. Thus, the concept of transformations links the general volcaniclastic facies (*multiple event facies*) to flow processes (*single event facies*) within a space–time framework from source (the volcano) to final deposition in marine or non-marine basins (proximal to distal facies). However, because of erosion, the lateral facies changes are commonly truncated. For example, in volcanic arc environments, the proximal source is commonly missing, with only distal facies (turbidite, submarine fan) and intermediate facies (fluvial, lahar to delta and shelf with or without submarine lahars) being present (Smith, 1988b).

As shown by the 1980 Mount St Helens eruptions, one facies lineage, linked by flow transformations, is as follows (Scott, 1988): eruption of pyroclastic surge or flow > lahar > hyperconcentrated flood flow > normal fluvial transport (in the Columbia River). Another lineage is fallout ash from vertical eruption plumes > initial large-scale debris avalanches > stop-gap storage of sediment on submarine shelves or slopes > submarine landslides > sub-aqueous lahars > turbidity currents.

Stratovolcanoes

Ruapehu volcano, New Zealand, is divided into two parts: a composite cone of volume $110\,km^3$ surrounded by a ring plain (Hackett & Houghton, 1989). Complementary parts of the volcano history are preserved in these two environments. Cone-forming sequences are dominated by sheet- and auto-brecciated-lava flows, that seldom reach the ring plain. The ring plain is built from the products of

explosive volcanism including distal primary pyroclastic deposits and reworked material eroded from the cone. Much of the material of the ring plain is deposited as lahars directly resulting from eruption processes or triggered by high intensity rain storms on volcano flanks. Deposits of the ring plain are further reworked and carried farther into alluvial systems and depositional basins immediately following eruptions or more gradually in the longer intervals between eruptions.

Thus, on present-day stratovolcanoes, major volcaniclastic facies associations can be divided on the basis of distance and geographic location — cone-forming sequences surrounded by voluminous ring plains corresponding to proximal and medial facies as presented above. Distal facies are far-travelled ash blankets that may be physiographically separated from the other deposits of the source volcano (Fisher & Schmincke, 1984).

Calderas

Calderas, as well as stratovolcanoes, can produce enormous amounts of volcaniclastic debris. Unlike stratovolcanoes, calderas have large-diameter craters generally without high-standing edifices, with correspondingly lower rates of erosional reworking of deposits. Commonly they form in backarc and other extensional tectonic regions such as rifts and grabens, therefore the chance for preservation of caldera fills, rim sequences and marginal caldera faults is greater than high-standing stratovolcanoes that can be rapidly worn down (Francis, 1983).

Other volcanoes commonly associated with each of the large volcano forms — stratovolcanoes and calderas — are domes, scoria cones and maar volcanoes, with chances of survival within sedimentary basin sequences dependent upon whether or not they occur as satellites on the slopes of the larger volcanoes, on highlands or in basins.

10.6.5 Volcaniclastic sedimentation and plate margins

Volcaniclastic sedimentation is charactistic of convergent plate margins in marine forearc sequences (Dickinson, 1976; Davies *et al.*, 1978; Ingersoll, 1978; Kuenzi *et al.*, 1979; Vessel & Davies, 1981; Miller, 1989), in marine to non-marine intra-arc grabens (Busby-Spera, 1986, 1988b), and in marine and non-marine environments of backarc or interarc areas (Van Houten, 1976; Mathisen & Vondra, 1983; Smith, 1987a,b, 1988a,b; Busby-Spera, 1988a; Turbeville *et al.*, 1989). The forearc region between the volcanic arc and the down-going subducting crustal slab, above which lies the trench, can be up to 300 km wide and can form a large forearc basin. Sedimentary environments along the shoreline include beach–shelf–slope–rise with fan-deltas, deltas, submarine canyons and submarine fans. The volcaniclastic component within the sedimentary fill depends upon intensity of volcanic activity and the volume of debris that enters this environment. Within the forearc basin itself, sediments are largely turbidite-dominated, and clastics are epiclastic volcanic and reworked pyroclastites and hydroclastites. Pyroclastic and hydroclastic materials dominate during episodes of volcanism, whereas epiclastics dominate between volcanic episodes. Thin primary fallout tephra deposits may be interbedded depending upon prevailing winds during volcanic eruptions.

Facies analysis of volcaniclastic aprons surrounding oceanic islands has led to the definition of several overlapping stages in the evolution of oceanic islands: (1) deep water stage; (2) shallow water–shield stage, (3) mature island stage; and (4) regressive erosional stage, each with different clastic processes giving rise to characteristic clast types and mixtures (Schmincke, 1987, 1988) (Fig. 10.32).

The above described sedimentary environments also occur in marine backarc regions, and therefore

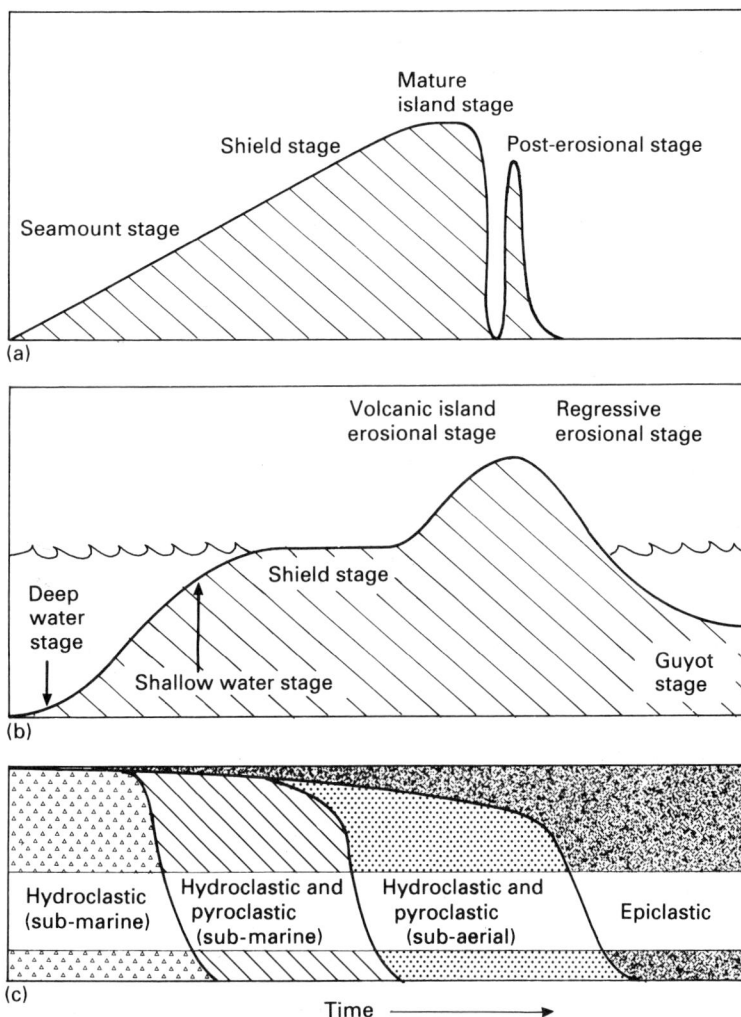

Fig. 10.32 Topographic and magma production rate evolution of volcanic ocean island as reflected in main types of volcaniclastic deposits. (a) Magma production rate; (b) topographic evolution (height); (c) dominant clastic processes. From Schmincke (1987).

are difficult to separate only on the basis of volcani-clastic lithology or type of transporting agent. Tectonic associations (extensional structures, grabens), chemical affinities (more alkaline in backarc) and rock associations (interbedding of cratonic sediments) may also be necessary to determine tectonic environment. In the North and South American Cordilleran environment, fallout tephra is much more common in backarc regions than in forearc environments whereas, in the Lesser Antilles, fallout tephra is far more common in the forearc region (Sigurdsson *et al.*, 1980; Sigurdsson & Carey, 1981).

There are several types of extensional environments, generally in backarc regions, both marine and non-marine, and in intra-arc regions (Smith *et al.*, 1987; Busby-Spera, 1988a,b). Chemical affinities of extensional volcanics are commonly alkaline. Although stratovolcanoes may grow within extensional environments to provide volcaniclastic debris, felsic ignimbrite deposits are characteristic products associated with caldera formation. Basaltic scoria cones (and maar volcanoes in water-rich environments) may be abundant. Extensional environments generally include graben structures that act as sediment traps. In lowland graben environments, basaltic volcanism is likely to give rise to abundant hydroclastics, and volcanic land forms can be partly covered and preserved within the sedimentary fill, which is likely to be epiclastic volcanics depending upon the intensity of volcanism. In addition to chemical evidence and rock type mentioned above, preservation of the volcano edifice signifies an extensional tectonic environment different from that of the convergent island arc stratovolcano that stands high and erodes away. Features of and evidence for arc graben depressions are briefly reviewed by Busby-Spera (1988b) for the early Mesozoic of the southwest Cordilleran United States. Important evidence is the great thickening of the depositional bodies, and the interbedding of quartz sandstone from the craton trapped in the graben and interbedded with pyroclastic and epiclastic materials, including ignimbrite. A modern arc graben analogue occurs in Central America (Burkart & Self, 1985). Smith *et al.* (1988) describe a late Miocene graben from the central Oregon High Cascades filled with volcaniclastics of the Cascade arc.

Volcaniclastic rocks of the arc are characteristically calc-alkaline andesite or basaltic andesite, but rhyolitic to dacitic ignimbrites also occur, but are more abundant and widespread in intra-arc grabens, back-arc regions and continental extensional tectonic zones.

10.7 References

Bagnold R.A. 1954. Experiments on a gravity-free dispersion of large solid spheres in a Newtonian fluid under shear. *Proc. R. Soc. Lond. A.*, **225**, 49–63.

Bagnold R.A. 1955. Some flume experiments on large grains but little denser than the transporting fluid and their implications. *Proc. Inst. Civ. Engnrs.*, **1**, (III), 174–205.

Ballance P.F. 1988. The Huriwai braidplain delta of New Zealand: A late Jurassic, coarse-grained, volcanic-fed depositional system in a Gondwana forearc basin. In: Nemec W. & Steel R.J. (eds) *Fan deltas: Sedimentology and Tectonic Settings*, pp. 430–444. Blackie & Son, Glasgow.

Barberi F., Innocenti F., Lirer L., Munno R., Pescatore T. & Santacroce R. 1978. The Campanian ignimbrite: A major prehistoric eruption in the Naples area (Italy). *Bull. Volcanol.*, **41**, 10–31.

Bitschene P.R. & Schmincke H.-U. 1990. Fallout tephra: Composition and Significance. In: Heling D., Rothe P., Förstner U., & Stoffers P. (eds) *Sediments and Environmental Geochemistry, Selected Aspects and Case Histories*, pp. 48–82. Springer-Verlag, Berlin.

Bogaard P.v.D. & Schmincke 1985. Laacher See tephra: A widespread isochronous late Quaternary ash layer in Central and Northern Europe. *Geol. Soc. Am. Bull.*, **96**, 1554–1571.

Bogaard P.v.D., Hall C.M., Schmincke H.-U. & York D. 1989. Precise single-grain ^{40}Ar/^{39}Ar dating of a cold to warm climate transition in Central Europe. *Nature*, **342**, 523–525.

Bogoyavlenskaya G.E., Braitseva O.A., Melekestsev I.V., Kiriyanof V.Yu. & Miller C.D. 1985. Catastrophic eruptions of the directed-blast type at Mount St Helens, Bezymianny and Shiveluch volcanoes. *J. Geodynamics*, **3**, 189–218.

Bramlette M.N. & Bradley W.H. 1942. Geology and biology of North Atlantic deep-sea cores between Newfoundland and Ireland. pt. 1, Lithology and geologic interpretations. *U.S Geol. Surv. Prof. Pap.*, **196A**, 1–55.

Burkart B. & Self S. 1985. Extension and rotation of crustal blocks in northern Central America and effect on the volcanic arc. *Geology*, **13**, 22–26.

Busby-Spera C.J. 1986. Depositional features of rhyolitic and andesitic volcaniclastic rocks of the Mineral King submarine caldera complex, Sierra Nevada, California. *J. Volcanol. Geotherm. Res.*, **27**, 43–76.

Busby-Spera C.J. 1988a. Development of fan-deltoid slope aprons in a convergent-margin tectonic setting: Mesozoic, Baja California, Mexico. In: Nemec W. & Steel R.J. (eds) *Fan Deltas: Sedimentology and Tectonic Settings*, pp. 419–429. Blackie and Son, Glasgow. 419–129.

Busby-Spera C.J. 1988b. Speculative tectonic model for the early Mesozoic arc of the southwest Cordilleran United States. *Geology*, **16**, 1121–1125.

Carey S.N & Sigurdsson H. 1980. The Roseau Ash: deep-sea tephra deposits from a major eruption on Dominica, Lesser Antilles Arc. *J. Volcanol. Geotherm. Res.*, **7**, 67–86.

Carey S.N & Sigurdsson H. 1982. Influence of particle aggregation on deposition of distal tephra from the May 18, 1980, eruption of Mount St Helens Volcano. *J. Geophys. Res.*, **87**, 7061–7072.

Carey S. & Sigurdsson H. 1986. The 1982 eruptions of El Chichon volcano, Mexico (2): Observations and numerical modelling of tephra-fall distribution. *Bull. Volcanol.*, **48**, 127–141.

Carey S. & Sparks R.S.J. 1986. Quantitative models of the fallout and dispersal of tephra-fall from volcanic eruption columns. *Bull. Volcanol.*, **48**, 109–125

Costa J.E. 1984. Physical geomorphology of debris flow. In: Costa J.E. & Fleischer P.J (eds) *Developments and Applications of Geomorphology*, pp. 268–317. Springer, Berlin.

Crandell D.R. 1971. Postglacial lahars from Mount Rainier volcano, Washington. *U.S. Geol. Surv. Prof. Pap.*, **677**, 1–75.

Crowe B.M. & Fisher R.V. 1973. Sedimentary structures in base-surge deposits with special reference to cross-bedding. Ubehebe Craters, Death Valley, California. *Bull. Geol. Soc. Am.*, **84**, 663–682.

Davies I.C., Querry M.W. & Bonis S.B. 1978. Glowing avalanches from the 1974 eruption of the volcano Fuego, Guatemala. *Bull. Geol. Soc. Am.*, **89**, 369–384.

Devine J.D., Sigurdsson H., Davis A.N. & Self S. 1984. Estimates of sulfur and chlorine yield to the atmosphere from volcanic eruptions and potential climatic effects. *J. Geophys. Res.*, **89**, 6309–6325.

Dickinson W.R. 1976. Sedimentary basins developed during evolution of Mesozoic–Cenozoic arc-trench systems in western North America. *Can. J. Earth Sci.*, **13**, 1268–1287.

Fisher R.V. 1961. Proposed classification of volcaniclastic sediments and rocks. *Bull. Geol. Soc. Am.*, **72**, 1409–1414.

Fisher R.V. 1966. Mechanism of deposition from pyroclastic flows. Am. J. Sci., **264**, 287–298.

Fisher R.V. 1971. Features of coarse-grained, high-concentration fluids and their deposits. *J. Sedim. Petrol.*, **41**, 916–927.

Fisher R.V. 1977. Erosion by volcanic base-surge density currents: U-shaped channels. *Bull. Geol. Soc. Am.*, **88**, 1287–1297.

Fisher R.V. 1983. Flow transformations in sediment gravity flows. *Geology*, **11**, 273–274.

Fisher R.V. 1984. Submarine volcaniclastic rocks. In: Kokelaar B.P. & Howells M.F. (eds) *Marginal Basin Geology: Volcanic and Associated Sedimentary and Tectonic Processes in Modern and Ancient Marginal Basins. Spec. Publ. Geol. Soc. London*, **16**, 5–27.

Fisher R.V. 1990. Transport and deposition of a pyroclastic surge across an area of high relief: The 18 May 1980 eruption of Mount St Helens, Washington. *Bull. Geol. Soc. Am.*, **102**, 1038–1054.

Fisher R.V. & Heiken G. 1982. Mt Pele, Martinique: May 8 and 20, 1902 pyroclastic flows and surges. *J. Volcanol. Geotherm. Res.*, **13**, 339–371.

Fisher R.V. & Mattinson J.M. 1968. Wheeler Gorge turbidite–conglomerate series: inverse grading. *J. Sedim. Petrol.*, **38**, 1013–1023.

Fisher R.V. & Rensberger J.M. 1973. Physical stratigraphy of the John Day Formation. *Univ. Calif. Publ. Geol. Sci.*, **101**, 1–45.

Fisher R.V. & Schmincke H.-U. 1984. *Pyroclastic Rocks.* 472 pp. Springer-Verlag, Berlin.

Fisher R.V., Glicken H.X. & Hoblitt R.P. 1987. May 18, 1980, Mount St Helens Deposits in South Coldwater Creek, Washington. *J. Geophys. Res.*, **92**, 10267–10283.

Fiske R.S. & Matsuda T. 1964. Submarine equivalents of ash flows in the Tokiwa Formation, Japan. *Am. J. Sci.*, **262**, 76–106.

Francis E.H. 1983. Magma and sediment — II. Problems of interpreting palaeovolcanics buried in the stratigraphic column. *J. Geol. Soc. Lond.*, **140**, 165–183.

Francis P.W., Roobol M.J., Walker G.P.L., Cobbold P.R. & Coward M. 1974. The San Pedro and San Pablo volcanoes and their hot avalanche deposits. *Geol. Rundschau*, **63**, 357–388.

Freundt A. & Schmincke H.-U. 1985. Hierarchy of facies of pyroclastic flow deposits generated by Laacher See-type eruptions. *Geology*, **13**, 278–281.

Freundt A. & Schmincke H.-U. 1986. Emplacement of small-volume pyroclastic flows at Laacher See (East-Eifel, Germany). *Bull. Volcanol.*, **48**, 39–59.

Glicken H. 1986. *Rockslide-debris Avalanche of May 18, 1980. Mount St Helens Volcano, Washington.* PhD dissertation, University of California, Santa Barbara, California.

Glicken H., Meyer W. & Sabol M.A. 1989. Geology and ground-water hydrology of Spirit Lake Blockage, Mount St Helens, Washington, with implications for lake retention. *U.S. Geol. Surv. Bull.*, **1789**, 1–33.

Guerrera F. & Veneri F. 1989. Neogene and pleistocene volcaniclastites of the Apennines (Italy). *Geol. Mijn.*, **68**, 381–390.

Hackett W.R. & Houghton B.F. 1989. A facies model for a Quaternary andesitic composite volcano: Ruapehu, New Zealand. *Bull. Volcanol.*, **51**, 51–68.

Hamilton W.B. 1988. Plate tectonics and island arcs. *Bull. Geol. Soc. Am.*, **100**, 1503–1527.

Harris D.M., Rose W.I., Jr, Roe R. & Thompson M.R. 1981. Radar observations of ash eruptions. In: Lipman P.W. & Mullineaux D.R. (eds). *The 1980 eruptions of Mount St Helens, Washington. U.S. Geol. Surv. Prof. Pap.*, **1250**, 323–333.

Hay R.L. 1959. Formation of the crystal-rich glowing avalanche deposits of St Vincent, BWI. *J. Geol.*, **67**, 540–562.

Heiken G.H. 1971. Tuff rings: examples from Fort Rock–Christmas Lake Valley, south-central Oregon. *J. Geophys. Res.*, **76**, 5615–5626.

Heiken G.H. 1974. An atlas of volcanic ash. *Smithsonian Contr. Earth. Sci.*, **12**, 1–101.

Heiken G.H. & Wohletz K. 1985. *Volcanic Ash.* University of California Press, Berkeley.

Hoblitt R.P. 1986. Observations of eruptions, July 22 and August 7, 1980, at Mount St Helens, Washington. *U.S. Geol. Surv. Prof. Pap.*, **1335**.

Hoblitt R.P. & Kellogg K.S. 1979. Emplacement temperatures of unsorted and unstratified deposits of volcanic rock debris as determined by paleomagnetic techniques. *Bull. Geol. J. Soc. Am.*, **90**, (I), 633–642.

Hoblitt R.P., Miller C.D. & Valance J.W. 1981. Origin and stratigraphy of the deposit produced by the May 18 directed blast. In: Lipman P.W. and Mullineaux D.R. (eds) *The 1980 Eruptions of Mount St Helens, Washington. U.S. Geol. Surv. Prof. Pap.*, **1250**, 401–419.

Houghton B.F. & Landis C.A. 1989. Sedimentation and volcanism in a Permian arc-related basin, southern New Zealand. *Bull. Volcanol.*, **51**, 403–450.

Ingersoll R.V. 1978. Petrofacies and petrologic evolution of the late Cretaceous fore-arc basin, northern and central California. *J. Geol.*, **86**, 335–352.

Johnson A.M. 1970. *Physical Processes in Geology.* Freeman, Cooper, San Francisco.

Kennett J.P. 1983. *Marine Geology.* Prentice Hall, Englewood Cliffs, NJ.

Kozu S. 1934. The great activity of Komagatake in 1929. *Tschermak's Mineral. Pet. Mitt.* **45**, 133–174.

Kuenzi W.D., Horst O.H. & McGehee R.V. 1979. Effect of volcanic activity on fluvial–deltaic sedimentation on a modern arc-trench gap, southwestern Guatemala. *Bull. Geol. Soc. Am.*, **90**, (I), 827–838.

Kuno H., Ishikawa T., Katsui Y., Yagi K., Yamasaki M. & Taneda S. 1964. Sorting of pumice and lithic fragments as a key to eruptive and emplacement mechanism. *Jap. J. Geol. Geog.*, **35**, 223–238.

Lacroix A. 1904. *La Montagne Pélé et Ses Eruptions.* Masson et Cie, Paris.

Lipman P.W. & Mullineaux D.R. (eds) 1981. The 1980 eruptions of Mount St Helens. *U.S. Geol. Surv. Prof. Pap.*, **1250**.

Lorenz V. 1974. Vesiculated tuffs and associated features. *Sedimentology*, **21**, 273–291.

Lowe D.R. 1982. Sediment gravity flows: II. Depositional models with special reference to the deposits of high density turbidity currents. *J. Sedim. Petrol.*, **52**, 279–297.

Lowe D.L., Williams S.N., Leigh H., Connor C.B., Gemmell J.B. & Stoiber R.E. 1986. Lahars initiated by the 13 November 1985 eruption of Nevado del Ruiz, Colombia. *Nature*, **324**, 51–53.

Major J.J. & Newhall C.G. 1989. Snow and ice perturbation during historical volcanic eruptions and the formation of lahars and floods. *Bull. Volcanol.*, **52**, 1–27.

Marshall J.R. (ed) 1987. *Clastic particles: Scanning Electron Microscopy and Shape Analysis of Sedimentary and Volcanic Clasts.* Van Nostrand Reinhold, New York.

Mathisen M.E. & Vondra C.F. 1983. The fluvial and pyroclastic deposits of the Cagayan Basin, Northern Luzon, Philippines — an example of non-marine volcaniclastic sedimentation in an inter-arc basin. *Sedimentology*, **30**, 369–392.

Mellors R.A., Waitt R.B. & Swanson D.A. 1988. Generation of pyroclastic flows and surges by hot-rock avalanches from the dome of Mount St Helens volcano, USA. *Bull. Volcanol.*, **50**, 14–25.

Middleton G.V. 1970. Experimental studies related to problems of flysch sedimentation. *Geol. Ass. Can. Spec. Pap.*, **7**, 253–272.

Miller M.M. 1989. Intra-arc sedimentation and tectonism: Late Paleozoic evolution of the eastern Klamath terrane, California. *Bull. Geol. Soc. Am.*, **101**, 170–187.

Mimura K. 1984. Imbrication, flow direction and possible source areas of the pumice-flow tuffs near Bend, Oregon U.S.A. *J. Volcanol. Geotherm. Res.*, **21**. 45–60.

Moore J.G. 1967. Base surge in recent volcanic eruptions. Bull. Volcanol., **30**, 337–363.

Nairn I.A. & Self S. 1978. Explosive eruptions and pyroclastic avalanches from Ngauruhoe in February, 1975. *J. Volcanol. Geothem. Res.*, **3**, 39–60.

Ninkovich D, Sparks R.S.J. & Ledbetter M.T. 1978. The exceptional magnitude and intensity of the Toba eruption, Sumatra: An example of the use of deep-sea tephra layers as a geological tool. *Bull, Volcanol.*, **41**, 286–298.

Osterkamp W.R., Hupp C.R. & Blodgett J.C. 1986. Magnitude and frequency of debris flows, and areas of hazard on Mount Shasta, Northern California. *U.S. Geol. Surv. Prof. Pap.*, **1396-C**, 1–21.

Paterne M., Guichard F. & Labeyrie J. 1988. Explosive activity of the south Italian volcanoes during the past 80 000 years as determined by marine tephrochronology. *J. Volcanol. Geotherm. Res.*, **34**, 153–172.

Perret F.A. 1937. The eruption of Mt Pele 1929–1932. *Carnegie Inst. Wash. Publ.*, **458**.

Pierson T.C. & Scott K.M. 1985. Downstream dilution of a lahar: transition from debris flow to hyperconcentrated streamflow. *Wat. Reso. Res.*, **21**, 1511–1524.

Pollack J.B., Toon O.B., Sagan C., Summers A., Baldwin B. & Camp W. Van 1976. Volcanic explosions and climate change: a theoretical assessment. *J. Geophys. Res.*, **81**, 1071–1083.

Pyle D.M. 1989. The thickness, volume and grain size of tephra fall deposits. *Bull. Volcanol.*, **51**, 1–15.

Rittmann A. 1958. Il meccanismo di formazione delle lave a pillows e dei cosidetti tufi palagonitici. *Atti Acc. Gioenia*, **4**, 310–317.

Ronov A.B. 1968. Probable changes in the composition of sea water during the course of geologic time. *Sedimentology*, **10**, 25–43.

Rose W.E., Jr, Pearson T. & Bonis S. 1977. Nuée ardente eruption from the foot of a dacite lava flow, Santiaguito Volcano, Guatemala. *Bull. Volcanol.*, **40**, 1–16.

Ross C.S. & Smith R.L. 1961. Ash-flow tuffs; their origin, geologic relations and identification. *U.S. Geol. Surv. Prof. Pap.*, **366**, 1–77.

Ruddiman W.F. & Glover L.K. 1972. Vertical mixing of ice-rafted volcanic ash in North Atlantic sediments. *Bull. Geol. Soc. Am.*, **83**, 2817–2836.

Schmid R. 1981. Descriptive nomenclature and classification of pyroclastic deposits and fragments: Recommendations of the IUGS Subcommission on the systematics of igneous rocks. *Geology*, **9**, 41–43.

Schmincke H.-U. 1967. Graded lahars in the type section of the Ellensburg Formation, south-central Washington. *J. Sedim. Petrol.*, **37**, 438–448.

Schmincke H.-U. 1982. Volcanic and chemical evolution of the Canary Islands. In: Rad U.V., Hinz K., Sarnthein M. and Seibold E. (eds) *Geology off the Northwest African Continental Margin*, pp. 720–785. Springer, Heidelberg.

Schmincke H.-U. 1987. Geological Field Guide of Gran Canaria, 2nd Edn. Pluto Press, Witten.

Schmincke H.-U 1988. Pyroklastische Gesteine. In: Füchtbauer H. (ed.) *Sedimentgesteine*, pp. 720–785. Schweizerbart'sche Verlagsbuchhandlung, Stuttgart.

Schmincke H.-U. & Bogaard P.v.D. 1991. Tephra layers and tephra events. In: Einsele G., Ricken W., Seilacher A. (eds) *Cycles and Events in Stratigraphy*, pp. 329–429. Springer, Heidelberg.

Schmincke H.-U., Fisher R.V and Waters A.C. 1973. Antidune and chute and pool structures in the base surge deposits of the Laacher See area, Germany. *Sedimentology*, **20**, 553–574.

Schumacher R. & Schmincke H.-U. 1990. The lateral facies of ignimbrites at Laacher See volcano. *Bull. Volcanol.*, **52**, 271–285.

Schumacher R. & Schmincke H.-U. 1991. Internal structure and occurrences of accretionary lapilli – a case-study at Laacher See volcano. *Bull. Volcanol.*, **53**, 612–634.

Scott K.M. 1988. Origins, behavior, and sedimentology of lahars and lahar-runout flows in the Toutle–Cowlitz system. *U.S. Geol. Surv. Prof. Pap.*, **1447-A**, 1–74.

Segerstrom K. 1950. Erosion studies at Paricutin, State of Michoacan, Mexico. *U.S. Geol. Survey Bull.*, **965-A**.

Self S. & Rampino M.R. 1981. The 1883 eruption of Krakatau. *Nature*, **294**, 669–704.

Self S., Kienle J. & Huot J.P. 1980. Ukinrek Maars, Alaska, II. Deposits and formation of the 1977 craters. *J. Volcanol. Geotherm. Res.*, **7**, 39–65.

Sheridan M.F. 1979. Emplacement of pyroclastic flows: A review. *Spec. Paper Geol. Soc. Am.*, **180**, 125–136.

Sheridan M.F. & Updike R.G., 1975. Sugarloaf Mountain tephra – a Pleistocene rhyolitic deposit of base-surge origin. *Bull. Geol. Soc. Am.*, **86**, 571–581.

Siebert L., Glicken H. & Ui T. 1987. Volcanic hazards from Bezymianny- and Bandai-type eruptions. *Bull. Volcanol.*, **49**, 435–459.

Sigurdsson H. & Carey S.N. 1981. Marine tephrochronology and Quaternary explosive volcanism in the Lesser Antilles Arc. In: Self S. & Sparks R.S.J. (eds) *Tephra Studies*, pp. 255–280. Reidel, Dordrecht.

Sigurdsson H. & Carey S. 1989. Plinian and co-ignimbrite tephra fall from the 1815 eruption of Tambora volcano. *Bull. Volcanol.*, **51**, 234–270.

Sigurdsson H., Carey S.N. & Fisher R.V. 1987. The 1982 eruptions of El Chichon volcano, Mexico (3): Physical properties of pyroclastic surges. *Bull. Volcanol.*, **49**, 467–488.

Sigurdsson H., Sparks R.S.J., Carey S.N. & Huang T.C. 1980. Volcanogenic sedimentation in the Lesser Antilles Arc. *J. Geol.*, **88**, 523–540.

Sisson T.W. 1982. *Sedimentary Characteristics of the Airfall Deposit Produced by the Major Pyroclastic Surge of May 18, 1980 at Mount St Helens, Washington.* MS Thesis, University of California at Santa Barbara.

Smith G.A. 1986. Coarse-grained nonmarine volcaniclastic sediment: Terminology and depositional process. *Bull. Geol. Soc. Am.*, **97**, 1–10.

Smith G.A. 1987a. Sedimentology of volcanism-induced aggradation in fluvial basins: Examples from the Pacific Northwest, U.S.A. In: Ettridge F.G., Flores R.M. & Harvey M.G. (eds) *Recent Developments in Fluvial Sedimentology, Spec. Publ. Soc. Econ. Palont. Mineral*, **39**, 217–228. SEPM, Tulsa.

Smith G.A. 1987b. The influence of explosive volcanism on fluvial sedimentation: the Deschutes Formation (Neogene) in central Oregon. *J. Sedim. Petrol.*, **57**, 613–629.

Smith G.A. 1988a. Sedimentology of proximal to distal volcaniclastics dispersed across an active foldbelt: Ellensburg Formation (late Miocene), central Washington. *Sedimentology*, **35**, 953–977.

Smith G.A. 1988b. The influence of explosive volcanism on fluvial sedimentation: The Deschutes Formation (Neogene) in central Oregon. *J. Sedim. Petrol.*, **57**, 613–629.

Smith G.A., Campbell N.P., Deacon M.W. & Shafiqullah M. 1988. Eruptive style and location of volcanic centers in the Miocene Washington Cascade Range: Reconstruction from the sedimentary record. *Geology*, **16**, 337–340.

Smith G.A., Snee L.W. & Taylor E.M. 1987. Stratigraphic, sedimentologic, and petrologic record of late Miocene subsidence of the central Oregon High Cascades. *Geology*, **15**, 389–392.

Smith R.L. 1960. Ash flows. *Bull. Geol. Soc. Am.*, **71**, 795–842.

Sparks R.S.J. 1976. Grain size variations in ignimbrites and implications for the transport of pyroclastic flows. *Sedimentology*, **23**, 147–188.

Sparks R.S.J. 1986. The dimensions and dynamics of eruption columns. *Bull. Volcanol.* **48**, 3–15.

Sparks R.S.J. & Wilson L. 1976. A model for the formation of ignimbrite by gravitational column collapse. *J. Geol. Soc. Lond.*, **132**, 441–451.

Sparks R.S.J., Moore J.G. & Rice C.J. 1986. The initial giant umbrella cloud of the May 18th, 1980, explosive eruption of Mount St Helens. *J. Volcanol. Geotherm. Res.*, **28**, 257–274.

Sparks R.S.J., Sigurdsson H. & Carey S.N. 1980. The entrance of pyroclastic flows into the sea, II. Theoretical considerations on subaqueous emplacement and welding. *J. Volcanol. Geotherm. Res.*, **7**, 97–105.

Sparks R.S.J., Wilson L. & Hulme G. 1978. Theoretical modeling of the generation movement and emplacement of pyroclastic flows by column collapse. *J. Geophys. Res.*, **83**, 1727–1739.

Staudigel H. & Schmincke H.-U. 1984. The Pliocene seamount series of La Palma (Canary Islands). *J. Geophys. Res.*, **89**, 11195–11215.

Suzuki T., Katsui Y. & Nakamura T. 1973. Size distribution of the Tarumai Ta-b pumice-fall deposit. *Bull. Volcanol. Soc. Japan*, **18**, 47–64.

Suzuki-Kamata K. 1988. The ground layer of Ata pyroclastic flow deposits, southwestern Japan—evidence for the capture of lithic fragments. *Bull. Volcanol.*, **50**, 119–129.

Taylor G.A. 1958. The 1951 eruption of Mount Lamington, Papua. *Austr. Bur. Min. Res. Geol. Geophys. Bull.*, **38**, 1–117.

Tokunaga T. & Yokoyama S. 1979. Mode of eruption and volcanic history of Jukaiyama Volcano, Nii-Jima. *Geog. Rev. Japan*, **52**, 111–125. (In Japanese)

Turbeville B.N., Waresback D.B. & Self S. 1989. Lava-dome growth and explosive volcanism in the Jemez Mountains, New Mexico. Evidence from the Plio-Pleistocene Puye alluvial fan. *J. Volcanol. Geotherm. Res.*, **36**, 267–291.

Valentine G.A. 1987. Stratified flow in pyroclastic surges. *Bull. Volcanol.*, **49**, 616–630.

Valentine G.A. & Wohletz K.H. 1989. Numerical models of Plinian eruption columns and pyroclastic flows. *J. Geophys. Res.*, **94**, 1867–1887.

Van Houten F.B. 1976. Late Cenozoic volcaniclastic Andean foredeep, Columbia. *Bull. Geol. Soc. Am.*, **87**, 481–495.

Vessel R.K. & Davies D.K. 1981. Nonmarine sedimentation in an active fore arc basin. *Soc. Econ. Paleont. Mineral. Publ.*, **31**, 31–45.

Walker G.P.L. 1984. Characteristics of dune-bedded pyroclastic surge bedsets. *J. Volcanol. Geotherm. Res.*, **20**, 281–296.

Walton A.W. 1979. Volcanic sediment apron in the Tascotal

Formation (Oligocene?), Trans-Pecos Texas. *J. Sedim. Petrol.*, **49**, 303–314.

Waters A.C. & Fisher R.V. 1971. Base surges and their deposits: Capelinhos and Taal Volcanoes. *J. Geophys. Res.*, **76**, 5596–5614.

Weirich F.H. 1989. The generation of turbidity currents by subaerial debris flows, California. *Bull. Geol. Soc. Am.*, **101**, 278–291.

Whitham A.G. 1989. The behavior of subaerially produced pyroclastic flows in a subaqueous environment: Evidence from the Roseau eruption, Dominica, West Indies. *Mar. Geol.*, **86**, 27–40.

Wilson C.J.N. 1980. The role of fluidization in the emplacement of pyroclastic flows: an experimental approach. *J. Volcanol. Geotherm. Res.*, **8**, 231–249.

Wilson C.J.N. 1984. The role of fluidization in the emplacement of pyroclastic flows 2: Experimental results and their interpretation. *J. Volcanol. Geotherm. Res.*, **20**, 55–84.

Wilson, C.J.N. 1985. The Taupo eruption, New Zealand II. The Taupo ignimbrite. *Phil. Trans. R. Soc. Lond. A*, **314**, 229–310.

Wilson L. & Head J.W., III 1981. Ascent and eruption of basaltic magma on the earth and moon. *J. Geophys. Res.*, **86**, 2971–3001.

Wilson L. & Walker G.P.L. 1987. Explosive volcanic eruptions — VI. Ejecta dispersal in plinian eruptions: the control of eruption conditions and atmospheric properties. *Geophys. J. R. Astr. Soc.*, **89**, 657–679.

Wilson, L., Sparks, R.S.J. & Walker, G.P.L. 1980. Explosive volcanic eruptions, IV. The control of magma properties and conduit geometry on eruption column behavior. *Geophys. J. R. Astr. Soc.*, **63**, 117–148.

Wilson L., Sparks R.S.J., Huang T.C. & Watkins N.D. 1978. The control of volcanic column eruption heights by eruption energetics and dynamics. J. Geophys. Res., **83**, 1829–1836.

Wohletz K.H. 1983. Mechanisms of hydrovolcanic pyroclastic formation: grain-size, scanning electron microscopy, and experimental studies. *J. Volcanol. Geotherm. Res.*, **17**, 31–63.

Wohletz K.H. 1986. Explosive magma–water interactions: Thermodynamics, explosion mechanisms, and field studies. *Bull. Volcanol.*, **48**, 245–264.

Wright J.V. 1979. *Formation, Transport and Deposition of Ignimbrites and Welded Tuffs*. PhD Thesis, Imperial College, London.

Wright J.V. & Mutti E. 1981. The Dali Ash, Island of Rhodes, Greece: a problem in interpreting submarine volcanigenic sediments. *Bull. Volcanol.*, **44–2**, 153–167.

Yokoyama S. 1974. Mode of movement and emplacement of Ito pyroclastic flow from Aira Caldera, Japan. *Sci, Repts Tokyo Kyoiku Daigaku, Sect. C (Geog. Geol. Mineral.)*, **12**, 17–62.

Index